Plant Mutation Breeding and Biotechnology

CABI is a trading name of CAB International

CABI
Nosworthy Way
Wallingford
Oxfordshire OX10 8DE
UK

CABI
875 Massachusetts Avenue
7th Floor
Cambridge, MA 02139
USA

Tel: +44 (0)1491 832111
Fax: +44 (0)1491 833508
E-mail: info@cabi.org
Website: www.cabi.org

T: +1 800 552 3083 (toll free)
T: +1 (0)617 395 4051
E-mail: cabi-nao@cabi.org

A catalogue record for this book is available from the British Library, London, UK.

Library of Congress Cataloging-in-Publication Data

Plant mutation breeding and biotechnology / edited by Q.Y. Shu, B.P.Forster, H. Nakagawa.
p. cm.
Includes bibliographical references.
ISBN 978-1-78064-085-3 (CAB International : alk. paper)
1. Plant mutation breeding. 2. Mutagenesis. 3. Plant biotechnology. I. Shu, Q. Y. II. Forster, Brian P. III. Nakagawa, H. (Hitoshi)

SB123.P5637 2012
631.5'3--dc23

2012009993

Published jointly by CAB International and FAO

ISBN-13: 978-925107-022-2 (FAO)
          978-178064-085-3 (CABI)

The designations employed and the presentation of material in this information product do not imply the expression of any opinion whatsoever on the part of the Food and Agriculture Organization of the United Nations (FAO) concerning the legal or development status of any country, territory, city or area or of its authorities, or concerning the delimitation of its frontiers or boundaries. The mention of specific companies or products of manufacturers, whether or not these have been patented, does not imply that these have been endorsed or recommended by FAO in preference to others of a similar nature that are not mentioned. The views expressed in this information product are those of the author(s) and do not necessarily reflect the views of FAO.

Applications for such permission should be addressed to:

Chief
Electronic Publishing Policy and Support Branch
Communication Division
FAO
Viale delle Terme di Caracalla, 00153 Rome, Italy

or by e-mail to:
copyright@fao.org

Commissioning editor: Nigel Farrer
Printed and bound in Malta by Gutenberg Press Ltd

Plant Mutation Breeding and Biotechnology

Edited by Q.Y. Shu, B.P.Forster, H.Nakagawa

Plant Breeding and Genetics Section
Joint FAO/IAEA Division of Nuclear Techniques in Food and Agriculture
International Atomic Energy Agency,
Vienna, Austria

www.cabi.org

Joint FAO/IAEA Programme
Nuclear Techniques in Food and Agriculture

Foreword

Up until the 20th century, spontaneous mutations were the only source of novel genetic diversity that mankind could exploit in selecting plants and animals suitable for domestication and breeding. A leap in plant breeding came when ionizing radiation was discovered to modify the genetic make-up of organisms. The pioneering work of L.J. Stadler in the late 1920s marks the beginning of plant mutation breeding, despite Stadler himself being less than optimistic about its real value. It was not until the establishment in 1964 of the Joint Food and Agricultural Organisation of the United Nations / International Atomic Energy Agency (FAO / IAEA) Division of Nuclear Techniques in Food and Agriculture, with its global coordinating and synergistic roles, that plant mutation breeding became a common tool available to plant breeders worldwide. Since these early days the Joint Division continues to play a considerable role in fostering the use of mutation techniques for crop improvement in FAO and IAEA member states. It does so by coordinating and supporting research, by promoting capacity building and technology transfer, by providing technical services and policy advice, and by collecting, analysing and disseminating information. By the end of 2009 the number of mutant varieties officially released worldwide had reached 3088, up from a mere 77 in 1964.

An early initiative of the Joint Division was the compilation of the *Mutation Breeding Manual*, published by the IAEA in 1975, with a second edition in 1977. The Manual was subsequently translated into several languages and has received wide acclaim as a reference book for plant breeders and as a text book at universities; it has played a pivotal role in educating several generations of plant breeders, including myself. But time does not stand still, and especially the past decade has seen a rapid emergence of new tools of relevance to plant breeders, including bio-informatics, genetic transformation and genomics. With these tools, and with the fast accruing knowledge of plant mutagenesis, the earlier perception of mutation induction as a random, uncontrolled process of empiric nature has also changed, and plant mutagenesis is now fully capitalizing on advances in molecular- and biotechnologies, such as Targetting Induced Local Lesions in Genomes (TILLING), and is an essential tool also in research on gene discovery and gene function. They have surely brought new vigor to plant mutation breeding, and have re-injected this discipline into the mainstream of science-based breeding.

Initially we planned to "merely" update the protocols in the earlier Manual, but quickly realized that this would neither do justice to the vast number of recent scientific and technological developments relevant to plant mutation breeding, nor would it come anywhere close to fulfilling the expectations of modern plant breeders and research scientists. In the book you are now reading we present contemporary knowledge of mutagenesis in plants, state-of-the-art technologies and methodologies and their underlying principles, and provide exemplary case studies on mutation induction, identification and utilization in plant breeding and research. I hope this book will meet your expectations and that you will find it a worthy successor to our earlier *Mutation Breeding Manual*. It is my sincere hope that it will help the global agricultural community to generate more and better crop varieties in its challenging effort to reach global food security and to minimize the currently widening gap between the rich, the poor and the famished.

Qu Liang
Director, Joint FAO/IAEA Division of Nuclear Techniques in Food and Agriculture
International Atomic Energy Agency

Preface

Evolution and practical breeding both depend on genetic variation. Over the years since Darwin, naturalists and a diversity of scientists have learned how to create, detect and utilize mutations. The development of genomics has more recently increased the power of plant mutagenesis in crop improvement. This unique book elegantly shows how biology, physics and chemistry all interplay to provide the nexus of theory and practice. Clearly written and illustrated, the book provides an up-to-date and comprehensive manual for understanding the application of mutagenesis and its scientific basis. Many inset boxes are included in the text that aid in the explanations of the approach or summarize studies related to the point being addressed. Definitions and glossaries are included and add clarity to the discussion. A sense of history is provided through the listing of milestones in the development of the technology. Useful reference lists and websites are provided in each chapter. Comments also are included at appropriate places relative to safety considerations.

Broad in its scope, many plant species from crops to ornamentals are covered in the book as well as mutant traits such as dwarfing genes, male sterility, disease resistance, chromosome pairing, fatty acid composition and many others. Examples are given that illustrate the need for additional breeding, usually by incorporation of genetic modifier genes, to somewhat change the original mutant phenotype in order for the mutation to become a significant breeding target. The reader is informed early in the book that today's plant mutagenesis not only includes "induced mutagenesis" via the traditional physical or chemical mutagenesis procedures, but also "insertional mutagenesis" and "site-directed mutagenesis". These two latter approaches – which will become even more commonplace in the future – allow greater certainty of obtaining the desired mutant phenotype; these approaches often require considerable molecular genetic information about the trait. Zinc finger nucleases against specific sequences are given as an excellent example of the achievable increased precision. This book explains pertinent molecular genetics aspects from promoters to enhancers, and different types of mutations from insertions/deletions to frameshift mutations. The importance of DNA repair in homologous recombination and mutagenesis also is discussed in detail. Relatively new molecular genetics techniques for detecting genetic variation are changing the precision and frequency of success; TILLING, de-TILLING, and eco-TILLING are discussed as efficient means of finding genetic variation. The next-generation DNA sequencing procedures are providing another leap forward.

Many points learned by experience via plant mutagenesis studies give the reader insights that earlier researchers had to learn the hard way. The outcome of any mutation experiment depends on many factors such as the type of mutagen, dose and dose rate, genotype, growth conditions, etc. Dose and dose responses and how these differ among methods are given in several instances. Success in achieving a high frequency of mutation is a sort of balancing act between maximizing the mutagenic effect and recovering viable/fertile plants; some discussion is allotted as to how to assess plant injury. Another valuable aspect of the book relates to the complications introduced through chimerism. Several mutagenesis methods may lead to a plant where the mutation is not in every cell of the plant. Proper interpretation of the results depends on knowledge of embryo development in that species relative to cell layers and number of primordial cells at the time of treatment.

Mutagenesis techniques – even the newer ones – have resulted in many impressive mutant varieties. The number of such varieties is high across economic plant species, including vegetatively propagated species. The reading of this book provides a lucid review of the basis of mutagenesis, gives many practical tips for the efficient production of mutant types, and portends an important future for such techniques in basic and applied biology. Anyone contemplating the use of mutagenesis as an approach to improving or modifying a trait or achieving basic understanding of a pathway for a trait will find this book an essential reference.

Ronald L. Phillips
Regents Professor Emeritus
University of Minnesota

Table of Contents

Section 1

Concepts, Historical Development and Genetic Basis

C01

Plant Mutagenesis in Crop Improvement: Basic Terms and Applications

B.P.Forster[a],* and Q.Y.Shu[b]

[a] BioHybrids International Limited, P.O. Box 2411, Earley, Reading RG6 5FY, UK
[b] Joint FAO/IAEA Division of Nuclear Techniques in Food and Agriculture, International Atomic Energy Agency, Wagramer Strasse 5, P.O. Box 100, A-1400 Vienna, Austria
Present: Institute of Nuclear Agricultural Sciences, Zhejiang University, Hangzhou 310029, China
*Corresponding author, E-MAIL: brianforster@biohybrids.co.uk

1. Introduction

Before the turn of the 21st century, experiments in plant mutagenesis were driven by the potential use of mutants in plant improvement. During the past ten years, genomics and molecular techniques have become part of plant mutagenesis research and induced mutants have become an established resource in genomics studies. Although plant-induced mutagenesis has been used widely as a tool in basic studies and practical breeding programmes, it is seldom considered to be an independent subject by plant scientists or plant breeders. There are only a very limited number of books or other publications with comprehensive treatments of this subject, particularly its principles and technologies, which sometime leads to ambiguous concepts and misuse of scientific terms. This book describes the underlying principles of plant experimental mutagenesis, its associated enabling technologies and its application to research and plant breeding. Examples and success stories are given to illustrate the practicality of methods. In order to understand these and subsequent subjects the reader must become acquainted with the common terminology of the discipline, which is set out below.

2. Definitions of Basic Terms

2.1. Mutagenesis and Experimental Mutagenesis

Mutagenesis is the process by which the genetic information of an organism is changed in a stable manner. This happens in nature as a result of errors in DNA repair (**see Chapter 5**). Mutagenesis is the process by which mutations are generated. Mutagenesis can be exploited experimentally (experimental mutagenesis) by physical, chemical and biological means (**Box 1.1**).

The following terms are used frequently:

- **Mutation**: This was originally defined in a series of articles by de Vries (1901, 1903 and 1905) as a sudden heritable change in the genetic material not caused by recombination or segregation. De Vries used the word "sudden" to differentiate between subtle changes that could be explained by the normal processes of recombination. "Sudden" changes (mutations) in plant forms (phenotypes) were obvious,

apparent and unusual and therefore of interest. Mutation however, especially at the gene sequence (genotype) level can lead to small and subtle changes in phenotype which may not become immediately apparent and these can now be detected using molecular techniques; thus the word "sudden" as applied to phenotype can be deleted from the definition.

- **Mutants**: individuals carrying a mutation that may be revealed using molecular means or identified by phenotyping tools. Different types of mutant can be generated using experimental mutagenesis (**Box 1.1**).

2.2. Mutation Genetics and Breeding

The term mutation breeding ("Mutationszüchtung") was first coined by Freisleben and Lein (1944) to refer to the deliberate induction and development of mutant lines for crop improvement. The term has also been used in a wider sense to include the exploitation of natural as well as spontaneous mutants, and in the development of any variety possessing a known mutation from whatever source. The argument is semantic as all genetic variation is *ipso facto* mutation, in the broad sense mutation breeding can be regarded simply as breeding. However, the term "mutation breeding" has become popular as it draws attention to deliberate efforts of breeders and the specific techniques they have used in creating and harnessing desired variation in developing elite breeding lines and cultivated varieties. Similarly the term mutant variety is simply a variety (var.), but draws attention to the fact that it carries an important trait controlled by a known mutant gene or it has been developed using mutation techniques (**for details see Chapter 24**). In some parts of the world the term cultivar (cv.) is used to describe a cultivated variety, for consistency, this book uses the word "variety". However it should be noted that the word "variety" is also used as a botanical taxonomic descriptor.

Although mutants can be produced using different kinds of experimental mutagenesis, mutation breeding is commonly restricted to the use of physically and chemically induced mutagenesis; other types of experimental mutagenesis are more commonly used in functional genomics studies.

Generation advancement is a key component in both genetic research and in breeding programmes. Here are some generally accepted terms in breeding. P_1, P_2, P_3, P_4, etc., denote parent one, two, three, four, etc. in a crossing programme. The product (progeny) of a cross between two parents is the first filial generation or F_1 generation. Subsequent generations are termed F_2, F_3, F_4, etc.

In line with this nomenclature the first mutated generation is termed the M_1. Plants that are produced directly from seeds (or gametes) treated with a mutagen are M_1 plants, the next generation is the M_2, followed by the M_3, M_4, etc. Seeds prior to mutagenic treatment are termed M_0 and after treatment referred to as M_1. Seeds that develop on the M_1 plants are therefore the M_2 generation which develop into M_2 plants. Embryos produced from crosses in which either the pollen or embryo sac has been treated with a mutagen also represent the first mutated generation and are therefore termed M_1. In vegetatively propagated crops (VPCs) the notation M_1V_1, M_1V_2, etc is used for consecutive vegetative (V) generations (it can also be simplified as MV_1, MV_2, etc.). Similarly, somaclonal variants are denoted as SV_1, SV_2, etc. Note that this terminology is different to that used in transformation studies, here the regenerated plants produced after transformation are denoted as T_0, and their progenies are T_1, T_2, etc. In the past, some groups also used X_1, X_2, etc. to define generations treated with X-rays, and SP_1, SP_2, etc., for materials exposed to space conditions.

The following terms are relevant to mutation genetics and breeding:
- **Mutant selection.** The process of identifying individuals with a target mutant phenotype; this includes two major steps: **mutant screening** and **mutant confirmation (or mutant verification)**. Mutant screening is a process of selecting out individuals from a large mutated population that meet the selection criterion. For example, M_2 plants flowering three days earlier than their

wild type parent(s) are screened as potential early flowering mutants, and plants without disease symptoms might be screened as potential disease-resistant mutants. Since flowering is dependent on both genotypic and environmental factors they can only be regarded as "**putative mutants**", which means they are not necessary "true mutants". This is the case for many traits including disease resistance as here non-infection may simply be the result of the absence of the pathogen.
- **Mutant confirmation** is the process of re-evaluating the putative mutants under replicated and stringent conditions, using larger sample sizes (usually the progenies of selected putative individuals, e.g. M_3 lines of selected M_2 plants). Many putative mutants of quantitative traits, e.g. growth duration, yield, quality, disease resistance, might be proven to be false mutants.

2.3. Genetic Features and Effects of Induced Mutations

Randomness of induced mutations. Physically or chemically induced mutations occur randomly across the whole genome and within any locus or gene. This is a very important feature of induced mutagenesis, because it not only provides the probability of generating mutations for any gene of interest, but enables the development of multiple mutations for any target gene in a predictive manner. Multiple mutant alleles are the source of genetic diversity for crop breeding as well as functional analysis of the targeted genes.

Dominance *versus* **recessiveness.** Most gene mutations produced by radiation effectively kill gene function as the gene is either knocked out or the mutation product is non-functional. Hence the vast majority of mutated genes are recessive. The frequency of recessive mutations has been reported to be in the range of 90–100% and that for dominant mutations 0–6%. However, as mutation techniques become more sophisticated, more subtle changes can be manufactured and detected. Point mutations induced by chemical mutagenesis for example are more likely to result in functional mutations than larger alterations in the genetic material, and these functional mutations have a greater chance of being dominant or co-dominant, i.e. they result in new functional alleles (**see Chapter 4**). However, dominant mutations can occasionally result from a DNA deletion event. For example, the dominant low glutelin content 1 (*Lgc 1*) mutation in rice is the result of a 3.5-kilobase (kb) deletion between two highly similar glutelin genes; the deletion causes a tail-to-tail inverted repeat, which might produce a double-stranded RNA molecule, a potent inducer of RNA silencing, and therefore silence the expression of key glutelin genes in a dominant manner (Kusaba *et al.*, 2003). Another example where a deletion leads to a dominant mutation which may be more common is that in most white-grained rice varieties carrying the mutant allele *rc* (a 14-base pair (bp) deletion) of the *Rc* gene, another 1-bp deletion in the *rc* gene reverses the frame shift and generates a pseudo-wild type red rice mutant (Brooks *et al.*, 2008).

Pleiotropic effects. Pleiotropy or a pleiotropic effect is the phenomenon whereby a gene influences multiple phenotypic traits; hence a mutation in a certain gene may have an effect on some or all traits simultaneously. This phenomenon is common in mutations of major genes such as those controlling plant height and flowering time (**see Section 3.4. below**). At the molecular level, pleiotropy can occur in transcriptional genes and genes upstream of a pathway. Downstream mutations in the starch biosynthesis, for example in the waxy gene, have no pleiotropic effects (**for more examples of pleiotropic effects see Chapter 24**).

Genetic and environmental effects on mutant gene expression. The expression of mutated genes can be influenced by the genetic background and environmental factors and in some cases these are conditional. Gene expression can vary as the genetic background changes, for example across successive generations of a breeding programme. In extreme cases the desired phenotype may disappear in advanced lines. This is particularly prevalent where the source of the mutation is in a genetic background distantly related to the target, commercial material. Effects of pleiotropy (the effect of one gene on the expression of another) may also temper mutant gene expression. Effects of the environment (biotic and abiotic) are easily observed when well characterized mutant populations, e.g. in the barley Bowman backcross lines (**see Chapter 25**) are grown over a range of environments and seasons. Obvious examples include mutations in genes controlling early maturity in response to photoperiod such as *Eam1* and *Eam6* (long day adapted) and *Eam5* and *eam9* (short day adapted); long day adapted material may fail to mature in short day conditions. The *uzu1* gene of barley is a good example where the environment controls the expression of plant height. In cool environments *uzu1* mutant lines appear near normal in height, but develop as extreme dwarfs in hot environments. In terms of yield, the smooth awn gene (*raw1*) and the unbranched style gene (*ubs4*) mutants show variable seed set responses to environmental stresses and different genetic backgrounds. Biotic stresses such as seed blight diseases can also restrict the development of seed pigments.

3. Mutation Techniques in Crop Breeding

3.1. A Brief Overview of Evolutionary Stages in Crop Development

Genetic change (mutation) has provided the natural variation (building blocks) for species evolution. Changes in species have not only been important for adaptation to the natural environment, they have also been exploited by man in the agricultural processes of species domestication and crop improvement. Before Mendel had developed his laws on inheritance plant breeding was regarded more as an art form than a science; it was a matter of selecting superior lines which arose by chance, and this included spontaneous mutants that caught the eye of the selector. Mendel's laws however allowed the development of plant breeding as a science based on simple mathematical principles; they provided a means of predicting progeny types from deliberate matings. Breeders therefore became more diligent in recording progeny performance data (weight and measure) in following inheritance patterns in plant pedigrees.

Figure 1.1 illustrates the generalized trend of increasing crop yields over the three major stages of crop evolution: 1) gathering from the wild, 2) domestication and 3) breeding of a plant species. No set times are given in **Figure 1.1** as the periods for domestication and the initiation of breeding vary enormously (from decades to centuries) among crops. Some of the earliest crops to be domesticated include sesame, fig, barley and wheat in the Near East and rice in East Asia some 10,000–8,000 BC, however crops such as blueberry, jatropha are as recent as 2000 AD. Before domestication yields from plants growing in the wild were seasonal, sporadic and totally dependent upon the prevailing natural environment. Yields varied unpredictably bringing years of feast and famine. Domestication represented a major event in human and crop evolution as it provided a degree of food security for domiciled people. Human interventions such as seed storage and plant cultivation mitagated against negative environmental factors such as poor or erratic rainfall and incidences of pests and diseases and promoted high, reliable yields. Before Mendel's laws and the development of structured plant breeding strategies, crop improvement from domestication onwards was largely governed by incremental improvements in agronomy such as watering, the application of fertilizers, weed control, crop rotation, disease control, soil preparation, etc. The genetic nature of the plants being cultivated changed very little except when major steps forward were made by the appearance of rare, but exceptional mutants. These occurred in many species and had major consequences for human as well as plant evolution, a notable example being bread wheat which evolved in cultivation to become a high yielding crop through the selection of mutations at the ploidy, genome, chromosome and gene levels. Without genetic improvement the impact of agronomy on a crop gradually reaches an optimum and plateaus off. However, after Mendel's revelations plant breeding became a test bed for new ideas in genetics, many of which resulted in breaking yield barriers (**illustrated as stepped improvements in Figure 1.1**). Ideas and theories tried and tested by plant breeders that have been successful include classical recombinant selection, induction of polyploidy, cytogenetic manipulations (chromosome engineering), doubled haploidy, F_1 hybrid production and transformation. All of which have met with spectacular successes in improving yield and yield stability in various crops: recombination breeding provided predictive methods of obtaining desired genotypes in sexually propagated

crops; polyploidy doubled the size of some vegetables; cytogenetics produced disease-resistant bread wheat; doubled haploidy produced pure, uniform and stable varieties in many crops, F_1 hybrids increased yields in maize six-fold and transformation increased yields massively through the incorporation of herbicide, pest and disease resistance genes in, for example cotton. The deployment of any, or a combination, of these biotechnologies becomes even more powerful when combined with genetic marker selection systems which can be deployed to accelerate the breeding process by genotypic rather than phenotypic based selection. These classic success stories are equalled if not surpassed by achievements in mutation breeding.

3.2. Milestones in Mutation Breeding

Improvements in plant breeding can only be made when sufficient variation for a given trait is available to the breeder. In the best case scenario the required variation (e.g. a disease resistant trait) is available within the elite gene pool of the crop. In many cases, however, the desired variation may exist but only present in material distantly related to elite lines, i.e. in out-dated varieties, old landraces or wild relatives. Retrieving such variation and developing it into a genetically finely tuned commercial variety is a protracted breeding process which has little appeal to plant breeders. The prospect of delivering such variation directly into elite material without recourse to extensive upgrading (crossing and selection) was immediately ceased once methods of

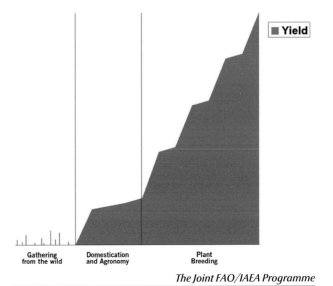

The Joint FAO/IAEA Programme

Figure 1.1 Trends in crop yields through the processes of gathering from the wild, domestication, agronomy and plant breeding.

induced mutation were discovered. In developing his mutation theory Hugo de Vries recognized the potential of induced mutation in plant and animal breeding and stated:

"We may search for mutable plants in nature, or we may hope for species to become mutable by artificial methods. The first promises to yield results most quickly, but the scope of the second is much greater and it may yield results of far more importance. Indeed, if it once should become possible to bring plants to mutate at our will and perhaps even in arbitrarily chosen directions, there is no limit to the power we may finally hope to gain over nature."

Initial successes in breeding were achieved by harnessing spontaneous (naturally occurring) mutants. The most famous example is the deployment of semi-dwarf mutants of wheat and rice in the "Green Revolution" (between 1965 and 1980). The genes for reduced height in wheat and rice were spontaneous mutants derived from the varieties "Norin 10" and "Dee-geo-woo-gen", respectively. These were used to breed short, stiff-strawed varieties that produced high grain yields in response to fertilizer inputs. Unprecedented yields were obtained which helped solve many problems in food security in developing countries at the time. India for example changed from being an importer to a net exporter of wheat. Other notable and early examples include: 1) the brown mid-rib mutant of maize (1920s), which is characterized by having a low lignin content making it highly digestibility as a fodder; 2) sweet lupin (palatable, non-toxic) spontaneous mutants selected by von Sengbusch as early as 1942 using mass screening of millions on seed; 3) bud sports in many flowers, ornamentals and fruit trees became a major and traditional source of the variation for crop improvement. For example, between 1942 and 1952 an estimated 25% of apples in North America had arisen from bud sports (spontaneous mutations). In the mid 1950s about 10% of all American fruit was derived from bud sports (cherries, nectarines, peaches, pears, plums and prunes) as were 30% of cut flowers in the Netherlands. Bud sports continue to be an important source of variation in these crops.

The first example of an induced mutant variety was a pale green tobacco, var. Chlorina, released in Indonesia in 1936. In the following decades a few other induced mutant varieties were released in a range of plant species, e.g. the tulip var. Faraday (1949, Netherlands), white mustard var. Primex (circa 1950, Sweden), bean var. Schaefer's Universal (circa 1950, Germany); rapeseed var. Regina II (1953, Sweden), fodder pea var. Weibull's Stråll (1957, Sweden), navy bean var. Sanilac (1957, USA), groundnut var. N.C.4 (1959, USA), oat var. Florad (1960, USA) and many un-named fruit trees and ornamentals. It was not however until radioisotopes became accessible during the Atomic energy programme that induced mutation, predominantly by gamma ray irradiation took off worldwide. **Box 1.2** shows a chronology of the major milestones in the history of mutation breeding.

3.3. The FAO/IAEA Division and the Mutant Variety and Genetic Stock Database

In 1964 the Joint FAO/IAEA Division of Nuclear Techniques in Food and Agriculture was established at the IAEA, Vienna with a Plant Breeding and Genetics Section. This promoted and stimulated worldwide application of mutation technologies in plant breeding providing services in information, irradiation and training, and developing collaborative international projects. According to FAO/IAEA figures 77 mutant varieties were released in 1969, which grew dramatically to 1,330 by 1989.

By 2009, about 3,100 mutant cultivated varieties in about 190 plant species were listed in the FAO/IAEA Database of Mutant Variety and Genetic Stock (http://mvgs.iaea.org). The largest numbers of mutant varieties are in the small grain cereal crops and reported in Asian countries (**Figure 1.2**).

The FAO/IAEA database can be interrogated by Latin or common species names, breeding methodology, location (country and continent) and by dates. Phenotypic descriptions, including photographs are provided. Over half the mutant varieties are in Asia (1,858, notably in India, Japan and China) followed by Europe (899), North America (202), Africa (62), Latin America (48) and the Australia/Pacific region (10). The database defines a mutant variety as a variety carrying a known mutant gene, and is not necessarily the direct product of a mutation treatment.

The FAO/IAEA figures for the numbers of mutant varieties are recognized as being a gross under-estimate for the following reasons:

1. Mutant varieties (particularly in ornamentals) are not always registered because such information is only collected from publicly available sources (mostly English) or voluntarily provided by breeders who have some connection with the FAO/IAEA programmes.

Box 1.2: Milestones in mutation breeding

300 BC	The ancient Chinese book "Lulan" provides the first documentation of mutant selection in plant breeding: maturity and other trait in cereals in China (Huang and Liang, 1980).
1590	The first verifiable (spontaneous) plant mutant described, "incisa" mutant of greater celandine.
1667	The first known description of a graft-chimera; Bizarria-orange, Florence, Italy.
1672	One of the oldest publications describing variability in trees, shrubs and herbaceous plants; Waare Oefeninge der Planten by A. Munting, see van Harten (1998).
1774 onwards	Descriptions of various mutants in wild and cultivated species by the taxonomist Linnaeus.
1865	Laws of inheritance published by Mendel.
1900	Mendal's work is accepted by the scientific establishment.
1901–1904	de Vries suggests and promotes radiation to induce mutations in plants and animals.
1907	Cramer publishes extensive examples of spontaneous mutants in crop plants.
1927	First proof of induced mutations in plants; radium ray treatment of *Datura stramonium* (Gager and Blakeslee, 1927).
1927	Muller working with *Drosophila* provides proof of mutation induction by X-rays. Muller champions induced mutation for animal and plant breeding and opens a new era in genetics and breeding.
1927	von Sengbusch invents the first mass selection method for mutants. Lupin is developed into a crop species by screening mutants for domestication and quality traits.
1928	Stadler publishes the first results of mutation induction in crop plants, barley, maize, wheat and oat, but is sceptical about the use of induced mutation for crop improvement.
1930s	Mutation breeding programmes are set up, notably in Sweden, Germany and USA.
1936	The first induced mutant variety is released, tobacco var. "Chlorina" using X-rays in Indonesia (then the Dutch East Indies).
1942	First report of induced disease resistance in a crop plant; X-ray-induced mildew resistance in barley (Freisleben and Lein, 1942).
1944	The term mutation breeding ("Mutationszüchtung") was coined by Freisleben and Lein.
1944/46	First reports of chemical induced mutation (Auerbach and Robson, 1944).
1949	First plant mutation experiments using ^{60}Co gamma ray installations. Cobalt-60 was chosen as a suitable radioisotope for continuous gamma irradiation as advantages included a long half-life, and it was relatively cheap, abundant and available. ^{60}Co became a standard tool in mutation induction of crop plants (Sparrow and Singleton, 1953).
1954	The first release of a mutant variety in a vegetatively propagated crop: tulip var. Faraday with an improved flower colour and pattern (see van Harten and Broertjes, 1989).
1964	The FAO/IAEA Joint Division was set up with a mandate to support and encourage the production of induced mutations (and related biotechnologies) for crop production particularly for food security issues in developing countries.
1966	First chemically induced mutant variety, Luther, of barley was released in the USA.
1972	The "Mutation Breeding Newsletter" is launched and published by the FAO/IAEA.
1993	Register of plant mutant varieties set up by the FAO/IAEA, which became the mutant variety genetic stock database (http://mvgs.iaea.org) in 2008.
2000–2009	Development of high-throughput genotyping and phenotyping using automated, robotic and computerized systems.
2000 onwards	Development of TILLING populations.

2. Most modern mutant varieties are not direct mutants, but a product of additional breeding. In general breeders are not particularly interested in the source of the variation and mutated lines are considered as basic, raw materials. Once a mutation for an important trait is captured and used over many years its novelty/origin is often lost, ignored or forgotten.

3. Politics and ownership. Mutation breeding is often a collaborative venture among institutions worldwide. Issues in international politics, protectionist policies, plant breeders' rights, ownership of materials, patents and profit sharing often contrive to promote individual interests and often cloud the source of the variation.

4. Economically important mutants can arise as bi-products of more basic research and not directly from a deliberate mutation breeding programme.

5. There is a concern in some countries that mutations may come under the umbrella of and be labelled as GMOs (genetically modified organisms, a term usually restricted to transformation events).

6. End-users are often reluctant to advertise the use of mutants because of vagaries of consumer perceptions.

3.4. Significant Mutant Genes and Mutant Varieties

Mutation breeding, particularly using gamma rays, took off worldwide in the 1960s and has resulted in spectacular successes, notably in seed propagated crops (SPCs) (**Box 1.3**). The widespread exploitation of mutant varieties has generated billions of additional incomes

for farmers and significantly promoted social-economic development of local communities (Ahloowalia *et al.*, 2004).

The short, stiffed-strawed barley varieties, Diamant (in Czechoslovakia) and Golden Promise (in UK) were direct mutants and carried the mutant dwarfing genes, *sdw1* and *ari.e.*GP, respectively. These varieties had a huge impact in Europe as they were better adapted to combine harvesting, they set new benchmarks for yield and quality and consequently were used extensively in subsequent variety development. In addition to Europe, semi-dwarf barley mutant varieties were induced and released in the USA (1981).

The first rice semi-dwarf induced mutant, var. Reimei was released in Japan in 1966. Reimei carried a mutation in the semi-dwarf gene *sd1*. As in the first semi-dwarf mutants in barley, Reimei was used extensively as a parent to breed many other successful varieties. Similarly, the first semi-dwarf rice mutant variety Calrose 76 developed in the USA in the late 1970s, was widely used in breeding programmes, which produced more than 20 new semi-dwarf varieties that are widely cultivated in the USA, Australia and Egypt (e.g. Giza 176). The *sd1* mutations in Reimei and Calrose 76 are allelic to the Chinese "Dee-geo-woo-gen" (DGWG) semi-dwarf mutation. In addition to short stature (and similar to the barley semi-dwarf mutants), *sd1* of DGWG provided useful pleiotropic effects on stiff straw, lodging resistance, day-length insensitivity, seed dormancy and high yield in response to fertilizer,

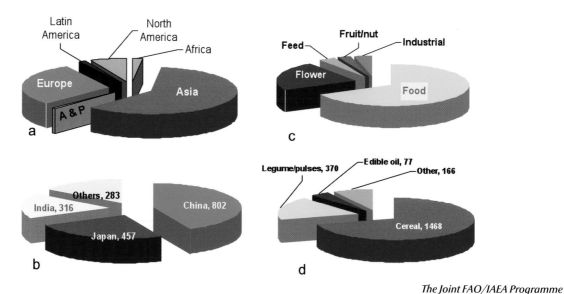

The Joint FAO/IAEA Programme

Figure 1.2 Distribution of mutant varieties (a) among continents (A&P: Australia and the Pacific); (b) in Asia; (c) among different types of end-uses; and (d) within food crops. Data source: FAO/IAEA Database of Mutant Variety and Genetic Stocks (http://mvgs.iaea.org, 2009).

Box 1.3: Some highlights of mutant varieties in the world

Bulgaria: Mutant durum varieties have occupied about 90% of the cultivating area since the 1980s.

China: Each of the following mutant varieties has cumulatively been grown on acreage of more than 10 million ha: rice varieties **Yuanfengzao**, **Zhefu 802** and **Yangdao No. 6**; wheat variety **Yangmai 156**; and the cotton variety **Lumian no.1**.

Costa Rica: Rice variety **Camago** occupied 30% of the cultivated area.

Europe: Many barley varieties widely grown in Europe are derived from mutant varieties **Diamant** or **Golden Promise**.

India: Mutant varieties are the prevailing varieties for pulses and legumes, for example the **TAU-1** mutant of blackgram has occupied 95% of the blackgram acreage in the State of Maharashtra and the groundnut varieties of the **TG** series (e.g. **TG24** and **TG37**) cover 40% of the groundnut acreage.

Italy: Durum wheat cultivation area was significantly expanded due to the cold tolerant mutant varieties (e.g. **Creso**).

Japan: Most rice varieties carry the *sd1* mutant allele from the variety **Reimei**. Japanese pear cultivation was rescued from extinction by the development of disease-resistant mutant varieties, **Gold Nijisseiki** and its derivatives.

Pakistan: The wheat mutant variety **Kiran 95** and the cotton mutant variety **NIAB-78** were planted on over 30% and 80% of the cultivation area for each crop, respectively.

USA: The rice mutant variety **Calrose 76** was the *sd1* donor for more than 10 successful varieties. **Star Ruby** and **Rio Red** are the two most important commercial grapefruit varieties (with the trade mark **Rio Star**).

Vietnam: VND and DT serial mutant rice varieties (e.g. **VND95-20** and **DT38**) have been cultivated on more than half million ha per year during the last decade, and DT serial mutant soybean varieties have been cultivated on more than 50% area with **DT84** being the leading variety in the past 10 years.

and was exploited in the development of the "Green Revolution" rice varieties.

With respect to VPCs there was a major expansion in the production of induced mutation in bulbs (e.g. hyacinths and tulips) and cut flowers (especially chrysanthemum, but also azalea, carnation, dahlia, rose, streptocarpus and others, **see Chapter 26**) that began in the Netherlands in the 1950s. In fruit crops, the first induced mutant variety was the sweet cherry var. Compact Lambert released in 1964 in Canada. The development of the disease-resistant Japanese pear variety Gold Nijisseiki saved this fruit as a crop in Japan (**Box 1.3, see also Chapter 26**). In Texas, USA, two mutant varieties Star Ruby and Rio Red are the preferred grapefruit varieties and sold under the trademark "Rio Star", which occupied 75% of the cultivation area.

3.5. New Developments in Plant Mutation Breeding

Mutation induction activities had peaks in the 1950s, 60s, 70s and 80s and enjoyed major successes in terms of mutant variety releases. However, mutation induction declined towards the end of the 20th century. There were various reasons for this, the negative arguments of Stadler still posed problems (**see Box 1.2 and Chapter 24**) and there was a general concern that mutation induction and mutation breeding were non-scientific. Another major

drawback was that all mutation detection and selection was done (painstakingly) at the phenotypic level.

The early years of the 21st century witnessed a resurgence in mutation technologies due to a rapid and greater understanding of mutagenesis and related disciplines, which led to more applications. The understanding of the molecular basis of mutagenesis (**DNA damage and repair; see Chapters 5 & 6**) transformed mutation induction from chance events into science-based techniques. The use of molecular and genomic tools for mutant screening and characterization also enabled mutation breeding to embrace and utilize the very recent findings and technological innovations in plant genomics and molecular biology research (**see Chapters 20 & 23**). And last but not the least, induced mutants are regarded as valuable tools in bridging the gap between phenotype and genotype, an important issue in plant breeding and plant genomics.

New species are emerging as being important in agriculture, but also to the environment, medicine and energy production. There is an urgent need to identify, develop and establish domestication traits for these crops. For species that have been in domestication for thousands of years many of these traits have been delivered by spontaneous mutations and were incorporated into the crop as they arose. The cereals: rice, wheat, barley, etc. provide good examples. In this respect, induced mutagenesis play a significant role. For crops new to agriculture, and also those that have experienced little in the way of genetic improvement through plant breeding, huge leaps forward can be made in a short period of time. For example, there has been a general worldwide trend to produce semi-dwarf crops of cereals and other crops. Semi-dwarfs tend to have a yield advantage as reserves normally used up by vegetative growth can be redirected towards the harvestable product, such as grain or fruits, thereby improving harvest index. As previously mentioned semi-dwarf mutations have played a major role in meeting targets for plant stature and yield. Semi-dwarf mutations are relatively common and can make up about a third of the phenotypic mutants in a mutated population (**see Chapter 25**), it would therefore be relatively simple to produce semi-dwarf types in species new to agriculture. Leaf colour and male sterile mutants are also relatively common mutant classes which are easily produced by induced mutagenesis and have been exploited in various species, e.g. pale green tobacco varieties. For mutant traits that are less common high-throughput screens may be required (**see Chapter 41**).

4. Experimental Mutagenesis in Crop Genomics

Gene identification and the functional analysis of allelic variation for a particular trait are two major fields in plant genomics research. The use of T-DNA insertion mutants and transposon mutants are considered to be the most direct means for filling the genotype–phenotype gap and directly connect the (disrupted) gene with the mutated traits. This is commonly referred to as reverse genetics, as it begins with an altered gene and works towards the effects on traits. However, these tools are not always applicable to all plant species, and they have inherent limitations, for example, most of the mutants produced by T-DNA insertion are either knock-out mutants or over-expression mutants (when the T-DNA with a strong promoter is inserted into the flanking sequence of a gene) and hence it is very difficult to study the effects of altered sequence variation (different alleles of a gene). In addition many crop plants are recalcitrant to transformation technologies and are excluded from such analyses. The use of T-DNA insertion mutants and transposition mutagenesis in functional genomics are discussed further in **Chapters 38 & 39**.

Mutants generated through induced mutagenesis have been used in genetics studies from the very beginning. In the (functional) genomics era, induced mutagenesis has become an important tool in plant genomic research, which can be explored by using both forward and reverse approaches (**Figure 1.3**), more details are available in **Chapters 35–37**.

4.1. Forward Genetics Approaches

Forward genetics works from traits (phenotypes) to genes and is the typical approach in plant breeding, genetics and genomics studies. Once a mutant phenotype is identified, the underlying mutated gene can be cloned through mapping, fine mapping and positional cloning.

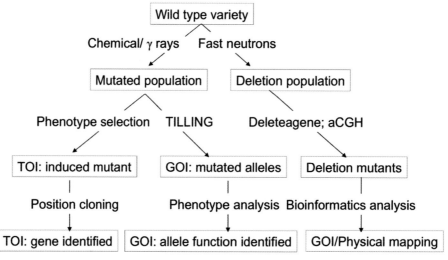

The Joint FAO/IAEA Programme

Figure 1.3 Induced mutagenesis in plant genomics research. TOI: traits of interest; GOI: gene of interest.

The detailed steps for cloning a mutated gene are given in **Chapter 23**. Although this is not as straightforward as the reverse genetics approach, it is the only valid method for identifying new genes that have no known gene homologues.

Positional cloning of mutated genes will become more and more efficient with the development of comprehensive genome sequencing that is ongoing in more and more plant species. Development of sequencing and microarray technologies will also facilitate the identification of mutated genes.

4.2. Reverse Genetics Approaches

Induced mutagenesis was almost irrelevant to reverse genetics before the development of "Targeting Induced Local Lesions in Genomes" (TILLING) and similar generic reverse genetics strategies. This approach can be used both for functional genomics and practical breeding. In plant genomics research, in addition to identifying the function of a particular gene in a given plant species, the effect of various mutant alleles can be assessed using TILLING technologies. Details of these and modified versions and their application are described in **Chapters 20–22**.

5. Future Prospects

The past decade has witnessed several advancements in high-throughput genotyping. The ability to interrogate the genetics of populations and individuals in detail resulted in a resurgence in experimental mutagenesis as it provided a means of determining gene function. Progress in genotyping has recently taken another giant leap forward with the increasing ease in obtaining sequence data, the "sequencing revolution". A range of "next generation" sequencing platforms exists: Roche's 454 GS FLX, Illumina's Genome Analyzer and Applied Biosystem's SOLiD system (Smith *et al.*, 2008), and new technologies are being developed in the quest for rapid sequence data generation. Today, sequencing factories boast the ability to sequence 20 human genomes in one day. The routine use of "next generation" platforms in the comprehensive sequencing of whole genomes will provide powerful tools in our understanding of genes and their functions *via* mutant characterization. The sequencing revolution will demand robust support

systems and there will be a need for standard reference genotypes of crop species. The increased rate of data acquisition will demand improved computer power and bioinformatics, programming and statistical expertise if the data are to be exploited fully. Since sequencing is a generic technology it may be applied to any organism and is expected to spread rapidly to encompass all crop species. Reference (wild type) genotypes for each crop will be needed to be agreed upon by the scientific community in order to identify mutant sequences by comparison with wild type.

The ability to determine gene function depends on associating genes with their end products, i.e. phenotypes. Our abilities to explore the genotype and build genomic data are at unprecedented levels and far exceed those of phenotyping. Indeed the unprecedented levels of information that the sequencing revolution is now generating widens the genotype–phenotype gap further. In plant breeding terms the phenotype is all important, i.e. the yield of a crop, its quality, responses to pests and diseases and abiotic stressors such as heat, cold, drought and salinity, and (in today's environmentally conscious context) the crop's carbon footprint and hydrogen yield. Linking genotype to phenotype remains a major challenge. The weakness in the scope and depth of phenotypic data has been recognized and several high-throughput phenomics facilities are under development around the world (**see Chapter 41**). These are generally based on artificial environment conditions and take non-destructive measurements of plants at various stages during their development and in response to imposed conditions, e.g. biotic and abiotic stressors (**see Chapter 41**). However, performance in the field remains the acid test for a crop.

To date the vast majority of experimental mutagenesis has focused on genetic effects, i.e. the alternation of the genetic code and the knock-on effects on phenotype. Scientists however are becoming increasing aware that phenotypic traits can be inherited by non-genetic means, i.e. *via* epi-genetic phenomena. Epi-genetics has been dubbed the "second code" whereby the genetic code is influenced by factors such as DNA methylation and gene silencing through DNA binding with small nuclear RNAs. Epi-genetics is a new and exciting discovery, currently there are few data on the effects of physical and chemical mutagens on epi-genetic factors though the mode of action would suggest these could be significant.

6. References

6.1. Cited References

Ahloowalia, B.S., Maluszynski, M. and Nicjterlein K. 2004. Global impact of mutation-derived varieties. *Euphytica*. 135: 187–204.

Auerbach, C. and Robson, J.M. 1944. Production of mutations by allyl isothiocyanate. *Nature*. 154: 81.

Brooks, S.A., Yan, W.G., Jackson, A.K. *et al.* 2008. A natural mutation in rc reverts white-rice-pericarp to red and results in a new, dominant, wild-type allele: Rc-g. *Theor Appl Genet*. 117: 575–580.

de Vries, H. 1901. Die Mutationstheorie I. Leipzig: Veit & Co.

de Vries, H. 1903. Die Mutationstheorie II. Leipzig: Veit & Co.

de Vries, H. 1905. Species and Varieties: Their Origin by Mutation. Chicago: The Open Court Publishing Company.

Freisleben, R.A. and Lein, A. 1942. Über die auffindung einer mehltauresistenten mutante nach röntgenbestrahlung einer anfälligen reinen linie von sommergerste. *Naturwissenschaften*. 30: 608.

Freisleben, R.A. and Lein, A. 1944. Möglichkeiten und praktische durchführung der mutationszüchtung. *Kühn-Arhiv*. 60: 211–22.

Gager, C.S. and Blakeslee, A.F. 1927. Chromosome and gene mutations in Datura following exposure to radium rays. *Proceedings of the National Academy of Sciences of the USA*. 13: 75–59.

Huang, C. and Liang, J. 1980. Plant breeding achievements in ancient China. *Agronomic History Research*. 1: 1-10 (in Chinese).

Kusaba, M., Miyahara, K., Lida, S. *et al.* 2003. Low glutelin content 1: A dominant mutation that suppresses the glutelin multigene family *via* RNA silencing in rice. *Plant Cell*. 15: 1455–1467.

Ríos, G., Naranjo, M.A., Iglesias, D.J. *et al.* 2008. Characterization of hemizygous deletions in Citrus using array-comparative genomic hybridyzation and microsynteny comparisons with the popular genome. *BMC Genomics*. 9: 381.

Smith, D.R., Quinlan, A.R., Peckham, H.E. *et al.* 2008. Rapid whole-genome mutational profiling using next-generation sequencing technologies. *Genome Research*. 18: 1638–1642.

Sparrow, A.H. and Singleton, W.R. 1953. The use of radiocobalt as a source of gamma rays and some effects of chronic irradiation on growing plants. *The American Naturalist*. 87: 29–48.

Stadler, L.J. 1928a. Genetic effects of X-rays in maize. *Academy of Sciences of the USA*. 14: 69–75.

Stadler, L.J. 1928b. Mutations in barley induced by X-rays and radium. *Science*. 68: 186–187.

van Harten, A.M. 1998. Mutation Breeding. Theory and Practical Applications. Cambridge: Cambridge University Press.

van Harten, A.M. and Broertjes, C. 1989. Induced mutations in vegetatively propagated crops. *Plant Breeding Review*. 6: 55–91.

von Sengbusch. 1927. Süsslupinen und öllupinen. *Landwirtschaftliches Jahrbuch*. 91: 723–880.

6.2. Websites

Australian Plant Phenomics Facility:
http://www.plantphenomics.org.au

The IAEA/FAO mutant variety genetic stock database:
http://mvgs.iaea.org

Lemna Tec high-throughput plant phenotyping systems:
http://www.lemnatec.com

6.3. Further Reading

Blakeslee, A.F. 1935. Hugo de Vries 1848–1935. *Science*. 81: 581–582.

Bronowski, J. 1973. The ascent of man. BBC.

Cramer, P.J.S. 1907. Kritische übersicht der bekannten fälle von knospenvariation. Natuurkundige verhandelingen der Hollandische maatschappij van wetenschappen, Haarlem, 3.6.

Darwin, C. 1859. The Origin of the Species by Means of Natural Selection. London: John Murray.

Darwin, C. 1868. The Variation of Animals and Plants under Domestication. In two volumes. 10th impr. of the 2nd edition (1921). London: John Murray.

Dubcovsky, J. and Dvorak, J. 2007. Genome plasticity a key factor in the success of polyploid wheat under domestication. *Science*. 316: 1862–1866.

Hagberg, A. and Åkerberg, E. 1962. Mutations and polyploidy in plant breeding. Stockholm Bokförlaget Bonniers.

Mendel, G. 1865. Versuche uber Pflanzen-Hybriden. Vesh. Naturforschung Ver. In *Brunn Verh*. 4: 3–47.

Muller, H.J. 1927. Artificial trans-mutation of the gene. *Science*. 66: 84–87.

A Brief History of Plant Mutagenesis

M.C. Kharkwal*

Division of Genetics, Indian Agricultural Research Institute, New Delhi 110 012, India
Present: C - 8 / 269 B, Keshav Puram, New Delhi – 110 035, India
* Corresponding author, E-MAIL: mckharkwal@gmail.com

1. Introduction

Mutation is a natural process that creates new variants (alleles) of genes. The variation so created is further amplified by recombination of alleles on homologous chromosomes and their independent assortment at meiosis. Mutation is the primary source of all genetic variations existing in any organism, including plants. Variation so created provides the raw material for natural selection and a driving force in evolution. Spontaneously arising mutations are very rare and random events in terms of the time of their occurrence and the gene in which they occur. In this way mutant forms showing both large and small effects on the phenotype arise for all kinds of traits. Many of the mutations may be deleterious making the organism less adapted to its environment and some may even be lethal. Some may be neutral in their effects and may confer no immediate advantage, but may help to generate a wide range of useful recombinant genotypes through the subsequent process of independent segregation and crossing over of genes. Others are of direct and immediate benefit to the plant. It is this variability, created through natural mutations and amplified by subsequent recombination of genes during sexual reproduction, on which natural selection operates to bring about evolution of new races and species.

In the book *Mutation Breeding – Theory and Practical Applications* van Harten (1998) has brought out the important historical developments that contributed to the conceptualization and maturation of mutation breeding as a scientific discipline as well as applications of plant mutation research. In this chapter, the history of plant mutation research and application to breeding is extended by adding a fifth period to van Harten's four-period scheme (**Box 2.1**); the historical background, the important discoveries and the major events that happened in the first half of the 20th century are described. Information about the technological development of biotechnologies and genomics that have transformed mutation research into the molecular paradigm is available in the relevant chapters of this book and hence is not elaborated here.

2. Distinct Periods of Plant Mutation Research and Application

The history of plant mutation, according to van Harten, could be traced back to 300 BC with reports of mutant crops in China. Based on featured research activities, important scientific discoveries and momentous events, van Harten grouped the history of mutation breeding into four distinct periods (**Periods I to IV, Box 2.1**).

van Harten wrote in his book: *"The 1990 Joint FAO/ IAEA symposium, in our opinion, marks the end of fourth period in which mutation breeding has clearly proved its worth, but also reached its peak as a subject of research"*. Although he was not able to speculate about the future of this subject, numerous landmark developments, both scientifically and technologically, have since then greatly advanced the subject, hence a fifth period featuring the integration of plant mutation with biotechnology and genomics has been added (**Box 2.1**).

In the late 19th century, while experimenting on the "rediscovery" of Mendel's laws of inheritance, Hugo de Vries (1889) found variation in evening primrose (*Oenothera lamarkiana*) and snapdragon (*Antirrhinum*) which did not follow Mendelian patterns of inheritance (3:1), but was nevertheless heritable. Mutation as a mechanism of creating variability was first identified by Hugo de Vries in 1901 and he considered them as heritable changes by mechanisms distinct from recombination and segregation. De Vries (1901, 1903), who is credited with the discovery of mutation, described these events as suddenly arising changes in organisms, which were inherited and produced relatively large effects on the phenotype. He coined the term "mutation" and presented an integrated concept concerning the occurrence of sudden, shock-like changes (leaps) of existing traits, which lead to the origin of new species and variation. He also clearly included "small effects" in this definition of mutation. In his experiments with evening primrose, de Vries observed many aberrant types, which he called "mutants". The notion of mutation which was used by de Vries to indicate sudden genetic changes as a major cause of evolution quickly became established after the publication of his great *"Mutationstheorie"*, 1901–1903. Because of de Vries's concept of mutation as the source of genetic variation and his early ideas about their potential value for plant breeding, his work around the turn of the 20th century may be marked as the starting point in the history of mutation techniques in the disciplines of plant breeding and genetics.

3. Classical Mutation Induction Experiments

Before the discovery of X-ray mutagenesis by Muller, there was little to be said about the causes of mutation except that they were rare, sudden and discrete events that cause "genes" as Johannsen (1909) called them, to pass from one stable state to another. In his experiments with common bean, Johannsen (1913) described not only spontaneous drastic mutations, but also slight mutations affecting the seed index (the ratio of seed width to seed length). This is a character which falls into the class of continuous variation, and hence Johannsen may be regarded as the first who really proved the existence of spontaneous mutations with small effects. Baur (1924) also emphasized repeatedly the importance of small mutations, which he called "Kleinmutationen", in evolution processes resulting from the accumulation of a large number of mutations, the majority of which exert only slight effects.

In his publications, De Vries (1901, 1903) suggested that the new types of radiations like X-rays and gamma rays discovered by Konrad von Roentgen (1895), Henry Becquerel (1896) and Pierre and Marie Curie (1897/1898)] might be greatly useful to induce mutations artificially. This was first realized through experiments with *Drosophila*, followed by plant experimental mutagenesis.

3.1. Induction of Mutations in Drosophila

The early extensive work on mutations in fruit fly (*Drosophila melanogaster*) especially that of Morgan (1910) on the basis of careful visual observations uncovered more than 100 physical abnormalities through mutations. He furnished several instances of the occurrence of new dominant genes, and many new sex-linked recessive mutants, in pedigree material where the event could be analysed in some detail. These examples confirmed that mutation can occur in a single gene in a single cell, and that it can occur at any stage of development. In fact, Herman Joseph Muller, a student of Morgan concluded that the frequency per cell is probably the same for each stage in the germ line. However, these results were merely qualitative, since the frequencies were too low for a quantitative study and were also strongly subjective. What was needed was an objective index and one that would recognize a class of mutations that was frequent enough to give significant numerical values. Both of these requirements were met in the elegant mutation experiments devised by H.J. Muller in the study of the newly discovered sex-linked lethals in *Drosophila*.

Muller's mutation technique depended on the study of the sex ratios from individual females that were heterozygous for sex-linked "marker" genes. This technique allowed unambiguous determination of the frequency of the sex-linked lethals, but it was very laborious, since detailed counts had to be made for each tested chromosome. Muller published a brief report in *Science* (Muller, 1927) to establish priority for his discovery of X-ray mutagenesis on fruit fly (*Drosophila melanogaster*) before leaving to attend the 5th International Congress of Genetics (ICG) to present his work on X-ray induced mutagenesis. The first paper on the use of mutation techniques and the discovery of induced mutagenesis presented at the 5th ICG, Berlin in 1927 was Muller's comprehensive report on his X-ray work on fruit fly, which appeared with the sensational title: "Artificial transmutation of the gene". This was an important discovery with far-reaching implications for genetics. Muller demonstrated an elegant technique that he had developed for scoring as well as determining rates of mutations in *Drosophila*. He reported that mutation rates of sex-linked recessive lethals could be greatly increased in *Drosophila* following treatment of sperm in male flies with high doses of X-rays. He observed that about one-seventh of the flies bred from these males contained individually detectable mutations in their X chromosome. His technique revealed that genes generally show a mutation rate of 10^{-5}–10^{-6} per locus per generation. This means that one out of 100,000 copies of the wild type gene can mutate in the course of one generation. He also introduced the concept of generation time while considering mutation rates in different organisms. Muller's discovery of rates of mutation has contributed enormously to our understanding of genes and their evolution.

Muller (1930) improved the mutation technique by using the "ClB" chromosome that he found in his experiments. This is an X chromosome with an inverted segment (*C*), which carries a recessive lethal (*l*) and the dominant mutant gene "*Bar*" for bar eye (*B*). The inverted segment (*C*) acts as a crossover suppressor and as a result, the *C*, *l* and *B* are always inherited together in this chromosome. This is now a classic text book example. The method made it possible to detect new sex-linked lethals without anaesthetizing the flies or making counts – which were replaced by simple and rapid examination of individual culture bottles. Through this technique one could test large number of

Box 2.1: History of plant mutation research and application[a]

Period I: Observation and documentation of early spontaneous mutants

300 BC	Early mutant crops in China.
1590	The "incisa" mutant of *Chelidonium majus*.
1672	Variability in plants.
17th century	"Imperial Rice" in China: a spontaneous mutant?
Late 17th century	Spontaneous mutant for the ornamental "morning glory" (*Ipomoea nil*) in Japan.
1741 and following years	Description of various mutants by Carl von Linné.
1859	*The Origin of Species* published by Charles Darwin.
1865	E.A. Carriére publishes his book: *Production et Fixation des Variétes dans les Végétaux*.
1894	W. Bateson published: *Materials for the Study of Variation, Treated with Special Regard to Discontinuity in the Origin of Species*.

Period II: Conceptualization of mutation and mutation breeding

1895–1900	The discovery of various kinds of radiation (X-rays, α, β and γ radiation).
1897–1908	Early work on irradiation of plants: mostly physiological effects and damage to nuclei and cell division.
1901	Hugo de Vries, coined the term "mutation" for sudden, shock-like changes of existing traits. *Die Mutationstherorie* of Hugo de Vries published. *Theory of Heterogenesis* published by S. Korschinsky.
1901 and 1911	First proof of mutations induced by chemicals in bacteria.
1904 and 1905	Hugo de Vries suggests artificial induction of mutations by radiation.
1907	P.J.S. Cramer's work on bud variations.
1909–1913	W. Johannsen describes spontaneous drastic mutations and slight mutations affecting seed index.
1910	Thomas Hunt Morgan: first mutation experiments with *Drosophila melanogaster*.
1920	N.I. Vavilov's "law of homologous series of variation".

Period III: Proof of induced mutations and release of the first commercial mutant varieties

1926	N.I. Vavilov's theory on gene diversity centres or "Centres of Origin".
1927	C. Stuart Gager and A. F. Blakeslee report on induction of mutations in *Datura stramonium*. Definite proof of mutation induction by X-rays by H.J. Muller, indicating the possibility of obtaining genetically superior plants, animals and man by applying X-radiation.
1928	Successful induction of mutations after irradiation of barley and maize by Lewis John Stadler.
1928–1934	Continued studies on mutation theory and practical applicability.
The 1930s	Start of the Swedish mutation research programme by Åke Gustafsson.
1934–1938	The first commercial mutant variety "Chlorina" obtained after X-radiation in tobacco by D. Tollenar released in Indonesia.
1934 and following year	The physical mutation theory – Hit and Target Theory – was established by N.W. Timoféeff-Ressovsky and co-workers.
1937	The chromosome doubling effect of colchicine on plant chromosome.
1941	Chemical mutagenesis: C. Auerbach, I.A. Rapoport, F. Oehlkers and others.
1942	First report on X-ray-induced resistance in barley.

Box 2.1: History of plant mutation research and application[a]

1951	Barbara McClintock reported controlling elements (later established as transposable genetic elements or transposons).
1953	The Watson–Crick model of the gene.
The early 1950s	Mutations induced by gamma rays were produced in gamma field by chronic irradiation, but the first evidence that gamma rays do induce mutation was not clear.
1956	E.R. Sears transfers resistance from *Aegilops* to wheat by radiation-induced translocation.
1958 and following year	Application of chemical mutagens on higher plants.

Period IV: Large-scale application of mutation breeding

1964	Establishment of the Joint FAO/IAEA Division of Nuclear Techniques in Food and Agriculture at Vienna, Austria; Internationally coordinated mutation breeding research programmes were launched.
1960s	Numerous national research institutes specialized in Nuclear Techniques in Food and Agriculture were established.
1969	The Pullman Symposium on Induced Mutations in Plant Breeding: The first classified list of mutant varieties published.
1981	First major symposium on the "Use of Induced Mutations as a Tool in Plant Research" organized by FAO/IAEA at Vienna, Austria.
1990	Joint FAO/IAEA symposium in Vienna, Austria to assess results of 25 years of applied mutation breeding.

Period V: Integration of plant mutation with biotechnology and genomics

1983	Four groups independently reported the production of first transgenic plants, laying down the basis of T-DNA insertion mutagenesis.
1983	The transposable controlling elements Ac and Ds were isolated, laying down the basis of transposon mutagenesis using Ac-Ds and modified genetic systems.
1997	Retrotransposons were re-activated *via* tissue culture in rice, which prompted the establishment of the large Tos 17 mutant collection.
2000	The first plant genome (*Arabidopsis* genome) was sequenced. Methodology for targeted screening of induced mutations established, which is now widely known as TILLING (Targeting Induced Local Lesion in Genomes).
2002–2005	Genomes of the *indica* and *japonica* rice subspecies were sequenced.
2005	Establishment of mutant populations for functional genomics studies, including TILLING and T-DNA insertion mutant populations in crop plants.
2008	International Symposium on Induced Mutations in Plants in Vienna, Austria to assess applications of induced mutation in plant mutation research and breeding in the genomics era.

[a]Adapted from A.M. van Harten (1998) for Period I – IV with addition/deletion of a few events.

X chromosomes and thus could obtain adequate data on the lethal frequencies.

When Muller applied his "*ClB*" technique to the study of gametes treated with X-rays, it was apparent that there was a marked increase in the frequency of newly arising lethals. This then is a clear example of the use of mutation for artificial induction of mutations. Muller concluded that high energy radiation is dangerous not only to the exposed individual, but also to their descendants.

3.2. Stadler's Pioneering Work on Mutation Induction in Plants

Induced mutations through radiation as a tool for generating novel genetic variability in plants took off only after the discovery of the mutagenic action of X-rays, demonstrated in maize, barley and wheat by Stadler in 1928 and 1930. Stadler had begun his studies with barley at about the same time when Muller began working with X-rays, but since he was using an annual plant species, his results lagged behind and were not available until after Muller's findings were published.

Stadler, widely recognized for his significant contributions in the field of mutation breeding, irradiated seeds of barley in the course of his experiments. At the time of treatment, the cells destined to give rise to different tillers were already differentiated in the embryo. Each of the several shoots of the developing plant was represented by a single diploid cell. He observed that radiation treatment can induce mutations in one tiller without affecting the others. The induced mutations were related to chlorophyll defects, resulting in white, yellow or virescent seedlings which could be scored readily. When the plant was mature, each shoot was separately self-pollinated and the seed planted. Any induced recessive mutant segregated in a 3:1 ratio and could be recognized as a new variant, since it would not appear from other shoots of the same plant. In all, Stadler found 48 mutations in distinct seedling characters from irradiated seeds, and none from a large control series. Some of the seeds were treated with X-rays while others were exposed to radium – the latter also proving to be mutagenic. Stadler (1941) went on to introduce a new mutagenic agent, ultraviolet light, into his experimental laboratory.

The three major discoveries made by Stadler's mutation work in 1928 with barley are:

1. Seeds soaked in water to initiate germination gave almost eight times more mutations than dry dormant seeds.
2. Mutation rate was independent of the temperature at the time of irradiation.
3. The relation of mutation rate to total dosage was linear – doubling the dose doubled the mutation rate.

The exhaustive results of the pioneering work of Muller and Stadler unambiguously proved that mutations can be artificially induced.

4. Establishment of Mutation Induction Techniques

After X-ray mutagenesis was established in both animals and plants, numerous new mutagens, both physical and chemical, were found to be effective in generating genetic variability.

4.1. Physical Mutagenesis

Hanson and Heys conducted detailed studies on the induction of lethal mutants in fruit fly by using radium (Hanson and Heys, 1929). They interposed lead shields of different thickness and recorded the ionization in each treatment. The curves for ionization and mutation rate could be superimposed leading to the conclusion that ionization is responsible for the mutations and that the relation is a simple, direct one – a "one hit" phenomenon. Oliver (1930) confirmed this conclusion by showing that varying duration of exposure to a constant X-ray source also gave a linear curve relating to mutation rate. It was also apparent that projecting this curve to a zero dosage did not give zero mutations. Muller showed in the same year that the amount of background radiation is far less than would be required to produce the normal "spontaneous" frequency of mutations. It came to be generally accepted that the total ionization is all that needs to be considered in connection with radiation-induced mutations, at least within a strain.

The mutagenic effect of ultraviolet light was discovered by Altenberg (1934) through irradiation of the polar cap cells of fruit fly eggs. The mutagenic potential of these rays has since been confirmed in many organisms in which germ tissue could be easily exposed to the low-penetrating ultraviolet light. With the wide establishment of Cobalt-60 irradiation facilities, gamma radiation has become a popular mutagen since the 1950s (**see Chapter 1**). Various forms of neutron had also been studied extensively for their use in mutagenesis in the 1960–70s. Though it has proved to be an effective mutagen, particularly for producing large DNA fragment deletions, the application of neutrons in induced mutagenesis has been limited.

During the past two decades, ion beams – either through implantation or irradiation have become a new type of physical mutagen (**see Chapters 9 and 10**). More recently, plant materials have been sent out into aerospace to study the intricacies of mutation induction in space. It has been speculated that the special environment of space flight, such as cosmic radiation, microgravity, weak geomagnetic field, etc. contains the potential agents of mutation induction. However, knowledge of the underlying genetics of aerospace mutagenesis is so far scarce.

4.2. Chemical Mutagenesis

Induction of mutations by chemical agents was attempted by many people over a long period, but there were no clear or convincing positive results until 1939 when Thom and Steinberger found that nitrous acid was effective in causing mutations in *Aspergillus*. Auerbach (1941) was the first to report that mustard gas (*1,5-dichloro-3-thiapentane*) had a mutagenic effect on fruit fly, which was similar to that of X-rays on plants. Auerbach and Robson (1946) later obtained clear evidence to show that mustard gas is mutagenic. Chemical mutagens were found to be highly effective in inducing true gene mutations and the specificity of action could be investigated through analysis of their reaction with different DNA bases. However, the question of whether chemical agents do indeed produce mutations with the same frequency as the physical mutagens (like the ionizing radiations) was settled after the first paper published by Auerbach and Robson (1946). They used the standard technique devised by Muller to score recessive and visible gene mutations in *Drosophila* following exposures of flies to a predetermined dose of the gas. Their most important observation was that mustard gas is highly mutagenic and capable of producing lethal and visible mutations at rates comparable with the effect of X-rays. Besides gene mutations, chromosomal aberrations in the form of deletions, inversions and translocations were produced. Through such extensive investigations mustard gas was found to be a very potent mutagen in *Drosophila*. Oehlker (1943), and Gustafsson and Mackey (1948) proved that mustard gas was mutagenic in barley.

Rapoport (1946, 1948) and others in Russia also discovered and demonstrated mutagenic effects of mustard gas and several other chemicals such as form-aldehyde, diethylsulphate, diazomethane and other compounds, and established that alkylating agents are the most important group of chemical mutagens. Since then hundreds of chemical agents belonging to several groups such as alkylating agents, nitroso compounds, base analogues, azide, acridine dyes, etc., have been found to produce mutagenic activity in a range of organisms (**see Chapters 12 – 13**).

4.3. Mutant Generation *via* Biotechnologies

In addition to mutations induced through physical and chemical mutagenesis, a few other approaches have been established for the production of mutants during the past two decades, which has become one of the major developments of the fifth period of plant mutation research. These methods were developed based on either new biotechnologies such as genetic transformation or discovery of genetic elements such as transposons and retrotransposons. The recently emerged target-selected mutagenesis or site-directed mutagenesis is a more sophisticated approach based on detailed insights on DNA repair mechanism. Detailed information about the history and recent development of these technologies is given in **Chapters 38–40**.

5. Understanding the Mechanisms of Mutation

With the mutagenic action of alkylating agents firmly established in the 1950s, the future direction of work on chemical mutagens was determined mainly by rapid advances in the understanding of gene structure and function following the demonstration of DNA as the genetic material. Watson and Crick while proposing their model for the structure of DNA in their celebrated 1953 paper pointed out that both replication and muta-tion of genes can be understood in terms of the new structure. It now became possible to plan studies with compounds which were found to react with and modify the DNA bases in specific ways. Thus, nitrous acid, based on chemical studies, was expected to convert cytocine to uracil and adenine to hypoxanthine, two of the known analogues of the nucleic acid bases. Among the various kinds of mutational changes at the molecular level are base substitutions, a term meaning simply nucleotide changes that involve substitution of one base for another. This can happen through mis-pairing of the base ana-logue in the treated DNA during replication, leading to mutation through transitions when exchanges occur either between purines (A–G) or between pyrimidines (T–C) and transversions when purines are exchanged for pyrimidines or *vice versa* (A, G–T, C). Freese (1959) reported extensive studies on reverse mutations resulting in wild type phenotype in *rII* mutants by two base ana-logues – 5-bromouracil (BU) and 2-amino-purine (AP). These base analogues were found to be highly mutagenic in viruses of the T4 series through their action in inducing A:T to G:C transitions. Transitions and transversions are the simplest kinds of base pair changes, but they may result in phenotypically visible mutations. There are no restrictions on the different kinds of sequence changes in the DNA of a gene following different types of misprints during replication. Another common error would be addition or deletion of a nucleotide base pair when one of the bases manages to pair with two bases or fails to pair at all. These kinds of sequence changes resulting in an alteration in the reading frame of the gene's DNA are known as frameshift mutations. They are more drastic in their effect as they may completely change the message of the gene starting with the point of deletion/addition. Some of the mutations occur from rearrangement of bases in the DNA. A small or large sequence of bases may be inverted as a result of chromosome breakage, and reunion of the broken ends may involve different DNA molecules in a reciprocal rearrangement or in loss of a fragment. Duplication of a DNA sequence is another common mechanism for change in the structure of a gene leading to gene mutation. Detailed information about radiobiology, DNA damage and repair is given in **Chapters 5, 6 and 11**.

6. Applications of Induced Mutation to Plant Breeding

In *Drosophila*, experimental mutagenesis has been used for genetic studies since the very beginning and for gene identification since the mid-20th century. In plants, although induced mutants have also contributed to the establishment of genetic maps or linkage groups of a few crops, its application had been mainly oriented for breeding new crop varieties till less than a decade ago. With the development of molecular biology and emergence of new genomic tools, understanding of the biological process of mutagenesis has greatly deepened. The applications of induced mutation have since been significantly expanded and its efficiency significantly increased.

Though Stadler (1928) was sceptical about the potential of induced mutations in crop breeding, extensive work following his work showed the practical potential of radiation as a plant breeding tool and resulted almost immediately in the recovery of some economically useful mutants in wheat (Sapehin, 1930 and Delaunay, 1931). Stubbe (1934) described "small mutations" for the first time in higher plants such as *Antirrhinum*, the snapdragon. Tollenaar (1934) was the first worker to isolate a light green *"Chlorina"* mutant in tobacco, which was released for commercial cultivation. Later, Freisleben and Lein (1942) reported the induction of mildew resistance in barley by X-irradiation.

A major stimulus for much of the subsequent work in further practical refinement of mutation technique was reported in barley by Swedish scientist A. Gustafsson (1947, 1963). In his study, a very large number of mutations, e.g. chlorophyll characters, short stature, stiffness of straw and dense ears were induced in different genotypes. The Swedish mutation programme led by him covered several crop species and generated a lot of information on successful use of mutation techniques in breeding and was influential in instigating mutation breeding in other crops (Lundqvist, 2009).

During 1950–70, several countries including China, Germany, India, the Netherlands, Sweden, UK, USA and Japan took up extensive crop improvement programmes through the use of induced mutagenesis and made spectacular accomplishments in evolving several superior mutant varieties in a large number of crop species (Kharkwal *et al.,* 2004).

The following selected examples amply substantiate the important role that mutation techniques have played and continue to play in plant breeding. The rapeseed variety "Regina II" was developed by mutation in Sweden and was released in Canada in 1953. "Redwood 65" flax (registered in 1965) was derived from a mutation programme at the University of Saskatchewan and is present in the pedigrees of many western Canadian flax varieties. The low linolenic traits in both flax and canola were more recently produced through mutation breeding techniques. "Pursuit Smart" canolas are also the product of mutagenesis. Elsewhere in the world, the FAO estimated in 1994 that almost 70% of the durum wheat in Italy was mutant varieties and that there were 400 rice varieties derived from mutagenesis programmes. In Europe two barley mutants, "Golden Promise" and "Diamant" provided dwarfing genes for most barley. The mutant gene, *mlo*, has provided resistance to mildew for many years in barley. The photoperiodic mutation in "Mona" barley allowed barley cultivation to be expanded into equatorial countries of short day length. "Michelite", an X-ray-induced white (common) bean mutant with altered plant type, is in the pedigrees of most of the white beans grown in North America. In summary, breeding programmes based on efficient mutation techniques have been widely used by plant breeders and many of our food crops are derived either directly or indirectly from such programmes (Kharkwal and Shu, 2009). A detailed description of major developments in plant mutation breeding and its application is given in **Chapters 1 and 24**.

During the early part of the era of induced mutagenesis, the technique was used as a tool for improvement of traditional traits like yield, resistance to disease and pests, etc. in various agricultural crops. Subsequently the emphasis shifted to more diversified uses of crop end-products, enhancing quality and nutritional value, tolerance to abiotic stresses, etc. During the recent decades, thanks to the tremendous progress happening in the research of plant molecular biology and biotechnology, particularly plant genomics, we are witnessing new impulses in plant mutation research, from fundamental studies of mutagenesis to reverse genetics. Breeders are now aware of the newer potentialities and far reaching implications of induced mutation techniques and are able to use them with more sophistication and efficiency than ever before dreamed possible.

7. References

7.1. Cited References

Altenberg, E. 1934. The artificial production of mutations by ultraviolet light. *Amer. Naturalist.* 68: 491–507.

Auerbach, C. 1941. The effect of sex on the spontaneous mutations rate in *Drosophila melanogaster. J. Genet.* 11: 255–265.

Auerbach, C. and Robson, J.M. 1946. Chemical production of mutations. *Nature.* 157: 302.

Baur, E. 1924. Untersuchungen ueber das Wesen, die Entstehung und die Vererbung von Rassenunterschieden bei *Antirrihnum majus. Bibl. Genet.* 4: 1–170.

De Vries, H. 1901. *Die Mutationstheorie.*I. Leipzig: Veit & Co. Leipzig, Germany (English translation, 1910. The Open Court, Chicago.)

De Vries, H. 1903. *Die Mutationstheorie.*II. Leipzig: Veit & Co. Leipzig, Germany (English translation, 1910. The Open Court, Chicago.)

Freisleben, R., and A. Lein. 1942. Uber die Auffindung einer mehltauresistenten mutante nach Roentgenbestrahlung einer anfaelligen reinen Linnie von Sommergerste. *Naturwissenschafen.* 30: 608.

Freese, E. 1959. On the molecular explanation of spontaneous and induced mutations. *Brookhaven Symp. Biol.* 12: 63-73.

Gustafsson, A. 1947. Mutations in agricultural plants. *Hereditas.* 33: 1-100.

Gustafsson, A. 1963. Productive mutations induced in barley by ionizing radiations and chemical mutagens. *Hereditas.* 50: 211-263.

Gustafsson, A. and Mackey, J. 1948. The genetic effects of mustard gas substances and neutrons. *Hereditas.* 34: 371-386.

Hanson, F.R. and Heys, F. 1929. An analysis of the effects of the different rays of radium in *Drosophila. Amer. Nat.* 63: 201-213.

Kharkwal, M.C. and Shu, Q.Y.. 2009. Role of induced mutations in world food security. pp. 33-38. *In*: Q.Y. Shu (ed.) Induced Mutations in the Genomic Era. Rome: Food and Agriculture Organization of the United Nations.

Kharkwal, M.C., Pandey, R.M. and Pawar, S.E. 2004. Mutation Breeding in Crop Improvement. pp. 601-645. *In*: H.K. Jain and M.C. Kharkwal (ed.) Plant Breeding – Mendelian to Molecular Approaches. New Delhi: Narosa Publishing House.

Lundqvist, U. 2009. Eighty years of Scandinavian Barley Mutation Research and Breeding. pp. 39–43. *In* : Q.Y. Shu (ed.) Induced Mutations in the Genomic Era. Rome: Food and Agriculture Organization of the United Nations.

Morgan, T.H. 1910. Sex limited inheritance in *Drosophila. Science.* 32: 120–122.

Muller, H.J. 1927. Artificial transmutation of gene. *Science.* 66: 84–87.

Muller, H.J. 1930. Types of visible variations induced by X-rays in *Drosophila. J. Genetics.* 22: 299–333.

Oehlker, F. 1943. Chromosome mutation in meiosis by chemicals. *In*: C. Auerbach (ed.) (1976), Mutation Research – Problems, Results and Prospects. London: Chapman and Hall.

Oliver, C.P. 1930. The effect of varying the duration of X-ray treatment upon the frequency of mutation. *Science.* 71: 44.

Rapoport, I.A. 1946. Carbonyl compounds and the chemical mechanism of mutation. *C.R. Doklady Acad. Sci. USSR.* 54: 65.

Rapoport, I.A. 1948. Alkylation of gene molecule. *C.R. Doklady Acad. Sci. USSR.* 59: 1183–1186.

Sapehin, A.A. 1930. Rontgen-mutationen beim Weizen (*Triticum vulgare*). *Der Zuchter*. 2: 257–259.

Stadler, L.J. 1928. Genetic effect of X-rays in maize. *Proc. Natl. Acad. Sci. USA*. 14: 69–75.

Stadler, L.J. 1930. Some genetic effects of X-rays in plants. *J. Hered*. 21: 3–19.

Stadler, L.J. 1941. Genetic studies with ultraviolet radiation. pp. 269–276. *In*: R.C. Punnet , Proc. Seventh Intl. Congress of Genetics. Cambridge: Cambridge University Press.

Stubbe, H. 1934. Einigekleinmutantionen von *Antirrhinum majus* L. Zuechter 6: 299–303.

Tollenaar, D. 1934. Untersuchungen ueber Mutation bei Tabak: I. Entstehungsweise und wesen kuenstlich erzeugter gene-mutanten. *Genetica*. 16: 111–152.

Van Harten, A.M. 1998. Mutation Breeding: Theory and Practical Applications. Cambridge: Cambridge University Press.

7.2. Websites

IAEA/FAO mutant variety genetic stock database: http://mvgs.iaea.org

7.3. Further Reading

Ahloowalia, B.S., Maluszynski, M. and Nichterlein, K. 2004. Impact of mutation-derived varieties. *Euphytica*. 135: 187–204.

Anonymous. 1977. Manual on Mutation Breeding (Second Edition), Technical Reports Series, No. 119. Vienna: Joint FAO/IAEA Division of Atomic Energy in Food and Agriculture, International Atomic Energy Agency.

Gaul, H. 1961. Use of induced mutations in seed propagated species. pp. 206-252. *In*: Mutation and Plant Breeding. Nat. Acad. Sci., Nat. Res. Council, USA.

Gaul, H., Ulonska, E., Winkel, C.Z. *et al*. 1969. Micromutations influencing yield in barley–studies over nine generations. *In*: Induced mutations in plants. Proc. Symp. Pullman, IAEA, Vienna, pp. 375–398.

Gottshalk, W. and Wolff, G. 1983. Induced Mutations in Plant Breeding. pp. 323-327. *In*: Monograph on Theoretical and Applied Genetics. Berlin: Springer-Verlag.

Gustafsson, A. and Mackey, J. 1948. Mutation work at Svalof. Svalof, 1886–1946: 338–355.

Kharkwal, M.C., Gopalakrishna, T., Pawar, S.E. *et al*. 2008. Mutation breeding for improvement of food legumes. pp. 194-221. *In* : M.C. Kharkwal (ed.), Food Legumes for Nutritional Security and Sustainable Agriculture. Vol. 1. Proc. Fourth International Food Legumes Research Conference, October 18–22, 2005, New Delhi. Indian Society of Genetics and Plant Breeding, New Delhi, India.

MacKey, J. 1956. Mutation Breeding in Europe. *Brookhaven Symposium in Biology*. 9: 141–156.

Maluszynski, M., Nichterlein, K., Van Zanten, L. *et al*. 2000. Officially released mutant varieties – the FAO/IAEA database. *Mutation Breeding Review*. 12: 1–84.

The Structure and Regulation of Genes and Consequences of Genetic Mutations

S.Lee, S.Costanzo and Y.Jia*

USDA-ARS Dale Bumpers National Rice Research Center, P.O. Box 1090, Stuttgart, Arkansas 72160, USA
* Corresponding author, E-MAIL: yulin.jia@ars.usda.gov

1. Introduction

The fundamental genetic unit of heredity, the gene, is a nucleic acid sequence characterized by three main parts: a promoter region where gene expression levels are controlled by transcription factors and other accessory proteins, the coding region (often interrupted by intervening sequences or introns), and a 3' untranslated region. Following transcription, the completed transcript, often referred to as "messenger RNA" (mRNA), is used as a template for the production of proteins. The order of amino acids encoded by mRNAs translates into specific proteins, which may be further processed by cells to attain their mature forms. Proteins are involved in diverse biological functions. The entire process of transcription and translation is precisely regulated in order to maintain normal cellular functionality. Perturbations to this highly sophisticated regulatory network can have drastic consequences, and result in

tinuous structures. The strands provide the essential information required for the production of cellular proteins. In this chapter, gene structure, genetic regulation and the consequences for gene expression of various sequence alterations by mutation are described.

2. Gene Structure and Regulation

2.1. Gene structure

A typical gene is associated with three types of sequences: a promoter, a coding region interrupted by intron sequences and untranslated regions. Introns, also sometimes referred to as "intervening sequences", are found in primary transcripts but are subsequently spliced out of the nascent mRNAs. Introns typically interrupt coding regions, but can also be located in untranslated regions of a gene, which are located on

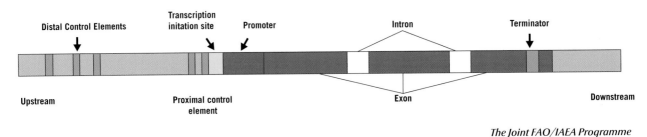

The Joint FAO/IAEA Programme

Figure 3.1 Typical gene structure. The coding regions (blue) contain the information used to define the sequence of amino acids in a protein. The untranslated regions (gray and red) are found in the mRNA but are not used to define the protein sequence; they are sometimes regulatory in nature. Introns (white) are found in the primary transcript but are subsequently spliced out of the mRNA.

phenotypic changes adversely affecting an organism's survival, and are a driving force in evolution.

Biological organisms have evolved and coexisted in various natural ecosystems for many millions of years. Phenotypes displayed by organisms are largely genetically programmed, and can be inherited over generations. The discovery of the gene as the fundamental unit of inheritance has led us to a better understanding of the molecular basis of phenotypes. It is now known that genetic information is stored in double helical polynucleotide chains of DNA (or RNA for some viruses). Two purines: adenine (A) and guanine (G) and two pyrimidines: thymine (T) and cytosine (C), or uracil (U), are linked by phosphate groups and sugars (deoxyribose in DNA, ribose in RNA) in these polynucleotide chains. The two strands of DNA are complementary in sequence to each other, and form stable and con-

either side of the coding region (**Figure 3.1**). The boundaries of a protein-encoding gene are defined as the points at which transcription begins and ends.

In general, the transcribed mRNAs from genes are to be used for producing proteins. In eukaryotes, mRNAs produced in the nuclei are translocated into the cytoplasm for protein synthesis. This process is called translation. The process of transcription and translation in a eukaryotic cell is described in **Figure 3.2**.

2.1.1. The Role of the Promoter Regions

The promoter region is a specific regulatory DNA sequence located upstream of a gene that is recognized and physically interacts with a class of proteins known as transcription factors. These proteins bind to determined regions within the promoter sequences, ultimately enabling the recruiting of the RNA polymerase enzyme

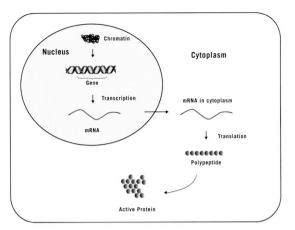

Figure 3.2 A diagram of DNA-RNA transcription and RNA-protein translation in a plant cell.

complex that synthesizes RNA based on the coding region of the gene. In prokaryotes, the RNA polymerase and an associated sigma factor are brought to the promoter DNA by an activator protein binding to its own specific DNA sequence. In eukaryotes, the complexity of this process is greater, and at least seven different factors are necessary for the transcription of an RNA polymerase II-associated promoter. Promoters also represent critical elements that can work in coordination with other regulatory regions (such as enhancers, silencers, insulators and other boundary elements) to control the level of transcription of a given gene. As promoters are generally located in close proximity, immediately upstream of a gene, positions in the promoter are designated relative to the transcriptional start site (TSS), where transcription of RNA begins for a particular gene.

The basal or core promoter is the minimal portion of contiguous DNA sequence required to initiate transcription properly by RNA polymerase II. It encompasses a region between –37 bp and +32 bp relative to the transcription start site. It may consist of various conserved motifs including a TATA box, an initiator (Inr) encompassing the transcription start site,

as well as general transcription factor binding sites. The TFII<u>B</u> (transcription factor) <u>R</u>ecognition <u>E</u>lement (BRE) is located immediately upstream of the TATA element and the TFIIB–BRE interaction facilitates the assembly of the complex, TFIIB, TATA-binding protein (TBP) and TATA box (**Figure 3.3**).

In addition to the core promoter, other conserved regions include a proximal or "upstream" promoter, and a distal promoter. Proximal promoter sequences upstream of genes tend to contain primary regulatory elements, and may extend as much as 200 bp upstream of the TSS, and contain specific transcription factor binding sites. The distal promoter sequences are found further upstream of genes, and may contain additional regulatory elements, often with weaker influences on gene expression than those of the proximal promoter.

Many distinct features separate prokaryotic from eukaryotic promoters. In prokaryotes, the promoter consists of two short sequences at positions –10 and –35 upstream from the transcription start site. Sigma factors are proteins that assist RNA polymerase in correctly binding to specific promoter regions, generally targeting functionally related genes participating in a given biological response. The sequence at –10 is also known as the Pribnow box, or the –10 element, and usually consists of the six nucleotides TATAAT. The Pribnow box is required for the initiation of transcription in prokaryotes. The other sequence at –35 (the –35 element) usually consists of the six nucleotides TTGACA, and is involved in enhancing gene transcription rates. Some prokaryotic promoters contain a so-called "extended –10 element" (consensus sequence 5'-TGNTATAAT-3'), and in these "extended –10" promoters the presence of the –35 element appears to be unimportant for transcription.

In eukaryotes, promoters are extremely diverse and are more difficult to characterize than their prokaryotic counterparts. As in prokaryotes, they typically lie

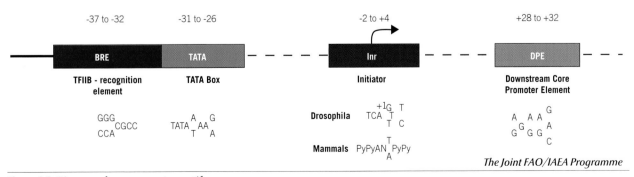

Figure 3.3 Diagram of core promoter motifs.

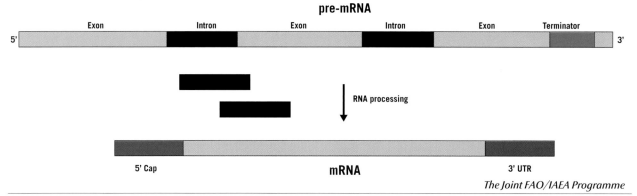

Figure 3.4 Diagram of pre-mRNA to mRNA splicing.

upstream of a gene and can also possess regulatory elements several kilobases away from the TSS. The transcriptional complex can also cause DNA to bend back on itself, facilitating the involvement of regulatory sequences located far from the actual site of transcription initiation. Some eukaryotic promoters also possess a TATA box (consensus TATAAA), which in turn binds a TATA-binding protein that assists in the formation of the RNA polymerase transcriptional complex. The TATA box typically lies within close proximity to the transcriptional start site, often within 50bp.

2.1.2. The Role of the Coding Regions

Protein-coding genes constitute the vast majority of genes in a genome. They are diverse in size and organization but generally maintain several conserved features. A protein-coding gene is defined by the extent of its primary transcript. The coding region contains the information used to define the amino acids in a given protein, containing the nucleotide sequence that is eventually translated into the correct sequence of amino acids. The coding region begins with the initiation codon (also known as the start codon), which is most frequently "ATG", encoding for the initiating amino acid methionine in a protein sequence. In prokaryotes, alternative start codons are also utilized, such as GUG and UUG, as well as several others. The coding region ends with one of three termination codons: TAA (ochre), TAG (amber) or TGA (opal).

2.1.3. The Role of Introns

Introns are non-coding DNA elements within a gene that are transcribed and still present in heterogeneous nuclear RNA (hnRNA) or pre-mRNA, but are spliced out within the nucleus to produce a mature mRNA (**Figure 3.4**). Based on known splicing mechanisms, introns

can be classified into four groups. Introns are common in eukaryotic RNAs, but are quite rare in prokaryotes. The exons (originally derived from the term expressed regions) are portions of a gene that are joined together to form the final functional mRNA. The number and length of introns varies widely among species and among genes within the same species. Genes in animals and plants often have numerous introns, which can be much longer than the adjacent exons.

Introns sometimes allow for alternative splicing of a gene, such that several different proteins sharing some common segments can be produced from a single gene. The control of mRNA splicing, and of which alternative splice variant is produced, is orchestrated by a wide variety of signalling molecules. Introns contain several short conserved sequences that are important for efficient and correct splicing. The exact mechanism for these intron-splicing enhancers is not well understood, but it is thought that they serve as binding sites on the transcript for proteins that stabilize the spliceosome complex. It is also possible that RNA secondary structures formed by intron sequences may have an effect on splicing, and in alternative splicing, an exon sequence in one splice variant is intron in another. Some introns, such as Group I and Group II introns are actually ribozymes that are capable of catalyzing their own splicing from the primary RNA transcript (Patel and Steitz, 2003). After removing introns from pre-mRNA, the transcribed exons spliced together to form the mRNA. However, if the mutation shifts the reading frame by one or two substitutions, the reading frame will shift to produce a new stop codon. This phenomenon is called exon skipping that is caused by non-sense mutation.

Nuclear or spliceosomal introns are spliced by the spliceosome and a series of snRNAs (small nuclear RNAs). Specific splice signals (or consensus sequences)

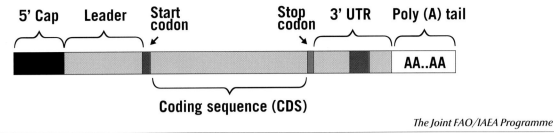

Figure 3.5 Diagram of a eukaryotic protein-coding mRNA.

within these introns assist in the splicing (or recognition) of these introns by the spliceosome. Group I, II and III introns are self-splicing introns and are relatively rare compared with spliceosomal introns. Group II and III introns are similar, share a conserved secondary structure, and use the so-called "lariat" pathway in their splicing mechanism. They perform functions similar to that of the spliceosome, and may be evolutionarily related to it. Group I introns are the only class of introns whose splicing requires a free guanine nucleoside, and they possess a secondary structure distinct from that of Group II and III introns. Group I introns are found in most bacteria and protozoa.

The initial discovery of intron sequences prompted investigators worldwide to examine their prevalence among diverse taxa. It was soon discovered that introns are widespread in eukaryotes, but quite rare in prokaryotic species. This has prompted much speculation with regards to the evolution of organisms in general, and the role introns may have played in that evolution. There are two main schools of thought concerning their potential roles: 1) Introns-early. Proponents of this school argue that introns were an essential feature of the earliest organisms. Their absence in bacteria, they argue, is due to the fact that the shorter division times of bacterial cells means that bacteria have had many more generations in which to evolve relative to eukaryotes. This rapid evolution has brought about the loss of nearly all ancestral introns. 2) Introns-late. Proponents of this school argue that the earliest organisms did not contain introns. They hypothesize that introns are a relatively recent arrival in the eukaryotic lineage, which have been necessary for the evolution of the diversity of regulatory mechanisms required to control gene expression in highly differentiated and multicellular organisms. In this view, prokaryotes do not have introns because they never had them in the first place.

2.1.4. The Role of 3' Untranslated Regions

The 3' untranslated region (3' UTR) is part of the mRNA situated downstream from the coding sequence (CDS), which is transcribed but not translated into protein. In eukaryotes, it immediately follows the termination codon of the CDS and directly precedes the polyA tail (**Figure 3.5**). In prokaryotes, 3' UTRs are also present in protein-coding mRNAs, although these transcripts lack polyA tails.

Several types of regulatory sequences can occur in eukaryotic 3' UTRs, including polyadenylation signals (usually AAUAAA or a slight variant thereof) that direct the site for cleavage and polyadenylation of the nascent mRNA approximately 30 bp downstream. Several hundred consecutive adenine residues are added at this cleavage position, comprising the poly-A tail of the mature mRNA. Additional regulatory sequences associated with 3' UTRs can include binding sites for proteins affecting mRNA stability or cellular location, as well as binding sites for microRNAs (miRNAs) (Mazumder *et al.,* 2003).

2.2. Gene Regulation

The regulation of gene expression is the primary cellular mechanism for controlling the abundance and timing of appearance of functional gene products. Although a functional gene product may be either RNA or a protein, the mechanisms regulating the expression of protein-coding genes have been the most intensively studied. Any step of gene expression may be modulated, from the initial transcription of RNA from a DNA template, through the post-translational modification or stability of a protein. Gene regulation gives cells the ability to control their ultimate structure and function, and is the basis for cellular differentiation, morphogenesis, and the versatility and adaptability of all organisms.

It is important to note that environments can influence the phenotypes. A phenotype is the result of interactions of both genetic and environmental factors. In addition, phenotypes can also be altered by epigenetic events. This is defined as a set of reversible heritable changes that occur without any changes of DNA sequences that can link gene expression to the environment.

2.2.1. Transcriptional Regulation

In the case of prokaryotic organisms, regulation of transcription is an essential mechanism for cells to quickly adapt/respond to an ever-changing environment. The availability of different types and quantities of nutrients affect various metabolic pathways, and ultimately determine which genes are expressed and their specific levels of expression. In prokaryotes, repressors bind to regions termed "operators" that are generally located downstream from, but proximal to the promoter. Activators bind to upstream promoter sequences, such as the CAP (Catabolite Activator Protein) binding region. In prokaryotes, a combination of activators, repressors and, in rare cases, enhancers determine whether a gene is to be transcribed.

In eukaryotes, transcriptional regulation tends to involve combinational interactions between several transcription factors, which allows sophisticated responses to complex environmental conditions. This complex regulation also allows control over spatial and temporal differences in gene expression. Eukaryotic genes extensively make use of enhancer elements, distal regions of DNA that can loop back and allow interactions with proteins bound to promoter sequences. A major difference between eukaryotes and prokaryotes is the fact that eukaryotes have a nuclear membrane that prevents simultaneous transcription and translation. In plants, the regulation of transcription plays a major role in determining the tissue- and developmental stage-specific activity of many genes, as well as responses to environmental cues and stresses, the regulation of metabolic pathways and the regulation of photosynthesis.

There are several strategies used by eukaryotes to control gene expression, including altering the rate of

transcription of a gene (**Box 3.1**), and some examples are discussed below:

1. Chemical modification of DNA by methylation is a common method of silencing gene expression. Methyltransferase enzymes typically methylate cytosine residues in DNA occurring within a CpG dinucleotide content (also called "CpG islands" that have a large number of cytosine and guanine adjacent to each other in the backbone of DNA). The "p" in CpG refers to the phosphodiester bond between the cytosine and guanine. Abnormal methylation patterns are thought to be involved in carcinogenesis.

2. Transcription of DNA is also determined by its three-dimensional structure. In general, the density of its packing is indicative of the frequency of transcription. Octameric protein complexes called nucleosomes (composed of proteins termed "histones") are responsible for the amount of supercoiling of DNA, and these complexes can be biochemically modified by processes such as phosphorylation, methylation and acetylation. Such modifications are considered to be responsible for more long-term or permanent changes to gene expression levels. Histone acetylation can also play a more direct role in transcriptional regulation, as histone acetyltransferase enzymes (HATs) such as the CREB-binding protein that helps dissociate DNA from nucleosomes, allowing transcription to proceed. Additionally, DNA methylation and histone acetylation can work together to promote gene silencing. The combination of the two seems to provide a signal for DNA to be packed more densely, resulting in reduced gene expression levels.

3. RNA polymerase activity can be regulated by at least four mechanisms: 1) specificity factors alter the specificity of RNA polymerase for a given promoter or set of promoters, making it more or less likely to bind to them (i.e. sigma factors used in prokaryotic transcription); 2) repressors bind to non-coding sequences on the DNA strand that

are close to, or overlapping the promoter region, impeding RNA polymerase's progress along the strand, thus impeding the expression of the gene; 3) basal factors position RNA polymerase at the start of a protein-coding sequence and then release the polymerase to transcribe the mRNA; 4) activators enhance the interaction between RNA polymerase and a particular promoter, stimulating the expression of the gene.

4. Regulatory proteins are involved in regulating gene expression (**Figure 3.6**), and usually bind to a regulatory binding site located near the promoter, although this is not always the case. Regulatory proteins often need to be bound to regulatory binding sites to switch a gene on (activators), or to shut off a gene (repressors). Generally, as organisms become more sophisticated, their cellular protein regulation becomes more complex. Post-transcriptionally, a second mechanism of gene regulation exists, represented by a growing class of 22-nucleotide-long non-coding RNAs, known as microRNAs (miRNAs), which function as repressors in all known animal and plant genomes. Two main types of gene regulation are recognized: inducible systems and repressible systems. An inducible system is inactive unless in the presence of specific molecules (referred to as inducers) that facilitate gene expression. The molecule is said to "induce expression", and the manner in which this occurrence is dependent on the existing control mechanisms, as well as differences between prokaryotic and eukaryotic cells.

A repressible system is constitutively active except in the presence of repressor molecules that turn off gene expression. The mechanism by which transcriptional repression occurs is also dependent on the existing control mechanisms as well as differences between prokaryotic and eukaryotic cells.

The transcription machinery can itself be regulated. In order for a gene to be expressed, several events must occur. First, there needs to be an initiating signal from either the cell itself or from the environment. This can be achieved through the binding of some ligands to a receptor. This signal gives rise to the activation of a protein called a transcription factor, which recruits other components of the "transcriptional apparatus." Transcription factors generally bind simultaneously to DNA, RNA polymerase, as well as other factors necessary for transcription (i.e. histone acetyltransferases, scaffolding proteins, etc.). Transcription factors, and their cofactors, can be regulated through reversible structural alterations such as phosphorylation, or inactivated through such mechanisms as proteolysis.

Transcription is initiated at the promoter's TSS, as increased amounts of active transcription factors occupy their target DNA sequences. Other proteins known as scaffolding proteins bind other cofactors and hold them in place. DNA sequences located relatively far from the point of initiation, known as enhancers, can aid in the assembly of the transcription apparatus. Frequently, extracellular signals induce the expression of transcription factors or components thereof, which

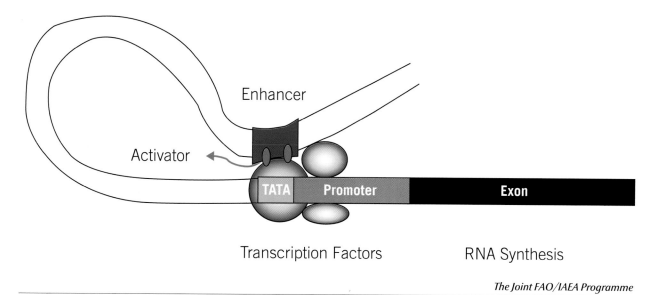

Enhancer

Activator

TATA   Promoter   Exon

Transcription Factors   RNA Synthesis

The Joint FAO/IAEA Programme

Figure 3.6 Diagram of the interaction between an enhancer and a promoter region for initiating transcription.

then influence the expression of other genes, or even their own expression.

2.2.2. Post-Transcriptional Modification

In the nucleus of eukaryotic cells, the primary transcript is converted into a mature mRNA. This post-transcriptional modification process allows mRNA to be exported out of the nucleus to be translated into proteins by ribosomes in the cell cytoplasm. In contrast, the majority of prokaryotic mRNAs do not require any further processing for translation. Three main types of modifications are performed on primary transcripts: 5′ capping, 3′ polyadenylation and intron splicing. Addition of the 5′ cap is the first step in pre-mRNA transcript processing. This modification occurs before transcription is complete, after the growing RNA strand has reached approximately 30 nucleotides. The process is catalyzed by a capping enzyme that associates with the carboxyl-terminal domain of RNA polymerase II, the main enzyme involved in mRNA transcription. The 5′ capping reaction replaces the triphosphate group at the 5′ end of the RNA chain with a special nucleotide (a methylated guanine) that is referred to as the 5′ cap. This process may serve to protect mRNA from the enzymatic activity of 5′ exonucleases, which degrades foreign RNA, and improves mRNA recognition by the ribosome during translation. The second step is the cleavage of the 3′ end of the primary transcript followed by the addition of a polyadenosine (poly-A) tail. This end of the pre-mRNA contains a sequence of around 50 nucleotides that acts as a signalling region recognized by a protein complex. The protein complex promotes association of other proteins, including those involved in transcript cleavage as well as polyadenylate polymerase (PAP), the enzyme responsible for the addition of the poly-A tail. PAP binding is required before transcript cleavage can occur, thus ensuring the tight coupling of the two events. RNA splicing is the process by which introns are removed from pre-mRNAs, and the remaining exons connected together to form a single continuous molecule. Although most RNA splicing occurs after the complete synthesis and end capping of the pre-mRNA, transcripts with many exons can be spliced while being transcribed. A large protein complex called the spliceosome catalyses the splicing reactions, guided by small nuclear RNA molecules that recognize splice sites in pre-mRNA sequences. Some pre-mRNAs can be spliced in multiple ways to produce different mature mRNAs that encode different protein sequences. This process is known as alternative splicing, and allows production of a large variety of proteins from a limited amount of DNA.

3. Mutations and their Consequences

3.1. Classes of Gene Mutations

Mutations can be classified based on the extent of the DNA sequence affected by the mutational event as either small-scale mutation involving one or a few nucleotides, or large-scale mutations where the chromosomal structure is affected. Small DNA changes may be further classified into point mutations, deletions and insertions.

Box 3.2: Examples of different types of mutations

Point mutations (substitution mutations)
Transition: A➔G, G➔A, C➔T, and T➔C
Transversion: A➔C, A➔T, C➔A, C➔G, T➔A, T➔C, G➔C and G➔T
Silent mutation: GT<u>C</u>(Val) ➔ GT<u>A</u>(Val)
Missense mutation: <u>G</u>TC(Val)➔<u>TT</u>C(Phe)
Non-sense mutation: <u>A</u>AG(Lys)➔<u>T</u>AG(Stop)

Insertion and Deletion (indels)
Insertion: ATATGTATAAAG➔ATATGT<u>CT</u>GATAAAG
Deletion: ATA<u>TG</u>TATAAAG➔ATAATAAAG
Frameshift mutation: <u>G</u>TC CTG TTA A···TAA(Stop)➔TCC TGT <u>TAA(Stop)</u>
                            GTC CTG TTA A···TAA(Stop) ➔ <u>GAG</u> TCC TGT <u>TAA(Stop)</u>

Inversion
5′ AG<u>GTTTGC</u>CTACTGG 3′ ➔ 5′ AG<u>CGTTTG</u>CCTACTGG 3′

Large-scale mutations, including inversions and gene duplications, will be discussed elsewhere in this book. In the present section, the mechanistic consequences of gene mutations will be described in detail.

3.1.1. Point Mutations

Also called single-base substitutions; these involve the replacement of a single nucleotide with a different one. This is the most common type of mutation. Point mutations can occur in two different ways, referred to as transitions and transversions. Transitions are events where a purine is converted to a purine (A to G or G to A), or a pyrimidine is converted to a pyrimidine (T to C or C to T). A transversion occurs when a purine is converted to a pyrimidine or a pyrimidine is converted to a purine. Transitions occur more frequently than transversions in mutational events. There are three types of point mutations: missense, non-sense and silent mutations, which are defined below. In missense mutations, a single nucleotide change alters a codon such that it encodes a completely different amino acid in the protein, also termed a nonsynonymous mutation. This type of mutation usually changes the conformation and activity of a protein. However, if an amino acid is replaced with another amino acid with similar chemical properties, then the function of the protein may be unaffected. Silent mutations result in no change to the amino acid sequence, and are also referred to as synonymous mutations. Synonymous changes are therefore not detectable at the amino acid level, and can be identified only in the DNA coding region. Although synonymous mutations are always silent, they may, in rare cases, create a new intron splice site that changes exons to introns, resulting in a different protein sequence. Non-sense mutations are identical to miss-sense mutations, except that the codon for an amino acid is replaced with a translation termination codon. As a result, the premature termination of translation leads to the production of a truncated protein, which is most likely non-functional.

3.1.2. Insertions and Deletions (Indels)

Insertions occur by adding one or more nucleotides and deletions result from eliminating one or more nucleotides from a DNA sequence. Deletions and insertions are collectively referred to as InDels because it is often difficult to differentiate between insertions and deletions when two orthologous sequences are compared. If Indels involve one or two base pairs within a coding region, a shift in the open reading frame results and is referred to as a frameshift mutation. Frameshift mutations typically create premature stop codons that produce shorter polypeptides than the original (non-mutated) versions. Indels occurring within coding regions may also alter intron splicing of pre-mRNAs. This type of mutation is called a splice site mutation at the intron region. Splice site mutation may also alter the sequence of the final translated protein.

Box 3.3: Examples of consequences of mutation in different genic regions in plants

Promoter mutations
- Reduced expression of pathogenesis related gene (PR1-a) (Buchel *et al.,* 1999).
- Increase disease resistance to bacterial blight in rice (xa13) (Chu *et al.,* 2006).
- Gain of virulence of the AVR-*Pita* by the Pot3 insertion (Kang *et al.,* 2001).

Coding region mutations
- The loss of the *avrRpt2* function for the disease resistance to *Pseudomonas syringae* pv. "tomato" (Lim and Kunkel, 2004).
- Gain of disease resistance to Flax rust (Dodds *et al.,* 2006).
- The loss of function of the homeobox gene OSH15 (Sato *et al.,* 1999).
- Abnormal tillering in rice (*moc1*) (Li *et al.,* 2003).
- Slender glume mutation by transposon element *mPing* in rice (Nakazaki *et al.,* 2003).

Intron mutations
- Exon skipping in *Arabidopsis* (Brown, 1996).
- The reduced expression of the splicing efficiency in *Arabidopsis* (Lewandowska *et al.,* 2004).

3' trailer mutations
- Altered flowering time in *Arabidopsis* (Swiezewski *et al.,* 2007).
- Natural variations in floral symmetry (Cubas *et al.,* 1999).

3.1.3. Inversions

The entire double-stranded DNA segment comprising two or more base pairs is reversed in an inversion mutation. This type of DNA rearrangement can occur *via* chromosome breakage and rejoining or by recombination occurring between homologous segments within the same chromosome. Inversion mutations may have little or no effect on an organism's phenotype unless a functional gene (or multiple genes) is altered.

3.2. Examples of Various Mutations which Affect Gene Expression

Mutations of different categories, as well as those found in different locations, can have distinct consequences for the function of a given gene. As described earlier, each gene consists of functionally distinguishable regions such as the promoter, untranslated regions and coding regions required for the production of functional gene products. Therefore, the effects of any gene mutation on an organism will vary, depending upon where the mutation occurs, and whether the function of an essential protein has been altered. In this section, we will focus on the consequences of mutations occurring in four different gene functional regions, including gene promoters, introns, coding regions and 3′ UTRs (**Box 3.3**).

3.2.1. Mutations in Promoter Regions

Mutations within the promoter region of a gene may cause deviations from normal gene expression patterns and/or levels, and may be either harmful or beneficial to an organism. There are numerous, well-characterized examples in the plant literature documenting some of these effects. For example, mutations in the pathogenesis-related gene *PR-1a* promoter reduce the level of inducible gene expression for a hypersensitive plant response and systemic acquired resistance (Buchel *et al.,* 1999). The *PR-1a* promoter region contains seven potential binding sites for GT-1, a complex that simultaneously interacts with multiple promoter regions. The experimental introduction of point mutations within GT-1 binding sites in the *Pr-1a* promoter results in significantly decreased formation of the GT-1 complex, leading to a reduction in salicylic acid (SA)-inducible gene expression. However, for promoter mutations having no effect on GT-1 binding, effects on SA-inducible gene expression are not observed.

Promoter mutations are also involved in the acquisition of disease resistance in rice (Chu *et al.,* 2006). The recessive allele *xa13*, whose locus represents an essential protein required for normal pollen development, confers resistance to bacterial blight caused by the pathogen *Xanthomonas oryzae* pv. *oryzae*. Sequence comparisons between the *Xa13* and *xa13* alleles reveal insertions and deletions within both the coding and the promoter regions. However, after examining *xa13* and *Xa13* sequences from different susceptible and resistant varieties, it was determined that observed amino acid changes resulting from nucleotide substitutions within the coding region of *xa13* do not contribute to xa13-mediated resistance. In contrast, it was determined that differences due to Indels within the promoter regions of various xa13 alleles led to the reduced accumulation of *Xa13* transcripts, and this suppression of *Xa13* expression increased whole-plant resistance to *X.oryzae*. Thus, mutations in the promoter regulatory sequences of *Xa13* resulted in the recessively inherited resistance against bacterial blight.

The avirulence gene *AVR-Pita* of *Magnaporthe oryzae* (formerly *Magnaporthe grisea*) triggers resistance mediated by the rice resistance gene Pi-ta. A mutation in the promoter of AVR-Pita caused by the insertion of a Pot3-type transposon results in the gain of virulence toward rice cultivar Yashiro-mochi, which possesses the Pi-ta gene. The insertion of a Pot3 transposon into the *AVR-Pita* promoter caused a spatial separation of the promoter sequences from the remainder of the gene, resulting in disruption of the *AVR-Pita* gene expression required for *Pi-ta*-mediated disease resistance to rice blast (Kang *et al.,* 2001). The transposon mutation in the rice blast pathogen may be one of the important mechanisms associated with race variation.

3.2.2. Mutations in Coding Regions

Point mutations or indels in the coding region of a gene can result in various types of mutations, such as amino acid-altering mutations and frame shift mutations. Thus, gene products (which control phenotypes) can be altered significantly by mutations in the coding region of a gene, which can potentially cause drastic changes in the function of essential proteins. For example, the Pi-ta gene in rice is a resistance gene against the rice blast pathogen *Magnaporthe oryzae* containing the *AVR-Pita* locus. *Pi-ta* encodes a predicted polypeptide of 928 amino acids associated with the nucleotide binding site

(NBS) class of resistance proteins. DNA sequences of various *Pi-ta* alleles from resistant (Yashiro-mochi) and susceptible rice varieties revealed a single amino acid difference (alanine to serine) resulting from a single nucleotide substitution in the *Pi-ta* coding region (Bryan *et al.*, 2000). Similarly, specific mutations in the coding region of *AVR-Pita* allow the pathogen *Magnaporthe oryzae* to circumvent *Pi-ta*-mediated resistance. In sequence comparisons between the wild type *AVR-Pita* allele and those of race-shift isolates, non-synonymous and synonymous substitutions were detected in the *AVR-Pita* coding region. These amino acids changes in the AVR-Pita protein cause a loss of pathogen recognition by the host, and may represent an adaptive evolution of the *AVR-Pita* gene under selective pressure from the major R gene in rice, *Pi-ta*. Protein–protein interactions between Pi-ta and AVR-Pita are critical for the expression of resistance. Single amino acid substitutions in either the leucine-rich domain of Pi-ta or in the AVR-Pita$_{176}$ protease motif disrupt physical interactions *in vitro*, and result in a loss of resistance to the pathogen (Jia *et al.*, 2000). The Pi-ta LRD region is important for the recognition of the AVR-Pita protein by the host rice plant, the first step leading toward the initiation of *Pi-ta*-mediated resistance (**Box 3.4**).

Plant development patterns can also be significantly altered by mutations in gene coding regions. For example, homeobox genes in plants are involved in meristem maintenance and the development of lateral organs. The insertion of a transposable element into the rice homeobox gene *OSH15* causes loss of gene function, resulting in plants exhibiting reduced internode lengths, particularly in internode 2 (Sato *et al.*, 1999). Sequence analysis of the entire coding region of mutant *d6-ID6* revealed deletions (~700 bp) spanning the entirety of exon 1 and a portion of the 5′ upstream region in *OSH15*, and normal growth was restored to mutants transformed with wild type *OSH15*. These findings demonstrated that the *d*6-type dwarf phenotype in rice was due to the loss of function of *OSH15*.

Rice tillering is an important determinant for grain yield, and this process can also be impaired by mutations in coding regions of essential genes. Rice mutants containing a mutated MONOCULM 1 (*Moc1*) gene completely lose the ability to initiate tillering. In *moc1* mutants, there is a failure in the formation of tiller buds along the non-elongated basal internodes. Sequence analysis of *Moc1* alleles determined that a retrotransposon insertion into the ORF of *moc1* resulted in a premature stop codon. *Moc1* is the first functionally characterized regulator of the tillering in rice.

Rice mutants exhibit a "slender glume" phenotype caused by the activation of a novel transposable DNA element. An active rice transposon *miniature Ping (mPing)* belonging to the Tourist-like superfamily of miniature inverted-repeat transposable elements (MITEs)

Box 3.4: Allelic diversity of the *Pi-ta* locus conferring rice blast resistance

Nucleotide polymorphisms between *Pi-ta* (resistant) and *pi-ta* (susceptible) alleles (Bryan *et al.*, 2000; Jia *et al.*, 2003)
Yashiro-mochi (resistant) / Tsuyuake (susceptible)
- A single nucleotide difference in 5′ leader region and the 3′ noncoding regions.
- Transversions in the first exon and transition/transversion in the second exon.
- High level of nucleotide polymorphisms in intron regions.
- Seven-bp substitutions in the coding region result in the five-amino acid differences.

Gene diversity of the *Pi-ta* alleles
Total four Pi-ta haplotypes
- *Pi-ta* mediate resistant rice cultivars: Yashiro-mochi, K1, Reiho, Tetep, Tadukan, Katy, and Drew.
- *pi-ta* containing indica rice cultivars: El Paso 144, Cica 9 and 93-11.
- *pi-ta* containing japonica rice cultivars: Tsuyuake, Nipponbare, and Sariceltik.
- C101A551.

Interactions of Pi-ta with AVR-Pita gene

Rice variety	0-137 AVR-Pita	CP3337	Pi-ta present
Yashiro-mochi, K1, Tadukan, Tetep, Reiho, Katy	AVR	Vir	Yes
C101A51, YT16	Vir	Vir	No

CP3337 is a spontaneous mutant of O-137 that has lost AVR-Pita.
AVR: avirulence and Vir: virulence.

was identified within exon 4 of the *slg* (<u>sl</u>ender glume) coding region (Nakazaki *et al.,* 2003). However, due to the mobility of mPing, this particular mutant slg allele is not stable, and occasionally it can revert back to the wild type allele due to the loss of the *mPing* element.

The avirulence gene *avrRpt2* from *Pseudomonas syringae* pv. "tomato" triggers resistance in *Arabidopsis* plants expressing the resistance gene RPS2. Mutations in the *avrRpt2* coding sequence introduced by site-directed mutagenesis abolish the RPS2-dependent hypersensitive response in *Arabidopsis*, suggesting a loss of avirulence activity. Consistent with this hypothesis is the observation that *Pseudomonas syringae* mutants lacking a functional avrRpt2 protein retain significant virulence activity on susceptible *Arabidopsis* plants (Lim and Kunkel, 2004).

The final example of the consequences of mutations to gene coding regions concerns the avirulence gene *AvrL567* of flax rust pathogen. Spontaneous mutations are known to occur in flax rust avirulence genes due to the selective pressure imposed by host R-genes. Non-synonymous substitutions have been identified within the coding regions of *AvrL567*, and the resulting amino acid substitutions in *AvrL567* correspond to residues playing an important role for the avirulence activity of flax R-genes. Finally, a Pot3 transposon element was identified in the *AVR-Pita* coding region of a *M.oryzae* field isolate which caused significant crop damage to a rice variety containing the corresponding resistance gene (Zhou *et al.,* 2007).

3.2.3. Mutations in Introns

Mutations in gene introns can significantly inhibit the normal splicing of pre-mRNAs, leading to the production of aberrant mRNAs. Moreover, the alteration of splice sites by mutation can alter essential gene products *via* the insertion or deletion of amino acids, and as mentioned, can also introduce premature stop codons which cause truncated gene products, a non-sense mutation. However, the formation of truncated proteins does not occur frequently *in vivo* by employing the non-sense-mediated mRNA decay pathway. Non-sense mediated decay is a cellular surveillance mechanism of mRNA that prevents the expression of truncated proteins. Examinations of mutations in intron splice sites in *Arabidopsis* have detailed the various mis-splicing patterns generated in plants harbouring these alleles. Mutations within 5′ and 3′ splice sites predominantly

cause loss of splicing and exon skipping in *Arabidopsis*. The observed exon skipping occurred by inhibiting the recognition or stable assembly of splicing factors for 3′ splice site mutations, and unstable U1snRNP association and interference with exon scanning and splicing factor assembly at upstream 3′ splice sites in 5′ splice site mutations (e.g. Jacobsen *et al.,* 1996).

U12-dependent introns represent one class of nuclear pre-mRNA introns found in eukaryotes, and mutated plant U12-dependent introns have been analyzed for splicing efficiency in *Arabidopsis*. Several introduced nucleotide substitutions resulted in reduced splicing efficiency, and additionally, insertions of two or three U-rich elements introduced into the mutated U12-dependent introns increased splicing efficiency. Thus point mutations and insertions within plant U12-dependent introns impact significantly on host gene expression.

Pi-ta is an effective resistance gene deployed in agricultural systems for many decades against blast disease in rice. The sequence diversity of this gene was studied recently in *Pi-ta* alleles isolated from resistant and susceptible rice varieties. *Pi-ta* is a single-copy gene in rice encoding a predicted cytoplasmic protein with a centrally located NBS, and a carboxyl terminal leucine-rich repeat (LRR) region (Bryan *et al.,* 2000). When the sequences of resistant (*Pi-ta*) and susceptible alleles (*pi-ta*) were compared, point mutations and indels were identified within the 5′ and 3′UTR regions, the first exon and within introns. The majority of the nucleotide polymorphisms that differentiated Pi-ta alleles derived from resistant vs. susceptible rice varieties were found within intron sequences, however their effect on *Pi-ta* splicing and activity remains to be determined.

3.2.4. Mutations in 3′ Untranslated Regions

The function of 3′ UTR regions is closely linked with the efficient translation of their associated proteins. The sequences that act as determinants of mRNA stability have been identified in 3′ UTR regions in plants. Two sequences such as the DST sequence and the AUUUA motif were found to destabilize transcripts in plants, and both sequences commonly present in many unstable plant transcripts. In mutagenesis experiments, the insertion of an AUUUA repeat into the 3′ UTR reporter genes caused the rapid degradation of the corresponding transcripts in stably transformed cells of tobacco. Also, in transgenic plants, the decrease of mRNA accumulation was observed in transgenic plants after the insertion of an AUUUA repeat into a globin reporter gene (Green, 1993). Therefore, mutations in this region can affect protein synthesis, which may in turn impact other critical cellular processes in organisms. For example, the *Arabidopsis* gene *FLOWERING LOCUS C (Flc)* is a central regulator involved in the control of flowering timing. In a recent study performed by Swiezewski *et al.* (2007), a T-DNA insertion mutation in the 3' UTR region of *Flc* caused mis-expression of *Flc*, and resulted in delayed flowering. On the other hand, two different mutants with T-DNA inserted either upstream or downstream from the 3' UTR of *Flc* showed early flowering. Significantly, alterations in the levels of *Flc* mRNA were observed for all three mutants.

3.2.5. Mutations in Regulatory Genes

The alteration of plant phenotypes can also be caused by mutations in genes encoding proteins such as transcription factors, which directly affect the expression of the target genes they regulate. Plant transcription factors are involved in diverse processes such as organ development, responses to environmental stimuli (including pathogen challenge) and the regulation of metabolic pathways. Mutations, which alter the activity or expression of transcription factors, are often involved in morphological and physiological changes associated

with the adaptation and domestication of crop species. The short stature of many cereal crops, for example, has been intentionally developed for improved grain yield and reducing crop loss due to wind and rain damage. In semi-dwarf wheat varieties, mutations in the Rht-1 gene, encoding a transcription factor responsible for gibberellin signaling, cause significantly reduced plant heights, and transformation of this same allele into transgenic rice results in a similar dwarf phenotype.

Mutations in regulatory proteins have also been demonstrated to affect seed colour in maize, such as the mutant allele *pale aleurone color1*, which causes pale aleurone pigmentation by reducing anthocyanin biosynthesis. Other important examples include mutations in various MADS box genes, a family of plant-specific transcription factors genes important in controlling floral and plant development. The important role of many of the genes of the MADS family and the action of MADS domain proteins in the plant life cycle have been widely described in *Arabidopsis*. For example mutation of OsMADS1, a MADS box gene from rice, results in early flowering (Jeon *et al.*, 2000).

Regulatory genes not only encode proteins, but may also produce regulatory RNAs. Post-transcriptional gene silencing (PTGS), or RNA interference (RNAi) refers to an important process involved in developmental regulation, maintenance of genome integrity and defense

Box 3.5: Examples of plant phenotype changes by regulatory gene mutations

Mutational consequences	Mutation	Gene	Plant	Reference
Reduced plant height by inhibited gibberellin signalling	Transcription factors	*Rht-B1/Rht-D1, dwarf-8*	Wheat, Rice, Maize	Peng *et al.*, 1999
Selection during the maize domestication	Regulatory gene regions	*tb1*	Maize	Wang *et al.*, 1999
Early flowering time	Transcription factors	*Leafy hull sterile 1*	Rice	Jeon *et al.*, 2000
Colour variations; pale aleurone colour	Transcription factors	*pac1*	Maize	Carey *et al.*, 2004
Increased seed weight	AGP regulatory gene	*Sh2*	Maize	Giroux *et al.*, 1996
Production of decaffeinated coffee plants	3' untranslated region (RNAi)	*CaMXMT1*	Coffee	Ogita *et al.*, 2003
A dominant high-lysine opaque maize variant	Regulatory gene regions (RNAi)	*O2*	Maize	Segal *et al.*, 2003
Defects in pollen grain development	Promoter region (RNAi)	*TCP16*	*Arabidopsis*	Takeda *et al.*, 2006

against foreign nucleic acids. In cells, gene regulation by RNAi occurs through the generation of small RNAs (smRNAs), derived from larger double-stranded non-coding transcripts by nucleolytic processing into small (21–26-bp) RNA molecules. smRNAs mediate the post-transcriptional silencing of protein-coding genes having complementary sequences by both targeted degradation as well as translational inhibition. Several important plant transcription factors involved in developmental pathways are known to be regulated by these silencing mechanisms. In several plant studies involving smRNAs (specifically a smRNA sub-class referred to as "miRNAs" or "micro-RNAs"), mutants with impaired miRNAs exhibit abnormal development. The discovery of these smRNA-mediated regulatory processes has in recent years provided biotechnologists with powerful new tools to perform targeted disruption of gene expression for a wide range of experimental organisms. The introduction of synthetic DNA or RNA molecules into cells that give rise to double-stranded RNAs *in vivo*, has proved to be a highly efficient means to silence the expression of a gene of interest, and these methods are now routinely performed by many laboratories studying various developmental processes. For example, the *Arabidopsis* TCP16 gene encodes a transcription factor playing an essential role in early pollen development, and transgenic plants expressing an RNAi transgene cassette designed against TCP16 produce inviable pollen grains due to defects in their development (Takeda *et al.,* 2006). In addition, RNAi technology has also been used to improve crops for nutritional value. For example, caffeine content was significantly reduced in coffee plants by RNAi-mediated suppression of the caffeine synthase gene (Ogita *et al.,* 2003), and maize plants with increased seed lysine content have been generated by inhibiting the expression of maize zein storage proteins with poor lysine content using RNAi technology (Segal *et al.,* 2003).

4. Acknowledgments

This material is based on work partially supported by the project 'Genomic characterization of rice germplasm' of USDA-ARS National Program NP301 in Dale Bumpers National Rice Research Center, Stuttgart, Arkansas, USA and by the National Science Foundation under Grant No. 0638820, USA. We thank Rebekah Clemons for her artwork on the graphs and Dr. Scott Baerson for critical review.

5. References

5.1. Cited References

Bryan, G. T., Wu, K.-S., Farrall, L. *et al.* **2000.** A single amino acid difference distinguishes resistant and susceptible alleles of the rice blast resistance gene *Pi-ta*. *The Plant Cell.* 12: 2033–2045.

Buchel, S.A., Brederode, F.T., Bol, J.F. *et al.* **1999.** Mutation of GT-1 binding sites in the Pr-1A promoter influences the level of inducible gene expression *in vivo*. Plant Molecular Biology. 40: 387–396.

Chu, Z., Yuan, M., Yao, J. *et al.* **2006.** Promoter mutations of an essential gene for pollen development result in disease resistance in rice. *Genes & Dev.* 20: 1250-1255.

Green, P.J. 1993. Control of mRNA stability in higher plants. *Plant Physiol.* 102: 1065–1070.

Jacobsen, S.E., Binkowski, K. and Olszewski, N.E. 1996. SPINDLY – a tetratricopeptide repeat protein involved in gibberellin signal transduction in *Arabidopsis*. *Proc. Natl Acad. Sci. USA.* 93: 9292–9296.

Jeon, J.-S., Jang, S., Lee, S. *et al.* **2000.** *leafy hull sterile1* is a homeotic mutation in a rice MADS box gene affecting rice flower development. *The Plant Cell.* 12: 871–884.

Jia, Y., McAdams, S.A., Bryan, G.T. *et al.* **2000.** Direct interaction of resistance gene and avirulence gene products confers rice blast resistance. The EMBO Journal. 19: 4004–4014.

Kang, S., Lebrun, M.H., Farrall, L. *et al.* **2001.** Gain of virulence caused by insertion of a Pot3 transposon in a *Magnaporthe grisea* avirulence gene. *Molecular Plant-Microbe Interactions.* 14: 671–674.

Lim, M.T.S. and Kunkel B.N. 2004. Mutations in the Pseudomonas syringae avrRpt2 gene that dissociate its virulence and avirulence activities lead to decreased efficiency in AvrRpt2-induced disappearance of RIN4. Molecular Plant-Microbe Interaction 17: 313–321.

Mazumder, B., Seshadri, V. and Fox, P.L. 2003. Translational control by the 3'-UTR: the ends specify the means. *Trends Biochem. Sci.* 28(2): 91–98.

Nakazaki, T., Okumoto, Y.A., Horibata, S. *et al.* **2003.** Mobilization of a transposon in the rice genome. *Nature.* 421: 170–172.

Ogita, S., Uefuji, H., Yamaguchi, Y. *et al.* **2003.** Producing decaffeinated coffee plants. *Nature.* 423: 823.

Patel, A.A. and Steitz, J.A. 2003. Splicing double: insights from the second spliceosome. *Nature Reviews Molecular Cell Biology.* 4(12): 960–970.

Peng, J., Richards, D.E., Hartley, N.M. *et al.* 1999. 'Green revolution' genes encode mutant gibberellin response modulators. *Nature.* 400: 256–261.

Sato, Y., Sentoku, N., Miura, Y. *et al.* 1999. Loss-of-function mutations in the rice homeobox gene *OSH15* affect the architecture of internodes resulting in dwarf plants. *The EMBO Journal.* 18: 992–1002.

Segal, G., Song, R. and Messing, J. 2003. A new opaque variant of maize by a single dominant RNA interference-inducing transgene. *Genetics.* 165: 387–387.

Swiezewski, S., Crevillen, P., Liu, F. *et al.* 2007. Small RNA-mediated chromatin silencing directed to the 3' region of the Arabidopsis gene encoding the developmental regulator, FLC. *Proc. Natl Acad. Sci. USA.* 14: 3633–3638.

Takeda, T., Amano, K., Ohto, M. *et al.* 2006. RNA interference of the *Arabidopsis* putative transcription factor TCP16 gene results in abortion of early pollen development. *Plant Mol. Bio*. 61: 165–177.

Zhou, E., Jia, Y., Correll, J. *et al.* 2007. Instability of the *Magnaporthe oryzae* avirulence gene *AVR-Pita* alters virulence. *Fungal Genetics and Biology* (doi:101016/jofgb.2007.02.03).

5.2. Further Reading

Bartel, D.P. 2004. MicroRNAs: genomics, biogenesis, mechanism, and function. *Cell.* 116: 281–297.

Jia, Y., Wang, Z. and Singh, P. 2002. Development of dominant rice blast resistance *Pi-ta* gene markers. *Crop Sci.* 42:2145–2149.

Lemon, B. and Tjian, R. 2000. Orchestrated response: a symphony of transcription factors for gene control. *Genes & Dev.* 14: 2551–2569.

Lodish, H., Berk, A., Matsudaira, P. *et al.* 2004. Molecular Cell Biology. New York: W. H. Freeman. 5th ed.

Riechmann, J.L. and Ratcliffe, O.J. 2000. A genomic perspective on plant transcription factors. Current *Opinion in Plant Biology.* 3(5): 423–434.

Scott, W.R. and Walter, G. 2006. The evolution of spliceosomal introns: patterns, puzzles and progress. *Nature Reviews Genetics.* 7: 211–221.

Watson, J.D., Baker, T.A., Bell, S.P. *et al.* 2004. Molecular Biology of the Gene, 5th ed. Long Island, NY: Cold Spring Harbor Laboratory Press.

C04

Mutation Categories

U.Lundqvist[a], J.D.Franckowiak[b] and B.P.Forster[c,*]

[a] Nordic Genetic Resource Center, P.O. Box 41, SE-230, 53 Alnarp, Sweden
[b] Agri-Science Queensland, Department of Employment, Economic Development and Innovation, Hermitage Research Station, Warwick, QLD 4370 Australia
[c] BioHybrids International Ltd, P.O. Box 2411, Reading RG6 5FY, UK
* Corresponding author, E-MAIL: brianforster@biohybrids.co.uk

1. Introduction

Mutation has been described as a sudden change in the genetic material of living cells. However, at the beginning of mutation research, over 100 years ago, these sudden changes were observed and detected by the phenotypes they affected. The "sudden phenotypic changes" were heritable and therefore had a genetic base and mutation detection and monitoring in research and plant breeding were based entirely on phenotyping. Later, from the mid-1950s onward, light microscopy provided a means of observing aberrations at the ploidy, karyotype and chromosome levels. Today mutations can be detected at the DNA sequence level. Details of gene structure and regulation at the molecular level and consequences of intra-genic mutations are described in **Chapter 3**.

Since mutation deals with genetic material, mutations can be found in all cellular constituents that carry DNA, i.e. the nucleus, the mitochondria and (for plants) plastids. Various mutation classifications have been proposed, which differ according to detection methods used and/or the bias of the classifier. For example:

- Stubbe (1938) divided mutations into three classes: 1) genome, 2) plastidom (all plastids in the cytoplasm of a cell) and 3) plasmon (the total cytoplasmic system of eukaryotes, mitochondria and plastids) with genome mutations being sub-divided into: genome, chromosome and gene classes.
- Gustafsson and Ekberg (1977) categorized mutations into: 1) genome mutations, 2) chromosome mutations and 3) extra-nuclear mutations.
- Auerbach (1976) classed mutations into: 1) changes in chromosome number, 2) inter-genic mutations and 3) intra-genic mutations.
- van Harten (1998) put forward a comprehensive catalogue of mutation classes: 1) nuclear and extra-nuclear mutations, 2) spontaneous and artificially induced mutations, 3) macro-mutations and micro-mutations and 4) phenotypic mutations.

The above classifications serve the purposes of the respective authors and although generally similar they can be ambiguous and confusing. A new, simple system is used here. This focuses on mutations of most interest to plant breeding. These are predominantly those in the DNA of the nucleus; mutations in cytoplasmic organelles are dealt with in **Chapter 17**. Nuclear muta-tions are divided into genome, chromosome and gene mutations and these can be sub-divided further based on changes in number and composition.

2. Classification at the Phenotypic Level

Variation in phenotype was the first means by which mutations were described and catalogued. The origins were in the variants described in taxonomic treatises of wild and cultivated plants by Linné, and in spontane-ous mutants (bud sports, etc.) of value and interest in domestication and crop improvement (**see Chapters 1–2 and 24**). When the first induced mutants were produced in plants and animals, these were described on the basis of their phenotype. Subsequent genetic studies revealed the controlling genes and these were named and coded based on the phenotypes they produced. Thus the gene for semi-dwarfism in barley has the symbol *sdw1* (see below for more on genetic nomenclature and mutant cataloging).

Phenotypic mutations are of particular interest to plant breeders as they describe traits that are of economic value, and since plant breeding has been traditionally a process based on phenotypic selection there is a natural link between breeding and mutation cataloging. Large mutation populations have been developed for research and plant breeding and classed according to phenotype. In cereals the most common mutation classes are for leaf colour, plant stature and sterility. As interest grew more phenotypic classes were developed. These cover all stages in development from seed traits to germination, to vegetative and generative growth and energy storage (**see Chapter 25**). In the case of barley, mutants have been collected from all sources (spontaneous and induced) worldwide and more than 600 are described in the Barley Genetic Stocks Ace Database: http://ace.untamo.net. Both Latin and English have been used to describe phenotypes and (subsequently) gene symbols.

3. Classification at the Genotypic Level

3.1. Genome Mutations

The genome consists of a basic set of chromosomes and is often assigned a species specific symbol. Diploid species are given a double symbol, e.g. for barley, *Hordeum vulgare* this is HH. The genome number of symbols correlates to the ploidy level thus H, HH, HHH and HHHH, etc. correspond to haploid, diploid, triploid, tetraploid barley, etc. In this example the same genome is multiplied to produce autoploids. Some polyploid species are composed of different genomes and known as alloploids, for example durum wheat is tetraploid with two diploid genomes, AABB, and bread wheat is hexaploid with three diploid genomes, AABBDD. Polyploidy has been a major evolutionary mechanism in plants and has been exploited in both plant domestication and crop improvement (Hagberg and Åkerberg, 1961).

3.1.1. Ploidy Mutations

Deficiencies. These include a reduction in genome number, e.g. from diploid barley (HH) to haploid barley (H). Haploid production and thereby doubled haploid production to produce homozygous lines is a valuable technology in plant breeding and genetics of many species. This is usually achieved by inducing embryogenesis in haploid (gametic) cell cultures and inducing genome doubling. In some species however, there are natural genetic mechanisms that promote a reduction in ploidy, e.g. the haploid inducer genes in maize and barley that increase the frequency of haploid embryos in seeds (Barrett *et al.,* 2008 and Finch, 1989, respectively). Recently, Ravi and Chan (2010) reported the production of haploids produced by centromere-mediated genome elimination. A mutant in the centromere protein gene (*cenh3*) was isolated from an *Arabidopsis thaliana* TILLING population, in which mutants were induced by ethylmethane sulphonate. When *cenh3* mutants are crossed with wild type the chromosomes of the mutant fail to attach to zygotic mitotic spindles. Mutant chromosomes are thereby lost and haploid embryos are formed. This method of targeting mutants causing perturbations in centromere function may be applied in producing haploids in a wide range of crops.

Irradiation of pollen is another valuable technique in producing haploid embryos. Here the irradiation treatment does not affect the pollen's ability to germinate, grow and penetrate the embryo sac, but does disable the sperm. Thus pollination occurs without subsequent fertilization, the method is used routinely in many fruit species (**see Chapter 30**).

Reductions in genomes commonly occur in interploidy crosses, natural or man-made. In genetics and plant breeding these can be manipulated in introgressing genes from one species to another, e.g. from a wild diploid wheat species into cultivated durum (tetraploid) and bread (hexaploid) wheats. These methods use cytogenetic methods and ultimately aim to restore the ploidy level of the crop species.

Additions. Polyploidy either by genome duplication (autoploids) or genome addition (alloploids) has occurred naturally in the evolution of many species and has also been induced for crop improvement. One effect of polyploidy is to increase the volume of the nucleus; this in turn increases cell size and tissue, organ and plant size. Alloploids have additional advantages as the different genomes contain different sets of genes and thereby enrich gene diversity (add new genes) and promote heterosis (hybrid vigour). Polyploids, spontaneous or induced have therefore been attractive for domestication and crop improvement. Crops domesticated as polyploids include:

- triploids: banana, watermelon and apple.
- tetraploids: cotton, brassicas, durum wheat, leek, potato and tobacco.
- hexaploids: bread wheat, oat, triticale and chrysanthemum.
- octaploids: dahlia, strawberry and pansies.

In some species the polyploidy event is so ancient that it cannot be recognized by cytological techniques and can only be detected at the molecular level, e.g. by discovering gene duplications. Such paleopolyploids include: maize, rice, rubber and soybean. Deliberate induction of polyploidy began in the mid 20th century and was successful in vegetable crops where organ size is of particular importance (Hagberg and Åkerberg, 1961). A more recent example is the development of the wheat/rye hybrid, triticale, which combines productivity of wheat with abiotic stress tolerance of rye. Triticale has been produced at the hexaploid (AABBRR) and octoploid (AABBDDRR) levels and the hexaploids are now established as a crop for marginal lands not suitable for wheat cultivation.

3.1.2. Genome Rearrangements

Genomes are not static and evolution has involved many forms of qualitative change. This is becoming increasing apparent with the discovery of paleopolyploids. Here the interactions, exchanges and fusions within and between genomes have produced a composite that effectively functions as a single genome. Intermediary stages, such as recent chromosome translocations, duplications and deletions can be more readily detected by cytological observations, e.g. by chromosome pairing at meiosis. However, a complication has been the co-evolution of pairing control mechanisms. For example hexaploid wheat is composed of three similar genomes (AA, BB and DD), each with seven pairs of chromosomes. One may therefore expect seven hexavalents to form at meiosis, but instead hexaploid wheat has been diploidized and produces 21 bivalents. The strict pairing of homologous chromosomes is enforced by the action of the *Ph1* gene. Mutations of this gene, e.g. *ph1b* have been produced and exploited to induce meiotic recombination between the genomes of wheat and related species, thus providing a means of transferring genes from one genome to another (**see Chapter 19**).

3.2. Chromosome Mutations

The chromosome is the body that links genes together, it is also known as a linkage group. Chromosomes have two arms, which may be of equal or unequal lengths; they are joined at the centromere (the location for the attachment of spindle fibres during mitotic and meiotic cell divisions). Each species normally has a standard set of chromosomes and this is known as the euploid complement.

3.2.1. Aneuploidy

Any change from the normal euploid complement is termed aneuploid and includes the addition and/or loss of one or more chromosomes or parts of chromosomes. Duplications and deletions, especially the loss of (or part of) a chromosome, cannot be tolerated by most diploid species. Some diploids can tolerate additional chromosomes, but these duplications are often lost during meiosis and in the subsequent generations the aneuploids revert to euploidy. Polyploids have more genetic buffering and viable aneuploid plants can be produced.

3.2.2. Chromosome Addition, Loss and Substitution

The development of aneuploid stocks has been a major asset in wheat genetics. Aneuploids can occur naturally, but are produced in abundance in progeny of crosses where parents contribute unequal numbers of genomes or chromosomes. This was exploited in wheat by Prof. Ernie Sears (University of Missouri, USA) to produce aneuploids for each of wheat's 21 chromosomes. Aneuploid stocks in wheat include: monosomics (deficient for one chromosome), nullisomics (deficient

for a pair of chromosomes), tetrasomics (containing an additional chromosome pair), nulli-tetrasomics (missing one pair of chromosomes but compensated for by an additional pair); ditelosomics (missing a pair of chromosome arms), etc. Aneuploids in wheat also include additional or substitutional chromosomes from other species (alien chromosome addition and substitution lines). Often aneuploid lines are unstable during meiosis and have to be checked and re-selected at every generation, otherwise they revert to euploidy. However, some alien chromosomes have the ability to ensure their transmission to the next generation and have been dubbed "cuckoo chromosomes". An example is the wheat/alien addition line for chromosome 4S of *Aegilops sharonensis*. Rather than reverting to euploidy, the transmission of the 4S chromosome to the next generation is guaranteed by developmental failure of pollen and/or eggs that lack the chromosome. The exploitation of aneuploids in wheat is discussed in Law *et al.*, (1987).

Random deletions in chromosomes can be induced by irradiation treatments (**see Chapter 11**).

3.2.3. Chromosome Rearrangements

Chromosome breaks produced by mutagenic treatments must be repaired if the cell is to survive such damage, but mistakes in the cellular "repair" system can lead to:

- **Inversions**, where a piece of chromosome is re-ligated in place but rotated through 180°, so that the linear order of genes is opposite to the wild type orientation. If the inversion involves the centromere it is known as a pericentric inversion, if not it is known as a paracentric inversion. Meiotic recombination events involving inversions can result in recombinants carrying duplications or deficiencies for parts of that chromosome.

- **Translocations**, where a piece of chromosome is relocated on the same chromosome (intrachromosomal) or transferred to a different (inter-chromosomal) location. Translocations do not involve loss of genetic material. Translocations are detected by cytological checks for multivalent formation at meiosis. They are an important means of transferring blocks of genes from one species/genome to another, especially in polyploid crops (**see Chapter 19**).

- **Reciprocal translocations** are the most common form of translocation and occur when two simultaneous chromosome breaks produce pieces that swap positions. These have been used successfully in plant breeding, particularly in polyploid crops. X-ray induced reciprocal translocation was used to transfer leaf rust resistance into wheat from a wild relative, *Aegilops umbellulata* (Sears, 1956). Other examples include the transfer of: disease resistance from rye and *Agropyron elongatum* into wheat; disease resistance from *Avena barbata* into oat; nematode resistance from *Beta patellaris* and *Beta procumbens* into sugar beet, and virus resistance from *Nicotiana glutinosa* into tobacco (**see Chapter 19**).

3.3. Gene Mutations

Gene structure and gene mutations are the topic of **Chapter 3** and only a brief description will be given here. Similar to genome and chromosome mutations these can be broadly divided into changes in gene number and composition.

3.3.1. Gene Copy Number Mutation

The copy number of a gene (paralogues) can increase and decrease by spontaneous or induced mutation. Duplications of a gene can occur at any site in the genome though they are most frequent in homologous chromosomes. Multiple copies of genes are found naturally for proteins produced in high amounts, e.g. the grain storage proteins of cereals. Mutations that affect the copy number may reduce or increase expression of these genes resulting in altered seed composition. Changes in copy number mutations are usually a product of mispairing during meiosis and recombination. However, the more common mutation event is a reduction in the effective copy number caused by point mutations.

3.3.2. Non-Structural Mutations: Changes in DNA Sequence

This category involves changes in the base pair sequence of genes. The DNA sequence is made up of purine (A and G) and pyrimidine (T and C) bases, which are recognized as codons (triplets of bases) that are eventually translated into gene products or amino acid sequences (proteins). Mutations are also caused by changes in non-coding regions of the genome e.g. promoters, introns, heterochromatin and repetitive DNA. Common non-structural mutations of genes include:

- **Point mutations**, these are the most common type of mutation induced by chemical mutagens. They involve single-base changes and may result in missense codons (codes for a different amino acid), silent (no change in amino acid coding) and non-sense (amino acid codon is changed to a translation stop codon). Point mutations are also known as single-base substitutions, which contribute to the build up of single nucleotide polymorphisms (SNPs).
- **Insertions and deletions** of one or more nucleotides. Collectively known as indels and may result in changes to codons and frameshifts in codon reading frames with qualitative and quantitative effects on the proteins produced.
- **Mutations** in the specific gene regions; such as promoter, coding, stop, intron and 3′ untranslated sequences.
- **Transposon insertion**, also known as insertional mutagenesis. Transposons are DNA sequences that can transpose, move in and out of chromosomes. They are often activated by environmental stress and their activity can knock out or activate genes by inserting into or exiting genes. Transposable elements and mutagenesis are discussed in detail in **Chapter 39**.

3.3.3. Mutation Mimics

Various physiological disorders can result in what appear to be "phenotypic mutants". However, these are not genetic and can be differentiated from true mutations by an absence of inheritance in progeny. Many epigenetic effects are poorly understood and the term is used as an umbrella to cover many physiological aberrations that are caused by factors other than genetic. The best understood of these is DNA methylation, which can interfere with normal gene expression. Methylation is a common disorder of DNA in tissue culture-generated plants. The condition may be transient in which case the regenerated plants grow out of the disorder and revert to a normal phenotype, in other cases the abnormal phenotype may persist for the whole life time of an individual. However, some epigenetic changes are heritable when methylation patterns are not completely reset in gametes and after fertilization.

4. Genetic Nomenclature

Standardized genetic nomenclature is essential for effective communication in any scientific discipline. This allows scientists to read and understand literature, talk to each other, compare data and request pertinent material, such as specific mutant resources. Names and symbols for genomes, chromosomes, loci, genes and alleles are normally debated, recommended and assigned by international elected individuals or committees for a given species or taxonomic group. The process is under continual review and often pragmatic in outcomes. For most species various systems have been used by various workers in various regions at various times. The incongruity is relatively easy to resolve within a scientific community working on a given species, for example barley. However, as syntenic relationships are established between species, e.g. between barley and other members of the small grain cereals (wheat, rye, etc.) and further to model species (*Arabidopsis*, *Brachypodium*, etc.) some standardization is needed at a more general level.

Barley, *Hordeum vulgare*, has relatively simple inheritance patterns; it is a diploid species with just seven pairs of chromosomes: and it is an inbreeding, annual crop. Barley has long been established as a model species for genetics and in particular for mutation research. Barley researchers have experienced and overcome many problems relating to genetic nomenclature, and therefore barley is a good example to illustrate the various issues involved.

4.1. Barley Genetic Nomenclature

The barley nuclear genome consists of seven distinct (heterologous) chromosomes. In early work barley cytologists numbered these 1 to 7 based on the length (longest to shortest with 6 and 7 having satellites) as observed when using a light microscope (Burnham and Hagberg, 1956). However, once mapping of genes began, it became apparent that the gene order found on a barley chromosome is similar to that found in homoeologous chromosomes of related species. Groups of homoeologous chromosomes (those with similar gene sets) were established in the Triticeae and as a consequence a new numbering system was developed and recommended for barley in 1996. Recommendations for the designation of barley chromosomes and their arms were made at a business meeting of the Seventh International Barley Genetics Symposium (University of Saskatchewan, Canada, 1996). The following has been adapted from the resolutions passed:

1. Each of the seven barley chromosomes is designated by a figure from 1 to 7 according to its homoeologous relationship with chromosomes of other Triticeae species. The figure is followed by the letter H; e.g. 1H, 2H, 3H, 4H, 5H, 6H and 7H.

2. The genome of barley is symbolized by the letter H. **Note:** The use of the genomic symbol H instead of I agrees with usage by most barley workers, but

---

Box 4.1: Locus, gene and allele nomenclature

The American system for gene symbols

Mdh	Three lettered code for the malate dehydrogenase gene
Mdh1	Locus number one
Mdh2	Locus number two
Mdh1.a	Allele a at locus 1
Mdh1.b	Allele b at locus 1
Mdh2.c	The third Mdh mutant allele assigned to the Mdh2 locus
Mdh.d	The fourth Mdh mutant, but not assigned to a locus
Mdh	Note that the first letter is upper case and indicates dominance.

Note that a full stop is used to separate the locus and allele portions of the gene symbol.

The Classic system for gene symbols

ari	Three lettered code for short awn (breviaristatum) genes
ari-e	The fifth locus having a breviaristatum gene
ari-e.1	The first breviaristatum mutant assigned to the fifth locus
ari-e.GP	The Golden Promise mutant at the ari-e locus
ari.137	The 137th short awned mutant, but not assigned to a locus
ari	Note the lower case first letter indicates recessiveness

goes against a proposal for the designation of all *Triticeae* genomes presented at the Proceedings of the Second International Triticeae Symposium.

3. The chromosome arms are designated by the letters S and L. The two arms of chromosome 1H are therefore: 1HS and 1HL, for 2H; 2HS and 2HL, etc. **Note:** The arm designations should now be regarded as "names" and do not refer to physical short and long arms.

Gene and allele nomenclature in barley, as in many other species, is not fully unified, though there are certain standard rules. There are two commonly used systems in barley, an American system and a Classic (Latin) system. The similarities and differences between the two systems (for locus, gene and allele nomenclatures) are also issues found in the genetic nomenclature of other plants. Barley can therefore act as a general guide. Both the American and Classic systems use a three letter italicized symbol for a locus. In the American system the three letters are derived from the English description of the phenotypic class whereas in the Classic system the letters derives from the Latin name (**see Box 4.1**).

The locus is the position at which a gene, or cluster of genes resides in the genome. Phenotypically similar genes (loci) are ordered numerically in the American system, but alphabetically in the Classic system. Alleles (mutants) in a phenotypic class are generally ordered alphabetically or numerically in the respective systems with permanent symbols assigned to a particular allele (mutant). As more information about a given mutant is obtained, the locus symbol may change, but the allele symbol is retained. **Box 4.1** gives examples of locus, gene and allele nomenclature using the American and Classic system, respectively. More details and guidelines for assigning symbols to barley genes are given in Franckowiak and Lundqvist (2009).

4.2. Allelism Testing

From **Box 4.2** it can be seen that the brachytic gene, *brh1*, has had many previous names, e.g. *br*, *ari-i* and *dx1*. Gene symbols are often ascribed to mutants once they are discovered with the gene symbol reflecting the trait. However, many traits, such as dwarfism (of which brachytic is one type) are controlled by various genes at several loci. Major issues are to separate the genetically different phenotypic classes of mutants and to group similar mutants that are alleles of the same gene (locus). This is done by allelism testing. Since the vast majority (over 90%) of mutant genes are recessives simple allelism testing is recommended. The standard practice in barley is as follows: a newly discovered mutant, for example a dwarf mutant, is crossed onto standard stocks for all or the most likely candidate mutant genes (loci). In cases where the new mutant and the standard stock match, the mutant phenotype will be expressed in the F_1 hybrid and allelism proven. In all other cases complementation between the mutant gene and its

---

Box 4.2: An excerpt of descriptions for BGS001 as an example of information available from the BGS database

Stock number:	BGS001
Locus name:	Brachytic 1
Locus symbol:	*brh1*
Previous nomenclature and gene symbolization:	Brachytic = *br*; Breviaristatum-i = *ari-i*; Dwarf x = *dx1*
Inheritance:	Monofactorial recessive. Located in chromosome 7HS in bin 7H-2, about 9.3 cM from the *fch12* locus.
Description:	Plants have short leaves, culms, spikes, awns and seeds. The seedling leaf is about 2/3 normal length, and a similar size reduction in the size of other organs is observed, but the awns are less than 1/3 normal length. Seeds are small and yields are low. The mutant phenotype is easy to classify at all stages of growth.
Origin of mutant:	A spontaneous mutant in Himalaya.
Mutational events:	*brh1.a* in Himalaya; *brh1.c* in Moravian; *ari-i.38* in Bonus; *brh1.e* in Aramir; *brh1.f* in Domen; *brh1.t* in Akashinriki; *brh1.x* in Volla; *brh1.z* in Aapo; *brh1.ae* in Steptoe.
Mutant used for description and seed stocks:	*brh1.a* in Himalaya; *brh1.a* in Bowman; *ari-i.38* in Bowman; *brh1.e* in Bowman; *brh1.t* in Bowman; *brh1.x* in Bowman.
References[a]:	Dahleen *et al.* (2005), Fedak *et al.* (1972); Franckowiak (1995); Gustafsson *et al.* (1969); Holm and Aastveit (1966); Kucera *et al.* (1975); Powers (1936); Swenson (1940); Szarejko and Maluszynski (1984); Tsuchiya (1974).
Prepared[a]:	Tsuchiya and Haus, 1971.
Revised[a]:	Tsuchiya, 1980; Franckowiak, 1997; Franckowiak and Dahleen, 2007.

[a] For references see BGS001 in Barley Genetic Stocks Ace Database: http://ace.untamo.net

wild type allele will produce normal offspring. Thus in **Box 4.2**, the mutation events isolated in the barley varieties 'Himalaya', 'Moravian', 'Bonus', 'Aramir', 'Domen', 'Akashinriki', 'Volla', 'Aapo' and 'Steptoe' were found to be allelic to *brh1.a* , and therefore added as alleles at the *brh1* locus.

When new mutations are discovered, for example where no allelism is found, crosses with various morphological mutant stocks can be used to generate segregating populations from which linkage with previously mapped loci can be determined.

4.3. Databases

A description of barley genes and genetic stocks is available at the GrainGenes website: http://wheat.pw.usda.gov/ggpages/germplasm.shtm by following the 'Barley' link to 'Barley Genetic Stocks'. A link to the Barley Genetic Stocks AceDB is also available which allows database searches for germplasm, genes, loci, alleles, and mutational events by various barley genotypes, genetic map locations, seed stocks, authors, compliers, revisions and references. Photographs of wild type and mutant phenotypes are also provided. The Barley Genetic Stocks (BGS) database houses information on more than 600 loci. As an example of the type of information contained an excerpt for the first entry, BGS001, is provided in **Box 4.2**.

5. References

5.1. Cited References

Auerbach, C. 1976. Mutation Research. Problems, Results and Perspectives. London: Chapman and Hall.

Barret, P., Brinkmann, M. and Beckert, M. 2008. A major locus expressed in the male gametophyte with incomplete penetrance is responsible for *in situ* gynogenesis in maize. *Theoretical and Applied Genetics*. 117: 581–594.

Burnham, C.R. and Hagberg, A. 1956. Cytogenetic notes on chromosomal interchanges in barley. *Hereditas*. 42: 467–482.

Finch, R.A. 1989. The *hap* gene causes facultative pseudogamy in barley. *Barley Genetics Newsletter*. 13: 4–6.

Franckowiak, J.D. and Lundqvist, U. 2009. Rules for nomenclature and gene symbolization in barley. *Barley Genetics Newsletter*. 39: 77–81.

Gustafsson, Å. and Ekberg, I. 1977. Types of mutations. Manual on mutation breeding. Technical reports no. 119. Vienna: IAEA, pp. 107–123.

Hagberg, A. and Åkerberg, E. 1961. Mutations and Polyploidy in Plant Breeding. Bonniers, Stockholm: Scandinavian University Books, p. 150.

Law, C.N., Snape, J.W. and Worland, A.J. 1987. Aneuploidy in wheat and its uses in genetic analysis. *In*: F.G.H. Lupton (ed.), Wheat Breeding. Its Scientific Basis, pp. 71–127, London and New York: Chapman and Hall.

Ravi, M. and Chan S.W.L. 2010. Haploid plants produced by centromere-mediated genome elimination. *Nature*. 464: 615–619.

Sears, E.R. 1956. The transfer of leaf-rust resistance from *Aegilops umbellulata* to wheat. *In*: Genetics in Plant Breeding. New York: Upton, pp. 1–22.

Stubbe, H. 1938. Genmutation I. Allgemeiner Teil. Handbuch der Ver-erbungswissenschaft. Band II F. Berlin: Verlag von Gebrüder Bornträger.

van Harten, A.M. 1998. Nature and Types of Mutations. Mutation Breeding. Theory and practical applications. Cambridge: Cambridge University Press, pp. 64–110.

5.2. Websites

Barley Genetic Stocks Ace Database:
http://ace.untamo.net
GrainGenes:
http://wheat.pw.usda.gov/ggpages/germplasm.shtm
Gramene:
http://www.gramene.org
ITMI (International Triticeae Mapping Initiative):
http://wheat.pw.usda.gov/ITMI/
NASC (European Arabidopsis Stock Centre):
http://arabidopsis.info/
Plant OntologyTM Consortium:
http://www.plantontology.org
MaizeGDB:
http://www.maizedb.org
Mutant Variety and Genetic Stocks:
http://mvgs.iaea.org
Oryzabase:
http://www.shigen.nig.ac.jp/rice/oryzabase
Solanaceae Genomics Network:
http://www.sgn.cornell.edu

TAIR (The Arabidopsis Information Resource):
http://www.arabidopsis.info
TIGR (The TIGR Arabidopsis thaliana Database):
http://www.tigr.org
USDA National Plant Germplasm System
http://www.ars-grin.gov/npgs/

5.3. Further Reading

Linde-Laursen, I. 1997. Recommendations for the designation of the barley chromosomes and their arms. *Barley Genetics Newsletter.* 26: 1–3.

Franckowiak, J., Lundqvist, U. and Konishi, T. 1997. New and revised names for barley genes. *Barley Genetics Newsletter.* 26: 4–8.

DNA Repair Pathways and Genes in Plant

M. Curtis*

Oregon State University, Department of Botany and Plant Pathology, 2050 Cordley Hall, Corvallis, OR 97331,USA
* Corresponding author, E-MAIL: curtism@science.oregonstate.edu

1. Introduction

DNA is frequently damaged by exogenous radiation and chemicals, and by endogenous oxidants and metabolites. DNA lesions vary in their toxicity and mutagenic ability. Lesions blocking transcription can directly interfere with cell function. Lesions blocking replication can elicit the DNA damage response (DDR), which delays cell division and sometimes kills the cell, and blocked replication forks are susceptible to recombination and subsequent chromosomal rearrangements. Postreplication repair (PRR) pathways promote replication past DNA lesions, but frequently mispair or misalign, creating mutagenic mismatches. Research on mutagenesis firmly demonstrated that organisms actively repair DNA when damaged by radiation, chemicals or replication errors, and in the absence of repair in unicellular organisms, viability declines sharply and surviving colonies are mutated.

Prokaryotes and eukaryotes have evolved conserved pathways to repair DNA lesions. While we have gained enormous knowledge of DNA damage, repair and mutagenesis in microbes and animals, the field of plant DNA repair is still in its infancy, with the possible subfield exceptions of direct repair of UV-photoproducts, and double-strand break repair (**see Chapter 6**). The origin and types of DNA lesions are briefly described and several DNA repair pathways are elaborated with links to plants whenever possible.

2. Origin and Types of DNA Lesions

DNA decays in the oxidative and aqueous cellular environment. Hydrolysis of the base–sugar bond yields apurinic/apyrimidinic sites (AP site), and deamination of cytosine or 5-methylcytosine yields uracil (U:G) or thymine (T:G), respectively. Metabolism creates reactive oxygen species (the hydroxyl radical ($^{\bullet}$OH) being the most reactive) and alkylating agents (S-adenosylmethionine, nitrosoamines (RN=NOH)) that oxidize or alkylate DNA. Both biotic and abiotic stress increases reactive oxygen, either purposefully, as for signaling molecules, or unintentionally as metabolic electron flow spills over onto oxygen. Secondary damages of ROS, such as lipid peroxidation, also cause DNA lesions. Exogenous chemicals, such as the commonly used mutagens, ethyl methanesulphonate (EMS), or N-methyl-N-nitrosourea, methyl chloride and some antibiotics increase alkylation of DNA bases. Exogenous gamma (γ)- and UV-radiation cause double-strand-breaks (**DSB, see Chapter 6**) and UV-photoproducts, respectively, and also induce some oxidative DNA damage. The mutagenicity of a DNA lesion depends on frequency, recognition by repair pathways, mode of lesion bypass (PRR) during DNA replication and types of lesion and mismatches.

The types of DNA lesion, their occurrence and effect, and the repair pathways recruited are given in **Table 5.1**.

3. DNA Damage Response and Repair

3.1. Response to DNA Damage

DNA lesions elicit cellular responses directly or indirectly. Different pathways directly recognize particular lesions and initiate repair, keeping steady-state levels of DNA damage low. Bulky DNA lesions block transcription, recruiting repair proteins to the damaged site. A diverse variety of DNA damages block replication, which signals responses that delay/arrest the cell-cycle, kill the cell and/or promote mutagenic DNA repair (**Figure 5.1**). Balancing responses to replication blocks is significant to mutagenesis and effects of DNA damage on tissue development. Consequences of DNA lesions depend on the affected cell type: quiescent, stem, amplifying (rapidly dividing) or mature. Arrest or death of quiescent or stem cells may switch meristem development from indeterminant to determinant, while cell-cycle delay of rapidly dividing cells may decrease growth rate and overall plant fitness (reduced fecundity i.e. reproductive success). Conversely, quiescent or stem cells that are progenitors of the germline may pass mutations to progeny.

3.2. Mutagenic DNA Repair

Mutation is an endpoint of mutagenic pathways that start with replication of undamaged or damaged DNA (**Figure 5.2**). Organisms have several general mechanisms that repair DNA lesions including base excision repair (BER), nucleotide excision repair (NER) and direct repair (DR). These repair systems keep steady-state levels of damaged DNA low, however, persistent DNA lesions block high-fidelity replicative DNA polymerases, increasing single-strand DNA (ssDNA) and the probabil-

Table 5.1: Common DNA lesions

DNA lesion	Occurrence	Base-pairing	Block to replication	Repair pathway
Oxidized pyrimidines				
Thymine glycol (Tg)	Low, spontaneous, gamma radiation	Tg:A>G	Yes	BER
Oxidized purines				
8-oxoguanine (8-oxoG)	High, spontaneous, stress, radiation	8-oxoG:A	No	BER
Formamidoguanine (FapyG)	Same as 8-oxoG	FapyG:A	Yes	BER
Alkylation				
O^6-methylguanine (O^6-MeG)	Spontaneous, chemicals	O^6-MeG:T	Yes	Direct
N^3-methyladenine (N^3-MeA)	Same as O^6-MeG		Yes	BER
1-N^6-ethenoadenine (εA)	Lipid peroxidation	εA:C, A or G	Yes	BER
Hydrolysis/deamination				
Apurinic/apyrimidinic (AP site)	Spontaneous, BER processing	No template base	Yes	AP endo.
Uracil (C →U)	Spontaneous, dUTP	U:G, U:A	?	BER
Thymine (^{5me}C→T)	Spontaneous	T:G	No	BER
Xanthine (A→X)	Spontaneous	X:G	No	BER
Hypoxanthine (G→HX)	Spontaneous	HX:A	No	BER
Other damages				
Cyclobutane pyrimidine dimers (CPDs)	UV radiation	3' C:T[a]	Yes	NER, direct
Pyrimidine [6-4] Pyrimidionine ([6-4]s)	UV radiation	3' T:G[a]	Yes	NER, direct
Double-strand breaks (DSB)	Gamma radiation, fork collapse	Rearrangements	DDR*	DSBR[c]
Single-strand breaks (SSB)	Oxidation, hydrolysis, AP and BER processing	Rearrangements	Fork collapse, DDR*	SSBR[c]
Cross-links	Radiation, oxidation		Yes	NER
Mismatches	DNA Replication	12 possible[b]	No	MMR
Base-loopouts	DNA Replication	No base-pairing	No	MMR

[a] The 3' base of UV-photoproduct dimers tends to be more mutagenic than the 5' base.
[b] G:T, T:G, G:A, A:G, G:G, C:T, T:C, C:A, A:C, C:C, T:T, A:A.
[c] DSBR = double-strand break repair; SSBR = single-strand break repair.
* DNA-damage response (DDR) is a complex network of pathways signalling cell-cycle delays or cell death.

ity of double-strand breaks (**DSBs, see Chapter 6**), both substrates for chromosomal rearrangements and elicitors of the DNA damage response (DDR). Organisms have several mechanisms to promote DNA replication past DNA damage, thereby avoiding cell-cycle delays and cell death that may perturb plant development.

DNA-lesion bypass is referred to by several names: damage tolerance, damage avoidance and post-

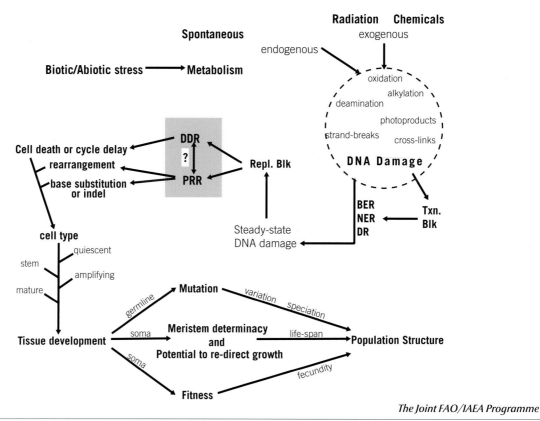

Figure 5.1 Responses to DNA damage in plants. The relationship (?) between post-replication repair (**PRR**) and the DNA damage response (**DDR**) influences DNA damage and effects on mutation and tissue development. See text for details. Base excision repair, **BER**; nucleotide excision repair, **NER**; direct repair, **DR**; replication block, **Repl. Blk**; transcription block, **Txn. Blk**; insertion/deletion, **indel**.

replication repair (PRR), and includes (*i*) specialized polymerases for trans-lesion synthesis (*TLS pols*), (*ii*) recombination or (*iii*) a template-switch. Bypass of damaged DNA by recombination or template-switch can be error-free but otherwise risks chromosomal rearrangements. Bypass by TLS pols is efficient but risks frequent (10^{-2}–10^{-3}) nucleotide mispairing or misalignment resulting in mismatches.

Mismatches also occur infrequently (10^{-7}–10^{-8}) during replication of undamaged DNA, but frequent enough to cause reduced fitness by mutation accumulation. Mismatch repair (MMR) directs excision of mutagenic mismatches, reducing the error-rate by 10- to 100-fold. The overall spontaneous mutation rate is therefore 10^{-9}–10^{-10}. Mismatches escaping MMR result in base substitution or insertion/deletion (indels), which along

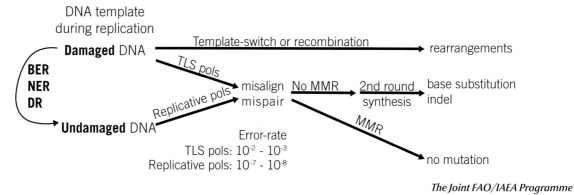

Figure 5.2 Mutagenic pathways. Replication of damaged DNA or undamaged DNA by trans-lesion synthesis polymerases (TLS pols) or replicative polymerases, respectively, can result in misaligned or mispaired DNA. Replication by TLS pols risks an error rate up to 100,000-fold greater than replication by replicative polymerases. Mismatch repair (MMR) prevents mutation. Damaged DNA may also be bypassed by a template-switch or recombination pathway that is error-free, but risks chromosomal rearrangements. Base excision repair (BER), nucleotide excision repair (NER) and direct repair (DR) limit the amount of damaged DNA.

(a) Base excision repair

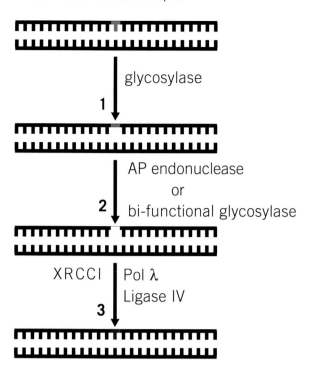

(b) Nucleotide excision repair

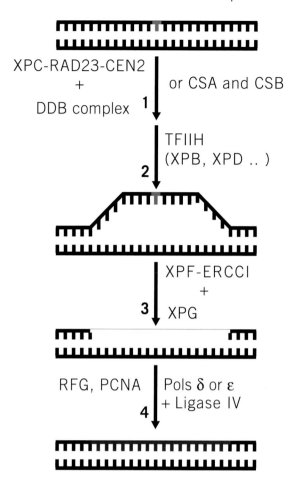

Figure 5.3 DNA excision repair mechanisms. (a) Base excision repair. (b) Nucleotide excision repair.

with chromosomal rearrangements that incur during genome replication are the endpoints of mutagenic pathways (**Figure 5.2**).

4. DNA Repair Pathways

4.1. Base Excision Repair (BER)

Base excision repair is initiated by DNA glycosylases that have specific or broad recognition of modified DNA bases (**step 1, Figure 5.3a**). Mono-functional glycosylases excise damaged bases leaving an AP site that is further processed by AP endonucleases, while bi-functional glycosylases have intrinsic endonuclease activity. Cleavage of the phosphate backbone by endonuclease activity (**step 2, Figure 5.3a**) releases the abasic deoxyribose

leaving a 3′ and 5′ end for DNA synthesis and ligation, respectively. If the 3′-end is not a hydroxyl and/or the 5′-end is not a phosphate, end modification is required for DNA repair synthesis (**step 3, Figure 5.3a**).

4.1.1. BER Mechanism

The challenge for BER is efficient recognition of damaged bases in a background of millions to billions of undamaged bases. Over-expression of DNA glycosylases can cause toxicity by excessive increases in AP sites. DNA glycosylases scan for damaged bases by trying to kink or bend the DNA helix and flip the target base into the glycosylase catalytic pocket. An aromatic amino acid that can base stack promotes the flipping out of the target base. Specificity for damaged bases is achieved at two steps. First, damaged bases destabilize base stacking and thus are more easily flipped out of the helix and act as flexible hinges for helix conformations that stabilize the glycosylase:DNA complex. Second, damaged bases have a distinct shape and distribution of electrostatic and hydrogen bonding potentials that are complementary to the catalytic pocket of DNA glycosylases. Once in the pocket, the N-glycosidic bond between the deoxyribose and base is cleaved.

4.1.2. Examples of BER in Plants

Plants have much 5-methylcytosine (5-MeC) at CG, CNG (N; any nucleotide) and CHH (H; A, C or T) sequences. Spontaneous deamination of 5-MeC to thymine causes C→T mutations. There are two plant-specific, bi-functional DNA glycosylases that excise 5-MeC initiating repair synthesis and insertion of unmethylated C. Mutants lacking the REPRESSOR OF SILENCING 1 (ROS1) show hyper-methylation of endogenous loci suggesting steady-state levels of DNA methylation depend on glycosylase activity. The DEMETER (DME) glycosylase gene is expressed in the two central cells of the embryo sac. DME demethylates three imprinted genes silenced by methylation, MEA, FIS2 and FWA. These genes are expressed in the central cells and from the maternal loci in the endosperm after fertilization, while the methylated paternal alleles remain silent.

The oxidized bases, 8-oxoguanine (8-oxoG) and formamidoguanine (FapyG), form stable base pairs with A, causing G→T mutations. Plants have the 'GO' system, first described in E. coli, that counters oxidative muta-genesis. MutM (FPG) and MutY recognize oxidized purines paired with C or A, respectively. MutM excises the oxidized purine, while MutY specifically excises A paired with oxidized purines. In addition, MutT hydrolyzes 8-oxo-dGTP, preventing its insertion across from A. The GO system therefore limits G→T and A→C transversions. A number of eukaryotes lack MutM, but have a different glycosylase that cleaves oxidized purines, called OGG1. OGG1 and FPG have overlapping substrate specificity. Plant OGG1 and FPG have been cloned and shown to cleave DNA with either 8-oxoG or FapyG (Murphy and George, 2005). Plant FPG is found in numerous forms due to alternative splicing.

Plants have many P450-family proteins associated with extensive profiles of secondary metabolites. Plants have multiple glycosylases for excising bases modified by alkylation: six copies of TAGA (expected to be specific for 3- MeA), two ALKA genes, one MPG and MAG1. ALKA, MPG and MAG1 are expected to have broad, overlapping activity for N- and O-methylated base lesions. This extensive repertoire for repairing alkyl-DNA damage suggests some secondary metabolites or intermediates of secondary metabolite detoxification are common mutagens in plants.

4.2. Nucleotide Excision Repair (NER)

Lesions that distort the DNA helix or block transcription elicit nucleotide excision repair (NER). Of par-

---

Box 5.1: Excision repair proteins

1. BER proteins

Oxidized pyrimidines: endonuclease III (NTH, 2*)
Oxidized purines: 8-oxoguanine glycosylase (OGG), formamidopyrimidine glycosylase (FPG/MMH/MutM), mutY-DNA glycosylase (MUTY/MYH)
Alkylated bases: methyl-binding domain glycosylase (MBD4), 3-methyladenine glycosylase (TAGA, 6*), methylpurine glycosylase (MPG/MAG/AAG), repressor of silencing-1 glycosylase (ROS1), demeter glycosylase (DME), alkyl glycosylase (AlkA, 2)
Deaminated bases: MBD4, MPG, uracil glycosylase (UNG)
Apurinic/apyrimidinic sites: AP endonuclease (APE/ARP, 3*)
DNA repair synthesis: poly(adenine)-ribose polymerase (PARP, 2*), flap endonuclease-1 (FEN-1, 2*), X-ray repair cross-complementing protein 1 (XRCC1), polymerase, ligase IV (Lig IV)

2. NER

Global recognition: xeroderma pigmentosum group C (XPC), radiation-sensitive group 23 (RAD23, 3*), centrin-2 (CEN2, 2*), damage-specific DNA-binding protein 1 (DDB1, 2*), xeroderma pigmentosum group E (XPE/DDB2), cullin4 (CUL4), ring-finger protein (RBX1)
Transcription-coupled recognition: cockyane syndrome group B (CSB), cockyane syndrome group A (CSA, 2*)
Core: transcription factor IIH (TFIIH), xeroderma pigmentosum group B, D, F and G (XPB/D/F/G, 2/1/1/1*), excision repair cross-complementing group 1 (ERCC1), replication repair protein A (RPA, 6*)

* Number of gene copies in plants

---

ticular interest are cylcobutane pyrimidines (CPDs) and pyrimidine (6-4) pyrimidionine ([6-4]s) dimers that are induced by ultraviolet radiation. NER excises a single strand of 25–30 nucleotides containing the damaged DNA (**Figure 5.3b**). NER functions in two modes, global genome repair (GGR) and transcription-coupled repair (TCR), which differ only in how DNA lesions are recognized. NER functions as a sequence of coordinated steps: lesion recognition, helix opening, dual incision 5′ and 3′ of the lesion and then DNA synthesis and ligation (**steps 1–4, Figure 5.3b**). Analyses of UV-sensitive mutants show plants depend on NER for repair of UV-photoproducts, in addition to having direct repair of UV-photoproducts (photoreactivation).

4.2.1. Lesion Recognition in Plants

The first step of global NER is recognition of helix-distorting lesions by the XPC-RAD23-CEN2 complex. The function of XPC and RAD23 remains uncharacterized in plants, however, Arabidopsis deficient in AtCEN2 show modest sensitivity to UV-C radiation and the cross-linking agent, cisplatin. Extracts from AtCEN2-deficient plants are reduced in efficiency for *in vitro* repair of UV or cisplatin-damaged plasmids. Deficiency in AtCEN2 also resulted in an increase in homologous recombination (HR). UV-B radiation increases the frequency of somatic HR. NER of UV-photoproducts may antagonize repair by recombination (Molinier *et al.*, 2004).

The DNA-damage-binding complex (DDB) enhances recognition of UV-photoproducts, which is particularly important for recognizing CPDs that are poorly recognized by XPC. The DDB complex forms an E3-ligase that targets specific E2-ubiquitin conjugating enzymes to several targets including DDB2 and possibly XPC. The DDB complex consists of DDB1, DDB2 (XPE), Cullin4 and Rbx1. Ubiquitination of XPC after UV-irradiation enhances its binding for damaged DNA. In contrast, DDB2-ubiquitin is targeted for degradation. Arabidopsis plants deficient in DDB2, CUL4 or DDB1 are sensitive to UV radiation. Plant extracts lacking CUL4 or DDB2 are reduced in *in vitro* repair of UV-damaged plasmids. Over-expression of either DDB1 or DDB2 increased tolerance to UV radiation, suggesting recognition of lesions by the DDB complex is a rate-limiting step to lesion repair (Molinier *et al.*, 2008).

In contrast to global NER, transcription-coupled NER requires the recruitment of CSA and CSB proteins to the DNA where RNA polymerase has stalled at the lesion.

Plants appear to have two orthologues of CSA and one of CSB, however they remain uncharacterized.

4.2.2. Helix Opening in Plants

After lesion recognition, CSB or XPC recruit the TFIIH complex (nine sub-units, including XPB and XPD) that opens the DNA helix around DNA lesions. The XPB 3′→5′ and XPD 5′→3′ helicase activities are necessary for opening the helix. Arabidopsis has two copies of XPB (AtXPB1 and AtXPB2) each with the helicase motifs conserved. The genes lie in a head-to-tail orientation, the result of a possible recent duplication (Morgante *et al.*, 2005). AtXPB1 complements the UV-sensitivity of the *S. cerevisiae rad25* mutants, however plants deficient in AtXPB1 do not have increased sensitivity to UV radiation, but are sensitive to the alkylating agent, MMS. Arabidopsis deficient in AtXPD have increased sensitivity to UV radiation, and show some mild growth defects. Removal of pyrimidne (6-4) pyrimidionine photoproducts (6-4s) is less efficient in *xpd* mutants. Complete loss of AtXPD-function is likely lethal (Liu *et al.*, 2003).

4.2.3. Dual Incision in Plants

Unwinding of the DNA helix exposes ssDNA, which is then coated with the RPA protein. In animals, RPA and XPA are necessary for opening the pre-incision complex fully, and XPA is necessary for XPF-ERCC1 recruitment and incision. Plants appear to lack an orthologue of XPA. An Arabidopsis mutant deficient in ERCC1 is sensitive to UV radiation, MMS and the inter-strand cross-linking agent, mitomycin C. This mutant is also gamma sensitive as is a mutant deficient in the plant orthologue of XPF. The XPF-ERCC1 is thought also to be involved in homologous recombination. XPF-ERCC1 incises the DNA 5′ of the lesion. XPG incises the DNA 3′ of the lesion. Mutants in AtXPG are sensitive to UV radiation as well as gamma radiation and H_2O_2. After incision, the RFC clamp loader loads PCNA at the 5′ incision site (3′ OH), and replicative polymerases fill in the gap.

4.3. Direct Repair in Plants

In mature leaves, the direct repair of UV-photoproducts may be the major repair pathway. Plants encode two distinct photolyases that directly repair both CPDs and [6-4] photoproducts. Photolyases have two cofactors: methenyltetrahydrofolate absorbs light (375–400nm) and transfers the excitation energy to the second

A. Translesion bypass at replication fork and lagging strand gap-filling

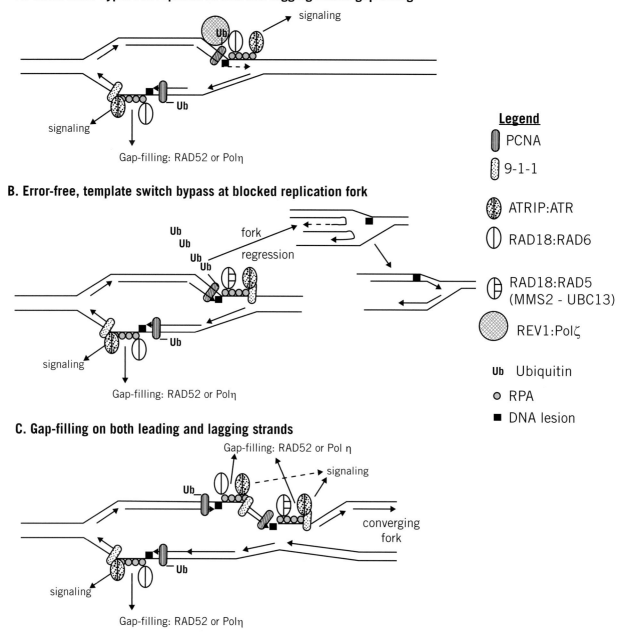

B. Error-free, template switch bypass at blocked replication fork

C. Gap-filling on both leading and lagging strands

Legend

- PCNA
- 9-1-1
- ATRIP:ATR
- RAD18:RAD6
- RAD18:RAD5 (MMS2 - UBC13)
- REV1:Polζ
- **Ub** Ubiquitin
- RPA
- DNA lesion

The Joint FAO/IAEA Programme

Figure 5.4 By-pass of DNA lesions by post-replication repair. (A–C) DNA lesions block maturation of Okawski fragments on the lagging strand, and block the replication fork on the leading strand. In both cases, ssDNA is coated by replication factor A (RPA), which recruits RAD18:RAD6 for mono-ubiquitination (**Ub**) of PCNA or recruits RAD18:RAD5 for poly-ubiquitination of PCNA. ATRIP-ATR is also recruited to ssDNA:RPA and signals the DNA damage response. The 9-1-1 clamp (RAD9-HUS1-RAD1) enhances ATR signalling. Lagging strand gaps are filled by either recombination (RAD52) or trans-lesion synthesis by Pol h. (A) Replication of the leading strand past the DNA lesion may be direct by trans-lesion synthesis (REV1-Polζ), promoted by mono-ubiquitinated PCNA, or (B) RAD5 poly-ubiquitinates PCNA and promotes the template switch pathway involving fork regression and the nascent lagging strand as the template for leading strand synthesis. (C) Replication restart or a converging fork downstream of a DNA lesion may create leading strand gaps, filled in by recombination or trans-lesion synthesis.

cofacter, FADH–, which then transfers an electron to the pyrimidine dimer, splitting it into two pyrimidines with concomitant transfer of an electron back to FADH°, regenerating FADH–. There are two classes of photolyases. Class I are microbial CPD photolyases that are not related to the class II CPD photolyases of eukaryotes. Plant CPD photolyases (PHR1) belong to the class II photolyases. The (6-4)-photolyase (UVR3) is more related to the microbial class I. An additional family of proteins sharing sequence orthology to class I photolyases are the cryptochromes (CRYI and CRY2), which regulate blue-light responses but do not repair damaged DNA. Rice and Arabidopsis deficient in CPD photolyase are more sensitive to UVB-irradiation, showing reduced growth. A quantitative trait locus (QTL) for sensitivity or tolerance to UV-B radiation in rice mapped to a CPD photolyase locus.

Some alkylated-bases, such as mutagenic and toxic O^6-MeG, are repaired without excision of the DNA backbone. In *E. coli*, the oxidative demethylase AlkB and alkylguanine-DNA alkyltransferase (ADA) directly reverse N1-methylpurines and N3-methylpyrimidines. Plants lack an ADA orthologue, but do have AlkB orthologues.

4.4. Post-Replication Repair (PRR)

BER, NER and direct repair keep steady-state lesions low, however, persisting DNA lesions block replication forks. Two critical networks essential for DNA replication of damaged DNA are the DNA damage response (DDR) and post-replication repair (PRR). The primary DDR response proteins do not involve DNA repair proteins and are not addressed here. PRR pathways employing trans-lesion synthesis polymerases (TLS pols) are responsible for damage-induced and spontaneous base and frameshift mutation. Alternatively, lesions can be by-passed by error-free template switch and recombinational mechanisms, but these may require more time and risk chromosomal rearrangements.

4.4.1. Replication of Damaged DNA

In UV-irradiated microbial and animal cells, newly synthesized DNA strands are shorter relative to nascent strands in unirradiated cells. These short nascent strands arise from UV-photoproducts blocking the replicative DNA polymerases. DNA lesions on the lagging strand block Okazaki fragment maturation, leaving a gap, but the replication fork is not blocked (**Figure 5.4A–C**). DNA damage blocking the leading strand polymerase blocks the replication fork and results in uncoupling of leading and lagging synthesis (**Fig 5.4A–C**). As the replicative helicase continues to unwind duplex DNA to template the lagging strand, ssDNA accumulates ahead of the block on the leading strand. RPA coats the ssDNA and stimulates both PRR and DDR. Gaps in the leading strand (**Figure 5.4C**) may arise in two ways. Studies in yeast and with reconstituted bacterial pro-

teins demonstrate that leading strand synthesis may be re-primed downstream of the damage, leaving a gap. Alternatively, firing of a proximal redundant origin or an already active fork may complete synthesis up to the blocked fork, again leaving a gap (Lehmann and Fuchs, 2006).

PRR acts to fill gaps or resume fork progression by promoting by-pass of DNA lesions (**Figure 5.3**). Because PRR does not repair DNA lesions, it is also called DNA damage avoidance or tolerance. There are three general pathways for by-passing DNA lesions: (*i*) specialized polymerases for trans-lesion synthesis (TLS *pols*), (*ii*) radiation-sensitive group 52 (RAD52)-dependent recombination or (*iii*) RAD5-dependent template-switch. All three pathways are partly regulated by the post-translational modification of PCNA. RPA-coated ssDNA binds the RAD18 protein that recruits RAD6 or RAD5. RAD6:RAD18 mono-ubiquitinates PCNA at lysine 165 (K165), while RAD5 in complex with methyl methanesulphonate-sensitive protein 2 (MMS2) and ubiquitin-conjugating enzyme 13 (UBC13) adds ubiquitin to mono-ubiquitinated PCNA, resulting in poly-ubiquitinated PCNA. Mono-ubiquitinated PCNA promotes translesion synthesis while polyubiquitination directs the error-free template switch pathway; fork-regression driving annealing of nascent leading and lagging strands so that the lagging strand serves as a temporary, lesion-free template for leading strand synthesis (**Figure 5.3B**). K165 of PCNA (and to a lesser extent K127) can also be modified with the small protein, SUMO. This modification recruits the SRS2 helicase that antagonizes RAD51-coated ssDNA filaments, thus suppressing recombination. Yeast *srs2 rad6* mutants are less UV-sensitive than *rad6* mutants, likely because RAD52-mediated recombination is not suppressed and can provide some PRR. How, when and where these independent lesion by-pass pathways are employed remains undefined, but essential to understanding regulation of replication through damaged DNA.

4.4.2. Trans-Lesion Synthesis Polymerases

TLS pols typically have DNA synthesis error-rates 5 orders of magnitude higher than replicative polymerases. TLS pols lack $3' \rightarrow 5'$ proofreading. TLS pols have open active sites and make few contacts with the replicating base pair, which relaxes base selection. Residues around the replicating base pair are not well conserved among Y-family polymerases and may influence their lesion specificity. TLS pols are thought to have evolved for relative error-free synthesis across specific lesions with reduced fidelity when incorporating nucleotides across non-cognate lesions. There is overlap among lesions by-passed by different TLS pols, and in some cases, two polymerases are required to complete lesion by-pass: one inserting a nucleotide at the lesion, and the other extending the primer beyond the lesion. The TLS pol, POLζ, is specialized for extension of mismatched primer termini, which likely accounts for its central role in mutagenesis; 90% and 50% of damage-induced and spontaneous mutation in yeast, respectively, is dependent on POLζ.

4.4.3. TLS Pols in Plants

Arabidopsis plants deficient in POLζ are sensitive to UVB-irradiation (Sakamoto *et al.*, 2003) and plant POL η complements UV-sensitivity in *rad30* mutant yeast. UVB-irradiation of Arabidopsis roots deficient in TLS pols show decreased root growth, apparent cell-cycle delays/arrest and specific killing of stem cells with increasing severity: wt < AtPOLη– < AtPOLζ– < AtPOLη– AtPOLζ–. Arabidopsis AtPOLζ– and AtPOLη– AtPOLζ– roots also show spontaneous death of stem cells, implying TLS pols are required for lesion bypass of some endogenous lesions. AtPOLη– AtPOLζ– roots do not recover normal growth until 5 to 6 days after UVB-irradiation, although the majority of CPDs are removed by 2 days. Division of quiescent cells 3 days after UV-B in AtPOLη– AtPOLζ– roots suggest the delay in growth may involve a collapse of the auxin gradient (Curtis and Hays, 2007). Thus, efficient by-pass of DNA lesions by TLS pols may be a balancing point between rates of mutation and avoiding perturbed tissue development. AtPolη may serve similar functions in the shoot meristem because over-expression of AtPOLη in Arabidopsis increases UV-tolerance of shoot growth. Deficiency of the TLS polymerase AtREV1 also shows modest sensitivity to UV-B radiation. High levels

of AtREV1 over expression appear to be toxic (Santiago *et al.*, 2008).

Three plant TLS pols have been purified from Arabidopsis. AtPOLκ is proficient in extension of primer-termini base-mismatches and inserts several nucleotides before dissociating from the template. Arabidopsis POLη was purified in parallel with yeast and human POLη. AtPOLη by-passed T[CPD]T, but was unable to by-pass T[6-4]T. By-pass efficiency of POLη was dependent on the template sequence (Hoffman and Hays, 2008). REV1 was purified and showed deoxy-cytidyl transferase activity, efficiently inserting dCMP opposite AP sites and undamaged bases of any type.

4.4.4. General Mechanism of Template Switch

RAD5 is a member of the SWI/SNF family of DNA-dependent ATPase, chromatin-remodeling proteins, and has conserved helicase motifs. *In vitro*, yeast RAD5 binds preferentially to structures mimicking replication forks and drives fork regression. RAD18 may recruit RAD5 to blocked replication forks. After PCNA is mono-ubiquitinated by RAD6:RAD18, RAD5 along with MMS2 and UBC13 poly-ubiquitinate PCNA. The function of poly-ubiquitinated PCNA has not been resolved, but may disrupt PCNA's interaction with a number of core replication factors causing partial disassembly of the replication fork to allow helicase-driven fork regression. Such fork remodelling may be timely and risk misalignment of nascent strands with homologous sequences other than the intended nascent lagging strand. It remains to be determined when cells choose between RAD5-template switch *versus* TLS pols for lesion by-pass.

4.4.5. Template Switch in Plants

Arabidopsis has two orthologues of UBC13 (AtUBC13A and AtUBC13B) and four E2 ubiquitin-conjugating enzymes, UBC/E2 variants (UEVs) that catalyze poly-ubiquitination of PCNA. UEVs and UBC13s are likely targeted to PCNA by the E3 ubiquitin ligase activity of

---

Box 5.2: Examples of DNA lesions bypassed by TLS pols

Polymerase η:
T[CPD]T:AA, 8-oxoG:C > A; O^6MeG:C ≈ T, Tg:A > C or G. Blocked by AP sites, εA and N^2-guanine adducts.

Polymerase ζ:
Efficient extension of mismatched primer termini at any DNA lesion, Tg:A.

Rev1: AP site:C, N^2-guanine:C.

RAD5. Both AtUBC13A and B physically interact with the animal or yeast UEV-protein, MMS2, which is required for error-free lesion by-pass. AtUBC13A or AtUBC13B complement the spontaneous mutagenesis and MMS sensitivity of yeast *ubc13* mutants. All four Arabidopsis UEV orthologues physically interact with AtUBC13, and suppress increased spontaneous mutations in *mms2* mutant yeast. Finally, Arabidopsis deficient in AtRAD5a are sensitive to cross-linking agents (mitomycin C and cis-platin) and MMS. Arabidopsis has two orthologues of yeast RAD5, but deficiency of AtRAD5b does not increase sensitivity to these DNA-damaging agents (Chen *et al.*, 2008). Arabidopsis also has an orthologue of the animal RAD5-orthologue, SHPRH, but it has not yet been characterized.

4.5. Mismatch Repair (MMR)

Mismatches are mutagenic lesions that occur during the replication of undamaged DNA (10^{-7} – 10^{-8}) and at a higher frequency during replication of damaged DNA (10^{-2}–10^{-3}). MMR recognizes mismatches and couples recognition to incision on the nascent strand. Exonucleases (5′-3′ or 3′-5′) then access the strand break and hydrolyze phosphodiester bonds past the mismatch. Alternatively, exonucleases may use a proximal strand break, if available, without necessity for an endonuclease. Re-synthesis by POLδ fills in the gap and ligase seals the backbone. MMR decreases error rates by 10 to 100-fold.

4.5.1. Lesion Recognition (Types of Mismatches)

Plants encode seven eukaryotic MutS orthologues (AtMSH1-AtMSH7), with MSH7 unique to plants (Hays, 2002). MSH proteins act in heterodimers to recognize mismatches. Mismatch recognition in plants is similar to that of animals, with MSH2:MSH3 failing to recognize base-pair mismatches, but recognizing base loop-outs. MSH2:MSH6 recognizes base mismatches and single nucleotide loop-outs, but not loop-outs > 1 base. The unique MSH7 orthologue forms a heterodimer with MSH2, and recognizes only base-pair mismatches, not base loop-outs (Culligan and Hays, 2000). The MSH2:MSH7 dimer has greater recognition than MSH2:MSH6 of A/C, A/G, A/A and G/A mismatches relative to recognition of G/T. Arabidopsis deficient in AtMSH2 show increased microsatelite instability and accumulate mutations during seed to seed propaga-

tion. MMR therefore is necessary for maintaining germ line genome stability in plants. Arabidopsis also has an orthologue of the yeast MSH1. AtMSH1 has a putative mitochondrial targeting signal that must be cleaved for AtMSH1 to form a homodimer. Mutations in AtMSH1 (CHM gene) alter mitochondrial genome stability.

4.5.2. Incision on the Nascent Strand

Mismatch recognition is coupled to strand incision in eukaryotes by several orthologues of the prokaryotic MutL protein: MLH1, PMS2 (PMS1 in yeast), MLH2 and MLH3. These proteins form three distinct heterodimers with MLH1 common to all. MLH1-PMS2 (MutLα) couples mismatch recognition by either MutS heterodimer to strand incision and repair synthesis. The function of the MLH1-MLH2 (MutLβ) is not well understood, but does suppress indels in yeast. MLH1-MLH3 (MutLγ) functions in meiosis, but also suppresses indels in yeast. During meiosis, MutLγ functions with MSH4:MSH5 in promoting cross-over during meiosis. In *E. coli*, recognition by MutS dimers couples MutL dimers to the MutH endonuclease that nicks the nascent strand by interpreting hemi-methylated Dam Methylase restriction sites (GATC). There is a delay in methylation at GATC sites on nascent strands. The mismatch is then removed by an exonuclease. Eukaryotes lack MutH orthologues, however, MutLα heterodimers have an intrinsic endonuclease activity. Repair synthesis in eukoryotes requires 5′ → 3′ exonuclease activity (likely EXO1) and/or 3′ → 5′ exonuclease activity intrinsic to POL and POL to excise the nascent strand with the mis-incorporated nucleotide.

4.5.3. MMR Role in Recombination

Plant MMR proteins are involved in recombination (Hays, 2002). Recombination between sequences showing increasing divergence (particularly during inter-species crosses) results in mismatches in recombination intermediates. AtMSH2 suppresses meiotic recombination during crosses between Arabidopsis ecotypes, and somatic recombination between direct repeat sequences containing 10 mismatches. AtMSH2 also suppresses meiotic recombination between homoeologous sequences. The MSH2 partner responsible for suppressing homoeologous recombination remains undefined, but could be AtMSH7, which shows specificity for base-pair mismatches. An EMS-mutant of wheat (*ph2a*) shows increased meiotic recombination in hybrids of diverged species. Wheat *TaMSH7* is located at the *ph2* locus. Specific silencing of *TaMSH7* (TaMSH6 transcripts not silenced) resulted in reduced fertility implicating a role for MSH7 in recombination (Hays, 2002). Arabidopsis deficient in AtMSH5 show reduced fertility associated with a reduction in chiasma (successful crossover) frequency, with residual chiasma randomly distributed. Similarly, deficiency in AtMSH4 reduces fertility and chiasma frequency (Higgins *et al.*, 2004). Thus, as with other organisms, MSH4:MSH5 is required for homologous recombination and crossover in plants. The AtMLH1:AtMLH3 heterodimer appears to function with MSH4:MSH5 in promoting resolution of meiotic holiday junctions into crossovers.

5. DNA Repair and Mutagenesis

Excision and direct repair keep steady-state levels of DNA lesions low, keeping mutagenic lesions from serving as templates for DNA replication. Replication of DNA across lesions by post-replication repair pathways greatly increases the risk of mutagenic mismatches or rearrangements. Some mutagenic mismatches form during replication of undamaged templates. Mismatch repair recognizes and repairs a variety of mismatches, further lowering the risk of mutation. Plants deficient in DNA repair enzymes or mismatch repair are expected to show elevated rates of mutation, with the mutagenic pathway that is altered depending on the DNA lesion(s) recognized by the respective repair protein. Deficiency in PRR also may elevate mutation rate or alter mutation spectra. The recent availability of mutation reporter genes for all single base substitution pathways and others for insertion/deletion events provide a means to evaluate the mutagenic consequences for plants deficient in specific repair proteins.

6. References

6.1. Cited References

Chen, I.P., Mannuss, A., Orel, N. *et al.* **2008.** A homolog of ScRAD5 is involved in DNA repair and homologous recombination in *Arabidopsis*. *Plant Physiol.* 146: 1786–1796.

Curtis, M.J. and Hays, J.B. 2007. Tolerance of dividing cells to replication stress in UVB-irradiated Arabidopsis roots: requirements for DNA translesion polymerases eta and zeta. *DNA Repair (Amst).* 6: 1341–1358.

Culligan, K.M. and Hays, J.B. 2000. *Arabidopsis* MutS homologs-AtMSH2, AtMSH3, AtMSH6, and a novel AtMSH7-form three distinct protein heterodimers with different specificities for mismatched DNA. *Plant Cell.* 12: 991–1002.

Hays, J.B. 2002. *Arabidopsis thaliana*, a versatile model system for study of eukaryotic genome-maintenance functions. *DNA Repair (Amst).* 1: 579–600.

Higgins, J.D., Armstrong, S.J., Franklin, F.C. *et al.* **2004.** The *Arabidopsis* MutS homolog AtMSH4 functions at an early step in recombination: evidence for two classes of recombination in *Arabidopsis*. *Genes Dev.* 18: 2557–2570.

Hoffman, P.D., Curtis, M.J., Iwai, S. *et al.* **2008.** Biochemical evolution of DNA polymerase eta: properties of plant, human, and yeast proteins. *Biochemistry.* 47:4583-4596.

Lehmann, A.R. and Fuchs, R.P. 2006. Gaps and forks in DNA replication: rediscovering old models. *DNA Repair (Amst).* 5: 1495–1498.

Liu, Z., Hong, S.W., Escobar, M. *et al.* **2003.** *Arabidopsis* UVH6, a homolog of human XPD and yeast RAD3 DNA repair genes, functions in DNA repair and is essential for plant growth. *Plant Physiol.* 132: 1405–1414.

Molinier, J., Ramos, C., Fritsch, O. *et al.* **2004.** CENTRIN2 modulates homologous recombination and nucleotide excision repair in *Arabidopsis*. *Plant Cell.* 16: 1633–1643.

Molinier, J., Lechner, E., Dumbliauskas, E. *et al.* **2008.** Regulation and role of *Arabidopsis* CUL4-DDB1A-DDB2 in maintaining genome integrity upon UV stress. *PLoS Genet.* 4:e1000093.

Morgante, P.G., Berra, C.M., Nakabashi, M. *et al.* **2005.** Functional XPB/RAD25 redundancy in *Arabidopsis* genome: characterization of AtXPB2 and expression analysis. Gene. 344: 93–103.

Murphy, T.M. and George, A. 2005. A comparison of two DNA base excision repair glycosylases from *Arabidopsis thaliana*. *Biochem Biophys Res Commun.* 329: 869–872.

Sakamoto, A., Lan, V.T., Hase, Y. *et al.* **2003.** Disruption of the AtREV3 gene causes hypersensitivity to ultraviolet B light and gamma-rays in *Arabidopsis*: implication of the presence of a translesion synthesis mechanism in plants. *Plant Cell*. 15: 2042–2057.

Santiago, M.J., Alejandre-Duran, E., Munoz-Serrano, A. *et al.* 2008. Two translesion synthesis DNA polymerase genes, AtPOLH and AtREV1, are involved in development and UV light resistance in *Arabidopsis*. *J. Plant Physiol*. 165: 1582–1591.

6.2. Websites

Arabidopsis DNA repair proteins:
http://www.uea.ac.uk/~b270/repair.htm
Human DNA repair proteins:
http://www.cgal.icnet.uk/DNA_Repair_Genes.html

6.3. Further Reading

Branzei, D. and Foiani, M. 2007. Template switching: from replication fork repair to genome rearrangements. *Cell*. 131: 1228–1230.

Bray, C.M. and West, C.E. 2005. DNA repair mechanisms in plants: crucial sensors and effectors for the maintenance of genome integrity. *New Phytol*: 168:511-528.

Britt, A.B. 1999. Molecular genetics of DNA repair in higher plants. *Trends Plant Sci*. 4: 20–25.

Friedberg, E., Walker, G., Siede, W. *et al.* 2006. DNA Repair and Mutagenesis. Second Edition. Washington, DC: ASM Press.

Friedberg, E.C., Wagner, R. and Radman, M. 2002. Specialized DNA polymerases, cellular survival, and the genesis of mutations. *Science*. 296: 1627–1630.

Lawrence, C.W. 2007. Following the RAD6 pathway. *DNA Repair (Amst)*. 6: 676–686.

Double-Stranded DNA Break, Repair and Associated Mutations

K.Osakabe, M.Endo and S.Toki *

Division of Plant Sciences, National Institute of Agrobiological Sciences, Kannondai, Tsukuba, Ibaraki 305-8602, Japan
* Corresponding author, E-MAIL: stoki@affrc.go.jp

1. Introduction

Throughout evolution the DNA of living organisms has been exposed to risks from a wide range of damaging factors such as UV irradiation and free radicals. DNA damage that is left unrepaired, or is repaired incorrectly, causes genetic alterations and, in the worst case, could lead to cell death. Double-stranded DNA breaks (DSBs) are the most threatening type of DNA damages in living cells. For the maintenance of genome integrity and cell survival, it is critical that cells repair such breaks accurately and faithfully. Cells possess several DNA repair pathways to deal with DSBs. Genetic and biochemical studies have provided considerable data on the genes and their encoded proteins involved in these repair systems and their functions. Although plants have long played a key role in building our understanding of genetics, relatively little is known about DNA repair in plants. The completion of genome sequencing of Arabidopsis and several major crops has greatly facilitated the search and study of genes involved in DNA repair. A number of homologous genes involved in DSB repair in yeast and vertebrates have been identified in higher plants (**Table 6.1**). In addition, the huge collection of T-DNA/transposon tagging mutants in Arabidopsis has helped identify the function of some of these genes in plants. Several Arabidopsis mutants that are deficient in homologous recombination (HR, **Box 6.1**) and non-homologous end joining (NHEJ, **Box 6.1**) have been identified; and appeared to be hypersensitive to DSBs-inducing agents. However, with a few exceptions, these mutants do not show embryonic lethality or severe growth defects, which is in contrast to cases in vertebrate animals. This may represent a distinct advantage of plant systems for the study of DSB repair and recombination. Here the current understanding of DSB repair and its role in mutagenesis in higher plants is described.

2. DNA Damage and Double-Stranded DNA Break Repair

DSBs are not always due to DNA damaging agents, they often represent intermediate molecules of active cellular metabolism. Physical mutagens such as ionizing radiations and UV (especially UV-C) irradiation, and a few chemical mutagens often lead to DSBs (**see Chapter 5**). Cellular processes such as DNA replication or repair of other types of DNA lesions also give rise to DSBs. For example, DNA repair by base excision repair or nucleotide excision repair involves endonucleases that introduce single-strand nicks, which can lead to DSBs (**see Chapter 5**). A single-strand nick or gap upstream of a replication fork can also be processed into a DSB during DNA replication. On the other hand, the formation of DSBs constitutes an intermediate stage in a number of cellular processes, including meiosis, the production of the variable regions of vertebrate immunoglobulin heavy and light chains, and mating type switching in yeast. Studies of the mechanisms of genetic recombination have shown that DSB formation is the critical event for the initiation of recombination.

There are two fundamental mechanisms for the repair of DSBs: 1) NHEJ, and 2) HR (**Box 6.1, Figure 6.1**). While HR is the predominant DSB repair pathway in bacteria and yeast, NHEJ is believed to be the principal DSB repair pathway in higher eukaryotes, including higher plants. Recent studies have revealed that both HR and NHEJ contribute to DSB repair in higher eukaryotes, with the relative contribution of the two repair pathways depending on the phase of the cell cycle and the developmental stage. HR is important during the late S and G2 phases of the cell cycle, when sister chromatids are available as templates, and during early stages of development. Thus, a deficiency in either repair pathway is highly toxic for higher eukaryotes.

3. Non-Homologous End Joining

NHEJ is a straightforward pathway, largely independent of the sequences involved, and acts simply to rejoin the two ends of the break. As a consequence, NHEJ is a relatively inaccurate process and is frequently accompanied by insertion or deletion of DNA sequences. This repair pathway is known to function throughout the cell cycle. Recently it has become clear that NHEJ comprises several different pathways (McVey and Lee, 2008).

3.1. Molecular Processes of NHEJ Repair Pathways

The most well-studied NHEJ pathway is the classical NHEJ (C-NHEJ), which is dependent on DNA-dependent protein kinase (DNA-PK) complex (Ku70, Ku80 and DNA-PK catalytic subunit (DNA-PKcs), XRCC4 and DNA ligase IV (LigIV).

Box 6.1: Glossaries of Double-stranded DNA break repair

Double-stranded DNA break (DSB)	The two complementary strands of the DNA double helix are broken simultaneously at sites that are sufficiently close to one another that base-pairing and chromatin structure are insufficient to keep the two DNA ends juxtaposed.
Non-homologous end joining (NHEJ) repair	NHEJ is a pathway that can be used to repair DSBs. NHEJ is referred to as "non-homologous" because the broken ends are directly ligated without the need for a homologous template.
Homologous recombination (HR)	HR is the process by which a strand of DNA is broken and joined to the end of a different DNA molecule. HR commonly occurs during meiosis as chromosomal crossover between paired chromosomes in eukaryotes.
Nucleotide excision repair (NER)	NER is a pathway to repair DNA damage affecting longer strands of 2–30 bases. This pathway system recognizes bulky distortions on the DNA helix caused by chemicals and UV as well as single-strand breaks. Recognition of these distortions leads to the removal of a short single-stranded DNA segment including the DNA lesion, creating a single-strand gap in the DNA, which is subsequently filled in by DNA polymerase.
V(D)J recombination	V(D)J recombination is a somatic recombination of immunoglobulins involving the generation of a unique immunoglobulin variable region. This recombination generates diverse patterns of T cell receptor and immunoglobulin proteins that are necessary for the recognition of diverse antigens.
Rad52 epistasis group	A definition of their genetic inter-relationship in HR repair pathway. This group includes Rad50, Rad51, Rad52, Rad54, Rad55 Rad57, Rad59, Mre11 and Xrs2. They were originally identified in a genetic screen for X-ray-sensitive mutants in budding yeast.
Holliday Junction (HJ)	After a DSB occurs a modified break end forming a single-stranded nucleofilament starts strand invasion for searching homologous template DNA. A displacement loop (D-loop) is formed during strand invasion between the invading 3' overhang strand and the homologous DNA. After strand invasion, a DNA polymerase extends the invading 3' strand, changing the D-loop to a more prominently cruciform structure. This four-stranded branched DNA structure is called a Holliday junction, and it is an intermediate in crossover or gene conversion.
Crossover and non-crossover	These are outcomes of meiotic recombination. In a paired homologous chromosome, meiotic DSBs occur and the DSBs are re-joined. When chromosome arms on opposite sides of the recombination initiation site exchange partners the event is referred to as a reciprocal crossover. If the original configuration of chromosome arms is retained, the event is referred to as a non-crossover.
Gene conversion	Gene conversion is an event in genetic recombination. This process results in the non-reciprocal exchange of a small fragment of genetic material from a donor DNA sequence to a homologous acceptor DNA sequence. It can occur between sister chromatids, homologous chromosomes or homologous sequences on either the same chromatid or different chromosomes.
Sister chromatid	Sister chromatids are identical copies of a chromosome; they contain the same genes and same alleles.

The first step of C-NHEJ is the binding of the Ku heterodimer proteins to the two DNA ends. The Ku heterodimer is composed of a 73 kDa subunit (Ku70) and an 86 kDa subunit (Ku80). Ku70 and Ku80 have a similar three-dimentional structure. The crystal structure of the Ku heterodimer with DNA showed that the heterodimer forms a toroidal structure with a central hole, which is large enough to accommodate duplex DNA. This ring shape of the Ku heterodimer indicates that Ku is able to form a bridge between the two broken ends, suggesting that Ku also contributes to the protection of DSB ends. Subsequently, DNA-PKcs is recruited by Ku. DNA-PKcs is a serine/threonine protein kinase with specificity for serine–glutamine or threonine–glutamine sequences in target proteins. The recruitment of DNA-PKcs to DNA breaks by Ku results in activation of its kinase function. The other NHEJ factors are hence phosphorylated by DNA-PKcs, which is important for the juxtaposition of DSB ends. Cells lacking Ku exhibit radiosensitivity and are defective in DSB repair. In animals, individuals lacking either of the Ku subunits show higher radiosensitivity, immune deficiency and defective DSB repair, similar to DNA-PKcs null animals; Ku null animals show growth defects and pre-immune senescence, which together

Table 6.1: HR and NHEJ factors in budding yeast, vertebrates and Arabidopsis

Budding yeast	Vertebrates	Arabidopsis	Function
Non-homologous end joining (NHEJ) DNA repair			
Ku70 (Hdf1)	Ku70	AtKu70	DNA end binding
Ku80 (Hdf2)	Ku80	AtKu80	DNA end binding
Not found	DNA-PKcs	Not found	DNA-dependent protein kinase
Not found	Artemis	Not found	DSB end processing
Dnl4	DNA ligase IV	AtLigIV	Ligation of DNA ends
Lif1	XRCC4	AtXRCC4	Complex with LigIV
Homologous recombination (HR)/NHEJ DNA repair			
Mre11	Mre11	AtMre11	DSB end processing
Rad50	Rad50	AtRad50	DSB end processing
Xrs2	NBS1	AtNBS1	DSB end processing
HR DNA repair			
Rad51	Rad51	AtRad51	Strand invasion
Dmc1	Dmc1	AtDmc1	Meiosis specific strand invasion
Sae2	CtIP	AtCom1/AtGR1	5′-3′ resection
Exo1	EXO1	AtEXO1	5′-3′ resection
Dna2	DNA2	AtDNA2	5′-3′ resection
Sgs1	BLM	AtRecQL4	5′-3′ resection HJ resolution
Rad52	Rad52	Not found	Recombination mediator
Rad51 paralogs			
Rad55	Rad51B	AtRad51B	Recombination mediator
Rad57	Rad51C	AtRad51C	Recombination mediator
	Rad51D	AtRad51D	Recombination mediator
	XRCC2	AtXRCC2	Recombination mediator
	XRCC3	AtXRCC3	Recombination mediator
Not found	BRCA2	AtBRCA2a, AtBRCA2b	Recombination mediator Rad51/ssDNA nucleofilament assembly
Rad54	Rad54	AtRad54	Chromatin remodelling Recombination mediator
Top3	TopIII	AtTop3	HJ resolution
Mus81	Mus81	AtMus81	HJ resolution
Mms4	Eme1	AtEme1	HJ resolution

indicate that Ku and DNA-PKcs have distinct, as well as overlapping functions.

If the juxtaposed DNA ends can be ligated, the repair would require only a ligation reaction. However, DSBs generated by exposure to DNA damaging agents rarely have ligatable DNA ends. Therefore, a second step to process the DNA ends to create suitable substrates for DNA ligase is required. The Mre11/Rad50/NBS1 (MRN) complex (The budding yeast homolog of NBS1 is Xrs2, and thus it is referred to as the MRX complex), which also plays an important role in the early stages of HR (see next section), is involved in this step in yeast. This complex functions as a structure-specific nuclease, and also plays a role as the end-bridging factor. The other structure-specific nuclease, Artemis was identified on the basis of its mutation in patients with human radiosensitive-severe combined immunodeficiency syndrome due to a defect in V(D)J recombination. The defect of V(D)J recombination (**Box 6.1**) in cells lacking Artemis is caused by a failure to open hairpins occurring at joint formation that would be required for subsequent end-joining during NHEJ. Although it is unclear whether Artemis participates directly in DNA damage-induced DSB repair, it is likely that Artemis is required to repair a subset of DSBs during NHEJ.

The final step is a ligase reaction and filling of the small gap by DNA polymerase. The ligation occurs in the LigIV/XRCC4 complex. LigIV is an ATP-dependent DNA ligase with an amino-terminal catalytic domain and carboxy-terminal BRCT domain. The BRCT domain is a conserved domain similar to the C-terminal portion of the BRCA1, and thought to function in protein-protein interactions. The interaction of LigIV (*via* this BRCT domain) with XRCC4 stimulates ligase activity, and the LigIV/XRCC4

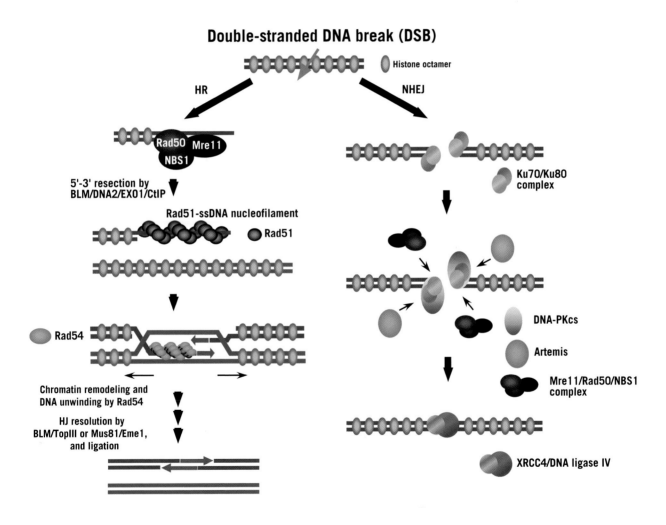

The Joint FAO/IAEA Programme

Figure 6.1 Schematic representation of two DSB repair pathways, HR and NHEJ.

complex is thought to be involved in DNA alignment or gap-filling prior to ligation. XRCC4 can also interact with DNA-PKcs, Ku and DNA polymerase μ. This might indicate that XRCC4 mediates to connect the sequential reactions that occur during NHEJ.

Recent studies have defined end-joining repair pathways other than C-NHEJ, which are independent of Ku70/Ku80/DNA-PKcs/XRCC4/LigIV. These pathways are so called 'alternative-NHEJ' (A-NHEJ) pathways. One of the well-defined A-NHEJ pathways is microhomology-mediated end joining (MMEJ; McVey and Lee, 2008). The most distinguished feature of MMEJ is the use of 5 to 25 bp microhomologous sequences for alignment of break ends during re-joining. This results in deletions flanking the original break, and thus, MMEJ is an error-prone repair pathway. Although the molecular process of MMEJ is less clear compared with that of C-NHEJ, recent studies predict the factors involved in MMEJ; Mre11/Rad50/Xrs2 and EXO1 for resection, poly(ADP-ribose) polymerase-1 (PARP1), XRCC1 and Ligase I (LigI) and III (LigIII) for ligation (McVey and Lee, 2008).

3.2. NHEJ in Higher Plants

Many of the factors involved in C-NHEJ mentioned above, except DNA-PKcs and Artemis, have been identified in the Arabidopsis genome, and their mutants have been characterized (**Table 5.1**). Characterization of C-NHEJ-deficient mutant plants demonstrated that these proteins are important for DSB repair in Arabidopsis (see review, Bleuyard *et al.*, 2006). Although genes involved in A-NHEJ pathway, such as *PARP-1, XRCC1* and *LigI* are found in the arabidopsis and rice genomes, the molecular process of A-NHEJ is less pronounced in higher plants.

C-NHEJ is thought to be critical for the integration of transforming plasmids or T-DNAs into the genomes of higher plants. Li *et al.* (2005) reported a decreased frequency of T-DNA integration in *atku80* plants during the transformation in somatic cells in a root transformation assay. In addition, overexpression of Ku80 in transgenic plants increased the rate of T-DNA integration (Li *et al.*, 2005). In contrast, Gallego *et al.* (2003) reported that the frequency of T-DNA integration in *atku80* plants was comparable with that in wild type plants during the transformation of germ-line cells by *in planta* transformation. These results indicate that the NHEJ pathway used for T-DNA integration is partly Ku-dependent, at least at certain developmental stages and/or in certain tissues of somatic cells.

4. Homologous Recombination

HR is generally an accurate pathway, and ensures the repair of DSBs without any loss of genetic information, but it requires the presence of the undamaged sister chromatid or homologous chromosome as a template. HR is a required repair mechanism for the progression of meiosis. The repair of DSBs by HR involves numerous steps and factors, including recognition of the homologous sequences, strand invasion, DNA synthesis and resolution of complex structures.

4.1. Molecular Process of HR

HR involves sequential steps that are catalyzed by multiple enzymes (**Figure 6.1**). In eukaryotes, these sequential steps are carried out by members of the Rad52 epistasis group (**Box 6.1**).

After DSBs occur, in yeast, the ends of duplex at the break are extensively processed by the MRN complex comprising Mre11, Rad50 and NSB1 to produce single-stranded 3' overhangs. The next key step is a further resection of the processed ends by the cooperation of CtIP (The budding yeast homolog is Sae2), EXO1 and DNA2 to produce long 3'-single-stranded DNA tails, which are required to form nucleoprotein filaments and important for homology search and strand invasion processes.

Facilitated by Rad52, Rad51 paralogs (**see Table 6.1**), a recombinase protein, Rad51 binds the resected ends of the break to form a Rad51/single-stranded DNA (ssDNA) complex (Rad51-nucleoprotein filament). In higher eukaryotes, BRCA2 also joins in this step to stabilize the Rad51-nucleoprotein filament. The resulting Rad51-nucleoprotein filament invades into a homologous double-stranded DNA (dsDNA) molecule, and starts DNA synthesis to extend the end of the invading strand. During this step, Rad54 plays several important roles. Rad54 is a member of the SWI2/SNF2 family of ATP-dependent chromatin remodelling factors. In the pre-synaptic phase, Rad54 combines with the Rad51-nucleoprotein filament, stabilizing the latter. In the synaptic phase, Rad54 protein serves as a constituent factor of the Rad51-nucleoprotein filament and remodels chromatin during a strand-transfer event, which, together with Rad51 nucleoprotein filament, leads to the creation of the D-loop. In the post-synaptic phase, Rad54 increases the rate and extent of heteroduplex extension and disassembles Rad51 nucleoprotein filaments on dsDNA.

By recapture of this strand, a co-joined molecule that contains the double Holliday junction (HJ, **Box 6.1**) is generated (**Figure 6.1**). The inter-linked molecules are then processed by branch migration, resolution of the HJ and DNA ligation (**Figure 6.1**). Genetical and biochemical studies revealed that two protein comlexes are important for HJ resolution. One is the BLM/TopIII (Sgs1/Top3 in budding yeast) complex and the other is the Mus81/Eme1 (Mus81/Mms4 in budding yeast) complex. While the Sgs1/Top3 complex produces non-crossover products, the Mus81/Eme1 complex is thought to function for crossover products. In meiosis, meiotic recombination-specific factors play important roles in recombination.

During meiosis, HR reaction starts at DSBs introduced by the Spo11 protein, which cleaves DNA *via* a topoisomerase-like reaction to generate covalent protein-DNA linkages to the 5' DNA ends on either side (Li and Ma, 2006). Spo11 is found in many organisms, including budding and fission yeasts, *Drosophila melanogaster*, mice, humans and plants, and Spo11 deficiency leads to sterility with very little residual meiotic recombination. The other factor is Dmc1, which is a meiosis-specific Rad51-like protein (Li and Ma, 2006). Although Dmc1 could form nucleoprotein filaments similar to Rad51, the behavior of complex formation between the two is different. Rad51 complex formation is required for Dmc1 association, presumably to promote recombination events, whereas Rad51 forms complexes in meiotic nuclei independently of Dmc1. Thus Dmc1 and Rad51 apparently have different functions in meiosis.

4.2. HR in Higher Plants

The arabidopsis genome possesses a set of genes involved in HR found in yeast and vertebrates, although some are not found or not identified yet. For example, higher plants are likely to lack the *Rad52* gene. Rad52 is an important factor in the assembly and stabilization of Rad51 nucleoprotein filaments in yeast. In contrast, the importance of Rad52 is less pronounced in vertebrates. In addition, the Arabidopsis genome possesses

a set of five Rad51 paralog genes found in mammals as an assembly factor of Rad51 to make nucleoprotein filament (**Table 6.1**). Arabidopsis Rad51 paralogues play an important role in the assembly and stabilization of Rad51 nucleoprotein filament in place of Rad52.

Spo11 is encoded by a single gene in most eukaryotes. The arabidopsis and rice genomes each possess three *SPO11*-like genes (*AtSPO11-1, AtSPO11-2* and *AtSPO11-3* in Arabidopsis), sharing 20–30% identity with each other (Stacey *et al.*, 2006). Arabidopsis mutants deficient in either AtSpo11-1 or AtSpo11-2 show severe meiotic defects. The apparently non-overlapping roles of AtSpo11-1 and AtSpo11-2 suggest a plant-specific metabolism for DSB induction in meiosis.

According to characterization of mutants deficient HR genes in *Arabidopsis*, mutant plants deficient in HR genes show higher sensitivity to DNA damaging agents, being especially hypersensitive to DNA cross-linking agents. In addition, these mutants show reduced frequencies of recombination events in somatic cells. Thus, each sequential step of HR is required for efficient HR repair in arabidopsis. In contrast, the contribution of the HR genes in meiosis is different in Arabidopsis. While Spo11-1, Spo11-2, Mre11, Rad50, Rad51, Dmc1, Rad51C and XRCC3 are required in meiosis, NBS1, Rad51B, Rad51D, XRCC2 and Rad54 are dispensable for the progression of meiosis.

Finally, recent studies have closed up the genetic link between HR repair and organization of the programmed cell cycle of shoot apical meristems in higher plants. Plants with deficiencies of AtMre11, AtBRCA2 and AtRad51C displayed fasciation and abnormal phyllotaxy phenotypes (Abe *et al.*, 2009). These phenotypes were not increased in mutants deficient for NHEJ-specific genes compared to wild type. These results suggest that HR repair during S to G2 of cell cycle is important to maintain cell-cycle control for shoot apical meristems.

5. DSB Repair and Chromatin Remodeling

The packaging of eukaryotic genomes into nucleosomes reduces access to sites of DSBs, impairing not only the detection of lesions but also their repair. Overcoming the barriers imposed by chromatin structure is important if HR is to be enhanced.

The chromatin assembly factor 1 (CAF-1) consists of three subunits of p150, p60 and p48 proteins, and is involved in nucleosome assembly following DNA replication and NER (Green and Almouzni, 2002). CAF-1 functions as a molecular chaperone of histone proteins to load onto DNA. Defects in CAF-1 might cause the temporary presence of open chromatin structures due to delayed chromatin assembly. Indeed, Costa and Shaw (2006) showed, *via* three-dimensional fluorescence *in situ* hybridization on intact root epidermal tissue, that most nuclei of root epidermal cells in *caf-1* mutants were in an open chromatin state. Arabidopsis *caf-1* mutants showed a hyper-recombination phenotype in *planta* (Endo *et al.*, 2006). Late S–G2 phase is suitable for HR because sister chromatids are available. G2 retardation might make the chromatin structure of CAF-1 mutants relatively open, allowing enhanced HR.

When DSBs occur, chromatin remodellers could also theoretically influence DNA repair directly by targeting action at repair sites to enhance repair protein access, or indirectly as general chromatin fluidizers, increasing site exposure throughout the genome to enhance damage detection and repair. One of the earliest detectable events in cells treated with ionizing irradiation is the rapid and extensive phosphorylation of the histone variant H2AX. ATM and ATR are the major kinases responsible for histone H2AX phosphorylation following replication blocks and DNA damage (Green and Almouzni, 2002). Phosphorylation could alter chromatin structure to permit the access or action of repair factors, or it may function as a signal amplification event, perhaps helping to recruit not only repair or downstream signaling factors, but also additional chromatin modulating activities. In addition to covalent modification, ATP-dependent chromatin remodeling has also been implicated in DSB repair. Ino80, which belongs to the SWI2/SNF2 chromatin remodeling gene family, has been shown to be important for efficient somatic HR in Arabidopsis (Fritsch *et al.*, 2004); HR frequency of *atino80* allelic mutants was 15% that of wild type plants. As described in the previous section, Rad54 also belongs to the SWI2/SNF2 chromatin remodeling gene family, and overexpression of yeast Rad54 enhances the gene targeting frequency in Arabidopsis (Shaked *et al.*, 2005). Furthermore, HR frequency was stimulated in an overexpressor of the Arabidopsis gene encoding MIM, which is one of the factors responsible for the structural maintenance of chromosomes (Hanin *et al.*, 2000).

The above discussion serves to highlight candidate proteins that might create an open chromatin structure during the process of DSB repair. Processive DNA events, such as RNA transcription and DNA replication, which result in increased accessibility to DNA, open further opportunities for DNA repair proteins to locate and access damaged sites. Recent rapid technological advances in fluorescent marker proteins and real time imaging techniques make it likely there will be new insights into this fascinating aspect of the cellular DNA damage response.

6. Implications and Perspectives

6.1. DSB Repair and Mutation Induction

The most widely used mutagens in plant genetics are chemicals, ionizing irradiation and T-DNAs/transposons. Of these, ionizing irradiation- and T-DNA/transposon insertion-mediated mutagenesis relies, at least in part, on the mechanism of DSB repair. Ionizing irradiation has long been used as a plant mutagen in forward genetic studies, and such mutants have proved valuable in the fields of genetics and breeding. Ionizing irradiation induces several types of mutations, including deletion, insertion and base substitution, and frequently causes rearrangement (**see Chapter 19**). Although the molecular mechanisms of ionizing irradiation-induced mutation are unclear, the NHEJ repair pathway participates in the induction of some of these mutations, since all NHEJ-deficient plants show higher sensitivity to ionizing irradiation. In the context of the cell cycle, NHEJ is thought to be the predominant pathway for DSB repair in seeds shortly after imbibition. In addition, aged seeds are generally thought to contain relatively high levels of DNA damage, and a prolonged G1 arrest is required during germination so that DNA repair can occur after imbibition (Whittle *et al.*, 2001). If the error-prone repair pathway functions during the prolonged phase of the cell cycle, the mutation rate in the seeds might be increased.

Currently, T-DNA/transposon insertion mutagenesis has become a common method for introducing mutations into plant genomes (**see Chapter 38_39**). T-DNA insertions require NHEJ activity and the situation is likely to be similar in the case of transposon insertions. In contrast to ionizing irradiation-mediated mutation, insertions of T-DNAs/transposons and retrotransposons are not always random. Rather, insertions of T-DNAs are preferentially found in 5'-gene regulatory regions, poly-adenylation site regions and A+T-rich regions (Gelvin and Kim, 2007). Similary, non-random distributions of T-DNAs and retrotransposons were observed in rice (**see Chapter 39**). These results suggest that insertions of T-DNA/transposons/retrotransposons would be biased dependent on loci, and further suggest that it is difficult to establish saturation of mutation of a genome *via* T-DNA/transposon/retrotransposon insertion, although the number of insertion mutations is still growing.

Recently, an NHEJ-based targeted-mutagenesis strategy was developed in several organisms by using synthetic zinc finger nucleases (ZFNs) to introduce a DSB at a specific site in the genome (**see Chapter 40**). Subsequent repair of the DSB by NHEJ produces deletions and/or insertions at the re-joining site. A large number of DNA-binding modules are currently available and the method of constructing ZFNs is established. Thus, if a cleavage site in a gene of interest can be identified and a method for introduction of ZFNs into the host plant cell is available, ZFN-mediated targeted mutagenesis represents a powerful tool for saturation mutagenesis of the genome.

6.2. Future Perspectives

The recent studies with Arabidopsis mutants described here have expanded the possibilities of investigating the molecular mechanisms of DSB repair in higher plants. Comparative studies of DSB repair in higher plants and vertebrates have revealed many similarities. On the other hand, several important differences between these two living systems have also emerged.

One striking difference is the viability of DSB repair-deficient mutants in Arabidopsis. Although the DNA damage-checkpoint system in Arabidopsis is similar to that of vertebrates, mechanisms of downstream processing, for example selection for repair or cell death, might differ between higher plants and vertebrates. One relevant factor in this context is that the germlines of plants are created at a late stage in development. Genomic changes in somatic cells, especially in shoot apical meristem cells, potentially risk transmitting genomic changes to the next generation. Hence the genome integrity of shoot apical meristem cells should be maintained by error-free DNA repair pathways. To date, HR repair has been found to be less pronounced in

somatic cells, but fasciation phenotypes in HR-deficient mutant plants may compensate for the requirement for HR repair in somatic cells. Investigation into the genetic interactions of HR and other DNA repair pathways (**see Chapter 5**) will certainly prove interesting.

One aim of the study of the molecular mechanisms of HR in plants is to be able to establish efficient target mutagenesis by gene targeting and to upregulate HR. It remains unclear how a particular repair pathway, whether HR or NHEJ, is chosen, activated and regulated when DSBs occur. Recent studies have shown that the frequency of HR is influenced by growth stage, day-length and temperature (Boyko *et al.*, 2006). HR is also increased by induction of DSBs. Recently, it was found that ZFN can be used to introduce DSBs at specific loci, and that such treatment drastically enhanced gene targeting in several organisms (**see Chapter 40**). Gene targeting of rice has now been established (**see Chapter 40**), and gene targeting coupled with ZFN in rice will allow efficient, well-designed targeted-mutagenesis of rice to be established.

7. Acknowledgements

We thank our colleagues for many stimulating discussions. We acknowledge the financial support of a PROBRAIN (Program for Promotion of Basic Research Activities for Innovative Biosciences) grant to S.T. from the Bio-Oriented Technology Research Advancement Institution (BRAIN) of Japan, and grants from the Ministry of Agriculture, Forestry and Fishery of Japan and budget for Nuclear Research from the Ministry of Education, Culture, Sports, and Technology of Japan. M.E. was also supported by a fellowship from Grant-in-Aid for JSPS (Japan Society for the Promotion of Science).

8. References

8.1. Cited References

Abe, K., Osakabe, K., Ishikawa, Y. *et al.* 2009. Inefficient double-strand DNA break repair is associated with increased fasciation in *Arabidopsis BRCA2* mutants. *Journal of Experimental Botany.* 60: 2751–2761.

Bleuyard, J.Y., Gallego, M.E. and White, C.I. 2006. Recent advances in understanding of the DNA double-strand break repair machinery of plants. *DNA Repair.* 5: 1–12.

Boyko, A., Zemp, F., Filkowski, J. *et al.* **2006.** Double-strand break repair in plants is developmentally regulated. *Plant Physiology.* 141: 488–497.

Costa, S. and Shaw, P. **2006.** Chromatin organization and cell fate switch respond to positional information in *Arabidopsis. Nature.* 439: 493–496.

Endo, M., Ishikawa, Y., Osakabe, K. *et al.* **2006.** Increased frequency of homologous recombination and T-DNA integration in *Arabidopsis* CAF-1 mutants. *EMBO Journal.* 25: 5579–5590.

Fritsch, O., Benvenuto, G., Bowler, C. *et al.* **2004.** The INO80 protein controls homologous recombination in *Arabidopsis thaliana. Molecular Cell.* 16: 479–485.

Gallego, M. E., Bleuyard, J. Y., Daoudal-Cotterell, S. *et al.* **2003.** Ku80 plays a role in non-homologous recombination but is not required for T-DNA integration in *Arabidopsis. The Plant Journal.* 35: 557–565.

Gelvin, S.B. and Kim, S.I. **2007.** Effect of chromatin upon *Agrobacterium* T-DNA integration and transgene expression. *Biochemistry Biophysics Acta.* 5-6: 409–420.

Green, C. M. and Almouzni, G. **2002.** When repair meets chromatin. First in series on chromatin dynamics. *EMBO Reports.* 3: 28–33.

Hanin, M., Mengiste, T., Bogucki, A. *et al.* **2000.** Elevated levels of intrachromosomal homologous recombination in *Arabidopsis* overexpressing the MIM gene. *The Plant Journal.* 24: 183-189.

Li, J., Vaidya, M., White, C. *et al.* **2005.** Involvement of *KU80* in T-DNA integration in plant cells. *Proceedings of the National Academy of Sciences USA.* 102: 19231–19236.

Li, W. and Ma, H. **2006.** Double-stranded DNA breaks and gene functions in recombination and meiosis. *Cell Research.* 16: 402–412.

McVey, M. and Lee, S.E. **2008.** MMEJ repair of double-strand breaks (director's cut): deleted sequences and alternative endings. *Trends in Genetics.* 24: 529–538.

Stacey, N. J., Kuromori, T., Azumi, Y. *et al.* **2006.** *Arabidopsis* SPO11-2 functions with SPO11-1 in meiotic recombination. *The Plant Journal.* 48: 206–216.

Shaked, H., Melamed-Bessudo, C. and Levy, A.A. **2005.** High-frequency gene targeting in *Arabidopsis* plants expressing the yeast *RAD54* gene. *Proceedings of the National Academy of Sciences USA.* 102: 12265–12269.

Whittle, C.A., Beardmore, T. and Johnston, M.O. **2001.** Is G1 arrest in plant seeds induced by a p53-related pathway? *Trends in Plant Science.* 6: 248–251.

8.2. Further Reading

Bennardo, N., Cheng, A., Huang, N. *et al.* **2008.** Alternative-NHEJ is a mechanistically distinct pathway of mammalian chromosome break repair. *ProS Genetics.* 4: e1000110.

Britt, A.B. **1996.** DNA damage and repair in plants. *Annual Review of Plant Biology.* 47: 75–100.

Cann, K.L. and Hicks, G.G. **2007.** Regulation of the cellular DNA double-strand break response. *Biochemical Cellular Biology.* 85: 663–674.

Hamant, O., Ma, H. and Cande, W.Z. **2006.** Genetics of meiotic prophase I in plants. *Annual Review of Plant Biology.* 57: 267–302.

Li, X. and Heyer, W.D. **2008.** Homologous recombination in DNA repair and DNA damage tolerance. *Cell Research.* 18: 99–113.

Ma, Y., Lu, H., Schwarz, K. *et al.* **2005.** Repair of double-strand DNA breaks by the human nonhomologous DNA end joining pathway: the iterative processing model. *Cell Cycle.* 4: 1193–1200.

Roth, D.B. **2003.** Restraining the V(D)J recombinase. *Nature Review Immunology.* 3: 656–666.

Shrivastav, M., De Haro, L.P. and Nickoloff, J.A. **2008.** Regulation of DNA double-strand break repair pathway choice. *Cell Research.* 18: 134–147.

Section 2

Mutagens and Induced Mutagenesis

C07

Mutagenic Radiations: X-Rays, Ionizing Particles and Ultraviolet

C.Mba[a,*,#], R.Afza[a] and Q.Y.Shu[b]

[a] Plant Breeding and Genetics Laboratory, Joint FAO/IAEA Division of Nuclear Techniques in Food and Agriculture, IAEA Laboratories Seibersdorf, International Atomic Energy Agency, Vienna International Centre, P.O. Box 100, Vienna, Austria
[#] Present Address: Plant Production and Protection Division, Food and Agriculture Organization of the United Nations, Viale delle Terme di Caracalla, 00153 Rome, Italy
[b] Joint FAO/IAEA Division of Nuclear Techniques in Food and Agriculture, International Atomic Energy Agency, Wagramer Strasse 5, P.O. Box 100, A-1400 Vienna, Austria
 Present: Institute of Nuclear Agricultural Sciences, Zhejiang University, Hangzhou 310029, China
* Corresponding author, E-MAIL: Chikelu.Mba@fao.org

1. Introduction

The physical agents that cause damage to DNA molecules of a living organism are referred to as physical mutagens or mutagenic radiations. Early in the 20th century, the 'rediscovery' of Gregor Mendel's work, now known as Mendelian genetics, was followed closely by theories that mutations, heritable change to the genetic makeup of an individual, could be induced, and induced mutations would mimic spontaneous mutations, the drivers for evolution and speciation (**see Chaptera 1 and 2**). The discoveries around the turn of the 19th century of X-rays (Roentgen, 1895), radioactivity (Becquerel, 1896), and radioactive elements (Marie and Pierre Curie, 1898) provided sound foundations for experimental mutagenesis. Muller and Städler's pioneering discoveries of the mutagenic effects of X-rays on fruit flies and maize, respectively, transcended induced mutagenesis to a reality in the late 1920s (**see Chapter 2** for more information on pioneering research). In the past 80 years, physical mutagens, mostly ionizing radiations, have been used widely for inducing hereditary aberrations and more than 70% of mutant varieties were developed using physical mutagenesis.

2. Types and Mutagenic Effect of Radiation

2.1. Types and Features

Radiation is defined as energy travelling through a distance in the form of waves or particles. It is broadly divided into two main categories, corpuscular and electromagnetic radiations, to distinguish between particulate and wave types of radiations, respectively (**Box 7.1**).

Electromagnetic radiations are classified further into different types, based on their source, but the most basic distinction lies in the amount of energy that is involved, their frequency and source. **Table 7.1** provides the characteristics of common types of radiation.

2.2. Ionizing and Non-Ionizing Radiations

Based on their capacity for producing ions, radiations are divided into ionizing and non-ionizing radiations. Non-ionizing radiation can be strong enough to influence the atoms it comes into contact with although not strong enough to affect their structure. By contrast, ionizing radiation has enough energy to directly affect the structure of atoms of impacted materials, including living plant and animal tissues. The term 'ionizing' arises from the fact that when these forms of radiation pass through a tissue, there is always the tendency to dislodge an electron from its orbit around the nucleus thereby producing an ion as the corresponding proton becomes positively charged (ionized).

3. Ionizing Radiation

3.1. Atomic Structure and Isotopes

In order to understand the modes of action of different physical mutagens, an appreciation of the central role of the atom is required. All substances are composed of atoms, which are mainly composed of space. The three particles of an atom are protons, neutrons, and electrons (except hydrogen which has no neutrons). At the centre of the atom are the protons and neutrons which are tightly bound together in the positively charged

Box 7.1: Types and features of radiation

1. Electromagnetic radiation is energy that travels through free space or through a material medium in the form of electromagnetic waves, such as radio waves, infrared radiation, visible light, ultraviolet radiation, X-rays and gamma rays. It has wave-like properties such as reflection, refraction, diffraction and interference, as well as particle-like properties in that its energy occurs in discrete packets, or quanta. All types of electromagnetic radiation travel at the same speed, but vary in frequency and wavelength, and interact with matter differently.

2. Corpuscular radiation is made up of sub-atomic particles, such as electrons, protons, neutrons or alpha particles, that travel in streams at various velocities. All the particles have definite masses and travel at various speeds; while their mass is determined when they are emitted from isotopes, their speeds can be further manipulated in various types of accelerators.

3. Ionizing radiation consists of highly energetic particles or *electromagnetic* radiations that can detach (ionize) at least one electron from an atom or molecule. Ionizing ability depends on the energy of the impinging individual particles or waves, and not on their number.

4. Non-ionizing radiation refers to any type of *electromagnetic* radiation that does not carry enough energy per quantum to ionize atoms or molecules — that is, to completely remove an electron from an atom or molecule.

Table 7.1: Classification of radiations over the range of the electromagnetic spectrum [a]

Frequency (Hz)	Wavelength	Nomenclature	Typical source
Mutagenic radiations			
10^{23}	3×10^{-15}	Cosmic photons	Astronomical
10^{22}	3×10^{-14}	γ-rays	Radioactive nuclei
10^{21}	3×10^{-13}	γ-rays, X-rays	
10^{20}	3×10^{-12}	X-rays	Atomic inner shell
10^{19}	3×10^{-11}	Soft X-rays	Electron impact on a solid
10^{18}	3×10^{-10}	Ultraviolet, X-rays	Atoms in sparks
10^{17}	3×10^{-9}	Ultraviolet	Atoms in sparks and arcs
10^{16}	3×10^{-8}	Ultraviolet	Atoms in sparks and arcs
Non-mutagenic radiations			
10^{15}	3×10^{-7}	Visible spectrum	Atoms, hot bodies, molecules
$10^{13} \sim 10^{14}$	3×10^{-5} ~ 3×10^{-6}	Infrared	Hot bodies, molecules
10^{12}	3×10^{-4}	Far-infrared	Hot bodies, molecules
$10^{4} \sim 10^{11}$	3×10^{-3} ~ 3×10^{3}	Radio, radar, and microwaves	Electronic devices

[a] Adapted from *McGraw-Hill Encyclopedia of Science and Technology* (10th edition, Parker, 2007).

nucleus while negatively charged electrons orbit the nucleus. The number of protons in the nucleus is unique and determines what material (element) the atom is. For example, if the nucleus contains 8 protons, the atom is oxygen, and if it contains 17 protons, the atom is chlorine. The number of neutrons in the nucleus of all atoms of the same element is not fixed. Atoms of the same element with different numbers of neutrons are called isotopes. For example, all atoms of the element carbon have 6 protons, but while most carbon atoms have 6 neutrons, some have 7 or 8. An isotope therefore refers to one of two or more atoms that have the same number of protons but different numbers of neutrons in their nuclei.

An isotope is also referred to as a nuclide. A nuclide is any particular atomic nucleus with a specific atomic number, Z and mass number A; it is equivalent to an atomic nucleus with a specific number of protons and neutrons. The term isotope is more appropriately used when referring to several different nuclides of the same element; nuclide is more generic and is used when referring to only one nucleus or several nuclei of different elements.

According to the IUPAC (International Union of Pure and Applied Chemistry) nomenclature, isotopes are named by giving the name of the element followed by a hyphen and the mass number – the sum of the neutrons and protons, or number of nucleons, in the isotope's nucleus, e.g. helium-3, carbon-12, carbon-13, iodine-131 and uranium-238. In symbolic form, the number of nucleons is denoted as a superscripted prefix to the chemical symbol, e.g. ^{3}He, ^{12}C, ^{13}C, ^{131}I and ^{238}U.

When the nuclear forces holding the protons and neutrons together at the nucleus of an atom are sufficient to overcome electrical energy that tries to push the protons apart, such an atom is said to be stable. An atom becomes unstable or radioactive when the number of neutrons is above a certain limit, and is referred to as a radionuclide, radioactive isotope or radioisotope. For example, carbon-14 is radioactive while carbon-12 and carbon-13 are stable isotopes. In the unstable state, some of the atom's excess energy begins to escape in the form of radiation. This may be in the form of alpha (α) particles, beta (β) particles, or gamma (γ) rays or combinations of all three. The process of emitting the radiation is called radioactive decay. When the nucleus of a radioactive isotope decays, emitting ionizing radiation, the nucleus is altered. It is transformed into another isotope which in many cases is a different element. This new isotope may be stable or unstable. If it is stable, the new isotope is not radioactive. If it is unstable, it also will decay, transforming its nucleus and emitting more ionizing radiation. Several decays may be required before a stable isotope is produced.

3.2. Types and Sources

While most ionizing radiations are emitted from naturally decaying isotopes, they can also be produced artificially in reactors and further manipulated through accelerators; the latter have also been used a lot in induced mutagenesis. A summary of ionizing radiations that have been used in plant mutation induction is presented in **Table 7.2**.

3.3. Ionizing Radiations Applied in Plant Mutagenesis

In general, ionizing radiations have the common property of being able to cause the release of energy – called ionizations or ion pairs – as they pass through matter. However, they vary in their ionization capacities and have unique properties (**Table 7.2**). In plant mutagenesis, the energy and penetration ability are the two key technical parameters that affect the effectiveness of a mutagen, other factors, e.g. the availability of and accessibility to the source, the suitability for treating a

Table 7.2: Types, sources and properties of various ionizing radiations [a]

Types of radiation	Source	Description	Energy	Hazard	Necessary shielding	Penetration into plant tissue[b]
X-rays	X-ray machine	Electromagnetic radiation	Commonly 50–300 keV	Dangerous, penetrating	A few mm of lead except high-energy installations	A few mm to many cm
Gamma rays	Radioisotopes and nuclear reaction	Electromagnetic radiation similar to X-rays	Up to several MeV	Dangerous, very penetrating	Heavy shielding, e.g. decimetres of lead or metres of concrete	Through whole part
Neutron (fast, slow and thermal)	Nuclear reactors or accelerators	Uncharged particle, slightly heavier than proton, observable only through interaction with nuclei	From less than 1 eV to several MeV	Very hazardous	Thick shielding of light elements, such as concrete	Many cm
Beta particles, fast electrons or cathode rays	Radioactive isotopes or accelerators	An electron (– or +) ionize much less densely than alpha particles	Up to several MeV	May be dangerous	Thick sheet of cardboard	Up to several cm
Alpha particles	Radioisotopes	A helium-nucleus, ionizing heavily	2–9 MeV	Very dangerous	Thin sheet of paper sufficient	Small fraction of a mm
Protons or deuterons	Nuclear reactors or accelerators	Nucleus of hydrogen	Up to several GeV	Very dangerous	Many cm of water or paraffin	Up to many cm
Low-energy ion beams	Particle accelerators	Ionized nucleus of various elements	Dozens of keV	Dangerous	Instrumental	A fraction of a mm
High-energy ion beams	Particle accelerator	Ionized nucleus of various elements	Up to GeV	Dangerous	Instrumental	Up to a few cm

[a] Adapted (with modifications) from IAEA (1977).
[b] Penetration is dependent upon many variables but it is assumed here that penetration is into an ordinary plant tissue of average density.

particular type of plant tissue, the safety for both treatment and post-treatment management, the cost of treatment, can all affect the usefulness of a particular mutagen.

Since the 1960s, gamma rays have become the most commonly used mutagenic radiation in plant breeding (**see Chapter 2**); during the past two decades, ion beam radiation has also emerged as an effective and unique mutagen (**see Chapters 9 and 10**). Other types of mutagenic radiation, e.g. X-rays, α- and β-particles, neutrons and UV light, have also demonstrated usefulness in plant mutation induction, either for particular types of material, or for particular purposes (e.g. fast neutrons in inducing deletion mutations), and they are described in the following sections.

3.3.1. X-Ray Irradiation

The X-ray was first discovered by Roentgen in 1895; this was achieved courtesy of a device called a Crooke's tube, essentially a glass envelope subjected to a high vacuum, with a wire element at one end forming the cathode, and a heavy copper target at the other end forming the anode. By applying a high voltage to these two electrodes, electrons formed at the cathode are pulled towards the anode and in the process strike the copper at very high energy levels. Roentgen discovered that very penetrating radiations were produced from the anode, which he called X-rays.

It is now known that X-rays can be produced whenever high-energy electrons impact a heavy metal target, like tungsten, molybdenum or copper. On impact, some of the electrons will approach the nucleus of the metal atoms on account of electrostatic attraction between the negatively charged electrons and the nucleus that has a net positive charge. An electron approaching the nucleus comes under the influence of a nuclear force, which being stronger than the electrostatic attraction decelerates the electron and deflects it from its original course, a phenomenon known as Bremsstrahlung (German for 'braking radiation'). The deceleration and deflection results in the loss of kinetic energy; the lost energy is emitted in the form of photons of radiation, i.e. X-radiation. An electron may also approach the atom directly and be stopped completely by the electrostatic field; this type of collision results in all the energy of the electron being converted to a single photon of X-ray. An atom that is hit by a high-energy electron may also become ionized through the dislodging of an electron

from the atomic orbit. The energy is lost as another electron, usually from the outer shell at a higher energy level, sheds its excess energy in order to occupy the void left by the ejected electron, and is emitted in the form of a single photon of X-radiation.

Over the past century, various kinds of X-ray machines have been produced for different applications, from medical imaging to airport security to non-destructive phenotyping. Induced mutations were first generated in plants using X-rays by Städler (1928); however, no commercial X-irradiator was specifically and widely adapted for mutation induction in plants. With continuing global security and safety concerns, it is becoming increasingly difficult to transport radioactive materials, like gamma irradiators. This is impacting negatively on the procurement and establishment of new gamma irradiation facilities as well as on the replenishment of depleted sources. Consequently, the renewal of interest in X-irradiation as a viable alternative to gamma rays is emerging. Some X-ray machines designed for tissue irradiation seem to be potentially useful for mutation induction due to both ease of operation and excellent technical adaption.

The Rad Source X-rays RS series (Rad Source Technologies Inc., Alpharetta, GA, USA), for example, have proven to be equivalent to cesium-137 and cobalt-60 sources as a blood irradiator; additionally, they are now being used in various biological research assays including the induction of genetic changes. These RS X-ray irradiators are cabinet-housed and require neither nuclear-specific certification nor specialized personnel for operation. Also, no special precautions, such as lead-lined rooms or controlled access are required for this series of X-ray irradiators. The obviation of these security provisions as well as the absence of the need for the eventual disposal of the cobalt-60 or casium-137 sources makes the X-ray option attractive. Although there has been no report on the use of such equipment, there is real potential that such X-irradiators may become preferred equipment for mutation induction in plants.

3.3.2. Fast and Thermal Neutrons

Neutrons are uncharged atomic particles of varying levels of kinetic energy. A neutron emission can arise naturally by being ejected from the nucleus during the process of nuclear fission, i.e. the splitting of an atom, a process accompanied with the release of energy. The collision of the ejected neutron with another atom leads to another fission event resulting in the ejection of yet more neutrons. These in turn collide with other atoms leading to more fission reactions and the ejection of yet more neutrons. This process is repeated countless times over leading to a self-sustaining chain reaction. Radioactivity is produced in the process as the neutrons also, on account of the absence of a charge, impact on the nucleus rendering it unstable. In practice, not all the neutrons contribute to the sustenance of the chain reaction; some pass from the core of the reactor through outlet ports and so could be used for other purposes such as irradiation.

Neutrons can be mono-energetic and poly-energetic, and can be produced through nuclear fission and further processed into different types (**Box 7.2**). Neutrons produced through nuclear fission in nuclear reactors are termed fast neutrons (N$_f$). When fast neutrons pass through moderators such as carbon, hydrogen, paraffin or water, their energy is gradually lost. When a neutron has the energy of a gas molecule at room temperature, it is known as a thermal neutron (N$_{th}$). Thermal neutrons are the most common type of neutrons produced by the preponderant gas-cooled and light water reactor types, and typically have different reaction profiles than matter from fast neutrons.

The FAO/IAEA Neutron Seed Irradiation Programme (1966–1974) stimulated extensive international research on the use of fast neutrons in seed mutation induction (IAEA 1967, 1968, 1972). One of the outputs of the program is the development of a Standard Neutron Irradiation Facility (SNIF) for swimming-pool-type reactors (Burtscher and Casta, 1967). The SNIF for the now decommissioned ASTRA reactor of the Austrian Research Centre, Seibersdorf, Austria consisted of a 46-cm-long cylindrical tube of 43 cm diameter. Constructed with lead (which effectively reduced

gamma contamination), additionally the walls were coated with boron carbide (B$_4$C) to minimize thermal neutron contamination. With this facility for the swimming-pool-type reactor, contamination of gamma rays and thermal neutrons can be reduced effectively. An additional extension tube made of lead was used to minimize further gamma ray contamination as sample chambers were transported into and out of the SNIF. The SNIF was movable and accurate dose rates could be determined based on its distance from the core of the reactor.

Other similar facilities developed for neutron radiation include the Lucas Heights Laboratories of Australian Institute of Nuclear Science and Engineering, a dedicated leg for handling irradiation of biological samples was adapted on the 3 MeV Van de Graaff Accelerator. Similarly, the Uranium Shielded Irradiation Facility (USIF) was used for TRIGA-type reactors. A Standard Column Irradiation Facility (SCIF) was designed similarly and used in thermal columns for exposure of biological samples to thermal neutrons (Novak and Brunner, 1992).

The use of fast neutrons as a mutagen has so far been limited due to various reasons. First, there are always accompanying contamination of γ-rays and thermal neutrons in fast neutron irradiation, which cannot be easily quantified nor can the individual effects of these different types of radiation be measured; second, the accessible fast neutron facilities are limited and its operation is much more complicated and expertise- and labour-demanding and hence is less cost-effective than other type of radiations.

3.3.3. Alpha Particles

An alpha particle is composed of two protons and two neutrons; it is essentially the nucleus of a helium atom and hence positively charged and energetic. Alpha particles are emitted in the course of the radioactive decay of heavy radioactive elements such as uranium and radium. The energy of α-particles emitted varies, with higher energy α-particles being emitted from larger nuclei, but most α-particles have energies of between 3 and 7 MeV, corresponding to extremely long to extremely short half-lives of α-emitting nuclides, respectively. Because of their charge and large mass, α-particles are easily absorbed by materials, and they can travel only a few centimetres in air. They lose energy rapidly when passing through matter and hence their penetrability is relatively low compared with gamma and X-rays (**Table 7.1**). However, α-particles are strongly ionizing, and with large enough doses can cause any or all of the symptoms of radiation poisoning. It is estimated that chromosome damage from α-particles is about 100 times greater than that caused by an equivalent amount of other radiation.

The penetrability of α-particles into plant tissues is very low hence external radiation of α-particles is rarely used in mutation induction in crops. Recently, particle delivery mechanisms especially charged-particle microbeams have improved significantly in cancer therapy. This may also hold promise for alleviating the constraints to the use of α-particles in induced crop mutations. Indeed, α-particles when managed with instruments can be considered as a special type of ion beam radiation – helium ion beams (**see Chapter 9**).

3.3.4. Beta Particles

Beta particles are fast moving electrons emitted from the nuclei of many fission events, i.e. they are products of radioactive decay of unstable nuclei such as the isotopes, ^{32}P and ^{35}S. They can be either negatively or positively charged but regardless of the charge, they have a very small mass (approximately 1/2000 the mass of a neutron). As mutagens, they have the same effects as X or γ-rays except that like α-particles, however, their low penetrability compromises their efficacy.

To overcome the limitation of low penetrability of particles from radioactive isotopes, a few methods have been developed for plant mutagenesis. The simplest way to administer β-particles is by immersing plant materials (e.g. seeds or propagule) in ^{32}P solution. The absorbed ^{32}P is integrated into germline cells of M_1 plants and causes mutations. This method has been used successfully in the development of rice and cotton mutant varieties. However, since the plant materials

are contaminated with the source radioactive isotope, safety becomes a concern in growing these materials. For external radiation, β radiation facilities for the exposure of plant propagules have also been developed. Bores and Bottinot (1974) described the design of strontium-90 (^{90}Sr) – yttrium-90 (^{90}Y) β radiation facility suitable for inducing mutations in plants. By exposing plant propagules to β-particles, no radioactive contamination remains in the treated materials.

Today, β-particles can be generated and transformed into electron beams for sophisticated irradiations using electron beam technology. Electron beams can have energy from hundreds of thousands of eV to several MeV and can be used as mutagens (Shu *et al.*, 2002), however, their potential has not been explored widely.

4. Non-Ionizing Radiation – Ultraviolets

UV radiation, first described by Johann Wilhelm Ritter in 1801, is light energy emitted between the wavelengths of 100 and 400 nm, i.e. between the electromagnetic spectra of X-ray and visible light (**Table 7.1**). UV radiation is of relatively low energy (compared with γ- and X-rays) and is not ionizing, i.e. does not dislodge electrons.

UV radiation is further divided into different forms based on wavelength, i.e. ultraviolet A (UVA) 315–400 nm, ultraviolet B (UVB) 280–315 nm, ultraviolet C (UVC) 100–280 nm. UVC is the most energetic and biologically damaging, it is not found in sunlight as it is absorbed by the ozone layer; UVB is the major mutagenic fraction of sunlight; and UVA also has deleterious effects primarily because it creates oxygen radicals, which can indirectly damage DNA.

The mutagenic effect of UV derives from the fact that it can react with DNA and other biological molecules since UV wavelengths are preferentially absorbed by bases in DNA molecules and by aromatic amino acids of proteins. UVB and UVC produce pyrimidine dimers on interaction with DNA; UVA produces very few of these. These pyrimidine dimers, on account of their ability to form lesions that interfere with transcription and DNA replication, lead to mutations, chromosomal rearrangements and lethality.

UV is produced from arc welding equipment and for practical purposes UV is artificially produced by the excitation of mercury vapour at low and medium pressures. Instruments of UV radiation are relatively cheap and available in almost every plant tissue culture or molecular biology laboratory. However, due to the low penetrability, their use in plant mutagenesis has been limited mainly to irradiation of materials and equipment used in tissue culture.

5. References

5.1. Cited References

Bores, R.J. and Bottinot, P.J. 1974. Design and dosimetry of a strontium-90–yttrium-90 beta irradiation facility. *Health Phys.* 26: 99–101

Burger, G. and Broerse, J.J. 1974. Neutron sources available for radiobiological research. *In*: Anonymous (ed.) Biological Effects of Neutron Irradiation. Vienna: International Atomic Energy Agency, pp. 3–20.

Burtscher, A. and Casta, J. 1967. Facility for seed irradiation with fast neutrons in swimming-pool reactors: a design. Neutron Irradiation of Seeds I. Technical Reports Series No. 76. Vienna: IAEA, pp. 41–62.

IAEA 1967. Neutron Irradiation of Seeds I. Technical Reports Series No. 76. Vienna: IAEA.

IAEA 1968. Neutron Irradiation of Seeds II. Technical Reports Series No. 926. Vienna: IAEA.

IAEA 1972. Neutron Irradiation of Seeds III. Technical Reports Series No. 141. Vienna: IAEA.

Novak, F.J. and Brunner, H. 1992. Plant breeding: induced mutation technology for crop improvement. IAEA BULLETIN, 4/1992: 25–33.

Parker, S. 2007. McGraw Hill Encyclopaedia of Science and Technology, 10th Edition.

Shu, Q.Y., Li, W.H., Xia, Y.W. *et al.* **2002.** Biological effects of electron beam on M1 plants in rice. J. Nucl. Agric. Sci. 17: 162–164.

Websites
Discovery of X-rays:
http://nobelprize.org/nobel_prizes/physics/laureates/1901/index.html

Radioactivity:
http://nobelprize.org/nobel_prizes/physics/laureates/1903/becquerel-bio.html

Discovery of radioactive element:
http://nobelprize.org/nobel_prizes/physics/articles/curie/

X-ray irradiator:
http://www.radsource.com/products/

5.2. Further Reading

IAEA 1977. Manual on Mutation Breeding, Second Edition. Vienna: International Atomic Energy Agency.

L'Annunziata, M.F. 2007. Radioactivity - Introduction and History. Amsterdam: Eslevier.

C08

Gamma Irradiation

C.Mba[a,*,#] and Q.Y.Shu[b]

[a] Plant Breeding and Genetics Laboratory, Joint FAO/IAEA Division of Nuclear Techniques in Food and Agriculture,IAEA Laboratories Seibersdorf, International Atomic Energy Agency, Vienna International Centre, P.O. Box 100, Vienna, Austria
[#] Present Address: Plant Production and Protection Division, Food and Agriculture Organization of the United Nations, Viale delle Terme di Caracalla, 00153 Rome, Italy
[b] Joint FAO/IAEA Division of Nuclear Techniques in Food and Agriculture, International Atomic Energy Agency, Wagramer Strasse 5, P.O. Box 100, A-1400 Vienna, Austria
  Present: Institute of Nuclear Agricultural Sciences, Zhejiang University, Hangzhou 310029, China
* Corresponding author, E-MAIL: Chikelu.Mba@fao.org

1. Introduction

Among the physical mutagens, X- and γ-rays are the most commonly used mutagens in mutation breeding. Over the past 40 years, the use of γ-rays in mutation induction has become particularly prevalent, while the use of X-rays has been significantly reduced (**Figure 8.1**). The preference for the use of γ facilities for irradiation is probably on account of both its wide availability and versatility of use. During the past half century, gamma sources were installed in several types of irradiation facility. While numerous facilities have dual or multiple uses, e.g. for medical uses, mutation breeding and food irradiation, some specially designed facilities have been constructed specifically for plant mutation induction and breeding, e.g. the Controlled Environment Radiation Facility of the Brookhaven National Laboratory, Upton, NY, USA, which comprises gamma cells (consisting of lead-shielded gamma emitters, ^{60}Co and ^{137}Cs), gamma phytotrons and gamma fields in which plants can be grown and irradiated chronically.

2. Source of Gamma Rays

The naturally occurring isotope, potassium-40 (^{40}K), is one source of γ-rays in the environment. For practical purposes, mutation induction in plants is achieved by γ-rays produced by the disintegration of the radioisotopes cobalt-60 (^{60}Co) and casium-137 (^{137}Cs). The physical parameters of ^{60}Co and ^{137}Cs make them suitable as gamma sources (**Table 8.1**).

The decision on which type of gamma emitter is procured is usually based on the consideration of the relative merits of these parameters for the purposes for which the source is being installed. For instance, ^{137}Cs has a much longer half-life than ^{60}Co, which probably explains why it is the more commonly used in γ irradiation facilities (both for mutation induction and for the sterilization of food products). In selecting between these two types of sources, it should also be borne in mind that ^{137}Cs and ^{60}Co have different properties. For instance, 13 times more ^{137}Cs is required to generate the same activity as ^{60}Co; 1 curie (Ci) of activity is generated by 0.88 mg and 11.5 mg of ^{60}Co and ^{137}Cs, respectively. Also, more than four times more activity of ^{137}Cs is needed to produce the same dose rate as ^{60}Co. Probably counterbalancing the foregoing is the fact that ^{60}Co decays about six times faster than ^{137}Cs. With half-lives of 5.3 and 30 years, respectively, ^{60}Co decays 12.3% per year as compared with 2.3% for the same period for ^{137}Cs. Also, less shielding is needed with ^{137}Cs than with ^{60}Co as indicated by their respective tenth-value layer (TVL, the thickness of an absorber required to attenuate a beam of radiation to one-tenth). Another important feature of ^{60}Co sources is that the radiations are composed of two gamma rays which occur in the decay of ^{60}Co to nickel-60 (^{60}Ni) and are present in equal numbers. Beta (β) particles are emitted from these two gamma emitters but the encapsulation and self-absorption of the source imply that the β-particles are of minimal importance biologically (detailed consideration of β-particles as irradiation agents is treated separately in **Chapter 7**).

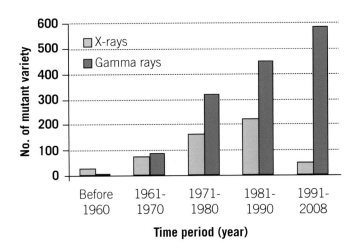

The Joint FAO/IAEA Programme

Figure 8.1 The global development of mutant crop varieties by using X- and gamma rays (source: http://mvgs.iaea.org accessed on 25 August 2009).

Table 8.1: Physical parameters of ^{60}Co and ^{137}Cs [a]

Parameters	^{60}Co	^{137}Cs
Half-life (year)	5.3	30
Energy	γ_1 = 1.33 MeV γ_2 = 1.17 Mev	γ_1 = 0.66 MeV
Specific γ-ray emission	1.29 R/h•Ci@1m	0.3 R/h•Ci@1m
Specific activity	1–400 Ci/g	1-25 Ci/g
Shield (TVL)	~ 5 cm lead	~ 2 cm lead
Chemical form	Metal element	Chloride or sulphate salt
β-emission	0.31 MeV (100%)	0.52 MeV (92%) 1.18 MeV (8%)
Applications	Commercial gamma irradiation facility for sterilization of foods and other items; gamma field, gamma cell	Gamma cell, phytotron, glasshouse

[a] adapted from Briggs and Konzak (1977); TVL: tenth value layer.

Box 8.1: The ionizing mechanism of gamma rays

1. The photoelectric effect: A gamma photon on interacting with an electron in the atom transfers all its energy to that recipient electron. The recipient electron is ejected from the atom (knocked out of orbit) to form a photoelectron whose kinetic energy is equal to that of the incident gamma photon minus the binding energy of the electron. The photoelectric effect is the dominant energy transfer mechanism for gamma ray photons with energies below 50 keV, but it is much less important at higher energies.

2. Compton scattering: This describes the photon–electron interaction in which an incident gamma photon loses enough energy to an atomic electron to cause the ejection of the latter from the atom. The remnant energy retained by the incident photon is emitted as a new, lower energy gamma photon with an emission direction different from that of the incident gamma photon. There is an inverse relationship between the probability of Compton scatter and photon energy. Compton scattering is thought to be the principal absorption mechanism for gamma rays in the intermediate energy range of 100 keV to 10 MeV, an energy spectrum.

3. Pair production: The gamma photon on encountering the nucleus causes the production (ejection) of an electron and additionally the electron's anti-matter, the positron leading to the so-called electron–positron pair. The positron has the same mass as an electron, but it has a positive charge equal in strength to the negative charge of an electron. Energy in excess of the equivalent rest mass of the two particles (1.02 MeV) appears as the kinetic energy of the pair and the recoil nucleus. The positron has a very short lifetime (if immersed in matter) (about 10-8 seconds). At the end of its range, it combines with a free electron. The entire mass of these two particles is then converted into two gamma photons of 0.51 MeV energy each. When an electron and a positron interact, they will be annihilated, a term describing their 'conversion' to two gamma rays. It is plausible to infer that when a positron enters any normal matter, it will find an abundant supply of electrons with which to annihilate.

The secondary electrons (or positrons) produced in any of these three processes frequently have enough energy to produce many ionizations (knocking electrons from their orbit) up to the end of range.

3. Mechanism of Ionization

As indicated above, ^{60}Co produces greater radioactivity than the same quantity of ^{137}Cs. This should be borne in mind with regard to dosimetry considerations in radiation-induced mutagenesis when using these two different sources of gamma rays. In general, as gamma rays pass through a tissue, the ionization process occurs through three main mechanisms: 1) the photoelectric effect; 2) Compton scattering and 3) pair production (**Box 8.1**).

4. Gamma Facilities

Gamma irradiation sources are commonly available in many countries. The IAEA has supported its Member States in Africa, Asia, Europe and South and Central America in building dozens of gamma facilities through its Technical Cooperation programme. Also, the IAEA, through its Plant Breeding and Genetics Laboratory of the FAO/IAEA Division of Nuclear Techniques in Food and Agriculture situated at the IAEA Laboratories in Seibersdorf, Austria provides gamma irradiation services for its Member States.

The Joint FAO/IAEA Programme

Figure 8.2 A Gamma cell. **A:** A cobalt-60 gamma source with a raised loading stage; **B:** Close-up of the raised loading stage of a cobalt-60 gamma source showing rice grains in a Petri dish.

Gamma instruments are divided into two main types for mutation induction, acute and chronic irradiation facilities. Acute irradiation, implying the one-time exposure of plant parts (usually propagules) to irradiation, is most conveniently carried out in a purpose-built laboratory making use of gamma cells and other dual-use irradiation facilities. On the other hand, gamma irradiation is also amenable for prolonged exposure of plant materials; when this is desired, the gamma source is placed in a glasshouse, field or in a chamber under controlled environmental conditions so that plants can be irradiated as they grow over extended periods of time.

The different adaptations of gamma sources for radiation-induced mutations are described below. A typical gamma cell irradiator is described as a source for acute irradiation while a gamma glasshouse in Malaysia, a gamma field in Japan and a gamma phytotron in Korea are described as examples of facilities providing chronic irradiation for induced plant mutations.

4.1. Gamma Cell Irradiator

The gamma cell is the most commonly used irradiator as it can be adapted for several functions including radiation-induced mutagenesis and sterilization of foods (**Figure 8.2**). The irradiators are suitable for seed irradiation as long as the size of the irradiation space is sufficient and the dose rate allows practical irradiation times. Gamma cell irradiators, designed for both vertical and horizontal loading of samples provide a uniform field for the irradiation of samples. Gamma sources are commonly provided as a line source, similar in shape to

a pencil. Within a sample chamber, this shape enables the gamma exposure to be provided as a uniform dose over a larger distance than is the case for a point source. The point source on the other hand projects the dose differently, as radiation coming from a single small point. When the distance from a line source is increased (about three times the length of the source) the line source acts as a point source. The exposure to the operator standing about 30 cm from the source, for instance, would be considered a point source due to the distance from the source itself.

4.2. Gamma Phytotron

A gamma phytotron is a completely closed chamber in which plants are grown under controlled environmental conditions and a gamma source is built for chronic irradiation. One such facility was built at the Advanced Radiation Technology Institute (ARTI), Korea Atomic Energy Research Institute (KAERI), Jeongeup, Jeonbuk, Korea. This chamber has a ^{60}Co ionizing source with radiation activity of about 400 Ci. The source is stored 1 m underground and remotely raised for irradiation in the operation room (Kang *et al.*, 2010).

The phytotron at ARTI has functionality for housing a range of plant materials from potted plants to cultured callus. The materials destined for chronic irradiation are exposed to low doses of γ-rays in the irradiation room (**Figure 8.3**). Immediately adjacent to this room is the non-irradiation control room where the same types of plant materials are grown under the same conditions of light, temperature and humidity; as control samples, gamma irradiation is excluded. Additionally, the facility

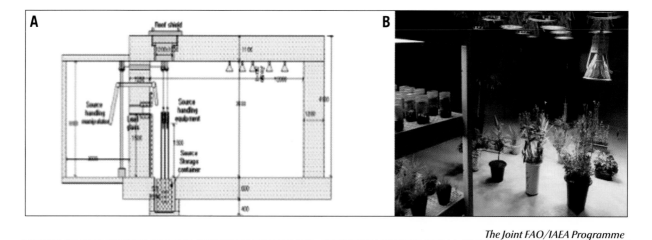

A   B

The Joint FAO/IAEA Programme

Figure 8.3 The gamma phytotron at the Advanced Radiation Technology Institute, Korea Atomic Energy Research Institute, Jeongeup, Jeonbuk, Korea. **A:** Diagram of the vertical section of the phytotron showing the location of the radiation source relative to the roof shield, the irradiation paraphernalia and lighting; and **B:** Photograph of the interior of the gamma phytotron showing potted and *in vitro* cultured plants arranged for chronic gamma irradiation; light irradiance is controlled by adjusting the lengths of the cables suspending the electric bulbs from the roof of the facility (courtesy of Dr. S.Y. Kang).

consists of a glasshouse for the acclimatization of irradiated plants before transferring into the field, an operations room where the panels for controlling and monitoring the conditions in the irradiation room are located and an office. The total floor area of the irradiation room is about 104.16 m².

The target plant materials for γ-ray irradiation are arranged in a circle around the ^{60}Co source starting from a distance of 2 m from the central source where dose rate was estimated at 612.9 mGy/h to a maximum distance of 7 m (dose rate, 60.1 mGy/h). The safety features of the phytotron include a 1.2-m-thick concrete wall surrounding the irradiation room and a two-fold lead shielding door between the operation and irradiation rooms. Also, the irradiation room is equipped with two charge-coupled device (CCD) camera systems in which all the features inside can be monitored from a remote operating room. The irradiation and control rooms have automated environmental control systems that permit temperature ranges of 15 to 35°C, and humidity of 50 to 80%. The light irradiance has a maximum range of approximately 417 µE/m²/s. All parameters are adjustable to suit desired growth conditions of targeted plants.

4.3. Gamma House

A gamma house is a special type of glasshouse which is similar to a gamma phytotron in overall design, but is much bigger in size and directly uses sunlight as a light source. Such facilities have been built in several countries such as Japan, Malaysia and Thailand.

The newly built gamma house at the Malaysian Nuclear Agency (Nuclear Malaysia) is a recent example (Ibrahim, 2010). Its design consists of concentric circles with an irradiation area of 30-m-diameter in the centre and is covered with a translucent polycarbonate roof (**Figure 8.4**). This central portion, where the double-encapsulated 800 Ci RSL6050 Caesium^{137} Source Pencil is located, is covered with a skyshine hood which protects the source from the elements while permitting the diffusion of the radiation from the gamma source through the air. The skyshine hood incorporates a water deluge system, which is connected to the fire detection system. This 30-m-diameter irradiation area is protected by a partial concrete wall fitted with an entry maze. This security feature is enhanced by an interlock system that permits the exposure of the source only after all the prerequisite safety conditions are met; it automatically returns the source to the safe storage position if any safety device is compromised. The source, shielded by lead and encased in fire deterrent steel jacket, is stored above the floor surface and is raised remotely when irradiation is desired.

The main irradiation area, the 30-m-diameter at the centre, is further encircled by a 300-m-diameter exclusion zone, also fitted with a safety interlock circuit. The source is only exposed when the entire 300-m-diameter site is free of personnel. To ensure maximum safety for personnel, the entry-exit maze is fitted with an override system, which allows a person trapped within the maze to open the gate. When activated, this override functionality automatically breaks the safety interlock

The Joint FAO/IAEA Programme

Figure 8.4 The gamma greenhouse at the Malaysian Nuclear Agency. **A:** An aerial view of the facility and immediate environs; the markings of the different security perimeters are overlaid on the picture. **B:** A close-up of the aerial view of the gamma greenhouse (courtesy of Dr. R. Ibrahim).

circuit causing the source to return to the safe storage position. Additionally, there is an emergency refuge area consisting of high density concrete of 1-m in thickness which is located at the 15-m-diameter point close to the normal entry-exit gate. This provides a safe area, free of radiation, for any person accidentally trapped in the facility when the source is activated.

The estimated dose rates are <0.011734 μSv per hour on the outside face of the concrete biological shield (30-m-diameter point); <0.004813 μSv per hour immediately above the skyshine roof; and <0.000718 μSv per hour at the 150-m-iameter fence. Currently, the irradiator operates 20 hours daily (from noon to 8 am). It is shut down for maintenance and sample preparations for 4 hours and activated for another 20-hour irradiation.

4.4. Gamma Field

A few large-scale irradiation facilities – gamma fields have been constructed around the world for chronic irradiation treatment of growing plants. For example, Kawara (1969) estimated that more than 20 gamma fields, including the largest one at the time, the Brookhaven National Laboratory (Upton, NY, USA) had been established all over the world in 1969. Some of these are now decommissioned, such as the one at the Brookhaven National Laboratory and the Gamma Field in Sichuan Agricultural Academy of Agricultural Science, Chengdu, China. However, the world's largest radiation facility for plant breeding – the gamma field at

the Institute of Radiation Breeding (IRB), Ohmiya, Japan has remained operational since it was commissioned in 1962 (**Figure 8.5;** Nakagawa *et al.*, 2010).

The IRB gamma field is a circular field of 100 m radius (an area of 31,400 m²) with 88.8 TBq ^{60}Co source at the center. The gamma field is surrounded by a shielding dyke 8 m high **(Figure 8.4)**, which prevents primary radiation from reaching the outside area. The ^{60}Co source has been renewed every 2 years since 1962. At the start of the operation, 20 hours (from 12 am to 8 am of the following day) daily irradiation has been conducted. Now, the daily irradiation time has been shortened to 8 hours (from 12 am to 8 pm) with extended restricted areas following the Recommendation of the International Commission on Radiological Protection (ICRP, 2001). Currently, the irradiation dose of the gamma ray at the nearest point of the field (10 m from the ^{60}Co source) is approximately 2 Gy/day. This diminishes progressively with increasing distance from the source, reducing by 150 times at a distance of 100 m from the source.

The facility is especially suited for the irradiation of large live plants *in situ*, and is utilized for chronic irradiation for various crops. Plants such as fruit trees, mulberry tree, tea tree, flowers, upland crops, rice, etc. are grown year round in the immediate vicinity of this facility; the nearest plants are at a distance of 10 m radius while the farthest are at a displacement of 100 m radius from the centrally located source. The facility is particularly useful for the irradiation of large plants.

The Joint FAO/IAEA Programme

Figure 8.5 The gamma field at the Institute of Radiation Breeding, Ohmiya, Japan. **A:** An aerial view of the facility showing concentric rings of crops grown in terraces around the irradiation source located at the centre of the field; **B:** A close-up image of the facility showing the irradiation tower, crops being exposed to chronic irradiation and freshly prepared field for planting; **C:** A close-up image of the facility showing the reinforced steel casing housing the ^{60}Co source (courtesy of Dr. H. Nakagawa).

Box 8.2: Safety considerations in the use of a gamma irradiation facility (from Dalhousie University, Canada)

System operation for gamma cell irradiator
- Verify that the source is in the "OFF" position before attempting to open the chamber door. After placing the sample into the chamber check the interlocks before placing the source in the active position.
- After using the irradiator, document use in the log book.
- The operator should stay outside the room when the unit has the source in the active position. If the sample is left in the irradiator for an extended period of time while the operator is not attending the system, the area must be locked.
- In the event of a problem with the system, the Radiation Safety Officer (RSO) should be contacted.

Emergency procedures for gamma cell irradiator
- In the event of malfunction during loading or unloading of the irradiator, the unit is to be taken out of service immediately. Evidence of malfunction includes binding or moving parts, the presence of metal shavings or chips, etc.
- Log and describe any abnormal occurrences in the log book.
- Should the "Release Source" fail, leave/secure the room immediately and contact the RSO.
- If at any time it is possible to open the cavity door without pressing the door release button, the interlock assembly is malfunctioning. **Do not use the unit!** Leave/secure the room and contact the RSO.
- If at any time it is impossible to raise the source with the door closed or to open the door with the source in the "OFF" position, either the interlock switches or interlock solenoids are malfunctioning. **Do not use the unit!** Leave/secure the room and contact the RSO.

Responsibilities in the use of a gamma cell irradiator
The principal investigator must:
- Maintain the irradiator in a clean and mechanically functional condition.
- Ensure that designated users receive training as required.
- Ensure that designated users wear whole body dosimeters when operating the unit.
- List and certify designated users.
- Ensure physical security of the key to the unit and prevent unauthorized use of the irradiator.
- Notify the RSO immediately of any malfunctions or problems with the irradiator.
- Arrange for repairs or maintenance of the unit by appropriate persons.

Designated users must:
- Operate the unit in accordance with the established procedures at all times.
- Wear a whole body dosimeter when operating the irradiator.
- Notify the principal investigator and the RSO of any malfunctions or problems with the irradiator.
- Ensure that the key is returned to secure storage following irradiation.

Radiation Safety Personnel must
- Maintain the licence/permit issued to the facility by the Canadian Nuclear Safety Commission (CNSC) and the Dalhousie University for operation of the irradiator.
- Conduct leak tests annually.
- Provide appropriate training.

Gamma cells are commonly used for acute irradiation while gamma rooms and gamma fields are most suited for chronic irradiation assays.

5. Safety Considerations

The containment vessel of an irradiator is normally heavily shielded, typically with lead. The operation of a gamma irradiation facility should in all cases be designated to a properly trained (and in many countries, duly certified) operator. Recognizing the potential hazards of unintended exposure to radiation, gamma irradiators are usually set up for operation by remote control. The inherent safety provisions, including interlocks and other safety features, are built into all types of gamma irradiator, all aiming at minimizing any potential exposure to operators. These are designed to move the source into the safe position if the chamber is opened or other situations arise. Administrative controls are established to make sure the area is clear and ensure that the operators have the appropriate training. Control of access to the facility is usually enforced through strict regulations on the handling of the key to the system. Quite importantly, workers using the system must be monitored for whole body gamma exposures as well in accordance with the statutory provisions of the country. More detailed guides on safe usage of gamma irradiation facilities are provided by the National Standards Handbook No. 73 (1960) while on-line resources include the websites of the Dalhousie University, Canada. An adaptation of the latter is provided in **Box 8.2**.

6. References

6.1. Cited References

Briggs, R.W. and Konzak, C.F. 1977. Radiation types and sources. In: Anonymous (ed.) Manual of Mutation Breeding. Vienna: International Atomic Energy Agency.

Ibrahim, R. 2010. The Malaysian Nuclear Agency Gamma Greenhouse. *Plant Mutation Reports*. 2: 50–51.

Kang, S.Y., Kim, J.B., Lee, G.J. *et al.* **2010.** Construction of "gamma phytotron": new chronic irradiation facility using ^{60}C gamma-ray. *Plant Mutation Reports*. 2: 52–54.

Kawara, K. 1969. Structural characteristics and the exposure rates in the Japan's Gamma Field. *Gamma Field Symposia*. 6: 1–18.

Nakagawa, H. 2010. Gamma radiation facilities at the Institute of Radiation Breeding. *Plant Mutation Reports*. 2: 55–56.

6.2. Websites

FAO/IAEA Database of Mutant Variety and Genetic Stock: http://mvgs.iaea.org

Radiation Safety Manual: http://environmentalhealthandsafetyoffice.dal.ca/files/Radiation_Safety_Manual_-_2008_revision.pdf

Ion Beam Radiation Mutagenesis

T.Abe*, H.Ryuto and N.Fukunishi

RIKEN Nishina Center, 2-1 Hirosawa, Wako, Saitama 351-0198, Japan
* Corresponding author, E-MAIL: tomoabe@riken.jp

1. Introduction

Ion beams consist of particles traveling along a path. The particles can vary in mass from a simple proton to a uranium atom. Alpha rays for example can be considered as a helium ion beam, though usually ion beams are generated through particle accelerators.

The history of ion beam research and its applications can be traced back to the establishment of accelerator facilities at the Lawrence Berkeley Laboratory (LBL, known as Bevalac) in the United States, in the 1970s. Before the Bevalac, no particle accelerator could produce ion beams with sufficient energy to pass through biological tissues. Relative biological effectiveness (RBE) values were determined for particles with relatively high energies and for relatively heavy ions at the LBL, and other facitlities in the 1980s. Biological effects depend on the local ionization density of the DNA. One of the difficult challenges of the 1990s was to understand the damage induced by ionizing radiation at the cellular and molecular (DNA) levels, and its relationship to biological effects.

The RIKEN RI-Beam Factory (RIBF, Japan) is a multidisciplinary facility for high-energy ion beam research. The light ion beam, up to 135 MeV/nucleon (about half the speed of light) has been established since 1986. Shortly after it was used in cancer therapy, collaboration was established among radiation oncologists, physicists and biologists. This positive experience encouraged plant scientists to use the RIBF for radiation biology research, which began in 1989. Plant breeding trials began in the mid-1990s and the work on induced mutations in plants is described below.

2. Ion Beam Radiation and Facilities

2.1. Nature and Characteristics of Ion Beam Radiation

There are two significant differences between photons, gamma rays and X-rays and ions: 1) an ion has a mass and 2) an ion has an electrical charge. By contrast photons, gamma and X-rays have neither. As a result, ion beams provide a different linear energy transfer (LET). For example this is 0.2–2 keV/µm for gamma-rays and X-rays, but is much larger and widely varying for ion beams, e.g. 23 keV/µm for C ions and 640 keV/µm for Fe ions.

High-LET radiation, such as ion beams, causes more localized, dense ionization within cells than low-LET radiation. As shown in **Figure 9.1**, the LET of ion beams changes dramatically with penetration distance. The peak of LET is achieved at the stopping point and is known as the Bragg peak (BP) (**Figure 9.1**). In human

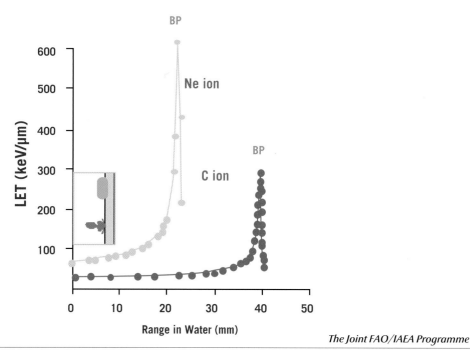

The Joint FAO/IAEA Programme

Figure 9.1 LET curves for Ne and C ion beam penetration in water. Water can be used as surrogate for soft tissue penetration in plants. A beam of high energy penetrates a plantlet and/or plant tissue with rather low and uniform LET.

cancer treatment, the BP is used to target malignant cells. However, a uniform dose distribution is important for even and reproducible mutagenic treatment, therefore plant cells and organs are treated using ions with stable LET (**Figure 9.1**).

The choice of ion beam depends on the characteristics of the ion with respect to electrical charge and velocity. When a high LET is required a heavier, highly charged ion is selected with a low velocity. Dose (in Gy) is proportional to the LET (in keV/µm) and the number of particles. For example, the number of particles (N) passing an area of 1 µm² in the case of 1 Gy dose deposition is given as,

$$N = 6.24/\alpha$$

where α is the LET value in keV/µm unit.

2.2. Ion Beam Facilities for Plant Mutation

The feature of ion beam generators that is probably most suitable for mutation breeding is that the accelerator can provide more than 0.2 pnA of C-ion beam at 100 MeV/nucleon. For plant mutation induction it is preferable for the accelerator facility to have an exclusive beam line with a beam spreading apparatus. There are a number of ion beam accelerators set up for medical use all over the world, and several large heavy-ion accelerator facilities have been constructed. However, most accelerators for medical use provide proton beams that have relatively low LET values, and the large heavy-ion accelerators are sometimes fully devoted to basic science (nuclear physics). Currently therefore, the number of accelerator facilities that are used for plant mutation induction and breeding is limited.

Table 9.1: Parameters of ion beam accelerator facilities for plant breeding

Facility	Ion beam	Energy (MeV/nucleon)	LET (keV/µm)	Range in water (mm)
RIBF (RIKEN, Japan)	C	135	23	40
	N	135	31	34
	Ne	135	62	23
	Ar	95	280	8
	Fe	90	640	4
TIARA (JAEA, Japan)	He	25	9	6.2
	C	26.7	86	2.2
	Ne	17.5	441	0.6
W-MAST (WERC, Japan)	H	200	0.5	256
	C	41.7	52	5.3
IMP (CAS, China)	C	80	31	17
	N	80	43	15
	O	70	62	10
	Ne	70	96	8
	Ar	69	327	5
LNS (INFN, Italy)	C	80	31	17

Note that the LET and ranges for W-MAST, IMP and LNS are the calculated values using the SRIM code (Ziegler *et al.*, 2008) at the kinetic energy written in the table.

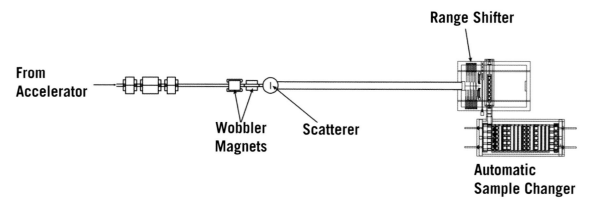

The Joint FAO/IAEA Programme

Figure 9.2 Schematic view of E5B beam line.

Table 9.1 shows a list of accelerator facilities in the world that have been used for plant mutation breeding. These include the RIKEN RI-Beam Factory (RIBF, Japan), the Takasaki Ion Accelerators for Advanced Radiation Application (TIARA, Japan), the Wakasawan Energy Research Center Multipurpose Accelerator System with Synchrotron and Tandem (W-MAST, Japan), Institute of Modern Physics (IMP, China) and the Laboratori Nazionali del Sud (LNS, Italy).

2.3. Ion Beam Preparation

The RIBF consists of five cyclotrons and an injector linac. An AVF cyclotron and a ring cyclotron, RRC, are used to obtain the beams listed in **Table 9.1**. The beam accelerated by the cyclotrons is transported to the E5B beam line, which is used exclusively for radiation of biology experiments. **Figure 9.2** shows the schematic view of the E5B beam line. The approximately 5-mm-diameter beam is transported from the accelerator and transformed into an approximately 8-cm-diameter beam with a uniform intensity distribution while passing through the E5B beam line. There are various methods used for spreading the beam, such as the raster scan method, the double scattering method and the wobbler scattering method (Chu *et al.*, 1985); the wobbler scattering method is used in the RIBF.

2.4. Automatic irradiation system

Ion beam irradiates samples *via* an automatic irradiation system, which consists of a range shifter and an automatic sample changer. The range shifter consists of

twelve 12-cm-diameter aluminium energy absorbers attached to air cylinders (from 0.02 to 20 mm thick). The kinetic energy of the beam is absorbed by the energy absorbers while passing through the range shifter, and the decreasing kinetic energy of the beam results in increasing the LET. Therefore, the LET of the beam incident on the samples can be adjusted using the range shifter. For example, the LET values from 22 to 285 keV/μm of C ion and from 60 to 700 keV/μm of Ne ion can be selected.

After the selection of the LET, the beam is used to irradiate samples, which is automatically positioned into the beam area using the automatic sample changer.

The Joint FAO/IAEA Programme

Figure 9.3 Photograph of cassettes filled with samples containers.

The automatic sample changer consists of a movable table with six stages for cassettes. The sample containers carrying the biological samples are placed into the cassettes, and the cassettes are automatically moved to the beam position. Typical sample containers are: a plastic box (5 × 7.5 × 1.25 cm) for dry seeds and scions, plastic Petri dishes (3, 6 and 9 cm diameter) for imbibed seed and tissue culture material and plant box for cultured plantlets (**Figure 9.3**). The maximum number of sample containers that can be automatically irradiated is more than 500 (Ryuto *et al.,* 2008).

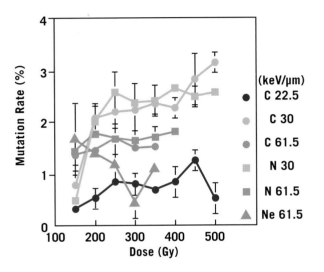

Figure 9.4 Effects of heavy-ion beams on mutation induction in *Arabidopsis*.

3. Biological Effects of Ion Beams

Ion beam irradiation is expected to produce double-stranded DNA breaks. Mutations induced by ion beam irradiation at the molecular level have been most extensively studied in mammalian cells. The frequency of deletion is higher for ion beam irradiation than for gamma rays (e.g. Suzuki *et al.,* 1996). On the other hand, ion beams can induce point-like mutants in yeast (e.g. Matsuo *et al.,* 2006) and red alga (*Cyanidioschyzone merolae*, Ohnuma *et al.,* 2007). These differences may be the result of different DNA repair systems between animals and micro-organisms. Non-homologous end-joining (NHEJ, **see Chapter 6**) leads to the rejoining of DSBs after mammalian cells are exposed to radiation, whereas HR is the major DSB repair pathway in micro-

organisms. In the case of *Arabidopsis* irradiated with C ions in TIARA, half of the mutants produced had point-like mutations and the other half large DNA alterations such as inversions, insertions, translocations and large deletions (5.4–233 kbp, Shikazono *et al.,* 2005). Several molecular analyses of C, N and Ne ion induced mutations in plants have been reported with deletion sizes ranging from 26 bp to 18 kbp in RIBF. Carbon ions at 23, 40, 60 keV/µm and Fe ions (640 keV/µm) induce point, insertion and deletion mutants in *Mesorhizobium loti*, in which HR is the major repair pathway (Ichida *et al.,* 2008). Insertion and deletion mutation rates for Fe ions were higher than those of C ions. The size of the deletion seems to be dependent on the LET; 43–203 bp for C ions and 119–641 bp for Fe ions. An LET of 30 keV/µm was found to be the most effective for mutation induction in *Arabidopsis* (**Figure 9.4**). In the case of rice, the highest mutation was observed in LET from 61 to 74 keV/µm with C and Ne ions (Hayashi *et al.,* 2008a). Thus, the LET of ion beams seems to be an important factor affecting mutagenesis.

4. Application in Plant Breeding

Initial studies showed that ion beams are highly effective in inducing mutations using developing seed embryos at a particular stage after fertilization without damage to other plant tissues. There are many types of mutants including albino, sectorial chimera, periclinal chimera, herbicide- and salt-tolerant phenotypes in M_1 plants in tobacco (Abe *et al.,* 2000). Exposure of plant materials to an ion beam for a few seconds to minutes triggered various genetic changes. The sensitivity of plant materials to ion beam irradiation decreased in the following order: callus, developing embryos and stem nodes, imbibed seeds and scions, and finally dry seeds (**Table 9.2**). Advantages of ion beam mutagenesis include: low dose with high survival rates, induction of high mutation rates and a wide range of variation. The use ion beam often alters only a single characteristic. Thus a new crop variety could be obtained by selecting a mutant with a modification to a target trait while retaining the existing valuable ones. The approach has been particularly successful in flower breeding.

In 2002, the sterile Vervena mutant, "Temari Bright Pink" became the first flower developed and marketed using ion beam mutagenesis. The development period

Table 9.2: Mutant lines developed in various crops using RIBF

Mutant phenotype/ plants (generation)	Variety, strain/plant material treated	Treatment ion/LET (keV/μm) dose (Gy)	Survival rate (%)/ mutation rate (%) [a]
Sterile			
Verbena	Temari Coral Pink/ stem node containing lateral buds	N/30/5, 10	84–90/0.9–2.8
	Temari Sakura/ stem node containing two lateral buds	N/30/5	81/1.0
Pelargonium	Splendide/ stem node containing a single axillary bud	C/23/5, 20	80/2.5–5
Asparagus (M_1)	Mary Washington 500W/ germinated seeds	N/30/20	80/8.3
Triticum monococcum	Dry seeds	N/28.5/50	ND/0.052–0.075
Eucalyptus	Shoot primordia	C/23/20	80/9.3
Chlorophyll deficient			
Tobacco (M_1)	BY-4/ovary containing fertilization embryos	N/28.5/10	75/0.6
Rice (M_2)	Koshihikari/imbibed seeds	Ne/63/10, 20	74-80/6.1–11.6
	Nipponbare/imbibed seeds	Ne/63/10	75/8.6
		C/23/20, 40	49-79/4.8–7.1
Flower colour and shape			
Petunia (M_1)	Ovary containing fertilization embryos	Ne/63/5	ND/1.0
Dahlia	Miharu/shoots	N/30/5, 10	NE/20.3–50.1
Rose	Bridal Fantasy/ dormant	Ne/61 /15	ND/51.7
Chrysanthemum	Chiyo/stem	C/23/10	93.7/14
	Beach ball/stem	C/23/7	44.8/4.5
Torenia	Summer Wave/leaf tissues and stem internodes	N/30/50	NE/1.9
		Ne/63/20	NE/1.6
Variegation			
Tobacco (M_1)	Xanthi/dry seeds	N/30/200	ND/0.08
Petunia Hybrida	Surfinia Blue Vein/stem node containing lateral buds	N/30/5	ND/1.8
Sandersonia (M_1)	Dry seeds	N/28.5/20	60/0.5
Apple	Fuji/dormant scions	Ne /84.1/7.5	50/4.2
Dwarf			
Barley	Kantohadaka 77/dry seeds	N/30/50, 100	ND/2.6
	Kantohadaka 77/imbibed seeds	N/30/5	ND/0.9
Millet	Mojappe/dry seeds	C/23/20	ND/0.1
Pepper (M_1)	California Wonder/dry seeds	Ne/61/10	80/1.3
Buckwheat	Shinano natsu soba/dry seeds	C/23/100	NE/0.6
	Rotundatiem/dry seeds	C/23/40	NE/1.1

[a] ND: no data, NE: no effect.

of the new variety took only three years. Similar successes were demonstrated by the new colour of Dahlia var. World (in 2002), the new sterile Verbena vars. Temari Sakura Pink (2003) and Temari Momo (2006), the new colour Petunia var. Surfinia Rose Veined (2003), the new colour Dianthus var. Olivia Pure White (2008), the new colour cherry blossoms var. Nishina Zao (2008) and the dwarf Delosperma vars. Reiko Rose and Reiko Pink Ring (2009). In addition three mutants of sweet pepper: two dwarf plants and a yellow pepper have been developed through ion beam irradiation (Honda *et al.*, 2006). These mutant phenotypes are recessive and due to single gene mutations. The isolation of monogenic homozygous recessive mutants in the M_1 generation have not normally been described in other mutational studies as there are complications with chimeras (**see Chapter 15**). However, ion beam treatment can induce homozygous recessive mutations in nuclear genes that allow detection of mutants even in M_1 plants (Abe *et al.*, 2000 and Imanishi *et al.*, 2007). Hence, the time needed for developing a new variety using ion beam radiation can be significantly shortened.

An international heavy-ion plant research consortium has been organized with 125 national user groups and 11 international institutes. The consortium includes agricultural experimental stations, universities, seed and horticulture companies. Recent results from the consortium involve studies in plant physiology, genetics, botany and agriculture using mutants of many species in the plant kingdom. Of note are four lines of salt-resistant rice, one of which has been grown in a saline paddy field with good yield (Hayashi *et al.*, 2007, 2008b). Other examples include: the development of semi-dwarf buckwheat that is shorter and sturdier than normal varieties (e.g. Morishita *et al.*, 2008); and the development of mutants that enable the molecular understanding of the mechanisms of flowering in sterile eucalyptus and wheat (Shitsukawa *et al.*, 2006).

The ion beam irradiation technique induces mutations at a high rate without severely inhibiting growth at relatively low doses. The irradiation treatment given to the various plant materials is short, only seconds or a few minutes, but is enough to induce mutation. Thus, the ion beam is an excellent tool for mutation breeding to improve horticultural and agricultural crops with high efficiency.

5. References

5.1. Cited References

Abe, T., Bae, C.H., Ozaki, T. *et al.* **2000.** Stress-tolerant mutants induced by heavy-ion beams. *Gamma Field Symp.* 39: 45–54.

Chu, W.T., Curis, S.B., Llacer, J. *et al.* **1985.** IEEE Trans. *Nucl. Sci.* 32: 3321.

Hayashi, Y., Takehisa, H., Kazama, Y. *et al.* **2007.** Isolation of salt-tolerant mutants of rice induced by heavy-ion irradiation. *RIKEN Accel. Prog. Rep.* 40: 253.

Hayashi Y., Takehisa H., Kazama, Y. *et al.* **2008a.** Effects of ion beam irradiation on mutation induction in rice. *Cyclotoron and their applications.* 2007: 237–239.

Hayashi Y., Takehisa H., Kazama, Y. *et al.* **2008b.** Characterizaion of salt-tolerant mutants of rice induced by heavy-ion irradiation. *RIKEN Accel. Prog. Rep.* 41: 234.

Honda, I., Kikuchi, K., Matsuo, S. *et al.* **2006.** Heavy-ion-induced mutants in sweet pepper isolated by M_1 plant selection. *Euphytica.* 152: 61–66.

Ichida, H., Matsuyama, T., Ryuto, H. *et al.* **2008.** Molecular characterization of microbial mutations induced by ion beam irradiation. *Mut. Res.* 639: 101–107.

Imanishi, S., Noguchi, A., Nagata, M. *et al.* **2007.** Construction of libraries of 'Micro-Tom' tomato mutations induced by heavy-ion bombardment. *Acta Horticulturae.* 745: 485–489.

Matsuo, Y., Nishinima, S., Hase, Y. *et al.* **2006.** Specificity of mutations induced by carbon ions in budding yeast *Saccharomyces cerevisiae. Mut. Res.* 602: 7–13.

Morishita, T., Miyazawa, Y., Saito, H. *et al.* **2008.** Semidwarf mutants of tartary buckwheat induced by heavy-ion beam irradiation. *RIKEN Accel. Prog. Rep.* 41: 232.

Ohnuma, M., Yokoyama, T., Inoue, T. *et al.* **2007.** Mutation induction by heavy-ion irradiation in Cyanidioschyzon merolae 10D. *RIKEN Accel. Prog. Rep.* 40: 252.

Ryuto, H., Abe, T., Fukunishi, N. *et al.* **2008.** Heavy-ion beam irradiation facility for biological samples in RIKEN. *Plant Biotech.* 25: 119–122.

Shikazono, N., Suzuki, C., Kitamura, S. *et al.* **2005.** Analysis of mutations induced by carbon ions in *Arabidopsis thaliana. J. Exp. Bot.* 56: 587–596.

Shitsukawa, N., Ikai, C., Shimada, S. *et al.* 2007. The einkorn wheat (*Triticum monococcum*) mutant, maintained vegetative phase, is caused by a deletion in the VRN1 gene. *Genes and Genetic Systems.* 82: 167–170.

Suzuki, M., Watanabe, M., Kanai, T. *et al.* 1996. LET dependence of cell death, mutation induction and chromatin damage in human cells irradiated with accelerated carbon ions. *Adv. Space Res.* 18: 127–136.

Ziegler, J.F., Biersack, J.P. and Ziegler, M.D. 2008. SRIM – The Stopping and Range of Ions in Matter. SRIM Co. ISBN 0-9654207-1-X. http://www.srim.org/.

5.2. Websites

http://www.riken.jp/engn/r-world/research/lab/ nishina/radia/index.html

http://www.rikenresearch.riken.jp/frontline/413/

http://wwwsoc.nii.ac.jp/ibbs/

5.3. Further Reading

Bae, C.H., Abe, T., Matsuyama, T. *et al.* 2001. Regulation of chloroplast gene expression is affected in *ali*, a novel tobacco albino mutant. *Ann. Bot.* 88: 545–553.

Goodhead, D.T. 1994. Initial events in the cellular effects of ionizing radiations: clustered damage in DNA. *Int. J. Radiat. Biol.* 65: 7–17.

Kazama, Y., Saito, H., Yamamoto, Y.Y. *et al.* 2008. LET-dependent effects of heavy-ion beam irradiation in *Arabidopsis thaliana*. *Plant Biotech.* 25: 113–117.

Kikuchi, S., Saito, Y., Ryuto, H. *et al.* 2009. Effects of heavy-ion beams on chromosomes of common wheat, *Triticum aestivum. Mut Res.* doi:10.1016/j.mrfmmm.2009.05.001.

Miyazaki, K., Suzuki, K., Iwaki, K. *et al.* 2006. Flower pigment mutations induced by heavy ion beam irradiation in an interspecific hybrid of Torenia. *Plant Biotech.* 23: 163–167.

Morishita, T., Yamaguchi, H., Degi, K. *et al.* 2003. Dose response and mutation induction by ion beam irradiation in buckwheat. *Nucl. Instrum. Methods Phys. Res.* B206: 565–569.

Niwa, K., Hayashi, Y., Abe, T. *et al.* 2009. Induction and isolation of pigmentation mutants of *Porphyra yezoensis* (Bangiales, Rhodophyta) by heavy-ion beam irradiation. *Physol. Res.* 57: 194–202.

Sasaki, K, Aida, R., Niki, T. *et al.* 2008. High-efficiency improvement of transgenic torenia flowers by ion beam irradiation. *Plant Biotech.* 25: 81–89.

Shimada,,S., Ogawa, T., Kitagawa, S. *et al.* 2009. A genetic network of flowering-time genes in wheat leaves, in which an APETALA1/FRUITFULL-like gene, VRN1, is upstream of FLOWERING LOCUS T. *Plant J.* 58: 668–681.

C10

Ion Implantation Mutagenesis

H.Y.Feng and Z.L.Yu*

Key Laboratory of Ion Beam Bioengineering, Chinese Academy of Sciences & Anhui Province, Hefei Institutes of Physical Science, CAS, P.O. BOX 1126, 230031 Hefei, China
* Corresponding author, E-MAIL: zlyu@ipp.ac.cn

1. Introduction

As presented in Chapter 9, heavy ion mutagenesis involves an ion beam that is generated through a heavy ion accelerator and has an ionic mass number (A) greater than 4 (A > 4), and an ionic energy of more than a million electron volts (MeV; Tsuji *et al.*, 1998). A heavy ion accelerator consists of an ion source, ion implantor (low energy accelerator), main accelerator and terminal. In general, ionic energy extracted from an ion implantor is in the range of thousands of electron volts (keV) in a heavy ion accelerator. Ion beams with the range of keV energy are commonly used for other purposes than mutagenesis, for example, ion implantation into materials and ion beam analysis.

Ion implantation is a method of implanting impurities below the surface of a solid. It is usually used for modifying surface properties of bombarded materials (Ziegler and Manoyan, 1988). In the mid-1980s, authors of this chapter discovered the bio-effects induced by ion implantation into plant seeds. The energies of ions used in these experiments (~keV) were 3 to 5 orders of magnitude lower than those typically used in heavy ion beam mutagenesis. It had generally been assumed that ~keV ions could penetrate into water with a depth that was far too small to induce biological aberrations. However, when the implantation dose (number of ions implanted) was sufficiently high, yellow stripes were observed on the leaves of rice, maize and wheat M_1 plants grown from seeds implanted with 30 keV nitrogen ions. This was similar to the observation on maize leaves in the Apollo-Soyuz space flight experiment, and these yellow characteristics could be stably inherited to later generations (Yu *et al.*, 1991).

Little attention had been paid until recently to interactions between low-energy ions and matter compared with high-energy ions; there had been almost no study on the effects of low-energy ions on complex organisms such as plant seeds. At present, it is generally accepted that molecular damage induced by high-energy ionizing radiation is actually the combined result of a multitude of low-energy (a few to several hundred eV) events, including secondary particles (electrons, free radicals, ions and segment molecules) caused by ionizing radiation in an ion track (Michael and O'Neil, 2000). The discovery of the mutation effect of ion implantation has opened a new branch in plant mutagenesis. Since the ion implantation technique is relatively simple and does

not need expensive equipment, it has been proven to be efficient in inducing important mutations and it has become an important new mutation method for plants and microbes. The underlying principles, techniques and methods of ion implantation mutagenesis, and its application in plant mutation breeding are described below.

2. Underlying Principles

Radiation damage to organisms induced by ionizing radiation is often considered to be primarily from the energy transferred along the ray path. It is now clear that the particle mass after losing all its energy can also have an impact. The term "damage" usually refers to DNA structural changes. Compared with high-energy ionizing radiation, the molecular damage resulting from ion implantation is characterized by the combined effect of energy deposition, mass deposition and charge exchange between ions and the various targeted compounds in an organism.

2.1. Energy Loss

From an atomic collision point of view once collisions occur the energy loss laws remain the same, regardless of the atomic complexity of the biological target and regardless of when and where the target atoms are hit. When ions are implanted into cells, they transfer some of their energy to a target atom through nuclear or electron collision energy loss (**Box 10.1**).

For a solid film, e.g. metal or a semi-conductor film, the distribution of the atomic volume density is continuous and homogeneous; the defects can be described at an atomic level. In other words its mass-thickness in the direction of incident ions is the same everywhere, so the α-particle transmission energy spectra (i.e. particle energy–lose energy) in a solid film should consist of a single energy peak. For biological materials, such as seeds, since the distribution of the atomic volume density is much more complicated, there are many voids such as channels or holes in the specimens, hence the energy transfer is different compared with solid film. It is thought that fluctuations in the incident ion energy loss along the direction of the ion trajectory is characterized by a beaded energy deposition due to the voids in which there is no energy loss when ions traverse

Box 10.1: Glossaries of ion implantation mutagenesis

1. Nuclear collision energy loss: an ion transfers some of its energy to a target atom and is deflected through a large angle, the energy transferred causes the target atom to be displaced; this is a nuclear collision process involving an interaction between an energetic ion and the atom hit. The energy transferred from the ion to the hit atom is termed 'nuclear collision energy loss'.

2. Electron collision energy loss: The moving ion transfers some of its energy to the electrons of the atoms it collides with causing ionization or excitation or electron capture. The energy loss in each collision in this case is relatively small and the ion undergoes only a small deflection. This is an electron collision process involving an interaction between the energetic ion and atoms hit. The energy transferred to the electrons of atom hit is termed 'electron collision energy loss'.

3. Ionization and excitation: The outer-shell of electrons of an atom absorb radiation energy in atomic physics, these electrons gain energy and jump to a higher energy state, known as an 'excited state'. The process is called 'excitation'. If the absorbed energy is sufficient the outer-shell electrons can break away from atom, this process is termed 'ionization'.

4. Alpha particule transmission energy spectrum: An energy spectrum is a distribution of energy among a large assemblage of particles. For any given value of energy, it determines how many of the particles have that much energy. The particles may be electrons, photons, alpha or a flux of elementary particles. Alpha particle transmission energy spectrum is the energy spectrum resulting from alpha particule penetration into matter.

5. Mass-thickness: Mass per unit area, g/cm².
Energy deposition: energy deposition is the energy transferred from the energetic particles to the matter hit; energy absorption is the energy absorbed by the hit matter. The radiation energy absorbed per unit mass is defined as the radiation dose (absorption dose).

these voids. **Figure 10.1** shows the α-particle transmission energy spectra in a piece of tomato skin (apparent thickness: ~50 μm) before and after ion implantation. Before ion implantation, the energy spectrum of the experimental system has a very broad distribution. This indicates a heterogeneous distribution of the spatial structure of the tomato skin, where areas with very small mass-thickness (mass each unit area, g/cm²) exist.

The energy spectrum given in **Figure 10.1** obviously shifts to the high-energy side, equivalent to the apparent thickness of the tomato skin being thinned by 7.5–9.1 μm. The thinnest of mass-thickness is almost zero, there appears to be a completely free α-particle peak at channel 933 in the spectrum (5.484 MeV). This indicates that the low-energy ion beam can etch away the tomato fruit coat and dig paths that can connect the voids (that

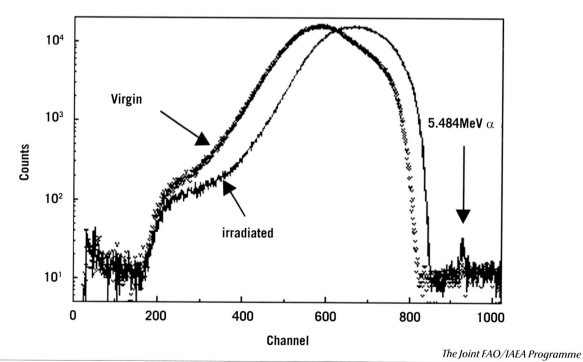

Figure 10.1 α-particle transmission energy spectrum of tomato fruit skin before and after N-ion implantation at a dose of 1 × 10¹⁷ ions/cm². X-Axis "Channel" can convert into particle energy, for example, Channel counts 934 equal to 5.484 MeV of α-particle transmission energy in this experiment.

in a natural situation are isolated from each other) in the direction of ion incidence. Thus, in further treatments this allows deeper penetration.

2.2. Mass Depositing Reaction

The ion concentration–depth distribution in the implanted samples can be detected by nuclear analysis. **Figure 10.2** shows the ion concentration–depth distribution in a peanut seed implanted with 200 keV vanadium (V) ions (Zhu *et al.,* 2001). In this experiment, a dry peanut seed was fixed in a specially made mould and put onto a slicer to remove part of the seed. The seed was placed in the target chamber with the cut side facing the ion beam and then implanted. After ion implantation, the sample was packed with paraffin, put onto the slicer again at the original position and sliced into 15-μm pieces. Each piece was analyzed using the proton induced X-ray emission (PIXE) technique for V-ion concentration. Scanning electron microscopy was used to investigate the V concentration distribution in more detail for the first piece (0–15 μm). The V peak in the first 0–15 μm was very high and about 98% of V is deposited in this region. The peak rapidly decreased in the second slice (16–30-μm) and continued to decrease in subsequent slices. In the region of 121–135 μm, the V concentration was close to the limits of PIXE detection. Since the element vanadium does not naturally occur in peanut (or in the cutter blade in PIXE analysis) the vanadium detected in this experiment can be only a result of implanted vanadium ions. These results demonstrated that the range of 200 keV V ions implanted in the peanut seed is of about three orders of magnitude greater than that in water, though water has almost the same mass thickness as peanut seed.

When ions are implanted into organs, tissues or cells, the deposited ions may react with nearby atoms and molecules. The reaction products may be stable or unstable (excited). The unstable products can participate in further reactions and eventually a balance is reached. The most simplified demonstration of this is when ice water was implanted with 30 keV $^{13}C^+$ at absolute temperature scale 16K, H_2O_2, $^{13}CO_2$ and ^{13}CO were formed (Strazzulla *et al.,* 2003). Similarly, nitrogen ion implantation into NAA (naphthyl acetic acid, with MW of 186, **Figure 10.3a**) can produce new products and derivatives (Huang *et al.,* 1996). It should be noted that there is no nitrogen atom in the molecular structure of NAA. After nitrogen implantation, new molecular structures in the implanted NAA sample were detected through GC-MS analysis. Its mass spectrum suggests a molecular weight of 267 g/mol. This new molecule is expected to contain a nitrogen atom according to chemical valence theory (**Figure 10.3**). These experiments demonstrated that ion implantation into organic molecules can result in the formation of new molecules.

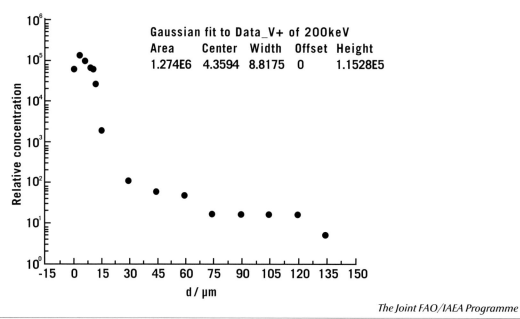

Figure 10.2 The relative concentration–depth distribution of 200 keV vanadium ions implanted in peanut seed measured by the PIXE technique.

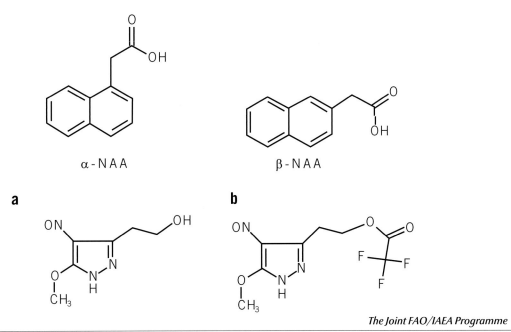

α-NAA

β-NAA

a

b

The Joint FAO/IAEA Programme

Figure 10.3 The molecular structure of NAA (above), and the structure of main product after nitrogen ion implantation into –NAA. (a) product; (b) derivative of product.

2.3. Charge Exchange

The surface of a biological sample, such as a cell, contains negative ion groups that can be observed to move to the anode during electrophoresis. When subject to incident ionizing radiation, the number of ionized groups on the surface of implanted samples is changed. Dissociation and topography change the cell–cell interaction, the interaction between Ca^+ ions and the cell membrane, the ion permeability of the cell membrane, and other physiological changes occur. In the case of interactions between energetic ions and biological organisms, energy deposition plays a role in ionizing radiation in damaging the ionized groups on the surface of the biological target. The important factor here is the charge exchange in the electro-characteristics of the sample.

Since biological organisms are not good electrical conductors, the accumulated surface charge is not immediately released. Instead, the charge is maintained

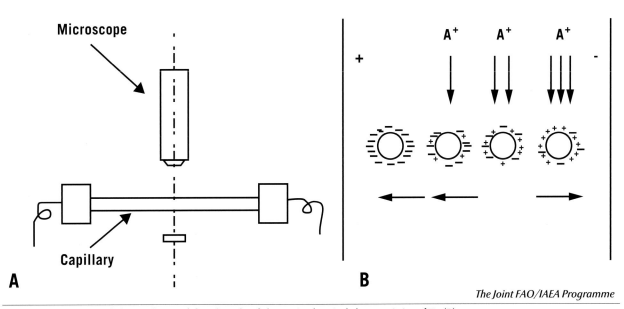

The Joint FAO/IAEA Programme

Figure 10.4 Schematic of electrophoresis (A) and mode of change in electrical characteristics of Tyr (B).

long enough to change the electrical characteristics of the sample surface, which can be examined using capillary-electrophoresis. For example, when positive ions are implanted in minute biological molecules, with the increase of implantation doses, the electrical characteristics of implanted molecules change accordingly. **Figure 10.4** shows the electrophoresis behaviour in the capillaries of the tyrosine (Tyr) crystal after ion implantation (Shao *et al.,* 1997). Tyr is an electro-negative molecule and moves to the anode during electrophoresis. When implanted with positive ions, it will slow down, stop and move towards the cathode as the implantation dose increases (**Figure 10.4**).

2.4. Dose and Dose Response

In radiobiology, biological effects appear when a biological organism absorbs radiation energy over a critical value. Therefore, the absorption dose, defined as the energy absorbed per unit mass, is the only major parameter that is used for characterizing radiation effects.

It should be noted that, for ion implantation mutagenesis, it is not sufficient only to consider energy absorption. Factors related to ion implantation-induced biological effects, as distinct from those due to radiation, include not only energy deposition (and momentum transfer), but also mass deposition and charge exchange. In brief, in the case of ion implantation the influencing factors are threefold: 1) the simultaneous input of energy, 2) substance and 3) charge of the biological organism. Parameters quantifying energy deposition, mass deposition and charge exchange should be used in assessing biological effects of ion implantation mutagenesis. However, for the convenience, the noun "dose" is still used in ion implantation, the unit is ions/cm^2.

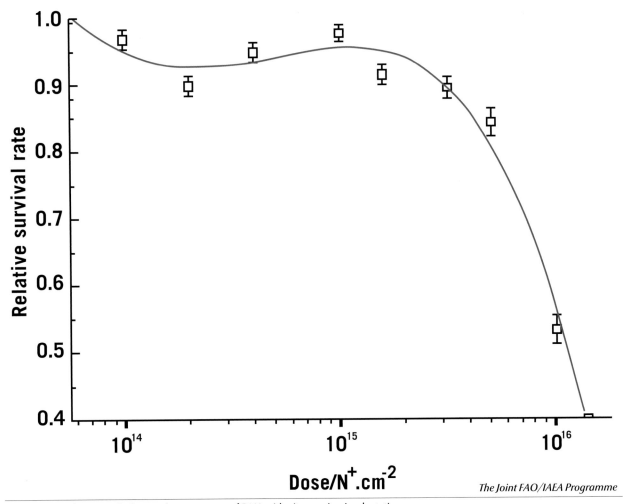

Figure 10.5 Relative survival rate – dose response of DNA with nitrogen ion implantation.

Since there are intrinsic differences between ion implantation and radiation, it is reasonable that the responses of biological organisms to ion implantation are different from other ionizing radiations. A typical dose–response curve is presented in **Figure 10.5**; it is worth noticing that there are oscillations, while the survival rate goes down with dose strength, i.e. the survival rate experiences a process of slowly dropping – slightly ascending–rapidly decreasing (Joiner *et al.,* 2001). This phenomenon of survival-dose response is universal in ion implantation mutagenesis of plants.

3. Technical Aspects of Ion Implantation Mutagenesis

3.1. Facility

A large quantity of samples, such as seeds, propagules, *in vitro* cultures, etc., need to be treated to establish M_1 populations of reasonable size. In ion implantation mutagenesis, because of the limited penetration ability of ion beams, each seed must be directly exposed to ion beam bombardment. Therefore, the diameter of the ion beam should be as large and homogeneous as possible. Although processing biological samples calls for some special requirements, the basic structure of an ion beam facility used for mutation induction is not greatly different from that of an industrial ion implanter (and in some respects simpler).

Figure 10.6 is a schematic diagram of a device used in mutation research. In the ion source part (1), an electrical discharge in gas or vapour produces plasma. Energy of ions extracted from the plasma by the extraction electrodes can generally reach ~50 keV. The ion beam is focused by a focal magnet (2) to run towards a magnet analyser (3). Ions with the same mass to charge ratio (m/z) are sorted by the magnet analyser to the acceleration section (4), here ions can be accelerated and energy of ions can be reached in the range of hundreds of keV. The accelerated ions enter into a target chamber (5) and are implanted into the seeds on a sample disc (6). For monitoring the ion beam parameters, various measurement systems are installed (7). Some of the extracted ions in passage between the ion source and the mass analyser are partly neutralized by

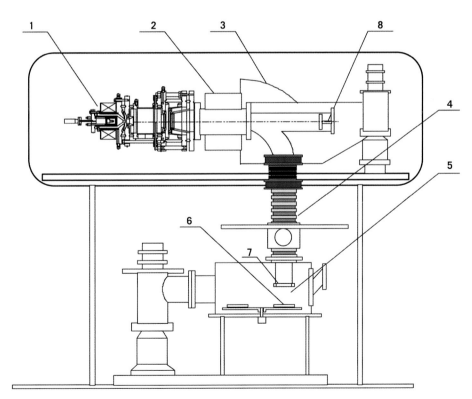

The Joint FAO/IAEA Programme

Figure 10.6 Schematic diagram of an ion beam line. (1) Ion source, (2) focusing lens, (3) magnet analyser, (4) accelerating tube, (5) target chamber, (6) sample disc, (7) measurement system and (8) all-pass target chamber.

charge exchange of ions and gas molecules; they are not deflected by the magnet analyzer, and hit directly a sample disc in a all-pass target chamber (8). Here the bio-effects induced by the implanted particles without charges could be studied.

The ion source consists of two basic parts: the plasma generator and the ion extraction system. The extraction system is usually composed of two, three or four electrodes. It is possible for three- or four-electrode extraction systems to form beams of ions with energy up to a hundred keV. If ionic energy from a ion source is further raised, to several hundred keV on request, ions will be accelerated in an accelerating tube. There are several aligned ion extraction holes or slits on each electrode.

Table 10.1: New mutant varieties bred using ion implantation

Varieties	Main characteristics	Ion with 30keV/dose range	Year of release
Tomato			
Lufanqie No. 7	High yielding, big fruit	Nitrogen ion/ 5×10^{15} ions/cm²	1997
Corn (maize)			
Wanyu No. 10	High yielding, high sugar	Nitrogen ion / $\sim 1.0 \times 10^{17}$ ions/cm²	1999
Wanyu No. 14	High yielding, opposite leaves	Nitrogen ion/ $0.8–1.2 \times 10^{17}$ ions/cm²	2000
Wanyu No. 15	High yielding, high sugar, opposite leaves	Nitrogen ion/ $1.5–1.8 \times 10^{17}$ ions/cm²	2003
Wheat			
Wanmai No. 32	High yielding, high disease resistance	Nitrogen ion/ $4.0–6.0 \times 10^{16}$ ions/cm²	1997
Wanmai No. 43	High yielding, resistant to smut	Nitrogen ion/ 3.0×10^{16} ions/cm²	1999
Wanmai No. 54	High yielding, resistant disease	Nitrogen ion/ 2.8×10^{16} ions/cm²	2003
Rice			
Wandao No. 20 (S9042)	High yielding, short stature period, resistant to disease	Nitrogen ion/ $\sim 2.6 \times 10^{16}$ ions/cm²	1994
Wandao No. 42 (D9055)	High yielding, resistant to rice blast	Nitrogen ion/ $\sim 3.0 \times 10^{16}$ ions/cm²	1994
Wandao No. 44	High yielding, high quality	Nitrogen ion/ $2.4–2.8 \times 10^{16}$ ions/cm²	1999
Wandao No. 45	High yielding, resistant to disease	Nitrogen ion/ $1.5–3.5 \times 10^{16}$ ions/cm²	1996
Wandao No. 48	High yielding, resistant to rice fulgorid	Nitrogen ion/ $4.0–5.0 \times 10^{16}$ ions/cm²	1999
Wandao No. 60	High yielding, high quality, resistant to rice stripe disease	Nitrogen ion/ $2.0–2.5 \times 10^{16}$ ions/cm²	2000
Wandao No. 71	High yielding, cold tolerance	Nitrogen ion/ $\sim 6.0 \times 10^{16}$ ions/cm²	1999
Wandao No. 96	High yielding, high quality, direct sowing	Nitrogen ion/ $\sim 30 \times 10^{15}$ ions/cm²	2006
Wandao No. 143	High yielding, for direct sowing	Nitrogen ion/ $\sim 65 \times 10^{15}$ ions/cm²	4005
Fengyou 293	High yielding, long-term storage	Hydrogen ion/ $1.2–1.5 \times 10^{16}$ ions/cm²	2006
Zhongyou 292	High yielding, long-term storage	Hydrogen ion/ $\sim 150 \times 10^{15}$ ions/cm²	2006
Wandao No. 179	High yielding, high quality, short growth period	Nitrogen ion/ $8.0–8.5 \times 10^{16}$ ions/cm²	2006
Soybean			
Jindou No. 28	High yielding, large grain	Argon ion/ $3.0–5.0 \times 10^{16}$ ions/cm²	2004
Fengdou 103	High yielding, large grain, black	Nitrogen ion/ $6.0–8.0 \times 10^{16}$ ions/cm²	2005

A wide ion beam can be obtained using multi-hole extraction electrodes.

In practice a device for ion implantation mutagenesis could be simpler than the one described above. The focusing system, magnet analyser and acceleration section may not be necessary for simplified devices. **Figure 10.7** shows a schematic of such a simplified device. The ion source on this device is similar to the one shown in **Figure 10.6** but installed vertically. The extraction system consists of three electrodes, each with 19 holes; the energy of extracted ions can reach to 50 keV. Ion beams of 15 cm in diameter are directly targeted, without being analysed or further accelerated, onto the seed sample disc on the turnplate 80 cm in diameter below the extraction electrode. The big target chamber is 1.0 m in height and 90 cm in diameter. The centre-line of the ion beam is 35 cm away from the centre-line of the big target chamber. The device has another small chamber underneath the big chamber, separated by a valve, for treatment of samples with high water content, such as microbes, cells, calli, etc.

When the valve is opened, cell or callus samples in the small chamber are suddenly exposed to a vacuum in the big chamber, evaporation of water removes considerable heat and the temperature of the cell or callus rapidly decreases. When a phase balance is reached, the temperature at the interface of the two phases is below 0°C. Thus the water at the cell or callus surface will freeze to form an ice shell and prevents the water inside the samples from evaporating further. For fixed vacuum pressure (i.e. the vapour pressure remains stable), the temperature of the ice shell surface is constant. Such a low temperature is harmful to living cells with no anti-freezing protection.

3.2. Methods in Sample Preparation and Treatment

3.2.1. Sample Preparation

Because the ion penetration range is limited, each seed must be directly bombarded by ion beam. Proper preparation of samples is helpful and sometimes essential for increasing the efficiency of ion implantation mutagenesis. Since the ion beam is produced in a vacuum, it is essential to protect samples with a high water content to attain a reasonable survival rate after bombardment.

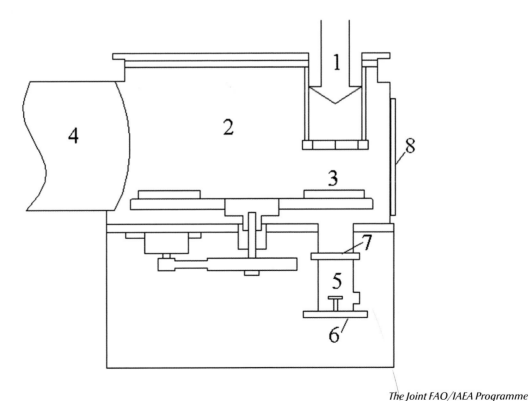

Figure 10.7 A schematic diagram of a simplified ion beam line. (1) ion beam from ion source, (2) big target chamber, (3) turnplate with sample discs, (4) vacuum system, (5) small target chamber, (6) in and out sample for small chamber, (7) vacuum valve connecting the two chambers, (8) in and out sample for big chamber.

The Joint FAO/IAEA Programme

Figure 10.8 Device for ion implantation mutagenesis.

3.2.1.1. Seed

The devices shown in **Figures 10.7 and 10.8** produce ion beams of about 15 cm in diameter and with dose homogeneity of 90%. Six sample discs can be placed on the turnplate. For dry plant seeds such as rice and wheat, each sample disc can hold about 1,000 grains. For investigation of the mutation effect of seeds with bourgeoning, germinating seeds, samples should be somewhat dried before bombardment. It is not always necessary, but it is helpful if the hulls of cereal grains are removed before treatment. For plant seeds with hard shells, such as gingko and cotton, the seed shell or part of it should be removed.

Table 10.2: Types and rates of aneuploid plants obtained by N-ion implantation in wheat and rye

Tested material	No. of euploid chromosomes	Ion dose (10^{16} cm^{-2})	No. of individuals	No. of chromosomes	Type of aneuploidy	Rate (%)
Wheat var.	42	3	38	41	2n–1	5.26
Premebi		4	22	41	2n–1	4.55
AR1 rye	18 (14R+4B)	4	23	19 (15R+4B)	2n+1	4.53
Tetraploid rye	28	1	27	27	2n–1	7.41
		2	29	29	2n–1+2t	3.45
		2	29	27	2n–1	6.89
		3	36	27	2n–2+1t	2.78
		3	36	27	2n–1	5.56
		3	36	26	2n–2	2.78
		4	33	26	2n–2	6.06
		4	33	27	2n–1	3.03
		4	33	29	2n+1	3.03

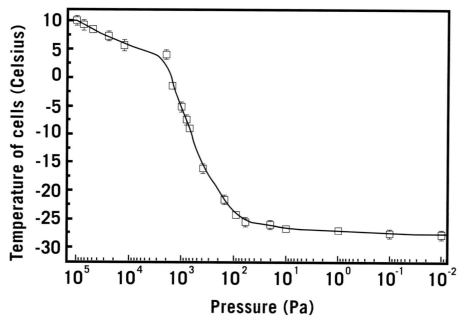

The Joint FAO/IAEA Programme

Figure 10.9 Temperature profile of cells during evacuation with the temperatures measured inside the cell pellet recorded during pumpdown.

3.2.1.2 Vegetatively Propagated Crops

For VPCs, bud tips should be used for ion implantation. In sample preparation, it is not advisable to take only young buds; a piece of stem or a segment of twig should also be included. The stem piece (such as for sweet potato) and twig (such as from poplar or fruit trees) should be wrapped with plastic film and smeared with a low melting point wax, so that only the bud tip is exposed to the ion beam. The prepared stem piece or twig is fixed on the sample support with bud tips oriented towards the ion beam.

3.2.1.3. Cells and Calli

Cell and callus samples will undergo pressure changes when loaded into the vacuum chamber. Cells and calli contain a large quantity of water, and their status in a vacuum is completely different from that of crop seeds. For simplicity, a cell can be considered to be a water droplet. When a group of cells is placed under vacuum, water evaporation can cause a precipitous drop of temperature in the cells, and the lower the pressure in the chamber, the lower the cell temperature. When the liquid–vapour equilibrium is reached, the temperature at the interface of the two phases is constant. According to the pressure–temperature equilibrium theory, the temperature of pure water would be below –70°C when the pressure reaches 1.33 Pa. Thus, cells begin

to freeze from the surface upon evaporation and an ice shell is formed. Although the ice shell surface can in theory prevent more water from evaporating, it has been observed in experiments that after cells have been pumped for 20 min, the entire cell pellet is transformed into icy dust. It is also reasonable to assume that all the ice in the cells will be evaporated eventually.

A real-time measurement indicates that cell temperature decreases gradually from room temperature to near 0°C when the pressure goes down from 1 atm to 1000 Pa (pre-pumping in the small chamber for ~12s) (**Figure 10.9**). After the isolation valve is opened, the cell temperature drops to –28 °C in about 100 s as the pressure diminishes from 1000 to 100 Pa. When the pressure continues to decrease, the temperature fall becomes slower implying either complete dehydration or a frozen ice solid. This rapid cooling rate and low temperature deactivate the cells (Feng *et al.*, 2004).

The poor vacuum tolerance of living cells and calli indeed presents a big obstacle to ion beam mutation. It is thus important to find means to mitigate the effects. A set of "glycerol and freezing, then vacuum" protocols can protect the treated cells and calli in a vacuum, for example, high glycerol concentrations (15% and 20%) have better cell protection effects against vacuum stress, and pretreatment with 20% glycerol increases cell survival rates from less than 1% to greater than 12%.

3.2.2. Ion Implantation and Dose Control

The ion beam mutation facility at the Institute of Plasma Physics, Chinese Academy of Sciences is used here as an example for illustrating the process of ion implantation (**Figures 10.7 and 10.8**). Crop seeds are inserted one by one on the sample discs/holders in advance. On the day of implantation, before applying vacuum pumping, the sample discs are placed on the turnplate in the big chamber (**Figure 10.10**). Six to ten sample discs can be simultaneously placed on the circumference of the turnplate. It normally takes about 1 hour for the chamber pressure to reach 0.001 Pa which is appropriate to start implantation. After ion implantation is completed for one sample disc, another sample disc will be rotated to the treatment position.

Before ion implantation, the dose rate should be measured. Ion implantation of biological samples usually makes use of an intermittent (repetitively pulsed) implantation mode. The number of ions implanted in each implantation pulse time is known and thus the total ion implantation dose can be calculated from the total implantation pulse times. For example, samples in a sample holder are implanted one at a time with an actual implantation on-time of 5 s and the implantation dose rate is 5×10^{14} ions/cm^2 per implantation time of 5 seconds; thus in a total of 50 implantation times the total dose received by the samples is $50 \times 5 \times 10^{14}$ ions/cm^2 = 2.5×10^{16} ions/cm^2.

The dose rate should be measured not only before implantation but also during ion implantation if a high dose (and hence a long implantation duration) is required. This is because the ion beam parameters are very sensitive to the discharge gas pressure and the electric field and magnetic field configuration in the beam line, and they may not remain constant throughout a long time run. The measurement is carried out when the beam is interrupted, and thus the time is not counted towards the applied dose.

The small target chamber is designed particularly for treatment of high-water content samples, such as cells, calli and small stems. The volume ratio between the big chamber and the small chamber is 600:1. The small chamber has a sterile environment (**Figure 10.11**). When samples are moved into and out of the small chamber, sterile air is blown in to ensure that the samples are not contaminated.

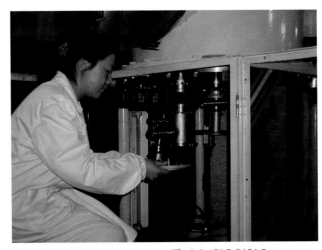

The Joint FAO/IAEA Programme

Figure 10.11 A culture plate with calli is loaded into the small chamber, sterile air is blown towards the laboratory assistant.

The Joint FAO/IAEA Programme

Figure 10.10 A lab assistant has a holder containing rice seeds to be loaded into the chamber.

4. Applications in Plant Breeding and Genetics

Since the mid-1980s, ion implantation technology has been studied extensively in China for its usefulness and uniqueness in plant breeding and genetics. At present, 19 devices (such as illustrated in **Figure 10.8**) are established in research institutes of several provinces. Since the mid-1990s, ion beam mutagenesis has also been developed in some other countries. As a mutation technology, it shares common features with mutation breeding and genetics; however, it also has some unique characteristics, and/or advantages over other mutation techniques.

4.1. Uniqueness of Ion Implantation Mutagenesis

- Safety and security: there is no requirement for radioactive isotopes (compared with gamma sources) and there is no toxic chemical waste (chemical mutagen), hence it is safe, secure and user friendly.
- Cost-effectiveness: it is cheaper and easier to build such a facility, and has a low running cost.
- Low physiological damage to M_1 plants means that the population size of M_1 plants can be reduced relative to other mutagenic methods.

4.2. Breeding of New Varieties

Since the discovery of its mutation effect, ion implantation has been used in breeding of various plant species. In addition to main crops, it has been proven to be useful for other plant species, such as *Ephedraceae*, clover, ginkgo and Chinese chestnut. As a result, 22 new varieties were bred by the authors' laboratory, and released during 1994–2006 in China (**Table 10.1**).

Among the released mutant varieties, a few varieties have already been widely grown. The wheat variety Wanmai 54 (also known as Wan 9926, **Table 10.1**) showed excellent performance in trials and farmers' fields, e.g. high yielding and resistance to head blight, its planting area has expanded to regions around the Yangtze River in China. In national new variety trials, Wanmai 54 exhibited yield advantages of 7% more than the standard variety Yangmai 158; it ranked first among all varieties in 2005–2006 production trials of winter wheat. Wanmai 54 also has excellent disease resistance and quality traits.

4.3. Generation of Novel Germplasm

Most induced mutants are not used directly as new varieties, either because the induced mutation is of no immediate agronomic value or it is linked with other inferior characteristics, or the parent variety has already lost its competitiveness to new varieties. For example, a total of ~13,000 mutant lines were developed through ion implantation by the end of 1994 by more than 60 breeding groups in China, of these only ~1.0% were used further in breeding programmes (e.g. Wanmai 54). However, they are still useful as materials for genetic research and gene identification.

Several mutants with novel characteristics were induced by ion implantation mutagenesis. The maize mutant with opposite leaves was identified in progenies of the inbred line Yellow 48 after treatment of 30 keV N^+ implantation; it is characterized by decussate phyllotaxis and fruits throughout the life cycle (**Figure 10.12**). The leaf number and total leaf area of this mutant is increased by almost a factor of two compared with other maize. This mutant might be used for studies on the genetics, phylogeny, phytomer modeling, plant physiology, plant taxonomy and breeding of maize.

In another example, ion implantation into seeds of perennial woody plants, such as gingko, Chinese chestnut and sea buckthorn, could be used to curtail their juvenile stage. **Figure 10.13** shows gingko in the first year of the juvenile stage when the ion implanted seeds have grown.

The Joint FAO/IAEA Programme

Figure 10.12 Maize plants that show opposite (left) or alternate (right) pairs of leaves and fruits.

Starting from a seedling it takes gingko 6–8 years to set fruit. After ion implantation treatment, a mutant with a significantly shortened juvenile stage was identified (**Figure 10.13**). Although the mechanism of curtailing the juvenile stage for perennial woody plants induced by ion implantation is not clear, it may provide materials for early breeding and for studies of plant physiology.

4.4. Application in Chromosome Manipulation

A combination of energy absorption, mass deposition and charge exchange of ion implantation into plant seeds can result in many types of chromosome damage. Taking wheat var. Premebi as an example,

Figure 10.13 Gingko fruiting in the first year of its juvenile stage.

chromosomal structure variations, such as centromere gaps, chromosomal terminal deletions, one-arm deletions of chromatids and chromosome fragments, all existed in the M_1 on metaphases of mitotic root meristematic cells. In meiosis of M_1, ion implantation resulted in monovalent or even bivalent lagging, and terminal centromere chromosome lagging, ring bivalent formation decreases, rod bivalents increased, as did unpaired monovalents. In mitosis of M_2 plants, a fraction of the root meristematic cells maintained the normal, euploid (2n) chromosome number, but a few were aneuploid. At the same time, in some metaphase divisions, chromosome segments, terminal deletions, lagging chromosomes, chromosome bridges and unequal segregation were observed.

Cytological abnormalities induced by ion beam treatment could be exploited using chromosome manipulation techniques. Aneuploids are basic materials used in plant genetics and breeding, particularly of polyploidy crops such as wheat. **Table 10.2** shows the types and rates of aneuploids induced in a hexaploid wheat, a diploid rye with supernumerary (B) chromosomes and a tetraploid rye by ion implantation. Aneuploidy was obtained in M_2 generations for all three tested materials. The majority of aneuploids were accounted for by a loss of one (2n–1) or more chromosomes. In some cases the loss of a chromosome arm resulted in the production of telomeric (t) chromosomes.

5. References

5.1. Cited References

Feng, H.Y., Wu, L.J., Xu, A. *et al.* 2004. Survival of mammalian cells under high vacuum condition for ion bombardment. *Cryobiology.* 49: 241–249.

Huang, W., Yu, Z., Jian, X. *et al.* 1996. Study of mass and energy deposit effects of ion implantation of NAA. *Radiation Physics Chemistry.* 48: 319–323.

Joiner, M.C., Marples, B., Lambin, P. *et al.* 2001. Low dose hypersensitivity: current status and possible mechanisms. *Int J Radiat Oncol Biol Phys.* 49: 379–389.

Michael, B.D. and O'Neill, P. 2000. A sting in the tail of electron tracks, *Science,* 287: 1603–1604.

Shao, C.L., Xu, A. and Yu, Z.L. 1997. Charge exchange effect of ion implantation to biomolecules. *Nuclear Techniques.* 20: 70–73 (in Chinese).

Strazzulla, G., Leto, G., Gomis, O. *et al.* 2003. Implantation of carbon and nitrogen ions in water ice. *Icarus.* 164:163–169.

Tsuji, H., Satoh, H., Ikeda, S. *et al.* 1998. Contact angle lowering of polystyrene surface by silver-negative-ion implantation for improving biocompatibility and introduced atomic bond evaluation by XPS. *Nuclear Instruments and Methods in Physics Research Section B.* 141: 197–201.

Yu, Z.L., Deng, J.G., He, J.J. *et al.* **1991.** Mutation breeding by ion implantation. *Nuclear Instruments and Methods in Physics Research Section B*. 59/60: 705–708.

Zhu, G.H., Zhou, H.Y., Wang, X.F. *et al.* **2001.** Measurement of low energy ion implantation profiling in seeds by PIXE and SEM with slicing-up technique. *Nuclear Techniques.* 24: 456–460 (in Chinese).

Ziegler, J.F. and Manoyan, J.M. **1988.** The stopping of ions in compounds. *Nuclear Instruments and Methods in Physics Research Section B*. 35: 215–228.

5.2. Websites

Key Laboratory of Ion Beam Biology, Chinese Academy of Sciences: http://www.ipp.ac.cn/sszy

Key Laboratory of Ion Beam Biology of Henan Province (Zhengzhou University): http://www4.zzu.edu.cn/wuli/newsite/show.aspx?id=100&cid=10

5.3. Further Reading

Feng, H.Y., Yu, Z.L. and Chu, P.K. **2006.** Ion implantation of organisms. *Materials Science & Engineering*. R54: 49–120.

Nastasi, M., Mayer, J.W. and Hirvonen, J. **1996.** Ion-Solid Interactions: Fundamentals and Applications. Cambridge: Cambridge University Press

Yu, Z.L. **1998.** Introduction to Ion Beam Biotechnology, Hefei: Anhui Sci. & Tech. Press (Chinese edition).

Yu, Z.L., Yu, L.D. and Brown, I.G. **2006.** Introduction to Ion Beam Biotechnology. NewYork: Springer-Verlag.

Effects of Radiation on Living Cells and Plants

P.J.L.Lagoda*

Joint FAO/IAEA Division of Nuclear Techniques in Food and Agriculture, International Atomic Energy Agency, Vienna International Centre, A1400, Vienna, Austria
* Corresponding author, E-MAIL: p.j.l.lagoda@iaea.org

1. Introduction

Radiobiology studies the effects of radiation on living organisms. This includes radiation across the electromagnetic spectrum including X-rays, UV radiation, visible light, microwaves, radio waves, emissions due to radioactive decay, low-frequency radiation such as ultrasound, heat waves and related modalities. Radiation biology is an interdisciplinary science that combines biology and radiation physics. The history of radiobiology began with the discovery of X-rays by Wilhelm Conrad Roentgen in 1895 (for more about the milestones in mutational genetics, **see Chapter 2**). In 1896, Henri Becquerel observed the emission of 'mysterious' rays from substances containing uranium, which Marie Curie named radioactivity. In 1898, Pierre and Marie Curie discovered radium and were awarded the Nobel Prize for Physics in 1903 jointly with Becquerel. In 1927, Hermann Joseph Muller published the first paper on the mutagenic effects of radiation on *Drosophila*. His discovery of the role of radiation in mutation induction initiated a new scientific research field of induced mutagenesis and earned him the 1946 Nobel Prize in Physiology and Medicine. Almost in parallel, Lewis John Stadler studied the mutagenic effects of radiation on higher plants and laid the scientific foundations of a systematic and reasoned extension of induced mutations in crop improvement (Stadler, 1928). Although he remained sceptical about its usefulness and economic applicability to crop genetics, his work nevertheless founded the new field of mutation breeding.

In order to enhance the efficiency of mutation induction for crop improvement, initial radiobiology studies focused on radiation effects on living, multicellular plant organisms. An impressive share of early development in this area took place in the Radiation Laboratory of the University of California (USA), beginning with the invention of the cyclotron by Ernest Lawrence in 1930. Studies with radioactive α-particle sources showed that radiations with high LET capacity were biologically more reactive than X-rays in eukaryotes. The generation of high-energy particles, with increased penetration ability, from accelerators made it possible to carry out *in vivo* radiobiological studies in the 1950s and 1960s. These particles became valuable tools for a range of investigations on the LET, radiation survival response, the dependence on relative biological effectiveness (RBE) and the oxygen enhancement ratio (OER). The OER refers to the enhancement of therapeutic or detrimental effects of ionizing radiation due to the presence of oxygen. The effects of low-LET ionizing radiations on cell survival are generally larger by about a factor of 3 in well oxygenated cells than in hypoxic cells. The influence of oxygen on the frequency of single- and double-strand breaks, measured as the OER, in different mammalian cells by electromagnetic radiation and fast electrons ranges from a 1.7- to 5.4-fold increase in the former and from a 3.6- to 3.8-fold increase in the latter (Tisljar-Lentulis *et al.*, 1983). While most radio-biological studies have focused on human health and food safety, work on crop improvement has been relatively limited. However, the underlying mechanisms and biological consequences of radiation are expected to be similar in animals and plants.

The process of radiation affecting plants involves physical, chemical and biological stages (**Table 11.1**). The biological consequences of radiation depend on its effect on cellular components such as proteins and organelles but more importantly on the plant's genetic material.

2. Physical and Chemical Effects

Radiation acts on living cells by the releasing of energy raising electrons to a higher energy state (excitation), ejecting electrons from target molecules (ionization) and other reactions (such as mass deposition and charge exchange). It involves both physical and chemical processes as detailed below.

2.1. Physical Stage

The timeframe of the first physical stage lies in the 10^{-15} seconds range; it mainly involves energy transfer, absorption and the subsequent ionization of plant molecules. Ionizing radiations generate ions in living material, producing highly reactive molecules. For low energy, short wavelength electromagnetic radiation such as visible and UV light, the photo-electric effect predominates. This includes molecular oscillation and excitation of atomic valence electrons, including their ejection called Compton scatter. For X-rays (usually with low energy), the photo-electric effect predominates, producing photo-electrons that transfer their energy in the same manner as Compton electrons. The Compton electrons scatter and travel through the target material,

Table 11.1: The different stages of radiation affecting living cells and plants

Stage	Duration	Process
Physical	10^{-15} s	Energy absorption, ionization, mass deposition, charge exchange.
Physico-chemical	10^{-6} s	Interaction with molecules, formation of free radicals.
Chemical and biochemical	Seconds to minutes	Interaction of free radicals with molecules, chromatins, proteins, DNA repair.
Biological	Minutes to years	DNA fragmentation, cell death, mutation, and reduced growth and reproduction.

colliding with atoms and thereby releasing packets of energy. Higher energy electromagnetic radiation, (e.g. γ-rays) can cause excitation and dissociation of atomic nuclei and creation of particle–antiparticle pairs. Neutrons with energies between 10 keV and 10 MeV transfer energy mainly by elastic scattering, i.e. billiard-ball-type collisions of atomic nuclei in the target material. In this process the nucleus is torn free of some or all of the orbital electrons because of its greater velocity. The recoiling atomic nucleus behaves as a positively charged particle. As the mass of the neutron is nearly the same as that of a hydrogen atom, hydrogenous materials are most affected by this energy transfer.

UV radiation induces a broad spectrum of events in the DNA and RNA of prokaryotes and eukaryotes. The scientific consensus is that X-rays and gamma radiation trigger initial physical processes, nourishing chemical reactions, whereas effects of UV radiation are of a chemical nature. In fact, the study of these is considered as a separate discipline, photobiology.

Ion beams as mutagens are different from other physical mutagens such as γ-rays or X-rays in that they not only involve energy transfer, but also mass deposition, charge exchange (**see Chapters 9 and 10**) and momentum transmission. Ions are heavier than electrons and they can gain a high momentum, i.e. for the same energy, the momentum of the ion is several orders of magnitude larger. According to the laws of conservation of energy and momentum, implanted ions interacting with cellular molecules will transmit their momentum which will then have a disruptive effect on the target. On the other hand, α-particles cause only a few ionizations, losing most of their ionization energy in the cytoplasm. However, the small ionization caused by the recoil of the parent nucleus during the α decay is potentially significant. Although this energy of the recoil nucleus is typically only about 2% of the energy of the α-particle, the limited 2–3 Å range of the recoil nucleus (**Figure**

11.1), renders it extremely potent due to its high electric charge and high mass. Thus, all of the ionization energy is deposited in an extremely small volume wherever the parent nucleus happens to be located. In the vicinity of chromosomes, the effect would be to 'wipe out' that region of the chromosome.

Investigations on the implantation of ^{7}Be and ^{7}Li ion beams into maize and wheat seeds have been conducted (Guo *et al.*, 2007). It is hypothesized, that ^{7}Li ion beam implantation in the irradiated seeds may lead to the production of ^{1}H (^{7}Li, ^{7}Be) following the electron-capture decay of ^{7}Be to the 478 keV first excited state in ^{7}Li. More studies are needed to consolidate this hypothesis, but if verified, this might open up a new range of physical mutation effects.

2.2. Physico-Chemical Stage

The timeframe of the second, physico-chemical, stage lies in the range of 10^{-6} seconds. This leads to macromolecular radicals that react with water and other cellular molecules and radicals until chemical stability is restored (direct chemical effect) and formation of reactive products from radiolysis of water (indirect chemical effect).

The target biological molecules in a living cell are mostly DNA, storage proteins and enzymes. The physico-chemical stage of radiation effects include the direct disruption of chemical bonds of molecules via the transfer of excitation and ionization energy through radiation or through mass deposition and charge exchange by ion beam implantation (**see Chapter 10**). All life forms maintain a reducing environment within their cells. A very efficient, but complex interplay of enzymes, powered by a constant input of metabolic energy buffers and equilibrates the internal reduced state. When this normal redox state is challenged and disturbed too far from its equilibrium, e.g. by ionizing

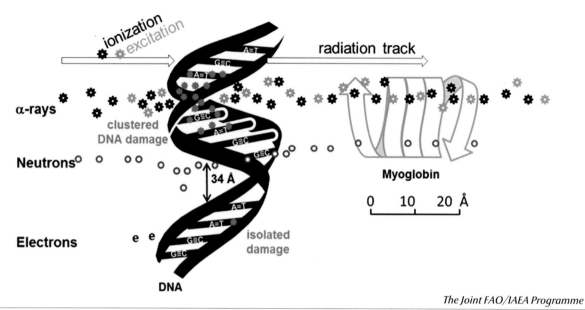

Figure 11.1 Illustration of DNA damage caused by ionization or excitation and separation of ionizations in relation to the size of a protein molecule (β chain of myoglobin) and a double helix when tissue is irradiated with α-rays, fast neutrons (recoil nuclei) and γ-rays (electrons). Note that the next ionization for the γ-rays in the electron track will come close to 60cm outside of the picture.

radiation (**Equation 1**), toxic effects arise through the production of reactive oxygen species that damage all components of the cell, including proteins, lipids, RNA and DNA. This process is called oxidative stress (Evans and Cooke, 2004).

$$H_2O \xrightarrow{\text{ionizing radiation}} \bullet OH, e_{aq}^-, H\bullet, H_2O_2, H_2$$

Equation 1. Interaction of ionizing radiation with cellular water
e_{aq} = hydrated electron (k ≈ 1.3–1.7 × 10^{10} M^{-1}s^{-1});
$\bullet H$ = hydrogen atom (k ≈ 1–5 × 10^8 M^{-1}s^{-1});
$\bullet OH$ = hydroxyl radical (k ≈ 5–10 × 10^9 M^{-1}s^{-1})
where k = diffusion constant

Thus, oxidative stress is caused by an imbalance between the production of reactive oxygen and an organism's ability to readily de-toxify the reactive intermediates and/or easily repair the resulting damage. Chemically speaking, oxidative stress is a large drop in the cellular reducing capacity or a large increase in the cellular reduction potential. The effects of oxidative stress depend upon the size of these changes, with a cell being able to overcome small perturbations and regain its original state. However, moderate to severe oxidative stress can trigger apoptosis and cause cell death, while more intense stress may cause necrosis, a form of cell death that results from acute tissue injury and provokes an inflammatory response. Chemical interactions, particularly of destructive reactive oxygen

species such as free radicals and peroxides cause extensive cellular damage.

2.3. Chemical Stage

The major portion of long term effects on the living cell is inflicted by damage to DNA. More than half of the biological effect of low LET ionizing radiation (for example, X-rays and γ-rays) is the result of indirect radiation. About 90% of this is due to the action of the hydroxyl radical (•OH) alone. Direct action predominates for high LET radiation.

2.3.1. Direct Radiation Effects

Conventional direct damage to DNA results in transversions (purines to pyrimidines and vice versa) and transitions (purine to purine and pyrimidine to pyrimidine) through the activities of different DNA repair mechanisms. Single- and double-strand breaks (SSBs and DSBs) by ionizing radiation are believed to be induced by electrons with sufficient energy to ionize DNA and produce chemically reactive macromolecules that interact with the cellular molecular environment. More recent findings (Boudaïffa *et al.,* 2000) challenge this conventional view by suggesting that damage to the genome by ionizing radiation is only induced by electrons with sufficient energy to ionize DNA. The mechanisms of direct damage by low-energy electrons are less well known, but a substantial proportion of

DNA strand breaks are thought to be initiated by single energy absorption events (**Figures 11.2a and 11.2b**). Low-energy electrons that are slowed down to energies too low to induce ionization of DNA (small transfers of energy in the order of 10 eV or less), undergo resonant attachment to DNA bases or to the sugar–phosphate backbone and produce complex DNA damage: a transient molecular anion is formed that reacts further to break one or both strands of the DNA. In this way, one electron can produce multiple lesions, thus amplifying the clustering of damage induced in DNA by a single radiation track.

2.3.2. Indirect Radiation Effects

A third of the *in vivo* damage to the genome caused by light ionizing radiation, such as X-rays, can be attributed to energy deposited directly in the DNA and its closely bound water molecules. The remaining two thirds can be indirectly attributed to free radicals produced by energy deposited in water molecules and other biomolecules surrounding the DNA. Since the reducing counterparts of $\bullet OH$, principally the hydrated electron, e_{aq^-} are relatively ineffective (**see Equation 1**) especially at inducing DNA strand breaks, almost all of the indirect damage to DNA is due to attacks by the highly reactive hydroxyl radical $\bullet OH$. Thus, the induction of a double strand break by a single burst of radiation is the result of a localized attack by two or more $\bullet OH$ radicals. This presents a challenge to the DNA repair machinery, and has vital consequences for biological effects of the cell, such as the development of a complex lesion through concomitant damage in close proximity by $\bullet OH$ or by direct effects on the DNA. One strand could be attacked by an $\bullet OH$ radical whereas the other strand may sustain direct damage within a 10 bp range of the $\bullet OH$ attack. Clusters of such hybrid damage are known to be generated by the closely spaced depositions of energy along the radiation tracks. Low-energy electrons might induce single strand breaks on each strand of the DNA as a single electron interaction. This is similar to the induction of a double strand break by a single $\bullet OH$ radical, in itself a minor contributor to indirect damage attributed to the effects of radical transfer between DNA strands. Thus, transfer of radicals or energy between DNA strands may play an important part in amplifying the complexity of DNA lesions over and above the level set by the physical clustering of events along the radiation tracks (**Figure 11.2c**).

Studies on radiation-induced DNA damage as a function of hydration, by gas chromatography/mass spectrometry (GC/MS) using an isotope-dilution technique, has enabled the cataloging of biochemical products of

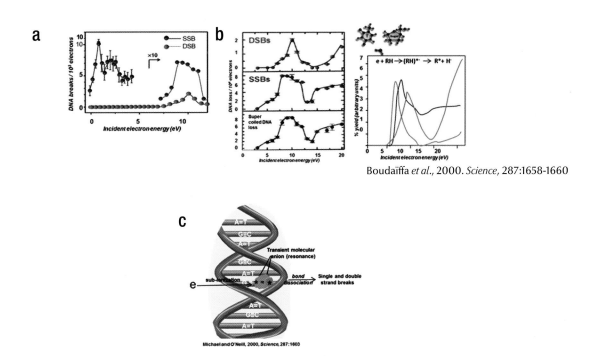

Boudaïffa *et al.*, 2000. *Science*, 287:1658-1660

The Joint FAO/IAEA Programme

Figure 11.2 DNA damage by electrons measured as a) strand breaks and b) DNA loss, quantum and energy dependent desorption yields. c) The mechanism of DNA damage showing anion formation which leads to strand breakage.

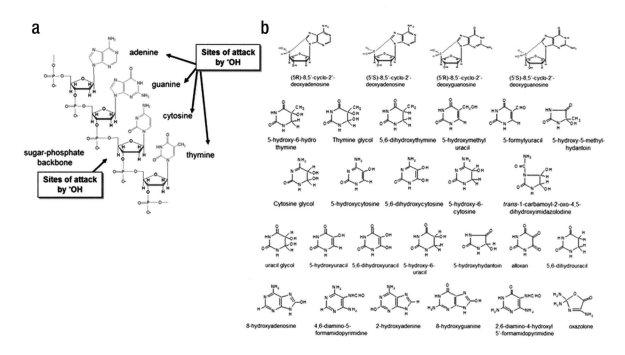

a

adenine

guanine

Sites of attack by ˙OH

cytosine

thymine

sugar-phosphate backbone

Sites of attack by ˙OH

b

(5'R)-8,5'-cyclo-2'-deoxyadenosine   (5'S)-8,5'-cyclo-2'-deoxyadenosine   (5'R)-8,5'-cyclo-2'-deoxyguanosine   (5'S)-8,5'-cyclo-2'-deoxyguanosine

5-hydroxy-6-hydro thymine   Thymine glycol   5,6-dihydroxythymine   5-hydroxymethyl uracil   5-formyluracil   5-hydroxy-5-methyl-hydantoin

Cytosine glycol   5-hydroxycytosine   5,6-dihydroxycytosine   5-hydroxy-6-cytosine   trans-1-carbamoyl-2-oxo-4,5-dihydroxyimidazolidine

uracil glycol   5-hydroxyuracil   5,6-dihydroxyuracil   5-hydroxy-6-uracil   5-hydroxyhydantoin   alloxan   5,6-dihydrouracil

8-hydroxyadenosine   4,6-diamino-5-formamidopyrimidine   2-hydroxyadenine   8-hydroxyguanine   2,6-diamino-4-hydroxyl 5'-formamidopyrimidine   oxazolone

The Joint FAO/IAEA Programme

Figure 11.3 DNA damage caused by oxidative stress. a) target sites for DNA oxidative damage, b) comprehensive catalogue of products from damaged DNA (courtesy of D. Dizdar).

DNA damaged by oxidative stress and ionizing radiation (**Figure 11.3**). Relative comparison has revealed that the products of DNA damage by direct ionizing radiation are essentially the same as those formed by reactions of the hydroxyl radical with DNA bases. The yields depend on the extent of DNA hydration.

DNA damage can be repaired. The timeframe of these biochemical processes (enzymatic repair of damaged DNA) lies in the range of minutes. The result of the action of the nuclear repair mechanisms could be error-free or error-prone (**see Chapters 5 and 6**). SSBs can be repaired quickly and with a low error rate. DSBs are more difficult to handle for enzymatic repair mechanisms and their error-prone repair leads to mutations, loss of reproductive capacity of the affected cell and ultimately to cell death. In some cases, the DNA might be irreparable, and the damaged genomic parts may be lost. Clustered lesions are difficult for the cell to repair and are therefore likely to lead to permanent damage to the genome.

2.3.3. Effects on Proteins (Enzymes)

On their path through the cells of a living organism, γ-rays randomly ionize molecules along their trajectories rupturing chemical bonds and releasing radiolytic products. As water is the most abundant molecule in living cells, the principal radiation products are reactive oxygen species (ROS) that diffuse into the watery environment of the different cellular compartments (e.g. cytoplasm, plastids, nuclei) and the cytosol. ROS react with macro-molecules in other cellular compartments. This indirect action of radiation is responsible for 99.9% of the damage to proteins. This injury translates into changes in protein structure and often induces a decrease or loss of protein function; there is seldom a gain of function. Irradiation of plants using γ-rays induces stress in the surviving plants which significantly affects physiological and biochemical processes: perturbations to protein synthesis, enzyme activity, hormonal balance, leaf gas-exchange and water exchange are prominently cited as the effects of seed irradiation. Functional, structural and morphological changes are correlated to the strength and the duration of the irradiation stress. Up to a threshold, the adaptive capacity of the plant is preserved and the observed changes are reversible. The plant responds to the radiation-induced stress in the same way it responds to environmental stress. An essential part of this response is the initiation of protein breakdown and recycling of breakdown products, which in turn depend on the levels of proteolytic enzymes as cellular proteins may need to be rebuilt or synthesized *de novo*. Degradation of damaged, misfolded and potentially harmful proteins provides free amino acids for the synthesis of new proteins. Indeed, after γ and neutron irradiation of pea seeds, enhanced degrading enzyme activities have been reported, leading to an inverse correlation with growth (Bagi *et al.,* 1988).

Every macro-molecule that suffers a direct ionization event is destroyed. The only survivors are those that escape direct ionization. The same is true for frozen samples. Ionizations still occur randomly and water is the principle molecule affected, but diffusion of radiation products is limited. At very low temperatures (77°K), a cut-off value is reached, where essentially all the radiation damage is due to primary ionization events occurring directly in macro-molecules. Thus, proteins in frozen solutions are 10^3 to 10^4 times less sensitive to radiation as those in the liquid state. The survival of frozen proteins after irradiation is a direct measure of the mass of the active structures and independent of the presence of other proteins.

2.4. Radio Effects of Ultraviolet Light

UV light can cause photochemical damage. Cyclobutane pyrimidine dimers (CPDs) are the main photoproducts following exposure to UV-B and UV-C, and they can lead to cell death and pre-carcinogenic lesions. Other types of dimers are considered to be especially mutagenic. DNA-protein cross-links can occur after UV radiation and can be lethal. Biological effects can arise when UV radiation is absorbed. Absorption is dependent on the chemical bonds of the material, and is highly specific. Sunburn in human is a form of erythema produced by over-exposure to the UV-B portion of the solar spectrum. A rare but deadly form of skin cancer in humans, malignant melanoma, is induced by exposure to sunlight, with occurrences localized to most frequently exposed regions of the body. Because UV can penetrate only a few layers of cells, the effects of UV radiation on humans are restricted to exposed tissues of the skin and the eyes. Survival of UV irradiation is reduced as the dose of radiation is increased. The shapes of survival curves are similar to those for lethality from ionizing radiation; they are dependent on the presence or absence of repair systems. The four repair systems that enhance biological survival include photo-reactivation by splitting of CPDs in the DNA; DNA excision repair; DNA recombination repair; and an inducible repair system of bacteria known as SOS repair. The first experimental evidence of photo-reactivation of

UV-induced DNA damage in plant ribosomal genes was produced by Petersen and Small (2001). Most original data obtained to date demonstrate that in the lack of visible light, young barley leaves are unable to remove an enormous fraction of CPD from the whole genome including ribosomal genes. It appears that the only mechanism capable of rescuing the barley genome from the vast majority of UV-C induced CPD is photo-reactivation. The investigations highlight the fact that UV-C induced oxidative base lesions have a significant impact on the stability of the plant genome. In addition, response to irradiation appears to be genotype dependent. The results in barley show the importance of light-dependent repair pathways for the maintenance of genome integrity and underscore the need for further intensive investigations in the field (Manova *et al.,* 2009).

2.5. Visualization of Damaged DNA

The damaged DNA molecules can be detected in irradiated cells using methods such as the Comet assay and TUNEL test.

2.5.1. Comet Assay

The comet assay (Singh *et al.,* 1988) or the Single Cell Gel Electrophoresis assay (SCGE) is a rapid, simple, visual and sensitive technique for measuring DNA breakage in individual eukaryotic cells. It is inexpensive and popular as a standard evaluation technique for DNA damage, repair and bio-monitoring. It can be used in conditions where more complex assays are not operable.

In this technique eukaryotic cells are immobilized in a low-melting-point agarose suspension. The captured, suspended cells are lysed under alkaline (pH > 13) conditions and subjected to electrophoresis. The underlying concept is that undamaged DNA retains a highly organized association with matrix proteins in the nucleus. Lysis disrupts this framework. The individual strands of DNA lose their compact structure and unwind, expanding out of the nuclear matrix. DNA is negatively charged. Large, undamaged DNA strands remain confined in the agarose matrix, whereas the smaller fragmented DNA pieces migrate towards the anode. Therefore, the amount of DNA that leaks through the gel matrix towards the anode is a measure of the amount of DNA damage in the cell. Electrophoresis is followed by visual analysis with DNA staining and measuring dye fluorescence to determine the extent of

DNA damage. This can be performed by manual scoring or automatically by imaging software. The stronger the signal from the migrated DNA the more damage present. The overall gel picture commonly shows a circular head corresponding to the undamaged DNA and a tail of damaged DNA. It resembles a comet (hence "comet assay"); the brighter and longer the tail, the higher the level of damage (**Figure 11.4**). DNA damage is usually due to SSBs and DSBs, AP sites (abasic sites missing either a pyrimidine or purine nucleotide) and sites where excision repair is taking place.

2.5.2. TUNEL Test

The TUNEL test (terminal deoxynucleotidyl transferase dUTP nick end labelling) is a method for detecting DNA fragmentation by labelling the terminal end of nucleic acids (Gavrieli *et al.*, 1992). Originally, TUNEL was a labelling method for detecting DNA fragmentation that results from apoptotic signalling cascades. The assay relies on the presence of nicks in the DNA which can be identified by terminal deoxynucleotidyl transferase, an enzyme that will catalyze the addition of dUTPs tagged with a secondarily labelled marker (e.g. fluorescence). It is also used to label cells that have suffered severe DNA damage, e.g. through radiation (**Figure 11.6**).

3. Biological Consequences

The biological consequences of radiation can appear at various stages of development, i.e. abnormal cell division, cell death, mutation, tissue and organ failure and reduction of plant growth. The effect is dependent on the type and dose of radiation, physiological status and the genetic composition of the treated material.

3.1. Effects on Cell Division and Plant Growth

Plant cells can swiftly respond to radiation treatment and initiate pathways to cope with such genomic stresses (**see Chapter 5**). Most DNA damage caused by radiation needs to be repaired before cells can start dividing again (the canonical cell cycle is shown in **Figure 11.6**). Severe irreparable damage may cause cell death, while lighter damage may be correctly or incorrectly repaired and lead to retardation of cell division, cytological abnormalities and induced mutation. The destruction of many enzymes by radiation also contributes to the slowdown of cell division and plant growth.

3.1.1. Effects on Mitosis

Depending on dose, several types of cellular effects can be observed. The most probable is cell death. Cell death can occur through apoptosis in which radiation triggers the genetic process of programmed cell death, cytolysis (swelling of cells until they burst and disappear), protoplasmic coagulation (irreversible gelatin formation in both the nucleus and cytoplasm), karyolysis (nuclear swelling followed by chromatin loss), pyknosis (nuclear contraction and chromatin condensation) or karyorrhexi (nuclear fragmentation).

At non-lethal doses, changes in cellular function can occur, including delays in certain phases of the mitotic cycle, disrupted cell growth, permeability changes and changes in motility. Mitosis may be delayed or inhibited following radiation exposure causing major alterations in cell kinetic patterns resulting in depletion of the affected cell populations. Dose-dependent inhibition of mitosis (reduced mitotic index) is particularly common in actively proliferating cell systems, e.g. meristems. This inhibition occurs when the chromosomes begin to condense in

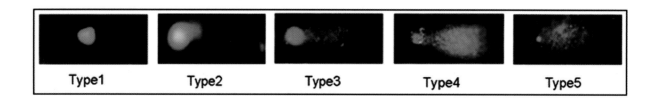

The Joint FAO/IAEA Programme

Figure 11.4 Barley comet assay. The comet assay was applied to nuclei isolated from root tips of germinated barley seedlings and used to monitor induction and repair of DNA damage produced in the barley genome *in vivo*. Experimental conditions were adapted to achieve an efficient detection of both single- and double-stranded DNA breaks. Five categories can be distinguished in the resulting comet populations, based on the % total DNA in the tail: Type1: <5%; Type2: 5–20%; Type3: 20–40%; Type4: 40–95% and Type5: >95% (courtesy of L. Stoilov).

early prophase of the mitotic cycle, but prior to the breakdown of the nuclear membrane. Subsequent irradiation after this transition point does not hold up mitosis.

Ionizing radiation produces cytological aberrations and defects in chromosome segregation. Cytological aberrations observed in mitosis include the production of micro-nuclei and chromosomal abnormalities. The micro-nucleus assay measures the proportion of micro-nuclei produced by a given radiation dose. It is a responsive test for investigations of cytogenetic damage produced by radiation in dividing cell systems *in vivo*. Chromosomal abnormalities in irradiated mitotic cells range from breaks, through exchanges, laggards and anaphase bridges, dicentric and centric ring formations, terminal fragments with telomeric signal at only one end and interstitial fragments that appear as double minutes without any telomeric signals (see review, Natarajan, 2005).

Defects in chromosome segregation play a critical role in producing genomic instability and aneuploidy. Merotelic kinetochore orientation is a major cause of lagging chromosomes during mitosis. Cells with mono-oriented chromosomes never enter anaphase and lagging chromosomes appear during anaphase after chromosome alignment occurs during metaphase. Cell

growth may also be retarded, usually after a latent period. This may be due to progressive formation of inhibitory metabolic products and/or alterations in the cell micro-environment. Irradiated cells may show both increased and decreased permeability. Radiation changes within the lipid bilayers of the membrane may alter ionic pumping. This may be due to changes in the viscosity of intracellular fluids associated with disruptions in the ratio of bound to unbound water. Such changes would result in an impairment of the ability of the cell to maintain metabolic equilibrium and could be very damaging even if the shift in equilibrium is small. The motility of a cell may be decreased following irradiation. However, the presence of normal motility does not imply the absence of radiation injury. Irradiated mammal spermatozoa, for example, may retain their motility and be capable of fertilization while carrying radiation-induced genetic changes which may affect subsequent embryogenesis.

3.1.2. Effects on Meiosis

Meiosis is an extremely complex process involving: DNA replication, cell division, genome reduction, chromosome recombination and chromosome re-assortment, and is easily disrupted by irradiation. An increase in radi-

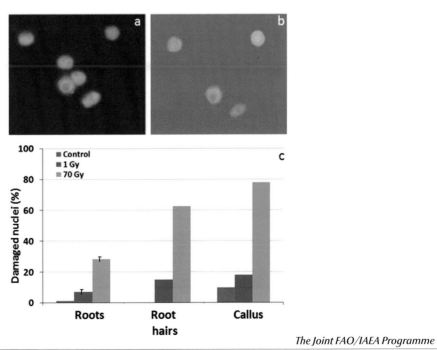

The Joint FAO/IAEA Programme

Figure 11.5 TUNEL test on *Crepis capillaris* cells (root and callus), challenged with γ radiation. Nuclei are revealed by fluorescence microscopy labeled with a) DAPI (4′,6-diamidino-2-phenylindole) and b) FITC (Fluorescein isothiocyanate). DAPI stains all types of nuclei whereas FITC is used to detect damaged nuclei. c) Frequency of damaged nuclei in tissues treated with different doses of γ-rays (courtesy of M. Maluszynski, 2009).

ation dosage is positively correlated to a decrease in the meiotic index. Mutated plants typically show reduced fertility, mainly caused by chromosomal rearrangements and genomic mutations during meiosis. The chromosomal aberrations involve reciprocal translocations, duplications, deficiencies (aneuploidy) and inversions.

3.1.3. Effects on Seed Germination, Plant Growth and Reproduction

Ionizing radiation significantly affects physiological and biochemical processes in plants. The irradiation of seeds disrupts protein synthesis, affects hormone balance, disturbs gas-exchange, water exchange and enzyme activity. The morphological, structural and functional changes depend on the strength and the duration of the radiation stress induced. In the case of moderate stress, the adaptability of plants is preserved and the observed changes are reversible. In general, seed germination, plant growth and reproduction are inversely correlated to irradiation dose.

3.1.4. Induced mutations

Although most damaged DNAs can be repaired, they do not necessarily regain their correct sequence or form. Various types of mutations in the DNA sequence can occur after exposure to mutagens. They include chromosomal aberrations (e.g. duplications, translocations), point mutations and short insertion/deletions (indels; **see Chapter 20**). Modern molecular high-throughput techniques such as TILLING (Colbert *et al.*, 2001; **see Chapters 21 and 22**) or pyrosequencing (Ronaghi *et al.*, 1998), allow for assaying SNPs and small INDELs in a high-throughput manner. In rice (*Oryza sativa*), the frequency of induced mutations varies with the developmental stage at the time of irradiation. The highest overall mutation frequency is observed when radiation

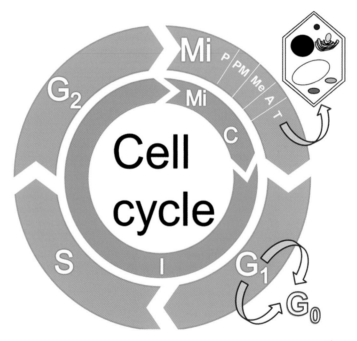

The Joint FAO/IAEA Programme

Figure 11.6 An illustration of the cell cycle:

Quiescent/senescent state
G$_0$ phase = Gap 0 (resting phase, the cell has left the cycle and has stopped dividing).

Interphase state
I = Interphase: (1) G$_1$ phase = Gap 1 (cell increases in size); (2) S phase = Synthesis (DNA replication), and (3) G$_2$ phase = Gap 2 (gap between DNA synthesis and mitosis, cell continues to grow).

Cell division state
M = M phase: (1) Mi = Mitosis (the cell's chromosomes are divided between the two daughter cells); (2) P = Prophase (chromatin in the nucleus condenses, mitotic spindle formation); (3) PM = Prometaphase (nuclear membrane dissolves, kinetochore formation); (4) Me = Metaphase (metaphase plate formation); (5) A = Anaphase (paired chromosomes separate to opposite sides of the cell); (6) T = Telophase (new membranes form around the daughter nuclei); (7) C = Cytokinesis (the cell's cytoplasm divides forming distinct cells, in plant cells, the rigid wall requires that a cell plate be synthesized between the two daughter cells).

is applied 10 days before anthesis, the late tetrad stage of microspores. This stage is more useful for radiation exposure than seeds and fertilized egg cells (Kowyama *et al.,* 1994). Pollen irradiation treatments can result in viable pollen tube growth but abnormal fertilization and is used as a standard treatment in the production of haploids in fruit crops (**see Chapter 30**) and in generating mutant germplasm in the next generation.

3.2. Radio Sensitivity and Relative Biological Effect

Various biological effects can result from the actions of ionizing radiation. Radio sensitivity measures the relative susceptibility of organisms, organs, tissues or cells to the damaging effects of ionizing radiation. Highly metabolically active cells, well nourished, quickly dividing cells and undifferentiated cells are most sensitive to radiation. Any given cells cycling through the different stages of the cell cycle are most sensitive at the M phase and gradually less sensitive during G_2, G_1 and least sensitive when in the S phase. Thus, as a rule of thumb, dividing tissues are radiosensitive (high mitotic rate), and non-dividing tissues are radio-resistant, thus reproductive cells/tissues are more affected than non-differentiated cells/tissues.

Data for one gymnosperm species (*Pinus strobus*) show that greater damage is produced by a 5-month chronic exposure during a period of active growth than by a comparable exposure during dormancy. The correlation between interphase chromosome volume (ICV) and radio-sensitivity was confirmed for these end points. The larger the ICV, the more sensitive the species. Thus sensitivity to chronic γ irradiation of untested woody plant species can be obtained if the ICV is known (Sparrow *et al.,* 1970). The radio-sensitivity of organisms varies greatly, being related to their intrinsic sensitivity to radio-biological damage and to their ability to repair the damage. Radiation doses resulting in 10% survival of a species range from 3 Gy (mouse and human cells) to greater than 1,000 Gy (the bacterium *Deinococcus radiodurans*). Developing embryos are relatively very radio-sensitive. Radio-sensitivity generally decreases as the embryonic development proceeds. The age-dependent change in radio-sensitivity may strongly depend on the degree of morphogenesis. In rice (*Oryza sativa*), the most radiosensitive stage (X-rays, 20 Gy), is the last pre-meiotic interphase.

The absorbed dose of ionizing radiation is measured in grays. One gray is the absorption of one joule of energy, in the form of ionizing radiation, by one kilogram of matter. 1 cm^3 of human tissue is usually composed of 10^9 cells. A 1 mGy dose indicates a hit rate of 1 in 1,000 or 10^6 cells. If 99.9% are repaired, 10^3 cells will result damaged by radiation. Assuming that 999 of the 1,000 damaged cells die, 1 cell may survive with damage and could be mutated.

The equivalent dose to a tissue, sievert, is found by multiplying the absorbed dose, in grays, by a dimensionless "Quality Factor" Q, a term that has replaced the relative biological effectiveness (RBE). RBE was a concept introduced in the 1950s by health physicists upon noting that different types of radiation might affect living organisms differently. Higher RBE correlates to greater biological damage for equivalent radiation exposure, dependent upon radiation type. Another dimensionless factor, N, depends upon the part of the body irradiated, the time and volume over which the dose was spread, and the species of the biological organism. Together, Q and N constitute the presently used radiation weighting factor, WR. For an organism composed of multiple tissue types a weighted sum or integral is often used. Although the sievert has the same dimensions as the gray (i.e. joules per kilogram), both measure different phenomena. For a given amount of radiation, measured in grays, the biological effect, measured in sieverts can vary considerably as a result of the radiation weighting factor WR. For example, temporary male sterility is induced by 0.15 Gy, whereas a four times higher (0.6 Gy) dose is needed for temporary female sterility (from http://rpop.iaea.org/). This variation in RBE is also attributed to the different LET of each type of radiation. Studies (Kazama *et al.,* 2008) have shown that the biological effect of ion beam radiation is dependent on absorption doses and LET values but independent of ion species, which means that the treatment of ^{12}C would produce similar biological effects on rice seeds as ^{20}Ne at the same dose (e.g. 50 Gy) and the same LET (e.g. 30 keV/μm).

4. References

4.1. Cited References

Bagi, G., Bornemisza-Pauspertl, P. and Hidvégi, E.J. 1988. Inverse correlation between growth and degrading enzyme activity of seedlings after gamma and neutron irradiation of pea seeds. *International Journal of Radiation Biology.* 53: 507–519.

Boudaïffa, B., Cloutier, P., Hunting, D. *et al.* 2000. Resonant formation of DNA strand breaks by low-energy (3 to 20 eV) electrons. *Science.* 287: 1658–1660.

Colbert, T., Till, B.J., Tompa, R. *et al.* 2001. High-throughput screening for induced point mutations. *Plant Physiol.* 126: 480-484.

Evans, M.D. and Cooke, M.S. 2004. Factors contributing to the outcome of oxidative damage to nucleic acids. *Bioessays.* 26: 533-542.

Gavrieli, Y. 1992. Identification of programmed cell death *in situ via* specific labeling of nuclear DNA fragmentation. *J Cell Biol.* 119: 493–501.

Guo, H.J, Liu, L.X., Han, W.B. *et al.* 2007. Biological effects of high energy ^{7}Li ion beams implantation on wheat. *Plant Mutation Reports.* 1: 31–35.

Kazama, Y., Saito, H., Yamamoto, Y.Y. *et al.* 2008. LET-dependent effects of heavy-ion beam irradiation in *Arabidopsis thaliana. Plant Biotechnol.* 25: 113–117.

Kowyama, Y., Saba, T., Tsuji, T. *et al.* 1994. Specific developmental stages of gametogenesis for radiosensitivity and mutagenesis in rice. *Euphytica.* 80: 27–38.

Maluszynski, M. 2009. Analysis of mutation types and frequencies induced in the barley genome by physical and chemical mutagens. CRP Effects of mutagenic agents on the DNA sequence in plants. D24011 (2003-2009), Vienna: IAEA.

Manova, V., Georgieva, M., Borisov, B. *et al.* 2009. Genomic and gene-specific induction and repair of DNA damage in barley. *In:* Q.Y. Shu (ed): Induced Plant Mutations in the Genomics Era. Rome: Food and Agriculture Organization of the United Nations, pp. 122–125.

Natarajan, A.T. 2005. Chromosome aberrations: Plants to human and feulgen to FISH. *Current Science.* 89: 335–340.

Petersen, J. L. and Small, G. D. 2001. A gene required for the novel activation of a class II DNA photolyase in Chlamydomonas. *Nucleic Acids Res.* 29: 4472-4481.

Ronaghi, M., Uhlén, M. and Nyrén, P. 1998. A sequencing method based on real-time pyrophosphate. *Science.* 281: 363–365.

Singh, N.P., McCoy, M.T., Tice, R.R. *et al.* 1988. A simple technique for quantitation of low levels of DNA damage in individual cells. *Experimental Cell Research.* 175: 184–191.

Sparrow, A.H., Schwemmer, S.S., Klug, E.E. *et al.* 1970. Radiosensitivity studies with woody plants: II. survival data for 13 species irradiated chronically for up to 8 years. *Radiation Research.* 44: 154–177.

Stadler, L.J. 1928. Mutations in barley induced by X-rays and radium. *Science.* 68: 186–187.

Stoilov, L. 2009. Genomic and gene-specific induction and repair of DNA damage in barley. *CRP Effects of Mutagenic Agents on the DNA Sequence in Plants.* D24011 (2003–2009), Vienna: IAEA.

Tisljar-Lentulis, G., Henneberg, P. and Feinendegen, L.E. 1983. The oxygen enhancement ratio for single- and double-strand breaks induced by tritium incorporated in DNA of cultured human T1 cells: Impact of the transmutation effect. *Radiation Research* 94: 41–50.

4.2. Websites

The different stages of radiation effect on living cells:
http://rpop.iaea.org/RPOP/RPoP/Content/Documents/TrainingRadiotherapy/Lectures/RT03_RadBiol1_RS_WEB.ppt

The DNA damage by low energy electrons:
http://http://www.isa.au.dk/networks/cost/radam07/talks/Denifl_WG1_RADAM07.pdf

The International Commission on Radiological Protection (ICRP):
http://www.icrp.org/

4.3. Further Reading

Anonymous 2009. ICRP Report 2006–2008 revision 1. ICRP Reference 90/547/09: 44.

Bedford, J.S. and Dewey, W.C. 2002. Radiation Research Society. 1952–2002. Historical and current highlights in radiation biology: has anything important been learned by irradiating cells? *Radiat. Res.* 158: 251–291.

Nias, A.H.W. 1998. An introduction to radiobiology, New York: Wiley, pp.38.

Skarsgard, L.D. 1998. Radiobiology with heavy charged particles: a historical review. *Phys Med.* 14(Suppl 1): 1–19.

C12

Chemical Mutagenesis

J.M.Leitão[*]

BioFIG, FCT, Universidade do Algarve, 8005-139 Campus de Gambelas, Faro, Portugal
* Corresponding author, E-MAIL: jleitao@ualg.pt, jleitao2@gmail.com

1. Introduction

The first attempts to induce mutations in biological systems using chemical compounds go back to the beginning of the past century. However, it was during World War II that the two most relevant names in chemical mutagenesis, Charlotte Auerbach and Iosif A. Rapoport, established the mutagenic properties of several chemical compounds (**Box 12.1**). A detailed review of these and other major moments in the history of plant chemical mutagenesis is given by van Harten (1998, **see Chapters 1 and 2**).

There is currently an enormous number of known chemical compounds able to induce mutations in prokaryotic and/or eukaryotic cells and this continues to increase. The continuous search and the synthesis of new mutagenic compounds is driven, not by the needs of experimental mutagenesis, but by the paradoxal fact that several mutagenic compounds, although carcinogenic, possess simultaneously anti-neoplastic properties and find application in anti-tumour therapy.

Despite the large number of mutagenic compounds, only a small number has been tested in plants. Among them, only a very restricted group of alkylating agents has found large application in plant experimental mutagenesis and plant mutation breeding. Over 80% of the registered new mutant plant varieties reported

in the IAEA database (http://mvgs.iaea.org/Search.aspx) obtained *via* chemical mutagenesis were induced by alkylating agents. Of these, three compounds are significant: ethyl methanesulphonate (EMS), 1-methyl-1-nitrosourea (MNU) and 1-ethyl-1-nitrosourea (ENU), which account for 64% of these varieties (**Figure 12.1**).

2. Alkylating Agents

2.1. Sources and Types of Alkylating Agents

Alkylating agents are strong mutagenic, carcinogenic and cytotoxic compounds (**Figure 12.2**). Paradoxically, the cytotoxic properties of some of the compounds are largely exploited in cancer therapy. Alkylating agents can be found among a large panoply of classes of compounds, including sulphur mustards, nitrogen mustards, epoxides, ethyleneimines and ethyleneimides, alkyl methanesulphonates, alkylnitrosoureas, alkylnitrosoamines, alkylnitrosoamides, alkyl halides, alkyl sulphates, alkyl phosphates, chloroethyl sulphides, chloroethylamines, diazoalkanes, etc.

Although most are synthetically produced, a few alkylating agents are of biological origin, e.g. the strong mutagenic glucosamine-nitrosourea (Streptozotocin) is produced by *Streptomyces achromogenes*.

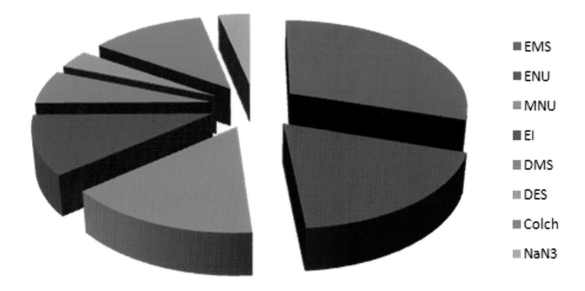

The Joint FAO/IAEA Programme

Figure 12.1 Relative number of released mutant varieties (direct and indirect) induced using the agents indicated. EMS – ethyl methanesulphonate; ENU – 1-ethyl-1-nitrosourea; MNU – 1-methyl-1-nitrosourea; EI – ethyleneimine; DMS – dimethyl sulphate; DES – diethyl sulphate; Colch – colchicine; NaN$_3$ – sodium azide.

Box 12.1: C. Auerbach and I. Rapport and their work on chemical mutagenesis

Charlotte Auerbach (1899–1994) was born to a Jewish family in Krefeld, Germany. She attended classes at the Universities of Berlin, Würzburg and Freiburg. After her "State examination" in 1924 and a short period of research in developmental biology she spent some years teaching at various schools in Berlin. After Hitler became the German Chancellor and new laws prohibited Jews from teaching in state schools she moved to the UK. There, through Professors H. Freundlich and G. Barger, she was introduced to F.A.E. Crew, head of the Institute of Animal Genetics in Edinburgh, where she started to work on Drosophila and was awarded a Ph.D. in 1935. In 1939 she acquired British citizenship, thus avoiding incarceration in internment camps during the second World War.

In 1938 Crew introduced Auerbach to H.J. Müller who stimulated her to test a number of chemical agents known to be carcinogenic for induction of mutations in Drosophila. After obtaining negative results with 1:2:5:6-dibenzanthracene, 9:10-dimethyl-1:2 dibenzanthracene and methyl-colanthrene, in 1940 Auerbach began to work in collaboration with J.M. Robson on the mutagenic effects of mustard gas.

Robson had at that time already observed the antimitotic effect of this compound in the vaginal epithelium of rats. During their collaboration most of the experimental treatments were carried out by Robson and his collaborators, while Auerbach performed the genetic analysis of treated Drosophila using the ClB method introduced by Müller, with whom she discussed experimental results.

In 1942, Auerbach and Robson sent several reports to the Ministry of Supply of the British Government where they described the induction of sex-linked lethal, chromosomes inversions and translocations by mustard gas (encrypted as substance "H"). They also reported mustard gas to act directly on the chromosomes while describing the induced visible mutations and discussing the similarities and dissimilarities of the effects of this chemical mutagen and those of X-rays.

In a letter to *Nature* in 1944, Auerbach and Robson mentioned that they had tested a number of chemical substances for their ability to produce mutations and that some of them were very effective, producing mutation rates of the same order as those with X-rays. However, it was not until 1946 in a new letter to *Nature* that the name of the mutagenic substance, dichloro-diethyl-sulphide or mustard gas, was disclosed. According to Auerbach the idea of testing mustard gas was suggested by the pharmacologist A.J. Clark who saw similarities between the long-lasting effects of this compound and that of X-rays, and hypothesized mustard gas to have effects on the genetic material.

Auerbach kept working in the field of mutagenesis, even long after she retired, publishing several articles and some major books, some of them translated into multiple languages. She was honoured with many prizes and distinctions including the Darwin Medal by the Royal Society in 1977.

Iosif Abramovich Rapoport (1912–1990) was born in Chernigov, Ukraine. In 1930 he started his studies at the Faculty of Biology at the Leningrad State University and in 1938 he was awarded a Ph.D. in biological sciences by the Institute of Genetics of the Academy of Sciences. A Doctor of Sciences degree was awarded in 1943 after he defended his thesis and while attending a rapid course for commanders at the Frunze Military Academy in Moscow.

Twice injured during the war (he lost an eye) Rapoport was awarded many major military medals of the Soviet Union and the USA Legion of Merit. After the war he resumed his research at the Institute of Cytology, Histology and Embryology in Moscow, working on chemical mutagenesis in Drosophila.

Although the first mutants induced by chemical mutagenesis were probably identified by V.V. Sacharov in 1932, the early work of Rapoport led to the identification of a number of mutagenic agents: formaldehyde, urotropine, acrolein, ethylene oxide, ethyleneimine, diethylsulphate, diazomethane and N-nitrosomethylurethane. As with Auerbach, Rapoport used Müller's ClB method to assess mutagenic effects in Drosophila. The first article on these discoveries: "Carbonyl compounds and chemical mechanisms of mutations" was published in 1946 and several other papers were published until 1948.

As an opponent to Lysenkoism and refusing to recognize his "error" Rapoport was excluded from the Communist Party and from 1949 to 1957 he worked as a paleontologist and stratigrapher. In 1957 (Stalin died in 1953) Rapoport joined the Institute of Chemical Physics of the Academy of Sciences in Moscow where he resumed his work on chemical mutagenesis.

Rapoport established the mutagenic properties of 55 chemical compounds including the nitrososureas, which he called "super mutagens" and still are largely used in plant chemical mutagenesis today. A centre for introduction of chemical mutagenesis in the biosynthetic (biotech) industry and agriculture was created in 1965. Headed up by Rapoport this centre had a tremendous impact in the utilization of the chemical mutagenesis in the Soviet Union, eastern European countries and other countries. Numerous mutant varieties were officially released in these countries and registered in the IAEA database; 383 chemically induced mutant varieties of major crop species were released in the former USSR alone by 1991.

Later in his life Rapoport was awarded the most prestigious social and academic distinctions of the USSR.

References:
Beale, G. (1993) The discovery of mustard gas mutagenesis by Auerbach and Robson in 1941. *Genetics.* 134: 393–399.

Some alkylating agents such as the methyl-donor S-adenosylmethionine (SAM), which in spite of being involved in about 40 metabolic reactions in mammalians is a weak methylating agent able to form adducts to DNA, are formed endogenously as natural products of organisms. Other alkylating agents such as chloromethane (CH_3Cl), formerly thought to be uniquely of ocean origin but today assumed to have a dominant terrestrial origin as result of the reaction of plant pectin with chloride, are naturally formed in the environment. A significant source of environmental exposure of humans to alkylating compounds, particularly to the carcinogenic 4-(methylnitrosamino)-1-(3-pyridyl)-1-butanone (NNK), 4-(methylnitrosamino)-1-(3-pyridyl)-1-butanol (NNAL) and N'-nitrosonornicotine (NNN), is tobacco smoke.

Alkylating agents are electrophilic compounds with affinity for nucleophilic centres in organic macromolecules to which they bind covalently. In DNA these compounds form covalently linked alkyl adducts to the bases and to the phosphodiesters.

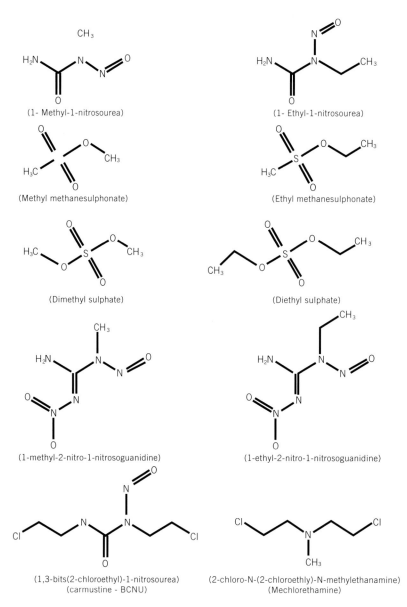

(1- Methyl-1-nitrosourea)

(1- Ethyl-1-nitrosourea)

(Methyl methanesulphonate)

(Ethyl methanesulphonate)

(Dimethyl sulphate)

(Diethyl sulphate)

(1-methyl-2-nitro-1-nitrosoguanidine)

(1-ethyl-2-nitro-1-nitrosoguanidine)

(1,3-bits(2-chloroethyl)-1-nitrosourea)
(carmustine - BCNU)

(2-chloro-N-(2-chloroethly)-N-methylethanamine)
(Mechlorethamine)

The Joint FAO/IAEA Programme

Figure 12.2 Molecular structure of commonly used alkylating agents in plant mutagenesis, and carmustine and mechlorethamine, two bi-functional alkylating agents used in anti-neoplastic clinical practice.

Despite some controversy, alkylating agents are usually classified as S_N1-type or S_N2-type according to their kinetic properties. Compounds for which the rate determining (slow) step in the alkylation reaction is a first-order kinetics formation of reactive intermediates independent of the substrate (in our case–DNA) are designated S_N1-type alkylating agents. Those compounds, where the rate-determining step is a second-order nucleophilic substitution reaction involving both the compound and the substrate (DNA) are designated S_N2-type alkylating agents (**Figure 12.3**).

The very reactive electrophilic species produced by the S_N1-type alkylating agents are generated independently of the substrate and are relatively unselective towards the nucleophile they alkylate. Consequently, these compounds alkylate nitrogen, oxygen and phosphate group sites in DNA. The S_N2 compounds, where the substrate (DNA) participates directly in the generation of the reactive species, react primarily with the most nucleophilic sites on DNA, i.e. the nitrogens N7 and N3 of guanine and N3 of adenine. (**Figure 12.4, Table 12.1**)

Although most of the commonly used alkylating agents in plant experimental mutagenesis produce similar spectra of alkylation lesions in DNA (**Figure 12.4**), differences in alkylation mechanisms (S_N1 vs S_N2) give rise to differences in the proportions of lesions produced.

Table 12.1: Alkylation products of major representative S_N1 (MNU, ENU) and S_N2 (MMS) compounds [a]

Alkylation products	MNU	ENU	MMS
O^2-alkylcytosine	<1	<1	<1
N^3-alkylcytosine	<1	<1	<1
N^3-alkylguanine	<1	<1	<1
O^6-alkylguanine	6	~7	0.2
N^7-alkylguanine	65	~13	83
O^2-alkylthymine	<1	~5	<0.1
N^3-alkylthymine	<1	~1	<1
O^4-alkylthymine	<1	~5	<0.1
N^1-alkyladenine	2	~1	1
N^3-alkyladenine	7	~6	11
N^7-alkyladenine	2	~1	2
Phosphotriesters	20	~65	1

[a] Data kindly provided by Prof. A. E. Pegg, Dep. of Cellular and Molecular Physiology, Milton s. Hershey Medical center, Pennsylvania State University College of Medicine, Hershey Pennsylvania 17033, USA

For example, 1-methyl-1-nitrosourea (MNU) and 1-ethyl-1-nitrosourea (ENU) react *via* an S_N1 mechanism and efficiently alkylate both nitrogens and oxygens in DNA. Methyl methanesulphonate (MMS), which reacts *via* an S_N2 mechanism, predominantly alkylates the nitrogens at the DNA bases and produces little alkylation of the oxygens in DNA bases and in the sugar–phosphate backbone (**Figure 12.4; Table 12.1**).

Alkylating agents are commonly divided into two classes: mono-functional and bi-functional. The alkylating agents usually used in plant experimental mutagenesis and mutation breeding are mono-functional.

Bi-functional alkylating agents (**Figure 12.2**) such as the chloroethylating agents (e.g. 1,3-bis(2-chloroethyl)-1-nitrosourea–BCNU), the nitrogen mustards or mitomycin C, are characterized by their ability to induce DNA strand cross-links and are utilized as anti-neoplastic compounds. This last class of alkylating agents has not been utilized in plant mutation breeding experiments.

2.2. Properties of the Alkylating Agents

The physical and chemical properties of the alkylating agents and of innumerable other chemicals, as well as

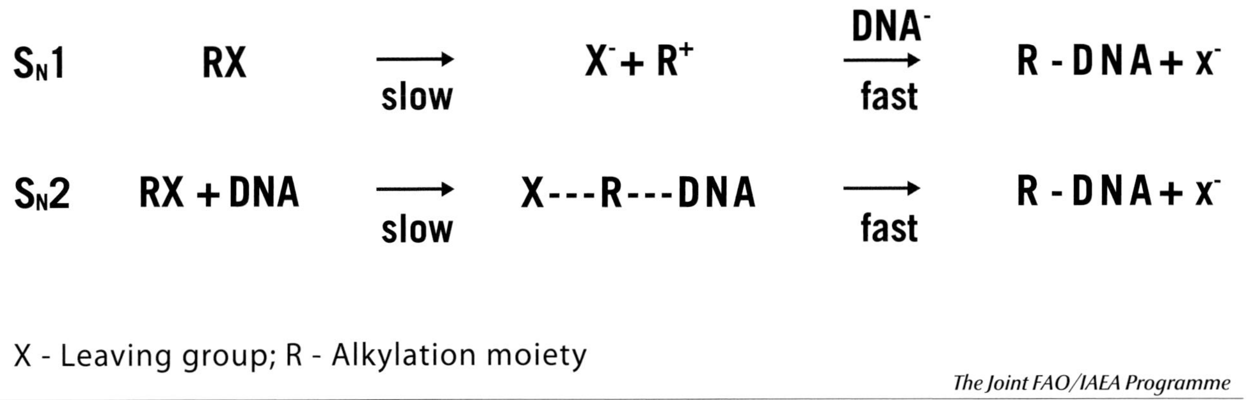

S_N1 $RX \xrightarrow{\text{slow}} X^- + R^+ \xrightarrow[\text{fast}]{DNA^-} R\text{-}DNA + x^-$

S_N2 $RX + DNA \xrightarrow{\text{slow}} X\text{-}{-}{-}R\text{-}{-}{-}DNA \xrightarrow{\text{fast}} R\text{-}DNA + x^-$

X - Leaving group; R - Alkylation moiety

The Joint FAO/IAEA Programme

Figure 12.3 The S_N1 and S_N2 mechanisms of alkylation.

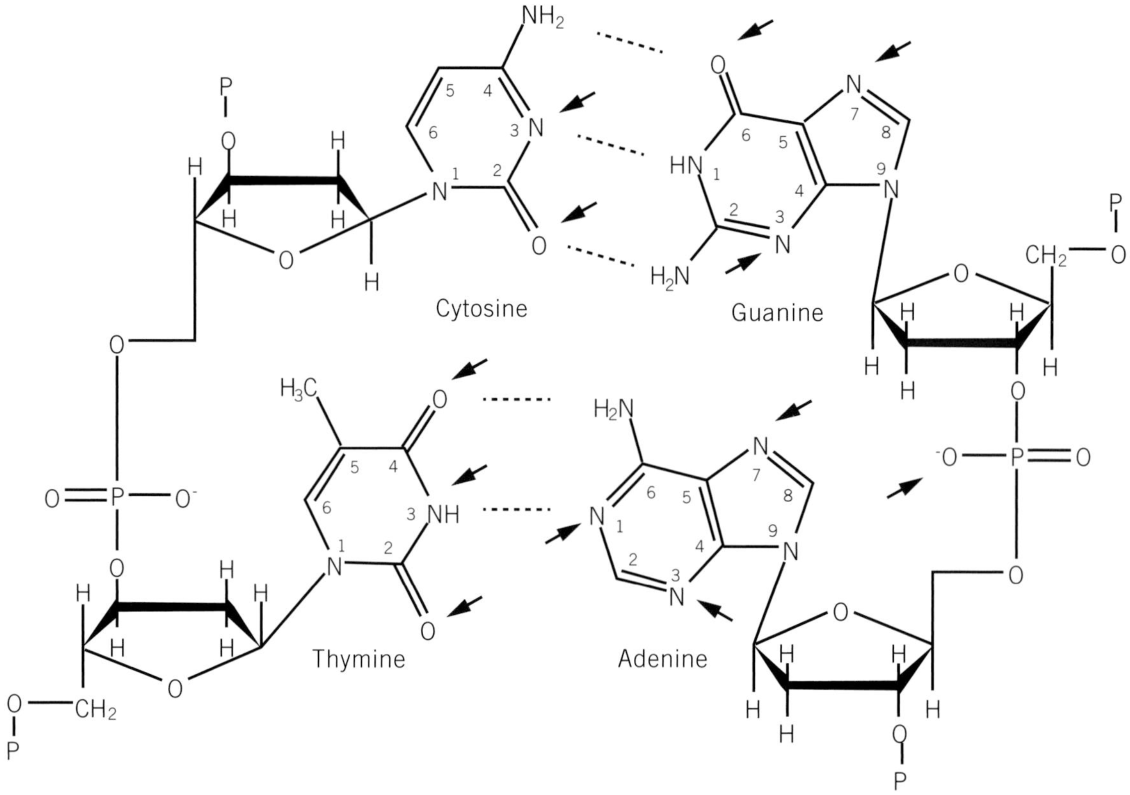

The Joint FAO/IAEA Programme

Figure 12.4 Most frequently alkylated sites in DNA.

information regarding their biological activity, chemical structure, and other properties, can be retrieved from the Pubchem database (http://pubchem.ncbi.nlm. nih.gov). The Substance Identity and the Compound Identity codes for some alkylating agents are given in **Table 12.2**. Both codes can be used interchangeably after accessing the database provided that the respective links "Substance" or "Compound" are previously selected.

Many other databases are available, inter-linked and inter-active. For example, the Toxicology Data Network (http://toxnet. nlm.nih.gov) links over a dozen different databases (including Pubchem). Broad and regularly updated and peer-reviewed information regarding multiple aspects of the compounds, from physical and chemical properties to biological activity and risk, handling, clean-up, disposal, etc., can be found in these databases. **Box 12.2** shows a partial example of the information that can be retrieved from Toxnet – in this case showing information regarding the inactivation and disposal of solid waste and mutagenic solutions after 1-methyl-1-nitrosourea treatments. Some general rules for inactivation of alkylating compounds are provided simultaneously.

2.3. Major Alkylation Adducts and their Repair

Eleven sites in the four bases and the phosphodiester groups constitute the 12 most common targets for the alkylating agents in DNA (**Figure 12.4**). Nevertheless, additional minor sites can be identified, such as the exocyclic 2-amino group of guanine in which alkylation by the bi-functional agent mitomycin C was recently determined.

N^7alkylguanine – the most nucleophilic site in DNA, the N^7 position of the guanine is the primary alkylated site in DNA. Although this represents the bulk of the DNA alkylation by most alkylating compounds, this altered base is, apparently, non-mutagenic.

O^6alkylguanine – a major characteristic product of the S_N1-type alkylating agents O^6-alkylguanine is strongly mutagenic since it mispairs with thymine and gives rise to G:C – A:T transitions. The repair of this lesion is particularly crucial in humans, given that besides the mutagenic implications there is a very well established strong correlation between the formation of this adduct and carcinogenesis. This DNA lesion is repaired in many prokaryotes and eukaryotes through a direct

Table 12.2: Commonly used alkylating agents in plant mutagenesis and mutation breeding

Compound IUPAC name	Acronym(s)	Substance[a] identity (SID)	Compound[b] identity (CID)
1-methyl-1-nitrosourea	NMU (MNU)	24897498	12699
1-ethyl-1-nitrosourea	NEU (ENU)	24897681	12967
methyl methanesulphonate	MMS	49815676	4156
ethyl methanesulphonate	EMS	24896575	6113
dimethyl sulphate	DMS	24893559	6497
diethyl sulphate	DES	24859256	6163
1-methyl-2-nitro-1-nitrosoguanidine	MNNG	49855726	9562060
1-ethyl-2-nitro-1-nitrosoguanidine	ENNG	77654587	5359974
N,N-dimethylnitrous amide	NDMA	24897656	6124
N,N-diethylnitrous amide	NDEA	49855498	5921

[a, b] Use the SID or the CID code to find the mutagen in the Pubchem database, respectively under Substance or Compound, to retrieve information regarding chemical structure, properties, safety, and other information about the compounds.

reversal and suicidal mechanism by O^6alkylguanine-DNA-alkyltransferases, which transfer the alkyl group from DNA to a cystein of the active site, which results in the inactivation of the repair protein.

The recent sequencing of the *Arabidopsis* genome revealed the absence of O^6alkylguanine alkyltransferase gene homologues in this plant species. The absence of similar gene homologues in other plants may explain the negative experimental results obtained regarding this repair activity in some plants (and in *Chlamydomonas*). Additional and alternative mechanisms for the repair of this adduct have been recently identified (**Figure 12.5**). One of the mechanisms implicates the product of alkyltransferase like (ATL) genes in the repair of this lesion *via* nucleotide excision repair (NER).

O^4*alkylthymine* – usually a minor lesion, even for SN1 compounds, this altered base can mispair with guanine and lead to A:T-G:C transitions. Directly repaired by *Ogt* and *Ada* (two different O^6-alkylguanine-DNA-alkyltansferases) gene products in bacteria this lesion is very inefficiently repaired by the mammalian O^6-alkylguanine-DNA-alkyltansferase protein (known as MGMT or AGT). It should be noted that 1-ethyl-1-nitrosourea tends to induce more A:T-G:C transitions than the related compound 1-methyl-1-nitrosourea (**Table 12.3**), which is in accordance to the higher alkylation rate of the O^4 position of the thymine by the ethylating (vs. methylating) compounds (**Table 12.1**).

N^3*alkyladenine* – although induced by all alkylating agents this altered base is a major product of the S_N2-

Table 12.3: Mutation spectrum of the alkylating agents at the molecular level

Compound	G:C – A:T (%)	A:T – G:C (%)	Transvertions (%)	Other
MNU	~100	-	-	-
ENU	72	21	6	-
MMS	20	14	66	-
EMS	93	1	2	4
NDMA	90	< 5	-	-
NDEA	60	21	9	10
MNNG	98	< 2	-	-
ENNG	95	5	-	-
DMS	74	3	20	3

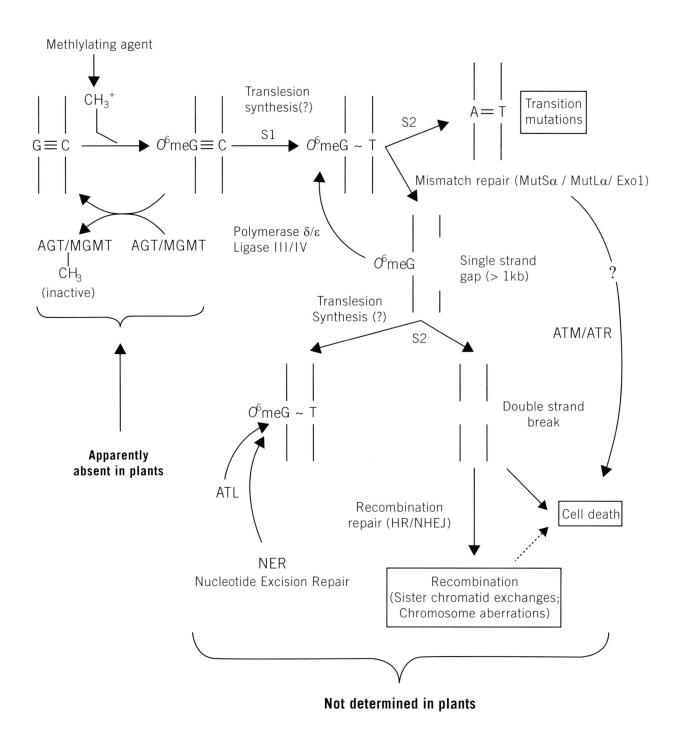

The Joint FAO/IAEA Programme

Figure 12.5 During DNA replication, thymine can be incorporated opposite O^6-alkylG if this lesion is not repaired by the direct reversal mechanism *via* MGMT/AGT proteins. The O^6-methylG:T mispair can be recognized by the mismatch repair (MMR) machinery, but the excised thymine is replaced again by thymine resulting in a "futile" repair cycle which can lead, on a further round of DNA replication, to the induction of DNA DSBs. DSBs can be repaired by the HR or non-homologous end joining (NHEJ) pathways, but can also result in clastogeny, SCE formation and/or apoptosis. Nucleotide excision repair (NER), a possible mechanism for repair of larger adducts, can, in some organisms, be recruited by alkyltransferase like (ATL) proteins to repair O^6-methylG, possibly even before S1. S1 and S2: first and second rounds of DNA replication after alkylation. Adapted from: Margison *et al.* (2002) Mutagenesis. 17: 483–487.

type compounds. This lesion and that of N³alkylguanine is repaired by the base excision repair (BER) mechanism, which initiates the removal of the altered base by a glycosylase activity that generates an abasic site in the double helix. Methyl methanesulphonate (MMS), an inducer of these two lesions, has been used for the study of this DNA repair mechanism in the various biological systems.

N¹alkyladenine and N³alkylcytosine – these two minor cytotoxic lesions are also preferentially induced by S_N2 agents, which could explain the increased frequency of transversion mutations after MMS treatments. These two adducts were recently shown to be repaired by a direct reversal mechanism involving oxidative DNA demethylases (Alk-B in E. coli). In contrast to the O⁶alkylguanine-DNA-alkyltransferase, homologues to the Alk-B gene have been already identified in Arabidopsis thaliana.

Alkylphosphotriesters – the alkylation of phosphodiesters in DNA results in alkylphosphotriesters, which in E. coli are repaired by the Ada-protein while in mammals they are assumed to be very poorly or even not repaired by the alkyltransferase protein (MGMT/AGT). It is noteworthy that for the S_N1 compounds MNU and ENU the alkylphosphotriesters account for approximately 20% and 70% of all alkylation adducts, respectively, and that there is an almost negligible amount of this lesion induced by the S_N2 compound MMS (**Table 12.1**).

The relative frequency of induction of a specific pre-mutagenic lesion in DNA depends mostly on the properties of the mutagen, e.g. S_N1 vs. S_N2 compounds or methylating vs. ethylating agents. The ultimate mutagenic effect of a chemical agent, however, depends on the lesions initially induced in the DNA, the lesions that remain unrepaired and the mutagenic effect of the repair mechanisms themselves.

When not repaired, the pre-mutagenic adducts tend to give rise to fixed mutations. The alkylated O⁶guanine tends to produce G:C to A:T transitions. The alkylated N³adenine gives rise to A:T to T:A transversions, while the non-repaired alkylated N³cytosine can result in C:G to T:A transitions and C:G to A:T and C:G to G:C transversions.

The elucidation of the fate of the alkylation products in plants constitutes one of the main challenges for plant experimental mutagenesis. The clarification of these processes will allow a better understanding of the mutational processes that determine success in plant mutation breeding.

3. Mutagenesis of Alkylating Agents in Plants

3.1. Mutagenicity

3.1.1. DNA Breakage and Clastogenicity

Despite the relative absence of experimental data on the fate of the formed DNA adducts, the mutagenicity and the clastogenicity of alkylating agents in plants has been documented for more than 40 years. The study of the clastogenic effects of alkylating agents and other chemical compounds in plants had, and still has, two main goals: 1) the assessment of several aspects of plant cell biology related to the plant response to chemical compounds, and 2) the study of the biological activity of chemical compounds using plants as biological assay in order to draw conclusions concerning the risk they represent to the environment and human health (e.g. risk of carcinogenicity). The sensitivity and reliability of the plant bioassays have been recognized by various prestigious international organizations such as the World Health Organization, which recommend their use for mutation screening and for detection of genotoxins in the environment.

Among the multitude of plant species used for such purposes, Crepis capillaris, Lycopersicon esculentum, Pisum sativum and Zea mays were the subjects of comprehensive and informative revisions (e.g. Grant and Owens, 2001). From these and other sources it is clear that all major alkylating agents (EMS, ENU, MNU, DES, etc.) show a positive clastogenic effect in all tested plant species. Effects can be seen at the cytological level and recorded as bridges and fragments in anaphase and telophase cells, as translocations, deletions, insertions, inversions, ring chromosomes, etc. in metaphase C cells, and as micro-nuclei in interphase and mitotic cells.

Regarding the clastogeny induced by alkylating agents in plants, it was found that prior treatment of root meristems with low doses of these compounds (or with environmental stress factors as heat-shock or heavy metal salts) induced a protective (reducing) effect against subsequent treatments with higher doses of alkylating agents. By analogy with the bacterial "adaptive response" to alkylating agents, this phenomenon was named "clastogenic adaptation".

The molecular basis of "clastogenic adaptation" in plants was studied in correlation with the formation and removal of O^6alkylguanine (Baranczewski *et al.,* 1997). Nevertheless, the apparent absence of the repair protein O^6alkylguanine-DNA-alkyltransferases in plants requires all former experimental data on "clastogenic adaptation" to be re-thought in the light of this new information. New explanations, including the role of other O^6alkylguanine repair mechanisms, are required.

DNA breakage is assumed to be the phenomenon that underlies the formation of chromosome aberrations and chromosome rearrangements. During the last two decades the assessment of genotoxic and clastogenic effects of chemical compounds in plants has been complemented with the direct assessment of DNA breakage, primarily using the single cell gel electrophoresis (SCGE) technique also known as the "Comet assay" (**see Chapter 11**).

While simultaneously evaluating DNA damage and somatic mutations in leaves of *Nicotiana tabacum,* Gichner *et al.* (1999) ranked the mutagenic potency of the four main alkylating agents as: MNU > ENU ≈ MMS > EMS. With respect to DNA damaging activity the rank order was slightly different: MNU > MMS > ENU > EMS. The reliability of the Comet assay in assessing DNA damage in different plant species is well established and a clear correlation between the extent of DNA damage and the concentration of the mutagen used has been observed in a wide range of species, including sugar beet, alfalfa, tobacco, lentil, maize, potato, durum wheat and bread wheat (Gichner *et al.,* 2003). A correlation between "clastogenic adaptation" expressed as a reduction of chromatid type aberrations, micro-nuclei and aneuploid cells, and the "clastogenic adaptation" expressed as a reduction of damaged DNA assessed by the "comet assay", was observed after the treatment with non-toxic doses of cadmium chloride prior to the challenge treatment with a high dose (5 mM, 2 h) of MNU.

3.1.2. Nature of Induced Mutations
Chemical mutagens in the early studies were termed "radiomimetic" due to the similarity of their effects compared with the radiation effects on biological systems. To the best of our knowledge, an extensive comparative analysis of the total spectra of mutations induced by radiation vs. chemical agents has never been carried out in plants. Nevertheless, in spite of differences in genome organization or cell cycle phase-dependent differences in chromatin packing that could determine differences in the accessibility of chemical compounds

to DNA, the ultimate DNA lesions induced by chemical mutagens: transitions, transversions, deletions, insertions, inversions, DNA single- and double-strand breaks and DNA recombination are similar, though in different proportion, to those induced by radiation. The spectrum of mutations induced by chemical and physical mutagens, in particular the so called "visible", "macro" or phenotypic mutations (chlorophyll, morphological, physiological) is expected to be similar.

Some mutations such as those conferring increased (plastome encoded) herbicide or antibiotic resistance seem to be more frequent in chemical (alkylating compounds) mutagenesis. However, the identification of these mutations may be simply the result of the applied selection procedures.

A major difference between chemical and physical mutagenesis is the possibility of achieving higher mutations rates with minimal effects on survival and fertility of M_1 plants by chemical mutagens, a situation that does not seem possible with acute treatments of physical mutagens.

3.2. Plant Materials and Methods of Treatment

3.2.1. Types of Plant Material
Chemical mutagenesis can be performed with all types of plant materials, from whole plants (usually seedlings) to *in vitro* cultured cells. Nevertheless, the most commonly used plant material is seed. Multiple forms of plant propagules such as bulbs, tubers, corms and rhizomes and explants used for plant vegetative propagation such as vegetative cuttings, scions, or *in vitro* cultured tissues as leaf and stem explants, anthers, *calli,* cell cultures, microspores, ovules, protoplasts, etc., are also used. Gametes, usually inside the inflorescences, are also targeted to mutagenic treatments (immersion of spikes, tassels, etc. into mutagenic solutions, **see Chapter 14**).

3.2.2. Types of Treatment
Mutagenic treatments are usually performed on plant tissues and result in multiple different mutations induced in a large number of cells, most of which normally do not develop into new plants. *In vitro* culture methods provide an exception in the sense that, for species where regeneration protocols are established, any mutated cell has the potential to regenerate into a new plant and thereby transmit its mutations to the next, sexual or vegetative, generation.

Box 12.2: Precautions for (and inactivation of) alkylating, arylating and acylating compounds

Precautions for "Carcinogens": Carcinogens that are alkylating, arylating or acylating agents per se can be destroyed by reaction with appropriate nucleophiles, such as water, hydroxyl ions, ammonia, thiols and thiosulphate. The reactivity of various alkylating agents varies greatly … and is influenced by solubility of the agent in the reaction medium. To facilitate the complete reaction, it is suggested that the agents be dissolved in ethanol or similar solvents. … No method should be applied … until it has been thoroughly tested for its effectiveness and safety on material to be inactivated. For example, in case of destruction of alkylating agents, it is possible to detect residual compounds by reaction with 4(4-nitrobenzyl)-pyridine.

1. Oxidation by potassium permanganate in sulphuric acid ($KMnO_4$ in H_2SO_4). The products of the reaction have not been determined. Degradation efficiency was >99.5%.

2. Reaction with sulphamic acid in hydrochloric acid solution (HCl). The strong hydrochloric acid causes displacement of the nitroso group. The nitrosyl chloride formed reacts with the sulphamic acid to form nitrogen and H_2SO_4. This reaction prevents any reformation of the nitrosamide. The products of the reaction are the corresponding amides produced by simple removal of the nitroso group. Degradation efficiency was >99.5%.

3. Reaction with iron filings in HCl solution. The strong HCl causes displacement of the nitroso group. The nitrosyl chloride formed is reduced by the iron filings in the acid to ammonia. This reaction prevents any reformation of the nitrosamide. The products of the reaction are the corresponding amides produced by simple removal of the nitroso group except for N-methyl-N'-nitro-N-nitrosoguanidine and N-ethyl-N'-nitro-N-nitrosoguanidine, where reductive removal of the nitro group causes the major products to be methylguanidine and ethylguanidine, respectively. Degradation efficiency was >99%.

4. Reaction with sodium bicarbonate solution ($NaHCO_3$). This weak base causes a slow, base-mediated decomposition. The rate of reaction is sufficiently slow so that any diazoalkanes that are formed react with the solvent before escaping from the solution. The products of the reaction have not been definitely identified. Degradation efficiency was >99.99% for N-methyl-N-nitrosourea, N-ethyl-N-nitrosourea, N-methyl-N-nitrosourethane and N-ethyl-N-nitrosourethane. The method is not suitable for N-methyl-N'-nitro-N-nitrosoguanidine, N-ethyl-N'-nitro-N-nitrosoguanidine, or N-methyl-N-nitroso-p-toluenesulphonamide.

5. Reaction with $NaHCO_3$ solution, then nickel-aluminum (Ni-Al) alloy and sodium carbonate (Na_2CO_3) solution, then potassium hydroxide (KOH) solution. The slow increase in pH of the solution produced by sequential addition of the bases causes a slow degradation of the nitrosamide. The degradation rate is sufficiently slow so that any diazoalkanes that are formed have time to react with the solvent before escaping from the solution. Degradation efficiency is >99.9%.

References

Montesano, R., Bartsch, H., Boyland, E., Della Porta, G., Fishbein, L., Griesemer, R.A., Swan, A.B., Tomatis, L. and Davis W. (eds.). 1979. Handling Chemical Carcinogens in the Laboratory: Problems of Safety. IARC Scientific Publications No. 33. Lyon, France: International Agency for Research on Cancer, p.17.
Nitrosamides. Lunn, G. and Sansone, E.B. 1994. Destruction of Hazardous Chemicals in the Laboratory. New York, NY: John Wiley & Sons, Inc., p. 279.

Table 12.4: Half-life of some alkylating agents in water solution at different temperatures and pH [Unit: $T_{0.5}$(h)]

Mutagens	Temp (°C)	pH6	pH7	pH8
MNU	20 (22)	24	(2.3)	
ENU	20(22)	31	(2.4)	
EMS	20 (25)		93.2 (7.8)	
MNNG	22			2.5
ENNG	22			7.0

One of the practical issues that concerns researchers and breeders is the optimization of the mutagenic treatments of the generative cells that will transmit mutations to the next (M_2) generation *via* sexual reproduction while lowering the injury of other tissues which will decrease plant survival or result in higher sterility in the M_1 generation.

During chemical mutagenic treatments, time (usually a few hours) is needed for the mutagen to reach the apical and/or axillary meristems in seed embryos, propagules, buds, etc., which are protected, respectively, by the seed testa and cotyledons and/or by primordial and adult leaves. In addition, during the mutagenic treatments the chemical mutagens undergo spontaneous degradation (**Table 12.4**) and part of the reactive chemical species will be lost in reactions (e.g. alkylation reactions) with the contents of tissues and

Table 12.5: Examples of mutagenic treatments with the most commonly used alkylating agents

Alkylating agent	Plant species	Plant material	Concentration (mM)	Exposure	References
MNU (MW 103.08)	Begonia	Leaf explants (in vitro)	0.2–10	1 h	Bouman and De Klerk (2001) Theor Appl Genet, 102: 111–117
	Lathyrus sativus	Pre-soaked seeds (12 h)	0.5–1.4	3 h	Rybinski (2003) Lathyrus Lathyrism Newsletter, 3: 27–31
	Lens culinaris	Seeds	0.49–3.88	6 h	Sharma and Sharma (1986) Theor Appl Genet, 71: 820–825
	Nicotiana glauca	Pre-soaked seeds (16 h)	1	2 h	Marcotrigiano and Hackett (2000) Annals of Botany 86: 293–298
	Oryza sativa	Panicles	1	45 min	Suzuki et al. (2008) Mol Genet Genomics, 279: 213–223
	Pisum sativum	Seedlings	1	1–4 h	Pereira and Leitão (2010) Euphytica 171: 345–354
ENU (MW 117.11)	Brassica oleracea				
	var. botrytis	Pieces of curd (in vitro)	0.3	Days	Deane et al. (1995) Euphytica, 85: 329–334
	Nicotiana plumbaginilfolia	Protoplasts	0.1	Days	Rey et al. (1990) Plant Cell Reports, 9 (5): 241–244
	Phaseolus vulgaris	Seeds	1.5–6.2	8 h	Svetleva (2004) J. Central European Agriculture, 5 (2): 85–90
	Pisum sativum	Seedlings	5	1–4 h	Pereira and Leitão (2010) Euphytica 171: 345–354
	Zea mays	Callus	1–30	4–8 h	Moustafa et al. (1989) Plant Cell and Tissue and Organ Culture, 17: 121–132
EMS (MW: 124.16)	Glycine max	Embryogenic cultures	1–30	4 h	Hofmann et al. (2004) Biologia Plantarum, 48 (2): 173–177
	Glycine max	Seeds	18	24 h	Wilcox et al. (2000) Crop Sci. 40: 1601–1605
	Helianthus annus	Pre-soaked seeds (4h)	80	5–12h	Nehnevajova et al. (2007) International Journal of Phytoremediation, 9: 149–165
	Oryza sativa	Panicles	94.2	(injected)	Lee et al. (2003) Genetics and Genomics, 22: 218–223
	Phaseolus vulgaris	Seeds	6.2–25	8 h	Svetleva (2004) J. Central European Agriculture, 5 (2): 85–90
	Trigonella foenum-graecum	Pre-soaked seeds (4h)	10–300	2–24 h	Saikat et al. (2008) Euphytica, 160: 249–258

cells other than the target meristematic cells, which will be reached some hours later by a depleted mutagenic solution.

To overcome this problem, the renewal – at least partial – of the mutagenic solution during the treatments has been suggested, which implies the repeated manipulation of the mutagen and the need for additional mutagenic solution. Another option is to pre-soak seeds (or other type of plant material) in water or buffer for some hours prior to mutagenic treatment. This allows the mutagen to diffuse more rapidly to the tissues of interest (meristems). Depending on the species and the experimental design diverse times of pre-soaking, from shorter periods of 4–5 hours to longer periods of 12–16 hours, may be used (**Table 12.5**). A third option is to pre-germinate seeds and to treat seedlings (**Table 12.5**). This procedure facilitates the access of the mutagens to the apical and axillary meristems and reduces the time of exposure to 1–2 hours. This has advantages in providing almost full survival and fertility among the M_1 plants and very high mutation rates in the M_2 generation. A range of other tissues can be pre-soaked or pre-germinated for use *in vivo* and *in vitro* mutagenesis (bulbs, corms, tubers, etc.).

In certain specific circumstances, other types of treatment may be used in chemical mutagenesis, including dipping inflorescences such as spikes or tassels in mutagenic solutions, or injection of mutagenic solutions in stems or culms, etc. There are also some reports of *in vitro* treatments, in which plant tissues are cultivated in the continuous presence of low concentrations of mutagenic agents (**Table 12.5**).

3.2.3. Concentration of Mutagens, pH and Exposure Time

It is advisable that the concentrations of the mutagens to be used is established and reported in molar units. Molar units refer to the number of molecules of the mutagen in the solution and facilitates the comparison of the biological effects of different mutagenic agents on an equimolar basis. In many earlier works mutagen concentrations are reported as percentage units, (v/v) or (w/v).

Some general conclusions can be reached by comparing the most used alkylating agents EMS, ENU and MNU. In acute treatments, EMS is frequently used at ranges of 10–100s milli-molar concentrations (e.g. 10–100 mM and over); ENU is generally used at concentrations an order of magnitude lower (e.g. about 5–6 mM) while MNU is usually used at much lower concentrations (e.g. 0.2–1.0 mM). The most important limiting factor regarding the mutagenic concentration is the toxicity of the compound, which rapidly increases with concentration and shows a clear negative effect on the survival and fertility of the M_1 progeny.

When used as a solid substance, the alkylating agents need first to be dissolved in appropriate solvents such as ethanol or DMSO and then added to buffers or water. The pH of the buffers is usually neutral or slightly acid (pH 6–7), this helps to minimize degradation of the mutagens (**Box 12.2**). Commonly used buffers include the phosphate buffers, in particular the Sorensen's phosphate buffer, at pH 6.8–7.0. Distilled water is slightly acidic and is also commonly used.

Exposure time varies substantially depending on the type of plant material and concentration, but usually ranges from 1 to 6–12 hours.

3.3. Handling M_1 and M_2 Generations

As for physical mutagenic agents, the level of injuries induced by the alkylating agents in M_1 plants can be assessed as germination and/or field emergence rates and parameters such as survival–percentage of plants that attain the adult phase; sterility–percentage of adult plants that do not produce progeny; plant morphoses, chlorophyll chimeras, plant height, number of inflorescences, number of fruits, fertile branches, fertile nodes, number of seeds, etc. (**see Chapter 14**). Several parameters can be conceptualized and analysed in different ways, for example: sterility can be assessed establishing multiple levels of expression from less sterile plants to totally sterile plants.

Much work has been dedicated to correlating injuries to the M_1 plants with mutagenic rates in the M_2 progeny. This has been done in order to establish predictive rules in generating the required mutant variation in the subsequent M_2 generation. Lethal dose thresholds have also been investigated. It became commonly accepted that doses inducing 25 to 50% lethality (LD_{25}–LD_{50}) among the M_1 plants will result in the highest mutations rates. Most of these parameters were developed from treatments of dry seeds which required relatively long treatment times, and this resulted in relatively high lethal injury to embryonic tissues. However, the use of protocols that minimize the injury to M_1 plants (e.g. pre-soaking seeds or carrying out short mutagenic treatments on germinated tissues) allow very high mutation rates in the M_2 generation to be reached while significantly reducing the lethality among the M_1 plants.

The normal, general mutation breeding methods apply to chemical mutagenesis. The pedigree method where M_1 plants are individually harvested and individually sown as M_2 families and the bulk method where all M_1 plants are bulk harvested and the bulked M_2 progeny is sown are commonly used. These two main breeding methods contrast in the way mutation rates are calculated: 1) Pedigree method – as percentage of M_2 families with mutations and 2) bulk method – as percentage of mutated M_2 plants. The first method however, permits a better correlation to be established between the M_1 and M_2 generations, as well as providing a better comparison between biological effects of different mutagens.

In a broad sense, and often in practise, an M_2 family is taken to be the progeny of an M_1 plant. However, this concept can be further constrained and an M_2 family can be formed from the progeny of a plant branch, individual spike, etc.

4. Other Chemical Mutagens

Paraphrasing the title of an article published by Ferguson and Denny (1995) it can be said that multiple groups of chemical compounds can be classed as an "underutilized resource" in plant chemical mutagenesis. Some of the promising compounds are discussed below.

4.1. Nitrous Acid and Nitric Oxide

4.1.1. Properties

The mutagenic effect of nitrous acid (HNO_2) in virus, bacteria, fungi and yeast was documented more than five decades ago. More recently nitric oxide (NO) was found to exert, through similar pathways, similar effects on biological systems. Both compounds induce two major types of mutations: transitions and DNA inter-strand cross-links.

Although hydrolytic deamination of the bases in DNA occurs spontaneously, nitrous acid and nitric oxide increase deamination rate, in particular of guanine. Both compounds induce nitrosamine deamination of adenine to hypoxanthine (Hx), of cytosine to uracil and of guanine to xanthine (Xan) and oxanine (Oxa) at a molar ratio of 3:1 (**Figure 12.6**).

Hypoxanthine pairs with cytosine lead to AT→GC transitions, while uracil pairs with adenine induce CG→TA transitions. Xanthine and oxanine can pair with cytosine or with thymine, in the latter case leading to GC→AT transitions.

Inter-strand cross-links are formed preferentially in 5'CG sequences where the two guanines on opposite strands remain covalently linked through a shared exocyclic amino group (**Figure 12.7**).

Uracil and hypoxanthine are repaired by the BER pathways, whereas an alternative repair pathway is deployed for hypoxanthine involving endonuclease V (recently identified in *E. coli*). Oxanthine seems to be repaired less efficiently than xanthine, *via* BER or NER pathways. The fate of the nitrous acid and nitric oxide-induced DNA inter-strand cross-links as well as the consequences on the ability of oxanine to form cross-link adducts with aminoacids, polyamines and proteins are currently very poorly understood.

In plants, nitric oxide is assumed to play an important role in plant growth and development and to be a ubiquitous signal involved in the induction of cell death and of defence genes, and in the interaction with reactive oxygen species during defence against pathogens or in response to the plant hormone ABA. Bearing in mind that nitric oxide over-production in chronically inflamed tissues has been implicated in carcinogenesis and that NO releasing compounds are potent growth inhibitors of cancer cell lines and prevent colon and pancreatic cancer in animal models, the study of nitrous acid and nitric oxide effects in biological systems is expected to expand.

4.1.2. Nitrous Acid and Nitric Oxide in Plant Mutagenesis

Although, nitrous acid mutagenesis is used for genetic improvement of virus, fungi and bacteria for multiple biotechnological purposes, its exploitation in plant genetic improvement has been lacking. Since nitrous acid is quite unstable, the mutagenic treatments with nitrous acid need to be quick, e.g. not longer than a few minutes, which makes the treatment of dry seeds ineffective. Young seedlings and especially *in vitro* cultured plant cells, tissues and other vegetative explants offer more potential as targets for nitrous acid mutagenesis. The low pH at which nitrous acid needs to be maintained is an additional stress factor that needs to be considered in any nitrous acid treatment of plant materials.

Protocols for nitrous acid (HNO_2) mutagenic treatments: nitrous acid needs to be prepared fresh at low temperatures (0°C) before each treatment. Usually potassium or sodium nitrites are solubilized in acetate buffer at pH 4.0–5.5. Alternatively, identical volumes of solutions with equi-molar concentrations of sulphuric acid and barium nitrite are mixed and the barium sulphate removed by centrifugation. This last solution has a very low pH (1.5). Mutagenic treatments are usually performed with 0.02 to 0.1 M solutions.

4.2. Base Analogues and Related Compounds

4.2.1. Types and Effects

Under specific conditions some base analogues (**Figure 12.8**), and their ribosides and deoxyribosides, can be more mutagenic to specific organisms than the alkylating agents. The analogue of thymine 5-bromo-uracil

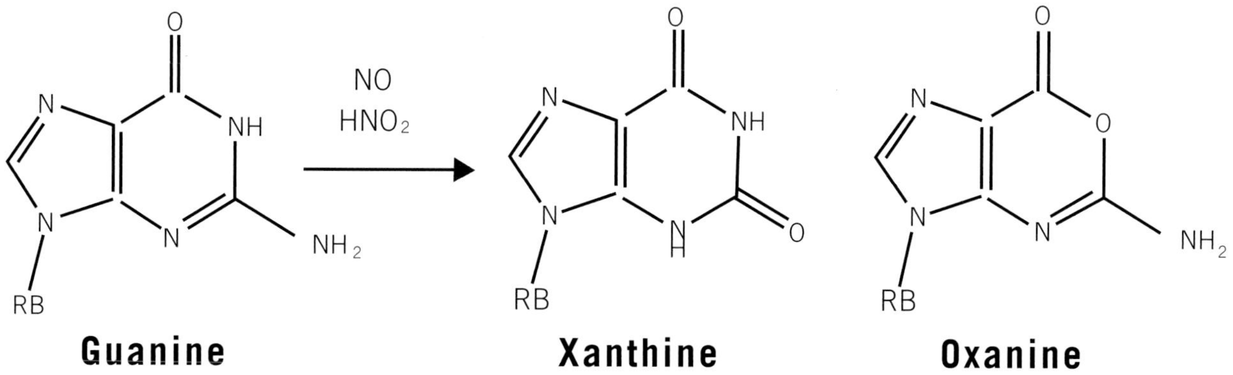

Figure 12.6 Nitrous acid and nitric oxide induce the nitrosamine deamination of guanine to xanthine and oxanine.

The Joint FAO/IAEA Programme

(BU) – and its deoxyriboside 5-bromo-2'-deoxyuridine (BUdR) – can incorporate into DNA and induce GC→AT and AT→GC transitions, and AT→TA and CG→AT transversions, as well as small indels that cause frameshift mutations, both *in vivo* and *in vitro*.

The ability of BUdR to incorporate into newly synthesized DNA strands is exploited in sister chromatid exchange analyses, and the anti-proliferative and radiosensitizing properties of this compound are being tested for anti-neoplastic treatments.

Other base analogues such as 2-aminopurine (2AP), 2,6-diaminopurine (2,6DAP), 6-N-hydroxylaminopurine (HAP) and 2-amino-6N-hydroxylaminopurine (AHAP) can function as analogues of adenine or guanine and can be incorporated into DNA where they induce high rates of C:G to T:A and T:A to C:G transitions and frameshift mutations. 2AP and 2,6DAP are very effective mutagens to phages and bacteria but weak mutagens to eukaryotic cells and weak carcinogens in mammalian systems. In contrast, HAP and AHAP are strongly mutagenic to eukaryotic cells and relatively carcinogenic. The clastogenic effect of base analogues in plants, including related compounds such as the alkylated oxypurines: 8-ethoxycaffeine and 1,3,7,9-tetramethyluric acid, was established over three decades ago almost simultaneously with the determination of the ability of some of these compounds to induce visible mutations in plants. Nevertheless, base analogues have not been tested intensively for the induction of mutants of interest in plants, and in this regard, the release of the malt-

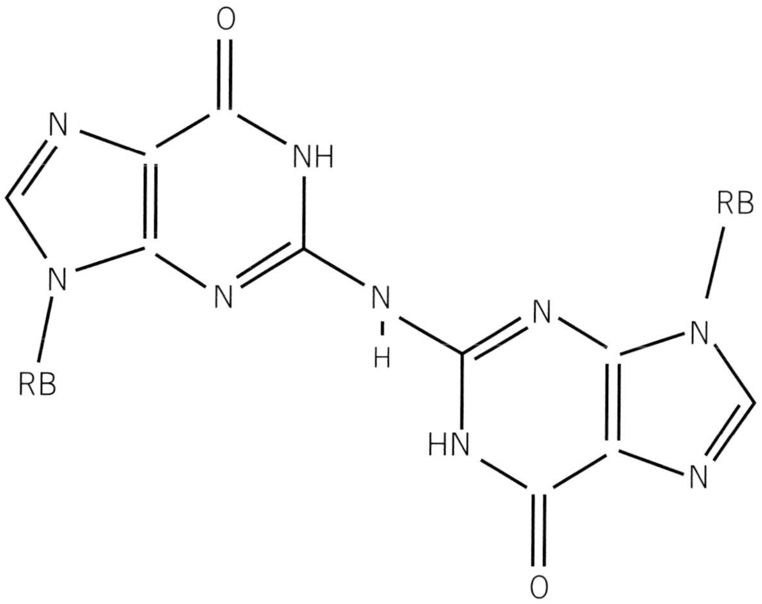

The Joint FAO/IAEA Programme

Figure 12.7 Covalent linkage of two guanines, on opposite strands, induced by nitrous acid and nitric oxide.

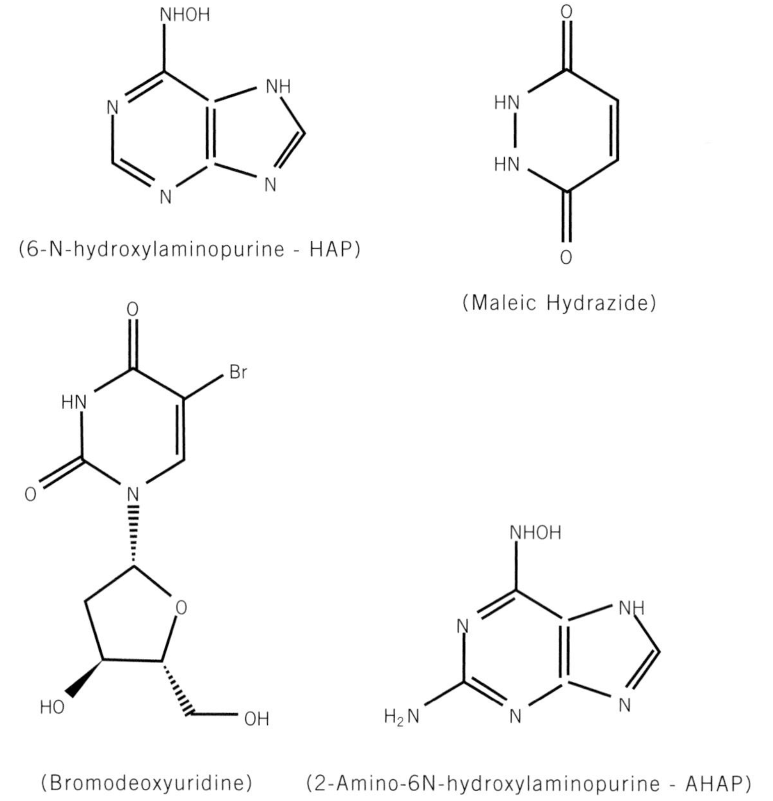
NHOH
NH
N
N
N
(6-N-hydroxylaminopurine - HAP)
O
HN
HN
O
(Maleic Hydrazide)
O
Br
HN
O
N
O
HO
OH
NHOH
NH
N
H_2N
N
N
(Bromodeoxyuridine)
(2-Amino-6N-hydroxylaminopurine - AHAP)

The Joint FAO/IAEA Programme

Figure 12.8 Molecular structures of several base analogues.

ing barley commercial variety "Fuji Nijo II" induced by BUdR + gamma rays treatment in Japan is exceptional.

Maleic hydrazide (1,2-dihydro-3,6-pyridazinedione, MH) is a structural isomer of uracil with plant growth regulator properties and is commonly used as a herbicide and sprout inhibitor. The genotoxic effects of this base analogue, particularly in plants, are rather intriguing and deserve further enlightenment. MH shows low mutagenic activity in bacteria, fungi and animal cells with seemingly no carcinogenic effects. However, this compound exhibits high mutagenic, clastogenic and recombinational activity in plants, frequently stronger than that of the most powerful alkylating agents (Gichner, 2003). In an experiment carried out simultaneously in two different laboratories the results of the comparative assessment of the genotoxicity of MH *versus* methyl methanesulphonate (MMS) in onion (*Allium cepa*) confirmed the much stronger clastogenic effect of MH: 24.0 and 46.4% cells with chromosome aberrations for concentrations of 5 mg/l and 10 mg/l, respectively, *versus* 19.1% aberrant cells for 10mg/l MMS (Rank *et al.*, 2002).

MH effects revealed by comet assay were contrary to the expected, since no DNA damage was observed (Gichner, 2003). However, Juchimiuk *et al.* (2006) also used the comet assay and registered extensive DNA damage both in *N. tabacum* and human leucocytes when MH treatments were performed on previously isolated cell nuclei. The comparison of the contradictory results of these experiments raises the question of the role of *in vivo* DNA repair mechanisms in the mutagenic effect of MH. Intriguing results were also observed when MH was used in combined mutagenic treatments. As expected, a synergistic effect was observed when MH was combined with X-rays in inducting somatic mutations in *Trandescantia* stamen hairs; but, conversely, an antagonistic mutagenic effect was registered when MH was combined with EMS.

4.2.2. Base Analogues and Plant Mutation Breeding

Apart from a very few exceptions, and despite the well documented genotoxic effects of MH in plants, reports on the use of this compound in plant mutation breeding are lacking. The relative low toxicity and carcinogenicity of most of the base analogues, which are relatively inert and non-volatile compounds, are properties expected to encourage the utilization of base analogues in plant mutation breeding. MH and other base analogues have been used as water or buffer (e.g. Tris HCl) solutions at 5 to 10 mg/l.

4.3. Antibiotics

4.3.1. Types and Effects

Antibiotics are defined functionally on the basis of their anti-microbial activity. This group of compounds includes a multitude of natural and synthetically synthesized substances which, according to their molecular structure, fall into very different classes of compounds. Some antibiotics (**Figure 12.9**), such as streptozotocin, mitomycin C or azaserine, can also be included in the group of alkylating compounds.

Streptozotocin (STZ) is a naturally synthesized broad spectrum antibiotic and a potent mutagen and carcinogen used as a diabetogenic and anti-neoplastic agent. DNA-specific sequence analyses showed that over 98% of the STZ-induced mutations were G:C to A:T transitions with a few A:T to G:C transitions. However STZ also produces DNA strand breaks, alkali-labile sites, unscheduled DNA synthesis, DNA adducts, chromosomal aberrations, micro-nuclei, sister chromatid exchanges and cell death. Although the ability of streptozotocin to induce visible mutations in plants, in particular chlorophyll mutations, was reported four decades ago this mutagen has not been taken up in plant breeding experiments.

Mitomycin C (MMC) is an antineoplastic antibiotic isolated from *Streptomyces caespitosus* that inhibits DNA, RNA and protein synthesis and induces apoptosis in mammalian cells, and intra-chromosomal recombination in plant somatic cells. MMC is a bi-functional alkylating agent that reacts with guanine residues to form DNA inter-strand cross-links at the 5-CG-3 sequences and six types of guanine adducts, four of them arising from the direct alkylation of DNA by MMC and two other resulting from the previous formation of 2,7-diaminomitosene (2,7-DAM), which then alkylates DNA (Paloma *et al.,* 2001).

Azaserine is a naturally occurring serine derivative with antineoplastic properties that functions as a purine antagonist and a glutamine amidotransferase inhibitor. Azaserine spontaneously decomposes to diazoacetate which carboxymethylates DNA, forming O^6CmG, N^7CmG and N^3CmA as major products and O^6meG and N^7 and N^3 methylpurines as minor products.

The fluoroquinolone ciprofloxacin is a bacterial gyrase inhibitor that causes DNA DSBs and induces a wide pattern of mutations including different kinds of base pair substitutions and 3 or 6 base pair insertions/deletions that result in frameshift mutations in bacteria. At high concentrations ciprofloxacin inhibits the eukaryotic topoisomerase-II and induces genotoxic effects in mammalian cells in *in vitro* studies, whereas no genotoxic or carcinogenic effects were observed when used in *in vivo* tests with rodents.

Some antibiotics are radiomimetic and induce predominantly SSBs and DSBs in DNA. A major representative group of these antibiotics are the bleomycins, a group of natural glycopeptides produced by *Streptomyces verticillus*. It has potent antitumor activity, which is commonly assigned to its strong induction of DNA breakage (Ferguson and Denny, 1995).

Many other antibiotics, such as fumagillin, amoxicillin and amoxicillin-related antibiotics, such as ampicillin, are also mutagenic to bacteria and mammalian cells, while others such as actinomycin D are potent apoptotic agents.

4.3.2. Antibiotic Mutagenicity Assays in Plants

The assessment of genotoxic effects of some antibiotics, in particular the study of the induced DNA lesions and respective repair pathways, has been carried out in plants, though at a much lower scale compared with the number of similar studies performed in bacteria and mammals.

The clastogenic properties of bleomycin have been determined in faba beans (*Vicia faba*), barley (*Hordeum vulgare*) and *Crepis capillaris* root cells. The formation and repair of bleomycin-induced DNA breaks in faba beans was assessed by neutral elution and by the comet assay. The very fast repair of bleomycin-induced strand breaks was first documented in *Arabidopsis* and more recently confirmed in barley (Georgieva and Stoilov, 2008).

Among the non-alkylating mutagenic compounds antibiotics have been the most extensively used in plant mutation breeding, with particular success in the induction of male sterile mutants in a number of plant species. Twenty-two cytoplasmic male sterile (CMS) and seven nuclear male sterile (NMS) mutants were selected in

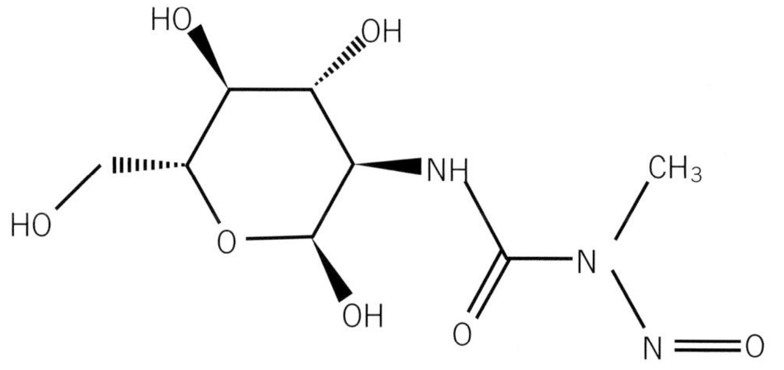

(Streptozotocin) 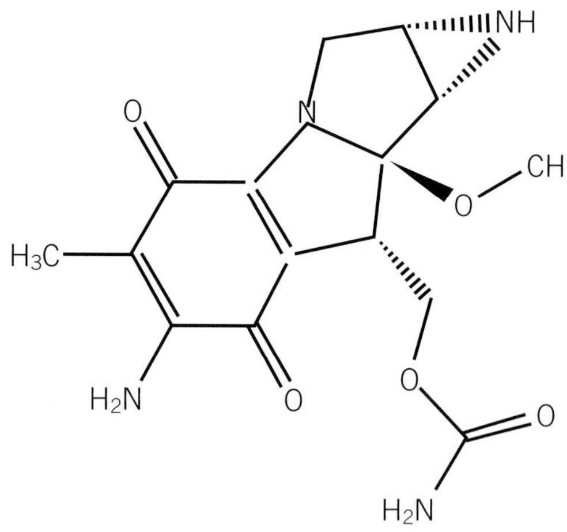   (Mitomycin C)

The Joint FAO/IAEA Programme

Figure 12.9 Molecular structures of two antibiotic/alkylating agents.

sunflower (*Helianthus annus*) after treatments of seeds with mitomycin C and streptomycin. Streptomycin proved to be more effective in the induction of male sterility mutations as 18 of the CMS mutants were induced by this antibiotic. Six of these CMS lines have been released by the USDA-ARS and the North Dakota AES (Jan and Vick, 2006). The efficiency of streptomycin to induce male sterile mutants was also proved in sugar beet, sorghum and pearl millet. In this last species male sterile mutants were also induced with mitomycin C. Streptomycin, penicillin, rifampicin, erythromycin and tetracycline were also tested for inducting male sterile mutants in *Linum usitatissimum*.

Despite the very limited use of antibiotics in plant mutation breeding and the very few published studies on the genotoxicity of these compounds in plants, it is worth mentioning that among the new plant genomic tools stands the publicly available microarray data from the AtGenExpress initiative (http://www.arabidopsis.org/info/expression/ATGenExpress.jsp). This includes the transcriptional response of 16-day old WT (Col) *Arabidopsis thaliana* seedlings to the genotoxic treatment with (1.5μg/ml) bleomycin and (22 μg/ml) mitomycin C for different (up to 24 hours) exposure times. Treatment conditions with antibiotics can vary substantially. Treatments with 50 to 200 μg/ml bleomycin, 5 to 500 μg/ml mitomycin C, 5 to 5000 μg/ml streptomycin and 1–5 mg/ml azaserine have been used.

4.4. Intercalating Agents and Topoisomerase Inhibitors and Poisons

4.4.1. Intercalating Agents

Intercalating agents can reversibly intercalate with double-stranded DNA, but they do not covalently interact with it. Classical intercalating compounds such as acridinium salts, ethidium bromide and propidium iodide are fused-ring aromatic molecules with positive charges on an attached side chain and/or on the ring system itself. Other compounds, such as DAPI, contain unfused aromatic systems with terminal basic functions and are classified as non-classical intercalators.

Initially used as disinfectants and anti-parasitic compounds, acridines and acridine derivatives constitute one of the biggest groups of the intercalating agents. These compounds have light absorbing properties and photo-enhanced cytotoxicity and mutagenicity and are exploited as dyes in biological and biochemical assays (e.g. acridine orange) and in clinics.

The most specific DNA lesion induced by the acridine compounds are frameshift mutations. The exploitation of these particular mutagenic properties allowed Francis Crick and co-authors to demonstrate the triplet nature of the genetic code using the acridine dye proflavine. The genotoxic activity of acridines and related molecules differ from compound to compound. While simpler molecules, such as 9 amino-acridine and quinacrine, induce

±1 frameshift mutations in bacteriophages and bacteria and are weak clastogens and mutagens to mammalian cells, other acridines such as quinacrine, amsacrine and anthracyclines are severe clastogens and carcinogens.

Acridines bearing additional fused aromatic rings (benzacridines) show little activity as frameshift mutagens, but following metabolic activation interact covalently with DNA inducing predominantly base pair substitution mutations (Ferguson and Denny, 2007).

Nitroacridines have been shown to be mutagenic and clastogenic to mammalians. Some of these compounds, such as nitracrine and the 3-nitroacridine Entozon, exhibit a predominant induction of -2 frameshift mutations. Like the nitroacridines, the acridine mustards and aflatoxin B1 can induce frameshift mutations and base pair substitution.

Acridine mustards, in which an alkylating mustard is attached to an intercalating acridine chromophore, can be 100-fold more cytotoxic than the free mustard. Some acridine mustards such as ICR-191 and C20 are known to form adducts to the N^7 position of guanine, but analogues with longer linker chains, such as C5, can form adducts almost exclusively at the N^1 position of adenine.

The recombinogenic and mutagenic activities of some mutagens, e.g. bleomycin, can be enhanced and synergistic effects can be observed with pre-treatments with aminoacridines, nitroacridines or acridine mustards.

4.4.2. Topoisomerase Poisons

Topoisomerases play critical roles in primary DNA processes such as replication and recombination as well as in chromosome segregation, condensation and decondensation. While type I topoisomerases remove super-helical torsions in DNA generating temporary single breaks in one strand, type II topoisomerases alleviate DNA over-winds and resolve DNA knots and tangles by a similar mechanism but generate transient DSBs. In their function, topoisomerases establish the so called cleavage complexes, constituted by transient covalent attachment between the tyrosyl residues of their active site and the terminal DNA phosphates generated during the cleavage reaction. Compounds that stimulate the formation or increase the persistence of these topoisomerase-DNA cleavage complexes are referred to as "topoisomerase poisons". Although there is some evidence for the involvement of topoisomerases

in the induction of frameshift mutations, e.g. a functional T4 topoisomerase is required for 9-aminoacridine mutagenesis in T4 bacteriophage, the strong clastogenicity of several intercalating agents has been ascribed to their topoisomerase II poisoning properties.

Topoisomerase II poisons are strong cytoxic and/ or clastogenic compounds, since stabilized cleavage complexes can inhibit DNA replication and cause DSBs. Multiple anti-cancer drugs such as etoposide and doxorubicin and antibiotic compounds such as ciprofloxacin and levofloxacin function as topoisomerase II poisons.

A second group of drugs that function by inhibiting the DNA binding and/or the catalytic cycle of topoisomerases are referred to as topoisomerase catalytic inhibitors. They do not induce DSBs and exhibit lower cytoxicity. Examples of eukaryotic (human) catalytic inhibitors are aclarubicin and merbarone; the former prevents the binding of topoisomerase II to DNA and the latter inhibits the DNA strands cleavage activity of this enzyme (e.g. McClendon and Osheroff, 2007).

Compounds that exhibit topoisomerase II poisoning activity can be found either among intercalating or non-intercalating agents. Genistein (a bioflavonoid), quinolones (e.g. CP-115,953), etoposide and teniposide are non-intercalating topoisomerase II poisons, while amsacrine, doxorubicin, mitoxantrone, proflavine and auramine are examples of intercalating topoisomerase II poisons (**Figure 12.10**). Quinones, which act as topoisomerase II poisons in the presence of cleavage complexes and as topoisomerase catalytic inhibitors when previously incubated with the enzyme in the absence of DNA, and benzene are other examples of strong topoisomerase poisons. Topoisomerase II catalytic inhibitors can also be classed as intercalating agents, such as 9-aminoacrine, chloroquine, tacrine and ethidium bromide or non-intercalating as merbarone.

Interestingly, topoisomerase II catalytical inhibitors can antagonize topo II poison activities. The attenuation of the DNA cleavage-enhancing properties of teniposide, amsacrine and etoposide by merbarone, or the inhibition of the etoposide induced micro-nuclei formation by chloroquine, A-74932, 9-aminoacridine and ethidium bromide, are examples of such antagonistic interaction.

Topoisomerase II does not become a genotoxic enzyme uniquely by the direct poisoning effect of chemical compounds. Poisoning effect on topoisomerase II

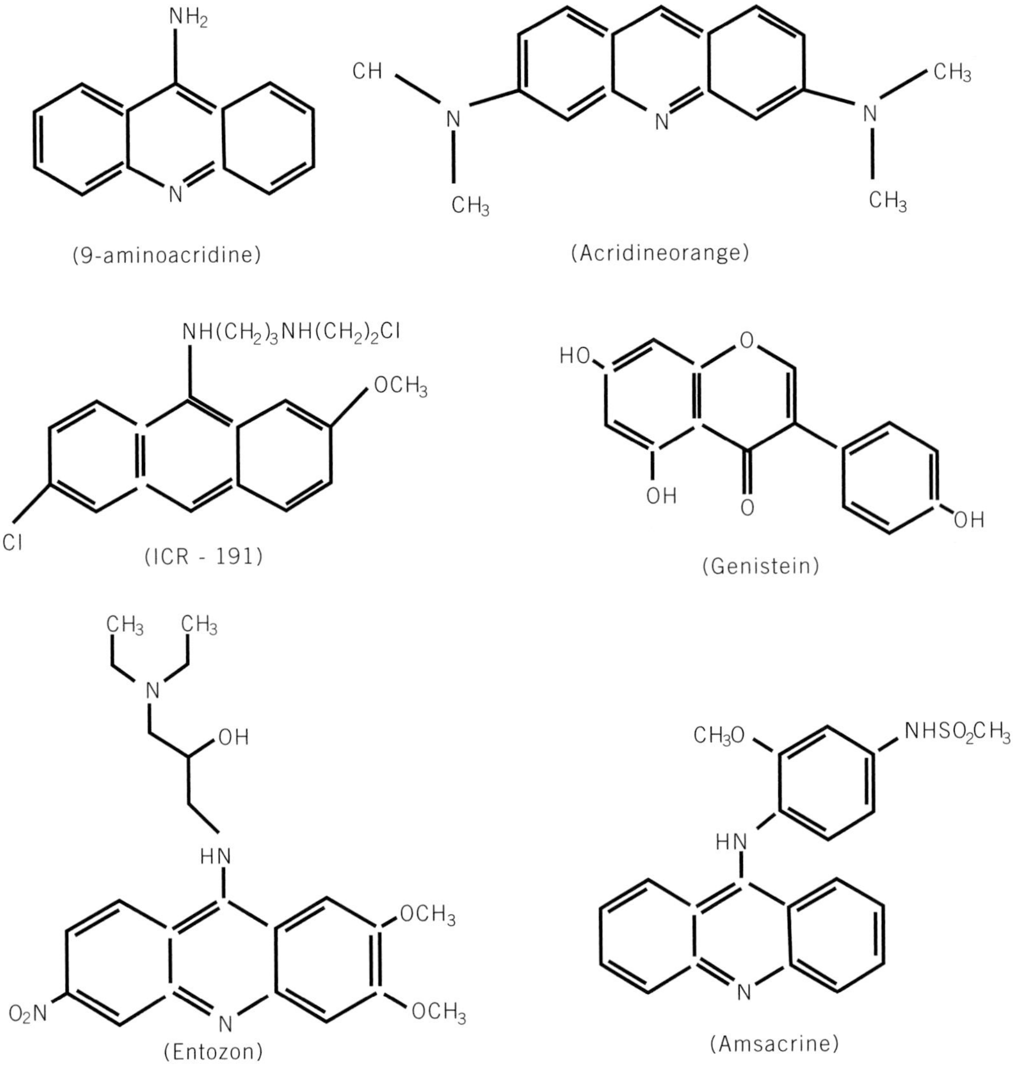

Figure 12.10 Molecular structure of several intercalating agents, topoisomerase catalytic inhibitors and topoisomerase poisons.

can be produced by double helix distortions caused by the presence of abasic sites generated by spontaneous hydrolysis, by DNA damaging agents or by base excision repair pathways (McClendon and Osheroff, 2007). In spite of the well demonstrated mutagenic activity of the intercalating agents and/or topoisomerase II catalytic inhibitors and topoisomerase II poisons in multiple prokaryotic and eukaryotic systems, and despite the well documented clastogenic and cytotoxic effects of acridines and their amino derivatives (Rank *et al.*, 2002), so far these compounds have been rarely tested in plants.

Intercalating agents and/or topoisomerase inhibitors and topoisomerase poisons have been used to induce mutants in bacteria, blue-green algae, *Chlamydomonas*, fish, animal and human cell lines, *Oenothera* chloro-

plasts, etc. However, to the best of our knowledge (except for a report of male sterile mutants induced by acriflavine and ethidium bromide in sugar beet) no other major publication has been produced reporting the use of this kind of compound for plant mutagenesis. The strong ability to induce frameshift mutations, DSBs and other types of mutational events, in some cases mediated by adduct formation, provides compelling reasons for more studies in plant systems. They might also become a new type of mutagen that can induce unique mutations for plant molecular biologists, e.g. for TILLING experiments, and for plant breeders.

Treatments with these mutagens are usually performed with low micro-molar solutions prepared from 5–10 mg/ml stock solutions in dimethylsulphoxide (DMSO), 10 mM HCl, 10mM KOH or in water. Optimal concentrations and exposure times for plant mutation breeding purposes need to be determined. Although plant mutagenic treatments using these compounds can be performed on seeds, it is to be expected that compounds with mutagenic effects associated with topoisomerase poisoning or topoisomerase inhibition activities will be more effective on vegetatively growing tissues, particularly on seedlings and *in vivo* and *in vitro* explants.

5. The Resurgence of Chemical Mutagenesis and Practical Tips

5.1. Resurgence of Chemical Mutagenesis

The mutagenicity of alkylating agents was largely determined by I. Rapoport and co-workers, and readers able to read Russian can find an enormous amount of relatively old literature reporting a panoply of induced mutations in a wide range of biological subjects, including plants. However, the most important sources of literature on plant chemical mutagenesis are the discontinued *Mutation Breeding Newsletter* and its successor, the *Plant Breeding & Genetics Newsletter*, as well as the proceedings of meetings and other literature edited by the IAEA. Another good source of information on chemically induce mutants are the "newsletters" edited by International Societies for specific crops, e.g. pea and barley, as well as web pages and information published by germplasm centres and curators, e.g. John Innes Institute, UK.

In recent years, experimental mutagenesis has re-emerged as an important tool for plant reverse and forward genetics. Large mutagenized plant populations are produced and used to screen both for induced phenotypically expressed mutations and for mutational changes in specific genomic sequences using a procedure commonly called TILLING (targeting induced local lesions in genomes, see **Chapters 21 and 22**). Although physical mutagenic agents such as gamma radiation and fast neutrons and other mutagens such as sodium azide (NaN$_3$) have been used to produce mutant populations for TILLING analyses, the alkylating agents, in particular EMS but also MNU, are the most frequently utilized mutagens for that purpose. As a consequence there has been a resurgence of mainstream articles and references to plant chemical mutagenesis.

5.2. Practical Tips for Plant Mutagenesis Experiments

The guide lines below were developed for working with strong mutagens and carcinogens such as alkylating agents, but may be followed when using mutagens assumed to be less toxic, less mutagenic or less carcinogenic.

1. Alkylating agents are very toxic and very strong mutagens and carcinogens. This should always be kept in mind when handling these substances, health and safety should be the first consideration.

2. Handling chemical mutagens is relatively safe if safety rules are strictly observed. It is worth mentioning that handling chemical mutagens with excessive fear can be dangerous. Fear overcomes rational behaviour and the risk of accidents (usually minor) increases.

3. Lab coats and gloves are absolutely mandatory when handling chemical mutagens. Protective eye covers are also mandatory in all operations not performed behind the protection of a fume hood front glass or other protected cabinet.

4. Vials containing alkylating agents should always be opened in a fume hood or other specialized containment facility due the possibility of toxic gas formation, sometimes under pressure.

5. Gloves do not protect completely from mutagenic solutions and should never be in contact with the compound, either solid or in solution. Gloves should be changed frequently and disposed of in a specific designated bag.

6. The weighing of solid alkylating agents should be done as quickly as possible, but not in a rush. The required amount should be calculated. As soon as a quantity close to the needed amount is reached weighing must be stopped. The final concentration of the solution can be corrected adding the appropriate amount of water or buffer.

7. Weighing can be avoided using pre-weighed mutagens in sealed bottles. Solid mutagens such as MNU or ENU can be acquired in ISOPAC vials to which solvents (buffer, DMSO, ethanol) can be added *via* injection through a resealable cap.

8. Solutions are safer to handle than substances in a powder/solid form. Liquid spills are also much easier to locate than small powder particles. When

weighing an alkylating agent it is advisable to add some millilitres of the solvent (DMSO, ethanol) immediately after weighing out the material – this will prevent further dispersal of the powder and will retain any liberated gas. It is also advisable that any tool (e.g. spatula) that has been in contact with the mutagen is immersed immediately in a solvent.

9. All surfaces, over which mutagens are supposed to be manipulated (space around balances, fume hood, etc.), can be covered with filter paper fixed with tape or other means. At the end of the treatment the paper can be carefully removed trapping any spilled particles or drops and disposed of in a specific designated disposal bag.

10. After the mutagen is dissolved in an adequate amount of buffer or water, the solution can be poured over the plant material to be treated. Plant materials that can cause splashes (seeds, tubers, corms, etc.) should first be placed into the container used for treatment (e.g. a simple glass beaker). Alternatively, the plant material can be immersed in the final volume of water or buffer solution to which the mutagen dissolved in a solvent (DMSO, ethanol) can be added with careful stirring (e.g. with a glass rod). During the mutagenic treatment containers, even in a fume hood, should be covered, e.g. by parafilm.

11. After pre-soaking, if the remaining water or buffer is removed, the amount of liquid taken up by the plant materials should be taken into account if an exact calculation of the final concentration of the mutagen is required.

12. At the end of the treatment period the mutagenic solution should be decanted or pipetted out, or the treated material removed from the mutagenic solution (easier for cuttings, tubers, corms, etc.). In both cases treated plant material should be washed immediately with water to remove the remaining mutagenic solution. After this, three or more changes of water should be used for at least one hour in order to remove as much of the mutagen as possible.

13. Immersion of the plant material in 10% sodium thiosulphate solution for a few minutes can help to inactivate the mutagenic agent.

14. Mutagenized plant material to be carried forward for *in vitro* culture needs to be surface sterilized after the mutagenic treatment.

15. Workers engaged in sowing, planting, grafting or *in vitro* culture of mutagenized material should be fully informed about the material they are manipulating. Careful supervision of workers is required to ensure that they wear appropriate protective clothing and gloves and that mutagenized materials are not manipulated directly by hand (even when wearing gloves). Direct contact with the treated material should be avoided; forceps can be used when sowing and planting and additional protection is necessary when preparing scions to reduce the risk of contact with the mutagen still in the plant material.

16. Chemical mutagenic treatments should always be performed by, or under the directed supervision of a specially trained officer to avoid contaminations and/or non-declared and non-eliminated contaminations.

17. Treatments can be performed in specialized laboratories or in common laboratories. In the last case, access to treatment rooms should be restricted. Formation of toxic gases (e.g. diazoalkanes from nitroso compounds at high pH) and their transportation to other laboratories *via* fume hood conduits should be avoided completely. Is advisable to perform mutagenic treatments out of regular working time.

18. Mutagenic solutions, including waste waters, should be inactivated and glassware and metallic instruments (spatula, forceps, etc.) chemically decontaminated. Plastics, paper, gloves, etc. should be incinerated (**Box 12.2**). Most commonly used alkylating agents can be inactivated by adding sodium thiosulphate solution (10% final concentration) to the mutagenic solution and mutagenic waste liquids, and by slowly increasing the pH using a sodium hydroxide solution or sodium hydroxide pellets (to 1% final concentration) and kept at least overnight to decompose. This will assure the complete inactivation of the mutagens.

A solution of 10% sodium thiosulphate and 1% sodium hydroxide, should be used for decontamination of glassware and disposable labware. Information regarding inactivation and disposal of specific mutagens can be found at http://toxnet.nlm.nih.gov/.

6. References

6.1. Cited References

Baranczewski, P., Nehls, P., Rieger, R. *et al.* 1997. Removal of O^6-methylguanine from plant DNA *in vivo* is accelerated under conditions of clastogenic adaptation. *Environmental and Molecular Mutagenesis.* 29(4): 400–405.

Ferguson, L.R. and Denny, W.A. 1995. Anticancer drugs: an underestimated risk or an underutilized resource in mutagenesis? *Mutat Res-Fund Mol M.* 331: 1–26.

Ferguson, L.R. and Denny, W.A. 2007. Genotoxicity of non-covalent interactions: DNA intercalators. *Mutat Res.* 623: 14–23.

Georgieva, M. and Stoilov, L. 2008. Assessment of DNA strand breaks induced by bleomycin in barley by the comet assay. *Environmental and Molecular Mutagenesis.* 49: 381–387.

Gichner, T., Ptáek, O., Stavreva, D.A. *et al.* 1999. Comparison of DNA damage in plants as measured by single cell gel electrophoresis and somatic leaf mutations induced by monofunctional alkylating agents. *Environmental and Molecular Mutagenesis.* 33: 279–286.

Gichner, T. 2003. Differential genotoxicity of ethyl methanesulphonate, N-ethyl-N-nitrosourea and maleic hydrazide in tobacco seedlings based on data of the Comet assay and two recombination assays. *Mutat Res.* 538(1-2): 171–179.

Gichner, T., Patková, Z. and Kim, J.K. 2003. DNA damage measured by the comet assay in eight agronomic plants. *Biologia Plantarum.* 47 (2): 185–188.

Grant, W.F. and Owens, E.T. 2001. Chromosome aberration assays in Pisum for the study of environmental mutagens. *Mutat Res.* 488: 93–118.

Hodgdon, A.L., Arenaz, P. and Nilan, R.A. 1981. Results of acridine dye treatment of Steptoe barley. *Barley Genetics Newsletter.* 11: 69–71.

Jan, C.C. and Vick, B.A. 2006. Registration of seven cytoplasmic male-sterile and four fertility restoration sunflower germplasms. *Crop Sci.* 46: 1829–1830.

Juchimiuk, J., Gnys, A. and Maluszynska, J. 2006. DNA damage induced by mutagens in plant and human cell nuclei in acellular comet assay. *Folia Histochem. Cytobiol.* 44 (2): 127–131.

McClendon, A.K. and Osheroff, N. 2007. DNA topoisomerase II, genotoxicity, and cancer. *Mutat Res.* 623(1-2): 83–97.

Paloma, Y., Belcourt, M.F., Tang, L.-Q. *et al.* 2001. Bioreductive metabolism of mitomycin C in EMT6 mouse mammary tumor cells: cytotoxic and non-cytotoxic pathways, leading to different types of DNA adducts. *The effect of dicumarol. Biocheml Pharmacol.* 61(12): 1517–1529.

Rank, J., Lopez, L.C., Nielsen, M.H. *et al.* 2002. Genotoxicity of maleic hydrazide, acridine and DEHP in Allium cepa root cells performed by two different laboratories. *Hereditas.* 136: 13–18.

van Harten, A.M. 1998. Mutation Breeding. Theory and Practical Applications. Cambridge: Cambridge University Press. pp. 353.

6.2. Websites

IAEA database:
http://mvgs.iaea.org/Search.aspx
PubChem:
http://pubchem.ncbi.nlm.nih.gov
Toxnet:
http://toxnet. nlm.nih.gov
AtGenExpress:
http://www.arabidopsis.org/info/expression/ATGenExpress.jsp

6.3. Further Reading

Bignold, L.P. 2009. Mechanisms of clastogen-induced chromosomal aberrations: a critical review and description of a model based on failures of tethering of DNA strand ends to strand-breaking enzymes. *Mutat. Res.* 681: 271–298.

Drabløs, F., Feyzi, E., Aas, P.A. *et al.* 2004. Alkylation damage in DNA and RNA-repair mechanisms and medical significance. *DNA Repair.* 3(11): 1389–407.

Hamilton, J.T.G., McRoberts, W.C., Keppler, F. *et al.* 2003. Chloride methylation by plant pectin: an efficient environmentally significant process. *Science.* 301: 206–209.

Loechler, E.L. 1994. A violation of the Swain-Scott principle, and not SN1 *versus* SN2 reaction mechanisms, explains why carcinogenic alkylating agents can form different proportionvs of adducts at oxygen *versus* nitrogen in DNA. *Chem Res Toxicol.* 7(3): 277–280.

Mishina, Y., Duguid, E.M. and He, C. 2006. Direct reversal of DNA alkylation damage. *Chem Rev.* 106(2): 215–232.

Palavan-Unsal, N. and Arisan, D. 2009. Nitric oxide signalling in plants. *Bot. Rev.* 75: 203–229.

Robertson, A.B., Klungland, A., Rognes, T. *et al.* 2009. DNA base excision repair: the long and short of it. *Cell Mol Life Sci.* 66(6): 981–993.

Strekowski, L. and Wilson, B. 2007. Noncovalent interactions with DNA: an overview. *Mutat Res.* 23(1-2): 3–13.

Tubbs, J.L., Latypov, V., Kanugula, S. *et al.* 2009. Flipping of alkylated DNA damage bridges base and nucleotide excision repair. *Nature.* 459: 808–813.

Sodium Azide as a Mutagen

D.Gruszka*, I.Szarejko and M.Maluszynski

Department of Genetics, Faculty of Biology and Environment Protection, University of Silesia, 40-032 Katowice, Jagiellonska 28, Poland
* Corresponding author, E-MAIL: damian.gruszka@us.edu.pl

1. Introduction

Sodium azide (NaN$_3$; SA), a common bactericide, pesticide and industrial nitrogen gas generator is known to be highly mutagenic in several organisms including plants and animals. SA is a potent mutagen in microorganisms and a very efficient mutagen in barley as well as in some other crop species, however it is marginally mutagenic in mammalian systems and not mutagenic in *Neurospora, Drosophila* or in *Arabidopsis thaliana* – a model plant species. Mutagenicity of SA is mediated through the production of an organic metabolite of azide and is highly dependent on acidic pH. The frequency of chromosome breakage caused by SA is relatively very low. The issues concerning metabolism, activity, cytotoxic and mutagenic effects of SA and comparisons with other mutagens are presented in this chapter.

2. Metabolism and Activity of SA

2.1. Effects on Seed Germination and Plant Growth

SA is a well-known inhibitor of heavy-metal enzymes with influences on metabolism and respiration of living cells. It is metabolized *in vivo* to a powerful chemical mutagen in many plant species, including barley, rice, maize and soybean (e.g. Owais and Kleinhofs, 1988; Szarejko and Maluszynski, 1999). The mutagenic effect of SA greatly depends on the pH of the treatment solution and similarly to N-methyl-N-nitrosourea (MNU), can be increased further by pre-germination of seeds prior to NaN$_3$ treatment. This mutagen generates M$_1$ sterility and high frequency of M$_2$ chlorophyll mutations in barley. The high frequency of chlorophyll and morphological mutations induced by SA is comparable with the

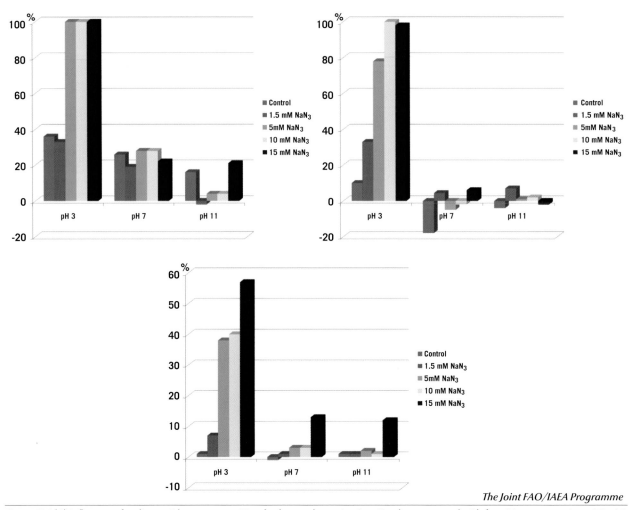

The Joint FAO/IAEA Programme

Figure 13.1 (A) Influence of sodium azide treatment pH on barley seed germination. Seed were treated with four SA concentrations: 1.5 mM, 5 mM, 10 mM and 15 mM for 4 h. (B) Influence of sodium azide treatment pH on barley seedling emergence reduction. Seed were treated with four SA concentrations: 1.5 mM, 5 mM, 10 mM and 15 mM for 4 h. (C) Influence of sodium azide treatment pH on barley seedling height. Seed were treated with four SA concentrations: 1.5 mM, 5 mM, 10 mM and 15 mM for 4 h.

frequency produced by alkylating agents. Influence of SA on barley seeds germination (**Figure 13.1A**), seedlings emergence reduction (**Figure 13.1B**) and seedlings height (**Figure 13.1C**) is greatly dependent on acidity of treatment solution, while in alkaline solution it is completely ineffective.

2.2. Metabolism of SA in Plant Cells

Since the pI (isoelectric point) of SA is at pH = 4.8, the predominant compound at pH = 3 is hydrogen azide (HN_3). The improved mutagenic effectiveness of SA in the acid form is due to facilitated penetration through the cell membrane by the uncharged HN_3 molecule. The mutagenic effect of SA is mediated through the synthesis of an organic metabolite, which was identified in bacteria and barley as an amino acid analogue L-azidoalanine [N_3-CH_2-CH(-NH_2)-COOH]. A free amino acid group is essential for mutagenic activity, when compared with the carboxyl group. Synthetic D-azidoalanine displays very low mutagenic activity, indicating that a stereo-selective process is involved in azidoalanine mutagenicity. The production of this metabolite is dependent on the enzyme O-acetylserine sulphhydrylase [E.C.4.2.99.8] inhibited *in vitro* by cysteine. This enzyme catalyses the addition of azide (N^{3-}) or sulphide (S^{2-}) to O-acetylserine, which leads to the synthesis of azidoalanine or L-cysteine, respectively. Alpha-methyl substitution blocks the mutagenicity of azidoalanine with alpha-methyl-azidoalanine being nearly devoid of mutagenic activity. Homologation of azidoalanine to yield 2-amino-4-azidobutanoic acid markedly increases the molar mutagenic potency. As in the case of azidoalanine, the mutagenic activity of this homologue is associated with the L-isomer. The lack of SA mutagenic effect on adult fruit flies and *Arabidopsis* seeds is attributed to the absence of cellular components conducting enzymatic conversion of SA to the metabolite azidoalanine (Sadiq and Owais, 2000).

2.3. Physiological Effects

SA is an inhibitor of the terminal segment of the electron transport chain. The physiological effect of SA is the inhibition of catalase, peroxidase and cytochrome oxidase, thus influencing respiratory processes. Proton-translocating ATPase complex (F_0F_1), which catalyses the terminal step of photo- or oxidative phosphorylation in chloroplasts, mitochondria and bacterial membranes is particularly prone to inhibition by SA. This leads to a reduction in ATP production and, as a consequence, DNA, RNA and protein synthesis. This inhibition is dose-dependent and is potent even at the lowest dose used for mutation induction in barley embryos. Recovery from this reduction was observed in the case of DNA replication, whereas there was only marginal recovery of protein synthesis. Due to the fact that SA affects DNA synthesis, it is most effective when it is applied at the S-phase of the cell cycle. SA acts as an inhibitor of the proton pump and alters the mitochondrial membrane potential. Additionally, in *E. coli* the primary site of the inhibitory action of SA is SecA protein, an essential component of the protein export system, displaying ATPase activity. Mutations causing resistance to SA in *E. coli* occur mainly in the *sec*A gene (Fortin *et al.*, 1990).

3. Cytotoxic and Mutagenic Effects

3.1. Influence on the Cell Cycle and Metabolism

A dose-dependent decline in mitotic index was observed when SA at concentrations 0.1–0.5% were applied to the seed of *Trigonella foenum-graecum*. At the highest concentration the reduction in mitotic index was 50%. The inhibitory influence of SA dose (concentration and duration of treatment) on the mitotic index was also observed in barley seedlings, which corroborated a cytotoxic effect of this mutagen (Ilbas *et al.*, 2005), and confirmed the inhibitory effect of SA on the cell cycle. The ATP demand of a dividing cell is much higher compared with a non-proliferating cell. The ATP deficiency caused by SA may be one of the reasons for the decrease in mitotic index. SA was found to decrease the cellular level of calmodulin, which is a calcium binding protein participating in signal transduction and cell division. SA at a concentration 1 mM caused a remarkable reduction of cell divisions in barley anthers (Castillo *et al.*, 2001).

The organization and movement of spindle fibres during the cell division is an ATP-dependent process. Due to the reduced synthesis and availability of ATP in SA treated cells, the spindle fibre organization is affected, which may in turn influence the organization of chromosomes at the metaphase plate and their migration during anaphase. Metabolic mechanisms that are the targets of SA activity are shown in **Figure 13.2**.

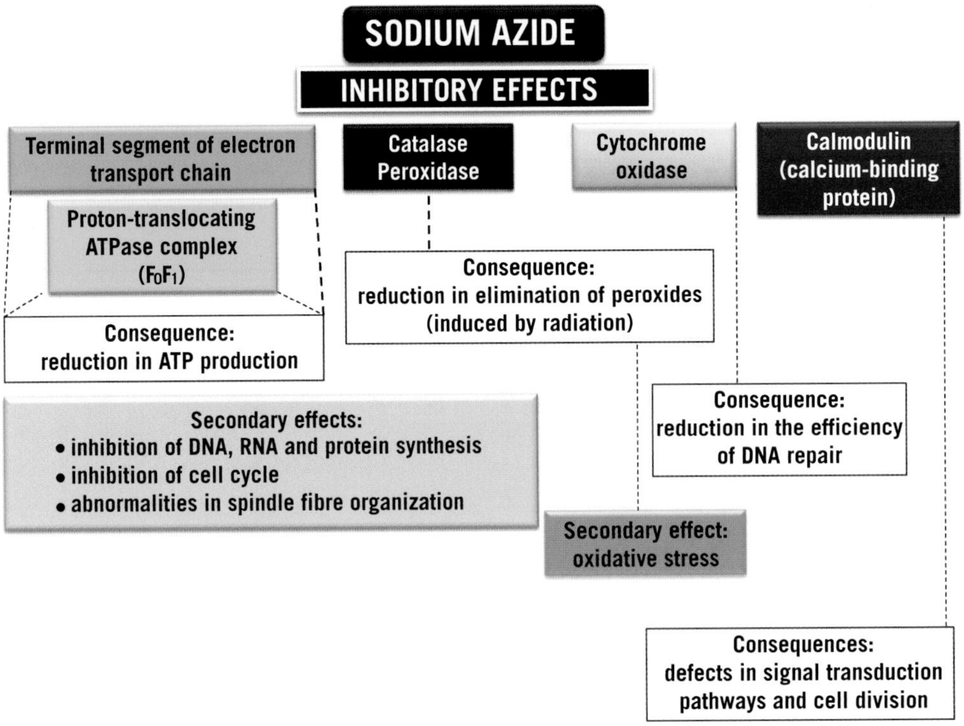

Figure 13.2 Targets and consequences of sodium azide activity in plant metabolism.

Apart from being a suppressor of cell divisions, positive effects of SA on pollen embryogenesis have been demonstrated in *Solanum nigrum* and *Hordeum vulgare* (Kopecky and Vagera, 2005; and Vagera *et al.*, 2004, respectively).

3.2. Mutagenic Effect

It is well known that SA induces chromosomal aberrations only at a very low rate compared with other mutagenic treatments. Is has been shown that L-azidoalanine does not interact with DNA directly under *in vitro* conditions. Mutagenic activity of this compound was attenuated by a deficiency in the excision of UV-like DNA damage in both plants and bacteria, therefore it seems that a lesion recognizable by the excision-repair mechanisms must be formed to evoke the effect. Mutagenesis proceeds from this by 'direct mispairing' (Owais and Kleinhofs, 1988).

SA was used for the first time as a mutagen by Nilan *et al.* (1973) in barley, when an increase in the frequency of chlorophyll mutations was observed in a dose-dependent manner, at the concentration range 1–4 mM at pH3, with highest frequency, 17.3%, recorded for 4 mM. A high frequency of chlorophyll mutations was also observed in barley variety Aramir after combined treatment with 1 mM NaN_3 (3 hours, pH = 3) followed by 0.7 mM MNU with a 6-hour inter-incubation (germination) period between treatments. Using this combination 6.4% of chlorophyll mutations were obtained in the subsequent M_2 population (**Figure 13.3A**), with a fertility and height reduction of M_1 plants lower than after the double treatment with MNU (**Figure 13.3B**).

Similarly, a very high level of point mutations was observed for other barley genotypes after combined treatment with NaN_3 and MNU. On average 5.6% chlorophyll seedlings were found in the M_2 derived from this treatment for six barley varieties tested and about 30–50% of M_1 plants carried a chlorophyll mutation. The reduction in M_1 plants fertility did not exceed 55%, which makes this combination a very efficient treatment for inducing point mutations in barley. Combined treatment of NaN_3 and MNU yielded a wide spectrum of gene mutations in many barley genotypes, leading to dwarf and semi-dwarf characters or changes in root system development and structure (Szarejko and Maluszynski, 1999). The protocol recommended as a highly efficient method of mutagenic treatment in barley is given in **Box 13.1**. In treatment of dormant rice seeds, the germination rate and seedling growth were reduced in

The Joint FAO/IAEA Programme

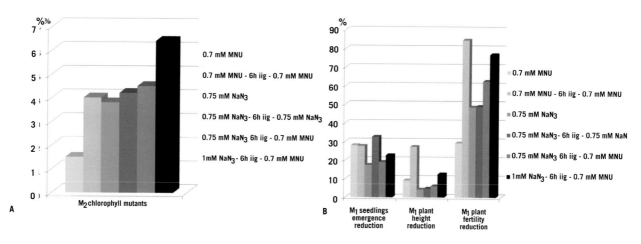

The Joint FAO/IAEA Programme

Figure 13.3 (A) Mutagenic effect of single, double and combined treatments with SA and MNU on chlorophyll mutation frequency in the M_2. The mutagens were applied on barley seed var. Aramir. **(B)** Mutagenic effect of single, double and combined treatments with SA and MNU on reduction in seedlings emergence, plant height and fertility of M_1 plants.

proportion to both 1) the increase in SA concentration and 2) the duration of treatment. SA alone also delayed seed germination. Given that the physiological effect of SA is inhibition of catalase, peroxidase and cytochrome

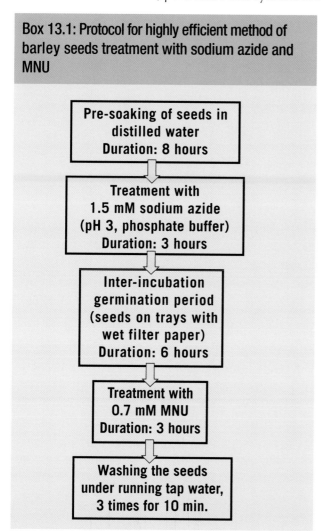

Box 13.1: Protocol for highly efficient method of barley seeds treatment with sodium azide and MNU

Pre-soaking of seeds in distilled water
Duration: 8 hours

⬇

Treatment with 1.5 mM sodium azide (pH 3, phosphate buffer)
Duration: 3 hours

⬇

Inter-incubation germination period (seeds on trays with wet filter paper)
Duration: 6 hours

⬇

Treatment with 0.7 mM MNU
Duration: 3 hours

⬇

Washing the seeds under running tap water, 3 times for 10 min.

oxidase, delay in germination may be explained by inhibition of the metabolic activity necessary for germination. The fact that SA has a limited effect on the fertility reduction of the M_1 supports the observation that somatic damages and mutation induced by this mutagen are not accompanied by chromosomal aberrations. The percentage sterility of barley M_1 ovules after SA treatment can be described by a second-degree parabola in which higher values of sterility are obtained at intermediate concentration of the mutagen (Prina and Favret, 1983). SA has been used successfully in the induction of mutations in plants, including numerous crop species (**Table 13.1**). The mutagenicity of SA can be affected by cellular metabolism during germination. SA, when applied at the DNA synthesis stage of the first cell division of germination process, is more effective in the induction of mutation. In rice, when applied at this stage at a dose of 10 mM for six hours, SA yielded the maximum percentage of mutations 11.1% and 1.22% based on M_1 panicles and M_2 seedlings, respectively (Hasegawa and Inoue, 1980).

SA proved to delay the *in vitro* differentiation of callus structures derived from leaf segments of sugar cane in a concentration-dependent manner (0.07 to 0.7 mM). The number of regenerated plantlets in SA-treated callus was reduced. Additionally, at a concentration of 0.9 mM necrosis in meristematic sugar cane callus was observed (Gonzalez *et al.*, 1990).

3.3. Types and Frequencies of Induced Mutation

The frequency and type of point mutations induced by this mutagen at a concentration of 1 mM were deter-

Table 13.1: The use of sodium azide for mutation induction in plant species

Species	Common name	Recommended treatment method (and outcomes)	Material
Brachypodium distachyon	Purple false brome	1.5 mM for 2 hours (estimation of mutagenic effects)	Seed
Corchorus capsularis	Jute mallow	20 mM (treatment period not available; development of new varieties)	Seed
Hordeum vulgare	Barley	1.5 mM for 3 hours (estimation of cytotoxic and mutagenic effects; in combination with MNU)	Seed
		1 mM for 3 hours (estimation of cytotoxic and mutagenic effects)	Seed
		1 mM for 2 hours (determination of mutations types)	Seed
		0.1; 1 mM for 6 hours (production of androgenic doubled haploid mutants)	Anther
		0.1; 0,5; 1; 5 mM for 20 hours (estimation of concentrations' influence on mutagenic effect)	Seed
		1 mM for 3 hours (determination of mutations frequencies; in combination with 15 mM MNU)	Seed
Oryza sativa	Rice	0.1; 1; 5; 10; 50 mM (estimation of cytotoxic and mutagenic effects)	Seed
Phaseolus vulgaris	Bean	0.04; 0.12; 0.36; 1.08 mM for 3 hours (estimation of mutagenic effects)	Seed
Pisum sativum	Pea	2 mM for 3 hours (development of nitrogen fixation mutants)	Seed
Saccharum officinarum	Sugarcane	0.07; 0.15; 0.46; 0.77 mM for 10 min (estimation of mutagenic effects)	Leaf callus
Solanum nigrum	Black nightshade	10; 20 mM for 24 hours (increase in the efficiency of androgenic production)	Seed
Sorghum bicolor	Sorghum	0.5; 1; 2; 4 mM for 4 hours (estimation of mutagenic effects)	Seed
Tradescantia hirsutiflora	Spiderwort	0.2 mM for 6 hours (estimation of cytological effect)	Inflorescence
Trigonella foenum-graecum	Fenugreek	0.1; 0.2; 0.3; 0.4; 0.5% for 6 hours (estimation of the rate of cytogenetic changes)	Seed
Triticum aestivum	Wheat	5 mM for 6 hours (development of thermo-tolerant mutants)	Seed
Zea mays	Maize	0.1; 1; 10 mM for 1 hour (mutation induction)	Immature embryo

mined by sequence analysis of the barley gene *Ant18* encoding dihydroflavonol 4-reductase, which catalyses the last step in the flavonoid synthesis pathway. SA, when applied at the above concentration, generated 21 base substitutions within the analysed sequence, which corresponds to 0.17% of the 12,704 nucleotides sequenced. Transitions made up 86%, whereas transversions constituted 14% of identified substitutions. The frequency of A-T→G-C transitions was about three times higher than G-C→A-T. Deletions and mutation hot spots were not found. Mutations induced by SA do not show a tendency for clustering. The absence of G-C→C-G transversions was observed in the analysed sequences. This kind of transversion is induced by ^{60}Co γ-rays or Fe^{2+} ions in the DNA through the action of oxygen free radicals, and absence of these transversions rules out the option that oxygen free radicals from azide-derived peroxide accumulation are the secondary mutagens, responsible for the identified substitutions (Olsen *et al.*, 1993).

4. Interactions with other Mutagenic Agents

Apart from being an independent mutagen, SA proved to display interesting interactions in terms of cytotoxic and mutagenic activity with other chemical and physical mutagens. A beneficial influence of simultaneous treatment of rice seed with SA and MNU was noted in terms of obtaining a high density of mutations with relatively low toxicity to treated seed. Concentrations of MNU above 15 mM cause a large drop in germination and viability of germinated plants. Therefore, a combination of 1 mM SA plus 15 mM MNU was applied and the density of induced mutations was determined as 1/265 kb. This observed mutation frequency is satisfactory for high-throughput TILLING strategy and proved to be about three times higher than has been determined for treatments with lower concentrations (Till *et al.,* 2007).

When applied at a concentration of 10 mM to barley seed, after γ-irradiation (160 Gy) in anaerobic conditions with a 4-hour treatment, SA considerably raises the frequency of chromosomal aberrations from 26 to 64%. SA is an inhibitor of catalase, peroxidase and terminal cytochrome oxidase. Catalase and peroxidase are important in the elimination of the peroxides, which are induced by radiation, whereas the inhibition of cytochrome oxidase interferes with the efficiency of the DNA repair process at the chromosome level, especially by reducing ATP production, which is required during the damage repair process. The synergistic relationship between these two mutagenic agents was noted only during acidic (pH 3) treatment. The interaction between SA and ionizing radiation and its association with a decreased catalase, proxidase and respiratory activity is explained by enhanced formation of peroxy radicals in irradiated tissues and the demand for ATP during the repair of the induced lesions. Additional corroboration comes from the fact that the induction of SA activity was also observed by pre-soaking of the seeds in either O_2^- or N_2^--bubbled water prior to and during the treatment.

SA treatment modifies the extent of the lesions and efficiency of the repair processes. The strong inhibition of catalase and peroxidase activity, induced by SA, results in the reduced scavenging of peroxide, leading to an increase in radiation-induced damage. Inhibition of respiratory activity brings about reduction of the rate of oxidative phosphorylation and thus a reduced supply of ATP to the affected cells. SA disrupts the cascade of oxidative phosphorylation re-directing the electron stream to flavoproteins and increasing the production of peroxide, furthermore accumulation of this compound is accelerated by the synchronous inhibition of catalase.

Taking into account that SA mutagenesis (also in combination with other mutagens) produces relatively high frequencies of mutation with mild effect on plants fertility the use of this mutagen in the future may be of significant importance in functional genomics when applied along with high-throughput technologies, like TILLING, especially for crop species.

5. References

5.1. Cited References

Castillo, A.M., Cistue, L., Valles, M.P. *et al.* 2001. Efficient production of androgenic double-haploid mutants in barley by the application of SA to anther and microspore cultures. *Plant Cell Reports.* 20: 105–111.

Fortin, Y., Phoenix, P. and Drapeau, G.R. 1990. Mutations conferring resistance to azide in *Escherichia coli* occur primarily in the *secA* gene. *Journal of Bacteriology.* 172: 6607–6610.

Gonzalez, G., Perez, M., Santana, I. *et al.* 1990. Mutagenic activity of 3-azido-1,2-propanediol and SA applied to sugar cane callus cells. *Biologia Plantarum.* 32: 388–390.

Hasegawa, H. and Inoue, M. 1980. Effects of SA on seedling injury and chlorophyll mutation in rice. *Japan Journal of Breeding.* 30: 301–308.

Ilbas, A.I., Eroglu, Y. and Eroglu, H. E. 2005. Effects of the application of different concentrations of NaN_3 for different times on the morphological and cytogenetic characteristics of barley (*Hordeum vulgare* L.) seedlings. *Journal of Integrative Plant Biology.* 47: 1101–1106.

Kopecky, D. and Vagera, J. 2005. The use of mutagens to increase the efficiency of the androgenic progeny production in *Solanum nigrum. Biologia Plantarum.* 49: 181–186.

Nilan, R.A., Sideris, E.G., Kleinhofs, A. *et al.* 1973. Azide – a potent mutagen. *Mutation Research.* 17: 142–144.

Olsen, O., Wang, X. and von Wettstein, D. 1993. SA mutagenesis: preferential generation of A-T→G-C transitions in the barley *Ant18* gene. *Proceedings of National Academy of Sciences of the U.S.A.* 90: 8043–8047.

Owais, W.M. and Kleinhofs, A. 1988. Metabolic activation of the mutagen azide in biological systems. *Mutation Research*. 197: 313–323.

Prina, A.R. and Favret, E.A. 1983. Parabolic effect in SA mutagenesis in barley. *Hereditas*. 98: 89–94.

Sadiq, M.F. and Owais, W.M. 2000. Mutagenicity of SA and its metabolite azidoalanine in *Drosophila melanogaster*. *Mutation Research*. 469: 253–257.

Szarejko, I. and Maluszynski, M. 1999. High frequency of mutations after mutagenic treatment of barley seeds with NaN3 and MNH with application of inter-incubation germination period. Mutation Breeding Newsletter. 44: 28–30.

Till, B.J., Cooper, J., Tai, T.H. *et al.* 2007. Discovery of chemically induced mutations in rice by TILLING. *BMC Plant Biology*. 7: 1–12.

Vagera, J., Novotny, J. and Ohnoutkova, L. 2004. Induced androgenesis *in vitro* in mutated populations of barley, *Hordeum vulgare*. *Plant Cell. Tissue and Organ Culture*. 77: 55–61.

5.2. Further Reading

Engvild, K.C. 1987. Nodulation and nitrogen fixation mutants of pea, *Pisum sativum*. *Theoretical and Applied Genetics*. 74: 711–713.

Engvild, K.C. 2005. Mutagenesis of the model grass *Brachypodium distachyon* with SA. Riso National Laboratory, Riso-R-1510: 1–8.

Falistocco, E., Torricelli, R., Feretti, D. *et al.* 2000. Enhancement of micronuclei frequency in the *Tradescantia*/micronuclei test using a long recovery time. *Hereditas*. 133: 171–174.

Hibberd, K.A. and Green, C.E. 1982. Inheritance and expression of lysine plus threonine resistance selected in maize tissue culture. *Proceedings of National Academy of Sciences of the U.S.A.* 79: 559–563.

Kleinhofs, A., Sander, C., Nilan, R.A. *et al.* 1974. Azide mutagenicity – mechanism and nature of mutants produced. *In*: Polyploidy and Induced Mutations in Plant Breeding. Vienna: IAEA, pp. 195–199.

Mullarkey, M. and Jones, P. 2000. Isolation and analysis of thermotolerant mutants of wheat. *Journal of Experimental Botany*. 51: 139–146.

Seetharami-Reddi, T.V.V. and Prabhakar, G. 1983. Azide induced chlorophyll mutants in grain *Sorghum* varieties. *Theoretical and Applied Genetics*. 64: 147–149.

Shamsuzzaman, K.M., Saha, C.S., Bhuya, A.D. *et al.* 1999. Development of a new jute (*Corchorus capsularis*) variety 'Binadeshipat-2' through SA mutagenesis. *Mutation Breeding Newsletter*. 44: 9–10.

Silva, E.G. and Barbosa, H.M. 1996. Mutagenicity of SA in *Phaseolus vulgaris*. *Brazilian Journal of Genetics*. 19: 319–322.

Section 3

Mutation Induction and Mutant Development

Methodology for Physical a
Mutagen

A.Kodym[a,b], R.Afza[a], B.P.Forster[a,c], Y.Ukai[d], H.Nakagawa[e] and C.Mba[a,*,#]

[a] Plant Breeding and Genetics Laboratory, Joint FAO/IAEA Division of Nuclear Techniques in Food and Agriculture,IAEA Laboratories Seibersdorf, International Atomic Energy Agency, Vienna International Centre, P.O. Box 100, Vienna, Austria

[#] Present Address: Plant Production and Protection Division, Food and Agriculture Organization of the United Nations, Viale delle Terme di Caracalla, 00153 Rome, Italy

[b] Melbourne School of Land and Environment, The University of Melbourne, Richmond, VIC 3121, Australia

[c] Biohybrids International Ltd, P.O. Box 2411, Earley, Reading RG6 5FY, UK

[d] Formerly, Graduate School of Agricultural and Life Sciences, The University of Tokyo, 1-1-1 Yayoi, Bunkyo, Tokyo 113-8657 Japan

[e] Institute of Radiation Breeding, National Institute of Agrobiological Sciences, P.O. Box 3, Kami-Murata, Hitachi-Ohmiya, Ibaraki 319-2293, Japan

* Corresponding author, E-MAIL: Chikelu.Mba@fao.org

on

[...]come of any mutation induction exercise is [...]y dependent on the appropriateness of the strate-[...]es adopted. The critical factors that could determine the spectrum and rate of induced mutations include the type of mutagen, the dose and dose rate administered and the method of treatment including choice of materials, pre- and post-treatment handling. These need to be carefully assessed and consequently executed according to project objectives and available resources. The standard procedures for inducing mutations using both physical and chemical mutagens are outlined in the following sections. The sensitivity of plant materials to various mutagens has been investigated and is summarized for many plant species. However, these should be considered as a good guide, not treated as fixed numbers, particularly for plant species that have not been studied extensively, since there is a significant genotypic effect in sensitivity to mutagenic treatments in plants. It is therefore recommended that preliminary assays to determine appropriate doses are carried out.

2. Effects of Mutagens in the M_1 Generation

Chemical and physical mutagens can directly or indirectly damage cellular components including DNA and proteins. For example, mutagenic radiations can directly cause DNA DSBs (**see Chapter 6**), while chemical mutagens on the other hand commonly interact with DNA molecules causing various kinds of DNA lesions (**see Chapter 12**). Such damage can cause cell arrest, abnormal cell division, and chromosomal and gene mutations. Toxic effects of mutagens can also result in physiological damage to the treated propagule, plant part or intact plant. A combination of the effects of the mutagen on the hereditary materials (at the gene and chromosome levels) and the primary injury results in reductions in growth and fertility and can lead to lethality in the first mutant population (the M_1 plants). Therefore, both physical and chemical mutagens can cause two types of effects on M_1 plants: (1) physiological damage (primary injury); (2) DNA damage (at the sequence to chromosome levels). The latter is the ultimate subject of any mutation programme.

2.1. Plant Injury in the M_1 Generation

Most of the observed effects in the M_1 generation are physiological. Plant injuries in the M_1 generation are indicative of the degree of the effects of mutagens on plants and can be determined quantitatively in various ways (**Box 14.1**). Physical injury is commonly measured using such parameters as reductions in germinability of seeds, growth rates of seedlings, vigour, sterility and even lethality of plants (**Box 14. 1**). These can be used as a surrogate to set threshold values of mutagen doses in acquiring the required mutation induction.

2.1.1. Seedling Height and Root Length

Seedlings are particularly sensitive to mutagens and provide an easy means of measuring treatment effects. Seedling height and root length are simple parameters. Seedling height is typically used as an indicator of genotype response to a mutagen and various methods can be devised depending on the species. **Figure 14. 1** shows the reduced growth of seedlings with the increase in dose of gamma irradiation in two genotypes of rice with varying levels of tolerance to radiation; The improved rice variety, IR29, has relatively scantier germination and seedling establishment compared

---

Box 14.1: Parameters used to estimate the degree of plant injury in the M_1 generation [a]

1. Seedling height, determined at a particular stage soon after germination. Normally carried out under controlled environmental conditions or in glasshouses.
2. Root length, determined soon after germination in controlled environment or glasshouse conditions.
3. Emergence under field conditions, or germination under controlled environment conditions.
4. Survival under field or controlled environment conditions.
5. Number of florets, flowers or inflorescences per plant.
6. Number of florets or flower parts per inflorescence.
7. Number of seed set.
8. Fruits and/or seeds per plant.

[a] Adapted from *Manual on Mutation Breeding*, International Atomic Energy Agency, Technical Report Series No. 119.

The Joint FAO/IAEA Programme

Figure 14.1 Seedlings from gamma ray-irradiated seeds showing seedling height and density decreasing progressively with increase in dose. (A) Improved rice variety, IR29. (B) Pokkali, an uncultivated wild relative of rice. (C) A Namibian variety of bambara groundnut.

with Pokkali, an uncultivated wild relative of rice, when they are treated at the same high dose of gamma rays (e.f. 500 Gy). indicating that Pokkali is more radio tolerant. **Figure 14.2** shows an easy method for measuring seedling height and root length: seedlings of barley are placed on a sheet of graph paper on which lengths had been previously marked.

2.1.2. Lethality and Sterility

Mutagens can completely prevent the germination of seeds at high doses. Lethality, when seeds germinate but are unable to grow and subsequently wither and die off, is also common at high doses. High mutagen doses can also reduce the fertility of M_1 plants and in extreme cases can result in total sterility. Since there are direct relationships between plant responses and mutagen dose, germination and survival rates as well as

reductions in fertility can be used in determining optimal doses of mutagens for treating plant propagules. Whereas germination rates are computed at an appropriate interval shortly after sowing, survival rates are determined at maturity (at harvest time) of the M_1 population. Survivors of the treatment may be defined as those plants that complete their life cycle and produce at least one floret, flower or inflorescence, regardless of whether seeds are produced. Actual plant death may occur at any time between the onset of germination and ripening. Survival rate determined in controlled environments can be significantly different from field tests, particularly if adverse conditions occur.

Fertility reduction in M_1 plants can be manifest in various forms (**Box 14.2**). The degree of M_1 sterility varies greatly from plant to plant and from inflorescence to inflorescence even within the population treated with

Box 14.2: Sterility of M_1 plants [a]

Manifestations of reduced fertility/sterility in M_1 plants
1. Severe stunting or growth inhibition which prevents flowering.
2. Flowers are formed but lack reproductive structures.
3. Reproductive structures are present but pollen is not viable (the most common occurrence).
4. Fertilization occurs but embryos abort before maturity.
5. Seeds form but fail to germinate properly or die after germination.

Causes of sterility in M_1 plants
1. Chromosome mutations (the major origin, particularly for physical mutagens).
2. Gene mutations.
3. Cytoplasmic mutations.
4. Physiological effects.

[a] Modified from *Manual on Mutation Breeding*, International Atomic Energy Agency, Technical Report Series No. 119.

the same dose. Seed set is the most commonly used criterion in quantifying sterility. Due to the wide range of variations among plants, sufficient number of inflorescences should be used to estimate this parameter.

2.2. Methods in Determining Dose Effects

It is well documented that effects of mutagens are dose dependent. However, the dose effect is not linear. With an increase in mutagen dose, the effect on plant growth and reproduction increases. Low doses can have a stimulatory effect on pollen and seed germination rate, seedling height and root length and on *in vitro* cultures. Prior to initiating any mutagen treatment, the most important step is to determine the effective dose. In the absence of detailed data, it is sometimes necessary to estimate the optimum dose in the M_1 population by estimating the LD_{50} or by RD_{50} *via* a small scale test. The LD_{50} is the dose that results in a 50% reduction in germinating seeds or viable plants. The RD_{50} represents the dose which reduces the growth and seed production of an M_1 population by 50%. As mentioned above, seedling height and root length are commonly used to determine dose effects of mutagens. The following controlled environment and glasshouse methods allow the determination of germination/emergence, seedling height and leaf spotting, all indices of the extent of damage occasioned by radiation or chemical mutagens.

It should be noted that the relationship between the effect on an M_1 trait and mutation rate in the subsequent M_2 varies between chemical mutagens and between radiations, especially between sparsely and densely ionizing radiations. For instance, the relative biological effectiveness (RBE) of neutrons often differs with M_1 traits.

2.2.1. Sowing Methods

Flat method. This is widely used for seeds of beans, cereals and other similar sized seed. The seeds are normally sown in a glasshouse in trays with holes for drainage containing heat- or steam-sterilized soil, and with adequate moisture to ensure germination. Seeds are planted in spaced rows keeping in mind the requirements of the species. Trays used at the Plant Breeding Unit of the IAEA have a size of 400 × 600 × 120 mm and accommodate 4–7 rows depending on the plant species and orientation of the tray. The seeds are planted in order of increasing dose with replications sown in different trays. This planting technique allows simple and rapid visual assessment of the various treatments (**Figure 14.1**), however it is limited to young seedlings. Alternatively, seeds can be sown in pots or individual cells of compartmentalized trays (commercially available) in which later stages can also be studied. The environmental conditions, as well as water and nutrient supply, and sowing depth, should be uniform for all treatments.

Petri dish method. This is particularly effective for seeds of cereals or smaller size and for seeds that require light for germination. Seeds are placed on wet filter paper in a Petri dish. The paper is kept moist at all times as germination is strongly influenced by water uptake. Fungal contamination may spoil the experimental set up, and can be controlled by using sterile filter paper, sterile Petri dishes, water and disinfected seeds (e.g. surface sterilization in 20% Clorox bleach, 5.25% w/v NaOCl active ingredient, for 20 min).

Sandwich blotter method. Seeds are pre-soaked and then placed between two wet blotter papers which are pressed together and supported vertically in racks. The racks are placed in plastic trays with water. This is a labour-saving and accurate technique that was especially designed for cereal seeds. However, it requires specific equipment such as a plastic film-covered growth cabinet and a humidifier.

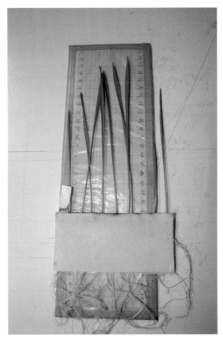

The Joint FAO/IAEA Programme

Figure 14.2 Seedlings of barley are placed on a sheet of graph paper on which lengths had been previously marked making it easy to measure the heights of the seedlings.

With any of the three methods, the plants should be grown in a climatically controlled environment, or in a uniform section of a glasshouse. The experimental set-up normally requires about 100 seeds per dose without replications; 20-25 seeds per dose with 3 to 4 replications and an equal number of control seeds is used for comparison.

2.2.2. Data Collection

Seedling height is determined by cell division as well as cell extension and as such provides an early simple means of determining treatment effect. It is not possible to distinguish between delayed germination and seedling height reduction in early seedling growth. Therefore, measurements are taken when the first true leaves in the control seedlings have stopped growing. It is therefore advisable to determine the time at which the first true leaf in control plants ceases to expand, in order to determine the best time for data collection.

It should be noted that at low doses, seedling heights of irradiated materials may exceed control values. There are differences in the morphogenesis of monocot and dicot seedlings, therefore different procedures are used for collecting data on plant injuries for the two types of plants.

In monocots, seedling height is measured from the soil level to the tip of the first or secondary leaf. In the Petri dish and sandwich blotter methods, measurement starts from the embryo axis. In cereal seedlings, the leaf which emerges through the coleoptile is the first true leaf. Seedlings where only the coleoptile develops are not considered for seedling height measurements.

In cereals, the reduction in seedling height, as a function of dose, emerges more clearly when control seedlings shoots are about 11-20 cm long. In barley measurements are taken usually after 10 to 14 days, when seedling height of the control is 16-20 cm.

In epigeous germination of dicot plants, the cotyledons are pulled upward through the soil during germination, e.g. *Phaseolus vulgaris*, Here the epicotyl length is measured between the points of attachment of the cotyledons to the tip of the primary leaves or to the stem apex. The hypocotyl region is relatively insensitive to radiation as growth here is determined mainly by cell elongation rather than cell division. The epicotyl exhibits more mitotic activity and is therefore a better measure of mutagen effects. Seedling height can also be taken from soil level to the tip of the primary leaves, or to the stem apex.

In hypogeous germination of dicot plants, the cotyledons remain below the soil surface, e.g. *Pisum sativum*, measurements are taken from soil level to the tip of the primary leaves (longest leaf) or to the stem apex.

2.2.3. Determination of LD_{50} and RD_{50}

After a radiosensitivity assay, a graph of the reductions in the measured parameter (usually expressed as a percentage of the control sample) against the corresponding dose of the mutagen is plotted. The indices, LD_{50} and RD_{50} (or any other percentage reduction), can be easily read off the gradient of the graph (the 0.50 position in **Figure 14.3**). Determining these values through substitutions in a straight line equation of the graph leads to the same results.

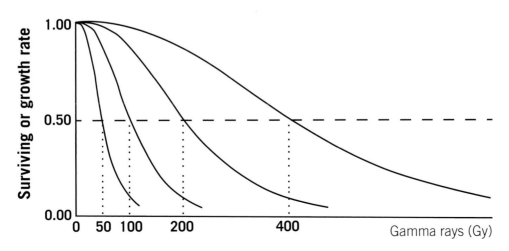

Figure 14.3 Model curves illustrating the effects of increasing dose of irradiation on physiological damage.

2.3. Factors Influencing Mutagenic Effects

Plant genera and species and, to a lesser extent, genotypes and varieties differ in their radio-sensitivity. These differences are accounted for by genetic, physiological, morphological and other biological modifying factors (such as ontology). These combine with environmental factors (such as oxygen and water content) to assert marked influences on the response of seeds (and other plant propagules) to ionizing radiation and chemical mutagens.

2.3.1. Genetic factors

The wide range of radio-sensitivities typically observed across plant species and even among breeding lines have been ascribed to differences in nuclear index, i.e. the interphase chromosome volume (ICV) of interphase cells of shoot meristems, defined as the nuclear volume divided by the chromosome number, in DNA content, and ploidy level. The higher the ICV with the same ploidy level, the higher the cellular radio-sensitivity. In irradiation of growing plants, radio-sensitivity as expressed by growth depression and/or seed sterility in the M_1 shows a close negative linear regression on ICV on a log–log plot, with a slope of –1. In seed irradiation, however, the correlation between radio-sensitivity and ICV is not so clear, though significant in comparisons of species with wide taxonomic diversity.

Another factor is the ploidy level. The homoeologous genes of polyploid plants have complementary effects, and as a result polyploid species are more tolerant to DNA damage than diploid species. In some genera, for instance, chrysanthemum, the higher the ploidy level, the higher the tolerance to radiations in the range of 2x to 22x.

Radio-sensitivity is also dependent on the DNA repair systems of the plant (**see Chapter 5**), a further indication of the influence of genetic backgrounds on the plants' responses to irradiation. A notable example of the striking variation in radio-sensitivity is that tree species are generally 2 to 2.5 times more sensitive to radiation than grasses.

An example of intraspecific genetic variation in radio-sensitivities was documented in the work of Takagi (1969) on soybeans which showed that the RD_{50} of acute gamma-irradiated seeds ranged from 60 Gy to 320 Gy for the high radiosensitive variety, Lexington, and the low radio-sensitive variety, Virginia bean, respectively. Similarly, about six-fold differences in radio-sensitivity to chronic irradiation (performed in a gamma field) were recorded between the highly sensitive soybean variety, Lexington with RD_{50} of 0.17 Gy/day and a low sensitive wild form Madara-ooba-tsurumame with RD_{50} of 1.05 Gy/day.

The acute radio-sensitivity differences between the high and low-sensitive varieties were identified to be controlled by two single recessive genes, *rs1* and *rs2*, while chronic gamma ray irradiation sensitivity was influenced by the single recessive gene, *rs1*. The *rs2* gene, which was discovered in the variety Goishi-shirobana (*RS1RS1rs2rs2*), increases the level of radio-sensitivity only during acute irradiation. Individuals possessing the double recessive genotype (*rs1rs1rs2rs2*) are known to exhibit significantly higher levels of radio-sensitivity and serious damage than

Table 14.1: Differences in radiosensitivity of soybean varieties exposed to acute and chronic gamma irradiations

Variety	Chronic irradiation		Acute irradiation		Genotype
	RD_{50}	Sensitivity	RD_{50}	Sensitivity	
Lexington	0.17	Highest	60	Highest	$rs_1rs_1\ rs_2rs_2$
Shin-mejito	0.18	High	82	High	$rs_1rs_1\ Rs_2Rs_2$
Goishi-shirobana	0.48	Low	82	high	$Rs_1Rs_1\ rs_2rs_2$
Tachi-suzunari	0.66	Low	180	Low	$Rs_1Rs_1Rs_2Rs_2$
Iyo-aogari	0.72	Low	205	Low	
Virginia bean	0.54	Low	320	Lowest	
Madara-ooba-tsurumame	1.05	Lowest	NT	NT	

the individuals possessing the *rs1rs1++* genotype (**Table 14.1**). A similar major gene *rs1,* which controls gamma-ray sensitivity in seed irradiation has been reported in barley (Ukai, 1986). In other species (for instance, rice) in which major genes have not been reported, varietal differences in radiosensitivity is generally polygenic.

2.3.2. Environmental factors

Oxygen is the main environmental modifying factor, while water content, temperature and storage conditions are secondary factors. The presence of oxygen can increase biological radiation damage hundred-fold (**see Chapter 11**). The moisture (or water) content of resting seeds, on account of its role in respiration and gas transport, also influences the seed's radio-sensitivity though the mechanism of this effect is not fully understood. In resting barley seeds, for instance, water content below 14% leads to increased sensitivity to both gamma and X-ray irradiations. Van Harten (1998) hypothesized that this may be attributable to the ability of drier seeds to retain more of the radicals produced from irradiation which are implicated in inducing the mutations; wetter seeds on the other hand have higher respiration rates. For practical purposes, the plant breeder should always determine and note the moisture contents of the seeds to be irradiated in order to achieve reproducible results. The effects of modifying factors can be minimized by controlling the seed water content and the reproducibility of parameters of primary damage should be made possible within practical limits. Thus, prior to gamma irradiation the seed water content is adjusted to 12–14% by holding seeds in a vacuum desiccator. The critical tissues are those of the embryo, but it can be assumed that the water content of the seed and the embryo of most species will be similar. Environmental factors are less important with densely ionizing radiation, consequently no seed moisture adjustment is necessary for fast neutron treatment. In thermal neutron irradiations, moisture content of seeds do not change radiosensitivity.

Compared with oxygen and water content, ambient temperature plays a less important role in influencing the outcome of radiation-induced mutagenesis, in seed irradiation. In growing plant irradiation, low temperatures increases M_1 damage and mutation rate. For chemical mutagenesis however, temperature affects the reactivity of the process and should therefore be monitored appropriately. The age of the seeds seems to affect the rate of mutations, with older seeds (e.g. five-year-old seed of annual cereals) having been shown to house a higher density of spontaneous mutations compared with freshly harvested seed. Seed ageing and the conditions of seed storage post-irradiation are also critical for the density of induced lesions. It is advisable to sow seeds soon after irradiation and if storage is desirable, the irradiated seeds must be stored dry, at low temperatures, in the dark and in oxygen conditions. Seed storage in small air-tight tinfoil bags or in sealed boxes has proven effective.

In general, botanical seeds are the most convenient materials for mutagenic treatment, especially for seed-propagated plants. Dry, quiescent seeds, i.e. seeds ready for germination are ideal, as differences in radio-sensitivity may occur as a result of the specific ontogenetic or physiological state of other plant materials. Also, actively dividing cells, as is the case in seeds about to germinate, are more sensitive to radiation than resting or dormant seed (van Harten, 1998).

2.3.3. Radio-Sensitivity of Plant Species

Over the past half century, a huge amount of data on the radio-sensitivity of various plant species have been accumulated. However, data available in the literature might vary even for the same species; such variations might be attributed to genotypic differences as well as the physiological status of the materials. In spite of these variations, it is recommended that these data are used as a basis for the optimization of irradiation doses prior to bulk irradiation of samples. The radio-sensitivities, shown as estimated LD_{50} or RD_{50}, of various plant species to fast neutrons and to both acute and chronic gamma irradiation are summarized in **Appendices 1, 2 and 3**. In some species, the suggested doses for practical application are also provided.

3. Methods and Techniques for Mutagenic Treatment

3.1. Acute, Chronic, Recurrent Treatment and Dose Fractionation

Irradiation, for the purposes of inducing mutations in plants, is normally carried out in one of three ways:

- Continuous exposure, usually of growing plants, over extended periods of time, ranging from weeks to months, to relatively low doses of irradiation; this is known as chronic irradiation. Gamma glasshouses and fields are particularly suitable for this extended exposure of growing plants to irradiation.

- The single exposure of plant propagules at higher doses over a short period of time (minutes) is known as acute irradiation.
- Recurrent irradiation involves the exposure to radiation of progeny of plants that had been irradiated in previous generation(s).
- Dose fractionation or split dose irradiation, somewhat akin to chronic irradiation, refers to the practice of exposing the plant propagule to more than one regimen of irradiation with the treatments separated by time intervals.

It used to be believed that acute irradiation resulted in relatively greater mutation frequencies. This does not appear to be supported by empirical data, but currently, most mutation induction exercises are of the acute type.

It is important to note that the degree and frequency of radiation-induced injuries depend not only on the total dose delivered but are also functions of the radiation dose absorbed per unit time, a parameter known as dose rate. The results from studies aimed at establishing relationships between dose rates and mutations have seemed inconsistent. While some studies indicate a direct relationship between dose rates and effects, i.e. higher dose rates leading to larger radiation damage and mutation rates, the inverse or a lack of a definite trend have also been reported. It seems that there is a critical point in dose rates above which the effectiveness of radiation increases with increasing dose rate and below which effectiveness decreases with increasing dose rate. More studies in this area are therefore clearly required.

3.1.1. Dose Rate Dependency of M_1 Damage

Acute and chronic irradiations correspond to higher and lower dose rates, respectively, although the threshold of dose rate between the two types of irradiations is not clear in plants. Increased radiation effects per unit dose (effectiveness) with increasing dose rate are a general phenomenon and have been reported in a range of plant species as well as bacteria and animals.

It is generally believed that repair mechanisms are responsible for decreased effects of radiations when the dose is delivered over a longer period of time. Repair of radiation damage takes place mainly during DNA replication (S stage) at interphase of the cell cycle. In this

context a radiobiological hypothesis has been proposed that states it is the dose per cell cycle that determines the magnitude of radiation effects produced in a population of dividing cells exposed continuously to radiations. According to this hypothesis the shorter the cell cycle or the longer the irradiation period, the less the effects of a given dose. In other words, lower dose-rates (chronic irradiations) are expected to result in fewer effects to cells as compared with higher dose-rates (acute irradiations).

Hall et al. (1966) irradiated seedling roots of Vicia faba with gamma rays at two different temperatures 19°C and 12°C. Since it was known that the average inter-mitotic period of the root meristem cells of Vicia cultured at 12°C is about twice that for roots grown at 19°C, it was expected that for a given dose rate roots at the lower temperature would accumulate twice the dose per cell cycle as roots at 19°C. They found that roots at 19°C required about twice the dose rate to produce a given reduced growth pattern compared with roots exposed at 12°C and concluded that the result was consistent with the hypothesis. In another experiment, Hall and Bedford (1964) avoided the influence of cell cycle by using roots of Vicia faba chilled to 3.5°C during irradiation treatment. Root growth and cell-division in the meristem was halted by the low temperature, but recovery of sub-lethal damage after acute doses took place, though slowly. They irradiated roots at 3.5°C with dose rates of 8.9, 15.2, 84.0 and 73,200 mGy/h and found that the radiation damage resulting from a given dose of radiation is diminished when the dose rate is reduced from 73,200 mGy/h to 84 mGy/h and from 84 mGy/h to 15.2 mGy/h, but that the radiation injury is independent of dose rate when the dose rate is as low as 15.2 mGy/h or less. However, Matsumura (1965) found greater effectiveness of chronic over acute exposures in terms of height reduction and in killing of various cereals using dry, quiescent seeds. These results suggest that although dose rate effects on dividing cells are mainly explained by cell cycle, there are other factors concerned.

Most observations on dose rate effects are based on, at most, a few dose rates and the range of dose rates investigated is somewhat limited. Bottino et al. (1975) irradiated germinating seeds of barley with ^{137}Cs gamma rays at various combinations (73 cases) of total dose (4–32 Gy) and dose rate (0.3–240 Gy/h) and measured seedling height 5 days after irradiation. The doses required to produce 20% and 35% growth inhibition at each exposure rate were determined. For both levels of growth inhibition, as dose rate increased, the total dose required to produce a given effect decreased (effectiveness increased). A linear relationship on a double logarithmic plot was observed in the range of dose rates between 0.3 to 15 Gy/h. At around 15 Gy/h, a change in slope occurred and a slight but consistent increase of required total dose (decrease of effectiveness) was observed from this critical dose rate area upwards 240 Gy/h. Conversion of exposure rate to exposure time demonstrated this point of change in effectiveness occurred at 0.3 to 0.4 h, which is well within one mitotic cycle (approximately 10 hours) and diminution of effects caused by protraction of exposure over more than one mitotic cycle could be eliminated as causative of the reversal. They suggested that the critical point may differ for other species and may depend on chromosome size and/or mitotic cycle time characteristics of a species.

An oxygen requirement has been reported for the expression of dose-rate effects (Dewey, 1969). It was suggested that this was due to the presence of repair mechanisms which require oxygen and function most efficiently at low dose rates.

3.1.2. Dose Rate Dependency of Mutation Rate

Like radiation damage, increased mutation rate with increased dose rate was found for many organisms, plants, animals and microbes.

Fujii (1962) reported the result of irradiation on air-dried seeds of einkorn wheat (*Triticum monococcum*) with 100 and 200 Gy of X-rays and gamma-rays at dose rates of 0.45 to 600 Gy/h. He found increased growth inhibition and seed sterility with increased dose rate. Chlorophyll mutation frequency per spike progeny also increased with increasing dose rate. When he irradiated seeds of an upland rice variety with 0.05 to 300 Gy of gamma-rays at dose rates of 0.1, 3 and 100 Gy/h, he found chlorophyll mutants to be more frequent in the highest treatment (100 Gy/h), however growth inhibition was most conspicuous at the lowest dose rate 0.1 Gy/h.

Using barley seeds with very low moisture content Natarajan and Maric (1961) reported that a low dose rate was more effective as regards seedling height, chromosome aberrations and the frequency of chlorophyll mutants segregating in progenies compared with a high dose rate. They considered the dose rate effect to be due to storage effects accumulated during the period of irradiation.

After irradiation of *Saintpaulia* with X-rays at dose rates ranging from 0.02 to 250 Gy/min, Broertjes (1972)

demonstrated a similar phenomenon to that of Bottino *et al.* (1975): the mutation frequency per Gy increased with increasing dose rate until a maximum mutation rate was obtained. Further increases in dose rate, however, resulted in decreases in mutation rate.

3.1.3. Recurrent Treatment

The strategy of irradiating the progeny of previously irradiated materials is known as recurrent treatment. Deriving from the notion that recurrent treatment would enhance both the density and spectra of mutation events, recurrent irradiation treatment was the subject of robust investigations from the 1940s through to the 1960s. It was expected that recurrent treatments would further broaden the induced genetic variability exploitable in crop improvement programmes and increase the chances of obtaining target mutant phenotypes. Investigators studied this strategy in both seed and vegetatively propagated plants using a wide range of mutagens including X- and gamma rays, thermal neutrons and the chemical mutagen, EMS; the alternation of EMS with irradiation was also studied. The results of these experiments did not bear out the expected results and were at best mixed. In most cases, radio-sensitivity, mutation density and spectra remained unaffected with repeated irradiation of subsequent generations. A telling result was the accumulation of so many deleterious alleles in barley after five generations of recurrent irradiation as to render the induced mutants useless as breeding materials. It is plausible to surmise that the possible increase in mutation density and spectrum might not justify the added expense and time invested in recurrent mutagen treatment. It is most practical to restrict the induction of mutations to either acute treatment (for plant propagules) or chronic treatment (for whole plants) and concentrate on producing sufficient mutants and developing high-throughput screens to increase the chances of obtaining the desired range and number of specific mutants in a wild type genetic background.

3.1.4. Dose Fractionation

The irradiation strategy involving the splitting of the predetermined acute dose over two or more irradiation treatments interspersed with variable time interval(s), a practice that received a lot of attention in the 1960s, is known as dose fractionation or split dose irradiation. The practice involved initiating a radiation treatment by administering a relatively small initial dose which was thought to either confer some protection against the subsequent higher dose or aid the plant in recovery from biological damage. In general, the studies seemed to suggest that the net biological damage from a split dose irradiation was less than from a single acute irradiation at a dose equalling the sum of the doses administered separately. Again, it is plausible to surmise that whatever advantages are derivable (in terms of reduced biological damage in dose fractionation), are nullified by its more cumbersome procedure. For practical purposes, especially in plant breeding, the administration of a single acute dose will remain the strategy of choice for inducing mutations.

3.2. Combined Physical and Chemical Treatments

Since chemical and physical mutagens have different modes of reaction with plant materials, some investigators held the notion that the induction of mutations could be enhanced, in terms of mutation rate and spectrum, by their combined use in a single treatment. The strategy of combined treatments usually involved the applications of a physical mutagen followed by a chemical mutagen. It was thought that the chemical mutagen would inhibit the DNA repair mechanism leading to a greater mutation rate and wider spectrum of mutation events. Instances of the serial exposures of the target materials to multiples of the same type of mutagen (physical or chemical) are rare in literature. The results from combined treatments have been largely ambivalent with the aforementioned expectations of enhanced mutagenicity never conclusively realized. This probably explains the significantly low number of officially released mutant crop varieties ascribable to a combination of physical and chemical mutagenic treatments (or multiple treatments with the same type of mutagen); these, in fact, account for approximately only 1% of all officially released crop mutants. In practical terms also, it is arguable whether increases in mutation rates beyond those currently routinely produced using single mutagenic treatments are useful in breeding; the unintended mutation events housed in the putative mutants would encumber the breeder with unnecessary extra efforts to get rid of the deleterious mutations.

3.3. Determination of Optimal Dose

The effectiveness and efficiency of any induced mutagenesis experiment (estimates of mutation density and ability of the mutagen to induce desirable changes with minimal unintended effects) are direct results of the choice of appropriate mutagen dosage. It is standard practice therefore to precede any bulk mutagen treatment with a preliminary determination of the dosage that would yield the greatest amount of desirable mutation per unit dose while at the same time producing the least density of undesirable effects. When using physical mutagens, this is known as radiation sensitivity testing, but more commonly referred to by the shortened form, radiosensitivity tests. A determination of the mutagen dose that results in a 50% reduction in seed germination or seedling viability, known as LD_{50}, has been used to predict accurately the most effective and efficient mutagen dose. Equally demonstrably useful in predicting such a mutagen dose is the determination of the mutagen dose that results in the reduction of average seedling growth by 50%; this is known as RD_{50}. In practice, mutagen doses leading to 30% lethality and 30% growth reduction, LD_{30} and RD_{30}, respectively, are also used in induced mutations aimed at generating mutant populations for crop improvement. The determinations of these parameters are fairly straightforward and involve comparisons between the untreated (control) samples with the treated samples (according to mutagen dose) for the effects the mutagen treatment. These comparisons are based on estimates of plant injury using one, or a combination of the phenotypic characters listed in **Box 14.1**.

In addition to empirically determined radio-sensitivity (providing the reference thresholds for mutation induction in plants), nature has also imposed inherent limitations on how many induced lesions a plant can reasonably tolerate. Beyond a certain threshold, the combination of biological and genetic damage can lead to sterile offspring and even lethality. The latter manifests in low or complete loss of germination of the treated materials; even when seeds germinate after exposure to extremely high doses of mutagens, the seedlings fail to grow and die soon afterwards.

3.4. Considerations for Induced Mutagenesis: Pre- and Post-Treatment Handling of Materials

3.4.1. Choice of Target Material: Seed Propagated Plants

Mutations are induced in plants through the exposure of their propagules to mutagens. For plants that are propagated through seeds, the procedure is more straightforward as botanical seeds are easily exposed to the mutagens. Seeds are easy to store and transport and generally permit the exercise of more controls over the dose of mutagens delivered. In order to expedite the recovery of the mutation events in homozygous states, procedures for inducting mutations using haploid gametic cells and using doubled haploidy to generate fertile diploid plants have been developed for a number of crops e.g. rice, barley and maize (see also **Chapter 29**).

When seeds are used as starting materials, dry, quiescent seed samples are used and further standardized by discarding any injured, abnormal or diseased seed. Seeds should be representative of the genotype and have a high germination capacity. In certain plants breeding materials may have distinct seed coverings, for example in oil palm thin-shelled commercial tenera palms are produced by crossing thick-shelled dura with shell-less pisifera types. The presence/absence and thickness of the shell will affect the mutation rate and treatments may need to be adjusted accordingly, or alternatively the shell may be removed and/or the embryo excised and the mutagenic treatment applied to naked (usually cultured) embryos. Procedures for breaking seed dormancy, such as cold treatment (prechilling), exposure to heat, breaking of hard seed coat (scarification) or hormonal treatment, precede irradiation of dormant seeds.

3.4.2. Choice of Target Material: Vegetatively Propagated Plants

In plants where sexual reproduction is not feasible, i.e. vegetatively propagated plants, other plant propagules are used, e.g. buds, tubers, stem cuttings, vines, twigs and *in vitro* nodal segments. With advances in cell and tissue culture, experiments are ongoing with a view to develop reproducible protocols for the use of different types of aseptic cultures including cell suspensions, somatic embryos and micropropagation systems as targets for inducing mutations. These types of cultures permit massive vegetative production in a short time frame and in relatively very little space (laboratories as compared with extensive hectarage in the field).

In general, depending on the type of propagule used, due considerations are given to the appropriate strategies for mitigating the effects of chimerisms (most pronounced with vegetatively propagated plants) and in recovering the sought after mutation events in homozygous states and in

suitable genetic backgrounds, usually with as few unintended mutation events as possible complicating selection on account of linkage drag. The strategies for handling mutagenic populations through subsequent generations as the putative mutants are evaluated, segregations are monitored and mutation events are fixed in homozygous states also vary between inbreeding and outcrossing crop species. Further consideration of induced mutation in vegetatively propagated crops is given in **Chapter 26**.

3.4.3. Handling of Seeds Before Exposure to Mutagens

Prior to gamma irradiation, the most commonly used form of physical mutagen, the moisture content of the seeds is equilibrated prior to exposure to irradiation, this is known as pre-conditioning. Pre-conditioning is done by packing the seeds loosely in water-permeable paper bags or mesh bags in lots of desired amounts per dose and labelled with information on species, name of variety or genotype, date and the treatment dose. The size of the bags must be suitable for the seeds and should fit into the irradiation compartment.

For seed moisture adjustment the seed coat and/or shell covering must be water permeable or else removed or mechanically or chemically modified. Seed coats can be rubbed with sandpaper, nicked with a knife or filed with a metal file. The seed sample is placed in a vacuum desiccator above a 60% glycerol-distilled water mixture; the desiccator is maintained at room temperature, with an internal relative humidity of about 73% (monitored with a hygrometer). Samples ranging from 10–500 g of seeds are normally handled and adjusted in a desiccator with a capacity for 1,000 ml of the glycerol/water mixture.

At the FAO/IAEA Seibersdorf laboratory, a routine pre-conditioning treatment for a minimum of 7 days with 60% glycerol is used for equilibrating the moisture contents for small grain cereals (such as rice, wheat and barley). This is extended for up to 14 days for greater quantities, larger size seeds (e.g. beans) and/or seed with thick seed coats. Different species may not equilibrate to the same water content at a particular relative humidity. In general it is sufficient to follow the desiccation procedure described above and to have a standardized method without the need for determining the seed water content. However, if uncertainties arise, the seed water content can be measured according to the rules of the International Seed Testing Association (ISTA, 2003).

Seeds are brought out from the desiccators just before radiation treatment. If the moisture equilibration cannot be followed immediately by radiation treatment, the seeds should be packed in air-tight containers during transport to the gamma source.

Environmental factors have minimal effects on the outcome of fast neutron irradiation; no seed moisture adjustment is therefore necessary prior to exposing seeds to fast neutrons. The above procedure described for seeds is designed for exposure to gamma radiation and is therefore omitted for far neutron treatments. The seeds are merely packed in lots of desired amounts per dose and labelled accordingly as above. Again, the size of package must be suitable for the seeds and should fit into the irradiation containers. Depending on the type of reactor it may be advisable to use plastic bags or plastic vials in order to prevent any potential spoilage by cooling water inside the reactor.

4. References

4.1. Cited References

Bottino, P.J., Sparrow, A.H., Schwemmer, S.S. *et al.* 1975. Interrelation of exposure and exposure rate in germination seeds of barley and its concurrence with dose-rate theory. *Radiation Botany.* 15: 17–27.

Broertjes, C. 1972. Use in plant breeding of acute, chronic or fractionated doses of x-rays or fast neutrons as illustrated with leaves of *Saintpaulia*. Thesis; Centr. Agric. Publ. Doc., Wageningen, Agr. Res. Report. (Verlag Landbouwk. Onderzoek.) 776, ISBN9022003884, 74pp. (cited by Bottino *et al.* 1975).

Dewey, D.L. 1969. An oxygen dependent X-ray dose rate effect in *Serratia marcescens*. *Radiation Research.* 38: 167–474.

Fujii, T. 1962. Comparison of the effects of acute and chronic irradiations. *Recent Advance of Plant Breeding (Ikushugaku Saikin no Sinpo).* 4: 60–69.

Hall, E.J. and Bedford, J.S. 1964. A comparison of the effects of acute and protracted gamma-radiation on the growth of seedlings of *Vicia faba*. Part I. Experimental observations. *Int. J. Rad. Biol.* 8: 467–474.

Hall, E.J., Oliver, R., Shepstone, B.J. *et al.* 1966. On the population kinetics of the root meristem of *Vicia faba* exposed to continuous irradiation. *Radiation Research.* 27: 597–603.

ISTA 2003. International Rules for seed testing, Edition 2003, International Seed Testing Association, Zuerich, Switzerland.

Matsumura, S. 1965. Relation between radiation effect and dose rates of X and gamma rays in cereals. Japan. *J. Genetics Suppl.* 40: 1–11.

Natarajan, A.T. and Maric, M.M. 1961. The time-intensity factor in dry seed irradiation. *Radiation Botany.* 1:1-9.

Takagi, Y. 1969. The second type of gamma-ray sensitive gene RS2 in soybean *Glycine max* (L.) Merrill. *Gamma Field Symposia.* 8: 83–94.

Ukai, Y. 1986. Development of various irradiation techniques for frequency enhancement of radiation breeding. *Gamma Field Symposia.* 25: 55–70.

van Harten, A.M. 1998. Mutation Beeding. Theory and practical Applications. Cambridge, New York: Cambridge University Press, p. 353.

4.2. Websites

NIAS (accessed May, 2004). Gamma field. Institute of Radiation Breeding, NIAS, MAFF, Japan: http://www.irb/affrc.go.jp/gf-l.gif

4.3. Further Reading

Brunner, H. 1985. Radiosensivity of a Number of Crop Species to Gamma and Fast Neutron Radiation. Standards for Laboratory Operations Involving Chemical Mutagens. A Training Manual. Plant Breeding Unit, Seibersdorf: FAO/IAEA Laboratories.

Nilan, R.A., Konzak, C.F., Wagner, J. *et al.* 1965. Effectiveness and efficiency of radiations for inducing genetic and cytogenetic changes. *Suppl. Rad. Bot.,* 5: 71–89.

C15

Chimeras and Mutant Gene Transmission

A.R.Prina*, A.M.Landau and M.G.Pacheco

Instituto de Genética "Ewald A. Favret", CICVyA-INTA. P.O. Box 25, B1712WAA, Castelar, Argentina
* Corresponding author, E-MAIL: aprina@cnia.inta.gov.ar

1. Chimeras

1.1. Definition and Types

In 1907, Winkler first used the word 'chimera' to describe a plant arising from grafting together two *Solanum* species (**Box 15.1**). The plant developed a shoot that longitudinally had two distinct halves, each belonging to a different species. A chimera is defined as an individual that has sectors made up of different cell genotypes. These cells differ in chromosome structures or in chromosome number (aneuploidy or ploidy) or even in only one nucleotide within a nuclear and/or an organelle genome. It is worth mentioning that in a broader sense some variegated phenotypes originating from epigenetic effects that are propagated through cell division and even across generations can be also considered to be chimeras.

Due to differences in genes affecting pigmentation plant chimeras can be revealed in clonal or cell lineage variegations (**Box 15.1**). The different patterns of clonal variegation are directly associated with the patterns of cell lineages, which differ markedly between monocots and dicots, particularly in leaf patterns. In both taxa, clonal variegations in the leaves are clear cut and are delimited by the leaf veins, showing up as elongated stripes in monocots and irregular patches in dicots (**Figure 15.1A and B**). These chimerical variegations are distinct from plants exhibiting positional variegation, as in the latter, the phenotypic differences among tissues of a leaf are due to differential gene expression and not due to genetic differences, and hence they are not chimeras (**Box 15.1**). Leaves with typical positional patterns, also called figurative patterns, are those presenting different pigmentations between the inter-venial spaces and the vein margins or between the upper and the lower part of the leaf blade (**Figure 15.1C and D**).

1.2. Chimeras Induced by Mutagenic Treatments

Multicellular organs, such as seed embryos or vegetative buds, are commonly subjected to mutagenic treatments for plant breeding purposes. Mutation rates are increased manyfold by mutagenic treatments; however, since every cell has multiple potential targets for mutation induction, any particular mutation can still be considered a rare event at the level of a single cell. Thus, it is very unlikely that a given mutational event

independently occurs in more than one of the cells that constitute a meristem and much less probable that it simultaneously affects both copies of any particular gene of a diploid cell. As a consequence the vast majority of mutations will be heterozygous. Each of these mutant cells, if able to multiply by mitosis, will form a unique heterozygous mutant cell lineage sector, which will grow surrounded by a normal background or by other different mutant sectors with variable sizes and patterns. Therefore, plants derived from seeds treated with a mutagen are complex chimeras.

Several factors can affect the formation of chimeras in M_1 plants: (1) the nature of the experimental material, such as the species ontogeny, the embryonic structural features and the developmental stage, specially the number of cells that will grow into different meristems; (2) the nature and dose of the applied mutagen, particularly its cell killing effects (**see Chapter 14**); and (3) the competitiveness of the mutant sector.

Most plants of the M_2 generation, are not chimerical, since every plant derives from a zygote by mitosis. The unicellular constitution of gametes and zygotes does not allow the transmission of chimeras based in nuclear genes from one to another generation. The same is not true for chimeras based on plastid or mitochondrial genes, since semi-autonomous organelles (plastids or mitochondria) are several per cell and they may be unevenly transmitted to daughter cells through mitosis.

When unicellular organs, such as those of the male or female gametophytes or zygotes, are treated with mutagens, chimeras based in nuclear genes can be avoided in M_1 plants, but organelle gene chimeras may still exist (**see Chapter 17**). Chimeras in M_2 plants or in further generations can also arise from genetically unstable mutant genotypes, *e.g.* those carrying mutator genes (**see Chapter 17**) or through the reactivation of transposons (Walbot, 1988), which can produce new variability recurrently.

1.3. Appearance of Chimeras in M_1 Plants

1.3.1. Somatic Chimeras

When mature seeds are subjected to mutagenic treatment, the number of leaf primordia cells is one of the factors determining the final size of the mutant sectors produced in fully grown leaves of the M_1 plants. In cereals the mature embryo is highly differentiated relative to that of dicots. For example, a mature barley embryo

Box 15.1: Key terms related to chimeras and mutant gene transmission

Chimera: from Greek *chimaira*: a mythological animal composed of parts of three different animals. It is traditionally applied in biology to define a plant (or an animal, an organ or a tissue) consisting of somatic sectors having different genetic constitutions. In modern biotechnology, it is also used to designate transgenic cells or organisms carrying DNA from more than one species. In plant mutation research, the former definition is adopted.

Cell-lineage or cell clone: cell population derived from a single cell by repeated mitosis.

Cell-lineage or clonal variegation: different pigmentations in a variegated plant are due to genetic differences between cell clones. This is an indication that the plant is chimerical. The patterns of variegation will depend on the ontogeny of the species.

Positional variegation: pigmentation differences among tissues of the same plant result from differential expression depending on the position of the cells and are not due to genetic differences. Plants carrying positional variegation are not chimeras.

Diplontic selection: competition occurring during the diplophase between somatic cells or cell clones carrying different genotypes in a chimerical organism. It is also called intra-individual or intra-somatic selection.

Haplontic selection: competition occurring during the haplophase. After overcoming diplontic selection in the soma of a chimerical plant, the mutant clones that reach M_1 inflorescences need to compete in gamete production and transmission to zygotes, in order to be transmitted into the second generation (M_2).

Ontogeny: refers to the developmental programme of an individual or organ.

Initial cells: cells capable of dividing to form a meristem. In a seed embryo, those mutations that affect initial cells, the originating cell lineages that eventually form reproductive tissues, are the only ones that may be transmitted to next generations.

The Joint FAO/IAEA Programme

Figure 15.1 Examples of clonal or cell lineage variegation (A and B) and positional variegation (C and D). A, barley; B, cotton; C, wheat; and D, cotton.

The Joint FAO/IAEA Programme

Figure 15.2 Chlorophyll somatic sectors of barley M₁ leaves. A, Plant showing two sectors of different colour, one *albina* on the fifth and the other *xantha* on the sixth leaf. B, *Albina* streak. C, Yellow streak on *viridis* background. D, *Viridis* streak on normal background. E, Normal green streak on a *viridis* heterozygous background. F, Yellow-*albina* double sector on the double heterozygous MC 136 (from Prina and Favret, 1988, Copyright 1988 by the American Genetic Association).

consists of the apical meristem, two auxiliary buds, three, sometimes four or exceptionally five, primordial leaves and several root initials (Mullenax and Osborne 1967). The first primordial leaf blade is composed of a greater number of cells than the second leaf, the third and so on. As a consequence the size of corresponding mutant sectors decreases in fully grown leaves: will consist of a few cells in the first leaf blade; but in the fourth or subsequent leaves they will appear as longitudinal stripes traversing the blade lengthwise (**Figure 15.2**).

Most induced mutant alleles are recessive and hence they are not expressed phenotypically in the heterozygous mutant cell-clones of M₁ plants. Even if dominant or semi-dominant mutant alleles are induced, most of them cannot be distinguished because they may affect characters that are not visual *e.g.* a mutant allele of root or grain quality traits. In addition, muta-

genic treatments can result in physiological disorders of M₁ plants, which can further make it difficult to identify mutated sectors. However, one type of chimera, which produces conspicuous changes in chlorophyll content, has been frequently observed and extensively studied in M₁ plants.

Since the early days of chimera research it has been noticed that leaves of some M₁ plants exhibit chlorophyll-deficient sectors, as longitudinal streaks in monocots and as irregular-shaped spots in dicots (**Figures 15.2 and 15.3**). When homozygous wild type barley seeds are treated with mutagens, either with chemicals or with X-rays, two types of chlorophyll deficiencies are often observed: light green (~70%) and albino (**see Chapter 25**). However, when heterozygous tester stocks are subjected to mutagenic treatments, the mutant spectrum of somatic sectors dramatically

The Joint FAO/IAEA Programme

Figure 15.3 Chlorophyll-deficient sectors on the first foliage-leaf in cotton after seed treatments with a) sodium azide or with b) X-ray treatments.

changes depending on both the gene, which is in a heterozygous condition and the type of mutagen. X-rays mutagenesis showed a more generalized effect than chemical treatments on different genes, including not only chlorophyll-deficient genes but also genes related to diseases resistance (Prina and Favret, 1988). In this respect, it is worth mentioning that in the case of vegetatively propagated species the experimental material usually carries several *loci* in a heterozygous condition, which can drastically influence the spectrum of mutant somatic sectors.

1.3.2. Cell Lineages and Chimeras in Reproductive Tissues

The chimerical nature of M_1 plants can be used to distinguish induced variability from that already present in the material before treatment, when M_1 spikes are individually harvested and grown, as proposed by Stadler (1930) at the dawn of experimental mutagenesis. (**Figure 15.4**). This is based on the premise that the reproductive tissues of each of the main spikes arise from different initial cells (**Box 15.1**). Newly induced mutations would segregate in only one (rarely in two) of the main M_1 spike progenies (**Figure 15.4**). On the other hand, phenotypic segregations observed in all or in most of the spike progenies correspond to variability already existing in a heterozygous condition, which could arise by spontaneous mutation.

Although most induced mutations are not expressed in somatic tissues of M_1 plants, they constitute hidden chimeras that play an important role in the development of mutated sectors that proceed into reproductive organs (male and/or female gametes). Knowledge of how different sectors grow and finally develop into reproductive organs is important for handling experimental material appropriately, *i.e.* how to grow and harvest M_1 plants, how to grow and select mutant plants in the M_2 and subsequent generations (**see Chapter 18**).

The initial cells of the mature embryo that develop into reproductive organs are called germ track cells. The number of germ track cells is termed the "genetically effective cell number" (GECN), and this must be estimated if mutation frequencies are expressed on a treated-cell basis.

Histological analyses that aim to determine the GECN are generally very difficult and often give contrasting results (see D'Amato, 1965). Interestingly, GECN can be estimated by analyzing the segregation ratios of chlorophyll mutants in M_1 plant inflorescence progenies. In this manner, the reproductive tissues in each of the main tillers that originate from more than one initial cell were determined in cereals (barley, rice and *Triticum durum*). According to Gaul (1964) 2-4 initial cells give rise to each of the main 4-5 spikes of barley in undisturbed development, but this number decreases after treatments that have high cell-killing effects (**see Chapter 14**). Considering the absence of deviations from selective forces and segregation of only recessive mutant alleles, it is estimated that rates near 1 mutant (M):3 normal (N) seedlings indicate that the reproductive tissues of the spike come from one initial cell; the rates will theoretically decrease to 1 M:7 N, 1 M:11 N and 1 M:15 N, for 2, 3 or 4 initial cells, respectively. In this manner, the segregation of 2 M:22 N observed in the spike-progeny represented in second place in **Figure 15.4A** suggests that the GECN in that spike was three. Obviously, spike progenies consisting of 20–30 M_2 seedlings, or as fewer

A. Spike progenies from the main tillers.
Segregation of recessive mutants induced by the treatment is expected to occur in only one M_1 spike progeny and usually in proportions below 3N:1M, because the reproductive tissues of the main tillers usually arise from more than one initial cell of the seed embryo.

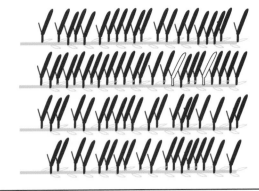

B. Spike progenies from lateral tillers. In lateral tillers mutant segregations take a bigger portion of the spike progeny. In plants with many lateral tillers it can take even more than one spike progeny, as in the example where two spike progenies originated in the same mutant initial cell.

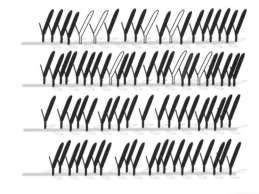

C. Spike progenies from a heterozygous plant.
Segregation of mutants is observed in all the spikes progenies indicating that the mother plant was a solid heterozygote. This means that the original mutation was induced previously to the treatment or came from pollen contamination.

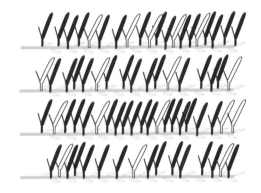

The Joint FAO/IAEA Programme

Figure 15.4 Three typical situations in the analysis of mutant segregation in individual M_1 spike progenies.

is commonly found after mutagenic treatments *e.g.* in barley, are very small for distinguishing situations with increasing GECNs. Another more precise method used for analysis of mutant sectors in M_1 barley spikes was achieved by the analysis of waxy pollen grains (*e.g.* Lindgren *et al.,* 1970). This character is determined by the pollen grain genotype and not by that of the mother plant and, therefore it is particularly useful for chimera analysis in the M_1 inflorescence. Such analyses indicate that mutant sectors present a wide variation in patterns and sizes, from one single anther to an entire spike.

It must be mentioned that lateral tillers can originate from cells still undetermined in the seed embryo, which indicates that distinction between germ or somatic lines is not strictly established in seed meristems and that some initial cells in the seed embryo can originate clones constituting both vegetative and reproductive tissues.

The variable number of initial cells forming reproductive tissues in main or lateral tillers of cereals and the variable size of the progenies in M_2 complicates the estimation of mutation rates. This matter prompted long discussions about the best scoring method for evaluating mutagenic effects. Conventionally, mutation frequency is estimated based on the number of chlorophyll mutations per hundred M_1 spikes. Gaul (1960) criticized this method considering that it may underestimate the actual mutation frequency in the case of high damaging treatments (see next section). He proposed to use the number of chlorophyll mutants per hundred M_2 seedlings as a general method. Several modifications for scoring mutation frequencies have been proposed, however, after theoretical comparisons of various methods it was concluded that none of them was invariably superior to others and that effectiveness of each method depends on the conditions and aims of the experiment (Yonezawa and Yamagata, 1975).

2. Competition of Mutated Versus Non-Mutated Cells

In M_1 plants, genetically different cell clones grow together in the same soma, which may bring about competitions for survival and growth. In this respect, there are two main sieves that a newborn mutant allele needs to pass through to be transmitted to the second generation. The first sieve acts at the level of the soma of the M_1 plant and depends on both the capability of the mutant cell to divide and generate (by mitosis) a mutant clone, and the fitness of that clone to compete with others (mutants or wild type) during M_1 plant development, and finally end up in reproductive tissues. This competition among somatic cells during the diplophase was called "diplontic selection" by Gaul (1964) and intra-individual or intra-somatic selection by Kaplan (see D'Amato, 1965). The second sieve occurs when the mutant clones, after overcoming diplontic selection, reach the reproductive organs of the M_1 plant. In order to get into the second generation (M_2) the mutant allele needs to compete in gamete production and in zygote formation. As this competition occurs during the haplophase, it is called "haplontic selection".

2.1. Diplontic Selection

Evidence of diplontic selection was observed in barley, rice and wheat, where the number of sectors constituting the M_1 inflorescences decreases with increasing radiation dosage (e.g. Gaul, 1964). This fact is explained by the high cell-killing effects of high-radiation doses that induce extensive chromosome disturbances and render some cells unable to multiply by mitosis. Accordingly, chemical mutagens have a slighter cell-killing effect than ionizing radiations and show a higher

number, but of smaller mutant sectors in M_1 inflorescences. It must be taken into account that favourable alleles can also be eliminated by diplontic selection when they are located in a cell that is affected by deleterious changes.

It was hypothesized that different levels of competition exist among clones arising from initial cells already pre-determined in dormant seeds and those clones derived from cells still undetermined, and hence M_1 inflorescences of later-forming tillers produce M_2 progenies with lower mutation rates (Gaul, 1964). According to this hypothesis, non-differentiated cells do not generally take over the function of those initial cells already pre-determined and so, competition for constituting the generative tissues in the main spikes is limited to only one to four cells in each. The initial cells not yet differentiated at the time of treatment, such as those of the lateral-tiller spikes, will be involved in a larger cell complex competing to form the reproductive tissues. On a practical level, differences in mutation rates between main and lateral tillers of barley may be magnified if the high plant lethality caused by high mutagenic doses is not compensated for by space planting M_1 seeds closely, in order to avoid having sparse plots with excessive tiller production.

It is worth mentioning that a distinction must be made with respect to competition among diploid cells grown *in vitro*, which occurs in a quite different context (**see Chapter 16**). Finally, it must be remarked that somatic competition is more complex in the case of induced mutations in organelle genes (**see Chapter 17**).

2.2. Haplontic Selection

During gametogenesis, recessive mutants are exposed to selection for the first time. In the case of mutant alleles expressed in a heterozygous condition, sporophyte and gametophyte functions can differ drastically and allele fitness values might differ from one phase to another.

In barley, seed set reduction of M_1 spikes has been largely used in estimating female gamete failure, even though other causes of seed set reduction, such as pollen abortion or early zygotic lethality cannot be ruled out. In the first decades of experimental mutagenesis, when the use of ionizing radiations was common, chromosome aberrations were assumed as the main source of M_1 sterility. Later, it was evident that other factors were involved, e.g. point mutations, small deficiencies, as well as physiological damages (Ekberg, 1969). Selection against most chromosomal deficiencies is usually much more intense among male gametes than females. However, as the population size in the male gametophyte (pollen production) largely exceeds that of the female counterpart (ovules) the effects of seed treatments on the male gametes have a much lower impact on M_1 seed-set.

Estimation of gametic competition by analysing M_2 mutant segregation is usually difficult because of deviations due to the chimerical constitution of M_1 inflorescences and diplontic selection. Evidence for moderate gametic competition has been obtained by analysing progenies of solid heterozygous plants. Extensive investigations carried in barley, wheat and pea showed a significant deficit in recessive mutants of about 20% in some cases, but an excess of recessive mutants was rarely found. It is generally agreed that deviating segregation frequencies in induced chlorophyll mutants are mainly due to reduced transmission of the mutant allele through the male gametophyte. Gametophytic lethal alleles have no transmission at all through the male gamete and represent an extreme case of (none) allele transmission. The pollen-irradiation method, used early on in maize and more recently in *Arabidopsis* (Naito *et al.*, 2005), allowed determination of sexual transmissibility of many mutations, including large deletions, induced by pollen irradiation.

It is obvious that gamete competition is only an issue for sexually propagated species. Therefore, some alleles having detrimental effects in gametes can be maintained in vegetative propagated species. In addition, lack of recombination in mitosis allows the maintenance of mutants in a heterozygous condition.

2.3. M_2/M_1 Mutant Frequencies Relationship

Blixt (1972) found a close correlation between M_2 chlorophyll mutants and M_1 leaf spotting induced by chemical mutagens in pea. However, it was observed in barley that the relationship between the frequency of M_2 mutant seedlings and the frequency of chlorophyll-deficient sectors observed in M_1 plants was much lower with X-rays than with chemical mutagens. Thus, after X-ray treatments, the ratio between M_2 chlorophyll mutations and M_1 chlorophyll sectors was as low as 0.4 and the opposite trend was observed after sodium azide treatments, which showed a $M_2:M_1$ ratio as high as 6.3. Meanwhile,

EMS showed an intermediate value of 2.5. In addition, as mentioned earlier, the expression of somatic mutations can drastically change depending on the genotype and, consequently, $M_2:M_1$ relationships can be markedly affected. Different trends in $M_2:M_1$ relationships were also observed when these mutagens were used in hexaploid wheat as compared with diploid barley. All the aforementioned indicates that the relationship between $M_2:M_1$ observations markedly depends on the particular mutagen–genotype combination and, therefore, making quick estimations of expected M_2 mutations based on the frequency of M_1 chlorophyll deficient sectors is only reliable if the relationship for the particular combination is already known.

Another interesting point to be considered is the relationship between the sterility of M_1 inflorescences and the mutation frequency in M_2. In several crop plants (barley, rice, pea and tomato) no correlation was found after seed irradiation. However, chemical mutagenic treatments on barley seed resulted in a high correlation between M_1 sterility and M_2 mutant frequencies, indicating that, at least in the case of some chemical treatments, elimination of highly sterile M_1 inflorescences as sometimes proposed in order to get rid of deleterious effects of the treatments, could contribute to a decrease in M_2 mutant frequencies (Prina *et al.*, 1986).

3. Acknowledgements

We wish to thank Graciela del Castaño, Mariela Trazar, Liliana Barchetta and Raúl Daniel Bassi for help provided during manuscript preparation.

4. References

4.1. Cited References

Blixt, S. 1972. Mutation genetics in *Pisum. Agr. Hort. Genet.* 30: 1–259.

D´Amato, F. 1965. Chimera formation in mutagen-treated seeds and diplontic selection. In: The use of induced mutations in Plant Breeding (Proc. Meet FAO/IAEA, Rome, 25th May–1st June 1964), pp. 3–16.

Ekberg, I. 1969. Different types of sterility induced in barley by ionizing radiations and chemical mutagens. *Hereditas.* 63: 255–278.

Gaul, H. 1964. Mutations in Plant Breeding. *Radiat. Bot.* 4: 155–232.

Lindgren, D., Eriksson, G. and Sulovska, K. 1970. The size and appearance of the mutated sector in barley spikes. *Hereditas.* 65: 107–132.

Mullenax, R.H. and Osborne, T.S. 1967. Normal and gamma-rayed resting plumule of barley. *Radiat. Bot.* 7: 273–282.

Naito, K., Kusaba, M., Shikazono, N. *et al.* 2005. Transmissible and nontransmissible mutations induced by irradiating *Arabidopsis thaliana* pollen with gamma-rays and carbon ions. *Genetics.* 169: 881–889.

Prina, A.R. and Favret, E.A. 1981. Comparative analysis of the somatic mutation process in barley. *In* : Barley Genetics IV (M.J.C. Asher (ed.), Edinburgh: Edinburgh University Press, pp. 886–891.

Prina, A.R. and Favret, E.A. 1988. Influence of marker genes on the expression of somatic mutations in barley. *J. Heredity.* 79: 371–376.

Prina, A.R., Hagberg, A. and Favret, E.A. 1986. Inheritable sterility induced by X-rays and sodium azide in barley. *Genetica Agraria.* 40: 309–320.

Stadler, L.J. 1930. Some genetic effects of X-rays in plants. *J. Hered.* 21(1): 3–19.

Walbot, V. 1988. Reactivation of the Mutator transposable element system following gamma irradiation of seeds. *Gen. Genet.* 212: 259–264.

4.2. Further Reading

Birky, C.W. Jr. 2001. The Inheritance of genes in mitochondria and chloroplasts: laws, mechanisms, and models. *Ann. Rev. Genet.* 35: 125–148.

Bengtsson, B.O. 1976. Two methods of estimating mutagenic effects after treatment of barley seeds. *Hereditas.* 83: 39–46.

Cecconi, F., Pugliesi, C. and Baroncelli, S. 1992. A clonal analysis in sunflower (*Helianthus annuus L*) – I. The capitulum development. *Env. Exp. Bot.* 32(4): 505–512.

Grant-Downton, R.T. and Dickinson, H.G. 2005. Epigenetics and its Implications for Plant Biology. 1. *The Epigenetic Network in Plants. Annals of Botany.* 96: 1143–1164.

Kirk, J.T.O. and Tilney-Bassett, R.A.E. 1978. The Plastids. Amsterdam : Elsevier, p. 960.

Koornneef, M. 2002. Classical mutagenesis in higher plants. *In*: Molecular Plant Biology Vol.1. P.M. Gilmarin and C. Bowler (eds.), Oxford: Oxford University Press, pp. 1–11.

Kurup, S., Runions, J., Köhler, U. *et al.* 2005. Marking cell lineages in living tissues. *The Plant Journal.* 42: 444–453.

Li, S.L. and Rédei, G.P. 1969. Estimation of mutation rate in autogamous diploids. *Radiation Botany.* 9: 125–131.

McDaniel, C.N. and Poethig, S. 1988. Cell-lineage patterns in the shoot apical meristem of the germinating maize embryo. *Planta.* 175: 13–22.

Bird, R.M.K. and Neuffer, M.G. 1987. Induced mutations in maize. *Plant Breeding Revs.* 5: 139–180.

Yu, F., Fu, A., Aluru, M. *et al.* 2007. Variegation mutants and mechanisms of chloroplast biogenesis. *Plant Cell and Environment.* 30: 350–365.

C16

Chimeras: Properties and Dissociation in Vegetatively Propagated Plants

T.Geier*

Fachgebiet Botanik, Forschungsanstalt Geisenheim, D-65366, Germany
* Corresponding author, E-MAIL: t.geier@fa-gm.de

1. Introduction

Chimeras are individuals composed of tissues of two or more idiotypes which are represented by their respective cell lines in the shoot apical meristem. In addition to 'chimeras' the term 'mosaics' is sometimes used synonymously. However, the latter is better applied to patterns resulting *e.g.* from permanent genetic instability which, unlike the 'true' chimeric patterns dealt with in this chapter, may be transmitted through single cells. Plant chimeras may be classified according to their origin, structure and the type of genetic difference in their component idiotypes. Regarding the origin, chimeras may be obtained as a result of grafting, sorting out from variegated seedlings, genetic transformation, as well as spontaneous or induced mutation. Only the latter is considered in more detail here.

In view of homohistic ('solid') mutant development in plant breeding, the classification of chimeras according to their structure is most important and hence will be considered in more detail. Mutagenic treatment of seeds and buds (tip/node cuttings), as commonly practised in mutation breeding programmes, leads to mutant M_1 plants exhibiting chimerism. This is because mutations occur in single cells and shoots are recovered from pre-existing multicellular meristems. Likewise, most of somatic mutations occurring spontaneously in nature, appear in a chimeric condition, which may persist during vegetative growth, sometimes unrecognized, over a long time. As will be detailed in the following paragraph, the mode of spreading and spatial arrangement of a mutant cell lineage results from the layered structure of the shoot apical meristem and ordered orientation of cell division planes, as well as the competitiveness of the mutant cells. A cell lineage derived from a mutated cell first spreads as a sector within the tissue layer where the mutation event had taken place. This kind of chimeric constitution called 'mericlinal' is unstable, the size and persistence of such sectors depending on the competitiveness of the mutated tissue. Lateral bud formation and sprouting within a mutated sector leads to a stable periclinal chimeric state in which mutated cells occupy a whole layer of the lateral shoot meristem.

In order to recover homohistic ('solid') mutants, the chimeric stucture of M_1 plants obtained after spontaneous or induced mutagenesis needs to be dissociated. In seed-propagated crops this is commonly done by selfing of M_1 plants and raising the M_2 progeny. In addition to chimera dissociation, selfing enables mutations to be brought in homozygous condition, which is required for true-to-type seed propagation. Chimera dissociation and mutant gene transmission in seed-propagated crops is described in detail in **Chapter 15** of this book. In the present chapter, main emphasis is placed on methods of chimera dissociation using plant regeneration *via* adventitious buds or somatic embryos. Unlike selfing, plant regeneration from somatic cells permits the genetic background of the treated varieties to be

---

Box 16.1: Types of chimeras

Plant chimeras can be classified:

1. According to the difference in their component idiotypes into:
- **Gene-differential chimeras**, which arise by somatic mutation of a nuclear gene. This type of chimera occurs spontaneously or may be induced by mutagenic treatments. To cause a chimeric phenotype, the mutation needs to be dominant, affect the dominant allele of a heterozygous genotype or occur in a haploid individual.
- **Organelle-differential chimeras**, which arise either by organelle mutation or by sorting out of two or more kinds of plastids or mitochondria from a mixed egg or zygote.
- **Chromosomal chimeras ('cytochimeras')**, the tissue layers of which differ in their chromosome number. This type of chimera is artificially produced by teatment with spindle poisons, e.g. colchicine, or sometimes arise spontaneously.
- **Transformation chimeras**, which arise by genetic transformation of a somatic cell and subsequent joint regeneration of the transformed cell and untransformed neighbours.
- **Graft chimeras ('species chimeras')**, produced artificially by grafting together different species or genera. At the graft union a callus tissue is formed and this gives rise to adventitious buds of chimeric structure.

2. According to their structure into:
- **Sectorial chimeras** (component idiotypes grow side by side and occupy distinct sectors of varying size).
  Sectorial chimeras are unstable and tend to segregate rapidly into pure types.
- **Mericlinal chimeras** (one of the component idiotypes occupies a sector of limited extent within a tissue layer of the plant body).
  Mericlinal chimeras are unstable and tend to lose their chimeric nature or develop into stable periclinal chimeras.
- **Periclinal chimeras** (component idiotypes occupy different peripheral tissue layers and core tissue of the plant body). Periclinal chimeras are fairly to very stable, provided the shoot apical structure is preserved.

retained, which is of particular importance in vegetatively propagated crops. Moreover, when mutagenic treatment is immediately applied to materials devoid of pre-existing meristems and this is followed by regeneration of adventitious buds, chimera formation can be largely avoided, because such buds often, though not always, originate from single cells. Employment of this so called 'adventitious bud technique' thus provides the possibility of immediate and efficient production of homohistic ('solid') mutants, but with the addition that this is true for nuclear mutations only. In case of plasmon mutations, cells with mixed mutated and wild type organelle populations are produced. Sorting out of organelles during subsequent regeneration is often incomplete and chimeras are frequently obtained, even if plant regeneration occurs from single cells.

2. Origin, Structure and Properties of Chimeras

2.1. Shoot Apical Structure and Plant Ontogenesis

Most spontaneous somatic mutations observed in nature are of mericlinal or periclinal chimeric condition. The different layered arrangements of component idiotypes in those chimeras (**Figure 16.2b–i**) result from the layered structure and highly ordered orientation of cell division planes within the shoot apical meristem (SAM), which was first described as 'tunica-corpus theory' by Schmidt (1924). In a typical angiosperm shoot tip (**Figure 16.1**), the core tissue (= corpus) is covered by two tunica cell layers. The number of tunica layers may vary (from only one to three or even more) among species and at different stages of development in the same species. Commonly, the layers are designated as L1, L2, L3 and so on (if there are more than the typical two tunica layers), starting from the outermost layer (= L1). All the following considerations are based on two tunica layers being present, L3 thus representing the corpus. Within the SAM proper, orientation of cell division planes in the

tunica layers is strictly anticlinal (i.e. perpendicular to the meristem surface), while both anticlinal and periclinal (i.e. parallel to the meristem surface) division planes occur in the corpus. The SAM layers are commonly termed 'histogenic layers', in order to denote that they produce independent cell lineages, which according to their position within the plant body differentiate into the various leaf and stem tissues. Thus, the epidermis is derived from L1 ("dermatogen"), whose cells, exept during trichome formation, do not divide perically. In contrast, cells of L2 origin, in addition to dividing anticlinally, may divide periclinally soon after they have left the SAM. Cells of L2 origin give rise to outer mesophyll as well as outermost cortical parenchyma and, notably, gametes, while L3 cells give rise to inner cortex, vascular tissues, roots and pith. Though this describes the usual pattern as observed in periclinal chimeras bearing pigment or ploidy mutations, the relative contribution of the histogenic layers to the different leaf and stem tissues, especially that of L2 and L3, is quite flexible and may vary a great deal even within an individual. Also, histogenesis in chimeras is strongly dependent on relative competitiveness of mutated and wild type layers. On the other hand, important to note is that the arrangement of histogenic layers in the SAM is carried over to lateral bud meristems, usually without being changed.

2.2. Types of Chimeras

Regarding the spatial arrangement of their component idiotypes, three main types of chimeras are distinguished:

2.2.1. Sectorial Chimeras
Sectorial chimeras, in which the component idiotypes occupy massive sectors of varying size, but reaching through more than one layer of the plant body (**Figure 16.2k**), are rarely seen in nature. They are formed, e.g. when an adventitious shoot meristem develops from a group of cells, one of which is mutated (**Figure 16.3c, d**).

Box 16.2: The 'tunica-corpus theory' of shoot apical organization

- In a typical angiosperm shoot apical meristem (SAM) a central core tissue named 'corpus' is covered by two, sometimes more, sometimes only one, tunica layer(s).
- While cell division planes are strictly anticlinal in the tunica layer(s), both anticlinal and periclinal divisons occur in the corpus.
- The layers, since they are the ancestors of the different shoot tissues, are also called 'histogenic layers', and are numbered starting with the outermost layer (L1).
- Normally, L1 gives rise to the epidermis only, L2 produces outer mesophyll, gametes and outermost stem cortex, L3 (usually the corpus) inner mesophyll, inner stem cortex, vascular tissues, roots and pith.

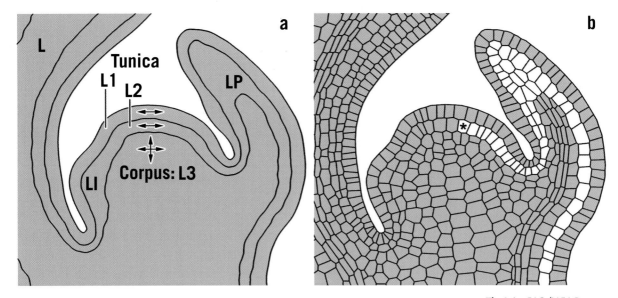

The Joint FAO/IAEA Programme

Figure 16.1 Longitudinal section of a typical angiosperm shoot tip with a two-layered tunica. a: Boundaries of histogenic layers and tissues derived from them. Within the SAM proper, cell division planes are strictly anticlinal in tunica layers L1 and L2, while anticlinal and periclinal divisions occur in the corpus (L3); arrows denote orientation of mitotic spindles; LI = area with periclinal divisions in L2 cells (cf. b), indicative of leaf initiation; LP = leaf primordium; L = older leaf. b: Cellular outlines; also shown is a mericlinal chimeric condition as depicted in Figure 16.2c. A mutation in a L2 cell at the position marked by an asterisk has spread by cell divisions within the sector marked white.

Sectorial chimeras are unstable and tend to segregate rapidly into pure types. By accident, they may attain a mericlinal chimeric state and finally stabilize as a periclinal chimera.

2.2.2. Mericlinal Chimeras

Due to the above described growth patterns a mutation occurring in a cell of one of the SAM layers results in the formation of a mericlinal chimeric constitution in which mutated cells occupy a sector of the tissue(s) derived from the respective histogenic layer (**Figure 16.1b, Figure 16.2b–d**). The fate of the mutated sector is governed by the growth pattern of the apex where the sector originates, the environmental conditions acting on the growth, and by diplontic selection, i.e. the selective value of the mutant idiotype relative to wild type. Mericlinal chimeras are unstable and tend to lose the mutated tissue or develop into stable periclinal chimeras. Because they may appear phenotypically similar, mericlinal chimeras are sometimes confused with sectorial chimeras.

2.2.3. Periclinal Chimeras

In periclinal chimeras mutated cells occupy one or more entire histogenic layer(s). This structure has first been visualized, and the independent behaviour of the histogenic layers in the SAM proven, in periclinal ploidy chimeras of *Datura*, in which the polyploid layers were 'marked'

by their increased nuclear and cell diameter (Satina *et al.*, 1940). The formation of a periclinal from a mericlinal chimera may happen through gradual driving out of wild type cells. However, this is to be expected only when the mutation is advantageous compared with wild type. More frequently, periclinal chimeras develop, when a lateral bud originates from within the sector bearing the mutated tissue layer. Since in this case the lateral meristem incorporates only mutated cells of the respective layer, sprouting of this bud will give rise to a periclinal chimeric shoot with the entire layer mutated. In this way, three primary types of periclinal chimeras (**Figure 16.2e–g**) may arise, depending on in which layer the mutation had occurred. The terminology for naming the different types of periclinal chimeras varies. In this chapter, a terminology is chosen referring to the layer(s) which is (are) mutated. Those chimeras bearing a mutated L1 are termed "ectochimeras", those with mutated L2 "mesochimeras" ("sandwich chimeras"), and those with mutated L3 "endochimeras". Trichimeras, bearing three distinct idiotypes are rare. They may originate e.g. by a further mutation event in one of the two wild type layers of a primary periclinal chimera, or by sorting out during plant recovery from a cell containing a mixed population of three or more different plastid types.

Commonly, different histogenic compositions are designated by letter combinations, where letters denote the idiotypes of the histogens starting with

L1. A mesochimera bearing a chlorophyll-deficient L2 (**Figure 16.2f**) is thus designated as GWG, using "G" for "green" (wild type) and "W" for "white" (chlorophyll-deficient). The two chimeric types in which one of two idiotypes resides in L1 and the other in L2+L3 (WGG and GWW; **Figure 16.2e, i**) appear phenotypically uniform, while the remaining (GWG, GGW and WWG) display variegation. The example of chlorophyll deficiency illustrates that even in case of a directly visible mutant trait, the chimeric nature or actual histogenic composition may be obscured due to cell type-specific expression of the mutant trait. Thus, because chlorophyll is not expressed in most epidermal cells, GGG and WGG phenotypes (**Figure 16.2a, e**) are superficially indistinguishable, as are GWG and WWG (**Figure 16.2f, h**), as well as GWW and WWW (**Figure 16.2i, j**). Distinction of these types requires microscopical inspection of stomatal guard cells or trichomes which normally express chlorophyll.

In the phenotypes depicted in **Figure 16.2**, cell-autonomous expression of the mutant trait is assumed. There are several cases, however, where chlorophyll-deficient mutated cells cause bleaching of chloroplasts in wild type cells of the neighbouring layer. For instance, endochimeras (GGW) of such type, instead of dark green leaves with a sharply delimited lighter centre (**Figure 16.2g**), possess leaves with a dark green margin and an indistinctly delimited white centre. The phenotype of periclinal chimeras may thus be modified by both cell type-specific expression and layer interactions. Further, chimeric patterns may be more or less obvious, depending on developmental stage-specific gene expression.

Periclinal chimeras can be fairly to very stable, which is exemplified by the many chimeric varieties of ornamen-

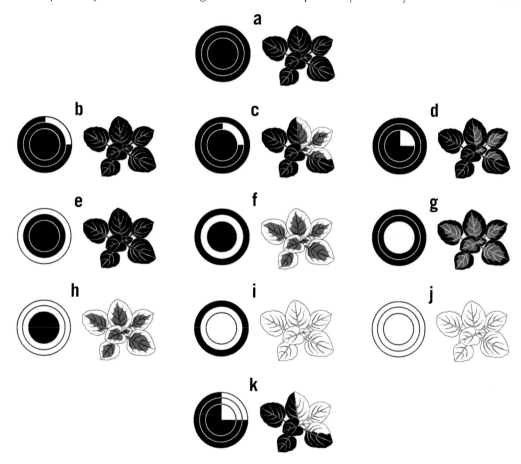

The Joint FAO/IAEA Programme

Figure 16.2 *Major chimera types, assuming a two-layered tunica in the shoot apical meristem.* Histogenic compositions (left; black = wild type, white = mutated) and phenotypes (right) are shown, taking chlorophyll deficiency as an example of a mutant trait (G = green, wild type; W = white, chlorophyll-deficient). **a**: Homohistic wild type – GGG. **b–d**: Mericlinal chimeras with mutated sector in L1 (b), L2 (c), and L3 (d), respectively. **e–g**: The three primary types of periclinal chimeras derived from the above mericlinal chimeras. **e**: Ectochimeric with mutated L1 – WGG. **f**: Mesochimeric ('sandwich') with mutated L2 – GWG. **g**: Endochimeric with mutated L3 – GGW. **h–j**: Three secondary types that may originate by histogenetic instability from the above periclinal chimeras. **h**: Ecto-mesochimeric ('diectochimeric') with mutated L1+L2 – WWG. **i**: Meso-endochimeric with mutated L2+L3 – GWW. **j**: Homohistic (solid) mutant – WWW. **k**: Sectorial chimera. Mutated tissue occupies a sector comprising all three histogenic layers. Partly adapted from Nybom (1961), and supplemented.

tals displaying variegated foliage with periclinal patterns (Marcotrigiano, 1997). These varieties are vegetatively propagated through tip and/or node cuttings which ensures the SAM structure remaining unchanged. Still, layer rearrangement may accidentally occur, urging on propagators to regularly remove off-types that have lost the desired pattern. Generally the extent of such variation is more prone in endochimeras than in mesochimeras. Causes of variegation other than periclinal chimerism are mentioned in **Chapter 15** of this book.

Ectochimeras usually are most stable. As the latter appear phenotypically uniform, their chimeric nature may remain unrecognized for long time and may only be revealed by rare layer rearrangements uncovering the mutant phenotype. Erroneously, this is often taken as a mutation event, but which, in fact, had taken place long before. In old clonal varieties, spontaneous somatic mutations, especially those which are weakly or not expressed in the epidermis, may accumulate in ectochimeric condition more or less unnoticed, grapevine being a good example (Franks *et al.*, 2002; Riaz *et al.*, 2002). This situation may not be undesirable, since in part, the chimeric nature may be responsible for certain unique characters of some cultivars/subclones, and, on the other hand, dissociation of the chimeric types may yield valuable new clones and thus lead to diversification within ancient varieties.

3. Dissociation of Chimeras

3.1. Histogenetic Instability

Different mechanisms, which can be summarized as histogenetic instability, may cause periclinal chimeric states to be changed and/or dissociated. Such processes have been studied in many species. They comprise rare periclinal ("illegitimate") divisions in one of the tunica layers within the SAM proper, resulting in replacement of the layers beneath. In this way, periclinal chimeras of the three primary types WGG, GWG and GGW (**Figure 16.2e–g**) may convert from one to the other type (GWG→GGW; **Figure 16.2f→g**), into secondary chimera types (e.g. WGG→WWG = ecto-mesochimeric or 'diectochimeric' – **Figure 16.2h**; GWG→GWW = meso-endochimeric – **Figure 16.2i**), or produce homohistic wild type (GGW→GGG; **Figure 16.2g→a**) and mutant (WGG→WWG→WWW; **Figure 16.2e→h→j**)

shoots, respectively. Another type of change is displacement of a layer by the one beneath (e.g. WGG→GGG, GWG→WGG, and so on). Combination of different mechanisms may even lead to layer reversal (GWG→WGW = ecto-endochimeric; not shown in **Figure 16.2**). Usually, all such kind of changes occur through transitional mericlinal states, i.e. the change first appears in a sector which, sooner or later, gives rise to a purely periclinal chimeric or homohistic shoot, e.g. through lateral bud formation within the changed sector.

Histogenetic instability may be enhanced and chimera dissociation favoured by conditions causing moderate damage to the SAM; notably radiation used for mutant induction may have this effect. Also, mechanical wounding and application of plant hormones can disturb the ordered growth of the SAM, thus accounting for higher percentages of off-types in plants obtained through meristem culture of periclinal chimeric varieties. Still, only partial chimera dissociation may be achieved in this way, as has been reported e.g. in a chimeric grapevine cultivar, where mostly chimeric plants were recovered from cultures of fragmented shoot apices (Skene and Barlass, 1983).

3.2. Dissociation Through Vegetative Propagation

As detailed before, mutagenic treatment of seeds and buds, respectively, first leads to an unstable mericlinal chimeric state. In order to achieve that the mutation is secured by conversion of the unstable mericlinal constitution into a more stable periclinal one, lateral bud sprouting should be forced. Buds located within mutated sectors will thus give rise to periclinal chimeric shoots in which loss of the mutation is less likely. In fruit trees, it has been proposed that before selection or rooting, shoots derived from original treated material should undergo several rounds of multiplication through branching, as grafted scions or *in vitro* shoot cultures (Predieri, 2001). Although some degree of layer rearrangement due to histogenetic instability may occur during repeated shoot sub-culture, and this may be favoured by *in vitro* shoot proliferation, it should be noted, however, that homohistic ('solid') mutants may not necessarily be obtained in this way. Thus, experiments conducted in cytochimeric banana, obtained after treatment with colchicine, showed that chimera dissociation was not completely achieved by micropropagation methods involving repeated axillary shoot proliferation (Roux *et al.,* 2001). In this connection it needs to be remembered of that ectochimeras as well as meso-endochimeras superficially appear uniform and thus are sometimes mistaken as being non-chimeric. Though rarely, their chimeric nature may give rise to variation through accidental layer rearrangements.

If complete chimera dissociation is intended, the only ways to achieve this with certainty are by selfing M_1 plants derived from treated material or by vegetative propagation *via* adventitious regeneration from single somatic cells. The latter is the method of choice in vegetatively propagated crops, since it enables the retention of the heterozygous genetic background of clonal varieties.

Adventitious buds may develop from wounded sites in *planta* or conventional cuttings devoid of preexisting buds (e.g. leaf or root cuttings), however, such regeneration capacity occurs rather erratically in a limited number of taxa. Compared with cuttings, explants cultivated *in vitro* are much more amenable to the induction adventitious regeneration. Based on the finding that shoot neoformation is promoted by application of cytokinins and auxins in appropriate concentration ratio, adventitious bud regeneration protocols have been developed for almost every angiosperm species of any commercial or scientific interest. Still, it needs to be mentioned that regeneration competence and culture requirements vary considerably among species and even idiotypes within a species, cases of regeneration recalcitrance demanding further improvement of the methods. In principle, adventitious shoots can be obtained from derivatives of all histogenic layers (e.g. Marcotrigiano, 1986), which is of particular importance for chimera dissociation. However, different cell types generally have different regeneration competence, and this may hardly be influenced by the culture medium composition or other factors. There are cases in which shoot regeneration almost exclusively

occurs from just one particular cell type (e.g. Hall *et al.,* 1995; Geier and Sangwan, 1996). In cases where adventitious buds are unavailable or more difficult to induce, chimeras may be dissociated through regeneration of somatic embryos as has been shown e.g. in grapevine (Franks *et al.,* 2002).

Most important for efficient chimera dissociation is the question of whether regenerants produced from multicellular explants originate from single cells or from a group of cells. Both has been observed, the type of origin seeming to be more or less species-specific. In cases where the origin is pluricellular, it is further important to know whether or not cells derived from more than one histogenic layer may be incorporated in the SAM of regenerants. Marcotrigiano (1986) has investigated this question, using a series of *Nicotiana tabacum* (T) + *N. glauca* (G) graft chimeras of four different histogenic compositions (TGG, GTT, TGT and TTG). Among a total of 658 regenerants he found 43 periclinal chimeric, 8 mericlinal chimeric and 607 non-chimeric with homohistic shoots of both species being recovered. This example nicely illustrates that adventitious regeneration quite effectively dissociates chimeras even in a system where regenerants are of pluricellular origin. A somewhat higher frequency of chimeric regenerants (10.8 %) was observed from petiole and internode explants of chimeric *Pelargonium zonale* bearing plastid defects in GGW and GWG histogenic composition (Li, 2005). Compared with explant and callus cultures, even more complete dissociation of chimeras may be achieved through shoot regeneration from cell suspension cultures.

3.3. Dissociation Through Seed Propagation

Seed propagation proceeds *via* unicellular gametes, and hence leads to complete dissociation of chimeras

---

Box 16.3: Chimera dissociation through vegetative propagation

- By forcing lateral bud sprouting in mericlinal chimeric M_1 plants, more stable periclinal chimeras can be obtained, and thus the risk of losing the mutation reduced.
- Histogenetic instability in periclinal chimeras may eventually yield homohistic ('solid') mutants through accidental layer rearrangements. Treatments changing the growth patterns of the shoot apical meristem, e.g. wounding and hormone application (meristem culture), lead to more frequent layer rearrangements.
- More reliable and rapid dissociation of periclinal chimeras is achieved *via* induction of *in vivo* or *in vitro* adventitious bud regeneration. Instead, somatic embryogenesis may be applied in species, where bud regeneration is unavailable, difficult to induce or embryogenesis is more efficient.
- For successful application of 'adventitious bud techniques', the most suitable type of cutting, explant or tissue culture system available in the respective species should be chosen, considering the regeneration efficiency, chimeric structure to be dissociated and cellular origin of regenerants.

bearing nuclear mutations. Selfing M_1 plants and raising the M_2 progeny further enables mutated alleles to be brought into homozygous condition, and thus to visualize recessive mutations. Moreover, homozygosity of mutated alleles is indispensable to true-to-type seed propagation. Mutant gene transmission in seed-propagated crops is described in detail in **Chapter 15**. Here, it should just be emphasized that seed propagation allows recovery of only the mutations residing in the germline, i.e. the cell lineage giving rise to gametes (usually L2), while those residing in tissue layers outside the germline are inevitably lost. In contrast, vegetative propagation *via* adventitous regeneration may enable the recovery of plants from any layer.

In vegetatively propagated crops, selfing is not usually applied to dissociate chimeras. This is for the following reasons: 1) Because of the general heterozygosity in clonal varieties, selfing inevitably results in complete loss of variety identity through recombination of many characters. For the same reason, buds (tip or node cuttings) or *in vitro* regeneration systems rather than seeds are subjected to mutagenic treatments. 2) Nuclear mutations affecting the dominant allele in heterozygous loci cause changed phenotypes which appear as

---

Box 16.4: Chimera dissociation through seed propagation

- Seed propagation proceeds through unicellular gametes and therefore leads to complete dissociation of chimeras bearing nuclear mutations.

- Selfing of M_1 plants is commonly practised in seed-propagated crops, however, only mutations residing in cells of the germline (usually L2) may be recovered in this way.

Table 16.1: Comparison of plant materials subjected to mutagenic treatment, with regard to the incidence of chimera formation, methods of chimera dissociation and other parameters

Material subjected to mutagenic treatment	Seeds	Tip and node cuttings	"Adventitious bud regeneration systems" [a]
Applicable to	Seed propagated crops	Vegetatively propagated crops	Seed- and vegetatively propagated crops
Constitution of M_1 plants	Usually mericlinal chimeric	Usually mericlinal chimeric	Homohistic (regeneration from single cells) or sectorial chimeric, rarely other constitutions (regeneration from more than one cell)
Methods of chimera dissociation	Selfing of M_1 plants, raising the M_2 progeny	1) Forcing of lateral bud sprouting and enhancing histogenetic instability 2) induction of adventitious bud or embryo regeneration	No dissociation required (homohistonts), or rapid spontaneous dissociation (sectorial chimeras)
Risk of loss of mutations due to diplontic selection	High (due to long duration from mutagenic treatment till flowering)	Medium (if dissociation of chimeras is effected soon after mutagenic treatment)	Low (due to rapid and independent development of regenerants)
Histogenic layer(s)/cells from which mutants may be recovered	Layer(s) giving rise to gamete formation (usually L2)	Derivatives of all histogenic layers; however, regeneration ability may vary between layers/cell types	All morphogenetically competent cells of the culture
Start of mutant selection	After chimera dissociation (in M_2)	Usually after plant regeneration, if applicable already during dissociation of chimeras	Usually after plant regeneration, if applicable *in vitro* during regeneration (selection for resistance/tolerance traits expressed at the cell level)

[a] Cuttings or *in vitro* explants devoid of pre-existing shoot meristems, tissue or cell cultures, capable of adventitious bud or embryo regeneration.

distinct sectors, already in M₁ plants. There is no need to achieve homozygosity of such mutant alleles, since vegetative propagation assures true-to-type reproduction. 3) Selfing may be impossible due to self-incompatibility, or clonal varieties may be sterile as e.g. in garlic and certain interspecific hybrids of ornamental species.

4. Single Cell Regeneration Systems for Avoiding Chimeras

In addition to their usefulness in dissociating existing chimeric structures, adventitious bud regeneration or somatic embryogenesis offer an efficient means of avoiding chimera formation. Thus homohistic mutants may be immediately obtained, when cuttings or explants capable of adventitious bud regeneration are subjected to mutagenic treatment and this is followed by regeneration from single cells (**Figure 16.3a, b**). More precisely, the SAM of regenerants needs to be derived from a single cell. There are several cases where adventitious shoots regenerate from single cells of conventional cuttings and many more such cases where this occurs *in vitro*. In other species, shoot regeneration normally occurs from a group of cells (Marcotrigiano, 1986). If more than one cell of the original explant contributes to the SAM of regenerants, and one of these cells is mutated, the regenerating bud will be sectorial chimeric (**Figure 16.3c, d; Figure 16.2.2k**). This state is unstable and its spontaneous dissociation leads to homohistic mutants more rapidly and freely than is the case after mutagenic treatment of preexisting buds. The formation of sectorial chimeric shoots after genetic transformation – like mutation, a single-cell event - has been illustrated by using appropriate marker genes (e.g. Domínguez *et al.*, 2004). As a rare exception, adventitious shoots may be even periclinal chimeric, as has been shown in transgenic tobacco (Schmülling and Schell, 1993).

While **Figure 16.3a–d** depicts situations in which adventitious buds arise directly from cells of the primary explant, a more or less extended phase of unorganized callus growth often precedes shoot regeneration. In this case, a mutated cell present in the primary explant multiplies to form a mutated callus sector. Buds regenerating from within such sectors are homohistic, even if their SAM is derived from more than one callus cell (**Figure 16.3e, f**). Buds developing from more than one cell at the border between mutated and wild type callus sectors may be sectorial chimeric. Primary somatic embryos are usually derived from single cells, while secondary embryos arising from the former, may

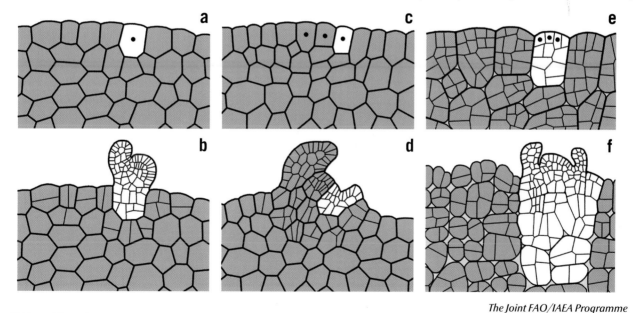

The Joint FAO/IAEA Programme

Figure 16.3 *Different modes of adventitious bud regeneration and their consequences for mutant production.* Cells contributing to shoot apical meristem formation are marked by dots, mutated cells are marked white. **a, b**: Direct regeneration from a single mutated cell results in a homohistic ('solid') mutant shoot bud. **c, d**: Direct regeneration from a group of cells. Three cells, one of which is mutated, participate in the formation of the apical meristem. The resulting bud has an unstable sectorial chimeric constitution. **e, f**: Indirect bud regeneration. Divisions of a single mutated cell have led to the formation of a mutated callus sector (e). A homohistic mutant bud has regenerated from a group of cells within the mutated callus sector (f). Sectorial chimeric buds may develop at the border between mutated and wild type callus sectors (not shown).

be of pluricellular origin. Consequently, some degree of chimerism may occur when mutagenic treatment is applied to somatic embryos, instead of explants or embryogenic callus.

Alternatively to multicellular explants, mutagenic treatment may be applied to single-cell cultures. Protoplast cultures, representing pure single-cell cultures, thus ensure complete avoidance of chimera formation and dissociation, respectively. They are, however, less easy to manage as the much more robust explant cultures, and reliable regeneration from protoplasts is available in fewer species. In any case, the type of plant material subjected to mutagenic treatment needs to be chosen considering not only the availability of regeneration systems in the treated species but also the ease of handling and expected overall efficiency. **Table 16.1** presents a comparison of plant materials subjected to mutagenic teatment, with regard to the incidence of chimera formation, methods of chimera dissociation and other parameters. It should be noted, however, that **Table 16.1** gives a generalized overview. The efficiency of different systems in terms of chimera dissociation and avoidance may vary, due to species- and tissue-specific regeneration ability and patterns. In addition to minimizing or avoiding chimera formation, *in vitro* adventitious bud regeneration or somatic embryogenesis systems may offer more advantages. Compared to seeds or cuttings, *in vitro* cultured tissues and cells, due to their small size and often more friable texture, may be easier to penetrate and hence more amenable to the application of chemical mutagens. Further, such systems may enable *in vitro* selection for increased resistance to abiotic and biotic stress, provided the resistance mechanism or components thereof are expressed at the cell level. *In vitro* selection methods are treated in detail in **Chapter 28** of this book.

Plastome mutations constitute a special case. They can be induced at high frequency by certain alkylating agents, N-nitroso-N-methyl urea (NMH) being particularly effective in this regard. Mutagenic treatment first gives rise to cells with a mixed population of mutated and wild type plastids. Regeneration from such heteroplastidic cells may lead to chimeric regenerants, even if these are derived from single cells. During bud regeneration and subsequent shoot growth some heteroplastidic cells may still be found, but sorting out of mutated and wild type plastids sooner or later leads to the establishment of stable periclinal chimeras of all possible histogenic layer compositions as well as homohistic shoots. NMH treatment of explants and subsequent shoot regeneration, compared with treatment of buds, can thus be an attractive method allowing rapid and easy production of different periclinal variegation patterns, desirable in ornamentals breeding.

One aspect that finally deserves attention is that different *in vitro* regeneration systems, to a greater or lesser extent, may involve increased frequencies of genetic changes summarized as 'somaclonal variation'. Consequently, if the aim is changing single traits through induced mutation without affecting the genetic background, *in vitro* culture procedures, applied as means of either chimera dissociation or avoidance of chimera formation, should be carefully evaluated in terms of their potential risk of somaclonal variation. On the other hand, somaclonal variation may be exploited as a potential source of useful mutations for crop improvement.

5. References

5.1. Cited References

Domínguez, A., Cervera, M., Pérez, R.M. *et al.* 2004. Characterisation of regenerants obtained under selective conditions after *Agrobacterium*-mediated transformation of citrus explants reveals production of silenced and chimeric plants at unexpected high frequencies. *Mol. Breed.* 14: 171–183.

Franks, T., Botta, R., Thomas, M.R. *et al.* 2002. Chimerism in grapevines: implications for cultivar identity, ancestry and genetic improvement. *Theor. Appl. Genet.* 104: 192–199.

Geier, T. and Sangwan, R.S. 1996. Histology and chimeral segregation reveal cell-specific differences in the competence for shoot regeneration and *Agrobacterium*-mediated transformation in *Kohleria* internode explants. *Plant Cell Rep.* 15: 386–390.

Hall, R.D., Verhoeven, H.A. and Krens, F.A. 1995. Computer-assisted identification of protoplasts responsible for rare division events reveals guard-cell totipotency. *Plant Physiol.* 107: 1379–1386.

Li, M.-Y. 2005. Observation of high-frequency occurrence of chimeral adventitious shoots in tissue culture from the chimeral tissues of *Pelargonium zonale*. *HortScience.* 40: 1461–1463.

Marcotrigiano, M. 1986. Origin of adventitious shoots regenerated from cultured tobacco leaf tissue. *Amer. J. Bot.* 73: 1541–1547.

Marcotrigiano, M. 1997. Chimeras and variegation: Patterns of deceit. *HortScience.* 32: 773–784.

Nybom, N. 1961. The use of induced mutations for the improvement of vegetatively propagated plants. *In*: Symp. Mutations and Plant Breeding. Ithaca, NY: Cornell University, NAS-NCR, Publ. 891: 252–294.

Predieri, S. 2001. Mutation induction and tissue culture in improving fruits. *Plant Cell Tissue Org. Cult.* 64: 185–210.

Riaz, S., Garrison, K.E., Dangl, G.S. *et al.* 2002. Genetic divergence and chimerism within ancient asexually propagated winegrape cultivars. *J. Amer. Soc. Hort. Sci.* 127: 462–710.

Roux N., Dolezel J., Swennen R. and Zapata-Arias F.J. 2001. Effectiveness of three micropropagation techniques to dissociate cytochimeras in *Musa* spp. *Plant Cell Tissue Org. Cult.* 66: 189–197.

Satina S., Blakeslee A.F., Avery A.G. 1940. Demonstration of the three germ layers in the shoot apex of Datura by means of induced polyploidy in periclinal chimeras. *Amer. J. Bot*. 27: 895–905.

Schmidt, A. 1924. Histologische studien an phanerogamen vegetationspunkten. *Bot. Arch.* 7/8: 345–404.

Schmülling, T. and Schell, J. 1993. Transgenic tobacco plants regenerated from leaf discs can be periclinal chimeras. *Plant Mol. Biol.* 21: 705–708.

Skene, K.G.M. and Barlass, M. 1983. Studies on the fragmented shoot apex of grapevine. IV. Separation of phenotypes in a periclinal chimera. *J. Exp. Bot.* 34: 1271–1280.

5.2. Further Reading

Broertjes, C. and van Harten, A.M. 1988. Applied mutation breeding for vegetatively propagated crops (Developments in Crop Science, Vol. 12). Amsterdam, New York: Elsevier.

Burge, G.K., Morgan, E.R. and Seelye, J.F. 2002. Opportunities for synthetic plant chimeral breeding: past and future. *Plant Cell Tissue Org. Cult.* 70: 13–21.

Pogany, M.F. and Lineberger, R.D. 1998. Plant chimeras in tissue culture: a review. http://aggie-horticulture.tamu.edu/tisscult/chimeras/chimera.html

Švábová, L. and Lebeda, A. 2005. *In vitro* selection for improved plant resistance to toxin-producing pathogens. *J. Phytopathol.* 153: 52–64.

Tilney-Bassett, R.A.E. 1986. Plant Chimeras. London, Baltimore: Edward Arnold.

Williams, E.G. and Maheswaran, G. 1986. Somatic embryogenesis: factors influencing coordinated behaviour of cells as an embryogenic group. *Ann. Bot.* 57: 443–462.

C17

Mutation Induction in Cytoplasmic Genomes

A.R.Prina*, M.G.Pacheco and A.M.Landau

Instituto de Genética "Ewald A. Favret", CICVyA, Instituto Nacional de Tecnología Agropecuaria (INTA), CC25, B1712WAA, Castelar, Buenos Aires, Argentina.
* Corresponding author, E-MAIL: aprina@cnia.inta.gov.ar

1. Introduction

Cytoplasmic genes are located in semi-autonomous cytoplasmic organelles, i.e. plastids and mitochondria. They are semi-autonomous in the sense that only a portion of the proteins they need are encoded by their own DNA, most are encoded in the nucleus, synthesized in the cytosol and finally imported into the organelles. It is widely accepted that mitochondria and plastids evolved from free-living prokaryotes and that during a long process of endosymbiosis they suffered a drastic loss of genetic material (90–95%). The lost genes were transferred to the nucleus of the host eukaryotic cell in which they originally invaded. In agreement with that idea both the mitochondrial and the plastid genomes share several similarities with bacterial genomes. During evolution these organelles became an essential component of the plant cell. The co-evolution of the three genomes involves an intensive and tightly coordinated cross-talk system. Many of the genes that still remain in these organelles encode components of their own genetic machinery and it can be inferred from information in bacteria that variability in these widely conserved genes not only includes diverse antibiotic resistances, but also differential abiotic stress responses.

Natural genetic variability residing in the cytoplasmic organelles is extremely narrow in comparison with that of the nuclear genome. The generation of variability using conventional tools is very difficult in most crop species due to the particular mode of inheritance, mostly mono-parental, and reduced or null recombination by means of traditional hybridization. Over the last

decades a huge amount of information on plastome genes was produced from experiments on site-directed mutagenesis using homologous recombination in *Chlamydomonas* and, more recently, in tobacco. However, in most crop plants species, plastid transformation is at present not easily performed and efficient procedures for plastome/chondriome mutagenesis are highly desirable.

Among the characters regulated by organelle genes, cytoplasmic male sterility (CMS) is by far the most commercially exploited, being a very useful tool for hybrid seed production in maize and several other crops. The CMS system has been determined in more than 150 species and it is usually associated with changes in mitochondrial genomes. Another cytoplasmic mutant commercially used in rapeseed varieties is a *psbA* gene allele that confers triazine tolerance. This plastid-encoded genotype has been found in several species as a spontaneous mutation.

2. Organelle Genomes and their Function

2.1. Plastid Genome (Plastome)

The plastid genome, or plastome, is a uni-circular DNA molecule of about 150 kbp encoding roughly 100–150 genes in the majority of higher plants. In addition to the components of the genetic machinery the plastome encodes most of the genes involved in photosynthesis (**Figure 17.1A**), including the reaction centre apoproteins, cytochromes and the large sub-unit of ribulose

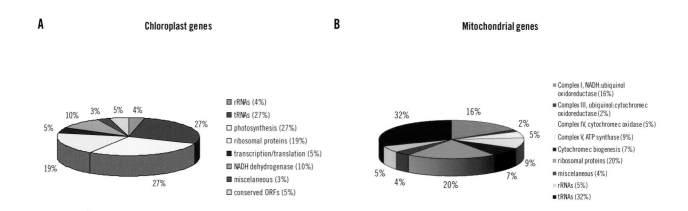

The Joint FAO/IAEA Programme

Figure 17.1 The gene composition of organelle genomes. A: The barley plastome (NC_008590). B: The wheat chondriome (NC_007579).

1,5-biphosphate carboxylase/oxygenase (RuBisCO). RubisCO is the key enzyme in photosynthesis and photorespiration, and probably the most abundant protein on earth. Variability in these genes has been observed to affect responses to abiotic stress and photosynthetic efficiency, and even a slight improvement in these characters can have a huge impact in plant productivity.

2.2. Mitochondria Genome (Chondriome)

In contrast to the plastome, the mitochondrial genome or chondriome of higher plants consists of a complex population of circular and linear DNA molecules with different sizes and sequences. In the chondriome internal recombination events, which can originate new arrangements, are common. Consequently, the gene order is not as strict as in the plastome. The size of the chondriome, ranging from 200 to 2,000 kbp, is much more heterogeneous among different plant species than the plastome. However, the number of genes encoding proteins does not necessarily correlate with the genome size. In mitochondria the genetic machinery is even more incompletely encoded than in plastids and thus they need to import not only proteins, but also some tRNAs from the cytosol. Several hundreds of proteins are necessary for a functional mitochondrion, but the chondriome is responsible for encoding and

synthesizing only a few of these, most of them involved in the electron transfer chain or belonging to the ATPase complex (**Figure 17.1B**).

3. Mutation Induction of Cytoplasmic Genes

It is widely accepted that mutation induction of cytoplasmic genes by treatments with typical mutagens, like ionizing radiations and chemical mutagens is much more difficult than that of nuclear genes. In some reports, the obtainment of cytoplasmic mutants was only assumed from observations of non-Mendelian transmission, sorting-out patterns of chlorophyll deficiencies and/or the presence of mixed cells, but they were not actually proven. On the other hand, it has been sometimes discussed whether these mutagens directly induced cytoplasmic mutations or if they induced nuclear mutants, which in turn produced cytoplasmic mutants. Some peculiarities of cytoplasmic genomes that are relevant for mutation induction are highlighted in **Box 17.1**.

3.1. Physical and Chemical Mutagenesis

3.1.1. Plastome

Among the typical mutagens, nitrosourea compounds have been revealed as promising agents for the induc-

---

Box 17.1: Cytoplasmic genomes and mutation induction

High number of copies of DNA molecules per cell: the cytoplasm of one cell contains numerous semi-autonomous organelles (plastids and mitochondria), each having multiple copies of DNA molecules. Thus, as mutations occur at the level of one of these copies, cytoplasmic chimeras cannot be avoided even when applying mutagenic treatments on unicellular organs (**see Chapter 15**).

Mixed cell: a cell carrying more than one type of allele in their plastids or mitochondria, also called heteroplastomic or heteroplastidic, when involving plastid DNA, and heteroplasmic, when involving any of these organelles. Cytoplasmic mutants always originate as mixed cells, they remain as mixed cells during several mitotic divisions and they can also persist as mixed cells after meiosis.

Somatic segregation or sorting out: a typical phenomenon of organelle genomes, which originates as a consequence of the so-called relaxed genome (see below). Thanks to somatic segregation an organelle mutant allele can differentially grow into some daughter organelles and daughter cells, segregating mutant vs. wild type homoplasmic cells without need of meiotic segregation.

Mode of inheritance and relaxed genome: in cultivated plants organelles have mostly a mono-parental mode of inheritance and they have the so-called relaxed genome, in contrast to the stringent nuclear one, which is copied only once per mitotic cell cycle. The more flexible mechanisms controlling replication and distribution of organelle-DNA molecules at cell division make the appearance of cytoplasmic mutants less predictable than that of nuclear ones (**see Chapter 15**). For nuclear genes, chimeras are usually limited to the soma of the M_1 plants, while segregation of solid mutants is expected to start in the M_2 generation. For cytoplasmic genes the appearance of chimeras is much more complex than for nuclear genes, they are not limited to the M_1 plants soma and the segregation of solid "homoplasmic" mutants is expected to occur in later generations.

Intracellular competition: as mentioned in **Chapter 15**, after mutagenic treatments applied on multicellular organs, competition between mutated vs. non-mutated cells can occur. This is true for both nuclear and cytoplasmic mutations, but in the last case, as different organelle-DNA molecules can replicate more often than others into a single cell, intracellular competition can also occur.

tion of plastome mutations in several higher plants species, including sunflower and tomato (Hagemann, 1982). Plastome-encoded antibiotics resistance was induced in *Solanaceae* species using N-nitroso-N-methylurea (NMU) (McCabe *et al.,* 1989) and atrazine tolerance in *Nicotiana* after treatment of N-ethyl-N-nitrosourea (ENU) (Rey *et al.,* 1990). Nitrosourea compounds seem to preferentially target the plastid DNA, but they also produce large amounts of nuclear mutations.

The availability of mutagenic treatments acting more specifically on the cytoplasmic DNA would be highly advantageous such that identification of cytoplasmic mutants can be easily performed in the absence of high amounts of nuclear gene mutants that occur when applying typical mutagens. In the uni-plastid *Chlamydomonas alga*, a model organism with several excellent attributes for plastid genetics studies, treatments with the antibiotic streptomycin induced plastome-resistant mutants and also cytoplasmically inherited mutants carrying photosynthetic defects. Specific increase in chloroplast gene mutations was also reported after growth of *Chlamydomonas* in the thymidylate synthase inhibitor 5-fluorodeoxyuridine (FUdr), a drug that causes the plastid to become thymidine starved (Würtz *et al.,* 1979). In this sense, some promising results came from experiments with the intercalating agent 9-aminoacridine hydrochloride in *Oenothera* (GuhaMajumdar *et al.,* 2004).

3.1.2. Chondriome

In the case of mitochondrial genes, cytoplasmic male sterile (CMS) genotypes have been the main target of mutation induction. Results of EMS mutagenesis for CMS has been so far contradictory. There are two reports of successful obtainment of CMS from ionizing radiation mutagenesis, one was in sugar beets with gamma rays (Kinoshita *et al.,* 1982) and the other in wheat with ion beam radiation (Wang and Li, 2005). Several chemicals known as effective inducers of cytoplasmic mutants in yeast have been successfully used in crop plants, *e.g.* acriflavine, streptomycin and ethidium bromide in sugar beets (Kinoshita *et al.,* 1982), and the last two drugs along with mitomycin in pearl millet (Burton and Hanna, 1982). The method of CMS induction with streptomycin in maize was patented in the US in 1971. Examples of mutation induction in cytoplasmic genes by artificial mutagenesis are listed in **Table 17.1**.

3.2. Cytoplasmic Mutators

Cytoplasmic variability can also be generated using genet-

Table 17.1: Some artificially induced cytoplasmic mutations reported in the literatures

Plant species	Genome	Phenotype	Mutagen
Oenothera	Plastome	Chlorotic sectors	9-amynoacridine hydrochloride
Chlamydomonas reinhardtii	Plastome	Non-photosynthetic or antibiotic resistant	5-fluorodeoxyuridine
Sunflower, Tomato and *Antirrhinum majus*	Plastome	Clonally variegated seedlings	N-ethyl-N-methylurea
Tomato	Plastome	Clonally variegated seedlings	N-ethyl-N-methylurea
Solanaceae	Plastome	Antibiotic resistant	N-ethyl-N-methylurea
Nicotiana plumbaginifolia	Plastome	Atrazine tolerant (*psbA* gene)	N-ethyl-N-nitrosourea
Chlamydomonas reinhardtii	Chondriome	Minute mutants	Acriflavine or ethidium bromide
Maize	Chondriome	CMS	Streptomycin
Sorghum	Chondriome	CMS	Colchicine
Pearl millet	Chondriome	CMS	Ethidium bromide, streptomycin or mitomycin.
Sugar beet	Chondriome	CMS	Gamma rays, acriflavine, streptomycin or ethidium bromide
Wheat	Chondriome	CMS	Electron beam

ically unstable genotypes, which probably are defective in genes that control organelle–genome integrity and are hence referred to as cytoplasmic mutators. Mutator lines produce maternally inherited changes and are therefore potential sources of organelle genetic variability.

3.2.1. Genetic Control of Genetic Instability

The genetic instability of mutator lines can be encoded by cytoplasmic or by nuclear genes, however, the latter are more useful for inducing cytoplasmic variability because they allow easy separation of both components by means of hybridization and backcrossing. Mutator genotypes already exist in nature and they can also be induced by mutagenic treatments.

3.2.2. Spectrum of Cytoplasmic Mutants Induced by Mutator Genotypes

The width of the mutant spectrum generated by cytoplasmic mutators can be either wide or narrow. The *iojap* maize and *albostrians* barley are typical examples of the narrow spectrum group. Other mutators reported in several dicots (*Arabidopsis, Epilobium, Oenothera, Nepeta* and *Petunia*) and in barley generate a wide spectrum of mutations. The mutator genotype of *Oenothera* increases 200–1000-fold the spontaneous appearance of pigment-deficient sectors. It was molecularly characterized as inducing deletions and duplications in the plastome through a replication slippage mechanism. Interestingly, when applying NMU on *Oenothera* plastome mutator genotypes a synergistic effect was observed suggesting an interaction between the mutator gene product and some repair systems in charge of correcting NMU-induced damage (GuhaMajumdar *et al.*, 2004).

The barley chloroplast mutator genotype induces a wide spectrum of chlorophyll mutants, which includes several viable and normal-vigour types. Mutator effects are slightly manifested in the M_2 generation (approximately 3 *striata* over 1,000 F_2 seedlings) and only subtle mutational changes have so far been detected in plastid DNA of mutator-induced mutants. Remarkably, the mutator induced plastome variability is easily observed because it occurs on a homogeneous nuclear genetic background and without the simultaneous induction of nuclear gene mutants. Most of the mutational changes so far observed consisted of several T/A – C/G transitions and only one insertion. They were observed in the plastid genes *inf*A, *psb*A and the *ycf*3 locus (Prina *et al.*, 2009). The *psb*A mutant was observed in families

selected for atrazine tolerance (Rios *et al.*, 2003). Both the wide spectrum of mutants and the subtle DNA changes induced suggest that this chloroplast mutator genotype can be an exceptionally valuable tool to explore the potential functionality of the otherwise highly conserved plastid genome.

Several mutants with chlorophyll-deficient sectors have been observed in association with mitochondrial anomalies. Thus, non-chromosomal stripe (NCS) maize mutants produce mitochondrial DNA rearrangements that cause deletion of essential genes; e.g. in the case of NCS 5 and NCS 6 mutants the mitochondrial *cox*2 (cytochrome oxigenase sub-unit 2) gene is affected and this finally impairs the development of normal chloroplasts showing yellow leaf stripes. Another chlorophyll-deficient mutant that alters the mitochondrial genome is the widely studied *iojap* maize, which has been observed to induce new cases of CMS (Lemke *et al.*, 1988). Another example is the maize nuclear genotype denominated P2 line that was reported as a natural mutagenesis system for mtDNA. It highly destabilizes the mitochondrial genome by increasing low copy-number sub-genomes, amplifying aberrant recombination products and causing loss of normal components (Kuzmin *et al.*, 2005). Finally, the *Arabidopsis chmchm* mutant, which confers maternally inherited leaf variegation and distortion and was previously classified as chloroplast mutator, has been observed to induce rearrangements in some mitochondrial genes for ribosomal proteins and to destabilize the mitochondrial genome. Furthermore, it causes accumulation of a rearranged subgenome, which was already present in the wild type at a very low level, by specific copy number amplification (substochiometric shifting), *e.g.* in the maternal distorted leaf (MDL) mutant a defective ribosomal protein gene *rps3* replaces the wild type copy. This mutator gene, renamed AtMSH1, encodes a homologue to MutS gene in *E. coli*, which is involved in DNA mismatch repair (Abdelnoor *et al.*, 2003).

4. References

4.1. Cited References

Abdelnoor R.V., Yule, R., Elo, A. *et al.* 2003. Substoichiometric shifting in the plant mitochondrial genome is influenced by a gene homologous to MutS. *PNAS.* 100(10): 5968–5973.

Burton, G.W. and Hanna, W.W. 1982. Stable cytoplasmic male-sterile mutants induced in Tift 23DB1 pearl millet with mitomycin and streptomycin. *Crop Sci.* 22: 651–652.

GuhaMajumdar, M., Baldwin, S. and Sears, B.B. 2004. Chloroplast mutations induced by 9-aminoacridine hydrochloride are independent of the plastome mutator in Oenothera. *Theor. Appl. Genet.* 108: 543–549.

Hagemann, R. 1982. Induction of plastome mutations by nitroso-urea-compounds. *In*: M. Edelman, R.B. Hallick and N.H. Chua (eds.) Methods in Chloroplast Molecular Biology, Amsterdam: Elsevier Biochemical Press, pp. 119–127.

McCabe, P.F., Timmons, A.M. and Dix, P.J. 1989. A simple procedure for the isolation of streptomycin resistant plants in Solanaceae. *Mol. Gen. Genet.* 216: 132–137.

Kinoshita, T., Takahashi, M. and Mikami, T. 1982. Cytoplasmic mutation of male sterility induced by chemical mutagens in sugar beets. *Proc. Japan Acad.* 58 Ser B: 319–322.

Kuzmin, E.V., Duvick, D.N. and Newton, K.J. 2005. A mitochondrial mutator system in maize. *Plant Physiol.* 137: 779–789.

Lemke, C.A., Gracen, V.E. and Everett, H.L. 1988. A second source of cytoplasmic male sterility in maize induced by the nuclear gene *iojap*. *J. Hered.* 79: 459–464.

Prina, A.R., Landau, A.M., Colombo, N. *et al.* 2009. Genetically unstable mutants as novel sources of genetic variability: The chloroplast mutator genotype in barley as a tool for exploring the plastid genome. *In*: Q.Y. Shu (ed.) Induced Plant Mutations in the Genomics Era. Rome: Food and Agriculture Organization of the United Nations. pp. 227–228.

Rey, P., Eymery, F. and Peltier, G. 1990. Atrazine- and diuron-resistant plants from photoautotrophic protoplast-derived cultures of *Nicotiana plumbaginifolia*. *Plant Cell Rep.* 9: 241–244.

Rios, R.D., Saione, H., Robredo, C. *et al.* 2003. Isolation and molecular characterization of atrazine tolerant barley mutants. *Theor. Appl. Genet.* 106: 696–702.

Wang, L.Q. and Li, G.Y. 2005. Role of mutation induction for wheat (*T. aestivum* L.) improvement. *In*: S.K.Datta (ed.) Role of Classical Mutation Breeding in Crop Improvement, Delhi: Daya Publishing House, pp. 156–185.

Würtz, E.A., Sears, B.B., Rabert, D.K. *et al.* 1979. A specific increase in chloroplast gene mutations following growth of Chlamydomonas in 5-fluorodeoxyuridine. *Mol. Gen. Genet.* 170: 235–242.

4.2. Further Reading

Birky, C.W. Jr. 2001. The inheritance of genes in mitochondria and chloroplasts: laws, mechanisms, and models. *Ann. Rev. Genet.* 35: 125–48.

Bock, R. and Khan, M.S. 2004. Taming plastids for a green future. *Trends in Biotechnology.* 22(6): 311–318.

Börner, T. and Sears, B. 1986. Plastome mutants. Review/ Genetics resources. *Plant Molecular Biology Reporter.* 4 (2): 69–92.

Gustafsson, A. and Ekberg, I. 1995. Extranuclear mutations. In: Manual on Mutation Breeding, second Edition. Vienna: Joint FAO/IAEA Div. of Atomic Energy in Food and Agriculture, pp. 114–117.

Hagemann, R. 1976. Plastid distribution and plastid competition in higher plants and the induction pf plastid mutations by nitroso-urea compounds. *In*: T. Bücher *et al.* (eds.) Genetics and Biogenesis of Chloroplasts and Mitochondria. Amsterdam: Elsevier/NorthHolland Biomedical Press, pp. 331–338.

Hagemann, R. 1986. A special type of nucleus-plastid-interactions: nuclear gene induced plastome mutations. *In*: G. Akoyunoglou and H. Senger (eds.) Regulation of Chloroplast Differentiation, New York: A.R. Liss, Inc., pp. 455–466.

Kirk, J.T.O. and Tilney-Bassett, R.A.E. 1978. The Plastids. Amsterdam: Elsevier, p. 960.

Woodson J.D. and Chory, J. 2008. Coordination of gene expression between organellar and nuclear genomes. *Nature Reviews/Genetics.* 9: 383–395.

Strategies and Approaches in Mutant Population Development for Mutant Selection in Seed Propagated Crops

Y.Ukai[a] and H.Nakagawa[b,*]

[a] Formerly, Graduate School of Agricultural and Life Sciences, The University of Tokyo, 1-1-1 Yayoi, Bunkyo, Tokyo 113-8657 Japan
[b] Institute of Radiation Breeding, National Institute of Agrobiological Sciences, P.O. Box 3, Kami-Murata, Hitachi-Ohmiya, Ibaraki 319-2293, Japan (Present address: Biomass Research & Development Center, National Agriculture and Food Research Organization, 3-1-1 Kannondai, Tsukuba 305-8517, Japan)
* Corresponding author, E-MAIL: luke154@jcom.home.ne.jp, ngene@affrc.go.jp

1. Introduction

The use of appropriate procedures for mutagenic treatment, mutant population development and management, and mutant selection are key elements in mutant exploitation. As described in **Chapter 14**, mutagenic treatments determine not only the mutation rate per cell in the treated material, but also cell development and genetic transmission. Once the treatment is completed, the next important step is to develop appropriate populations for screening and selecting mutants with desired traits. This includes the growing and harvesting of M_1 and M_2, and in some cases M_3 populations. It is also to be noted that rare traits can be the result of various combinations of mutated loci and these often do not show themselves until much later generations. This process is quite different from that of conventional cross-breeding and is one of the most critical steps in the process of mutant development. If the procedure is inappropriate, mutants cannot be detected efficiently even when they are generated at a high frequency.

The procedure of population development and managements can be different for self- and cross-pollinated crops, and for qualitative and quantitative traits. The development of a relevant procedure also needs to take into account ontology and chimerical nature of most M_1 plants, and the frequency and inheritance of induced mutations.

2. Population Management for Selecting Mutants in Self-Pollinated Crops

2.1. Genetic Features of M_1 and M_2 Populations

After mutagenic treatment, plant materials (seeds, tissues, organs, etc.) and plants grown from them are in the M_1 generation; the seeds harvested from M_1 plants and the plants grown from these seeds are the M_2 generation. M_1 and M_2 populations are populations that are composed of M_1 and M_2 plants. Their genetic structure is quite different from those of a traditional cross-breeding programme, i.e. F_1 and F_2 populations (**see Box 18.1**).

2.1.1. M_1 Population

After treatment with mutagenic radiations or chemical mutagens, mutations can occur in different genomic regions and different cells of the treated material.

The frequency of mutation occurrence at a certain gene at the time of treatment is defined as the mutation rate per gene. The frequency of mutation occurrence at one or more of the genes at a locus in a cell is called mutation rate per cell. Letting mutation rate of a gene of a particular locus be μ, the probability of occurrence of mutation at one or two genes of the loci is $2\mu(1-\mu) + \mu^2$ ($= 2\mu-\mu^2$), which is close to 2μ, since μ is usually much smaller than 1. Therefore, in diploid plants, the mutation rate per cell is double the mutation rate per gene.

Box 18.1: Comparison of genetic features between populations derived from mutagenic treatment and hybridization in self-pollinated crops

1. Typically, cross-breeding is initiated following the controlled hybridization of two different inbred or doubled haploid lines. Since the parents are highly or completely homozygous, the resulting F_1 plants are heterozygous and uniform in genotype. Therefore, a small number of F_1 plants are sufficient for developing the subsequent F_2 population. In contrast, the plants of an M_1 population are not genetically uniform, and exhibit a variety of genotypes due to the induction of various mutations. Since most induced mutations are usually recessive to the original allele, these mutations cannot be identified in the M_1, and phenotypically the plants may appear to be uniform. It is not advisable to grow a large number of M_2 plants from a small number of M_1 plants since this approach will reduce the efficiency of specific mutant identification. This is demonstrated mathematically in the following section.

2. The plants of an F_2 population are generated following self-pollination of F_1 plants. The various genotypes, as represented by the various loci provided by each parent, segregate in the typical Mendelian fashion, i.e. at any particular segregating locus, three genotypes are possible: *AA, Aa or aa*, where *AA* and *aa* are the genotypes of the two parental lines. Whether a locus segregates for genotypes or not depends on the combination of the genotypes of the parents used. The segregation of parental genotypic contribution provides a large spectrum of genotypes in the F_2 population. Similarly, in some plants of an M_2 population, the targeted locus may have three possible segregating genotypes, *AA, Aa'* and *a'a'* where *a'* is the mutant recessive allele at the locus. The differences are: 1) the mutation frequency at the targeted locus is very low in the whole population even following a mutagenic treatment (e.g. $<10^{-4}$); 2) within a single plant progeny-row of M_2 plants, the segregation of mutant (*a'a'*) and wild type (*AA* and *Aa'*) does not normally follow Mendelian ratio, and fewer mutant plants are expected due to the chimerical nature of the M_1 plants and diplontic and haplontic selection (see explanation in this chapter); 3) additional mutations can occur simultaneously at loci other than the targeted locus, in other words, each of the M_2 plants has the potential to have mutations at different loci.

The chance of a simultaneous occurrence of two or more mutations at the same locus of the two homologous chromosomes in diploid plants is rare and so is the chance of segregation of homozygous mutants in the M_1 population.

In seed propagated crops, only germline cells can transmit their genotype into subsequent progenies. Therefore in materials treated with a mutagen, only cells that develop into inflorescences ("initial cells" or "genetic effective cells" in a seed, **see Chapter 15**), can transmit mutated alleles into the M_2 generation. Other vegetative cell lines, e.g. those that develop into leaves and/or endosperms may also be mutated; however, these mutated cells cannot transmit mutated alleles to the next generation since they do not take part in meiosis and therefore do not give rise to male or female gametes.

When a mutation ($A \rightarrow a'$) is induced at one of the genes of a homozygous AA locus, the genotype becomes heterozygous (Aa'). Typically, the mutant allele is completely recessive to the original allele and the Aa' phenotype cannot be discriminated from the original AA phenotype. Hence, the vast majority of

induced mutants cannot be screened in the M_1 generation, though there are a few examples of dominant mutations.

The number of cells that compose the initial cells of the shoot apex in a seed is not one but several and since the mutation rate per gene is very small, the probability of the simultaneous induction of mutations at two or more initial cells is negligibly small, therefore any inflorescence that possesses the mutated allele is often in a chimeric state with both mutant and normal tissue sectors (**Figure 18.2, also see Chapter 15**). When an inflorescence (e.g. a spike of barley and wheat or a panicle of rice) or a part of the inflorescence is derived from a mutated initial cell, then this is called "a mutated inflorescence".

2.1.2. M_2 Population

The overall genetic feature of the M_2 generation varies with the method of how M_2 seed is harvested from M_1 plants and how they are grown up (see next section). Mutants that have homozygous mutant alleles ($a'a'$) will appear in the M_2 population as a result of self-pollination

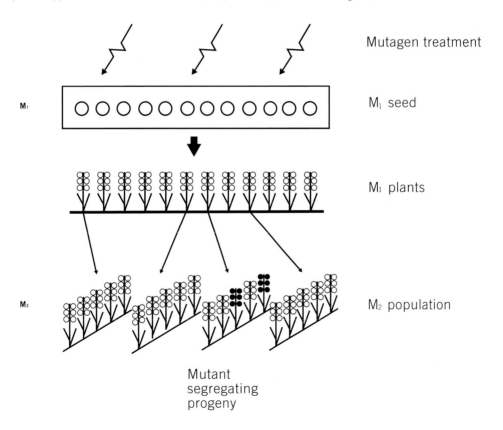

Mutagen treatment

M_1 M_1 seed

M_1 plants

M_2 M_2 population

Mutant
segregating
progeny

The Joint FAO/IAEA Programme

Figure 18.1 Population development for mutant selection by M1 spike progeny method in self-pollinated crops. Solid circles indicate homozygous recessive mutant seeds.

of M_1 plants (**Figure 18.1**). For example, early flowering, semi-dwarf and male sterile mutants can be visually recognized. However, due to the chimerical nature of M_1 plants, and subsequent diplontic selection in M_1 plants and haplontic selection during gameogenesis, as well as due to lethality of M_2 embryos or seeds with lethal homozygous mutant alleles, the proportion of mutant ($a'a'$) to wild type plants (AA and Aa') is usually less than those segregating in a Mendelian fashion. **Figure 18.2** shows different scenarios of the segregation of mutant plants in the M_2 generation when they are grown in single plant progenies. For example, if there are four initial cells in the treated seeds and only one of them is mutated after treatment, the segregation rate of mutant to wild type plants in the M_2 will be 1:15, compared to 1:3 in a typical (Mendelian) F_2 population.

The competition process that takes place between normal and mutant cells in vegetatively growing tissues is known as "diplontic selection". The survival rate of cells with the mutated allele in M_1 generation is lower than that of the normal cells. If such competition occurs at the reproductive stage such as meiotic division and fertilization, the process is known as "haplontic selection". Gametes with a mutated allele have a lower chance of completing double fertilization than normal ones. In both cases, the segregation ratio becomes lower than the expected ratio of 0.25.

2.2. Procedures for Selecting Mutants

2.2.1. Growing out the M_1 Population

In growing out the M_1 population, the following guidelines are helpful: 1) plants should be grown in suitable conditions; 2) plants should be isolated, physically or biologically, from other varieties or materials of the same species.

After mutagenic treatment, plant materials such as seeds often have a low level of viability. For example, the germination rate of M_1 seeds may be reduced to about 50% of control seeds in order to achieve a reasonable mutation frequency (**see Chapter 15**). This level of germination and subsequent emergence can be significantly reduced if the seeds are sown in adverse environments, e.g. in drought or saline fields. Therefore, even if the objective of a mutation project is to select mutants with enhanced tolerance to drought or salinity, the M_1 plants should be grown in a non-stressed condition; otherwise, there will be an insufficient number of plants generated in the M_2.

Mutagenic treatment not only causes germination reduction of M_1 seeds, but also leads to a decrease in pollen fertility of M_1 plants. Therefore, M_1 plants have a higher out-crossing tendency than non-treated plants. If there are plants of other genotypes (varieties or breeding materials) that are grown nearby and which flower at the same period as the M_1 plants (which is highly likely in seasonal crops), out-crossing can easily occur in M_1 plants, which will result in contaminations of M_2 seeds. Therefore, M_1 plants should be grown either at a reasonable distance from other varieties (physical isolation) or in a time period when no other plants would flower simultaneously (biological isolation).

2.2.2. Harvesting and Growing out the M_2 Seed

There are basically two methods of establishing an M_2 population in self-pollinated crops such as barley and wheat. In such monocots the inflorescence is known as a spike.

M_1-spike progeny method

This is a method in which M_2 seeds are harvested separately from each spike of M_1 plants. The method was developed by Stadler (1928) and has been used effectively by Swedish research groups guided by Å. Gustafsson (Lundqvist, 1991). Usually 10 to 20 plants are grown out from seed of individual M_1-spikes. The frequency of mutants per progeny is called the "segregation frequency" and is distinguished from the "segregation ratio", which refers to the ratio of a trait segregating according to its genetic control, e.g. a ratio of 3:1 in the progeny for a single gene controlled trait.

The segregation frequency depends on four factors: 1) segregation ratio; 2) existence of chimeras in a spike; 3) diplontic and haplontic selection; and 4) the number of plants per progeny. In typical diploid species, the segregation ratio of recessive mutants ($a'a'$) is 0.25; and in polyploids the segregation ratio is lower than 0.25. If a spike originates from a single mutated initial cell, and the number of plants per progeny is infinite, then the same is also expected for segregation frequency.

If there is a chimeric sector in a spike of an M_1 plant, the expected frequency of mutants per progeny is less than 0.25. Allowing the number of initial cells to be set as "k", the expected frequency of mutants in a progeny with an infinite number of plants is 0.25/k. The higher the number of initial cells, the lower the segregation frequency. In progeny derived from a spike of the main stem

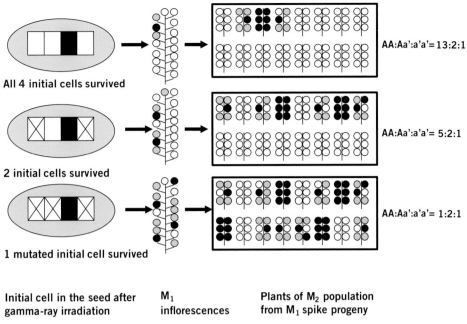

The Joint FAO/IAEA Programme

Figure 18.2 Chimeric inflorescences of M_1 plants and segregation of mutants in M_2 population when the number of initial cells is four. (Modified from Ukai (2003)). Solid squares indicate mutated initial cells; solid circles indicate homozygous recessive mutant seeds (a'a'); Grey squares indicate heterozygous seeds (Aa').

Table 18.1: The number of M_2 plants T_2(× 10,000) required in obtaining one or more mutants with the probability 0.95. The mutation rate per cell is assumed to be 10^{-4} and segregation ratio of the mutant 0.25 [a]

n[b]	Number of initial cells k				
	1[c]	2	3	4	6
1	**11.96**[d]	**11.96**	**11.96**	**11.96**	**11.96**
2	13.66	12.74	12.46	12.32	12.19
3	15.50	13.55	12.97	12.69	12.42
4	17.46	14.39	13.49	13.06	12.65
5	19.56	15.26	14.03	13.44	12.89
6	21.77	16.17	14.58	13.84	13.13
7	24.09	17.10	15.14	14.23	13.37
8	26.51	18.07	15.73	14.64	13.62
9	29.01	19.07	16.32	15.06	13.87
10	31.59	20.10	16.93	15.48	14.12
15	45.31	25.67	20.19	17.72	15.45
20	59.80	31.84	23.78	20.16	16.87

[a] Adopted from Ukai and Yamashita (1974).
[b] The number of plants per progeny.
[c] Absence of chimera.
[d] Figures in block letters are minimum values of total number of M_2 plants for a particular number of initial cells.

that develops from a higher number of initial cells, the segregation frequency is lower than in progeny derived from primary and secondary stems (tillers). If some of the initial cells are killed by mutagenic treatment of seeds (e.g. with high dose of radiation treatment), the segregation frequency of mutants becomes higher (**Figure 18.2**).

Even if progeny arise from a mutated spike, it is possible (by chance) that no mutant may be segregated out. When the number of plants per progeny is n, then, the probability that no mutant appears in the progeny by chance is $(1-0.25)^n$. If n is 10, for example, the probability is $(1-0.25)^{10} = 0.056$. The smaller the number of plants per progeny, the higher the probability that no mutant appears by chance. This is true in the absence of chimera, and/or diplontic and haplontic selection.

The segregation frequency of mutants per progeny is calculated by the total number of mutants divided by the total number of plants observed in the M_1-spike progenies in which one or more mutants were selected. In other words, the segregation frequency is calculated on the basis of the number of progeny in which one or more mutants actually segregated and not on the basis of progeny derived from a mutated spike. Some of the latter cannot be discriminated from normal progeny due to no segregation of mutants by chance. Hence, the segregation frequency is not consistent with the expected ratio of 0.25 even in the absence of chimerism, and diplontic and haplontic selection, unless n is indefinite. The smaller the value of n, the higher the segregation frequency with n = 1 being at the maximum.

If chimera and/or diplontic and haplontic selection occurs, then the segregation frequency is lowered and the probability that no mutant appears in the progeny by chance becomes higher.

One-plant-one-grain method
The method of constructing the M_2 population by planting only one grain (seed) from each plant was proposed as the "one-plant-one-grain method". In practice, the seeds are harvested not from each M_1 plant, but from each spike, hence this method is also known as the "one-spike-one-grain method". If a few seeds per spike are used for the establishment of the M_2 population, this approach is called the "one-spike-few-grain method".

In mutation breeding, a single target mutant from the entire M_2 population is sufficient for utilization and multiplication of the mutant for further selection and testing. Usually, two or more mutants of the same kind are

not necessary. The word "the same kind" here means the same change in DNA sequence in a strict sense, change in the same gene, or phenotypically same change. It depends on the aim of the experimenter and is subjective.

Based on this principle, the higher the probability of obtaining one or more mutants with the same total number of M_2 plants, the higher the efficiency of mutant development. The probability of obtaining one or more mutants under a given total number of M_2 plants is larger as the number of plants per progeny (n) becomes smaller, and is at the maximum when n = 1 (Freisleben and Lein, 1943). Details are shown in the following section.

2.3. Selecting of Qualitative Trait Mutants

It is important to determine the most efficient sizes of M_1 and M_2 populations to ensure a reasonable probability of identifying a desired mutant. As discussed above, factors such as chimerical nature, harvesting and growing practices, and the genetic nature of traits of interest can all significantly affect the population size and hence the efficiency in mutant development.

2.3.1. Determining M_1 and M_2 Population Size
Qualitative traits, including morphological traits such as semi-dwarf, and disease resistance are very often the target of a mutation programme. The appropriate population size can be determined through the following analysis according to Yoshida (1962):

Letting the number of M_2 lines be m and the number of plants per M_1 spike progeny n, the mutation rate per cell $p_1 (= 1-q_1)$, and the segregation ratio of the mutant trait $p_2 (=1-q_2)$, the probability (P) that one or more mutants are detected among the total number of M_2 plants $(T_2 = mn)$ is:

$$P = 1 - \psi^m \qquad (1)$$

where $\psi = q_1 + p_1 q_2^n$. In this equation, the segregation ratio is the expected ratio of mutants ($a'a'$) in progeny derived from a heterozygous mutant (Aa'). The segregation ratio is usually not equal to the segregation frequency. Hence, we have:

$$m = \frac{\log (1-P)}{\log (\psi)} \qquad (2)$$

Therefore, the total number of M_2 plants required is:

$$T_2 = mn = \frac{n \log (1-P)}{\log (\psi)} \qquad (3)$$

Setting P to be a constant (e.g. 0.95), T_2 is expressed as a function of n. A value of n that minimizes T_2 is the most efficient. The results of numerical calculations in the absence of chimera are shown in the second column ($k = 1$) of **Table 18.1**. T_2 is at minimum when $n = 1$, indicating that taking only one grain from each M_1 spike is the most effective approach. This is the theoretical basis of superiority of the one-spike-one-grain method in mutant selection. The minimum value of 11.96 in the table infers that under a mutation rate per cell of 10^{-4}, 119,600 plants are required to obtain one or more mutants with the probability of 0.95. If $n = 20$, T_2 is 598,000, indicating that the number of M_2 plants required increases to five times the number when $n = 1$.

In the one-spike-one-grain method, seed is pooled after a constant number of seed are obtained from each spike and sown in the M_2 as a bulk. Cost and labour required for growing the M_2 plants is usually smaller than in the M_1 spike-progeny method. This method, however, has some weakness. When a detected mutant is accompanied by severe or complete sterility, the mutant is lost. On the other hand, in the M_1-spike progeny method, another mutant of the same type may be found in the same progeny from which the mutant originally segregated. Moreover, sister-plants exhibiting a normal phenotype and heterozygous genotype (Aa') may be retained, and homozygous mutants ($a'a'$) may be recovered in the M_3. Another weakness is that a large number of plants must be grown in the M_1 as compared with the M_1 spike-progeny method. Dense planting, if possible, may be effective in reducing the space, cost and labour for growing out the M_1 plants. Obtaining a constant number of seed from each M_1-spike is laborious. In this instance, sometimes an equivalent number of seed are taken from a sample of bulked M_2 seed without separating the M_1-spikes. But in this approach, the high efficiency of detecting mutants by the one-spike-one-grain method is partly lost.

2.3.2. Effects of Chimeras in M_1 Reproductive Parts and Efficiency of Selection

The shoot primordium of a seed embryo has a multicellular structure, but only the genetic effective cells (GEC, or initial cells as described above) become germ-line cells which eventually produce progeny. Since the mutation rate is very low, mutation at the same locus

Table 18.2: The total number of M_1 and M_2 plants T_{1+2} ($\times$ 10,000) required in obtaining one or more mutants with the probability 0.95 [a]

n [b]	Number of initial cells k				
	1 [c]	2	3	4	6
1	23.93	23.93	23.93	23.93	23.93
2	**20.49** [d]	19.11	18.69	18.48	18.29
3	20.66	18.06	17.29	16.92	16.56
4	21.83	**17.98**	16.86	16.33	15.81
5	23.47	18.31	**16.83**	**16.13**	15.46
6	25.40	18.86	17.01	16.14	15.31
7	27.53	19.55	17.31	16.27	**15.28**
8	29.82	20.33	17.69	16.47	15.32
9	32.23	21.19	18.14	16.73	15.41
10	34.75	22.12	18.62	17.03	15.53
15	48.34	27.39	21.54	18.90	16.48
20	62.79	33.43	24.97	21.17	17.71

[a] Adopted from Ukai and Yamashita (1974).
[b] The number of plants per progeny.
[c] Absence of chimera.
[d] Figures in block letters are minimum values of total number of M_2 plants for a particular number of initial cells.

Table 18.3: Efficiency in the selection of mutants with a quantitative trait controlled by a single gene

Proportion of selection	SD	A[a]	B[b] %	C[c] %
Upper 1%	SD = 1/1d	1,008.1	9.1	0.90
	SD = 1/2d	1,036.1	37.1	3.58
	SD = 1/3d	1,073.8	74.8	6.97
	SD = 1/4d	1,094.2	95.2	8.70
Upper 0.1%	SD = 1/1d	101.7	1.8	1.77
	SD = 1/2d	113.6	13.7	12.06
	SD = 1/3d	146.3	46.4	31.72
	SD = 1/4d	181.7	81.8	45.02

[a] A: The number of plants selected.
[b] B: Proportion of selected mutants among the mutants segregated.
[c] C: Proportion of mutants among the plants selected.
It was assumed that the frequency of mutants in M_2 is 10^{-3}, and that the number of plants screened is 100,000.

usually occurs only in one of these GECs, which leads to mutant sectors or chimera in a spike. In other words, a spike is often composed of non-mutated and mutated cells. The larger the number of initial cells, the smaller the expected frequency of mutants per progeny. It is generally believed that the presence of chimera within spikes leads to a disadvantage in the screening of mutants. However, theoretical calculations have revealed that even in the presence of chimera, the one-spike-one-grain method is the best (**Table 18.1**). The superiority of the one-spike-one-grain method to other methods where two or more grains are obtained from each spike decreases with the increase in the number of initial cells.

If a densely planted M_1 plant population is not feasible, the cost required for growing the M_1 plants must be taken into account. The total number of M_1 and M_2 plants (T_{1+2}) required for detecting one or more mutants with a probability of 0.95 is shown by:

$$T_{1+2} = m + mn = \frac{(n+1)\log(1-P)}{\log(\psi)} \qquad (4)$$

The calculated results are shown in **Table 18.2**. If the cost per plant is equal between the M_1 and the M_2, the optimum number of plants per progeny is not 1 and becomes larger as the number of initial cells increases. If the number of initial cells is 1 (no chimera), 2, 3, 4 and 6, the total number of plants required are at the minimum when the number of plants per progeny are 2, 4, 5, 5 and 7, respectively.

2.3.3. Estimation of Mutation Rate

The estimation of mutation rate is important for determining the practical population sizes. There are two methods for estimating the mutation rate. The first method is based on 'M_2 mutant frequency', the second is based on 'mutant progeny frequency'.

The expected number of M_2 progenies that include one or more mutants (mutant segregating progenies) $E(M_{mut})$ is:

$$E(M_{mut}) = mp1(1-q_2^n) \qquad (5)$$

Here, $1-q_2^n$ represents the probability that one or more mutants are segregated out among the plants from an M_1 spike.

The expected number of mutants per progeny $E(N_{mut})$ for progenies in which one or more mutants are segregated is:

$$E(N_{mut}) = \frac{np_2}{1-q_2^n} \qquad (6)$$

Hence, the expected total number of mutants in the M_2 is:

$$E(T_{mut}) = E(M_{mut}) E(N_{mut}) = mnp_1p_2 \qquad (7)$$

From this, p_1 is estimated by:

$$p_1 = \frac{T_{mut}}{mnp_2} = \frac{T_{mut}}{T_2} \cdot \frac{1}{p_2} \qquad (8)$$

If p_2 is 0.25, four times the frequency of mutants in the M_2 is an estimate of the mutation rate p_1. This is the method proposed by Gaul (1960) which is based on 'M_2 mutant frequency'.

The frequency of mutant progenies (p_1^*) is often employed as an estimate of mutation rate that is sometimes called the 'mutations-per-spike' measure (Frydenberg, 1963). This is based on scoring individual mutational events, and represents a direct estimate of initial mutation frequency, but has a weakness as an estimate of mutation rate. In the absence of chimeras within spikes, p_1^* is represented as follows:

$$p_1^* = E(M_{mut})/m = p_1(1-q_2^n) \qquad (9)$$

If the number of plants per progeny (n) is not sufficiently large, the probability that no mutants segregate out in the progeny by chance is not negligible and p_1^* is smaller than p_1. Moreover, in the presence of a chimera and with a sufficiently large n, p_1^* is larger than p_1. When the number of initial cells in the meristem of a seed embryo is k, p_1^* is roughly k times p_1. The number of initial cells varies with several factors such as the dose of the mutagen, stage of plants treated, the species used as material and the type of stem (main or secondary) with which the spike is associated.

2.4. Selecting Quantitative Trait Mutants

2.4.1. Theoretical Considerations

A quantitative trait such as yield or quality of grains is generally controlled by many genes, and in addition, influenced by environmental factors. Since the mutation rate of a gene is very low, the chance of simultaneous occurrence of mutations at two or more genes is negligible. Hence the selection of mutants in mutation breeding is usually unsuccessful for a quantitative trait, particularly when the number of loci controlling the trait is many and the effect of each locus is small as compared with environmental variation. But many quantitative trait locus (QTL) studies have revealed that the genetic effect among the contributing loci is not equal. Often a few loci possess significantly higher genetic effects. Mutations at a gene at such loci with a large genetic effect can be selected after mutagenic treatment. For such mutations, too, the one-spike-one-grain method is most efficient for the selection of mutants. A quantitative trait mutant cannot be detected with a high level of confidence owing to its interaction with environmental factors, therefore, a different procedure is recommended for selecting such mutants.

Assume a quantitative trait of a phenotypic value which follows a normal distribution (A) with mean N and standard deviation s. In other words, the genotypic value of the normal plant is N, and its environmental variance is s^2. Suppose that a targeted mutation ($A \to a'$) is induced at a locus with a high genetic effect with a mutation rate per cell p_1. Assume that the phenotypic value of the homozygote $a'a'$ follows another normal distribution (B) of mean M (>N) with the same magnitude of standard deviation as a normal plant (σ). In other words, the mutant has a genotypic value of M and an environmental variance of σ^2. Assume also that the mutated allele is completely recessive to the original allele, and that the phenotypic value of heterozygote (Aa') follows the distribution A.

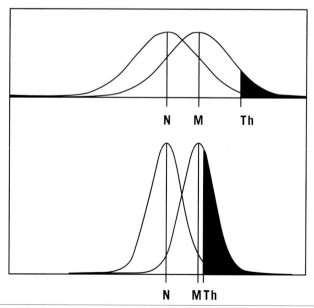

The Joint FAO/IAEA Programme

Figure 18.3 Selection of a quantitative mutant controlled by a single gene.

Phenotypic values of plants in the M_2 population follow a distribution which is a combination of two normal distributions A and B with a ratio of $1-0.25p_1$ and $0.25p_1$. Directional selection of M_2 plants is practised.

All of the plants with phenotypic values above a threshold point (Th) are selected. In this type of selection for a quantitative trait the determination of the value of Th in relation to the mutant genotype and environmental variation is important.

Computer simulations were performed for the selection of plants with phenotypes that are higher than Th (**Figure 18.3**). It was assumed that the frequency of mutants in M_2 is 10^{-3}, and that the number of plants screened is 100,000. The selection proportion was based only on the normal distribution of the normal plants, since in actual cases genetic value (M) of the targeted mutant is unknown, and was set at 1% and 0.1%. Standard deviation σ was set at $1d$, $\frac{1}{2}d$, $\frac{1}{3}d$ and $\frac{1}{4}d$, where $d = M-N$. The proportion of selected mutants among the mutants segregating in the population (B) and the proportion of selected mutants among the selected plants (C) were recorded. The results are shown in **Table 18.3** and indicate that (B) is sufficiently high when σ is $\frac{1}{3}d$ or $\frac{1}{4}d$, depending on the selection proportion. In the case of a 1% selection proportion, (B) is higher, however, (C) is lower than in the case of 0.1%. The decrease of environmental variation up to $\frac{1}{3}d$ or $\frac{1}{4}d$, if possible, is important so that one does not miss the segregating mutants. If this is possible, utilizing a higher selection proportion is better, but in this case the proportion of mutants among the selected plants is low and re-selection of mutants is required.

The normal curve on left and right hand show a distribution of phenotype of a quantitative trait of the original population and a mutant population, respectively. The latter actually has a very low frequency as compared with the original population. N and M stand for the phenotypic mean (genetic effect) of the original and a mutant population. Th is the selection threshold (critical). The solid area under a normal curve shows the proportion of selected plants, 1% for instance. The standard error of the normal distribution (= environmental variation) is set to be equal to the difference (d) of M and N $(d = M-N)$ and to half of d, in the top and bottom of the cases, respectively. Note that in the former case the proportion of mutants selected among the mutants induced are much lower than in the latter case, though the selection proportion is equal between the two cases.

2.4.2. Practical Example

For example, let the targeted mutation be a late-flowering mutation. Suppose that the original wild type plants flower on the 80th day after sowing, while targeted late mutants flower on the 90th day. $N = 80$ and $M = 90$, and hence $d = N-M = 10$. If the standard deviation of the phenotypic value is $0.5d$, namely 5 (days), and if the selection proportion is set at 1%, then the number of plants to be selected is approximately 1,036 out of the 100,000 plants, and the proportion of selected mutants among the mutants segregated is 37.1%, namely, 62.9% of the mutants included in the M_2 population fail to be detected. The proportion of mutants among the plants selected is only 3.58%, and hence re-selection for mutants is required. In order to increase the proportion of mutants detected among the segregating mutants, a decrease in the standard deviation i.e. environmental variations of the phenotypic value is required. Selection of mutants on the basis of the mean value of M_2-plant derived progeny in the M_3 instead of the value of the M_2 plant is a possible solution as shown in the next section.

2.4.3. Traits of Low Heritability

If the heritability of the targeted quantitative trait is low and is much influenced by environmental effects, mutants should be screened from the M_3 population based on the mean phenotypic value of progeny derived from M_2 spikes or plants. In such a case also, the results shown above may be helpful in determining the selection proportion if the phenotypic value of individual plant in M_2 is replaced by the mean phenotypic value of the progeny.

3. Population Management for Selecting Mutants in Cross-Pollinated Crops

Since the genetic structure of segregating generations in cross-pollinated crops is quite different from those in self-pollinated ones, the methods for selecting mutants in self-fertilizing species are not applicable for cross-fertilizing species.

3.1. Production and Genetic Structure of the M_2 Population

In some cross-fertilizing (allogamous) crop species (e.g. maize, melon, cucumber, oil palm), male and female flowers are spatially separated. In such species, It is possible to produce M_2 populations by artificial selfing of M_1 plants by pollinating the female flowers with pollens from male flowers of the same plant. However, homozygous mutants would not appear in such M_2 populations, since male and female gametes usually derive from different cells in the seed embryo. Since M_2 plants are free from chimerism, therefore homozygous mutant M_3 seeds are produced through selfing of heterozygous plants (*Aa'*) in M_2, and subsequently homozygous mutant plants are segregated out in M_3 populations. In many cross-fertilizing species, however, selfing is not successful due to self-incompatibility systems, and in other cases artificial selfing on a large scale is impractical due to very small size of flowers.

In an open pollinating plant population of an allogamous species in which a specific mutation (*A→a'*) was induced with a frequency per gene of μ, the frequency of the genotypes *AA*, *Aa'* and *aa'* in the population is expected to be $(1 - μ)^2$, $2μ(1 - μ)$ and $μ^2$, respectively, assuming random mating. Since the mutation rate is very low, the frequency of the plants possessing the genotype *a'a'* ($μ^2$) is negligible and it is practically impossible to obtain a mutant in the M_2. Thus the mutated genes induced by the treatment are mostly retained in the segregation population as heterozygotes *Aa'* which can rarely be discriminated from the normal genotype *AA* due to the recessive nature of the mutations. As a consequence, adopting a method

that forces a type of inbreeding is essential for obtaining mutants from the populations derived from mutagen treated seeds. The "Crossing-within Spike Progeny (CSP) method" developed by Ukai and Yamashita (1983) and Ukai (1990) was found to be very effective in Italian ryegrass (*Lolium multiflorum*), rye (*Secale cereale*) and other allogamous species.

3.2. Crossing-within Spike Progeny Method

The CSP method is composed of: 1) harvesting seeds separately from each spike of the M_1 plants or plants in a recurrently treated population (**see Chapter 14**); 2) sowing the seeds derived from each spike in a small hill-plot (plot of a small-area, e.g. 0.5 m × 0.5 m) in the next generation; 3) isolating each hill from the others by bagging all the plants of each hill-plot just prior to the start of flowering (**Figure 18.4**); and 4) harvesting seeds from each hill-plot and sowing them as a hill-plot progeny for the selection of mutants in the next generation.

This method is based on half-sib mating, a type of inbreeding that is achieved within each hill-plot from which homozygous mutants will segregate out in the following generation (M_3) with an expected frequency of 0.0625, if the M_1-spike is free from chimeras and the number of plants per hill and the number of plants per progeny in the selection generation are both infinite. In an M_1-spike derived from a mutated seed (*Aa'*) the fre-

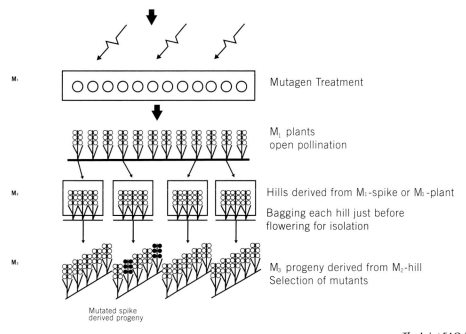

M₁   Mutagen Treatment

M₁ plants
open pollination

M₂   Hills derived from M₁-spike or M₁-plant

Bagging each hill just before flowering for isolation

M₃   M₃ progeny derived from M₂-hill
Selection of mutants

Mutated spike
derived progeny

The Joint FAO/IAEA Programme

Figure 18.4 Population development for mutant selection by crossing-within spike progeny method in cross-pollinated crops.

quency of the genotype A and a' of egg is 1/2 and 1/2, respectively, in the case of no chimeras. Under open pollination, the eggs are fertilized with pollen from the other plants. Due to very low mutation rates, nearly all pollen have the A genotype. Hence, the seed set on the M_1-spike have the genotype AA and Aa' both with an expected frequency of 1/2. Seed harvested from the mutated spike are sown in a hill-plot and the plants are bagged just before flowering to avoid fertilization with the pollen from the other hills. Fertilization is performed within each hill-plot. The frequency of gene A and a' is 3/4 and 1/4, respectively. So, under random mating within each hill, the frequency of the mutant seeds $(a'a')$ per progeny derived from the hill in the M_3 generation is expected to be $1/4 \times 1/4 = 1/16$.

3.3. Increasing the Frequency of Mutant Detection

When an M_1-spike is derived from k initial cells, the frequency of mutants per progeny in M_3 is expected to be $0.0625/k^2/(1-(1-p_3)^n)$, where n is the number of plants per progeny, and p_3 is the segregation frequency when n is infinitely large. The frequency of the progeny in which one or more mutants segregate is $kp_1(1-(1-p_3)^n)$, where p_1 is the mutation rate per cell. Hence, the frequency of mutants among the observed plants is $0.0625/k^2/(1-(1-p_3)^n) \times kp_1(1-(1-p_3)^n) = 0.0625\,p_1/k$. Thus, the frequency of the mutants in the M_3 population decreases with the inverse of k, the number of initial cells. Avoiding chimeric structure within M_1-spike is important in mutant selection in cross-pollinating crops.

Since the bagging of each hill-plot just before flowering time in M_2 is a laborious task, it is desirable to increase the frequency of the mutated gene in the population in the generation in which bagging is performed.

Recurrent mutagenic treatments for successive generations meet these two requirements. Unlike a self-pollinated species, seed sterility does not increase drastically after recurrent treatment in a cross-pollinated species, which was shown in Italian ryegrass after gamma-ray and chemical treatments (Ukai, 1990).

4. Other Considerations

Supplementary comments regarding the method of mutant development and selection are shown below.

1. In a cross-breeding programme, most important traits to be selected are quantitative and, in general, controlled by polygenes. The phenotype of a quantitative trait is influenced by environmental factors. Selections for a specific trait are made in the later generations when most of the loci governing trait are fixed. In a mutation breeding programme, the targeted traits are those usually governed by a single major gene and the selection of mutants is performed primarily in the M_2 generation. However, some breeders erroneously consider that the selection methods developed for cross-breeding can also be applied to mutant selection.

2. Seed sterility may be observed and is often caused by the mutagenic treatment. The treatments with radiations often induce chromosome translocations and inversions that lead to pollen and seed sterility. Sterility following chemical mutagenic treatments is often due to induction of sterility genes. The magnitude of sterility differs between plants and between spikes within plants. Either type of induced sterility is more or less inherited to the subsequent generations, hence, if the objective is not to develop mutants for fertility, inflorescences with normal seed fertility should be selected from the M_1 plants.

3. When the breeder identifies a mutant in field screening, a stake marker is often used and placed in the soil adjacent to the individual and a record (preferably electronic) is made in a field note which leads to a database describing the location and the characteristics of the targeted mutant(s). If the probability of identification of a mutant by any observer is one, as is the case with most of the chlorophyll and early maturity mutations, we identify these mutants as expressing "complete ascertainment". Usually semi-dwarf mutants and disease resistance mutants are generally not easily identified, and these types of mutants are defined as expressing "incomplete ascertainment".

4. In the selection of mutants in self-pollinating crops, the most important point in M_3 screening is to evaluate the mutations that have been selected at the M_2 generation. The mutation must be highly heritable and fixed in the following generation by selfing the M_2-spike progeny in the field. If the undesirable characteristics are associated with a mutant and cannot be removed by ordinary selection, it may be necessary to backcross (BC) the mutant to the origi-

nal variety and select a promising progeny from the BC population.

5. In instances where chronic irradiation is performed on growing plants instead of seeds, the screening of mutants differs with the developmental stages at which the mutation is induced. If the mutation occurred prior to pollen- and embryo sac development, the mutants can be selected at the M_2 generation as in the seed irradiation, although the number of initial cells within the spike is much higher than in seed irradiation If the mutation occurs at or after the pollen- and embryo-sac development, the mutants will appear at the M_3 generation.

6. Generation of doubled haploids from M_1, M_2, M_3, etc. generations is a valuable means of producing homozygous mutants and is particularly valuable in species that have long generation times e.g. perennial trees (**see Chapter 29**).

5. References

5.1. Cited References

Frydenberg, O. 1963. Some theoretical aspects of the scoring of mutation frequencies after mutagenic treatment of barley seeds. *Radiation Botany*. 3: 135–143.

Freisleben, R. and Lein, A. 1943. Vorarbeiten zur züchterischen Auswertung röntgeninduzierter Mutationen. II. Mutanten des Chlorophyllapparates als Testmutationen für die mutationsauslösende Wirkung der Bestrahlung bei Gerste. *Z. Pflanzenzücht.* 25: 255–283.

Gaul, H. 1960. Critical analysis of the methods for determining the mutation frequency after seed treatment with mutagens. *Genet. Agrar.* 12: 297–318.

Lundqvist, U. 1991. Swedish mutation research in barley with plant breeding aspects (a historical review) *In*: Plant Mutation Breeding for Crop Improvement (Proceedings FAO/IAEA Symposium, Vienna, 1990), Vol. 1, Vienna: IAEA, pp. 135–148.

Stadler, L.J. 1928. Mutations in barley induced by X-rays and radium. *Science*. 68: 186–187.

Ukai, Y. 1990 Application of a new method for selection of mutants in a cross-fertilizing species to recurrently mutagen-treated populations of Italian ryegrass. *Gamma Field Symposia*. 29: 55–89.

Ukai, Y. 2003 Plant Breeding, Tokyo: University of Tokyo Press, p. 455 (in Japanese).

Ukai, Y. and Yamashita, A. 1974. Theoretical consideration on the problem of screening of mutants I. Methods for selection of a mutant in the presence of chimera in M_1 spikes. *Bull. Inst. Radiation Breeding*. 3: 1-44.

Ukai, Y. and Yamashita, A. 1983. Crossing-within-Spike-Progeny method. An effective method for selection of mutants in cross-fertilizing plants. *Technical News of the Institute of Radiation Breeding*.

Yoshida, Y. 1962. Theoretical studies on the methodological procedures for radiation breeding. I. New methods in autogamous plants following seed irradiation. *Euphytica*. 11: 95–111.

5.2. Further Reading

Lida, S. and Amano, E. 1987. A method to obtain mutants in outcrossing crops. Induction of seedling mutants in cucumber using pollen irradiation. *Technical News of the Institute of Radiation Breeding*. 32.

Ukai, Y. 2010. Theoretical considerations on the efficiency for selection of mutants in an allogamous plant species. *Breeding Science*. 60: 267–278.

Irradiation – Facilitated Chromosomal Translocation: Wheat as an Example

H.-Y.Wang, Z.-H.Liu, P.-D.Chen* and X.-E.Wang*

State Key Laboratory of Crop Genetics and Germplasm Enhancement, Institute of Cytogenetics, Nanjing Agricultural University, Nanjing 210095, China
* Corresponding author, E-MAIL: Wang Xiu-e: xiuew@njau.edu.cn, Chen Pei-du: pdchen@njau.edu.cn

1. Introduction

1.1. Background

Bread wheat (*Triticum aestivum L.*) is one of the most widely grown food crops in the world. A continuous improvement in its production is a challenging task for wheat breeders to keep abreast of population growth and increasing consumption in certain regions such as in South Asia. As the genetic variability within cultivated wheat has been gradually eroded under modern agricultural systems, alien germplasm has become increasingly important as a source of new variation for wheat improvement. The related genera of wheat such as *Secale, Agropyron, Aegilops, Haynaldia, Elymus, Thinopyrum* and *Hordeum* represent an enormous pool of genetic variation that has great potential for wheat improvement. These species exhibit huge diversity in phenotype and adaptation to a wide range of environments, desirable traits such as disease resistance, drought tolerance, salt tolerance, winter hardiness and adaptability to poor soil. However, because of their genetic distance, it is difficult to introduce these useful genes into cultivated varieties by crossing, chromosome pairing and recombination between homoeologous chromosomes. The gene transfer can be achieved by chromosome manipulation, i.e. the development of amphiploids and alien addition, substitution and translocation lines. The amphiploid contains a complete set of the alien chromosomes, and addition or substitution lines contain a whole alien chromosome. These materials have little potential as varieties as many redundant genes would be introduced into the cultivated variety along with the target genes. However, they are extremely valuable as bridging lines in the production of translocation lines, especially interstitial translocation between a normal wheat chromosome and a small alien chromosome segment.

Chromosome translocations can be induced by ionizing radiation (**see Chapter 11**), tissue culture, homoeologous chromosome pairing, gametocidal genes from several *Aegilops* species, spontaneous wheat–alien chromosome translocation, or by centric breakage and fusion. Spontaneous alien translocation can occur as a result of occasional chromosome breakage and re-union. Due to the distant genetic relationship, homoeologous recombination between chromosomes of wheat and its wild relatives takes place at a low frequency due to the action of the homologous

chromosome pairing gene, *Ph*. Thus, the use of methods that can induce random chromosomal breakage at a relatively high frequency is useful in increasing the ratio of terminal translocations. Gametocidal chromosomes can induce extensive structural chromosome changes, but the variants obtained are often unstable.

Irradiation of wheat–alien hybrids, amphiploids and wheat/alien chromosome cytogenetic stocks can produce a mass of translocations, these can carry desired genes from alien species and are ready to be used in traditional breeding. Irradiation-facilitated translocation enables the transfer of genes from alien chromosomes that would rarely pair with those of wheat during meiosis. In addition, the alien chromosome segment containing the target gene could, theoretically, be inserted into a wheat chromosome without losing any wheat chromatin. A further advantage is that once the alien segment is introgressed into the wheat genome, it is genetically fixed as it will not recombine with wheat chromosomes in the presence of *Ph*. Therefore, this kind of irradiation technique is a powerful tool to introduce new and heritable variation into wheat for genetic improvement, and serves as an important complement to conventional breeding. So far, irradiation has been widely used for inducing the transfer of alien genes into wheat. Several wheat varieties have been developed using this approach, notably in Australia and the USA. In addition to their direct use in wheat breeding, translocation lines have been used widely in gene mapping.

1.2. Wheat Chromosome Nomenclature

Bread wheat, *Triticum aestivum* is a member of the subtribe *Triticeae*, which includes several important small grain cereals: durum wheat (*T. durum*), barley (*Hordeum vulgare*), rye (*Secale cereale*) and the wheat/rye hybrid (triticale) along with many wild grass species. Bread wheat is a hexaploid species comprising three genomes: AA, BB and DD. It is a cultivated species with no wild equivalent; however, each of the three genomes was contributed by separate ancestral wild diploid grasses. Triticeae genomes are composed of seven pairs (7") of chromosomes. Bread wheat has 42 chromosomes or 21 pairs of chromosomes. This is known as the euploid complement, and any deviation from 42 chromosomes is termed aneuploid. Basic chromosomal constitution is given by the sporophytic number (2n), the ploidy level (x) and the chromosome number and is thus 2n = 6x = 42. The chromosomes of

Table 19.1: Information on the species included in this chapter

Name used in this chapter	Species Synonyms	2n	Genome
Triticum aestivum L.		42	AABBDD
Triticum durum L.		28	AABB
Triticum umbellulatum (Zhuk.) Bowden	*Aegilops umbellulata* Zhuk; *Kiharapyrum umbellulatum* (Zhuk.) Á. Löve	14	UU
Secale cereale L.		14	RR
Agropyron elongatum (Host) P. Beauv.	*Thinopyrum elongatum* (Host) D. R. Dewey; *Elytrigia elongata* (Host) Nevski	14	EE
Agropyron intermedium (Host) P. Beauv.	*Thinopyrum intermedium* (Host) Barkworth & D. R. Dewey subsp. Intermedium; *Elytrigia intermedia* (Host) Nevski	42	JJJ^sJ^sSS
Thinopyrum ponticum (Podp.) Barkworth & D. R. Dewey	*Elytrigia pontica* (Podp.) Holub; *Lophopyrum ponticum* (Podp.) Löve	70	$JJJJJJJ^sJ^sJ^s$
Haynaldia villosa (L.) Schur	*Dasypyrum villosum* (L.) P. Candargy	14	VV
Hordeum vulgare L.		14	HH
Leymus racemosus (Lam.) Tzvelev	*Elymus giganteus* Vahl	28	$N_sN_sX_mX_m$

each genome are numbered 1 to 7: 1A, 2A, 3A, 4A, 5A, 6A and 7A for the A genome; 1B, 2B, 3B, 4B, 5B, 6B and 7B for the B genome; and 1D, 2D 3D, 4D, 5D, 6D and 7D for the D genome. These chromosomes are ordered according to relatedness, thus the group 1 chromosomes (1A, 1B and 1D) carry similar genes and are referred to as being homoeologous. The homoeologous relationship extends to all members of the Triticeae. Genomes and chromosomes not present in wheat, but which may be introduced from other species, are termed 'alien'. Further information on genetic nomenclature is given in **Chapter 4.**

2. Irradiation-Facilitated Chromosomal Translocation

2.1. Types of Chromosome Translocation Induced by Irradiation

Translocations are one type of chromosome aberration, which may result from the rejoining of broken chromosomes. Irradiation tends to induce chromosome breaks, which can rejoin at random, resulting in translocations. These are likely to be deleterious. Explanations for the different types of translocation are summarized in **Box 19.1.**

Terminal translocations (**Figure 19.1A**) are the most common type, these involve the distal segment of a chromosome which is replaced by a segment of another (e.g. an alien) chromosome. When breaks occur within the centromeric region, it may result in a whole arm translocation (**Figure 19.1B**), in which a whole arm of a chromosome replaces that of another. The normal divi-

Box 19.1: Homologous chromosomes, homoeologous chromosomes and translocation classes

Homologous chromosomes: These are chromosome pairs of a genome, one from each parent, they have the same genetic composition, e.g. 1A and 1A; 2A and 2A; 7D and 7D, etc.
Homoeologous chromosomes: These are related chromosomes with similar genetic constitutions, but from different genomes, e.g. 1A, 1B and 1D; 2A, 2B and 2D; 7A, 7B and 7D, etc.
Translocation: A chromosomal aberration resulting from the interchange of parts between non-homologous chromosomes.
Terminal translocation: Where the distal segment of a wheat chromosome is replaced by a segment of another (e.g. an alien) chromosome.
Whole arm translocation: A wheat chromosome arm is replaced by a whole arm of another (e.g. an alien) chromosome.
Reciprocal translocation: A type of chromosome rearrangement involving the exchange of chromosome segments between chromosomes that do not belong to the same pair of chromosomes.
Intercalary translocation: A chromosome segment inserted into another chromosome.
Compensating translocation: A chromosome segment replaces an equivalent segment of a homoeologous chromosome.

sion is either the separation of the two sister-chromatids during anaphase of mitosis and anaphase II of meiosis, or the separation of the pairing homologous chromosomes during the anaphase I of meiosis, however the mis-division is that a breakage occurs at the centromere region of a chromosome during anaphase and this will result in telo-, iso- or whole arm translocation chromosomes. Mis-division of univalent chromosome centromeres can only cause whole arm translocation. A reciprocal translocation (**Figure 19.1C**) is a type of chromosome rearrangement involving the exchange of chromosome segments between two chromosomes that do not belong to the same pair of homologous chromosomes. A compensating translocation, in which a desired alien segment replaces an equivalent segment of a homoeologous wheat chromosome, is more likely to be beneficial.

In an intercalary translocation (**Figure 19.1D**), a chromosome segment is inserted into another, however this seldom occurs as this requires several simultaneous breakage and reunion events. This type of translocation will be desirable when an alien segment containing desirable genes is inserted into a host chromosome without loss of host genes. This requires one break in the host chromosome and two breaks in the alien chromosome, with the desirable genes between the breaks. The excised alien segment has unstable ends, which can unite with the unstable ends of the host chromosome at the break point. If the inserted alien segment is quite short, it should not interfere with the pairing of the homologous host chromosomes, particularly if the chromosome with the insertion is made homozygous by selfing.

2.2. Steps to Develop Useful Translocation Lines

There are five steps essential for achieving useful translocations between chromosomes of wheat and an alien species (**Figure 19.2**). Firstly, identification of alien genotypes carrying desired genes required for wheat improvement; secondly, production of a wheat–alien hybrid, amphiploid, alien addition or substitution lines;

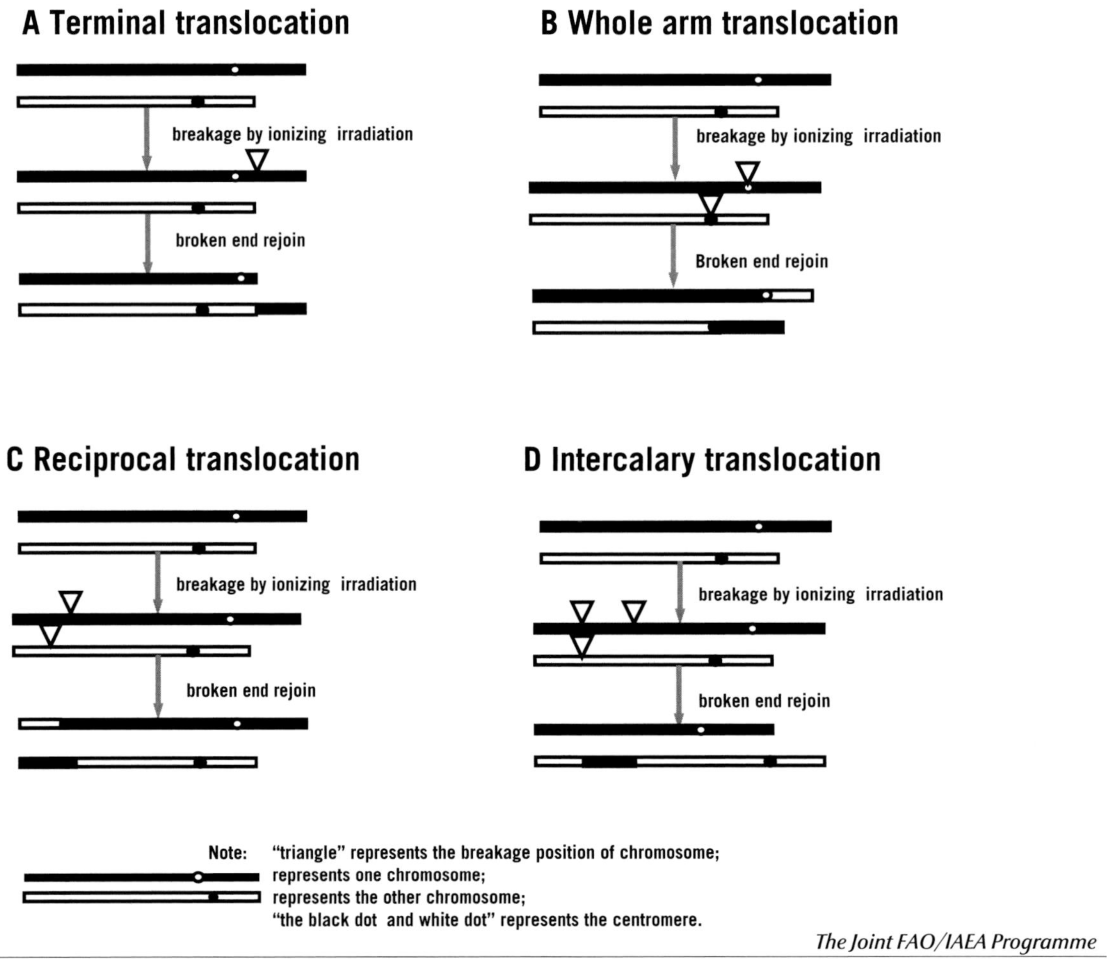

A Terminal translocation

breakage by ionizing irradiation

broken end rejoin

B Whole arm translocation

breakage by ionizing irradiation

Broken end rejoin

C Reciprocal translocation

breakage by ionizing irradiation

broken end rejoin

D Intercalary translocation

breakage by ionizing irradiation

broken end rejoin

Note: "triangle" represents the breakage position of chromosome;
        represents one chromosome;
        represents the other chromosome;
        "the black dot and white dot" represents the centromere.

The Joint FAO/IAEA Programme

Figure 19.1 Ionizing radiation-facilitated chromosome translocations.

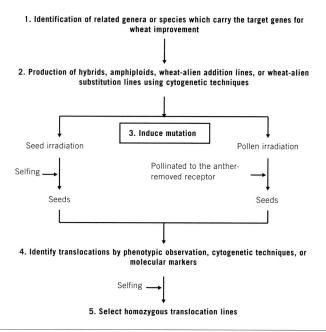

1. Identification of related genera or species which carry the target genes for wheat improvement

2. Production of hybrids, amphiploids, wheat-alien addition lines, or wheat-alien substitution lines using cytogenetic techniques

3. Induce mutation

Seed irradiation

Pollen irradiation

Selfing

Pollinated to the anther-removed receptor

Seeds

Seeds

4. Identify translocations by phenotypic observation, cytogenetic techniques, or molecular markers

Selfing

5. Select homozygous translocation lines

The Joint FAO/IAEA Programme

Figure 19.2 A generalized procedure of developing translocation lines by seed and pollen irradiation.

thirdly, irradiation of the hybrid, amphiploid or alien chromosome lines to induce translocations; fourthly, identification and selection of translocation events and finally, fixation of the translocated chromosome by producing homozygous lines *via* selfing. The first two steps (identifying genes for transfer and making wheat–alien hybrids), are the most difficult. In addition, the initial translocations obtained in this procedure may often be agronomically inferior to wheat because the alien segment may contain other genes with negative effects. In this case, an additional step is required to reduce the size of the transferred chromosome segment.

2.3. Radiation Mutagens Used to Induce Chromosome Breakage Events

Radiation treatments used for chromosome breakage–reunion induction include: X-rays, fast neutrons, gamma rays and ultraviolet. The choice of radiation is related to the type of materials to be treated, the availability of irradiation equipment and the expected/desired frequency and spectrum of mutations. Note that irradiation treatments differ only slightly with respect to their efficiency in producing desirable translocations. Generally, fast neutrons can induce relatively small segment deletions or translocations. X-rays and gamma rays are preferred because of their ease of use and safety in application, good penetration, high reproducibility, high translocation frequency and fewer disposal (radioactive) problems. X-rays were used most extensively in early studies, whereas gamma rays have become more widely used.

2.4. Genotypes Targeted for Irradiation Treatment

In general, four genotypes, i.e. wheat–alien amphiploids, alien chromosome addition lines, substitution

Box 19.2: Genotypes targeted for radiation treatment

Amphiploid: The chromosomally doubled product of a hybrid between two species. The chromosome number of an amphiploid is the sum of that of the two parents, and the genotype is entirely homozygous.
Wheat–alien addition line: A genotype that has a complete set of wheat chromosomes with additional chromosomes from another species.
Monosomic addition line: A genotype that has the normal complement of 21 pairs of wheat chromosomes plus a single alien chromosome.
Disomic addition line: A genotype that has 21 pairs of wheat chromosomes plus a pair of alien chromosomes.
Wheat–alien substitution line: A genotype that has a wheat genome in which wheat chromosomes are replaced by chromosomes from an alien species.
Disomic substitution line: A genotype that has 21 pairs of chromosomes, but in which a pair of alien chromosomes replaces a homoeologous pair of wheat chromosomes.
Monosomic substitution line: A genotype that has a wheat chromosome replaced by a single alien chromosome.

lines and whole arm translocation lines are used as initial materials for irradiation to induce chromosome translocations (**Box 19.2**). It is preferable that the desired genes of the alien species have been mapped to a chromosome or chromosome arm and their respective homoeologous relationships with wheat chromosomes clearly established. Furthermore, wheat–alien chromosome translocations resulting from these genotypes can be identified easily by cytological methods, by phenotypic analysis and/or by molecular markers specific to the alien chromosomes or target chromosomal regions.

Amphiploids (or amphidiploids) are usually the fertile products of spontaneous or induced chromosome doubling of sterile interspecific or intergeneric hybrids. The chromosome number of an amphiploid is the sum of the two parents, and the genotype will be entirely homozygous. Amphiploids provide a starting point for the production of wheat–alien chromosome addition lines or substitution lines and interspecific transfers.

Wheat–alien addition lines, including monosomic additions ($2n = 43$; $21_w'' + 1_a'$) and disomic additions ($2n = 44$; $21_w'' + 1_a''$), have complete wheat genomes with additional chromosomes from the alien species. These genotypes are used to identify alien chromosomes carrying useful genes and form the starting point for the cytogenetic transfer of alien chromatin into wheat.

Wheat–alien substitution lines, including monosomic substitutions and disomic substitutions, have a normal (euploid) wheat chromosome numbers ($2n = 42$), but a single chromosome, or a pair of wheat chromosomes are replaced by single (monosomic) or a pair of (disomic) chromosomes from an alien species. These genotypes have certain advantages over addition lines, although they are generally more difficult to obtain. By having an euploid chromosome complement they are more stable, allowing large-scale multiplication. They also permit the value of the genes carried by the alien chromosome to be evaluated against each of the three possible wheat homoeologous alleles in turn. Irradiation of disomic substitution or addition lines, which have two doses of an alien chromosome, has a greater chance of exchange with a host chromosome than when a single alien chromosome is present.

Wheat–alien disomic addition and substitution lines are usually used for translocation induction, because the target gene has been mapped to a specific alien chromosome and the identity of the translocation chromosome is easy to be determined. However, the

translocation lines produced only involve that specific alien chromosome. Furthermore, the alien translocation induction frequency is very low due to the low ratio of alien chromatin present. In contrast, a complete genome of the alien species is present in the wheat–alien amphiploids and could be used to induce a mass of translocation lines involving different chromosomes. Bie *et al.* (2007) obtained the large number and range of translocations by pollen irradiation of a *Triticum durum–Haynaldia villosa* amphiploid. This work produced a translocation line germplasm bank for further genetic studies, including physical mapping.

The use of available whole arm translocation for irradiation can improve the efficiency of creating interstitial translocations. Here a break in the alien chromosome can generate small fragment interstitial or terminal translocations, or deletions of the alien chromosome. Chen *et al.* (2008) observed a high frequency of both small fragment terminal translocations or deletions and intercalary translocations involving the short arm of chromosome arm 6V of *H. villosa* by irradiation of a whole arm translocation, 6VS/6AL.

2.5. Target Materials for Mutagenesis and Irradiation Dosage

Various parts of the same plant exhibit differences with respect to their radio-sensitivity. The occurrence of induced translocations may be affected by numerous factors such as genetic background (genotype), target tissue, water content in the target tissue, temperature, oxygen level and dose rate (**see Chapter 14**).

Seed

Seed is a convenient and common target for mutagenic treatment. Prior to treatment it is referred to as the M_0 generation, after treatment it becomes M_1. The water content of the seed is a critical factor affecting translocation-producing frequency (Smith, 1958). Compared to dry seeds, seeds soaked in water before treatment are more responsive to irradiation. Seeds with 12–14% water content usually give relatively higher mutation frequencies. Half-lethal dosage is widely used to obtain the maximum number of mutations, with 100 Gy being a reasonable dose. However, the dose and duration of application of a mutagen vary with genotype and should be determined through experimentation. It has been found that a 20 Gy dose (^{60}Co source) was sufficient for inducing a desirable level of wheat–rye chromosomal translocations in triticale seeds (Ahmad *et al.*, 2000). Rakhmatullina and Sanamyan (2007) estimated the efficiency of seed irradiation by thermal neutrons for inducing chromosomal aberration in M_2 of cotton, and they found that 15 and 25 Gy were the most efficient radiation doses for inducing chromosomal aberrations.

Chimeras are common in the first generation (M_1 plants) and, only with some exceptions (organelle genes, mutator genotypes) can be purged by selfing in the production of the M_2 generation. For crops like wheat, individual tillers (side branches) originate from different cells of the embryo of the treated seeds. If an aberration occurs in one of these cells, it will be carried in the tiller developed from that cell.

Adult plant shortly before meiosis

The transfer of rust resistance from *Triticum umbellulatum* (formerly, *Aegilops umbellulata*) into common wheat using irradiation by Sears (1956) is a classical example. In his experiment, adult plants were irradiated shortly before meiosis (with 15 Gy), and the irradiated pollen was pollinated onto spikes of normal, non-irradiated plants.

Irradiation of adult plants will be particularly advantageous if monosomic-substitution plants (being monosomic for an alien chromosome and also for a homoeologous wheat chromosome) and also deficient for chromosome 5B carrying the *Ph1* gene, are used. In the absence of the pairing homologous gene, *Ph1*, homoeologous chromosomes will associate at meiosis and recombine. Therefore, by irradiating plants that are not only doubly monosomic, but also lack *Ph1*, a substantial increase of the yield of desired translocations may be achieved. Alternatively, mutants carrying a non-functional *Ph1* gene, e.g. *ph1b* may be used. However, adult plants are not as easy to handle as seeds, especially if the radiation source is not readily available.

Pollen

Pollen is another relatively convenient target for irradiation treatment. The treatment dosage for pollen is less than that for seeds. In this treatment, spikes from tillers to be used for pollen irradiation are detached from the plant. After irradiation, they are maintained with their cut ends in water until flowering. The induced translocations in all parts or branches of the M_1 plant are present in a heterozygous state, and hence it is not necessary to harvest separately. The advantage of pollen irradiation is that the plants resulting from the fertilization of un-irradiated egg cells with irradiated pollen are usually free from chimeras and are heterozygous for any induced translocations. However, pollen irradiation is usually associated with problems such as non-availability of adequate quantities of pollen, poor pollen viability and the potential of inducing haploid embryo formation (**see Chapter 30**).

Mature female gametes

Chen *et al.* (2008) successfully enhanced the inducement frequency of translocations, especially interstitial translocations, by irradiation of mature female gametes of whole arm translocation lines. The use of mature female gametes has several advantages. First of all, female gametes are less lethal sensitive and can endure higher dosages and dosage rates. Second, female gametes are irradiated just before fertilization and they are pollinated with normal fresh pollen after irradiation. In this case, the structural aberrations have a greater chance of being involved in the fertilization process and thereby can be transmitted to the next generation. Third, these irradiated female gametes can be pollinated with mature and fresh pollen of normal wheat. This avoids the elimination of structurally aberrant chromosomes due to pollen tube growth and fertilization competition that occurs after pollination (normal pollen usually has a competitive advantage). Thus a higher proportion of the chromosome aberrations can be retained in the M_1 plants. M_2 seeds are obtained by backcrossing with normal pollen. Various structural changes (including interstitial translocation) observed in the M_1 can be recovered in the M_2. Backcrossing can improve the transmission frequency and vitality of the progenies. The fertility levels are generally enhanced with each progressive generation and hence the small fragment chromosome changes in the M_2 plants are easily transmitted to the next generation. The application of genomic *in situ* hybridisation (GISH) in the M_1 enabled the detection of more chromosome structural changes of alien segments in a relatively small population (**Box 19.3**).

2.6. Identification of Translocation Events

A number of well established methods for identifying translocations including phenotypic, cytological and molecular analysis are available for wheat geneticists and breeders (**Box 19.3**).

Box 19.3: Cytogenetic and molecular cytogenetic tools used to identify chromosome translocation

Chromosome C-banding: Method of defining chromosome structure by differential staining (banding) of constitutive (C) heterochromatin regions with Giemsa.

Genomic *in situ* hybridization (GISH): GISH is a genomic probing technique and a molecular cytogenetic technique, which uses genomic DNA from the alien species as a probe in combination with an excess of unlabelled wheat DNA in the hybridization solution to block cross hybridizations. GISH analysis allows physical determination of alien chromosome segments and the break points in these translocations and an estimation of the sizes of the transferred segments of the alien species.

Fluorescence *in situ* hybridization (FISH): A molecular cytogenetic technique in which a DNA probe is labelled with a fluorescent dye conjugate (that can be visualized under a fluorescence microscope) and then hybridized onto target DNA, usually chromosome preparations on a microscopic slide. FISH allows direct mapping of DNA sequences to chromosome, and has become an important technique in plant molecular cytogenetics research. It is used to map genes physically and precisely to a specific region of a chromosome and can enumerate chromosomes, and/or detect chromosomal deletions, translocations or gene amplifications in cells.

Molecular markers: A molecular marker (or genetic marker) is a specific fragment of DNA sequence that is associated with a part of the genome and can be identified within the whole genome. Molecular markers are used to 'flag' the position of a particular gene or the inheritance of a particular characteristic. In a genetic population, the characteristics of interest will usually stay linked with the molecular markers. In wide hybridization, the molecular markers specific for different species are used to trace the introduced alien chromatin in a background of cultivated species. In some cases the gene of interest can be monitored directly.

2.6.1. Phenotypic Markers

Morphological markers are most useful if the target genes have been located on specific chromosomes and have a distinct phenotypic effect when combined with the host genotype. A good example of this type of marker is a gene on rye chromosome 5R for pubescent peduncle (hairy neck), which shows clear expression in a wheat background.

2.6.2. Biochemical Markers

Certain biochemical markers such as proteins and isozymes are particularly useful for detecting alien chromatin. When crosses are made between distantly related species or genera, the presence in the progeny of isozyme markers from both parents confirms hybridization. A disadvantage is the relatively low level of polymorphism at loci identified by isozymes and proteins and these are now largely replaced by genome wide molecular markers.

2.6.3. Chromosome C-Banding

The development of chromosome banding techniques makes chromosome identification fast, reliable and economical. Segmental transfers can be detected by chromosome banding on mitotic chromosomes if the segments have a distinct banding pattern from that of the host chromosomes. If the alien-segment banding is not distinctive, meiotic observations of the pairing configurations can be combined with the use of *in situ* hybridization (ISH) and/or molecular markers analysis. However, chromosome banding techniques are uninformative if the alien

chromosome segments lack diagnostic bands. Banding polymorphism in different wheat genotypes sometimes also confuses the identification of the alien chromosome segments. This method requires skills in microscopy.

2.6.4. Genomic *in situ* Hybridization (GISH)

The ISH technique allows genes or DNA sequences to be localized directly on chromosomes in cytological preparations. The molecular cytogenetic method of ISH makes it possible to identify transfers of alien chromosomes or segments to host species more efficiently and, in the case of segmental transfers, with more precision than other methods. GISH uses genomic DNA from the alien species as probes in combination with an excess of unlabelled wheat DNA in the hybridization solution to block cross-hybridizations. Wheat–alien translocations and their breakpoints can be clearly identified by GISH. However, GISH only allows the detection of alien chromosome segments. Chromosome banding analysis is required to identify the wheat chromosomes involved in the translocations. The combination of chromosome banding and GISH techniques also allows the estimation of the size of the inserted alien chromosome segment and the missing wheat chromosome segment.

Recently, a sequential chromosome banding and GISH technique was developed. Using this technique, the size, position, breakage and reunion point of the alien chromosome segment along with the identity of the wheat and alien chromosomes involved in the translocation can be determined in a single experiment.

GISH was further improved with the advent of non-isotopic fluorescent reporter molecules for labelling of DNA in the 1980s leading to the development of fluorescence *in situ* hybridization (FISH) techniques. FISH methods have advantage over hybridization with isotope-based probes, including longer probe stability, speed, high sensitivity, spatial resolution and simultaneous detection of more than one probe. So, FISH is a useful tool in translocation identification and characterization at the physical level. When species-specific repetitive sequences or genomic clones (e.g. yeast or bacterial arti-ficial chromosomes, restricted fragment length polymorphisms clones) are used as probes, a specific genomic region can be targeted through FISH analysis. FISH and GISH require expertise in fluorescence microscopy.

2.6.5. Molecular Markers

Genome specific molecular markers with wide genome coverage such as restriction fragment-length polymorphisms (RFLPs) can provide a powerful tool in evaluating the amount and location of the alien chromatin in the translocation lines. They are especially valuable in deter-

Table 19.2: Wheat–alien translocations facilitated by irradiation

Target traits [a]	Alien species	Translocation line (wheat–alien translocation)
Resistance to leaf rust	*Triticum umbellulatum*	Transfer (T47) (T6BS.6BL-6U#1L); T40 (T6BL.6BS-6U#1L); T41 (T4BL.4BS-6U#1L); T44 (T2DS.2DL-6U#1L); T52 (T7BL.7BS-6U#1L)
Resistance to greenbug	*Aegilops speltoides*	CI17884 (T7AS-7S#1S. 7S#1L)
Resistance to leaf rust and stem rust	*Agropyron elongatum*	Agatha (T7DS.7DL-7Ae#1L)
Resistance to leaf rust and stem rust	*Agropyron elongatum*	Teewon (T1BL.1BS-3Ae#1L)
Resistance to stem rust	*Agropyron elongatum*	K2046 (T6AS.6AL-6Ae#1L)
Resistance to wheat streak mosaic	*Agropyron elongatum*	CI15322 (T4DS.4DL-1Ae#1L)
Resistance to wheat streak mosaic	*Agropyron intermedium*	WGRC27 (T4DL.4Ai#2S)
Resistance to leaf rust	*Agropyron intermedium*	T4 (T3DL.3DS-7Ai#2L); T7 (T6DS.6DL-7Ai#2L); T24 (T5AL.5AS-7Ai#2L); T25 (T1DS.1DL-7Ai#2L); T33 (T2AS.2AL-7Ai#2L)
Resistance to greenbug and powdery mildew	*Secale cereale*	Amigo (T1AS.1R#2S)
Resistance to leaf rust and stem rust	*Agropyron elongatum*	Amigo (T1BL.1BS-3Ae#1L)
Resistance to greenbug	*Secale cereale*	GRS1201 (T1AL.1R#3S); GRS1204 (T2AL.2AS-1R#3S; T2AS-1R#3S.1RL#3)
Resistance to leaf rust and powdery mildew	*Secale cereale*	Transec (T4BS.4BL-2R#1L)
Resistance to leaf rust	*Secale cereale*	ST-1 (T2AS-2R#3S.2R#3L)
Resistance to stem rust	*Secale cereale*	WRT238 (T3AS.3R#1S); 90M126-2 (T3AL.3R#1S); 90M126-9 (T3BL.3R#1S)
Resistance to Hessian fly	*Secale cereale*	88HF16 (T6BS.6BL-6R#1L); 88HF79 (T4BS.4BL-6R#1L); 88HF89 (T4AS.4AL-6R#1L-4AL)
Resistance to BYDV	*Thinopyrum intermedium*	Line-632-21 (7DS-7EL)
Resistance to FHB	*Leymus racemosus*	NAU618 (T1AS-Lr7S.Lr.7L); NAU611 (T4AL.Lr7S); NAU601 (T4BS.4BL-7Lr#1); NAU615 (T4BS.4BL-7Lr#1S-1); NAU614 (T6BL.6BS-5Lr#1)
Blue-grain	*Thinopyrum ponticum*	BT108 (1BS.1BL-4AgL); Line 9908 (2DL-4AgL); BT91 (4AgL-3AL)
Resistance to powdery mildew	*Haynaldia villosa*	92R137 (6VS/6AL)

[a] BYDV, Barley yellow dwarf virus; FHB, Fusarium head blight.
[b] Wheat–alien chromosome translocations induced by irradiation before 1996 were summarized by Friebe *et al.* (1996).

mining the relative position of translocation breakpoints. Furthermore, RFLP analysis can identify an interstitial translocation not detectable by GISH. The identification of markers tightly linked to the target gene allows marker-assisted selection. RFLP markers have been used for several alien introductions into wheat for pest or disease resistance: *H23* and *H24* genes for Hessian fly resistance (Ma *et al.,* 1993), the segment carrying *Pm13* for powdery mildew resistance (Donini *et al.,* 1995), *Yr15* for stripe rust resistance (Sun *et al.,* 1997), eyespot resistance from *Haynaldia villosa* (Yildirim *et al.,* 1998), and scab resistance from *Leymus racemosus* (Wang *et al.,* 2001).

Because of the lower costs and higher speed of analysis, PCR-based markers are now preferred in detecting alien chromosomes or chromosome segments. The co-dominance of simple sequence repeats (SSRs) and their high polymorphic information content often make these the markers of choice. Many wheat SSR markers have been developed, their primer sequences published, and their chromosomal locations determined. However, most SSR markers are genome-specific, and their transferability across related species is low.

The US wheat expressed sequence tag (EST) mapping project has mapped nearly 7,000 ESTs in chromosome bins covering all 21 wheat chromosomes, providing an excellent resource for marker development for specific chromosome regions. Alternatively, EST-based PCR products can be digested with frequent-cutting restriction enzymes to increase polymorphism. EST-SSR or EST-STS (sequence tagged sites) have now been applied to detect alien chromosomes and chromosome segments.

Comparative mapping studies have demonstrated extensive colinearity among different species and this colinearity is increased among closely related species. Many molecular markers have been mapped in particular homoeologous chromosomes or particular chromosome segments. In most cases, most markers were polymorphic in the region of the alien chromatin.

3. Exploitation of Translocation Lines in Wheat Improvement

3.1. Application of Translocation Lines in Breeding

Disease resistance gene cloning studies in model or other plant species have shown that resistance genes are often present as a gene cluster in a specific chromo-

some region. Alien translocation lines therefore provide a means of transferring blocks of such genes, translocation lines are genetically stable and their resistance is more durable compared to single gene transfer. The wheat–rye 1RS/1BL translocation has been utilized successfully in breeding programmes worldwide, one of the important reason is that several useful genes especially disease resistance genes are located in 1RS. More and more translocation lines, especially intercalary translocations with multiple useful traits are expected to be deployed in modern wheat breeding with the accelerated development of various translocation lines involving various different alien species.

A number of translocations have been induced by irradiation of amphiploids, addition lines and substitution lines between wheat and various alien species (**Table 19.1**). Although most of these translocations are either non-compensating translocations or contain undesirable genes, a number of improved strains carrying useful alien genes for resistance to various diseases of wheat have been developed using this technique. One of the most successful translocations involved the transfer of a segment from *Agropyron elongatum* chromosome 6E, carrying the stem rust resistance gene *Sr26*, to chromosome arm 6AL of wheat. Using this translocation line, stem rust resistance (*Sr26*) was successfully introduced into Australian wheats in 1971 and it has played an important role in protecting wheat crops from stem rust especially in northern New North Wales and Queensland. Kite, one of the first cultivated varieties with *Sr26*, was the highest yielding commercial variety in northern New North Wales for several years. A total of 12 registered Australian varieties (Eagle, Kite, Jabiru, Avocet, Bass, Blade, Flinders, Harrier, King, Quarrion, Sunelg and Takari) possess *Sr26*. In the US, Riley67, a commercial variety of soft winter wheat produced at Purdue University, possesses leaf rust resistance from *T. umbellulatum*, demonstrating further potential of this technique for crop improvement.

Using ionizing irradiation, a large number of translocations have been induced between chromosomes of wheat and related species such as *Secale cereale, T. umbellulatum, Leymus racemosus, Haynaldia villosa* and *Agropyron elongatum*, etc. (**Table 19.1**).

3.2. Application of Translocation Lines in Genetic Mapping

In contrast to their direct use in wheat breeding, translocation lines have been widely used in gene mapping. According to the data from identification and characterization of wheat–alien translocations, a large number of genes conferring disease resistance and other morphological traits have been mapped on chromosomes of wheat–alien species, thereby facilitating further directed chromosome engineering aimed at producing agronomically superior germplasm. As shown in **Table 19.1**, the majority of the genes located on wheat–alien translocation lines are related to disease and pest resistance. The original blue-grained wheat, Blue 58, was a substitution line derived from hybridization between common wheat and *Thinopyrum ponticum*, in which one pair of 4D chromosomes was replaced by a

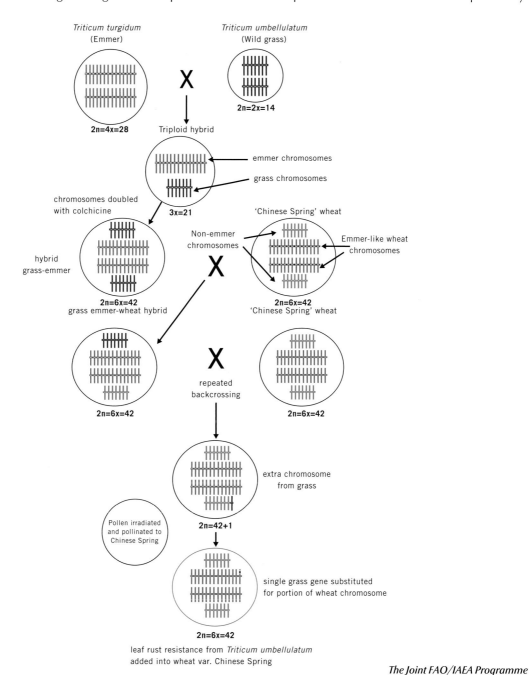

The Joint FAO/IAEA Programme

Figure 19.3 Irradiation-induced translocation between chromosomes of wheat and *T. umbellulatum*. The procedure by which a gene for resistance to leaf rust was transferred from *T. umbellulatum* to common wheat. An allohexaploid was made by combining the chromosomes of emmer (2n = 4x = 28, AABB) with those of *T. umbellulatum*. The resulting allohexaploid (2n = 6x = 42, AABBUU) was crossed with the wheat genetic stock Chinese Spring. Repeated backcrossing of rust-resistant plants to Chinese Spring gave rise to a plant with the chromosome content of common wheat, plus one chromosome from the wild grass, which carried the gene for rust resistance. X-ray irradiation of pollen induced chromosome rearrangement in which the gene for rust resistance was transferred to a wheat chromosome. The red shows the A and B genome; The blue shows the U genome; The green shows the D genome.

pair of alien 4Ag chromosomes (an unknown group 4 chromosome from *A. ponticum*). Blue aleurone located on chromosome 4Ag might be a useful cytological marker in chromosome engineering and wheat breeding. GISH analysis proved that 4Ag was a recombination chromosome, its centromeric and pericentromeric regions were from an E-genome chromosome, but the distal regions of its two arms were from an St-genome chromosome. Zheng *et al.* (2006) treated the translocation line with γ-rays to induce recombination between the 4Ag translocation and normal wheat chromosomes. By GISH and FISH analysis, a set of new translocation lines with different seed colours were identified and the gene(s) controlling the blue aleurone was located on the long arm of 4Ag. Further, the colour gene was physically mapped to the 0.71–0.80 regions (distance measured from the centromere of 4Ag).

3.3. Examples

Details of four examples of translocations for wheat improvement are given below.

3.3.1. "Transfer" – a Wheat/*Triticum umbellulatum* Translocation With Resistance to Leaf Rust

The first successful radiation-induced translocation was the wheat – *T. umbellulatum* translocation line with a dominant gene for leaf-rust resistance (*Lr9*). This study is a classical example of logic and careful planning, as is illustrated in **Figure 19.3**. The wild grass *T. umbellulatum* (2n = 2x = 14, genome UU) was first crossed to tetraploid emmer wheat, *T. turgidum* (2n = 4x = 28, genomes AABB), and chromosomes of the F$_1$ plant were doubled with colchicine to produce a fertile allohexaploid (2n = 6x = 42, genomes AABBUU). The experimental allohexaploid was then crossed to a hexaploid bread wheat genetic standard, Chinese Spring (2n = 6x = 42, genomes AABBDD). After the second backcross to Chinese Spring, leaf rust-resistant plants that contained 42 wheat chromosomes plus one additional chromosome were identified. The additional chromosome was postulated to be from *T. umbellulatum* and carrying the leaf rust resistance gene (*Lr9*). To induce a possible translocation and reduce the chromosomal load from *T. umbellulatum*, plants with 43 chromosomes (monosomic addition lines) were X-rayed shortly before meiosis with 15 Gy. An intact iso-chromosome, and probably the centric (non-translocated) part of the alien

chromosome with a wheat segment attached, behave as univalents at meiosis and were not included in most of the resultant gametes.

Crosses were made by pollinating pollen from the irradiated plants to untreated plants of Chinese Spring. The *Lr9* gene was traced through subsequent generations by screening progenies for resistance to the rust disease pathogen *Puccinia recondite*. Resistant plants were checked cytologically to select those that had an alien segment attached to a wheat chromosome while discarding those with an intact iso-chromosome. Out of 17 different translocations with an alien segment carrying *Lr9*, one was at first thought to be an insertion as described above. However, further study showed that a reciprocal exchange had occurred through a terminal transfer of the long arm of the *T. umbellulatum* chromosome 6U segment to the distal region of the long arm of chromosome 6B. This line was named "Transfer". It was later used as a source of resistance to leaf rust in the production of several commercial wheat varieties widely grown in the United States.

3.3.2. Translocations between Wheat and *Leymus racemosus* with Resistance to *Fusarium* Head Blight (Chen *et al.*, 2005)

Wheat scab (*Fusarium* Head Blight, FHB) is a destructive disease in warm and humid wheat growing areas of the world. Finding diverse sources of FHB resistance is critical for genetic improvement of resistance in the wheat breeding programme. *Leymus racemosus* is a wild perennial relative of wheat and is highly resistant to FHB. Three wheat – *L. racemosus* disomic addition (DA) lines DA5Lr, DA7Lr and DALr.7 resistant to FHB, were used to induce wheat – *L. racemosus* translocation lines through irradiation. Plants with scab resistance were irradiated at meiosis at the (pre-meiotic) boot stage by ^{60}Co γ-ray 5–11.25 Gy (0.75–1.0 Gy/min) or their mature pollen was irradiated at the early flowering stage by ^{60}Co γ-ray 10 Gy (1.0 Gy/min). The irradiated plants were then allowed to self-pollinate or used as pollen donors in crosses with susceptible varieties. The progenies were analysed by C-banding and GISH to identify intergenomic translocations involving wheat and *Leymus* chromosomes.

As shown in **Figure 19.4**, a total of five wheat–alien translocation lines with wheat scab resistance were identified by chromosome C-banding, GISH, telosomic pairing and RFLP analyses. In line NAU614, the long arm of 5Lr was translocated to wheat chromosome 6B.

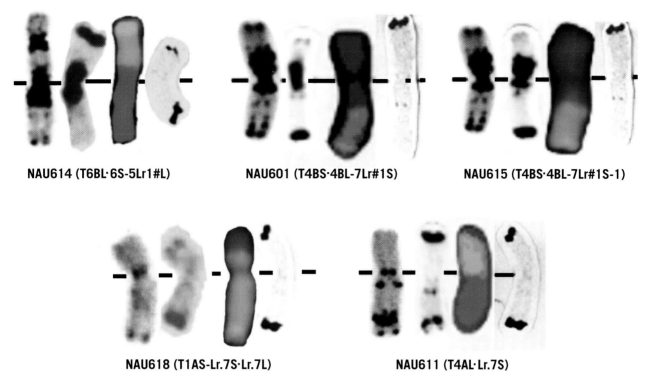

NAU614 (T6BL·6S-5Lr1#L)   NAU601 (T4BS·4BL-7Lr#1S)   NAU615 (T4BS·4BL-7Lr#1S-1)

NAU618 (T1AS-Lr.7S·Lr.7L)   NAU611 (T4AL·Lr.7S)

The Joint FAO/IAEA Programme

Figure 19.4 The C-banding and GISH analysis of wheat – *Leymus racemosus* translocation chromosomes. From left to right for each panel: C-banded intact wheat chromosome C-banded translocation chromosome, translocation chromosome after GISH, C-banded intact *L. racemosus* chromosome.

Two lines, NAU601 and NAU615, had a part of the short arm of 7Lr transferred to wheat chromosomes 4B. Two other lines, NAU611 and NAU618, contained translocations involving *Leymus* chromosome Lr.7 and different wheat chromosomes. The resistance level of the translocation lines with a single alien chromosome segment was higher than the susceptible wheat parent Chinese Spring, but lower than the alien resistant parent *L. racemosus* (**Figure 19.5**). At least three resistance genes in *L. racemosus* were identified. One was located on chromosome Lr.7, one was on the long arm of 5Lr and the other on short arm of 7Lr. These translocation lines will provide much-needed new sources of scab resistance for wheat improvement.

For irradiation, both dry seeds and spikes at pre-meiosis, meiosis or pollen stage can be treated. In this experiment, the frequency of translocations through irradiation of adult plants at meiosis or in spikes before pollination was much higher (11.8%) than that of irradiation of dry seeds. Although an irradiation treatment at meiosis has the advantage that chromosome translocations can occur in both male and female gametes, treating spikes just before flowering is much easier.

Monosomic addition lines were irradiated and used as male parents in crosses with susceptible lines. The plants of the M1 progeny with 2n = 42 chromosomes and good scab resistance were selected for further screening of translocations by chromosome C-banding and GISH. Because a monosomic alien chromosome will tend to be lost during meiosis and a male gametophyte (n+1) with a complete alien chromosome is at a disadvantage, meiosis and pollen competition favour preferential transmission of gametes with translocations.

3.3.3. Translocations Between Wheat and *Haynaldia villosa* Using Pollen Irradiation (Bie *et al.*, 2007)

Bie *et al.* (2007) used pollen irradiation treatment to induce chromosome translocations between *Triticum durum* and *Haynaldia villosa*. In their experiment, *T. durum* – *H. villosa* amphiploid pollen was first treated with 12 Gy γ-rays and then pollinated onto *T. aestivum*, 'Chinese Spring'. Ninety-eight intergeneric translocations between *T. durum* and *H. villosa* were detected by GISH in 44 of 61 M1 plants, with a higher translocation occurrence frequency of 72.1%. There were 26 whole arm translocations, 62 terminal translocations and 10

CS   wheat-L.racemosus   Sumai3   8545
translocation

The Joint FAO/IAEA Programme

Figure 19.5 Spikes of some translocation lines inoculated with *Fusarium graminearum*. Spikes from left to right: Chinese Spring (CS, susceptible wheat parent), NAU601, NAU611, NAU615, NAU618, NAU614, Sumai 3 (resistant check) and Mianyang 8545 (susceptible check). Disease symptoms only occurred rarely on one or two spikelets in the translocation lines, whereas the symptom occurred on half the spikes in the susceptible parent Chinese Spring and susceptible check Mianyang 8545.

intercalary translocations. The ratio of small alien segment terminal translocations was much higher than that of large alien segment terminal translocations. Pollen irradiation is therefore an effective method for rapid mass production of wheat–alien chromosomal translocations, especially terminal translocation will be more significant for wheat improvement.

3.3.4. Inducement of Chromosome Translocation with Small Alien Segments by Irradiating Mature Female Gametes of the Whole Arm Translocation Line (Chen *et al.*, 2008)

Haynaldia villosa has been proved to be an important genetic resource for wheat improvement. Currently, most of the reported *Triticum aestivum – Haynaldia villosa* translocation lines are involved in a whole arm or large alien fragments. Chen *et al.* (2008) reported a highly efficient approach for the creation of small chromosome segment translocation lines. In their experiment, the female gametes of a wheat – *H. villosa* 6VS/6AL translocation line were irradiated with ^{60}Co-rays of 16, 19.2, 22.4 Gy at 1.6 Gy/min before flowering. Anthers were removed from the irradiated florets on the same day and the florets were pollinated with normal fresh pollen of *T. aestivum* var. Chinese Spring after 2–3 days. GISH was used to detect the chromosome structural changes involving 6VS of *H. villosa* at mitosis metaphase of root-tip cell of M_1 plants. Among the 534 M_1 plants screened, 97 plants contained small segment

chromosome structural changes of 6VS, including 80 interstitial translocation chromosomes, 57 terminal translocation chromosomes and 55 deletion chromosomes. Irradiating mature female gametes of whole arm translocation is a new and highly efficient approach for creation of small segment chromosome structural changes, especially for interstitial translocations.

4. Future Developments

4.1. Combined Use of Irradiation with Induced-Homoeologous Pairing

Another approach for introducing alien genetic material to wheat is the induction of homoeologous pairing and recombination. In the absence of the *Ph1* gene, (e.g. by using nulli 5B lines or *ph* mutant lines), homoeologous pairing and recombination can occur among wheat and alien chromosomes of related species. One advantage of this technique is that all the resulting wheat–alien translocations occur between homoeologous (related) chromosomes. Therefore, the transferred alien chromatin can compensate for the loss of the missing wheat segments. This method complements the irradiation procedure as the exchanges are targeted between homoeologous chromosomes. However, it seems to work only for genes that are located in the distal half of chromosome arms, therefore the chance of transferring a proximally located alien gene is currently very low.

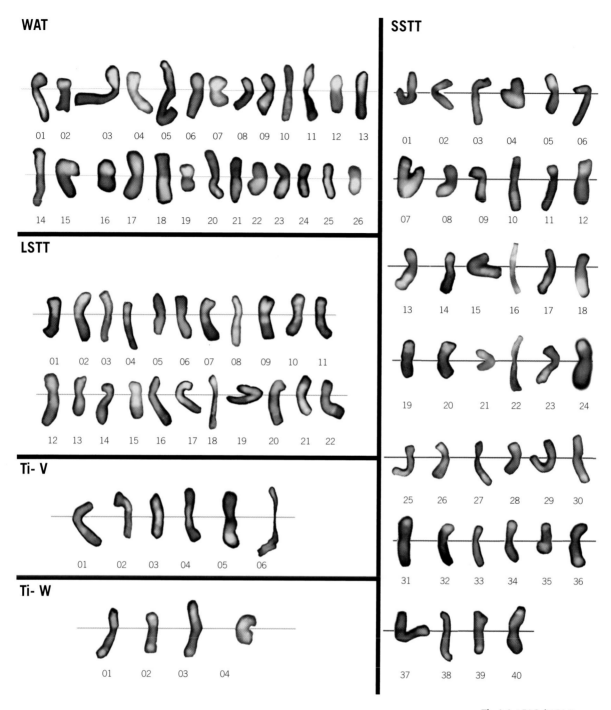

WAT

01 02 03 04 05 06 07 08 09 10 11 12 13

14 15 16 17 18 19 20 21 22 23 24 25 26

LSTT

01 02 03 04 05 06 07 08 09 10 11

12 13 14 15 16 17 18 19 20 21 22

Ti- V

01 02 03 04 05 06

Ti- W

01 02 03 04

SSTT

01 02 03 04 05 06

07 08 09 10 11 12

13 14 15 16 17 18

19 20 21 22 23 24

25 26 27 28 29 30

31 32 33 34 35 36

37 38 39 40

The Joint FAO/IAEA Programme

Figure 19.6 Ninety-eight γ-ray induced wheat (W) – *Haynaldia villosa* (V) translocations identified by GISH analysis. Ti-V, intercalary translocation type of 'W-V-W'; Ti-W, intercalary translocation type of 'V-W-V'; WAT, whole arm translocation 'W·V'; LSTT, large segment terminal translocation (W-V·V); SSTT, small segment terminal translocation (W·W-V). The yellow-green areas show hybridization of labelled *H. villosa* genomic DNA to the alien chromatin in the wheat background, and the red parts represent the chromosome fragment from common wheat. Total genomic DNA of *H.villosa* was labelled with Fluorescein-12-dUTP by nick translation and used as a probe.

In contrast, the irradiation method of obtaining wheat–alien exchanges has the advantage of high frequency of translocations induction. However, it has the disadvantage that the majority of the translocations are genetically imbalanced as they involve non-homoeologous chromosomes. Thus, a combination of these two methods may integrate both advantages of high-frequency and specificity. Presumably the low percentage of homoeologous translocations resulting from irradiation is due to the fact that broken ends of wheat and alien homoeologous chromosomes are rarely in close proximity in the irradiated metabolic nucleus. However, if chromosomes are associated in all interphase nuclei and not only pre-meiotic nuclei, relaxation of the *Ph* mechanism may result in wheat and alien chromosomes being somatically associated. Irradiation-induced broken ends of homoeologues would then be in close proximity for fusion of exchange ends. Therefore, irradiating plants or seeds that are 5BL-deficient may be the most efficient method for obtaining a high frequency of genetically balanced wheat–alien translocations.

4.2. Use of Molecular Markers in the Identification and Selection of Desired Translocations

Identifying target genes for transfer can be a difficult step in procedures for inducing translocations between chromosomes of wheat and alien species. Despite the enormous pool of alien genetic variation and the sophisticated techniques available for transfer to wheat, there still remains the problem of identifying, in the alien species, the characters required for transfer. So far the majority of the recorded wheat–alien translocations are related to resistance to fungal diseases and pests, as such translocations can be relatively readily identified by phenotypic screening. However, phenotype analysis is sometimes of low efficiency, especially where environmental effects are large as in many diseases. In the case of identification of γ-ray-induced wheat–alien translocation lines, Crasta *et al.* (2000) found that conventional phenotypic analysis only resulted in a low (4%) success rate of identifying barley yellow dwarf virus-resistant and susceptible translocation lines, whereas 58% of the susceptible progeny of this irradiated seed contained a *Thinopyrum intermedium* chromosome-specific repetitive sequence as identified by RFLP and GISH analysis.

With the increasing needs of improving wheat quality, both processing and nutritional, and of expanding wheat production into arid or semi-tropical areas where it has not been previously cultivated, wheat breeders wish to transfer other useful genes in regard to quality and adaptability from alien species to wheat. However, in comparison with those conferring to disease and pest resistance, these genes are difficult to identify by phenotype analysis such as physical and chemical determinations because of high genotype–environment interaction effects. In this situation, molecular markers linked to these new genes, together with GISH, C-banding and biochemical markers, would be helpful

for identifying and screening the desirable translocations induced by irradiation.

5. Acknowledgements

The authors wish to express their thanks to Ms Chunxia Yuan for her artwork in the graphics.

6. References

6.1. Cited References

Ahmad, F., Comeau, A., Chen, Q. *et al.* 2000. Radiation induced wheat-rye chromosomal translocations in Triticale: optimizing the dose using fluorescence *in situ* hybridization. *Cytologia.* 65(1): 1–6.

Bie, T.D., Cao, Y.P. and Chen, P.D. 2007. Mass production of intergeneric chromosomal translocations through pollen irradiation of *Triticum durum-Haynaldia villosa* amphiploid. *Journal of Integrative Plant Biology.* 49: 1619–1626.

Chen, P.D., Liu, W.X., Yuan, J.H. *et al.* 2005. Development and characterization of wheat- *Leymus racemosus* translocation lines with resistance to *Fusarium* Head Blight. *Theoretical and Applied Genetics.* 111(4): 941–948.

Chen, S.W., Chen, P.D. and Wang, X.E. 2008. Inducement of chromosome translocation with small alien segments by irradiating mature female gametes of the whole arm translocation line. *Science in China series C: Life Sciences.* 51(4): 346–352.

Crasta, O.R., Francki, M.G., Bucholtz, D.B. *et al.* 2000. Identification and characterization of wheat-wheatgrass translocation lines and localization of barley yellow dwarf virus resistance. *Genome.* 43: 698–706.

Donini, P., Koebner, R.M.D. and Ceoloni, C. 1995. Cytogenetic and molecular mapping of the wheat–*Aegilops* longissima chromatin breakpoints in powdery mildew-resistant introgression lines. *Theoretical Applied and Genetics.* 91: 738–743.

Friebe, B., Jiang, J., Raupp, W.J. *et al.* 1996. Characterization of wheat-alien translocation conferring resistance to disease and pests: current status. *Euphytica.* 91: 59–87.

Le, H.T., Armstrong, K.C. and Miki, B. 1989. Detection of rye DNA in wheat-rye hybrids and wheat translocation stocks using total genomic DNA as a probe. *Plant Molecular biology report.* 7: 150–158.

Ma, Z.Q., Gill B.S., Sorrells M.E. *et al.* **1993.** RFLP markers linked to two Hessian fly-resistance genes in wheat (*Triticum aestivum* L.) from *Triticum tauschii* (coss.) Schmal. *Theoretical and Applied Genetics.* 85(6-7): 750–754.

Rakhmatullina, E.M. and Sanamyan, M.F. 2007. Estimation of efficiency of seed irradiation by thermal neutrons for inducing chromosomal aberration in M_2 of cotton *Gossypium hirsutum* L. *Russian Journal of Genetics.* 43(5): 518–524.

Sears, E.R. 1956. The transfer of leaf rust resistance from *Triticum umbellulatum* to wheat. *Brookhaven Symp Biol.* 9: 1–21.

Sears, E.R. 1977. An induced mutant with homoeologous pairing in common wheat. *Canada Journal of Genetics Cytology.* 19: 585–593.

Sun, G.L., Fahima,T., Korol, A.B. *et al.* **1997.** Identification of molecular markers linked to the *Yr15* stripe rust resistance gene of wheat originated in wild emmer wheat, *Triticum dicoccoides. Theoretical Applied Genetics.* 95: 622–628.

Smith, H.H. 1958. Radiation in the production of useful mutations. *The Botanical Review.* XXIV (1):1-24

Wang, X.E., Chen, P.D., Zhou, B. *et al.* **2001.** RFLP analysis of wheat-L. *racemosus* translocation lines. *Acta Genetica Sinica.* 28 (12): 1142–1150.

Yildirim, A., Jones, S.S. and Murray, T.D. 1998. Mapping a gene conferring resistance to *Pseudocercosporella herpotrichoides* on chromosome 4V of *Dasypyrum villosum* in a wheat background. *Genome.* 41: 1–6.

6.2. Websites

Cytogenetic and Genome Research:
   http://content.karger.com/ProdukteDB/produkte.asp?Aktion=JournalHome&ProduktNr=224037

Cytogenetic Resources:
   http://www.kumc.edu/gec/prof/cytogene.html

The Wheat Genetic and Genomic Resources Centre, Kansas State University, USA:
   http://www.k-state.edu/wgrc/

6.3. Further Reading

Appels, R., Morris, R., Gill, B.S. *et al.* **1998.** Chromosome Biology. Boston, MA: Kluwer Academic Publishers.

Endo, T.R. 1994. Structural changes of rye chromosome 1R induced by gametocidal chromosome. *The Japanese Journal of Genetics.* 69: 11–19.

Lupton, F.G. 1987. Wheat Breeding: Its Scientific Basis. Lupton F G H. (ed.) London: Chapman and Hall.

Riley, R., Chapman, V. and Johnson, R. 1968. Introduction of yellow rust resistance of *Aegilops comosa* into wheat by genetically induced homoeologous recombination. *Nature.* 217: 383–384.

Molecular Techniques and Methods for Mutation Detection and Screening in Plants

Q.Y.Shu[a], K.Shirasawa[b,c], M.Hoffmann[d], J.Hurlebaus[d] and T.Nishio[b]*

[a] Joint FAO/IAEA Division of Nuclear Techniques in Food and Agriculture, International Atomic Energy Agency, Wagramer Strasse 5, P.O. Box 100, A-1400 Vienna, Austria
 Present: Institute of Nuclear Agricultural Sciences, Zhejiang University, Hangzhou 310029, China
[b] Graduate School of Agricultural Science, Tohoku University, Sendai 981-8555, Japan
[c] Present address: Kazusa DNA Research Institute, Kisarazu, Chiba 292-0812, Japan
[d] Roche Applied Science, Nonnenwald 2, 82372 Penzberg, Germany
* Corresponding author, E-MAIL: nishio@bios.tohoku.ac.jp

1. Introduction

Mutants are important genetic resources not only for breeding but also for basic studies of gene function. Before the turn of the 21st century, mutants were mostly selected by observing phenotypes of individual plants in mutated populations treated with chemical or physical mutagens. During the past 10 years, the genomes of a number of plant species including important crops (e.g. rice, maize, soybean, sorghum) have been sequenced or partially sequenced (e.g. wheat) and the function of many genes has been determined or annotated. Concurrently, various molecular and genomics tools have been developed for the detection of genetic variants including single nucleotide polymorphism (SNP). The integration of genomics information and molecular tools has resulted in the development of various molecular approaches that can be deployed in the screening for mutants in large populations derived from radiation and chemical mutagenesis. In this regard, the invention of Targeting Induced Local Lesions in Genomes (TILLING), **see Chapters 21 and 22**) has been significant in exploiting mutations in linking phenotype to genotype.

In this chapter, a brief overview is given for both the features and frequencies of mutations induced by various mutagens; several techniques for detection of point and small indel mutations are described, followed by high-throughput approaches to screen for mutants. This is a rapidly growing area and examples of some emerging platforms are included.

2. Induced Mutations: Molecular Features and Frequencies

During the past century, different types of physical and chemical mutagen have been explored for the induction of mutations in plants. It is well known that different mutagens have different effects on plants, resulting in different types and levels of damage, and consequently produce a range of induced mutations at different frequencies. There are also a few claims that certain mutagens can generate 'rare' mutations, for example, heavy ion beam mutagenesis produced several flower colour mutants that are seldom induced by using γ-rays (**see Chapter 9**). Such phenomena can only be explained by a good understanding of the genetic control of individual traits and the molecular features of induced mutations, and the latter is also critical for the proper design of approaches to screen for induced mutants.

2.1. Nature of DNA Lesions in Induced Mutants

2.1.1. Differences in DNA Lesions Observed in the M_1 and M_2 Generations

After mutagen treatment, cells with damaged DNA will either die or repair the DNA lesions. While most repairs are error-free, some are not (**see Chapters 5 and 6**). By irradiating *Arabidopsis* pollen with γ-rays (low LET) and carbon ions (high LET), Naito *et al.* (2005) revealed that M_1 plants, which were derived from seeds that were produced by pollinating non-treated plants with irradiated pollen, carried extremely large deletions of up to >6 Mbp (~5% of the whole genome). However, not all DNA lesions in M_1 plants are transmitted to the M_2 and higher generations. The transmission of different mutations depends on their effects on gamete development or survival; the majority of large deletions are not transmitted to M_2 progeny, while mutations containing 1- or 4-base pair deletions are generally transmitted normally. Further analysis of various transmission modes suggests that the non-transmissibility of the large deletions may be due to the deletion of a particular region that contains a gene or genes required for gamete development or survival. Since most chemical mutagen-induced mutations are point mutations, which have limited effects on gamete development and survival, hence DNA lesions observed in M_1 and M_2 generations are similar.

2.1.2. DNA Lesions Caused by Different Mutagens

A large amount of data on molecular features of induced mutations have become available during the last decade; they are either from TILLING analyses of M_2 (or M_3 in a few cases) plants or discovered after a mutated gene is cloned through other forward genetics schemes. In general, most mutations caused by chemical mutagens, particularly those of alkylating agents, are base pair substitutions (**see Chapter 12**). However, small deletions are also infrequently identified in progenies derived from chemical mutagenesis (**see Chapter 20**).

In contrast to the large amount of data available for chemical mutagenesis, molecular characterization of mutations induced by different types of radiation remains limited. Available data indicate that mutations induced by radiations consist of both deletions and base pair substitutions (**Table 20.1**). It has shown that carbon-ion irradiation can induce kilobase-scale deletions, as well as short deletions and rearrangements such as inversions, insertions and translocations. Fast neutrons are believed to result in kilobase-scale deletions, for example Li *et al.* (2001) identified deletions up to 12 kb.

Similarly, only one study analysed the molecular nature of mutations in the *Wx* gene in about two dozen rice mutants induced by sodium azide (Jeng *et al.*, 2009). Therefore, further studies are needed to have a thorough comprehension.

2.2. Mutation Frequencies

In the past, mutation frequencies have been estimated by observing phenotypic mutants, e.g. chlorophyll deficiency, plant height, sterility, etc., and expressed either by the percentage of individual M_2 mutants or by the percentage of mutated M_2 panicle rows, in which at least one mutant is observed. Therefore, mutation frequencies vary between different traits and among different M_2 populations derived from different muta-

genic treatments. In molecular studies, the estimation of mutation frequency is performed in a different way.

2.2.1. Method in Estimating Mutation Frequency

The frequency of induced mutation can now be estimated by assessing DNA lesions in selected DNA fragments of M_2 plants. If the total size of assessed DNA fragment is expressed as *i* kilo bases (kb), the total number of M_2 individuals as *j*, and the total number of DNA lesions identified as n, then the mutation frequency (*f*) is expressed as the number of mutations per 1,000 kb for ease of comparison:

$$f = (n \times 1000)/(i \times j)$$

This formula can be manipulated easily to determine mutation density, that is, one mutation per set number of DNA kb (**Table 20.2**).

To achieve an accurate estimation, false positive or negative mutations should be excluded. False negative mutations are DNA lesions that are not identified in the experiment due to technical inability. For example, Gady *et al.* (2009) reported that the false negative rate can be as high as 25% and 75%, respectively in 4× and 8× pooling, when the high resolution DNA melting curve analysis

Table 20.1: Types of DNA lesion resulting from treatment of different mutagens

Types of mutagen	Types of DNA lesion[a]	Observations	Reference or source [b]
Ethyl methanesulphonate	BPS	Mostly, G/C to A/T or *vice versa*	See Table 20.2 & Chapter 12
N-Nitroso-N-methylurea	BPS	Mostly, G/C to A/T or *vice versa*	See Table 20.2 & Chapter 12
Sodium azide	BPS; insertion/deletion	23-bp duplication; G to T change; microsatellite change	Jeng *et al.* (2009)
Gamma rays	Deletions and BPS	Small deletions (1 to a few bp) and large deletions (up to a few Mbp), as well as single nucleotide substitutions	Naito *et al.* (2005) & authors' unpublished data
Ion beams	Deletions and BPS	Small deletions (1 to a few bp) and large deletions (up to a few Mbp), as well as single nucleotide substitutions	Naito *et al.* (2005) & Chapter 9
Fast neutrons	Deletions	Deletions of 0.8–12 kb	Li *et al.* (2001)
Thermal neutrons	Deletions and BPS	A deletion of ~40 kb and an A to G change	Sato and Nishio (2003); Kawakami *et al.* (2009)

[a] BPS: base pair substitution.

(HRM) method was used for the detection of single nucleotide substitutions. In physical mutagenesis, since many mutants are expected to be large deletions, some of them can easily escape detection when a method designed for point mutation detection is applied. Conversely, when methods for screening deletions are applied, e.g. for fast neutron mutagenesis, it is not possible to identify point and small indel mutations. In mutagenesis, false positive mutations can also exist due to mechanical mixture and heterozygosity residue in the seeds used for mutagen treatment, and out-crossing of the M1 plants with pollen from a neighbouring variety. The existence of false negative and positive mutations can lead to the under- and over-estimation of mutation frequency.

2.2.2. Mutation Frequency and Influencing Factors

Mutation frequencies have been estimated in a number of TILLING programmes in various plant species (**Table 20.2**). They are mostly from chemical mutagenesis. The highest mutation frequency was reported in wheat, where ~40 mutations per 1000 kb DNA fragment were observed, while the lowest was only ~1 mutation per 1000 kb in barley (**Table 20.2**).

Table 20.2: Mutation frequency and estimated number of lesions per genome in M_2 populations of different plant species

Mutagen[a]	Mutation frequency	NLpG[b]	Reference
Arabidopsis: diploid; genome size: 125 Mb			
EMS	~1/170 kb	~700	Greene *et al.* (2003)
Barley: diploid; genome size: ~ 5.3 Gb			
EMS	~1/1,000 kb	~5,300	Caldwell *et al.* (2004)
Sodium azide (SA)	~1/374 kb	~15,000	Talame *et al.* (2008)
Maize, diploid, genome size: 2.5 Gb			
EMS	~1/485 kb	~5,100	Till *et al.* (2004)
Rice: diploid, genome size: 389 Mb			
EMS; SA + MNU	~1/300 kb	~1,400	Till *et al.* (2007)
MNU [c]	~1/135 kb	3,100	Suzuki *et al.* (2008)
gamma rays	~1/6,190 kb	63	Sato *et al.* (2006)
Sorghum: diploid, genome size: 735 Mb			
EMS	~1/526 kb	~1,400	Xin *et al.* (2008)
Soybean: paleopolyploid, genome size: 1.1 Gb			
EMS, MNU	1/550 – 1/140 kb	~2,000 – 8,000	Cooper *et al.* (2008)
Tomato: diploid, genome size ~950 Mb			
EMS [d]	1/730 kb	~1,300	Gady *et al.* (2009)
Wheat: hexaploid, genome size ~16 Gb			
EMS	1/35 – 1/24 kb	666,000	Slade *et al.* (2005); Uauy *et al.* (2009)
Durum wheat: tetraploid, genome size ~10.8 Gb			
EMS	~1/51 kb	~211,000	Uauy *et al.* (2009)

[a] EMS: Ethyl methanesulphonate; MNU: N-Nitroso-N-methylurea.
[b] NLpG: The estimated number of lesions per genome.
[c] M_2 plants were derived from partially sterile M_1 plants (fertile M_1 plants were not included in the production of M_2 population).
[d] Low seed set M_2 families (585 among a total of 8810) were excluded for the analysis.

Many factors can influence the mutation frequency in an experiment. First, physical mutagens generally produce relatively low frequencies, though a number of false negative mutations may exist as described above. Sato *et al.* (2006) screened a gamma ray-irradiated rice population by mismatch cleavage analysis using *Brassica* petiole extracts. Analyses of 25 gene fragments (1.23 kb length on average) of 2,130 M_2 plants revealed six mutants. The rate of mutation induced by gamma rays was estimated to be one mutation per 6,190 kb (**Table 20.2**). This is in sharp contrast to the mutation frequency of chemical mutagens; up to one mutation per 135 kb for MNU and roughly one per 300 kb for EMS (**Table 20.2**). Second, higher mutation densities are mostly reported in polyploid crops, such as hexaploid and tetraploid wheat in which multiple genomes (gene duplications) buffer mutation events. Third, the procedure of population development may affect the mutation frequency observed. For example in rice, Suzuki *et al.* (2008) observed a mutation frequency of seven mutations per 1000 kb in the MNU-derived M_2 population that was developed from low seed setting M_1 plants. This frequency is two times higher that reported by Till *et al.* (2007) in M_2 populations derived from randomly harvested M_1 plants after combined mutagenesis of MNU and sodium azide. This may also explain the relatively low mutation frequency observed in diploid tomato populations since in that experiment M_2 plants with low seed set were excluded (Gady *et al.*, 2009, **Table 20.2**). It should also be noted that the mutation frequency may vary significantly in different experiments even when the same mutagen, dose and variety are used (Xin *et al.,* 2008).

The different mutation densities in different crop species or in different populations induced by different mutagens at different doses result in different numbers of mutations. For example, there could be on average about 666,000 mutations at the DNA level in a single mutant line in wheat. This has significant implications for the utilization of induced mutants both in breeding and gene function analysis (**see Chapter 24**).

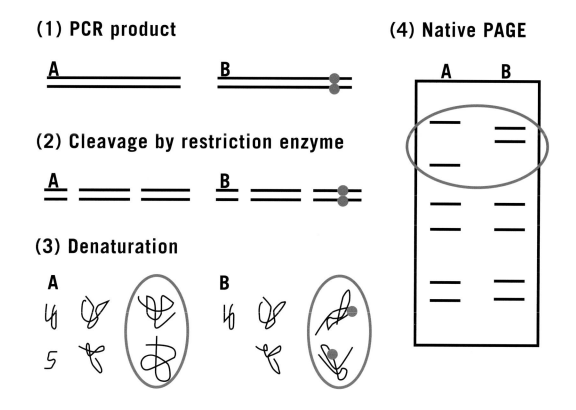

The Joint FAO/IAEA Programme

Figure 20.1 PCR-RF-SSCP analysis. PCR products are cleaved by restriction enzymes, denatured and analysed by electrophoresis in native polyacrylamide gel.

3. Techniques for Mutation Detection and Screening

Since many naturally occurring mutations are single nucleotide polymorphisms (SNPs), various technologies have been developed for SNP detection, most of which are also useful for the detection of small insertion/deletions (indels) (**Box 20.1**). Some are described below.

3.1. Single Strand Conformation Polymorphism

Single strand conformation polymorphism (SSCP) analysis can detect point mutations in DNA fragments (<500 bp) by the different motilities of mutant DNA and wild type DNA. Various modifications of the SSCP technique have also been developed. For example, in multiple fluorescence-based PCR-SSCP, PCR products labelled with fluorescent dUTPs are analysed by an automated DNA sequencer under SSCP conditions. PCR-RF-SSCP analysis is a combination of PCR-RFLP analysis and SSCP analysis, in which PCR products (~2 kb) are digested into short DNA fragments (<500 bp) by restriction enzymes and electrophoresed in polyacrylamide gels under SSCP conditions (**Figure 20.1**). This method is suitable for screening point mutations in longer DNA fragments than those used in normal SSCP analysis. In rice, point mutations in the *Wx* gene of many glutinous mutants have been detected efficiently and identified by the PCR-RF-SSCP technique (Sato and Nishio, 2003). However, this method is not applicable to the analysis of bulked DNA samples and thus cannot be used for screening mutations in large populations.

3.2. Temperature Gradient Gel Electrophoresis, Conformation-Sensitive Gel Electrophoresis and Conformation Sensitive Capillary Electrophoresis

3.2.1. Temperature Gradient Gel Electrophoresis

Temperature gradient gel electrophoresis (TGGE) is a method used for mutation detection. Here mutants are detected by differences in electrophoretic mobility in temperature gradient gels, due to differences in melting temperatures between the homo-duplex of wild type (WT) or mutant DNA, and the hetero-duplex of WT-mutant DNA (**Box 20.1**). DNA homo-duplexes are double-strand DNA fragments with perfect base complementation (complete strand matching), while hetero-duplexes contain mismatched bases between strands. When a WT DNA sample is mixed with mutant (M) DNA, three types of DNA duplexes can be formed by allowing them to dissociate then re-anneal (WT-WT, M-M and WT-M). Temperature gradient capillary electrophoresis

---

Box 20.1: Methods used for detecting DNA point and small indel mutations

Single-strand conformation polymorphism (SSCP): SSCP is revealed through electrophoretic analysis of denatured DNA fragments (<500 bp) in native polyacrylamide gel. It can be used for detection of unknown mutations, but not for analysis of pooled DNA samples.

PCR-restriction fragment length polymorphism (PCR-RFLP): PCR-RFLP is the electrophoretic analysis of PCR products cleaved by restriction endonuclease. A term "cleaved amplified polymorphic sequence (CAPS)" used by plant scientists is the same as PCR-RFLP. It is not suitable for the detection of unknown SNPs.

PCR-restriction fragment-SSCP (PCR-RF-SSCP): PCR-RF-SSCP is the analysis of PCR products cleaved by restriction endonuclease in native polyacrylamide gels. This method is highly useful for the identification of genes having unknown SNPs, but it cannot be used for pooled DNA samples.

Temperature gradient gel electrophoresis (TGGE): TGGE is the electrophoretic analysis of hetero-duplex DNA in temperature gradient gels. Differences in melting temperatures between the homo-duplex of wild type or mutant DNA and the hetero-duplex of such DNA result in the difference of mobility in gels. A modification using capillary electrophoresis is termed TGCE.

Conformation sensitive capillary electrophoresis (CSCE): The analysis is based on the different mobilities of DNA homo-duplexes and hetero-duplexes during electrophoresis in capillaries filled with CAP, a semi-denaturing polymer. It allows the identification of pools containing a mutation within the target fragment.

High resolution melting curve analysis (HRM): HRM is a non-enzymatic, robust and well-established method to characterize DNA fragments through post-PCR melting curve analysis. It makes use of special DNA-binding dyes whose binding/release characteristics allow them to be used and detected at high concentrations without inhibiting PCR. Due to their saturating, homogeneous staining of PCR products, such dyes give sharp, unique melting profiles that allow the differentiation between homo- and heterozygote samples, and, less frequently, even between homozygous wild types and mutants (whole amplicon melting).

Dot-blot-SNP: SNP analysis by hybridization of labelled oligonucleotides to dot-blotted PCR products, which was improved using competitive hybridization. This method is useful for genotyping mutated genes and for discrimination of a newly induced mutation from genetic variations, but cannot be used for analysis of unknown mutations.

Mismatch cleavage: In this method, wild type and mutated DNAs are mixed to form duplex DNAs; the hetero-duplex DNAs having a mismatched base pair are cleaved by a hetero-duplex cleavage enzyme such as Cel1 or ENDO1. The DNA fragments (cleaved or intact) are revealed by electrophoretic analysis. This method is amendable to pooled DNA samples and for the screening of unknown mutants.

For details **see Chapters 21 and 22.**

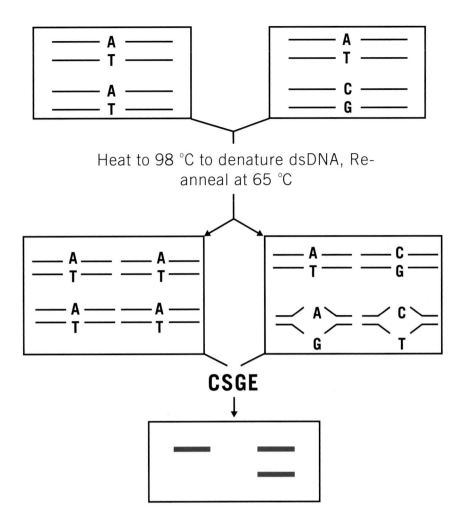

Heat to 98 °C to denature dsDNA, Re-anneal at 65 °C

CSGE

Figure 20.2 Illustration of homo-duplex and hetero-duplex generation, and band pattern seen in conformation-sensitive gel electrophoresis (CSGE). *Upper panel*: A PCR product from a mutant that is heterozygous (A/C) at a specific nucleotide position will contain two types of double stranded DNA (dsDNA). *Middle panel*: dsDNA is dissociated by heating to 98°C and re-annealed by incubation at 65°C and homo-duplex and hetero-duplex are formed. *Bottom panel*: homo-duplex and hetero-duplex are separated by CSGE.

(TGCE) has been developed for high-throughput screening of mutations, in which capillaries instead of slab gels are used for electrophoresis. There is no known report on the use of TGGE for mutation screening in plants, but it has been successfully applied in mice.

3.2.2. Conformation-Sensitive Gel Electrophoresis

Conformation-sensitive gel electrophoresis (CSGE) is based on the ability to distinguish between homo-duplex and hetero-duplex DNA fragments by electrophoresis under partial denaturing conditions (**Figure 20.2**).

Several factors can affect the sensitivity of CSGE for mutation detection, including amplicon size, nature and sequence context, and location of mismatches. In general, the optimal amplicon size is 200–500 bp. The nature of mismatch and the surrounding nucleotide sequence can also affect the detection, for example it

was reported that four hetero-duplexes (C:C, C:A, A:C and T:C) were indistinguishable from the homo-duplex band. Finally, the mismatches that are close to the end of the amplicon are less easily detected. CSGE is ideal for the detection of heterozygous single base changes and small indels, but large deletions might not be detected.

Conformation-sensitive capillary electrophoresis (CSCE) is similar to CSGE. The differences are that CSCE is performed by capillary electrophoresis and the amplicons are labelled using tagged PCR primers. CSCE is more amenable to high-throughput and has greater sensitivity than CSGE.

3.3. High resolution DNA melting analysis

High resolution DNA melting analysis (HRM) is a robust and well-established method to characterize gene

247

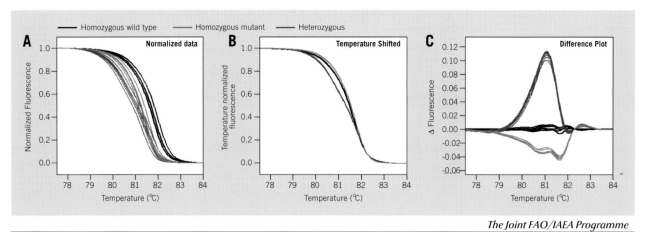

The Joint FAO/IAEA Programme

Figure 20.3 Identification of wild and mutant type DNA homo-duplex and their hetero-duplex using high resolution melting curve analysis (courtesy of Roche Diagnostics).

fragments through post-PCR melting curve analysis. High resolution analysis is achieved using special DNA-binding dyes and instruments with optical sensing devices. The special DNA-binding dyes have binding characteristics that allow them to be used at high concentrations without inhibiting PCR. Due to their saturating, homogeneous staining of PCR products, such dyes give sharp, unique melting profiles that allow the differentiation between homo- and heterozygote samples, and between homozygous wild types and mutants (whole amplicon melting).

In HRM analysis, unknown sequence variations become apparent in heterozygous samples due to the presence of hetero-duplex DNA. When amplified and during melting, these samples show fluorescence curves with significantly different profiles (shapes) than those derived from homozygous (wild type or mutant) samples. Software algorithms are available to analyse these differences by first normalizing the data and then temperature-shifting the curves such that differences of Tm between homozygotes with very similar curve shapes disappear, allowing the heterozygous samples to stand out (**Figure 20.3**). In cases where Tm differences between homozygotes are big enough, these can be more displayed by omitting the temperature shifting step. By finally plotting the difference in fluo-

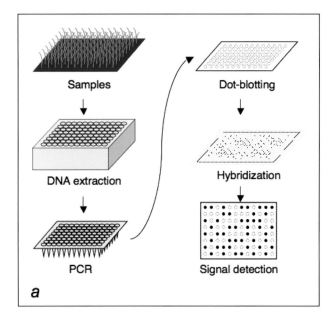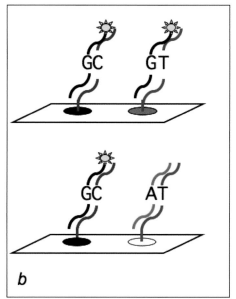

The Joint FAO/IAEA Programme

Figure 20.4 Dot-blot-SNP technique. **a:** Genomic DNAs are extracted, DNA fragments are amplified by PCR, and blotted by dot-blotting. Labelled oligonucleotides are hybridized and signals are detected. **b:** When only labelled oligonucleotides are hybridized, high background signals are detected because mismatched hybridization occurs (*upper panel*). By adding unlabelled competitive oligonucleotides to hybridization mixture, background signals are reduced (*bottom panel*).

rescence between each sample, the homozygotes and heterozygote samples can be easily identified (**Figure 20.3**). Machines that detect HRM are normally linked to a computer in which data can be analysed, stored or transferred to a database. A limitation of HRM and other methods that detect SNPs in a given length of DNA is that they cannot differentiate between two samples each with one SNP for the same base but at different locations, but this is not a big issue for detecting mutation in an artificially generated mutant population since the chance of two mutations existing in such a short DNA fragment is close to zero.

3.4. Dot-Blot SNP Analysis

Dot-blot SNP analysis is a high-throughput and simple method for detecting point mutations. In such analysis, a point mutation is detected by hybridization with an oligonucleotide probe and a competitive oligonucleotide (**Figure 20.4**). It can be used to identify a targeted mutation in a specific DNA sequence.

3.5. Mismatch Cleavage Analysis

It has been known for decades that DNA fragments with mismatches can be recognized and cleaved by certain enzymes such as S1 and Endo1. With this discovery and its application in analysis of SNPs and small indels, high-throughput mutation discovery platforms such as TILLING (**see Chapters 21 and 22**) have been established.

4. Methods for Molecular Mutation Screening

Different mutagens can cause different types of DNA damage, and consequently result in different types of DNA lesions in M_2 populations. Therefore, different technologies are needed to detect different types of induced mutation. Low mutation frequencies are problematic, for example the highest mutation rate ~40 mutations per 1000 kb DNA fragment in wheat (*via* chemical mutagenesis, **Table 20.2**) translates into only 40 plants carrying a mutation when 1,000 plants are analysed for a 1-kb target gene fragment. The number of plants required for screening increases significantly for other crops due to reduced mutation densities, particularly when populations are developed by radiation treatments (**Table 20.2**). Therefore, the screening needs to be high-throughput. Since the features and the loci of mutations are largely unknown, many methods based on sequence information are currently not applicable for mutation screening.

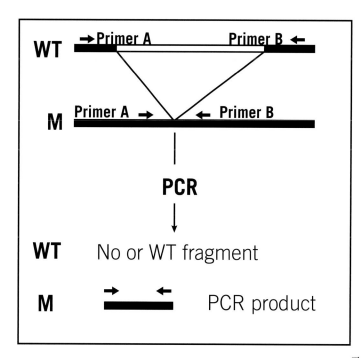

Figure 20.5 A simplified illustration of detecting mutant alleles using PCR. WT and M refer to wild and mutant alleles, respectively. The deleted fragment is shown as a white stripe, which can range from hundreds to hundreds of thousands of base pairs. When the deletion is large enough, e.g. >5 kb, there is no amplification of the WT allele.

The ability of several methods to detect mutant alleles in pooled DNA samples has, to some extent, provided a practical and feasible means of screening mutations in mutagenized populations using molecular tools (**see Chapters 21 and 22**). Although coined only in 2000, TILLING has proven to be a valuable methodology for finding induced mutations as well as those naturally existing in a number of plants. The principles and protocols of the typical TILLING technology are given in **Chapter 21 and 22**. The TILLING version using fast neutron mutagenesis and one simplified TILLING technique are briefly described below.

4.1. De-TILLING

4.1.1. Principles of De-TILLING

Fast neutron mutagenesis often results in kilobase-scale DNA deletions. As a new knock-out technique to obtain deletion mutants for target genes, a strategy to screen for rare deletion mutants in large fast neutron mutagenized populations was first developed by Li *et al.* (2001) and demonstrated in Arabidopsis and rice. It combines fast neutron mutagenesis and high-throughput PCR screening, known as "Deleteagene" (Delete-a-gene). This strat-

egy has been further developed and named deletion-TILLING, or de-TILLING (Rogers *et al.*, 2009).

The de-TILLING method includes three key technological aspects:

1. Fast neutron mutagenesis, which generates DNA deletions of different size.
2. A DNA pooling strategy to reduce the number of PCRs needed.
3. Technologies that allow a mutant allele, possessing an internal deletion, to be amplified in pools with excessive genomic target sequence (wild type allele).

As shown in **Figure 20.5**, when the deletion is sufficiently large, there is no amplification for the WT allele, while the mutant allele can be readily amplified. Li *et al.* (2001) observed that the deletion size of Arabidopsis mutations varied from 1 to 12 kb with half of them being about 2–4 kb. When a mutation is only about 1–2 kb, while the target region extends to only 2–3 kb, then the WT allele can be easily amplified. Since the WT amplicon is present tens of thousands times in excess over the mutant allele, there is a high risk that the mutant allele will not be amplified; hence special procedures are required to ensure the mutant allele is detected.

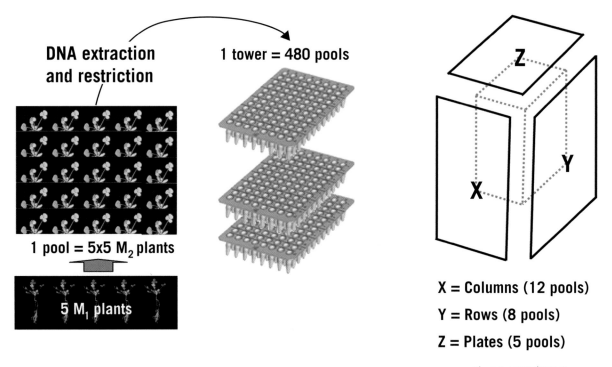

DNA extraction and restriction

1 tower = 480 pools

1 pool = 5x5 M₂ plants

5 M₁ plants

Z

Y

X

X = Columns (12 pools)

Y = Rows (8 pools)

Z = Plates (5 pools)

The Joint FAO/IAEA Programme

Figure 20.6 The 3D pooling strategy for de-TILLING. *Left panel*: five seeds of individual M₁ plants are harvested and grown up as M₂ plants. The DNA of the 25 M₂ plants is extracted and digested with restriction enzymes. Fragmented DNA is combined to produce a pool; a total of 480 pools becomes a tower; *Right panel*: the structure of one tower, which consists of pools of columns, rows and plates. Initial drawings courtesy of Dr T. Wang, John Innes Centre, UK.

4.1.2. Pooling Strategies

In de-TILLING, by preferentially amplifying the deletion mutant, it is possible to exploit deep poolings. However, by increasing the pooling factor, it is important to develop a strategy to de-convolute these pools, once a mutant has been detected in them. A multiple step pooling method was proposed by Li *et al.* (2001). Equal DNA samples of 18 M_2 lines were combined into the first level pool – sub-pool; 2 sub-pools make a pool, and 8 pools become a super-pool, and 9 super-pools create a mega-pool. The mega-pool, containing 2,592 M_2 lines, is used for detecting deletion mutants. Once a mega-pool is identified that contains a mutant locus,

all its super-pools are screened. The same procedure is applied for the pools and sub-pools until the individual mutant line is identified.

With the invention of techniques of preferentially amplifying mutant loci from an even larger pool, Rogers *et al.* (2009) were able to design a novel strategy for deeper pooling. In this strategy, the population is segregated into a towered structure consisting of 96-well plates of DNA extractions. Each tower is pooled to create a 3-dimensional (3D) pool matrix of rows, columns and plates (**Figure 20.6**).

In practice, the DNAs of about 10–15 towers (representing ~6,000 M_2 plants) are pooled further and used

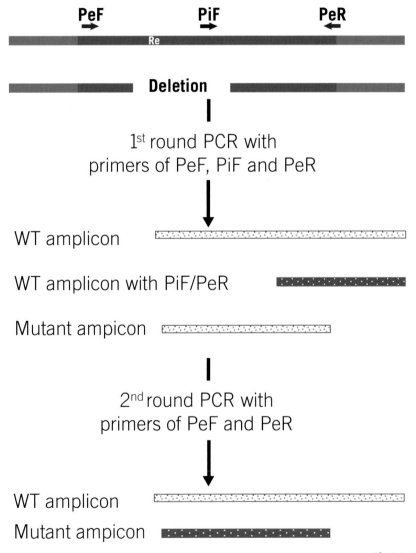

The Joint FAO/IAEA Programme

Figure 20.7 The de-TILLING strategy. Amplification from a vast excess of wild type sequences in DNA pools containing a deletion allele is suppressed by template restriction and production of the suppressor fragment from the poison primer. PeF and PeR: external forward and reverse primers; PiF: (internal) poison primer (forward). RE: Restriction site. Amplicons with red dotted white stripe are poorly amplified, while amplicons with white dotted red stripe are well amplified. Initial drawings provided by the courtesy of Dr T. Wang, John Innes Centre, UK.

for PCR in screening for mutants. When a mutation is identified within a PCR reaction, the compositional components of the 25 towers (X + Y + Z, **Figure 20.6**) are screened simultaneously to identify a single tower housing the mutant, in a single step. Similarly, once the mutant tower is identified, the 25 pools are screened to identify the pool that contains the mutant seed lot. This strategy has been deployed successfully in the identification of deletion mutants in *M. truncatula* (Rogers *et al.*, 2009).

4.1.3. Detection of Deletion Mutants

The identification of deletion mutants by PCR among wild types with great excess in quantity is a major challenge. In such cases, the amplification of wild type sequences should be suppressed efficiently, otherwise the mutant allele will not be amplified. Rogers *et al.* (2009) developed a strategy that allows a mutant allele, possessing an internal deletion, to be amplified in pools where the genomic target sequence is present at a 20,000-fold excess. This detection sensitivity is achieved by combining two approaches that suppress the amplification of the undeleted WT fragment – restriction suppression and a poison primer approach (Rogers *et al.*, 2009; **Figure 20.7**).

Restriction enzyme suppression relies upon the pre-digestion of highly complex DNA pools with one or a few selected restriction enzymes (**Figure 20.7**), which cut the target sequence. This prevents the WT sequence acting as a dominant PCR template. Since the deletion fragment does not contain the restriction sites for the enzymes used, the DNA template with the deletion is not affected.

WT sequences escaping the restriction digestion are subject to 'poison primer' suppression. The poison

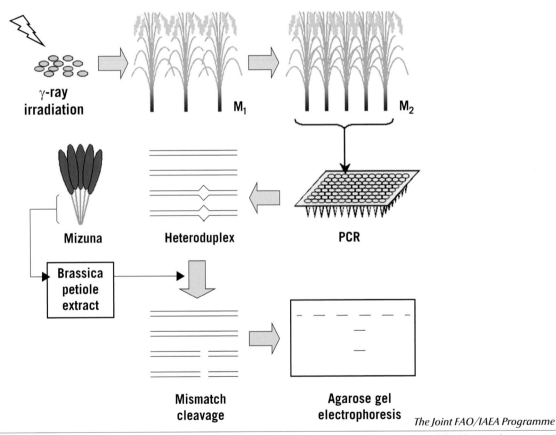

The Joint FAO/IAEA Programme

Figure 20.8 Schematic representation of mismatch cleavage analysis for mutant screening. DNA fragments amplified by PCR from genomic DNA of five M₂ plants are denatured by heating and then gradually cooled (3 min at 98°C, 1 min at 65°C, 1 min at 55°C, 1 min at 45°C, 1 min at 35°C and 1 min at 25°C with a pace of −0.3°C/sec) to form hetero-duplexes. The DNA in 5 μl PCR mixture is mixed with 6 μl of 5×buffer (50 mM KCl, 50 mM MgSO₄, 50 mM HEPES pH 7.5), 0.6 μl of 0.1% BSA, 0.6 μl of 0.1% Triton X100 and 1 μl of *Brassica* petiole extract (BPE) in a total volume of 30 μl, incubated at 45°C for 15 min and subjected to electrophoresis in 2% agarose gels containing SYBR Green I (Cambrex). BPE is extracted from 50 g powder of petioles of *Brassica rapa* L. var. *laciniifolia*, after being ground with a mortar and a pestle in liquid nitrogen, by 30 ml extraction buffer (100 mM Tris-HCl pH 7.5, 100 μM phenylmethylsulphonyl fluoride) at 4°C. Proteins are precipitated with 25% to 80% saturation of ammonium sulphate, dissolved in 2 ml extraction buffer, and dialysed against 100 mM Tris-HCl buffer (pH 7.5). This crude protein solution is used as an enzyme solution, after calibration.

primer is designed in the deletion region and together with the external reverse primer can amplify short fragments of the WT sequence. These short fragments, known as suppressor fragments, are produced more efficiently and act to suppress the amplification of the long fragments. Amplification of the deletion mutant allele produces a single fragment by the two external primers (**Figure 20.7**). Therefore, after the first round of PCR, the deletion mutant allele is enriched relative to the full WT allele. In the second round of PCR, only external primers are used and the mutant allele is preferentially amplified (**Figure 20.7**).

4.2. A Simplified TILLING Technique

In a typical TILLING analysis, target DNA fragments are amplified by PCR with fluorescent-labelled primers from bulked DNAs extracted from eight M_2 plants. Hetero-duplexes are formed between a wild type single strand and a mutant complementary strand and then cleaved by CelI endonuclease at a mismatch point. The CelI-treated PCR amplicons in the form of homo-duplex or hetero-duplex are then subjected to electrophoresis in denaturing polyacrylamide gels of the LI-COR gel system (**see Chapters 21 and 22**).

Several groups have developed simplified versions of the TILLING technology. In such systems, mismatched DNA are cleaved by crude extracts of the CelI enzyme (see ref. in Till *et al.,* 2007), and electrophoresis is performed in agarose or common polyacrylamide gels, hence expensive equipment is no longer necessary. For example, Sato *et al.* (2006) demonstrated that a crude extract prepared from mizuna petioles (*Brassica rapa*), which are easily ground to powder in liquid nitrogen, can be used as a mismatch-specific endonuclease. In their system, PCR amplicons of pooled DNA samples, after denaturing and re-annealing, are treated with *Brassica* petiole extract and subjected to agarose gel electrophoresis (**Figure 20.8**). By using this system, a single point mutation can be detected in a DNA mixture containing wild type and mutant DNAs at a ratio of 9:1, indicating that one heterozygous mutant among five plants can be identified. This technique has been applied successfully in screening for mutants induced by gamma irradiation (Sato *et al.,* 2006). Although the sensitivity is lower than that of the TILLING system using eight bulked M_2 plants, this system is much simpler as it neither involves an initial investment nor requires a high operational cost. In the agarose-gel-based system, one person can screen 4,320 M_2 plants in one day (3 runs/day) using one electrophoretic apparatus, in which 288 samples can be analysed at a time.

5. New Developments

Detection or scanning a mutation in targeted genome sequence in a limited number of individuals, or screening of induced mutations in a large mutagenized population are extremely useful tools for functional genomics and plant breeding. During the past decade, there have been numerous developments that are relevant to this subject.

5.1. Low Cost Sequencing

DNA sequencing is the ultimate step to verify a mutation in a gene both in reverse and forward genetics studies. Many molecular technologies have been designed as a preliminary screening tool to identify the individuals that carried a mutation in a large population (e.g. TILLING) or the gene fragment in which a target gene is located (e.g. fine gene mapping), so that the amount of gene sequencing could be significantly reduced. With the availability of Next Generation (NextGen) and Third Generation sequencing technologies, which produce gigabases of data at extremely low cost, the landscape has changed significantly allowing for a sequencing revolution. For example, a team has recently sequenced three human genomes for US$4,400 each – at least ten times less than that achieved with other technologies, and 20 human genomes can be sequenced in a day.

Such low cost sequencing has made re-sequencing of gene fragments of interest (e.g. the delimited region of a gene) and genes of interest in thousands of mutagenized inbred lines both feasible and inexpensive. This has led several laboratories to integrate TILLING projects with NextGen and Third Generation sequencing technology platforms, e.g. Seq-TILLING (Weil, 2009).

A major advantage to large scale Seq-TILLING is that DNA samples can be analysed in much deeper pools than with the CelI-based approach. For some sequencing instruments, pooling can be as high as 40- to 50-fold, compared to the 8-fold pooling in typical CelI TILLING. Similar to the de-TILLING described above, a 2- or 3D dimensional strategy can be applied. Additionally, a barcode system can be deployed by attaching barcode sequences to the fragments as primers used to amplify the sequence in preparation for loading into sequencing instruments (Weil, 2009).

5.2. New Platforms for TILLING

Typical or simplified TILLING systems are based on the hetero-duplex cleavage by endonucleases such as Cel1. As described above, there are instruments that can be used to differentiate hetero-duplex from homo-duplex, hence no cleavage is needed. The applicability of these instruments for TILLING is therefore determined by its throughput and cost-effectiveness. Gady *et al.* (2009) has recently applied two such high-throughput techniques – CSCE and HRM for detecting point mutations in large EMS-mutated populations (**Figure 20.9**). Using CSCE or HRM, the only step required is a simple PCR before either capillary electrophoresis or DNA melting curve analysis.

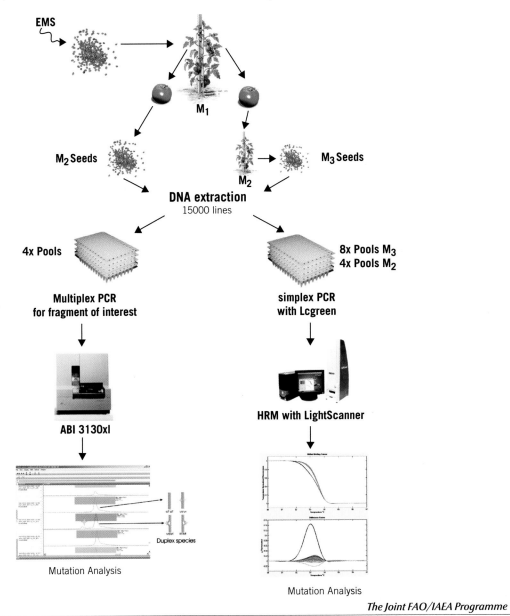

The Joint FAO/IAEA Programme

Figure 20.9 An example of mutant production and identification in tomato using the TILLING process. M_2 population, 10 seeds originating from the first M_1 fruit were ground and DNA isolated, the M_2 population comprises 8,225 lines. M_3 population, from the second M_1 fruit, consisting of 8,810 lines were grown and selfed, seeds were harvested for 7,030 lines and a seed lot subset (10 seeds) was used for DNA extraction. For both M_2 and M_3 population, DNA was pooled 4- or 8-fold, depending on the selected screening method: CSCE; after multiplex PCR amplification with fluorescent-labelled primers, samples are pooled directly and loaded into capillaries filled with CAP polymer. Pools containing a mutation are identified using Applied Maths' HDA peak analyser software. HRM; following PCR amplification in presence of LC-Green+™ or other dye, pools are analysed for their product melting temperatures. Adopted from *Plant Methods,* 2009, **5**: 13.

5.2.1. CSCE-Based TILLING

CSCE is a non-enzymatic differential DNA conformation technique for SNP discovery as described above (**Figure 20.2**). After PCR amplification, and the denaturing and re-annealing of amplicons, several duplex species are formed, e.g. homo-duplex of wild type (WT-WT) and mutant (M-M) and hetero-duplex of WT-M. Because of the mismatch formed in the hetero-duplex, it migrates at a different speed than the homo-duplex during electrophoresis in capillaries filled with CAP, a semi-denaturing polymer, thus allowing the identification of pools containing a mutation within the target fragment. Presence of hetero-duplex is identified as an altered peak shape as shown in **Figure 20.9**.

Experiments with different pooling depths demonstrate that an 8-fold pooling was feasible in diploid material using CSCE analysis, a 4-fold pooling is recommended for assurance of minimum false negative results. In comparison with the up to 1.5 kb of the amplicon in a typical TILLING, the target fragment length is optimal for 200–500 bp in CSCE. This reduction of efficiency is compensated for by multiplex PCR of up to four fragments and by using primers labelled with three different dyes for each product labelling. Combining multiplex PCR and multiple labelling with amplicons differing in size, it is possible to screen 12 fragments in one run by pooling all PCR products post-PCR reaction.

5.2.2. HRM-Based TILLING

Similar to CSCE, HRM is also a non-enzymatic mutation screening technique, it reveals sequence variants due to distinct patterns in DNA melting curve shape. It has been used recently in many ways as a novel approach to study genetic variation in many fields (human, animals, plants, fungi and microbes, and plant breeding), with applications ranging from qualitative SNP detection to semi-quantitative analysis of methylation. When using equipment and reagents such as the LightCycler® 480 System and its accompanying High Resolution Melting Mastermix, which contains a saturating fluorescent dye, real time PCR is carried out using a touchdown protocol, with annealing temperatures ranging from 70 to 60°C. HRM curve data are obtained at a rate of 25 acquisitions per °C.

In the study of Gady *et al.* (2009) the LCgreen Plus+™ molecule was used during PCR amplification, and pools containing a mutation were screened using a LightScanner® System (**Figure 20.9**). They demonstrated that it was possible to distinguish hetero-duplex formation even when diluted 32-fold, but they used an 8-fold flat DNA pooling in their study. The length of the amplicons is the same as in CSCE-TILLING.

6. References

6.1. Cited References

Cadwell, D.G., McCallum, N., Shaw, P. *et al.* **2004.** A structured mutant population for forward and reverse genetics in barley (*Hordeum vulgare* L.). *Plant J.* 40: 143–150.

Cooper, J.L., Till, B.J., Laport, R.G. *et al.* **2008.** TILLING to detect induced mutations in soybean. *BMC Plant Biology*. 8: 9.

Drmanac, R. 2009. Human genome sequencing using unchained base reads on self-assembling DNA nano-arrays. *Science*. DOI: 10.1126/science.1181498

Gady, A.L.F., Hermans, F.W.K., Van de Wal, M.H.B.J. *et al.* **2009.** Implementation of two high through-put techniques in a novel application: detecting point mutations in large EMS mutated plant populations. *Plant Methods*. 5: 13.

Greene, E.A., Codomo, C.A., Taylor, N.E. *et al.* **2003.** Spectrum of chemically induced mutations from a large-scale reverse-genetic screen in *Arabidospsis*. *Genetics*. 164: 731–740.

Jeng, T.L., Wang, C.S., Tseng, T.H. *et al.* **2009.** Nucleotide polymorphisms in the waxy gene of NaN₃-induced waxy rice mutants. *J Cereal Sci.* 49: 112–116.

Kawakami, S.I., Kadowaki, K.I., Morita, R. *et al.* **2009.** Induction of a large deletion including the waxy locus by thermal neutron irradiation in rice. *Breed Sci.* 57: 153–157.

Li, X., Song, Y., Century, K. *et al.* **2001.** Fast neutron deletion mutagenesis-based reverse genetics system for plants. *Plant J.* 27: 235–242.

Naito, K., Kusaba M., Shikazono, N. *et al.* **2005.** Transmissible and nontransmissible mutations induced by irradiating *Arabidopsis thaliana* pollen with gamma-rays and carbon ions. *Genetics*. 169: 881–889.

Rogers, C., Wen, J.Q., Chen, R.J. *et al.* **2009.** Deletion-based reverse genetics in *Medicago truncatula*. *Plant Physiol.* 151: 1077–1086.

Sato, Y. and Nishio, T. 2003. Mutation detection in rice waxy mutants by PCR-RF-SSCP. *Theor Appl Genet.* 107: 560–567.

Sato, Y., Shirasawa, K., Takahashi, Y. *et al.* 2006. Mutant selection of progeny of gamma-ray-irradiated rice by DNA hetero-duplex cleavage using *Brassica* petiole extract. *Breed Sci.* 56: 179–183.

Suzuki, T., Eiguchi, M., Kumamaru, T. *et al.* 2008. MNU-induced mutant pools and high performance TILLING enable finding of any gene mutation in rice. *Mol Genet & Genomics.* 279: 213–223.

Talame, V., Bovina, R., Sanguineti, M.C. *et al.* 2008. TILLMore, a resource for the discovery of chemically induced mutants in barley. *Plant Biotechnol J.* 6: 477–485.

Till, B.J., Cooper, J., Tai, T.H. *et al.* 2007. Discovery of chemically induced mutations in rice by TILLING. *BMC Plant Biology.* 7: 19.

Uauy, C, Paraiso, F., Colasuonno, P. *et al.* 2009. A modified TILLING approach to detect induced mutations in tetraploid and hexaploid wheat. *BMC Plant Biology.* 9: 115.

Weil, C.F. 2009. TILLING in grass species. *Plant Physiol.* 149: 158–164.

Xin, Z.G., Wang, M.L., Barkley, N.A. *et al.* 2008. Applying genotyping (TILLING) and phenotyping analyses to elucidate gene function in a chemically induced sorghum mutant population. *BMC Plant Biology.* 8: 103.

6.2. Websites

Conformation sensitive capillary electrophoresis (CSCE):
http://www.ngrl.org.uk/wessex/downloads/pdf/NGRLW_CSCE_3.1.pdf

LightCycler® 480 System & Mutation Detection Approaches:
http://www.lightcycler480.com.

6.3. Further Reading

Dujols, V., Kusukawa, N., McKinney, J.T. *et al.* 2006. High-resolution melting analysis for scanning and genotyping. *In*: D. Tevfik (ed.) Real Time PCR. Abingdon: Taylor and Francis, pp. 159–169.

Iwahana, H., Fujimura, M., Takahashi, Y. *et al.* 1996. Multiple fluorescence-based PCR-SSCP analysis using internal fluorescent labeling of PCR products. *Biotechniques.* 21: 510–519.

McCallum, C.M., Comail, L., Greene, E.A. *et al.* 2000. Targeting induced local lesions in genomes (TILLING) for plant functional genomes. *Plant Physiol.* 123: 439–442.

Muleo, R., Colao, M.C., Miano, D. *et al.* 2009. Mutation scanning and genotyping by high-resolution DNA melting analysis in olive germplasm. *Genome.* 52: 252–260.

C21

Discovery of Chemically Induced Mutations by TILLING

R.Bovina, V.Talamè, S.Salvi, M.-C.Sanguineti and R.Tuberosa*

Department of Agroenvironmental Sciences and Technology (DiSTA), University of Bologna, Viale Fanin 44, 40127 Bologna, Italy
* Corresponding author, E-MAIL: roberto.tuberosa@unibo.it

1. Introduction

The term functional genomics encompasses a number of different approaches aimed at determining gene function on a genome-wide scale. The application of these approaches is greatly facilitated by the utilization of new, high-throughput technologies applicable to almost any organism. As an example, sequence alignment-based comparisons are used to identify homologous sequences between and within species, transcriptional profiling to determine gene expression patterns and interaction analyses to help elucidate pathways, networks and protein complexes. However, although these analyses are extremely useful to extrapolate important features of a novel gene from a biochemical or a molecular point of view, they are not very informative in the context of the functional complexity of a living organism.

In order to overcome these limitations, different reverse genetics approaches have been conceived. Nonetheless, the tools for reverse genetics are not always transferable from one organism to another or from model species to non-model ones because in most cases the main drawback is the lack of efficient technical protocols exploitable for the majority of the plant species.

A novel, reverse genetics approach that combines the advantages of point mutations provided by chemical mutagenesis with the advantages of PCR-based mutational screening has been introduced with the name of TILLING (Targeting Induced Local Lesion in Genomes; McCallum *et al.,* 2000). From a technical standpoint, the first step of a TILLING assay is the PCR amplification of a target DNA fragment of interest from pooled DNAs of multiple mutant individuals. In sample pools, hetero-duplexes with a mismatched base pair are formed between wild type and mutated fragments by denaturing and re-annealing PCR products (**Figure 21.1**). Hetero-duplexes are cleaved by an endonuclease enzyme able to recognize the mismatch position. Cleaved products are then resolved using denaturing polyacrilamide gel or capillary electrophoresis. When a positive signal is identified, individual DNA samples of the pools are mixed in equal amounts with the wild type DNA and one-by-one re-analysed to identify the mutant individual plant; the induced mutations are eventually confirmed by sequencing. A detailed description of the technical aspects of the TILLING procedure is presented in the following section.

2. Technical Aspects of the TILLING Procedure

From a technical point of view, the TILLING protocol includes four main phases:
1. generation of a mutant population;
2. selection of target genes;
3. molecular screening;
4. recovery of mutants.

2.1. Generation of a Mutant Population

The first and most important prerequisite of TILLING is to create a proper mutant population. The accurate programming of the chemical mutagenesis step will influence all subsequent experimental choices and the quality of the TILLING results. Above all, it is of paramount importance to reach an appropriate frequency of mutations per genome maintaining a reasonably low percentage of lethality and plant sterility. Two fundamental aspects need to be considered before proceeding with the chemical treatment: 1) the chemical mutagen that works best with the target species and 2) the material to be mutagenized.

2.1.1. Mutagen and Mutagenic Treatment

An important feature to consider is the frequency of point mutations per genome induced by different chemical mutagens. The higher the number of mutations that are induced per genome, the more desirable the chemical mutagen will be. In fact, for TILLING purposes, the utilization of chemicals with a low point-mutation frequency requires the production of larger populations to allow for a sufficient number of mutant alleles per gene. This aspect should also be duly considered in terms of cost and labour efforts required to analyse a very large number of individuals. Among the chemical mutagens that have been used for producing point mutations (**see Chapters 12 and 20**), ethylmethansulphonate (EMS) has been favoured for several decades both in plants and in animal species because it produces a relatively high density of irreversible mutations per genome. An important advantage of using a standard mutagen, such as EMS, is that an extensive body of literature has confirmed its effectiveness in forward-genetics screens in a range of species. EMS has produced consistent results in different organisms; apparently similar levels of mutagenesis have been achieved in *Arabidopsis* seeds soaked in EMS (McCallum et al., 2000) and *Drosophila* males fed EMS (Bentley *et al.*, 2000). These features have made EMS the preferred mutagen for TILLING applications: EMS has been utilized in *Arabidopsis* (McCallum *et al.*, 2000), *Lotus japonicus* (Perry *et al.*, 2003), *Hordeum vulgare* (Caldwell *et al.*, 2004), *Zea mays* (Till *et al.*, 2004), *Drosophila melanogaster* (Winkler *et al.*, 2005), *Triticum* spp. (Slade *et al.*, 2005), *Brassica napus* (Wu *et al.*, 2007), *Oryza sativa* (Till *et al.*, 2007), *Medicago truncatula* (Porceddu *et al.* 2008) and *Sorghum bicolor* (Xin *et al.*, 2008). In other cases, less common mutagens have been used in particular species for which they exhibit high effectiveness. For example, an EMS-like alkylating agent called ethylnitrosurea (ENU) has been used efficiently on zebrafish (Wienholds *et al.*, 2003), sodium azide (NaN$_3$) was used to produce a barley mutant population for TILLING applications (Talamè *et al.*, 2008) and a combination of sodium azide plus methyl nitrosurea (NaN$_3$-MNU) was used to develop a rice TILLING population (Till *et al.*, 2007). Another crucial aspect to be considered to obtain the right level of mutations per genome is the intensity of the chemical treatment. The application of

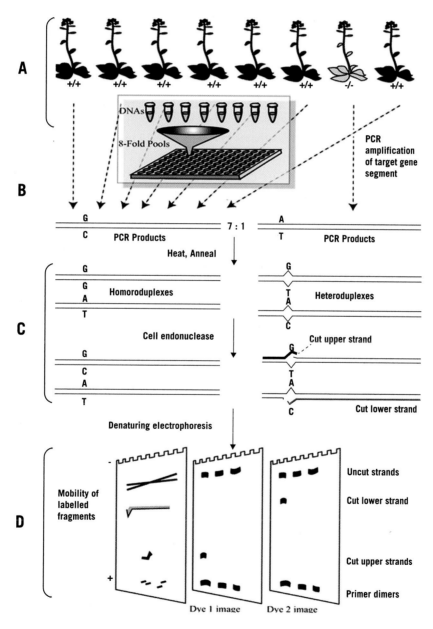

The Joint FAO/IAEA Programme

Figure 21.1 Technological steps of TILLING procedure. (A) DNA isolation and eight-fold pooling. (B) Amplification of a target locus in a pool containing one mutant sample. (C) Heteroduplex formation and cell digestion (D) Denaturing electrophoresis and mutants detection.

elevated mutagen doses increases the point mutation rate. However, there are good reasons to avoid using the highest possible mutagen dose because this will also lead to an increase in undesired mutations.

Commonly, it is of primary importance to test different mutagen dosages to determine the optimal dose especially while working with species for which no previous mutagenic information is available. It is known that each chemical mutagen can produce different effects in different species in terms of mutation rate and seed survival. Some parameters such as: embryo lethality, sterility rate and presence of pigment-defective sectors (chimerism) may provide an idea on the effectiveness of the mutagenesis treatment and on the optimal mutagen dosage identification (**see Chapter 14**).

2.1.2. Material to be Mutagenized

In plant species characterized by sexual reproduction, the mutagenic treatment can be applied to either seeds or pollen. Seed treatment is considered the preferred approach for TILLING applications. Seed mutagenesis, as compared to pollen mutagenesis, is technically easier because seeds do not have any temporal/developmental stage restrictions for treatment and because seeds are much more resistant to higher mutagen doses. As a negative counterpart, many of the mutations obtained in seed embryos are not preserved in subsequent generations (**see Chapter 15**). Individuals from the M_2 generation represent the first non-chimeric material which could be used for TILLING: almost all of the mutations identified in the M_2 generation will be recovered in the mutant lineage.

Pollen mutagenesis can be preferable under several points of view, particularly for the possibility of obtaining genetically homogeneous M_1 plants. In this case, M_1 plants can already be screened by TILLING because any mutation displayed in the molecular analysis is almost often sexually transmitted. So far, the pollen approach for TILLING purposes has been used only on maize, a monoecious species, with staminate flowers in the tassels separated from pistillate flowers on the ear. As a matter of fact, a substantial drawback of seed mutagenesis in species like maize is that male and female flowers almost invariably derive from different germline cells of the embryo. The selfed progeny of a mutagenized seed will not segregate for the recessive mutation because a recessive allele is only provided by one of the gametes. The following generation of selfing (M_3) will reveal the recessive mutants. To by-pass this chimerism problem, the pollen treatment approach appears to be a preferable alternative in monoecious plants as also shown in a study conducted in maize for discovering induced point mutations by TILLING.

2.2. Selection of Target Genes

Once a mutagenized population is made available, the next TILLING step is the selection of target genes. A typical reverse genetics screen starts with the identification of one or more genes of known sequence that are chosen because of their putative involvement in a particular biological or physiological process. In this context, the comparison of potential targets with functionally known paralogues or homologous genes, together with gene expression or protein interaction information can facilitate the identification of suitable target genes. Whenever possible, single-copy genes are considered the best choice for deploying TILLING because in this case, point mutations are more likely to generate detectable phenotypic alterations. For this purpose, whenever precise information on the gene in the target species is unavailable, the identification of a suitable gene is facilitated by information from other plant species and/or by further analysis (e.g. Southern blotting on the target sequence).

2.2.1. Selection of Target Amplicon

One of the main technical shortcomings of TILLING is that it is possible to screen for point mutations only in a portion (up to ~1,000 bp) of the target gene. The identification of the most appropriate gene window can be hindered by limited knowledge in terms of sequence information of the species of interest. If the genome sequence of a species becomes available, it is relatively simple to design specific primers for the target region; conversely, in species only partially sequenced, PCR amplification could result in spurious products due to paralogous and homoeologous sequences. A good solution to this problem has been shown by Slade *et al.* (2005) in wheat where gene-specific primers were designed for optimal TILLING of *waxy* loci.

If previous information on the position of the most important functional domains is not available, the choice of a gene window should take into account the gene model (exons and introns position), the effect of point mutations on codons and splice junctions (EMS alkylates

G residues, **see Chapter 3**), and sequence conservation of the encoded protein. The important task of designing primers suitable for TILLING is simplified by various software tools that utilize homology information to predict the effect of the mutations on the protein function. The free web-based tool, Codons Optimized to Detect Deleterious LEsions (CODDLE) was developed as a tool to facilitate the choice of gene regions more suitable for TILLING (http://www.proweb.org/coddle/). Once a genomic sequence and the corresponding coding sequence is available, CODDLE identifies regions where point mutations are most likely to cause deleterious effects on gene function. The programme exploits available information about the typology of induced mutation and plots the probability of misense and non-sense mutations along the coding sequence. Additionally, the programme searches the predicted protein sequence under study for the occurrence of conserved regions that are less likely to be functionally neutral. Finally, the programme identifies the gene portion most suitable for TILLING amplifications that will be used for primer design (Primer3 programme at: http://frodo.wi.mit.edu/). The selected amplicon and the effects of all possible polymorphisms induced by the mutagen are automatically tabulated by the programme.

2.3. Screening Mutant Pools

A detection system suitable for TILLING should be sensitive enough to discriminate between wild type and mutated sequences at the single nucleotide level. Key steps of the TILLING procedure can be summarized as: pooling, hetero-duplex formation and enzymatic cleavage, detection of generated fragments and phenotype recovery.

2.3.1. Pooling Strategies

Individuals from the M_2 or M_3 generation are used to prepare DNA samples that are bulked and arrayed on microtitre plates. An important aspect to be evaluated is the maximum number of DNA samples that can be pooled. In fact, as the pooling level increases, the proportion of hetero-duplexes decreases, with a consequent reduction in sensitivity. On the other hand, increasing the pooling level reduces the number of analyses required to screen the entire collection of mutants. The number of mutants that need to be screened to identify a suitable amount of mutant alleles can be very

high (in some cases, up to tens of thousands), making it necessary to pool as many individuals as possible before testing for the presence of a mutation in the target region. This aspect affects the throughput and the cost of the technique. A high-throughput TILLING protocol for consistent detection of heteroduplexes in eight-fold pools has been developed, thus enabling the detection of one genome in 16 in a diploid species (see endonuclease cleavage below). This implies that a single TILLING run can interrogate ~750,000 total bp (eight-fold pooling × 1,000 bp × 96 lanes) detecting mutations equivalent to ~3,000 sequencing lanes. The resulting savings, taken together with cost savings from pooling prior to PCR amplification, suggests that high-throughput TILLING is almost an order of magnitude more economical than full sequencing, even when accounting for the identification of the mutant in each pool and its sequencing.

There is also a trade-off between discovering a mutation in a pool and identifying the mutant individual. Ultimately, there is a break-even point, beyond which any further increase in pooling implies that the effort in finding the individual mutant within a positive pool exceeds that of screening the pools. The break-even point is proportional to the mutation rate and the size of the fragment because doubling the size of the fragment or doubling the mutation rate halves the number of assays required to identify the individual carrying the mutation. This situation is peculiar for wheat TILLING where, due to the high mutation frequency, the PCR screening was mainly performed in two-fold pools. Alternatively, DNA samples can be pooled according to a 2- or 3D gridding strategy, that increases the number of DNA pools for the PCR screens but allows for the identification of the mutant individual without proceeding with the analysis of the single components of the pool. As an example, a collection of 100 mutants can be pooled into a 2D-grid consisting of 10 rows and 10 columns, so that an individual line is identified by an address of one row and one column. When a mutation is discovered, it will appear in two separate gel lanes whose coordinates will identify the unique individual harbouring the mutation, thus reducing false-positive errors, as true signals will replicate in the relevant lanes. A 2D-pooling strategy was successfully utilized for the discovery of rare human polymorphisms by Ecotilling (Till *et al.*, 2006).

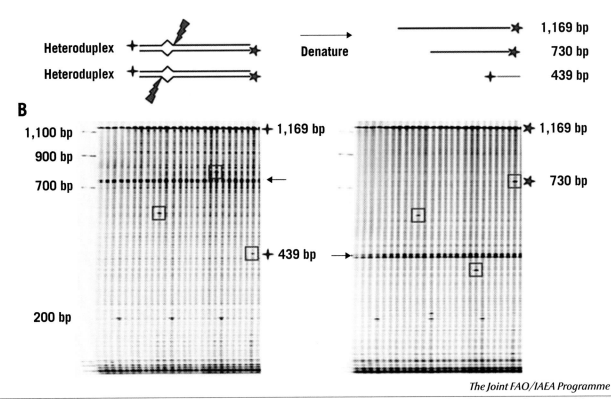

Figure 21.2 Schematic diagram of enzymatic mismatch cleavage (A) and typical production TILLING data from a 1,169-bp Drosophila gene target (B). The amplification is carried out with the forward primer labelled with IRDye 700 (blue four-point star) and the reverse with IRDye 800 (red five-point star) (A). Products are denatured (A) and fragments are size-fractionated using denaturing polyacrylamide gel electrophoresis and visualized using an LI-COR DNA analyser (B). The molecular weight of the cleaved IRDye 700 fragment (439 bp in this example) plus the IRDye 800 cleaved fragment (730 bp) equals the molecular weight of the full-length 1,169 PCR product (Till *et al.,* 2006).

2.3.2. PCR, Heteroduplex Formation and Enzymatic Cleavage

Sample pools are subjected to window-specific amplification by PCR using properly designed primers. The products of amplification are then denatured and re-annealed, generally by heating and cooling. If a point mutation occurs in the amplified region in at least one of the single DNA strands of the pool, it is likely that the mutant strand will re-anneal with the wild type strand so that a fraction of the pool will have a mismatch at the site of the mutation. A method that can distinguish the mismatched hetero-duplex identifies any pool containing a mutation (**see Chapter 22**). For high-throughput TILLING purposes, an efficient hetero-duplex detection method based on endonucleolytic cleavage and sizing on electrophoretic gels has been developed. Enzymes of the S1 nuclease family are characterized by the ability to recognize single nucleotide mismatches as a restriction site and to generate cleavages at a single strand exactly at the 3′-side of the mismatch. Consequently, incubating sample pools containing hetero-duplex molecules in the presence of an S1 nuclease will generate short fragments in addition to the uncleaved ampli-

con. Although the S1 nuclease can cleave mismatches, their detection can be limited and not always effective. A member of the S1 family purified from celery (*Cel*I) can reliably cleave mismatches on either strand on the 3′- side. The *Cel*I enzyme has rapidly become the most popular endonuclease for high-throughput TILLING applications although only one costly commercial formulation of the enzyme is currently available (Surveyor kit from Transgenomic Inc., NE, USA). More recently, it was shown that crude extracts from celery are sufficiently enriched in mismatch cleavage activity so that they can be used for high-throughput applications without further purification (**see Chapter 20**).

2.3.3. Detection of Cleaved Fragments

The first TILLING screenings were based on a denaturing high-pressure liquid chromatography (DHPLC) approach for hetero-duplex detection. The key point of using DHPLC is that the solid phase has a differential affinity for single- and double-stranded DNA based on their different hydrophobicity. In the case of duplex fragments, keeping the column temperature just below the melting temperature of the homo-duplexes causes mismatched fragments to begin melting out as the duplex passes through the column, decreasing its hydrophobicity and thus its retention time in the column. Consequently, even single mismatches in a duplex will cause a reduction in the affinity to the solid matrix and so the molecules appear as a leading peak because they exit the column sooner than perfectly matched fragments. The DHPLC approach can be easily automated as no labelling or purification of the DNA fragments is needed. The method is also relatively fast and highly specific. The requirement of scaling up the detection of the cleaved fragments generated by nuclease digestion has led to the development of an efficient detection method based on sizing by electrophoretic gels.

The approach implemented for the LI-COR sequencers (LI-COR Inc., 2006; www.licor.com) consists of denaturing polyacrylamide slab gel analysers. This electrophoretic instrument permits the visualization of fragments based on fluorescence detection. For this purpose, PCR amplification of the target fragments is conducted using infra-red dye-labelled primers. The forward primer is usually 5'-end labelled with a fluorescent dye IRDye700 (detected at ~700 nm) and the reverse primer is labelled with IRDye800 (detected at ~800 nm). For each run, primers fluorescence emissions are detected by LiCOR and two electronic image files are produced, one containing data from the 700-nm channel and the other from the 800-nm channel. In denaturing conditions, if a mutant allele is present, the cleaved fragments are displayed in the two channels so that the sum of their length will correspond to the complete amplicon size. The molecular weights of the fragments also provide the approximate position of the nucleotide change (**Figure 21.2**). Compared to the DHPLC, the electrophoretic TILLING method presents several important advantages: larger size fragments (~1,000 bp) can be explored, thus allowing for the identification of a higher number of point mutations per analysis. In addition, the *Cel*I-based method is feasible for any novel fragment without substantial protocol modifications while the DHPLC approach requires setting up new running conditions (i.e. the column temperature).

Although TILLING has so far extensively utilized the LI-COR gel analyser as a detection system, recently another mutation discovery strategy based on fluorescence detection has deployed the ABI capillary sequencer (Applied Biosystems, CA, USA). The method takes advantage of the automated loading procedure and the possibility to simultaneously analyse two different amplicons. Accordingly, the ABI prism developed a four-colour system that provides sufficient spectral separation to allow for the use of all four dye labels.

An additional method makes use of a commercial kit based on a mismatch cleavage by *Cel*I to identify mutations in double-stranded DNA fragments using an agarose-gel assay for fragment detection. This mutation detection kit can be applied particularly to small-scale projects, such as screening for mutations and polymorphisms in single genes. Unfortunately, the high cost of each analysis makes this approach unsuitable for extensive applications.

Whatever the selected detection system and once positive results are identified, a second round of TILLING is performed on the individuals from each positive pool. This step identifies the plant carrying the mutation and confirms the mutation. The ultimate validation of the results is obtained by sequencing the variant alleles that will allow for singling out the exact position and the typology of the mutation. Several kinds of single-base changes are expected to result from chemical mutagenesis.

Once a mutation is discovered, it is important to estimate the probability of protein-function alteration. As an example, truncation mutations are almost all clearly classified, while the effect of a missense mutation may not be so obvious. For this purpose, software tools based on protein homology information to predict damaging lesions are available. For example, PARSESNP (Project Aligned Related Sequences and Evaluate SNPs; http://www.proweb.org/parsesnp/) is a web-based tool for the analysis of polymorphisms in genes. The programme determines the translated amino acid sequence starting from a reference DNA sequence (genomic or cDNA) and a gene model, and the effects of the supplied polymorphisms on the expressed gene product. If a homology model is provided, predictions can be made as to the severity of missense changes. PARSESNP utilizes for this purpose conserved alignments in the Blocks database (the same information also used by CODDLE). Another similar bioinformatics tool developed for TILLING is SIFT (Sorting Intolerant From Tolerant http://blocks.fhcrc.org/sift/SIFT), a software able to predict whether an amino acid substitution affects protein function based on sequence homology and the physical properties of amino acids. The SIFT

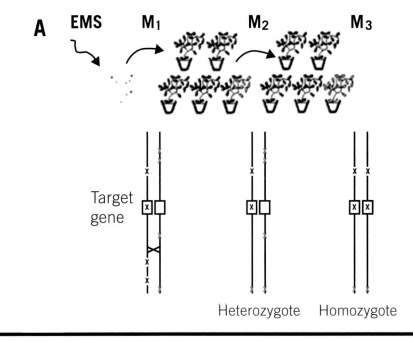

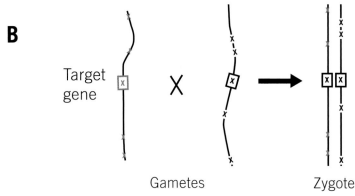

The Joint FAO/IAEA Programme

Figure 21.3 Phenotypic characterization of a target gene in a selfed TILLED line (A) or by crossing two independent TILLED lines (B).

programme uses PSI-BLAST searching of current databases to assess amino acid conservation.

2.3.4. Phenotype Recovery

When an allelic series of mutations is discovered through TILLING with some likely to affect the functionality of the encoded protein, a phenotypic analysis is required to verify the effect of the mutation. As previously illustrated, chemical mutagenesis introduces a large number of point mutations per genome (up to many thousands) in addition to those detected at the target locus. Consequently, in the standard procedure, the mutants are backcrossed or outcrossed in order to eliminate or at least reduce undesired effects caused by extra mutations that could interfere with the phenotypic appearance (**see Chapter 27**). However, when

advanced generations are scored for the phenotypic analysis, these procedures could be avoided. For example, when an M_2 plant heterozygous for the target mutation is considered, the subsequent M_3 family will show the 1:2:1 Mendelian segregation. Typing numerous plants for each Mendelian class should yield a perfect correlation between genotype and phenotype because crossovers scramble the background and will produce different subsets of homozygous background lesions in different M_3 plants (**Figure 21.3A**). A limitation is that the background mutations can be distinguished from the mutation in the gene of interest only if they are unlinked. When multiple mutant alleles for the same target gene are available, a complementation test offers a solution to this problem. In fact, two mutant lines could be crossed to make them heteroallelic at the target gene and to essentially make all background mutations heterozygous with the wild type (**Figure 21.3B**). Consequently, a phenotype attributable to the two non-complementing mutations will be found in every individual carrying both, whereas complementing background mutations will not show different phenotypes from the wild type.

It may happen that no phenotypic variation is detected even if a mutant line completely devoid of background mutations is available. Two main causes can be responsible for this finding: functional redundancy and the presence of a conditional phenotype (**Figure 21.4**). Functional redundancy among the members of a gene family or among homoeologous genes in polyploid species is a likely reason for the frequently observed lack of an identifiable phenotype associated with a mutation. Most plants in wheat mutant populations appeared to be similar to the wild type and only in rare cases (<0.5%) were abnormal phenotypes observed among field-grown plants. It is well known that the polyploidy nature of wheat provides buffering against the effect of mutations. To test for functional redundancy, genetic crosses between plants that bear mutations in different

redundant genes can be performed, resulting in the production of a multiple mutant.

Another possible reason for the lack of visible phenotypes is that individual gene-family members may have evolved to function only under specific physiological conditions (i.e. conditional phenotypes). Thus, only in such conditions will an effect on the phenotype be observed. To test for conditional phenotypes in a gene of unknown function, a broad range of growing conditions should be used (**see Chapter 25**).

Once a phenotype has been observed, additional steps are required to prove that the phenotypic characteristic is indeed controlled by the gene of interest (**Figure 21.4**). For example, in order to unambiguously assign a function to a gene, a complementation test with a functional transgene can be performed. However, this is still an expensive and time-consuming process in various species. The widely accepted alternative to complementation is the characterization of multiple alleles at the target locus. The observation of a similar phenotype in multiple, independent mutant lines of the same locus provides strong support that the locus is indeed controlling the phenotype.

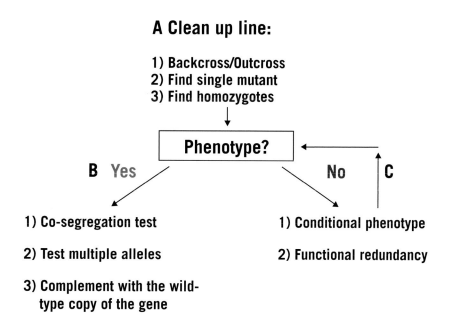

A Clean up line:

1) Backcross/Outcross
2) Find single mutant
3) Find homozygotes

Phenotype?

B Yes

No **C**

1) Co-segregation test

2) Test multiple alleles

3) Complement with the wild-type copy of the gene

1) Conditional phenotype

2) Functional redundancy

The Joint FAO/IAEA Programme

Figure 21.4 Flow chart of the main steps required for the recovery and confirmation of a mutant phenotype. (A) Procedure used for cleaning up the mutant lines. Mutants identified are usually backcrossed and/or outcrossed in order to generate plants carrying single-homozygous point mutations. Once an appropriate mutant line is realized the next step is to determinate the effect of the mutation on the phenotype (B). Common strategies utilized to finally demonstrate the correlation between the phenotypic characteristic and the mutation of interest. (C) Sometimes mutants have no readily identifiable phenotype (for example because of functional redundancy or low penetrance). In these cases, more steps could be required for the identification of a correlated phenotype.

Box 21.1

Project	Species	Website
Arabidopsis TILLING Project	*Arabidopsis thaliana*	http://tilling.fhcrc.org:9366/
Arcadia Biosciences	*Triticum aestivum, Triticum durum*	http://www.arcadiabiosciences.com/OurToolbox.htm
Canadian TILLING Initiative	*Arabidopsis thaliana, Brassica napus, Brassica oleracea*	http://www.botany.ubc.ca/can-till/
GABI TILLING	*Arabidopsis thaliana, Beta vulgaris, Hordeum vulgare, Solanum tuberosum*	http://www.gabi-till.de/project/project/gabi-till-project.html
LycoTill	*Lycopersicon esculentum*	http://www.agrobios.it/tilling/
Maize TILLING Project	*Zea mays*	http://genome.purdue.edu/maizetilling/
OPTIWHEAT	*Triticum durum*	http://www.rothamsted.bbsrc.ac.uk/cpi/optiwheat/indexcontent.html
RevGenUK	*Brassica rapa, Lotus japonicus, Medicago truncatula*	http://revgenuk.jic.ac.uk/
Rice TILLING Project	*Oryza sativa*	http://tilling.ucdavis.edu/index.php/Main_Page
SCRI Barley mutants	*Hordeum vulgare*	http://germinate.scri.sari.ac.uk/barley/mutants/
Shanghai RAPESEED Database	*Brassica napus*	http://rapeseed.plantsignal.cn/
Soybean Mutation Project	*Glycine max*	http://www.soybeantilling.org/
TILLMore	*Hordeum vulgare*	http://www.distagenomics.unibo.it/TILLMore/
UTILLdb	*Pisum sativum*	http://urgv.evry.inra.fr/UTILLdb

3. TILLING Applications

3.1. Genome Structure of the Target Species

Ideal targets for TILLING are species with a diploid and gene-rich genome where phenotypic alterations induced by mutagenesis are easy to identify because point mutations often occur in a functional region and they are not masked by gene redundancy. Unfortunately, eukaryotic genomes and plants in particular often do not present a simple organization, particularly for traits determined by gene families and quantitative trait loci (QTLs). Furthermore, the evolution of many crop genomes was characterized by complex modifications such as partial- or whole-genome duplications. Typical examples are the polyploid nature of cultivated wheat and the extensive segmental duplications in Arabidopsis. Despite these limitations, excellent TILLING results have been reported in both Arabidopsis (Till *et al.*, 2003) and wheat (Slade *et al.*, 2005).

The molecular analyses are more easily performed with highly homozygous, inbreeding species. Although heterozygosity and dioecy complicate the application of TILLING, in zebrafish TILLING resources have been successfully produced (Wienholds *et al.*, 2003) and in Drosophila (Winkler *et al.*, 2005), two highly heterozygous animal species. In these cases, it is preferable to perform mutagenesis on a few individuals (or preferably only one) in order to limit as much as possible the pre-existing variability within the population.

3.2. TILLING Projects and Public Services

In the plant kingdom, the TILLING technique was primarily set up in the model species Arabidopsis (McCallum *et al.*, 2000). The next goal was to provide the Arabidopsis community with a TILLING service known as the Arabidopsis TILLING Project (ATP), later on included in a larger project named the Seattle TILLING Project (STP) dealing with different species. Additional

TILLING projects have been extensively reported in the literature for both plant and animal species. **Box 21.1** reports the websites of a number of plant TILLING projects.

4. Advantages, Disadvantages and Prospects

TILLING provides a high-throughput, sensitive, cheap and rapid reverse genetics approach with several advantages.

Firstly, the utilization of chemical mutagens generally introduces point mutations which produce an allelic series for potentially any target gene. In fact, unlike techniques generating knock-outs, chemical mutagenesis causes destabilization of the proteins, ranging from not effective to severe. Ideally, using a TILLING approach, a target gene can be investigated for each domain or even each single amino acid.

The introduction of point mutations, and the consequent attenuation of peculiar genes, may have commercial applications in crop breeding. In some cases, for agricultural purposes, it could be useful for silencing altogether the expression of particular genes like those coding for protein required for pathogen multiplication or for the synthesis of toxic metabolites. TILLING provides an excellent opportunity for tailoring crop plants that display desirable traits but lack powerful genetic tools. Furthermore, an additional advantage of TILLING is that resulting crop varieties are not subject to the same regulatory approval requirements as transgenic crops. Another advantage of TILLING over other reverse genetics strategies centred mainly on model species, is its applicability to almost any species. This notwithstanding, TILLING could become obsolete when a general technology for easy modification of target genes becomes available. Recent improvements in these technologies, such as the selective enzymatic cleavage of chromosomal loci, may render them widely applicable also in recalcitrant systems (Urnov *et al.,* 2005).

A component of the success of the TILLING strategy refers to the ability to easily detect mismatches in DNA heteroduplexes using cleavage approaches. Additionally, in the near future the development of other techniques to discover and screen for single-nucleotide differences could improve and simplify the TILLING procedure. Advances in genomic technology, such as improved SNP detection or sequencing, will greatly improve the mutant detection. A good candidate for TILLING improvement is the direct sequencing that could facilitate homozygous/heterozygous discrimination, hence allowing one to analyse contemporary multiple sequences. Microarray technology could be very promising in a long-term view, although the high cost and uncertain reliability are issues that need to be resolved.

Finally, while TILLING is a fast procedure likely to be improved in its effectiveness, evaluating the phenotypic consequences of mutations is still a slow procedure. Furthermore, to produce a range of alleles rather than a single knock-out mutation requires an increased effort in downstream analysis. Probably the most challenging future efforts in functional genomics will be aimed at facilitating the downstream analysis of mutants, including sophisticated high-throughput phenotyping platforms and more attention to data management issues.

5. References

5.1. Cited References

Bentley, A., MacLennan, B., Calvo, J. and Dearolf, C. R. 2000. Targeted recovery of mutations in *Drosophila. Genetics.* 156: 1169–1173.

Caldwell, D. G., McCallum, N., Shaw, P., Muehlbauer, G. J., Marshall, D. F. and Waugh, R. 2004. A structured mutant population for forward and reverse genetics in barley (*Hordeum vulgare* L.). *Plant Journal.* 40: 143–150.

McCallum, C. M., Comai, L., Greene, E. A. and Henikoff, S. 2000. Targeted screening for induced mutations. *Nature Biotechnology.* 18: 455–457.

Perry, J. A., Wang, T. L., Welham, T. J., Gardner, S., Pike, J. M., Yoshida, S. and Parniske, M. 2003. A TILLING reverse genetics tool and a web-accessible collection of mutants of the legume *Lotus japonicus. Plant Physiology.* 131: 866–871.

Porceddu, A., Panara, F., Calderini, O., Molinari, L., Taviani, P., Lanfaloni, L., Scotti, C., Carelli, M., Scaramelli, L., Bruschi, G., Cosson, V., Ratet, P., de Larembergue, H., Duc, G., Piano, E. and Arcioni, S. 2008. An Italian functional genomic resource for *Medicago truncatula. BMC Research Notes.* 1: 129.

Slade, A. J., Fuerstenberg, S. I., Loeffler, D., Steine, M. N. and Facciotti, D. 2005. A reverse genetic, non-transgenic approach to wheat crop improvement by TILLING. *Nature Biotechnology*. 23: 75–81.

Talamè, V., Bovina, R., Sanguineti, M. C., Tuberosa, R., Lundqvist and U., Salvi, S. 2008. TILLMore, a resource for the discovery of chemically induced mutants in barley. *Plant Biotechnology Journal*. 6: 477–485.

Till, B. J., Cooper, J., Tai, T. H., Colowit, P., Greene, E. A., Henikoff, S. and Comai, L. 2007. Discovery of chemically induced mutations in rice by TILLING. *BMC Plant Biology*. 11: 7–19.

Till, B. J., Reynolds, S. H., Greene, E. A., Codomo, C. A., Enns, L. C., Johnson, J. E., Burtner, C., Odden, A. R., Young, K., Taylor, N. E., Henikoff, J. G., Comai, L. and Henikoff, S. 2003. Large-scale discovery of induced point mutations with high-throughput TILLING. *Genome Research*. 13: 524–530.

Till, B. J., Reynolds, S. H., Weil, C., Springer, N., Burtner, C., Young, K., Bowers, E., Codomo, C. A., Enns, L. C. and Odden, A. R. 2004. Discovery of induced point mutations in maize genes by TILLING. *BMC Plant Biology*. 28: 4–12.

Till, B. J., Zerr, T., Comai, L. and Henikoff, S. 2006. A protocol for TILLING and Ecotilling in plants and animals. *Nature Protocols*. 1: 2465–2477.

Urnov, F. D., Miller, J. C., Lee, Y. L., Beausejour, C. M., Rock, J. M., Augustus, S., Jamieson, A. C., Porteus, M. H., Gregory, P. D. and Holmes, M. C. 2005. Highly efficient endogenous human gene correction using designed zinc-finger nucleases. *Nature*. 435: 646–651.

Wienholds, E., van Eeden, F., Kosters, M., Mudde, J., Plasterk, R. H. and Cuppen, E. 2003. Efficient target-selected mutagenesis in zebrafish. *Genome Research*. 13: 2700–2707.

Winkler, S., Schwabedissen, A., Backasch, D., Bökel, C., Seidel, C., Bönisch, S., Fürthauer, M., Kuhrs, A., Cobreros, L., Brand, M. and González-Gaitán, M. 2005. Target-selected mutant screen by TILLING in *Drosophila*. *Genome Research*. 15: 718–723.

Wu, G. Z., Shi, Q. M., Niu, Y., Xing, M. Q. and Xue, H. W. 2007. Shanghai RAPESEED Database: a resource for functional genomics studies of seed development and fatty acid metabolism of *Brassica*. *Nucleic Acids Research*. 36: 1044–1047.

Xin, Z., Wang, M. L., Barkley, N. A., Burow, G., Franks, C., Pederson, G. and Burke J. 2008. Applying genotyping (TILLING) and phenotyping analyses to elucidate gene function in a chemically induced sorghum mutant population. *BMC Plant Biology*. 8: 103.

5.2. Websites

Codons to Optimize Discovery of Deleterious Lesions (CODDLE):
www.proweb.org/coddle/

Project Aligned Related Sequences and Evaluate SNPs (PARSESNP):
www.proweb.org/parsesnp/

Primer3 programme at:
http://frodo.wi.mit.edu/

Sorting Intolerant From Tolerant (SIFT):
http://blocks.fhcrc.org/sift/SIFT

5.3. Further Reading

Alonso, J. M. and Ecker, J. R. 2006. Moving forward in reverse: genetic technologies to enable genome-wide genomic screens in Arabidopsis. *Nature*. 7: 524–536.

Comai, L. and Henikoff, S. 2006. TILLING: practical single-nucleotide mutation discovery. *Plant Journal*. 45: 684–694.

Henikoff, S. and Comai, L. 2003. Single-nucleotide mutations for plant functional genomics. *Annual Review of Plant Biology*. 54: 375–401.

Ng, P. C. and Henikoff, S. 2003. SIFT: Predicting amino acid changes that affect protein function. *Nucleic Acids Research*. 31: 3812–3814.

C22

A Protocol for TILLING and Eco-TILLING

B.J.Till[a], T.Zerr[b], L.Comai[c] and S.Henikoff[d]*

[a] Basic Sciences Division, Fred Hutchinson Cancer Research Center, Seattle, Washington 98109, USA
  Present: Plant Breeding and Genetics Laboratory, Joint FAO/IAEA Division, International Atomic Energy Agency, Wagramer Strasse 5, P.O. Box 100, A-1400 Vienna, Austria
[b] Department of Genome Sciences, University of Washington, Seattle, Washington 98109, USA
[c] University of California Davis Genome Center, Davis, California 95616, USA
[d] Howard Hughes Medical Institute.
* Corresponding author, E-MAIL: steveh@fhcrc.org

Reprinted by permission from Macmillan Publishers Ltd:
[Nature Protocols] (B.J.Till, T.Zerr, L.Comai & S.Henikoff (2006) A Protocol for TILLING and Eco-TILLING. Nature Protocols 1, 2465–2477), copyright (2006)

1. Introduction

The acquisition of large-scale nucleotide sequence data has led to the introduction of methods that target the disruption of specific genes, also known as reverse genetics. Although potentially very powerful, reverse genetic approaches can be limited if the method used to induce the genetic change is not generally applicable to different organisms, as is often the case for transgenic strategies. However, chemicals and other agents have been used for several decades to induce mutations in a variety of organisms, making traditional mutagenesis an attractive strategy for reverse genetics. Chemical mutagenesis is especially valuable for agricultural applications because, unlike transgenic approaches, it is not hampered by restrictive regulatory and consumer issues. In addition, mutagenesis-based methods for reverse

genetic applications can provide more than just gene knock-outs; mutagens such as ethyl methanesulphonate primarily induce point mutations, providing allelic series that also include useful missense mutations (Greene *et al.,* 2003). Mutations can be induced by chemicals at a high density, allowing for the efficient recovery of multiple alleles in a small screening population. Targeting Induced Local Lesion in Genomes (TILLING) is a reverse genetic strategy that combines traditional mutagenesis with high-throughput discovery of single-nucleotide changes (McCallum *et al.,* 2000; Colbert *et al.,* 2001). The method is general and, following its original application to the model plant *Arabidopsis thaliana,* has been applied to a variety of plant and animal species including maize, lotus, barley, wheat, Drosophila and zebrafish (Till *et al.,* 2004a; Wienholds *et al.,* 2003; Winkler *et al.,* 2005; Perry *et al.,* 2003; Slade *et al.,* 2005; Caldwell *et al.,* 2004).

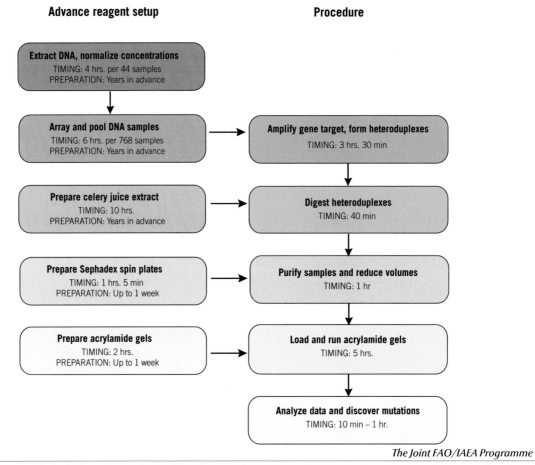

The Joint FAO/IAEA Programme

Figure 22.1 The major steps for high-throughput polymorphism discovery by TILLING and Eco-TILLING. Reagents can be prepared in advance, thus streamlining the high-throughput procedure. The approximate time required to prepare each reagent, along with an estimation of the length of time a reagent can be prepared in advance of screening is provided. The procedure for polymorphism discovery consists of (i) PCR amplification using fluorescently labelled gene-specific primers followed by the creation of hetero-duplexed molecules by denaturing and annealing polymorphic amplicons, (ii) enzymatic cleavage of mismatched regions in hetero-duplexed fragments, (iii) sample purification and volume reduction, (iv) polyacrylamide denaturing gel electrophoresis and (v) data analysis to discover polymorphic samples. The procedure can be completed in 1 day. Discovered polymorphisms can be sequenced to determine the exact nucleotide change. Sequencing is aided by the knowledge of the approximate position of the polymorphism acquired during gel electrophoresis.

1.1. Overview of TILLING

The TILLING procedure can be divided into a series of steps (**Figure 22.1**). DNA is first extracted from test samples. DNA aliquots are pooled (typically eight-fold) and arrayed into a multi-well microtiterplate containing 96 or 384 wells. Screening for mutations begins with PCR amplification of a target fragment of up to ~1.5 kilobases (kb) using gene-specific infrared dye-labelled primers. The forward primer is 5′-end labelled with a fluorescent dye that is detected at ~700 nm (IRDye 700) and the reverse primer is labelled with the IRDye 800,which is detected at ~800 nm. After PCR amplification, samples are denatured and annealed to form hetero-duplexes between mutant and wild type DNA strands. Samples are then incubated with a single-strand specific nuclease to digest mismatched basepairs (**Figure 22.2a**) (Till *et al.,* 2004b). After the reaction is stopped, DNA is purified from buffer components and sample volumes are reduced. A portion of each sample is then loaded onto a denaturing polyacrylamide slab gel. We use the LI-COR DNA analyser system to visualize fluorescently labelled DNA. Two electronic image files are produced per gel run; one containing data from the 700-nm channel and the other from the 800-nm channel (**Figure 22.2b**). Data analysis is aided by the use of GelBuddy, a freely available program for Macintosh and Windows PCs designed for the analysis of TILLING and Eco-TILLING gel data produced by the LI-COR analyser (Zerr and Henikoff, 2005). The exact nucleotide change is then determined using standard DNA sequencing methods. With eight-fold pooling, ~1,600 samples can be screened per day for mutations in a single ~1.5-kb gene target using a single LI-COR analyser. Higher throughput and economy of scale can be achieved by using multiple thermal cyclers and analysers.

Several alternative TILLING protocols are available. For example, TILLING organisms with large introns, such as zebrafish, requires short exon targets, leading to the development of alternative protocols for zebrafish

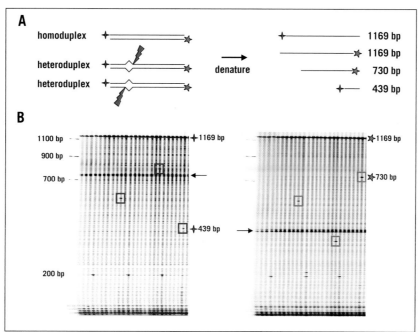

The Joint FAO/IAEA Programme

Figure 22.2 Mutation discovery by TILLING. Schematic diagram of enzymatic mismatch cleavage (A) and typical production TILLING data from a 1,169-bp Drosophila gene target (B). (A) PCR is performed using 5′ IRDye-labelled primers. The forward primer is labelled with IRDye 700 (blue four-point star) and the reverse with IRDye 800 (five-point star). Hetero-duplexed molecules are formed by denaturing and annealing amplified fragments. Mismatched regions are cleaved by treatment with a single-strand specific nuclease (green lightning bolts). Products are denatured and fragments are size-fractionated using denaturing polyacrylamide gel electrophoresis and visualized using an LI-COR DNA analyser. The molecular weight of the cleaved IRDye 700 fragment (439 base pairs (bp) in this example) plus the IRDye 800 cleaved fragment (730 bp) equals the molecular weight of the full-length PCR product (1,169 bp). The molecular weights of the fragments provide the position of the nucleotide change. (B) IRDye 700 (left) and IRDye 800 (right) images shown for the same 25 lanes of a 96-lane gel image. True mutations (boxed) produce a fragment in the IRDye 700 image (boxes) and in the IRDye800 image (boxes). The fragments illustrated in panel a are marked on the right of each image. To aid in gel analysis, 200-bp lane markers are added in every eighth lane, beginning in lane number 4. Molecular weight ladders are added to the end lanes of the gel to facilitate molecular weight calibration (sizes listed to the left of the left image). A strong common band (marked by an arrow) coincides with a stretch of 14 thymine residues starting at position 751. The band likely arises from breathing and subsequent cleavage of the duplex in this region.

TILLING (Draper *et al.*, 2004, Wienholds and Plasterk, 2004). In addition, TILLING can be performed using alternative readout platforms including slab gel and capillary systems (Perry *et al.*, 2003) (http://www.jic-genomelab.co.uk/services/mutation-detection/development.html). Although many groups use an enzymatic mismatch cleavage approach similar to that described in this protocol, other mutation discovery methods can potentially be substituted including denaturing HPLC and resequencing (e.g. McCallum *et al.*, 2000; Bentley *et al.*, 2000; Wienholds *et al.*, 2002). When choosing a single nucleotide polymorphism (SNP) discovery platform, it is important to consider criteria such as equipment cost, reagent cost, maintenance cost, automation and assay sensitivity.

1.2. High-Throughput Large-Scale Applications of TILLING

TILLING is a high-throughput, low-cost method that is suitable for a large-scale production operation. The first production-scale TILLING service to be offered was the Arabidopsis TILLING Project (ATP; http://tilling.fhcrc.org). Several computational tools were adapted to provide a completely web-based system for target selection, primer design and evaluation of sequence-verified mutations (Till *et al.*, 2003). The service began in August 2001 and after 5 years of operation had delivered more than 6,700 mutations to the Arabidopsis community (http://tilling.fhcrc.org/arab/status.html). Analysis of the first ~2,000 mutations showed the expected 2:1 Mendelian ratio of heterozygous to homozygous mutations in samples pooled eight-fold. Because heterozygous mutations will be present at half the concentration of homozygous mutations in an eight-fold pool (1/16th vs 1/8th), it is clear that TILLING is sensitive and robust on a production scale (Greene *et al.*, 2003). Since October 2005, ATP has recovered its operating expenses, including materials, labour and overheads, from user fees, which provide a realistic cost estimate for production-scale TILLING. Currently, the ATP charges $1,500 to screen ~3,000 lines, which includes DNA sequencing validation of mutations and a free replacement order should the initial screen fail. Approximately 90% of orders supplied by users are successful, with occasional failures usually attributable to primer design. ATP was used as a model to create maize (http://genome.purdue.edu/maizetilling/) (Till *et al.*, 2004) and Drosophila (Fly-TILL, http://tilling.fhcrc.org/fly/) TILLING services. There are also several independent TILLING services, including a service for lotus (http://www.lotusjaponicus.org/tillingpages/homepage.htm) and barley (http://www.scri.sari.ac.uk/programme1/BarleyTILLING.htm), which further illustrates the value of the method for production-scale reverse genetics.

1.3. Eco-TILLING

The TILLING protocol that we developed for reverse genetics has been adapted to survey natural variation within populations. Our first eco-TILLING project was accidental: we noticed bands in common for some individuals, which we realized were contaminants from a different Arabidopsis accession ("ecotype") from that used in the mutagenesis. A systematic survey of ~200 ecotypes indicated that the method is accurate and robust (Comai *et al.*, 2004). We found that multiple polymorphisms, including SNPs, small indels and variations in satellite repeat number could be discovered within a single target region. Detection of dozens of polymorphisms in a single fragment is possible because only a small fraction of hetero-duplexes are cleaved at any single position by the Cel1 nuclease used for TILLING (Till *et al.*, 2004b; Comai *et al.*, 2004). eco-TILLING can be used to discover and catalogue common polymorphisms without sample pooling, as was done for Arabidopsis and black cottonwood (*Populus trichocarpa*) (Gilchrist *et al.*, 2006). To detect homozygous polymorphisms, an equal amount of DNA from a reference sample is added to each assay well. However, when rare nucleotide polymorphisms are sought, samples can be pooled, as we have recently illustrated in using eco-TILLING to discover rare SNPs in the human genome (Till *et al.*, 2006).

1.4. Other Potential Applications

The protocol described here allows for the low-cost discovery of induced mutations and natural nucleotide polymorphisms on a production scale. The use of standard bench techniques and equipment makes TILLING and eco-TILLING additionally attractive for smaller scale projects. We can also foresee the extension of this protocol to other applications, where the rapid and low-cost discovery of SNPs, indels and other small nucleotide polymorphisms is desired, such as the identification of polymorphisms associated with quantitative traits and the screening for rare mutations in cancer.

Box 22.1: Preparation of celery juice extract

Timing ~10 h

1. Wash ~1 lb (one bunch) of celery in cool water. Remove any leafy material to aid in juicing. Pass the celery through a juicer. Approximately 400 ml of juice should be produced.

CRITICAL Perform all steps at 4°C.

2. Centrifuge the juice for 20 min at 2,600 g to pellet debris. Transfer the supernatant to a new tube.

3. Add 1 M Tris and phenylmethylsulphonyl fluoride (PMSF) to the cleared celery juice so that the final concentration of the solution is 0.1 M Tris-HCl, pH 7.7, 100 μM PMSF.

CAUTION! PMSF is hazardous. Wear gloves; avoid direct contact with skin or inhalation.

4. Bring the supernatant to 25% $(NH_4)_2SO_4$ by adding 144 g l^{-1} of solution. Mix gently at 4°C for 30 min. Centrifuge the supernatant at 13,000–16,200 g at 4°C for 40 min. Discard the pellet.

5. Bring the supernatant from 25 to 80% $(NH_4)_2SO_4$ by adding 390 g l^{-1} of solution. Mix gently at 4°C for 30 min. Spin at 13,000–16,200 g for 1.5 h. Save the pellet and discard the supernatant, being careful in decanting the supernatant.

PAUSE POINT The pellet from Step 5 can be stored at −80 °C for at least 2 weeks.

6. Suspend the pellets in ~ 1/10 the starting volume (e.g. 40 ml) with Tris/KCl/PMSF buffer, ensuring the pellet is completely dissolved.

7. Dialyse thoroughly against Tris/KCl/PMSF buffer at 4°C for 1 h. Use dialysis membrane with a 10,000 kDa cutoff size. Use at least 4 l of buffer per 20 ml of enzyme suspension. It may be easiest to split the suspension into several dialysis tubes. After 1 h, replace buffer with fresh Tris/KCl/PMSF buffer. Repeat two more times for a total of four buffer changes.

PAUSE POINT After two buffer changes, dialysis can be left overnight. Dialysis can proceed as long as 1 week without any apparent loss in enzyme activity. Four hours of dialysis can be taken to be the minimal amount of dialysis required.

8. If desired, clear the dialysed solution by spinning at 10,000 g for 30 min. Greater than 90% of the enzyme activity should be retained in the soluble fraction.

PAUSE POINT Aliquot into small volumes (~1 ml) and store at −80°C. This protein mixture does not require storage in glycerol and remains stable through multiple freeze–thaw cycles. When activity is determined, working stock volumes can be aliquoted (e.g. enough for 1 week) and stored at −20°C. One pound of celery should produce enough enzyme for approximately 500,000 reactions.

1.5. Limitations of TILLING

A successful TILLING project depends on the development of a densely mutagenized population and the preparation of DNA of suitable quality for PCR. Because protocols for chemical mutagenesis in different plant and animal species can vary dramatically, no single protocol can be considered general, and so we do not provide one here. Although many species have been successfully mutagenized, others have been less tractable for reasons that are not clear. In cases where mutagenesis is unfeasible, natural nucleotide diversity can nevertheless be discovered by eco-TILLING. In theory, mutation discovery may be limited in highly heterozygous species where the large number of bands from natural polymorphisms could potentially inhibit the detection of rare induced mutations. When encountering problems in obtaining consistent PCR amplification, alternative extraction methods or additives to the PCR mixture, or both, may be needed to ensure robust PCR (Wilson, 1997).

2. Materials

2.1. Reagents

- FastPrep DNA Kit (MP Biomedicals cat. no. 6540-400).
- Ribonuclease A (Sigma cat. no. R6513-10MG).
- Ammonium sulphate $((NH_4)_2SO_4$; Sigma cat. no. A2939).
- PMSF (Sigma cat. no. P7626-5G) **CAUTION!** PMSF is hazardous. Wear gloves; avoid direct contact with skin or inhalation.

- 1 M PMSF (stock in isopropanol). To prepare an aqueous solution of 100 μM PMSF, add 1 ml 0.1 M PMSF per litre of solution immediately before use.
- Tris/KCl/PMSF buffer: 0.1 M Tris-HCl pH 7.7, 0.5 M KCl and 100 μM PMSF in water.
- TE buffer: 10 mM Tris-HCl pH 7.5 and 1 mM EDTA prepared in water.
- Hot start Ex-Taq polymerase (Takara cat. no. RR006B).
- Ex-Taq buffer (supplied in kit with Ex-Taq polymerase).
- dNTPs (2.5 mM each, supplied in kit with Ex-Taq polymerase).
- $MgCl_2$ (25 mM in water).
- Unlabelled forward and reverse gene-specific primers (see REAGENT SETUP).
- IRDye 700 (forward) and IRDye 800 (reverse) labelled primers of the same sequence as the unlabelled primers (MWG Biotech). **CRITICAL** Limit exposure of IRDye-labelled primers to fluorescent light. Repeated freeze–thaw cycles can reduce fluorescent signal. Make multiple aliquots and store at –80°C. Working stocks can be kept at –20°C for months as long as thawing is limited.
- Celery juice extract (**prepared as described in Box 22.1**)
- 10× digestion buffer: 37.5 ml water, 5 μl 20 mg ml⁻¹ BSA, 100 μl 10% Triton X-100 (vol/vol), 2.5 ml 2 M KCl, 5 ml 1 M HEPES pH 7.5 and 5 ml 1 M $MgSO_4$. Prepare in advance and store aliquots at –20°C.
- Sephadex G-50 medium (Amersham Pharmacia cat. no. 17-0043-02).
- Formamide load buffer: 19.2 ml deionized formamide, 770 μl 0.25 M EDTA pH 8 and ~1 mg bromophenol blue. Store for long periods at –20°C. Working stocks can be held at 4°C for 1 week.
- Isopropyl alcohol (Sigma cat. no. 19030).

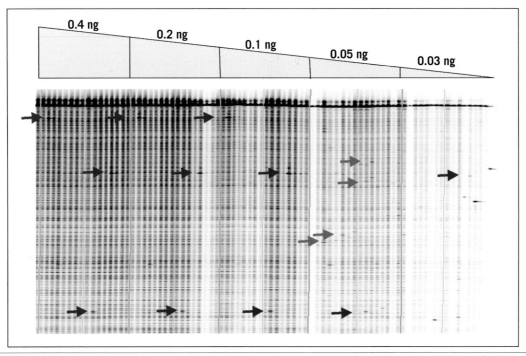

Figure 22.3 Gel image showing the effect of genomic DNA concentration on TILLING data. An LI-COR IRDye700 image shown for a ~1,000-bp Arabidopsis gene target. PCR was performed on 16 unique samples using different DNA concentrations (listed above the gel image). Vertical lines produced by the GelBuddy program bracket every eight lanes (a useful feature when using a 2D pooling strategy). Light blue lines bracket the set of 16 samples at different concentrations. Each sample contains DNA from a pool of eight different Arabidopsis plants. Blue arrows mark known mutations. At DNA concentrations of 0.05 and 0.03 ng, true mutations are lost and false-positive signals begin to appear (examples are marked by red arrows). Loss of positive signals at low concentrations may result from a failure to amplify from the small number of mutant genomic targets in the early rounds of PCR. False-positive signals likely arise from rare random Taq errors occurring on single molecules in the early rounds of PCR. When using low concentrations of genomic DNA, these induced errors represent a sufficient proportion of the total number of amplified molecules, such that they are visible as cleaved products on the LI-COR gel image. Errors of this type are not seen when using higher concentrations of genomic DNA because they are present on only one of thousands of PCR products produced during early rounds of amplification, and are thus beyond the limits of detection. Genomic DNA concentrations of 0.1 or 0.2 ng produce the best signal-to-noise ratio in this gel image. After testing more than five different gene targets, we determined that the target shown here produced a greater-than-average yield of PCR product, and ~0.4 ng of genomic template was chosen as the standard amount of DNA to use in all Arabidopsis PCR reactions. Single target optimization is impractical for high-throughput production.

Box 22.2: Extraction of genomic DNA and concentration normalization

Timming 4 h per 44 samples

1. Add the following components to each supplied reaction tube (prefilled with garnet particles and one ceramic bead).

CRITICAL STEP A variety of methods are available for the extraction of high-quality DNA from plant and animal tissues. We have used the FastPrep kit (MP Biomedicals) for all DNA produced at the Seattle TILLING Project. The method listed below is a modification of the manufacturer's protocol supplied with the kit. Unless stated otherwise, all solutions are supplied in the kit.

Component	Amount
Fresh (or lyophilized) tissue	100 mg (20–50 mg)
Cell lysis solution	800–1000 µl
Protein precipitating solution for plant tissue	200 µl

2. Place on shaker for 45 s at 4.5 m s^{-1}.

3. Centrifuge at room temperature at 14,000 g for up to 30 min.

4. Transfer as much as 800–900 µl to a new 2-ml tube.

5. Add an equal volume of binding matrix. Incubate for 5 min at room temperature with gentle agitation.

6. Centrifuge at 9,000–10,000 g for 3 min at room temperature. Discard the supernatant.

7. Add 500 µl wash buffer and vortex thoroughly to suspend pellet.

8. Centrifuge at room temperature at 9,000–10,000 g for 3 min. Discard the supernatant.

9. Air dry pellets for approximately 20 min.

10. Add 200 µl ultra-pure water and vortex thoroughly to suspend pellet. Incubate for 5 min at room temperature with gentle agitation.

11. Centrifuge at room temperature at 14,000 g for 5 min. Transfer the supernatant to a new microfuge tube. We use a tube with a rubber o-ring to reduce sample evaporation.

12. Add 20 µl of 10× TE buffer containing 3.2 µg ml^{-1} RNase A.

PAUSE POINT Store DNA at 4°C. In our experience, Arabidopsis DNA produced by this method has remained stable for over 4 years when stored in TE buffer at 4°C.

13. Normalize all samples to ensure that each individual will be equally represented in the pool. This can be done by performing agarose gel electrophoresis on aliquots and comparing ethidium bromide-stained fluorescence of the slow-migrating band to standards of known concentrations. The amount to load on a gel should be determined empirically. By using a 1.5% agarose gel, any degradation of the DNA can be detected, which should run as a single band of limiting mobility (**Figure 22.4**). We use software provided with a digital gel documentation (GelDoc) system to determine the DNA concentration of each sample, which we adjust to the desired concentration by adding TE buffer.

Figure 22.4 Examples of genomic DNA quality. High-quality (A), medium-quality (B) and low-quality (C) genomic DNA as assayed by agarose gel electrophoresis and ethidium bromide staining. With adequate quantification, medium-quality genomic DNA can be used to produce high-quality TILLING and eco-TILLING gel data. The use of low-quality genomic DNA can result in low-quality TILLING and eco-TILLING data, likely owing to the inhibitory effect of fragmented DNA on PCR amplification. Fragmented DNA can also lead to inaccurate quantification, which might lead to false-negative errors in pooled samples.

- Ficoll (Fisher cat. no. BP525-100). Prepare a 1% (wt/vol) solution in water. Aliquots can be stored at –20°C for months. Working stocks can be stored at 4°C for up to 2 weeks.
- APS ((NH_4)$_2S_2O_8$) (Sigma cat. no. A3678-25G).
- TEMED (N,N,N',N'-tetra methylethylene diamine) (Sigma cat. no. T9281).
- KB+ 6.5% acrylamide gel matrix (LI-COR cat. no. 827-05607). **CAUTION!** Acrylamide, TEMED and APS are hazardous. Avoid contact with skin and inhalation; wear protective clothing, gloves and goggles.
- 0.8× TBE running buffer: 89.2 g Tris-base, 45.8 g boric acid (H_3BO_3) and 68 ml 0.25 M EDTA pH 8; adjust the final volume to 10 l with water.

2.2. Equipment

- FastPrep Instrument for DNA extraction (MP Biomedicals cat. no. 6001-120) or equivalent upgrade.

- Microcentrifuge (Labnet International cat. no. C0233M-2) or equivalent.
- 20 and 200 μl variable volume multichannel pipettes (Rainin cat. nos. L8-20,L8-200) or equivalent.
- 300 μl variable volume electric multichannel pipette (Rainin, cat. no. E8-300) or equivalent (used for distributing PCR mix into multiwell plates).
- Multichannel electronic pipettor with expandable equal tip spacing (Matrixcat. no. 2139), used for transferring samples from tubes into 96- or 384-well sample plates.
- Vegetable juicer, L'EQUIPE Model 110.5 or equivalent.
- Stir plate (Fisher cat. no. 11-500-49SH).
- Nanopure (water treatment) (VWR (Barnstead) cat. no. 13500-866).
- pH meter (Fisher cat. no. 13-636-AR10).
- Centrifuge 5804 (Brinkman cat. no. 2262250-1) or equivalent.

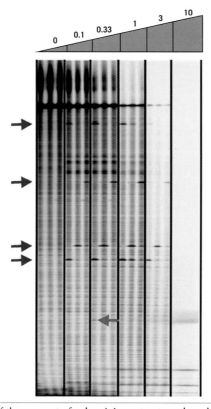

Figure 22.5 Gel image showing the effect of the amount of celery juice extract used to cleave mismatches. An LI-COR IRDye 700 image shown for an ~1,000-bp Arabidopsis gene target. Three samples containing known nucleotide polymorphisms were treated with increasing amounts of celery juice extract (listed above gel image). Units of celery juice activity are listed, with 1 unit defined as the amount providing the best signal-to-noise ratio. Blue arrows mark the location of true polymorphisms. The red arrow marks a band that is present in both image channels at the same position, likely resulting from a spurious mispriming event. Signal and noise decrease with increasing amounts of extract, presumably owing to cleavage of the end label as the duplex "breathes". This figure is modified from Till *et al.,* 2004b.

- Thermal cyclers, for example, dual 384-well GeneAmp PCR system 9700 (Applied Biosystems cat. no. N8050002).
- LI-COR 4300 S DNA Analyser (LI-COR cat. no. 4300-02).
- Vacuum sealer, for example, FoodSaver V900.
- Filter paper (Fisher cat. no. 09-806A).
- 100-tooth membrane combs (Gel Company cat. no. CAJ96).
- 25 cm glass plates, 25 mm spacers, 25 mm gel casting comb, gel rails and other accessories provided with the LI-COR DNA analyser. Additional materials can be purchased from LI-COR or the Gel Company.
- Skirted 96-well plates (Island Scientific cat. no. IS-800ARS).

- Millipore centrifuge alignment frame (Fisher cat. no. MACF09604).
- Millipore multiscreen 96-well separation plate (Fisher cat. no. MAHVN4550).
- Millipore multiscreen column loader (Fisher cat. no. MACL09645).
- Buffer Reservoir (Appogent Discoveries cat. no. 8095).
- 96- or 384-well reaction plates for PCR (Island Scientific cat. no. IS-800ARSor Abgene cat. no. TF-0384).
- Plate sealing adhesive tape (Applied Biosystems cat. no. 4306311).
- Spectra/Por* 7 Dialysis Membrane, 10,000 MWCO (VWR cat.no. 25223-800).
- Spectra/Por* Closures (VWR cat. no. 25224-100).

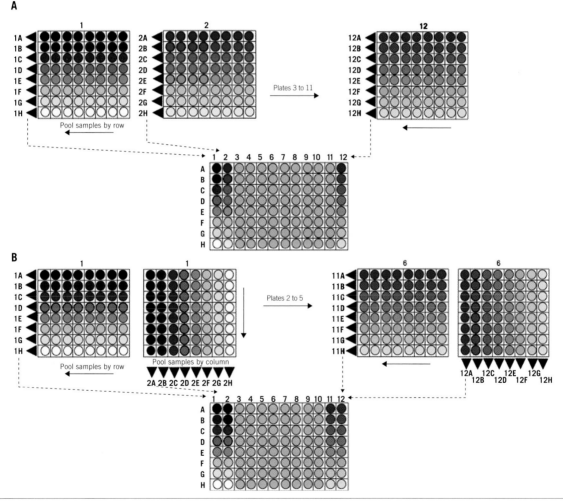

Figure 22.6 Pooling strategies used for TILLING and eco-TILLING. Samples are first arrayed in an 8×8 grid of 64 individual samples. (A) For one dimensional (1D) pooling, all eight samples in a row of the 8×8 plate are pooled together. Using an eight-channel pipettor, all 64 samples from an 8×8 plate can be pooled together with eight pipetting steps. The first 64 samples are deposited in column 1 of the 1D plate. Samples from the second 8×8 plate are deposited in column 2, and so on, for a total of 12 8×8 plates used to make a 1D plate containing 768 unique individuals. (B) For two dimensional (2D) pooling, samples from a common row on an 8×8 plate are first pooled as in (A). Samples from a common column are then pooled and deposited in the adjacent column of the 2D plate. A total of 384 unique samples are deposited in the 96-well 2D pool plate.

Box 22.3: Sample arraying and pooling

Timing 6 h per 768 samples

Arraying individual samples into a multiwell plate prior to pooling can reduce pipetting errors and the time required to construct a plate of pooled samples. We use standard 96-well microtiter plates for arraying because the spacing of the wells is compatible with multichannel pipettors and larger format pipetting devices.

One-dimensional pooling strategy

1. Array sixty-four samples into an 8 × 8 grid on a 96-well plate (leaving the last four columns empty).

2. Pool the samples using an eight channel pipettor; combine samples from each of the eight rows on the plate into a single column of the pooled plate, so that position A1 of the pool plate contains samples A1–8 of individual plate 1.

Using a one-dimensional pooling strategy, 12 pooled columns are produced from 12 plates of 64 individuals (**Figure 22.6A**). This arraying protocol allows the screening of 768 different individuals in a single assay. Tracking a putative mutation from a pool to an individual sample is straightforward. For instance, if a mutation is found in position B2 of the 96-well pool plate, the eight samples that contributed to this pool are found in row B of the second "individuals" plate of 64 samples used to create the pools.

Two-dimensional pooling strategy

1. Fill column one of the pooled plate as described above.

2. Combine all eight samples in a single column of the same individuals plate and deposit those samples into the adjacent column of the 96-well plate of pooled samples. Each individual is then represented in two unique pools.

Two-dimensional eight-fold pooling allows 384 unique samples to be screened per 96-well plate assay (**Figure 22.6B**). When a mutation is discovered, it will appear in two separate gel lanes, the coordinates of which will identify the unique individual harboring the mutation. Additionally, the strategy should reduce false positive errors, as true signals must replicate in the appropriate lanes. For projects where samples are not pooled, it may be useful to replicate samples within the assay to more easily define false negative and false positive signals. Once error rates are clearly defined, sample redundancy can be eliminated and screening throughput increased.

It is advantageous to prepare screening stock plates in a 96-well format. For 384 liquid handling, four 96-well stock plates are combined to create a 384-well assay plate. During Sephadex purification, the 384-well plate array is converted back into four 96-well plate arrays, which facilitates loading the 100 lanes of a LI-COR gel.

Prepare a master stock plate by arraying samples, diluted to the appropriate concentration (see REAGENT SETUP), into 96- or 384-well microtiterplates and pool as desired. Prior to PCR, transfer 5 µl from this plate to the assay plate. Assay plates can be prepared in advance and stored at –20°C for up to one week in a sealed container to limit evaporation. The master stock plate can be stored at 4°C for years. To limit evaporation, the plate is first sealed with adhesive sealing tape, and then vacuum sealed.

- 96- or 384-tip pipetting device (Apricot Designs cat. nos. PP550DS orPP125DS) (optional).
- 96-lane loading tray 1.5 mm spacing (Gel Company cat. no. TAY96) (optional).
- Comb-loading robot (Aviso USA, Ltd, cat. no. 8115-030201) (optional).

2.3. Reagent Setup

2.3.1. Primer Design

Primers should be designed to amplify a single target region. We have found that increasing the melting temperature of primers reduces low molecular weight background noise in TILLING assays, presumably arising from nonspecific primer hybridization. We use the web-based tool CODDLE (http://proweb.org/coddle) to select gene regions that have the highest density of potentially deleterious mutations caused by treatment with chemical mutagens (Till *et al.,* 2003; McCallum *et al.,* 2000). Primers are designed by CODDLE using the Primer 3 program (Rozen and Skaletsky, 2000) with melting temperatures from 67 to 73°C and of a length in the range of 20–30 nucleotides. A 100-µM solution of each primer is prepared in TE buffer and aliquots are stored at –80°C to avoid repeated freeze–thaw cycles that could reduce fluorescent activity.

2.3.2. Optimizing the Amount of Genomic DNA for PCR

Timing ~10 h

After DNA extraction and concentration normalization (**Box 22.2**), the optimal amount of genomic DNA to use for PCR should be determined empirically, by following the standard TILLING protocol with varying amounts of genomic DNA. For most organisms, we have determined the optimal amount of genomic DNA by using a molar ratio between primer and genomic DNA similar to that originally determined to be optimal for Arabidopsis. For example, 0.375 ng of Drosophila genomic DNA is used per PCR reaction. For maize, with a genome size approximately 14 times that of Drosophila, 4.5 ng of genomic DNA is used per amplification reaction. A concentration is chosen that provides the best signal-to-noise ratio when using a fixed amount of enzyme for hetero-duplex digestion (**Figure 22.3**). A higher concentration may be required when using lower quality genomic DNA samples (**Figure 22.4**).

2.3.3. Optimizing the Amount of Single-Strand-Specific Nuclease

Timing ~10 h

The optimal amount of single-strand-specific nuclease for hetero-duplex digestion should be determined empirically. We have found that the activity in celery juice extract remains stable for more than 1 year at −80°C, and therefore use the same amount of enzyme for all TILLING and eco-TILLING reactions performed in our laboratory. A celery juice extraction protocol is provided in **Box 22.1**. To determine the optimal amount of enzyme, perform TILLING reactions with varying amounts of enzyme and choose the amount that provides the best signal-to-noise ratio. We have measured similar activities from many independent celery juice extractions, and so the amount used in this protocol can serve as a guide. Using too little enzyme results in a very dark image where true bands are difficult to discover above the background of PCR failure products. Adding too much enzyme results in a blank image, presumably because the enzyme cleaves off the IRDye label from the 5′ DNA ends when the duplex "breathes" (**Figure 22.5**).

2.3.4. Determining the Number of Samples to Pool

Timing ~10 h

The number of samples to pool together before a screen can be determined empirically by testing multiple mutations or polymorphisms at different levels of sample pooling. For TILLING, only a small percentage of individuals will harbour a mutation in the PCR amplicon, so that doubling the number of individuals in a pool will

provide nearly a two-fold increase in the throughput of mutation discovery. For eco-TILLING, the expected frequency of nucleotide polymorphism will influence the level of desired pooling. When cataloguing common nucleotide changes, pooling is not appropriate because of the additional work needed to determine which individuals in a pool harbour the polymorphism (Comai *et al.*, 2004; Gilchrist *et al.*, 2006). However, for rare polymorphisms, pooling reduces effort but will obscure the identification of very common polymorphisms. For example, we discovered rare human SNPs much more efficiently in eight-fold pools than in unpooled samples, but common SNPs were in many pools, making the task of identifying the individuals harbouring very common SNPs more time consuming than if screening were performed on unpooled samples (Till *et al.*, 2006). Note that if sample pooling is not used, homozygous nucleotide changes will be discovered only if reference DNA is added to each sample. **See Box 22.3 and Figure 22.6** for recommended pooling strategies.

2.3.5. Preparation of Ladders and Lane Markers

Timing 3 h 30 min

IRDye ladders are commercially available; however, assay cost can be reduced by preparing custom ladders and lane markers. Design primers to produce PCR products of desired length. Follow the PCR amplification protocol (**Steps 1–6**) using genomic DNA of the corresponding species. After PCR amplification, purify the product through a Sephadex G50 spin plate, without the addition of formamide load dye or lane marker to the catch plate, and omitting the sample volume reduction step described in Step 19. Combine sample wells into a single tube and store at −20°C. This is the concentrated stock. Determine the amount to use empirically, typically ~0.13 ng per lane. Prepare a 200-bp fragment as a lane marker. To create a ladder, combine fragments of different molecular weights. This ladder should include a 700-bp fragment if using GelBuddy for image analysis.

2.4. Equipment Setup

2.4.1. Preparation of Sephadex Spin Plates

Timing ~1 h 5 min

Fill the wells of the Millipore column loader with dry Sephadex G-50 powder. Remove excess powder with

Box 22.4: PCR amplification program

Loop number	Cycle number	Denature	Anneal	Ramp	Extend	Hold
	1	95°C, 2 min				
1	2–9	94°C for 20 s	73°C for 30 s, reduce temperature at 1° C per cycle	72°C at 0.5 °C per s	72°C for 1 min	
2	10–54	94°C for 20 s	65°C for 30 s	72°C at 0.5 1C per s	72°C for 1 min	
	55				72°C for 5 min	
	56	99°C for 10 min				
3	57–126		70°C for 20 s, reduce temperature at 0.3°C per cycle			
	127					8°C

the supplied scraper. Lay a clean Millipore 96-well separation plate on top of the filled loader. Invert the plate and loader to deposit Sephadex into the plate. To ensure that all Sephadex enters the plate, tap the back of the loader with scissors. Remove the loader and clean off any excess Sephadex powder on top of the plate with a fine paintbrush. To hydrate the Sephadex, add 300 µl water to each well and let stand for 1 h before use. Plates can be stored for up to 1 week at 4°C in a sealed plastic bag containing a moist towel to prevent dehydration. Before use, gently rinse the bottom of the plate to remove excess Sephadex. When sample purification is complete, store the used plate for approximately 3 days until the Sephadex has dehydrated. Discard the Sephadex, rinse the plate thoroughly with water and allow it to completely dry before reuse (1–3 days depending on room temperature and humidity). Plates can be reused at least six times.

2.4.2. Preparation of Acrylamide Gels

Timing 2 h

Clean glass plates and assemble spacers and rails following the protocol provided with the LI-COR DNA analyser. Combine 20 ml of 6.5% acrylamide gel matrix, 15 µl TEMED and 150 µl 10% APS. Use a plastic syringe to deposit the acrylamide solution between glass plates. When done, insert the casting comb and apply the pressure plate. Pour excess acrylamide solution onto comb. Let it stand for 90 min before use. To store gels, apply a damp paper towel to each end of the gel and cover in plastic wrap. Gels can be stored for up to 1 week at 4°C. Before use, thoroughly wash both the front and back plate, remove the casting comb and remove all acrylamide from the well. Before placing the plates in the LI-COR DNA analyser, clean them with isopropanol, making sure that the laser detection region is clean. After the run is complete, the samples will have passed through the gel and new samples can be applied and the gel run again. Only two runs are suggested because the sample image becomes diffuse and hard to interpret during the third run. **CAUTION!** Acrylamide, TEMED and APS are hazardous. Avoid contact with skin and inhalation; wear protective clothing, gloves and goggles.

2.4.3. Reusing Membrane Combs

Timing ~40 min

A membrane comb can be reused many times until its teeth become bent and the comb is no longer easy to insert into the gel well. Clean a comb by soaking in a tub of deionized water for at least 30 min. If many combs are washed at once, care should be taken to avoid comb damage, and water should be exchanged several times to ensure the combs are thoroughly cleaned. Air-dry combs for several days before reusing.

3. Procedure

3.1. PCR Amplification of Target Regions and Hetero-duplex Formation

Timing ~3 h 30 min
1. Prepare genomic DNA as outlined in **Box 22.2** or by any other appropriate method. Pool samples and prepare assay plates containing 5 µl of genomic DNA per well, as detailed in **Box 22.3**. These can be prepared

fresh, or prepared in advance and frozen until required.

2. Make a fresh mixture of unlabelled and labelled primers. Mix 100 μM stock primers at a ratio of 3:2:4:1 (IRDye 700primer: unlabelled forward primer: IRDye 800 primer: unlabelled reverse primer). We do not store primers in mixed form, and therefore only make enough primer mixture for the number of reactions we are performing. To minimize pipetting errors, the minimal volume of primer mixture we prepare is 10 μl (enough for two 96-well reaction plates).

CRITICAL STEP Limit exposure of IRDye-labelled primers to fluorescent light.

3. Prepare the 96-well PCR master mix as outlined below (extra volume is made to avoid pipetting errors), adding Taq enzyme last. Vortex thoroughly.

Component	Amount (μl)
Water	360
25 mM MgCl$_2$	68
10× Ex-Taq buffer	57
2.5 mM each dNTP (the stock solution contains all four dNTPs)	92
Primer mix	4
5 U μl^{-1} hot start Ex-Taq	6
Total	587

CRITICAL STEP IRDye-labelled primers are sensitive to fluorescent light. When possible, lighting should be dimmed and samples shielded from direct light. Keep Taq polymerase on ice. The master mix components can be prepared at room temperature when using a hot-start Taq.

4. Add 5 μl of PCR master mix to each well of the assay plate and seal the plate with adhesive tape.
5. Spin the plate for 2 min at 1,300 g at room temperature (approximately 19–25°C).
6. Place the plate in the thermal cycler and run the amplification programme in **Box 22.4**.

PAUSE POINT When cycling is complete, the plate can be stored at –20 °C for weeks. Note that fluorescent signal can decay over time and it is best to process the assay plate as soon as possible.

CRITICAL STEP Alternatives: When screening a small number of samples, the cost of end-labelled IRDye primers can become prohibitive. Several universal primer labelling strategies have been described that can reduce primer cost (Wienholds *et al.*, 2003; Winkler *et al.*, 2005; Till *et al.*, 2006).

3.2. Hetero-duplex Digestion

Timing ~40 min

7. Prepare the celery juice extract reaction mix, as outlined below, and store on ice. The volumes indicated are sufficient for two 96-well plates. Details of how to prepare celery juice extract are provided in **Box 22.1**.

CRITICAL STEP We have tested a variety of enzymes and enzyme preparations for TILLING and eco-TILLING in addition to celery juice extract. These include commercially available Surveyor nuclease, mung bean nuclease and S1 nuclease (Till *et al.*, 2004b). When using commercial sources of S1 or mung bean nucleases, do not use the accompanying buffers because these are optimized for different applications.

CRITICAL STEP The nuclease concentration in each batch of celery juice extract must be determined empirically (see REAGENT SETUP). The amount listed here is for illustration purposes only.

Component	Amount
Water	4.8 ml
10× digestion buffer	840 μl
Celery juice extract	~7 μl
Total	~5647 μl

8. Spin the assay plate for 2 min at 1,300 g at room temperature.
9. Add 20 μl of celery juice extract reaction mix to each well of the plate. Mixing is optional, but is recommended when practical (e.g., using a multi-tip pipetting device).
10. Spin the assay plate for 2 min at 1,300 g at room temperature.
11. Incubate the plate for 15 min at 45°C.

12. Spin the assay plate for 2 min at 1,300 g at room temperature.
13. Stop the reaction by adding 5 µl of 0.225 M EDTA to each well of the assay plate.

PAUSE POINT The assay plate can be stored at −20 °C for weeks. Note that the fluorescent signal can decay over time and it is best to process the assay plate as soon as possible.

3.3. Sample Purification

Timing ~1 h
14. Prepare one catch plate for each 96-well reaction plate before performing Sephadex purification. Add 1.5 µl of formamide load buffer to each well of the catch plate. Add ~2 µl of 200-bp lane marker to the plate such that marker is deposited every eight lanes beginning with lane 4 (e.g. 4, 12, 20 and so on).

CRITICAL STEP Alternatives: In theory, any method that provides both salt and buffer removal in addition to volume reduction can be used in place of the Sephadex G50 column. A less expensive alternative would be alcohol precipitation (Draper *et al.*, 2004). When testing options, consider the time required, the percentage of sample recovery and the material cost of the assay.

15. Assemble a Sephadex G50 plate, an alignment frame and an empty 96-well waste plate to catch water. Centrifuge the assembly for 2 min at 440 g at room temperature.
16. Remove the waste plate and attach a sample catch plate containing formamide load dye and lane markers.
17. Apply the full sample volume from the assay plate onto the 96-well Sephadex plate. If using a 384-well assay plate, each quadrant of the plate is applied to a separate Sephadex plate. This is necessary because a maximum of 100 sample lanes can be loaded on an LI-COR gel.

CRITICAL STEP Apply the samples directly over the centres of the packed Sephadex columns and do not touch the columns with pipette tips. Make sure that the sample catch plate is properly labelled and that the samples maintain the same orientation between plates (position A1 of assay plate equals A1, and not H12, of sample catch plate).

18. Spin the Sephadex plate/catch plate assembly for 2 min at 440 g at room temperature.
19. Remove the catch plate containing the DNA solutions. Reduce sample volumes by incubating the plate for 45 min at 85°C, or until only 1.5 µl remains in each well.

PAUSE POINT After volume reduction, the plate can be stored at 4°C for several weeks. Note that fluorescent signal can decay over time and it is best to process the assay plate as soon as possible.

3.4. Loading Samples Onto 100-Tooth Membrane Combs

Timing 10–40 min
20. Heat the sample catch plate for 5 min at 90°C.
21. Store the sample on ice for at least 3 min before loading.
22. Apply ~0.5 µl of sample per membrane comb tooth starting with tooth number 4 and ending with tooth number 99. The bottom 1/3 of the tooth should be stained blue after loading.
23. Apply ~0.25 µl of molecular weight ladder to teeth 1, 3 and 100. This asymmetry ensures that sample 1 is never confused with sample 96 in the event that the comb is inverted before application to the gel. There are several options for loading samples onto membrane combs. The sample can be applied directly using a single-channel pipettor, can be preloaded into a comb-loading tray or the sample can be loaded directly using a comb-loading robot. Although the comb-loading robot is advantageous in that human error is reduced and loading is automated, using a comb-loading tray and a multichannel pipettor provides an inexpensive alternative.

PAUSE POINT The loaded membrane comb can be stored for an extended period of time before loading (at least 24 h). As the fluorescent signal can decay over time, it is advised to apply the comb to the gel within 2 h of loading.

CRITICAL STEP Alternatives: Samples can be loaded directly onto the gel using a shark's tooth or a square tooth comb. Although 96-tooth shark's tooth combs are available, loading can be tricky. Loading fewer than 96 samples on a single LI-COR gel reduces throughput without providing any clear benefit for data analysis.

Table 22.1: Troubleshooting table

Step	Symptom[a]	Probable causes	Tests and solutions
45	Weak or absent IRD	PCR amplification failure	Start by running some undigested PCR product on an agarose gel. If you observe less than ~5 ng μl^{-1} PCR product, inefficient PCR is likely the cause. Try alternative primers (Step 2) and test PCR solutions (Step 3). Check cycling conditions (Step 6). Validate the quality of the genomic DNA template (**Box 22.2**)
		Improper storage of IRD primers by (REAGENT SETUP, Step 2)	If adequate PCR product is observed by agarose gel electrophoretic analysis, load an aliquot of undigested PCR product on an LI-COR gel (~0.5 μl of unpurified product from the 10 μl PCR reaction is usually sufficient). If you can detect no signal, the fluorescent primers may have lost activity.
		LI-COR analyzer failure (Step 32)	If you can detect no signal, the fluorescent primers may have lost activity. If the ladders and lane markers appear, you can rule out an LI-COR gel analyzer failure
		DNA over-digestion	If adequate signal of undigested PCR product is observed on the LI-COR gel image, try varying the amount of single-strand-specific nuclease (Step 7) to determine if over-digestion is the problem. Make fresh EDTA stock. DNA will be over-digested if enzymatic reaction is not stopped (Step 13)
	Only the IRDye 800 signal is weak or missing	Unknown, possibilities include a failure during primer synthesis, or ambient laboratory conditions such as elevated levels of ozone	This is a common problem experienced at our facility. We have yet to obtain conclusive results indicating the cause of signal loss specific to the IRDye 800-channel image
			Perform reactions with primers giving a weak IRDye 800 signal in parallel with primers known to give a strong IRDye 800 signal. Record laboratory conditions such as temperature, humidity and ozone levels to determine if there is a correlation between laboratory conditions and signal loss
	IRD signal too strong	Too much sample loaded on the gel (Step 22)	Perform TILLING/eco-TILLING with positive controls and vary the amount of sample loaded on the gel (Step 22). If cleaved products are not present on the gel, perform an enzyme titration (Step 7) with new buffers and solutions, checking that proper incubation times and temperatures are followed
		Single-strand-specific nuclease for hetero-duplex digestion has reduced activity (Step 7)	
	Gel image smeary	Salts not removed from sample (Step 18)	If using Sephadex for sample purification, check centrifugation times and speed. Try purification using a new membrane plate. Perform the Sephadex purification step twice on samples. Try an alternative purification such as alcohol precipitation
		Sample not denatured before or during PAGE run (Steps 14, 20 and 32)	Check electrophoresis conditions. Perform reactions with positive controls using new reagents (including formamide buffer)
		PCR product degradation (Steps 7 and 14)	Skip hetero-duplex digestion with a subset of samples. Run these samples on the same gel as digested products. This tests for degradation during hetero-duplex digestion. A likely cause of sample degradation is the formamide load buffer

[a] All symptoms are observed during gel analysis.

3.5. Loading and Running Gels

Timing ~20 min for pre-run and comb application, 4 h 15 min for gel running

24. Remove a pre-poured gel from 4°C storage.
25. Clean the gel well and glass and insert the plate assembly into the LI-COR DNA analyser. Fill the upper and lower buffer tanks with 0.8× TBE buffer.
26. Pre-run the gel for 20 min, ensuring that the lasers focus properly (40 mA, 40 W, 1,500 V, 50°C, image width 1,028 and scan speed 2).
27. Rinse the gel well thoroughly using a plastic 10-ml syringe with no needle attached.
28. Remove buffer from the upper tank. Remove excess liquid from the gel well by inserting filter paper into the gel well.
29. Remove the filter paper and add 1 ml Ficoll solution to the gel well. A small bead should form at the edge of the front glass plate.

30. Insert the loaded membrane comb into the gel well at a 45° angle. Once the teeth touch the Ficoll solution, the comb should slide smoothly into the well. The comb should be gently inserted into the well and the teeth should not be forced into the acrylamide.

31. Gently fill the upper buffer tank with 0.8× TBE buffer and replace the top electrode.

32. Start a 4 h 15 min gel run (40 mA, 40 W, 1,500 V, 50°C, image width 1,028 and scan speed 2). Optional: After 10 min, remove the comb and gently rinse the Ficoll out of the gel well. Re-insert the top electrode and close the LI-COR door. The 4,200-series machines will automatically restart. If using a 4,300-series machine, the run will remain paused until the continue option is selected. We typically start runs at the end of the day and leave the machines running unattended.

CRITICAL STEP Excessive washing of the well can disturb the gel and result in lower quality gel data.

3.6. LI-COR Gel Analysis

Timing 10 min–1 h

33. Download the most recent version of GelBuddy from http://www.proweb.org/gelbuddy/. Installation instructions, technical documentation and descriptions of additional features are provided on the GelBuddy website.

34. Download IRDye 700 and IRDye 800 TIFF images from the LI-COR to your desktop.

35. Open GelBuddy. Import images by using the "File/Open 700 and 800 Channel Images" menu command.

36. Adjust the 700-channel image so that background bands become visible, using the slider bars at the top of the GelBuddy window.

37. Click the "Select Channel" toolbar button at the top of the window to view the 800-channel image. (Each button may be identified by the text that appears when the mouse cursor is placed over the button.) Adjust the 800-channel image as in **Step 36**.

38. Call lanes. Click the "Find Lanes" button. Set the number of sample lanes in the "Find Lanes" pop-up window (96 for a standard TILLING/eco-TILLING run). Unless one of the channels is very weak, use both channels for detecting lanes. After clicking the "OK" button, automatically generated lane tracks will be superimposed on the displayed gel image.

39. Edit lanes. Blue lane tracks should run through the 200-bp marker band found in every eighth lane starting with lane number 4. If the blue lane tracks do not coincide with the 200-bp markers, or if any lane tracks deviate from the sample lanes in the image, click the "Edit Lanes Mode" button. Click the mouse over the lane you wish to edit or delete, or over the lane adjacent to the area where you wish to add lanes. Use the "Edit/Insert Lanes" or "Edit/Delete Lane" menu command to insert or delete lanes. To change the path of a single lane track, drag the boxes on the selected lane horizontally to the desired location. When finished, click the "Show Lanes" button to hide the lane tracks.

40. Adjust the fragment length calibration ladder. The ladder is adjusted by dragging markers at the left margin of the image. Click the "Show Calibration Information" button. Curves representing the calibration ladder will be superimposed on the gel image. Place the mouse over the blue number 700 and drag the number vertically to place the corresponding calibration curve over the 700-bp-size standard band. Do the same with number 200, making sure the corresponding curve overlaps the 200-bp lane marker bands. Set the migration limit markers by dragging the red numbers to the bottom of the signal on the gel image (100%) and to the top of the full-length product (0%). When complete, click the "Show Calibration Information" button again to hide the calibration ladder.

41. Mark mutations. GelBuddy provides automatic (A) and manual (B) band marking tools. The automatic band marking tool is most appropriate for analysis of eco-TILLING images containing many cleavage product bands. The manual band marking tool can be used as the sole means to mark an image containing few cleavage products or to edit an automatically generated markup.

(A) Automatic band marking (optional)

42. Click the "Analyze" button. The "Edit Gel Information" pop-up window will appear.

43. Enter the size of the full-length product and click "OK". A log window will appear showing the progress of automated band detection. The total analysis time will appear when analysis is complete.

44. Close the log window to view the marked image.

(B) Manual band marking

45. Click the "Record Signals Mode" button. If you have not yet specified the size of the full-length product, you will be prompted to do so at this time. Enter the size of the full-length product and click "OK". Use the "Select Channel" button to switch between channels.

46. Click the mouse over a cleavage product band to mark it. Each mark in the 700 channel is indicated by a red box and each mark in the 800 channel is indicated by a blue box. To delete a mark, select the channel containing the marked band, place the cursor over the box and click the mouse while pressing the "Option" key (on Macintosh OS X) or "Alt" key (on Windows). If you are unsure of a mutation, note that when the mouse is over an image band, the inferred size of the cleavage product is indicated at the bottom of the window. For each polymorphism, the inferred cleavage product sizes for the blue and red boxes should sum to approximately the size of the full-length product.

47. Edit signal groups. GelBuddy automatically links comigrating bands to facilitate assignment of samples to haplotype groups in eco-TILLING applications. If you wish to review and edit grouping of comigrating bands, click the "Edit Signal Groups Mode" button. Each group will appear as a set of coloured boxes linked by horizontal lines. To select multiple bands, click on the first band you wish to select, then click on additional bands while pressing the "Shift" key. The "Edit/Group Selected Signals", "Edit/Merge Selected Signal Groups" and "Edit/Remove Selected Signals From Group" menu commands may then be used to alter signal grouping.

48. For two-dimensional (2D) pooling applications, the "Options/Show Pool Boundaries" menu command may be used to display lines delineating sets of row and column pools (vertical lines in **Figure 22.3**).

49. View report. A text report containing a list of paired 700- and 800-channel bands, lists of comigrating bands and other information may be viewed by selecting the "Report/View Concise Report" menu command. Use the "Report/Save Concise Report" command to save this report for future reference.

50. Save the gel markup file using the "File/Save Gel Markup" menu command. The "File/Open Gel Markup" command may be used during subsequent GelBuddy sessions to load the markup file for review or revision.

4. Troubleshooting

Timing

Approximate times for the major reagent preparation steps and the high-throughput screening procedure are listed in **Figure 22.1**.

Troubleshooting

Troubleshooting advice can be found in **Table 22.1**.

5. Anticipated Results

True polymorphisms will produce two cleaved fragments, one in the IRDye 700–channel image and another in the IRDye 800-channel image whose molecular weights add up to the molecular weight of the full-length PCR product. The full-length PCR product should produce a sharp and intense band toward the top of the gel image. Background bands arising from incomplete PCR should be of lower intensity (**Figure 22.2**). Assays can be optimized to increase the signal-to-noise ratio by altering the amount of genomic DNA used in PCR and the amount of enzyme used for mismatch cleavage (**Figures 22.3 and 22.5**). Bands that are scored on TILLING and eco-TILLING gels should represent true nucleotide polymorphisms. This can be verified by sequencing the putative polymorphic individual.

6. Acknowledgements

We thank the past members of the TILLING team for their input on protocol improvement and development. Steve Reynolds, Kim Young and Rob Laport developed modifications for extracting DNA using the FastPrep kit. We thank the current TILLING team – Jennifer Cooper, Christine Codomo, Aaron Holm, Margaret Darlow and Lindsay Soetaert – for their continued input on data quality and protocol improvement. We thank Takehito Furuyama and other members of the Henikoff lab for helpful discussions on the effect of low amounts of genomic DNA on PCR amplification and enzymatic mismatch cleavage. We thank Elizabeth Greene, Samson Kwong and Jorja Henikoff for developing the computational and informatics tools required to run a high-throughput production laboratory. This work was supported by grants 0234960 and 007777 from the National Science Foundation.

Published online at http://www.natureprotocols.com
Rights and permissions information is available online at http://npg.nature.com/reprintsandpermissions

7. References

Bentley, A., MacLennan, B., Calvo, J. et al. 2000. Targeted recovery of mutations in *Drosophila*. *Genetics*. 156: 1169–1173.

Caldwell, D.G. et al. 2004. A structured mutant population for forward and reverse genetics in Barley (*Hordeum vulgare* L.). *Plant J.* 40: 143–150.

Colbert, T. et al. 2001. High-throughput screening for induced point mutations. *Plant Physiol.* 126: 480–484.

Comai, L. et al. 2004. Efficient discovery of DNA polymorphisms in natural populations by Eco-TILLING. *Plant J.* 37: 778–786.

Draper, B.W., McCallum, C.M., Stout, J.L. et al. 2004. A high throughput method for identifying N-ethyl-N-nitrosourea (ENU)-induced point mutations in zebrafish. *Methods Cell Biol.* 77: 91–112.

Gilchrist, E.J. et al. 2006. Use of Eco-TILLING as an efficient SNP discovery tool to survey genetic variation in wild populations of *Populus trichocarpa*. *Mol. Ecol.* 15: 1367–1378.

Greene, E.A. et al. 2003. Spectrum of chemically induced mutations from a large-scale reverse-genetic screen in Arabidopsis. *Genetics*. 164: 731–740.

McCallum, C.M., Comai, L., Greene, E.A. et al. 2000. Targeting induced local lesions IN genomes (TILLING) for plant functional genomics. *Plant Physiol.* 123: 439–442.

McCallum, C.M., Comai, L., Greene, E.A. et al. 2000. Targeted screening for induced mutations. *Nat. Biotechnol.* 18: 455–457.

Perry, J.A. et al. 2003. A TILLING reverse genetics tool and a web-accessible collection of mutants of the legume *Lotus japonicus*. *Plant Physiol.* 131: 866–871.

Rozen, S. and Skaletsky, H. 2000. Primer3 on the WWW for general users and for biologist programmers. *Methods Mol. Biol.* 132: 365–386.

Slade, A.J., Fuerstenberg, S.I., Loeffler, D. et al. 2005. A reverse genetic, non transgenic approach to wheat crop improvement by TILLING. *Nat. Biotechnol.* 23: 75–81.

Till, B.J. et al. 2003. Large-scale discovery of induced point mutations with high throughput TILLING. *Genome Res.* 13: 524–530.

Till, B.J. et al. 2004a. Discovery of induced point mutations in maize genes by TILLING. *BMC Plant Biol.* 4: 12.

Till, B.J., Burtner, C., Comai, L. et al. 2004b. Mismatch cleavage by single-strand specific nucleases. *Nucleic Acids Res.* 32: 2632–2641.

Till, B.J. et al. 2006. High-throughput discovery of rare human nucleotide polymorphisms by Eco-TILLING. *Nucleic Acids Res.* 34: e99.

Wienholds, E. and Plasterk, R.H. 2004. Target-selected gene inactivation in zebrafish. *Methods Cell Biol.* 77:69–90.

Wienholds, E., Schulte-Merker, S., Walderich, B. et al. 2002. Target-selected inactivation of the zebrafish rag1 gene. *Science* 297: 99–102.

Wienholds, E. et al. 2003. Efficient target-selected mutagenesis in zebrafish. *Genome Res.* 13: 2700–2707.

Wilson, I.G. 1997. Inhibition and facilitation of nucleic acid amplification. *Appl. Environ. Microbiol.* 63:3741–3751.

Winkler, S. et al. 2005. Target-selected mutant screen by TILLING in *Drosophila*. *Genome Res.* 15: 718–723.

Zerr, T. and Henikoff, S. 2005. Automated band mapping in electrophoretic gel images using background information. *Nucleic Acids Res.* 33: 2806–2812.

Applications of DNA Marker Techniques in Plant Mutation Research

D.Wu[a], Q.Y.Shu[b] and C.Li[c],*

[a] State Key Lab of Rice Biology and Key Lab of the Ministry of Agriculture for Nuclear-Agricultural Sciences, Zhejiang University, Hangzhou 310029, P.R. China
[b] Joint FAO/IAEA Division of Nuclear Techniques in Food and Agriculture, International Atomic Energy Agency, Wagramer Strasse 5, P.O. Box 100, A-1400 Vienna, Austria
Present: Institute of Nuclear Agricultural Sciences, Zhejiang University, Hangzhou 310029, China
[c] Department of Agriculture, Government of Western Australia, 3 Baron-Hay Court, South Perth, WA6151, Australia
* Corresponding author, E-MAIL: chengdao.li@agric.wa.gov.au

1. Introduction

DNA-based molecular markers are polymorphic DNAs that may be anonymous or identified at specific locations within genome. They have been widely used for genotyping and diversity analysis of germplasm, mapping and tagging of genes. In the last two decades, DNA markers (often termed molecular markers) have been modified constantly to enhance their utility and to bring about automation and high-throughput in the process of genome analysis. A large number of economically important traits have been tagged using genetic markers in various crops. These developments have given new dimensions for plant breeding through marker-assisted selection (MAS) to develop better varieties faster.

In comparison to morphological markers, DNA markers can be used for evaluating multiple traits simultaneously, at any developmental stages, using DNA from all kinds of plant organ and tissues, and independent of the influence of environmental conditions. Since DNA can be analysed prior to sowing/planting in the field, genotyping (especially large scale screening) can be fitted into the off-season and selection made prior to field sowing/planting, e.g. in the case of seed genotyping. Therefore, DNA marker analysis is efficient in time and space, and thereby costs, and can accelerate the breeding processes by allowing rapid improvement of traits, particularly those that are difficult to improve by conventional methods. In a breeding context DNA markers become especially powerful if they are combined with pedigree information, thus markers and traits can be traced through pedigrees to discover the source of the initial variation/introduction and its retention in successful breeding lines and varieties. Genetic markers are also powerful when combined with other biotechnologies such as doubled haploid in producing homozygous lines carrying the trait (mutation) of interest.

DNA marker techniques can also be used widely in plant mutation breeding and genetic research for increasing both the efficiency and efficacy of mutation techniques. They can be used for tracing the pedigree of induced mutants and tagging important mutations. Consequently, closely linked markers of mutant traits can be used for MAS, pyramiding and cloning of mutant genes. However, the way they are used is somewhat different from conventional plant breeding programmes. Some biochemical markers such as proteins

and enzymes can be used as efficient markers in crop improvement, for example the grain storage proteins of wheat that contribute to quality characteristics and mutations in these genes can be screened for by altered protein profiles. Genetic markers are also important in the identification of radiation-facilitated chromosome translocations (**see Chapter 19**) .

2. Types and Characteristics of DNA Markers

2.1. Types of Molecular Markers

There is a wide assortment of genetic differences that can be revealed using various types of DNA analytical tools. The most common genetic variations in plant genomes involve: single nucleotide polymorphisms (SNPs), repetitive sequences, inversions and insertion/deletions (indels). Various technologies and strategies have been developed to uncover and utilize these genetic differences, which can be classified into four categories (**Box 23.1**).

Different DNA markers might have significantly different technical requirements, particularly for DNA quantity and quality, reproducibility and cost (**Table 23.1**). They may also differ in development and set up costs and amenability to automation, which is becoming more and more important for high-throughput genotyping. On the other hand, technological developments have dramatically increased the ability and efficiency of detecting DNA polymorphisms. For example, various technologies have been developed to detect SNPs (**see Chapter 20**), small indels can now be detected easily using sophisticated machines such as multi-capillary electrophoresis systems. Increasingly, the various processes are integrated to provide a seamless process from DNA extraction to DNA analysis to data capture.

2.2. Characteristics of Molecular Markers

Different DNA markers have different characteristics. Some DNA markers reveal genetic variation solely at a single position whereas others reveal variation simultaneously at multiple loci (**Table 23.1**). Certain DNA markers are co-dominant, allowing both alleles at a locus to be distinguished, while others are dominant, only indicating presence or absence of the marker;

Box 23.1: Classification of DNA markers based on the development and type of genetic variation

Genomic DNA restriction site-based markers

RFLP: Restricted fragment length polymorphisms. This is one of the earliest type of DNA marker. DNA is cleaved with restriction enzymes that recognize specific sequences in the DNA, the resulting fragments are separated by size using electrophoresis. Probes that are generated from specific regions of the genome are used to identify fragments corresponding to the target sequence.

PCR-based markers

RAPD: Randomly amplified polymorphic DNAs. Sections of DNA are randomly amplified from different regions of the genome using PCR primers of approximately 10 base pairs (bp) of arbitrary sequence to generate a range of markers, normally visualized as banding profiles in gels after electrophoresis.

SSR: Simple sequence repeats, also known as microsatellites. SSRs are tandemly repeated nucleotides of 1–4 bp in size, e.g. mono-nucleotide repeats, (A)n, (T)n, (G)n and (C)n; bi-nucleotide repeats such as (CT)n, (AG)n; tri-nucleotide repeats such as (TCT)n, (AGA)n, etc. Polymorphism is based on differences in the number of repeats (length of repeated sequence).

EST-SSR: SSR located in expressed sequence tags (ESTs). ESTs are directly related to gene expression and EST libraries can be developed for specific plant tissues (e.g. root ESTs) and responses to specific environments (e.g. heat stress). EST-SSRs are especially useful for comparative mapping and evolutionary studies because they have a high likelihood of being syntenic between species.

Indel: DNA insertion/deletion. Indels are polymorphisms of short insertions and deletions, which can be spread across the genome, and are therefore recognized as an abundant source of genetic markers, though not as common as SNPs. In most cases, indels are referred to as non-repetitive sequences.

ISSR: Inter simple sequence repeat. As the name implies ISSRs are genomic regions located between SSR sequences. These regions can be amplified by PCR using simple repeat sequences as primers. Various SSR sequences can be anchored at the 3′-end of the primer to increase their specificity (usually to reduce the number of bands on a gel to a manageable number). ISSRs are mostly dominant markers, though occasionally a few may exhibit co-dominance. An unlimited number of primers can be synthesized for various combinations of di-, tri-, tetra- and penta-nucleotides.

SNP: Single nucleotide polymorphisms. SNPs are single base changes in the DNA sequence that are generally bi-allelic at any particular site. SNPs can be revealed through various approaches (see Chapter 20), but sequencing methods are rapidly being developed that aid SNP detection.

SRAP: Sequence-related amplified polymorphism. SRAP is based on two-primer amplification. The primers are 17 or 18 nucleotides long and consist of the core sequences and selective sequences. The core sequences are 13 to 14 bases long, of which the first 10 or 11 bases at the 5′ end are sequences of no specific constitution (known as "filler" sequences), followed by the sequence CCGG in the forward primer and AATT in the reverse primer. The core is followed by three selective nucleotides at the 3′ end. The filler sequences of the forward and reverse primers must be different from each other and can be 10 or 11 bases long. In the PCR, the annealing temperature for the first five cycles is set at 35°C. The following 35 cycles are run at 50°C. The amplified DNA fragments are separated and detected by denaturing acrylamide gels.

ASAP: Allele-specific associated primers. Specific primers are designed according to sequence information of specific alleles and used for amplification of DNA to generate a single fragment at stringent annealing temperatures. They are also known as allele-specific or gene-specific markers.

VNTR: Variable number of tandem repeats. VNTRs are also known as minisatellites that are tandem repeats with a monomer repeat length of about 11–60 bp. The minisatellite loci contain tandem repeats that vary in the number of repeat units between genotypes and are referred to as variable number of tandem repeats or hypervariable regions (HVRs).

Combination of restriction and PCR

AFLP: Amplified fragment length polymorphism. This technique combines aspects of RFLP and PCR amplification. The technique is based on the detection of genomic restriction fragments by PCR amplification and can be used for DNAs of any origin or complexity without any prior knowledge of sequence, using a limited set of generic primers. The number of fragments detected in a single reaction can be 'tuned' by selection of specific primer sets.

CAPS: Cleaved amplified polymorphic sequence. This technique combines PCR amplification and restriction enzyme digestion to generate DNA polymorphisms. Polymorphic patterns are generated by restriction enzyme digestion of PCR products. PCR primers for this process can be synthesized based on the sequence information available in databases of genomic or cDNA sequences or cloned RAPD bands. These markers are co-dominant in nature.

STS: Sequence-tagged sites. RFLP probes specifically linked to a desired trait are sequenced and converted into PCR-based STS markers based on nucleotide sequence of the probe resulting in a polymorphic band pattern for the specific amplicon. In this technique the tedious hybridization procedures involved in RFLP analysis can be overcome.

Box 23.1: Classification of DNA markers based on the development and type of genetic variation

Array-based molecular markers

DarT: Diversity Array Technology. DarT is a hybridization-based method capable of generating a medium density scan of any plant genome, irrespective of genome sequence information. A single DarT assay simultaneously genotypes hundreds to thousands of SNPs and indels throughout the genome.

Microarray-based markers: A microarray-based method has been developed for scoring thousands of DNAs for a co-dominant marker on a glass slide. The approach was developed to detect insertion polymorphism of transposons and works well with NP markers. Biotin-terminated allele-specific PCR products are spotted unpurified onto streptavidin-coated glass slides and visualized by hybridization of fluorescent detector oligonucleotides to tags attached to the allele- specific PCR primers. Two tagged primer oligonucleotides are used per locus and each tag is detected by hybridization to a con-catameric DNA probe labelled with multiple fluorochromes.

BeadArray: This technology relies on laser fibre optics, which provides the possibility of testing >1500 loci in a single DNA sample in a single tube. The assay employs sets of locus specific oligodeoxynucleotides designed to interrogate each individual SNP. One of these is locus specific (Locus Specific Oligo, LSO), but non-discriminatory, and the other two are designed so that their 3'- base is complementary to one or other alternative allele sequences (Allele Specific Oligo's, ASO). After a complex series of reactions, the yield of a PCR product that contains one or other of the discriminating ASOs can be quantified and the SNP genotype at the specific locus ascertained. BeadArray is a genotyping system commercialized by Illumina and exploits an assay known as 'GoldenGate'. In this assay 1536 different SNP loci are interrogated by a series of extension and ligation reactions followed by PCR.

Table 23.1: Relative comparisons of some commonly used DNA markers in cereals

Features	RFLP	RAPD	AFLP	SSR	SNP
DNA required (µg)	10	0.02	0.5–1.0	0.05	0.05
DNA quality	Very high	Moderate	High	Moderate	High
PCR-based	No	Yes	Yes	Yes	Yes
Number of polymorphisms	Moderate	Moderate	High	Moderate	Low
Easy to use	Not easy	Easy	Easy	Easy	Easy
Amendable to automation	Low	Moderate	Moderate	High	High
Reproducibility	High	Unreliable	Moderate	High	High
Development cost	Low	Very low	Moderate	High	High
Cost per analysis	High	Low	Moderate	Low	Moderate
Inheritance	Co-dominant	Dominant	Dominant	Co-dominant	Both
Speed of analysis	Slow	Moderate	Slow	Fast	Variable

some DNA markers are developed based on the gene/allele sequence and hence are gene/allele specific while most others are not (**Table 23.1, Box 23.2**). The usefulness of a particular type of genetic marker also depends on its density, frequency, distribution and richness in polymorphism in the genome (evaluated by its polymorphic index, **Box 23.2**).

Simple sequence repeats (SSRs) are co-dominant and highly polymorphic. SSRs are not confined to non-coding regions of the genome, and many are found in coding sequences of genes, allowing the development

of EST-SSR markers (Varshney *et al.*, 2005). In contrast to SSRs, multiple regions of the genome can be targeted simultaneously with RAPDs. However, this is not a serious constraint as SSR analyses are easy to multiply. A down side to RAPD markers is they are dominant, not co-dominant. The RAPD technique is fairly simple however; it does not require restriction digestion, labelled probes, hybridization or prior knowledge of sequence and the use of hazardous detection chemicals. Small variations in experimental conditions often affect the reproducibility of the RAPD technique (e.g. between

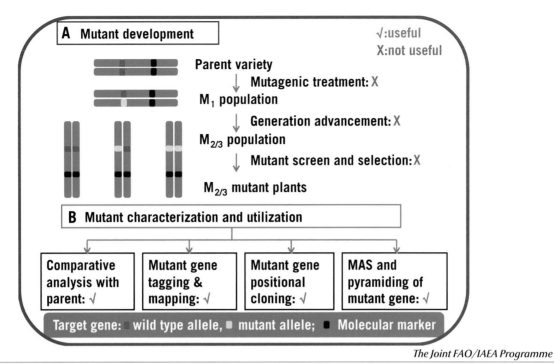

The Joint FAO/IAEA Programme

Figure 23.1 A schematic diagram of molecular marker techniques in plant mutation research

laboratories), which limits its use.

AFLP is a technique that combines aspects of both RFLP and DNA amplification using PCR; it is a more robust method than RAPD, but more technically demanding. AFLP typically allows the analysis of dozens of DNA markers simultaneously. Combinations of primers with alternate supplementary bases allow many different sub-sets of DNA marker combinations, or genomic representations, to be analysed, a feature that makes this technique very versatile. Like RAPD markers, AFLP markers are either present or absent. AFLP band intensities can be quantified and scored as co-dominant markers, especially between samples of closely related individuals such as backcross populations to discriminate heterozygous loci. AFLP is normally used as a genetic fingerprinting method or when large numbers of markers are required from which candidate markers linked to a target can be selected.

For sheer density of DNA markers, SNPs offer one of the richest sources. SNPs are increasingly used as DNA markers because techniques are becoming more widespread, cost effective and automated, and the genome sequence information of major crops are becoming increasingly available (**see Chapter 20**). With SNPs, as opposed to other types of DNA markers, marker density can be measured on the kilobase scale instead of the megabase scale. Over 37,000 SNPs were identified in a comparison of shotgun sequences between *Arabidopsis* accessions Columbia and *Landsberg erecta*. Because of their high density in the genome, SNPs are extremely well suited for high-resolution gene mapping, marker-assisted selection and studies assessing the level of diversity throughout the genome. SNP frequency is such that more than one SNP can be detected within a DNA fragment or across a region of interest, and the variation detected can be converted into haplotypes if required.

In addition to the above DNA markers, a number of other markers, i.e. CAPS, STS, SCAR, ISSR, SRAP, VNTR, DarT, and other derivative techniques have been developed and utilized during the past decades (the more common of these are described in **Box 23.1**).

3. Use of DNA Markers in Mutant Characterization

Plant mutation research projects can include either all steps, as shown in **Figure 23.1**, from mutagenic treatment to gene cloning and gene pyramiding, or involve only a part of them. DNA marker techniques can be used to facilitate some of investigations; however, they are usually not useful for screening and selection of induced mutants (**Figure 23.1**).

3.1. Most DNA Markers are Not Useful in Mutant Screening

During the past decade, numerous DNA markers have been identified for various traits of agronomic or quality importance. This has led to a commonly asked question: can DNA markers be used for screening or selecting an induced mutant? The answer is that most of them are not useful, with the exception of functional markers. Other commonly asked questions are given in **Box 23.3**.

As described earlier and shown in **Figure 23.1**, all DNA markers except gene specific markers are indirect markers and consist of a DNA fragment that is physically close to the gene of interest. Since plant mutagenesis induced by chemical or physical mutagens is a random process and occurs at an extremely low frequency, it is very unlikely that two mutations occur simultaneously, one at the marker locus, the other at the gene target; Furthermore, even if such an event occurs, it is almost impossible for the two mutations to occur in the same desired direction (e.g. the marker and target gene are mutated into previously reported desired alleles). Consequently, a mutant plant most probably has the wild type allele of the linked marker, and *vice versa*. Therefore, standard DNA markers that are only linked to a trait will not be useful for screening and selection of a mutated trait in mutant populations (M_2, M_3, etc).

Currently, the only exceptions are functional molecular markers, for example, SNPs responsible for herbicide tolerance and susceptibility (**see Chapter 31**), or

---

Box 23.2: Characteristics of DNA markers

1. **Dominant and co-dominant marker:** Dominant markers reveal only one allele, while co-dominant markers allowing each allele to be distinguished at a locus.
2. **Gene (allele)-specific marker:** These markers are developed based on the sequence of a gene or allele, and reveal nucleotide variation at the target locus.
3. **Functional marker:** DNA markers developed based on sequence differences that can affect a plant phenotype. As gene/allele specific markers, functional markers are superior to random DNA markers such as RFLPs, SSRs and AFLPs owing to absolute association with the locus, allele, mutation or trait.
4. **Polymorphic index (PI):** PI is an indicator of the degree of polymorphism of a maker among genotypes of a given population; this is normally done at the level of the allele.
5. **Marker density:** An indicator of the number of genetic markers per unit of genome region.

Box 23.3: Frequently asked questions relating to DNA markers in plant mutation research

Question 1: Can DNA markers be used for screening and selecting induced mutants?
Answer: *In general no, induced mutants cannot be screened or selected using most DNA marker systems. Functional markers can be used for a specific type of mutation.*

Question 2: To what degree can an induced mutant be different from its wild type parent for it to be distinguishable with molecular markers?
Answer: Theoretically, chemically induced mutants differ from their parents by only 5 per million base pairs and frequency is even less for those physically induced. Studies in cereals show that selected morphological mutant have identical (AFLP and SSR) genotypes to their parents. Therefore, if a mutant shows a high degree (e.g. >5%) of SSR variation compared to its parent, it might not be a true mutant of the said parent, and is more likely to be a contaminant.

Questions 3: Can a mutation be tagged or mapped using DNA markers?
Answer: Yes or no, it depends on the nature of the mutated trait. Yes, if the mutation happens to be a trait of qualitative nature and the mutated trait can be clearly identified in a segregating population (e.g. F_2, F_3) derived from a cross between the mutant and another variety that is sufficiently different in genetic constitution. Otherwise, it is not possible to tag or map the mutation.

Question 4: What kind of populations should be developed for tagging and mapping a mutant gene?
Answer: Mapping populations (F_2, BC_1, DHs, etc.) should be developed by crossing the mutant with wild type varieties that are genetically distant from the mutant (this provides a high level of polymorphism among the markers used to construct the map). The most common error for tagging a mutant gene is to use a segregating population derived from a cross between a mutant and its parent (there will be no marker segregation).

Question 5: What are the advantages of induced mutants compared with conventional germplasm in positional gene cloning?
Answer: Once a gene is delimited to a short genomic region (i.e. ~100 kb), in most cases there should be only one site of sequence difference between the mutant and its parent. The gene with this sequence difference is highly likely to be the target gene. This is often not the case for conventional germplasm, since more than one site of sequence difference is expected, for example, there is a SNP every 2kb of rice genome between the indica and japonica sub-species.

for tolerance or susceptibility to a particularly race of a disease (e.g. the rice blast resistance gene *Pita-2*, **see Chapter 3**). In such cases, a simple molecular marker can be developed and used for screening of a desired mutant; however, it will need to screen millions of M_2 plants, which is not cost effective for most traits. This may change with the dramatic reduction of sequencing cost owing to the new generations of sequencing facilities.

3.2. Excluding False Mutants Using DNA Markers

Due to reduced pollen fertility of M_1 plants (**see Chapter 14**) and because the mutant selection process (a process of selecting any plant that differs from its parent variety), there is plenty of opportunity for mixing within the mutant population and contamination from other sources, e.g. a segregant from an outcross of the parent with another variety. It is recommended that extreme care is taken to protect M_1 and M_2 plants from pollen and seed contamination through proper isolation, and avoid other sources of mixing. However, it is practically impossible to exclude all chances of contamination, especially when the material is grown out on a large scale in the field. A selected mutant is normally subject to scrutiny using the breeder's knowledge, expertise and experience, but further more stringent validation is required as such a process cannot always lead to a scientifically correct judgement due to potential pleiotropic effect of a mutation, which can make a mutant quite different from its parent at the phenotypic level.

On the other hand, it is not uncommon that the physiological, biochemical and agronomical effects of a mutated trait are evaluated through comparative analysis between a mutant and its parent. Recently, such comparisons have extended to gene expression and metabolism using various array and 'omics' techniques (Burow *et al.*, 2008; Zakhrabekova *et al.*, 2008). The scientific soundness of any conclusion derived from such comparative studies relies on the true origin of the mutant from the parent variety. Therefore, it has become increasingly important to develop an objective and efficient means to guarantee that the mutant is directly originated from the parent under study through induced mutagenesis (not a mixture, or a progeny from an outcross). In this regard, standard DNA markers can play a role in assuring the mutant being compared is truly a direct mutant of the parental variety.

As described above, SSRs are highly polymorphic markers that have been developed in many plant genotypes. If they have a mutation frequency similar to other genomic regions during chemical and physical mutagenesis, a mutant should have an almost identical SSR fingerprint to its parent. This is because 1) SSR markers are short DNA fragments, i.e. ~200 bp per SSR marker, therefore, only one mutation will be induced per 1,000 SSR markers (based on the observations of one mutation per 200 kb); 2) even such a mutation (mostly SNP) cannot be revealed since the polymorphism of SSR marker are based on the repeat number difference (at least 2 bp difference). It is known that SSRs mutate at a higher frequency than normal genomic regions, for example, it has found to be about five or more times higher in maize (Vigouroux *et al.*, 2002). Even so, a mutant would be expected to be identical to its parent when genotyped with 200 SSR markers. The chances of hitting a desired mutation with an SSR are negligible. Such theoretical postulation has recently been experimentally proven in rice.

Rice lines carrying a transgenic gene or a recessive mutant gene were used by Fu *et al.* (2008) to demonstrate out-cross derived contaminants, which were frequently selected as "mutant" in several rice populations. However, all seven true mutants have SSR haplotypes identical to their respective parents. They further demonstrated that false mutants can be identified easily using SSR analysis, and were able to correct the pedigree of a *xantha* mutant Huangyu B (Fu *et al.*, 2007). Therefore, SSR marker assays can be used as a simple method for assuring the genetic relationship of a mutant and its parent, and used to exclude any false mutant before performing comprehensive studies on the effects of the mutations.

3.3. Tagging, Mapping and Cloning a Mutant Gene

In the early days of mutagenesis, the existence of a gene was often first revealed when a mutant with a visible phenotype was recovered; from pea plants with wrinkled seeds to fruitflies with altered eye pigmentation. Mutants have long played a central role in genetic analysis, and in crop species like barley and rice and model species, genetic maps were established using spontaneous and induced mutations. For example in Arabidopsis the first comprehensive genetic map was established 20 years ago in which 76 genes associated with distinct phenotypes, from altered trichome morphology and seed coat pigmentation to reduced surface waxes and increased hypocotyl length were mapped (**see Chapter 4**). With the availability of various molecular markers, attempts to tag and map a mutant gene have become routine in molecular studies.

3.3.1. What Kind of Mutated Traits can be Tagged or Mapped?

Mutation can affect both qualitative (e.g. glutinous endosperm, extreme dwarfism) and quantitative (e.g. yield, maturity, grain protein content) traits. It is unlikely that two or more genes in the same pathway are mutated simultaneously in a single mutant plant (due to the very low frequency of mutation for any single gene). Therefore, the mutated trait of either qualitative or quantitative nature segregates as a single gene in segregating populations (F_2, BC_1, DHs, etc) derived from a cross between a mutant and its parent. However, when a quantitative trait mutant is crossed with another variety with a contrasting genotype, the mutated trait can no longer be identified in the resulting progeny, since other genes influence the trait (positively or negatively).

It should be noted that quantitative trait loci (QTLs) are regularly identified using genetic markers, but they are different from the mutated gene in a mutant line discussed above, and hence should not be confused. It should also be noted that, with the development of new generations of sequencing technologies, it might become possible to sequence large DNA fragments at reasonable cost and high-throughput; hence polymorphic DNA markers can be developed with much greater precision and used to tag mutated genes using populations derived from crosses between a mutant and its parent. Such an approach is highly desirable as it would preclude the development of developing mapping populations and thus save time.

3.3.2. Gene Tagging and Mapping

If a mutant trait is qualitative in nature it can be tagged, mapped and in some cases cloned (**Figure 23.2**). Currently, such work always starts with the development of a segregating population. It is recommended that two crosses are made between the mutant line and two wild type (WT) genotypes. The WT genotypes should be genetically distant from the mutant so a sufficient number of polymorphic markers can be assayed. In rice, for example, if the mutant is an *indica* rice, it is

recommended that mapping populations are developed by crossing to *japonica* genotypes. F_2, BC_1 and/or DH populations are commonly used for such purposes. The second population can be used as a validation population, which also provides a second option for tagging the mutant gene if the first population shows low polymorphisms for the available genetic markers. The second population is also very useful for delimiting the region in which the mutant gene is located (**Figure 23.2**). It should be noted that the larger the population the more accurate the marker-mutation linkage.

Bulked sergeant analysis (BSA, Michelmore *et al.,* 1991) is a popular method for tagging traits. In this method two DNA bulks are made from contrasting individuals at extremes ends of a frequency distribution for the trait. In the case of a disease resistance trait, one bulk with be composed of the DNA from say 10 individuals showing extreme resistance and the second bulk is made up of an equal number of extreme susceptible individuals. If the population is derived from a mutation programme the mutated traits will be absent in one bulk but present in the other. Therefore, after phenotyping 10 individuals with and 10 individuals without the mutant gene are selected and DNA extracted from each individual. The DNA is then bulked for each of the two groups. The two bulks and the parental lines are then analysed using various markers. If the location of the mutation is unknown a genome-wide marker system is required. If, however, the chromosome, chromosome arm or chromosomal region is known more targeted markers can be selected (for example, mutated traits that are assigned to a specific chromosome or linkage group through classical chromosome mapping). Any marker found in one bulk but not the other becomes a candidate tag for the mutant gene. When such markers are found their fidelity to the trait can be checked by analysing each individual that make up the bulks and confirming the marker–mutation association in the mutant parent.

These candidate tags are then further tested in the whole population in order to estimate the genetic distance between the markers and the mutant gene and to map the mutation. The process is the same as the common practice of gene tagging and mapping. For coarse mapping of a mutant gene, population size can vary from 100 to 400 plants depending on the plant species, for example, 100 mutant type plants would be typical for tagging a single gene in rice (**Figure 23.2**). Fine mapping requires larger mapping populations or

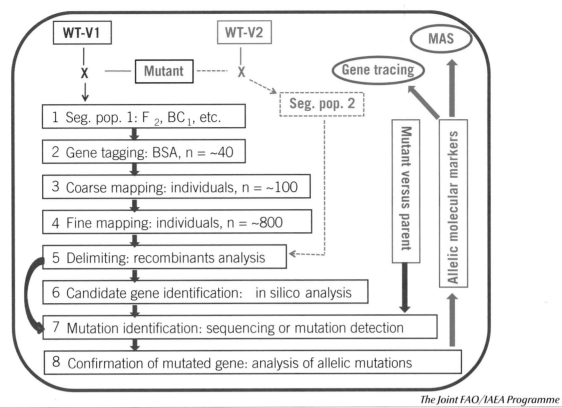

WT-V1 **WT-V2** **MAS**

Mutant **Gene tracing**

Seg. pop. 2

1 Seg. pop. 1: F$_2$, BC$_1$, etc.

2 Gene tagging: BSA, n = ~40

3 Coarse mapping: individuals, n = ~100

4 Fine mapping: individuals, n = ~800

5 Delimiting: recombinants analysis

6 Candidate gene identification: in silico analysis

7 Mutation identification: sequencing or mutation detection

8 Confirmation of mutated gene: analysis of allelic mutations

Mutant versus parent

Allelic molecular markers

The Joint FAO/IAEA Programme

Figure 23.2 A schematic diagram of tagging, mapping, cloning and tracing a mutated gene. WT-V1 and V2: wild type variety 1 and 2; Seg. pop.: segregating population; BSA: bulk segregants analysis; individuals are mutant type plants; MAS: marker-assisted selection.

specialist methods such as targeting recombination events around the locus of interest.

3.3.3. Fine Mapping, Delimiting and Cloning

Induced mutants have been purposely and systematically produced for gene isolation and gene analysis. The rice IR64 mutant collection at the International Rice Research Institute (IRRI) comprises more than 38,000 M$_4$ lines. Compared with the T-DNA mutant lines, induced mutations are unique and to some extent irreplaceable in gene discovery. For example, most T-DNA insertions cause gene knock-outs, and no mutant can be recovered if such genes are essential for plant survival. In contrast induced mutations are more "knock-down" mutations with some possessing unique functionality, they are therefore of great use in gene discovery and gene function studies. In many crops, it is not always possible to generate large numbers of insertion mutants, because of resilience to transformation or inefficient regeneration protocols, but it is possible to generate mutants in all crops through physical or chemical mutagenesis. Several examples of gene discovery are described in **Chapters 35–37**. Here the general procedure is discussed.

In the course of map-based cloning, mutant genes are first identified through linkage to a sufficiently small region by fine mapping and delimiting (**Figure 23.2**). A larger segregating population (e.g. 1,000 to 2,000 F$_2$ or DH plants) is grown, genotyped and phenotyped to determine fine-scale genetic linkage. The mutant can be delimited through analysis of recombinant plants based on reference genome sequence information which is available for most major crop species. At this stage, recombinant plants of a second population are always very useful since there may be additional recombinants that can be used for narrowing the fragment where the mutant gene is located and thereby developing more tightly linked markers (**Figure 23.2**).

When the mutant gene is delimited to a narrow region, the next step is to identify the most likely candidate gene through bioinformatics and *in silico* data searches (**Figure 23.2**). For example, it is possible to identify a gene that is homologous to a known gene in another species. For example, Kim *et al.* (2008) and Xu *et al.* (2009) identified the candidate genes responsible for two low phytic acid mutations through homologue data searches, and consequently demonstrated the mutations occurred in two genes. However, if the mutated trait is controlled by a gene for which no data are available, it is not possible to identify the candidate gene through such analysis. In such cases, mutation detection through sequencing or mismatch cleavage of the whole delimited region can be performed for the mutant and its parent, for example Zhao *et al.* (2008) identified a new low phytic rice gene using such techniques. Although this method is currently often limited to data searches in model species, the expansion of whole genome sequencing, particularly of major crop species will provide greater scope in the future as 'omics' data bases are developed in crops.

Induced mutants have a special advantage in identifying the target gene after it is delimited compared with conventional germplasm. Once a mutant gene is delimited to a short genomic region (e.g. ~100 kb) it can be recognized by its sequence variation. Sequence difference between a mutant and its parent has a frequency of about 1 per 200 kb for chemical mutagenesis and even lower for physical mutagenesis (**see Chapter 20**). Therefore, any gene with a sequence difference within the target region is highly likely to be the gene in question. This is not the case for conventional germplasm, since more than one site of sequence difference is expected, for example, there is a SNP every 2 kb of the rice genome between *indica* and *japonica* sub-species. If there are two or more allelic mutations for the same mutated trait, and all mutations occur in the same gene, this become compelling evidence for the gene discovery (**Figure 23.2**).

3.4. Marker-Assisted Selection, Pyramiding and Tracing Mutant Genes

3.4.1. Marker-Assisted Selection and Gene Pyramiding

After a mutant trait is tagged by one or more markers, it is possible to select the mutant trait using the markers; this is termed marker assisted selection (MAS). Marker assisted selection has become an important tool in plant breeding to increase breeding efficiency, allowing early generation selection, and reducing plant population size and in reducing redundant planting areas. In principle, there are no substantial differences between common gene and mutant gene tagging when MAS is applied to a breeding programme.

New varieties are required to be excellent for all important characters such as yield, desirable quality,

and resistance to pest and diseases. Therefore, a new variety should have the best combination of alleles for these traits. Gene pyramiding is the term used when two or more genes are simultaneously selected and bred into a single breeding line.

Gene pyramiding is especially useful for combining genes responsible for resistance to different races of a disease or biotypes of a pest, since resistances to multi races or multi biotypes are also considered to be more durable than resistance to a single race or single biotype. Gene pyramiding would be very difficult without genetic markers, since it is often impossible to distinguish resistance controlled by a single gene or by two or more genes at the phenotypic level. Genetic markers provide a powerful means for maximizing the utilization of mutant resources through MAS and gene pyramiding. It has already been successfully applied in several crop breeding programmes, and some new varieties and advanced lines possessing multiple attributes have been developed. A coordinated research project on 'Pyramiding of mutated genes contributing to crop quality' was initiated in 2004 by the International Atomic Energy Agency with the aim of developing elite breeding lines by combining mutant genes controlling nutritional value (amino acid composition, micronutrients, vitamins, secondary metabolites, nutraceuticals, etc.), end-user quality traits, and tolerance to biotic and abiotic stresses.

It is essential to develop multiplex marker technologies for efficient pyramiding of genes. PCR multiplexing and multi-pooling strategies can significantly reduce the cost of genotyping and increase genotyping throughput. These are available for many markers systems, e.g. a multiplex-ready PCR technology has been developed for genotyping with multiple SSR markers (Hayden *et al.*, 2008).

3.4.2. Allele-Specific Markers for Distinguishing and Tracing Mutant Genes

Once a mutant gene is identified and cloned, it is possible to develop genetic markers that differentiate the different alleles; these are known as allele-specific markers. For example, Zhao *et al.* (2008) developed allele-specific markers for the low phytic acid gene *Lpa1* in rice. After the KBNT *lpa1-1* mutation was identified to be a single nucleotide substitution of C/G to T/A, a CAPS marker (LPA_CAPS) was developed for distinguishing the mutant allele (*lpa1-1*) from the WT allele. Similarly, an indel marker (LPA1_Indel) was designed to discern the Os-*lpa*-XQZ-1 mutation (*lpa1-2*) from the WT allele. It is now possible to differentiate the two mutant alleles (*lpa1-1* and *lpa1-2*) and the WT allele (*Lpa1*) using these two markers. With allele-specific markers, MAS for the desired gene allele is 100% accurate and therefore of immense value in breeding.

With allele-specific markers, it also becomes possible to trace the origin of important gene alleles in modern varieties. The Green Revolution in rice was achieved with the development of semi-dwarf (SD) rice varieties during the 1960 and 1970s. The short stature of semi-dwarf varieties is due to mutations in the SD1 gene that encodes the GA20-oxidase-2 enzyme. It has been demonstrated that at least seven *sd1* alleles have been used in rice breeding in China, Japan and the USA. Apart from the IR8 *sd1* allele (a 383-bp deletion and hence null mutation, from the native variety Dee-geo-woo-gen, DGWG), two artificially induced *sd1* mutations, namely those of semi-dwarf mutant varieties Reimei (*sd1*-RM) and Calrose-76 (*sd1*-CR), have also been used widely in Japan and the USA. Both *sd1*-RM and *sd1*-CR are due to a single nucleotide substitution, which leads to amino acid changes, i.e. D349H and L266F, respectively (Asano *et al.*, 2007). Based on these findings, Kim *et al.* (2009) were able to design an allele specific marker and examined the nature and distribution of *sd1* alleles among US varieties.

4. References

4.1. Cited references

Asano, K., Takashi, T., Miura, K. *et al.* 2007. Genetic and molecular analysis of utility of *sd1* alleles in rice breeding. *Breed Sci.* 57: 53–58.

Burow, G.B., Franks, C.D. and Xin, Z. 2008. Genetic and physiological analysis of an irradiated bloomless mutant (epicuticular wax mutant) of sorghum. *Crop Sci.* 48: 41–48.

Fu, H.W., Li, Y.F. and Shu Q.Y. 2008. A revisit of mutation induction by gamma rays in rice (*Oryza sativa* L.): implications of microsatellite markers for quality control. *Mol. Breed.* 22(2): 281–288.

Fu, H.W., Wang, C.X., Shu, X.L. *et al.* 2007. Microsatellite analysis for revealing parentage of gamma ray-induced mutants in rice (*Oryza sativa* L.). Israel. *J. Plant Sci.* 55: 201–206.

Hayden, M.J., Nguyen, T.M., Waterman, A. *et al.* 2008. Application of multiplex-ready PCR for fluorescence-based SSR genotyping in barley and wheat. *Mol Breed.* 21: 271–281.

Kim, S.I., Andaya, C.B., Newman J.W. *et al.* 2008. Isolation and characterization of a low phytic acid rice mutant reveals a mutation in the rice orthologue of maize MIK. *Theor Appl Genet.* 117: 1291–1301.

Michelmore, R.W., Paran, I. and Kesseli, R.V. 1991. Identification of markers linked to disease-resistance genes by bulked segregant analysis: a rapid method to detect markers in specific genomic regions by using segregating populations. *Proc Natl Acad Sci USA.* 88: 9828–9832.

Varshney, R.K., Graner, A. and Sorrells, M.E. 2005. Genic microsatellite markers in plants: features and applications. *Trends Biotechnol.* 23: 48–55.

Vigouroux, Y., Jaqueth, J.S., Matsuoka, Y. *et al.* 2002. Rate and pattern of mutation at microsatellite loci in maize. *Mol Biol Evol.* 19: 1251–1260.

Xu, X.H., Zhao, H.J., Liu, Q.L. *et al.* 2009 Mutations of the multi-drug resistance-associated protein ABC transporter gene 5 result in reduction of phytic acid in rice seeds. *Theor Appl Genet.*

Zakhrabekova, S., Gough, S.P., Lundqvist, U. *et al.* 2008. Comparing two microarray platforms for identifying mutated genes in barley (*Hordeum vulgare* L.). *Plant Physiol Biochem.* 45: 617–622.

Zhao, H.J., Liu, Q.L., Ren, X.L. *et al.* 2008. Gene identification and allele-specific marker development for two allelic low phytic acid mutations in rice (*Oryza sativa* L.) . *Mol Breed.* 22: 603–612.

4.2. Websites

Coordinated Research Project on Gene Pyramiding: http://www.iaea.org/programmes/nafa/d2/crp/d2-crop-quality.html.

International Rice Research Institute: http://www.iris.irri.org

Rice Functional Genomics: http://signal.salk.edu/cgi-bin/RiceGE

List of mutant genes of *Arabidopsis*: http://www.nsf.gov/bio/pubs/reports/nsf9643/appx3.htm

DarT marker service: http://www.triticarte.com.au

GENica: http://www.genica.net.au.

4.3. Further reading

Guimarães, E.,Ruane, J., Scherf, B. *et al*. 2007. Marker-Assisted Selection-Current Status and Future Perspectives in Crops, Livestock, Forestry and Fish. Rome: Food and Agriculture Organization of the United Nations.

International Atomic Energy Agency 2002. Mutant germplasm characterization using molecular markers. *Training course series No.19*. Vienna: IAEA.

Lörz, H. and Wenzel, G. 2007. Molecular Marker Systems in Plant Breeding and Crop Improvement. Tokyo: Springer.

Servin, B., Martin, O.C., Mezard, M. *et al*. 2004. Toward a theory of marker-assisted gene pyramiding. *Genetics*. 168: 513–523.

Mutation Breeding

Principles and Applications of Plant Mutation Breeding

Q.Y.Shu[a,*], B.P.Forster[b] and H.Nakagawa[c]

[a] Joint FAO/IAEA Division of Nuclear Techniques in Food and Agriculture, International Atomic Energy Agency, Wagramer Strasse 5, P.O. Box 100, A-1400 Vienna, Austria
Present: Institute of Nuclear Agricultural Sciences, Zhejiang University, Hangzhou, China
[b] BioHybrids International Limited, P.O. Box 2411, Earley, Reading, Berkshire, RG6 5FY, UK
[c] Institute of Radiation Breeding, National Institute of Agrobiological Sciences, Kami-Murata, Hitachi-Ohmiya, Ibaraki 319-2293, Japan
* Corresponding author: Email: qyshu@zju.edu.cn

1. Introduction

Plant and animal breeding has been practised since the dawn of human civilization. Simple selection of desirable offspring was the first method of breeding and this utilized the occurrence of spontaneous mutations. The application of Mendel's laws of genetics at the beginning of the 20th century made a significant impact on plant breeding, which changed from empirical or experience-based practice to a science-based plant technology. Cross-breeding (or recombinant breeding), based on crossing of different genotypes followed by trait selection, has become the widely practised method in plant breeding. Later, the work on the induction of genetic alterations by X-rays, performed by the plant breeder Lweis John Stadler in the late 1920s and early 1930s, laid the foundation of another type of plant breeding – mutation breeding.

Mutation breeding involves the development of new varieties by generating and utilizing genetic variability through chemical and physical mutagenesis. The achievements of mutation breeding (**see Chapters 1 and 2**) contributed by various influential people (**Box 24.1**) in various species places mutation breeding on a strong foundation. It is now one of the three pillars of modern plant breeding (the other two are recombinant breeding and transgenic breeding). Although mutation breeding follows the same rules of traits being under genetic control, it differs from other breeding methods in that it generates new genetic variation. It also differs in the methods of screening and selection of desirable breeding lines, and in the potential to improve a new variety, e.g. by speeding up the development of desired phenotypes. As in other modern breeding schemes mutation breeding exploits advances in genomics in selecting desired lines by genotyping rather than phenotyping, this has been dubbed "molecular mutation breeding".

Mutation breeding is quite unique in several aspects, it is therefore of paramount importance to understand the advantages and limitations for its effectiveness in crop improvement. The principles and applications of mutation breeding in crop improvement, and the integration of genomics, molecular technologies and other related technologies into mutation breeding are briefly introduced in this chapter. Examples are given throughout. The chapter also acts as a guide to other chapters and further reading on specific topics.

2. Biological Basis of Mutation Breeding

A number of biological mechanisms underlie mutation breeding. They include the biology of DNA damage and repair, ontology and reproductive biology, gene and genome biology, and genetics and functional genomics.

2.1. DNA Damage and Repair

DNA damage caused by physical and chemical mutagens is the starting point of mutation breeding. As described in **Chapter 5**, DNA lesions are naturally generated in living cells in the process of metabolism and DNA replication. Treatments with chemical or physical mutagens can significantly increase the amount, and alter the profile, of DNA damage. Living cells can respond quickly to the DNA damages and initiate different mechanisms either by killing the damaged cell or by repairing DNA lesions; the consequences of these processes are directly linked to mutation breeding.

2.1.1. DNA Damage and Cell Survival

In general, the higher the dose of a mutagen, the more the DNA damage. Each cell can only tolerate certain amounts of DNA damage, above which the cell would be killed through processes such as apoptosis. At the plant level, or *in vitro* tissue culture, this is reflected by a reduction or cessation of growth after mutagen treatment above certain critical dose levels.

2.1.2. DNA Repair and Consequences

After mutagen treatment at non-lethal doses, viable biological mechanisms are initiated to repair the DNA lesions (**see Chapters 5 and 6**). This process is critical and has significant implications for mutation breeding since the consequences of this process are directly related to the type and frequency of mutations produced and hence it has an impact on the generation and frequency of desirable mutant plants.

Error-proof and error-prone DNA repair. If all DNA lesions are perfectly repaired, there would be no mutations. However, errors in DNA repair do happen and are inherent with certain mechanisms. For example, homologous recombination repair of DNA double strand breaks (DSBs) often results in restoration of the original DNA sequence, and hence produces relatively few mutations. Not all DNA lesions are repaired by solely one mechanism and some DNA repair mechanisms are.

Box 24.1: Influential people in the history of mutation breeding

Carl von Linné (1707–1778). Swedish botanist, the 'Father of taxonomy' and constructor of the system of naming, ranking and classifying organisms. From 1741 onwards described various examples of mutants and heritable variation in both wild and cultivated plants.

Gregor Johann Mendel (1822–1884). An Austrian monk who conducted studies on garden peas in the Augustinian Abbey of St Thomas at Brünn (now Brno, Czech Republic). In 1865 Mendel published studies on inheritance from which Mendelian genetics and Mendelian laws of inheritance were developed. Known as the 'Father of genetics'.

Wilhelm Conrad von Röntgen (1845–1923). German physicist and 'Father of diagnostic radiology' who discovered X-rays (1895). Nobel Prize in Physics laureate in 1901.

Antoine Henri Becquerel (1852–1908). French physicist who discovered spontaneous irradiation from uranium, referred to as Becquerel radiation (a mixture of α-, β- and γ-rays). Nobel Prize in Physics 1903 (shared with Pierre and Marie Curie).

Charles Robert Darwin (1809–1882). English naturalist who wrote the groundbreaking book *The Origin of the Species*. Darwin developed the evolutionary theory of natural selection through "survival of the fittest" in which heritable genetic changes (mutations) provide variation and the driving force for speciation. Darwin also defined bud variation (bud sport mutation) in 1868.

Hugo de Vries (1848–1935). Dutch scientist who developed the concepts of genes (pangenes) and mutation. A founding father of genetics. Suggested that newly discovered irradiation could be used to induce mutations artificially. Produced two major works: *The Mutation Theory* (published in two volumes: 1901 and 1903), and *Species and Varieties: Their Origin by Mutation* (1905).

Nikolai Ivanovich Vavilov (1887–1943). A Russian botanist, geneticist and plant breeder. Developed the concept of "Centres of diversity" where genetic variation for plant species, including the origins of crop species, occurred in certain geographic regions. According to Vavilov wild type (dominant) alleles dominate the centres with recessive alleles (the result of mutation) being more frequent on the periphery. In 1935 Vavilov famously stated that "plant breeding was evolution in human hands".

Herman Joseph Muller (1890–1967). An American geneticist who worked on mutation induction in Drosophila. Provided proof that X-rays could induce mutations in genes (1927). The ability to produce abundant mutants provided a foundation for many studies, e.g. in evolution, mechanisms of gene mutation and the properties of genes. Produced a method (CIB-method) to detect mutations on the X-chromosome of Drosophila. Muller was very optimistic about practical benefits of induced mutations. Nobel Prize in Physiology and Medicine, 1946.

Lewis John Stadler (1896–1954). A contemporary of Muller but worked on mutants in seed crops, particularly maize, wheat and barley. Stadler dramatically increased the mutation rate in crop plants. He demonstrated that mutant rates were proportional to dosage and that rates varied within and between species. In so doing, Stadler was the first to demonstrate the potential of induced mutagenesis in higher plants. Contrary to Muller, Stadler was critical about the benefits of induced mutation for breeding: 1) referring to most mutations as 'junk', 2) stating that sufficient variation was naturally available to breeders, 3) that mutations were invariably recessive and had deleterious effects and, 4) that rare worthwhile mutants would require the development of massive screening methods. Although Stadler was correct in his views, his pessimistic comments had a negative effect on deploying mutagenesis in plant breeding.

Herman Nilsson-Ehle (1879–1949). Swedish plant physiologist, geneticist and breeder. Along with co-worker Åke Gustafsson is credited to be the first to produce superior varieties (in barley) through induced mutation. He stressed the importance of combining theoretical and practical research. Worked at Svalöf, a Swedish plant breeding institute renowned for applying and testing emerging technologies and scientific principles to plant breeding.

Reinhold von Sengbusch (1898–1985). German scientist credited with the first example of efficient mass screening methods for mutant detection, "sweet", alkaloid-free lupin (begun in the 1920s). These mutants along with other spontaneous mutants for non-shattering and water-permeable seed led to the domestication of lupin and their development as a fodder crop.

Hans Stubbe (1902–1989). German pioneer of mutation breeding. Published extensively on fundamental issues in induced mutation (e.g. Stubbe, 1937). He recognized the importance of mutation in diversity and evolution and was among the first to study small mutation effects. He established the Institute of Plant Genetics and Crop Research with its world renowned Genebank at Gatersleben, Germany.

Åke Gustafson (1908–1998). A Swedish geneticist and plant breeder who was a keen advocate of mutation breeding. He debunked and addressed the criticisms of Stadler and stated that mutation induction could produce valuable mutations and that mutation induction was a useful tool in the hands of a plant breeder. He developed a large basic research programme to influence and improve methods for mutagenesis in breeding cultivated plants. Often referred to as the 'father of mutation breeding'.

Iosif Abramovich Rapoport (1912–1990). Russian pioneer of chemical-induced mutagenesis in plant breeding. Induced mutations in over 3,000 varieties of various crops from which 366 mutant varieties were produced.

Auerbach and Robson (1944–1946). Published the mutagenic effects of certain chemicals.

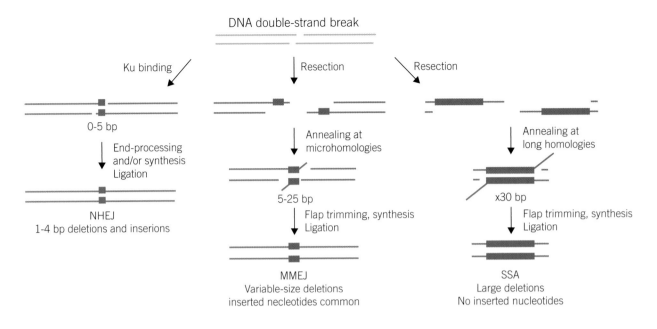

DNA double-strand break

Ku binding   Resection   Resection

0-5 bp

End-processing
and/or synthesis
Ligation

NHEJ
1-4 bp deletions and inserions

Annealing at
microhomologies

5-25 bp

Flap trimming, synthesis
Ligation

MMEJ
Variable-size deletions
inserted necleotides common

Annealing at
long homologies

x30 bp

Flap trimming, synthesis
Ligation

SSA
Large deletions
No inserted nucleotides

The Joint FAO/IAEA Programme

Figure 24.1 Comparison of non-homologous end joining (NHEJ), microhomology-mediated end joining (MMEJ) and single-strand annealing (SSA) pathways in *Sacchomyces cerevisiae*. During NHEJ repair of a DSB, the Ku70–Ku80 heterodimer binds, preventing DNA end resection. Repair proceeds by annealing at short micro-homologies (green boxes), fill-in by Pol4 and ligation using DNA ligase IV, resulting in small deletion and insertion products. Both MMEJ and SSA require end resection or DNA unwinding to reveal homologous sequences, although the length of homology required for MMEJ (5–25 bp) is shorter than for SSA. SSA and MMEJ also requires 30 flap cleavage before fill-in synthesis and ligation. Whereas MMEJ products can contain inserted nucleotides, these are never observed in SSA. McVey and Lee (2008) *Trends in Genetics*. 24: 529–538. Copyright permitted by Elsevier.

Mutation feature and spectrum. At the DNA level, mutations are either DNA base pair changes, or large or small DNA fragment insertions/deletions. At the phenotypic level, mutations can result in a wide spectrum of morphologies. Large DNA indels often cause truncation of encoded protein either due to deletion of a number of amino acids, or frameshift; small indels can also lead to truncated proteins due to frameshift that causes early stop coding. Therefore, most indels result in loss-of-function mutations and are recessive to wild type alleles. Large fragment deletions may involve more than one gene; hence at the biological level such mutations can have large effects on traits (particularly if it involves two or more genes belonging to the same gene family). On some rare occasions deletions may generate pseudo-wild type alleles when they restore the frame of the gene due to a further deletion (**see Chapters 1 and 4**), and at phenotype level, these mutations appear to be dominant.

Base pair changes, often induced by chemical mutagens, do not always have an impact on the gene product. However, some can result in either functional alteration of the gene product due to amino acid change, truncated proteins due to the generation of a premature stop code, or changed gene expression at the transcriptional level or by splicing modifications if the mutation occurs at the exon-intron interface.

Implications on mutation breeding. Since most deletions lead to gene knock-out or knock-down they are recessive to wild type alleles at the phenotype level. Therefore, they are most useful for improving recessive traits, a common example is semi-dwarfism. When a new gene function is needed, base pair changes are more likely to fulfil the objective, since they can result in either a gain of function or alteration of function, such as herbicide tolerance mutations (**see Chapter 31**). Therefore, mutation breeding is able to improve crop traits by both knock-out and knock-down of genes of interest or by gain of function or alteration of function through appropriate choice of mutagens.

Understanding the features of mutations at the DNA level is also important in designing molecular screening methods. For example, different strategies and methods are needed for screening point mutations and indels (**see Chapter 20**). The co-existence of error-proof and error-prone DNA repair mechanisms also provides possibilities to increase the mutation frequency through genetic engineering of the systems.

2.2. Ontology and Reproductive Biology

After DNA damage is repaired, the DNA lesions are either restored to the wild type, undamaged status or fixed as a mutation in the affected cells. They become useful in mutation breeding only when they are sourced by germline cells and pass through diplontic and haplontic selection (**see Chapters 15 and 16**). In this respect, a full understanding of cell ontology and plant reproductive biology is essential.

2.2.1. Ontology

In each propagule, such as a seed or a bud, there is usually more than one meristematic cell that develops to produce germline cells or new organs. The ontology of different plant species is very important in the management of M_1 and M_2 populations in seed crops and for the dissociation of chimeras in vegetatively propagated crops.

In seed crops, mature and dry seeds are often chosen for mutagenic treatment. Although the knowledge of the exact number of primordial cells in a seed may be unclear, it is commonly estimated, for example in cereals, that there are about four to six primordial cells that serve as the initials for germline cells. Therefore, the M_1 plants derived from mutagenized mature seeds are chimerical, and different tillers/branches carry different mutations. The chimerical nature of M_1 plants complicates the management of M_1 and M_2 populations, and precludes screening mutants in the M_1 generation. To avoid the formation of chimeras, pollen grains and fertilized eggs (zygotes) are sometimes targeted in mutagenic treatment (**see Chapter 14**).

In vegetatively propagated crops, the M_1V_1 plants generated by induced mutation are also chimerical. Here ontology knowledge is of particular importance in determining the number of sub-cultures needed in *in vitro* tissue culture, or for developing suitable *in vivo* methods to dissociate the chimeras. If insufficient rounds of *in vitro* sub-culture or *in vivo* cloning are performed, the plants will still possess chimeras, and the mutant may be lost in subsequent multiplications. On the other hand, if unnecessary sub-culture/cloning is carried out, the probability of having plants with the same genetic make-up increases; as a consequence this would necessitate increasing the size of the mutant population and hence the cost of mutant selection (**see Chapters 26 and 28**).

Despite the fact that ontology is a basic principle in plant development and has direct implications of mutation breeding, research in ontology has been limited. In this regard there is great potential in exploiting multiple recessive traits and mutant cell tracking (e.g. by molecular markers) to study ontology in greater depth.

2.2.2. Reproductive Biology

Reproductive modes of plants. Plants have evolved two distinct reproductive modes: asexual and sexual reproduction, each of which exhibits a rich diversification of mechanisms (**Figure 24.2**). Autogamy (self-fertilizing) includes self-pollinated species without self-incompatibility mechanisms; their progeny is homozygous with the same genotype as the parent plant. Allogamy (separate male and female parents) includes cross-pollinated species with high or low self-incompatibility systems to prevent autogamy. The genotypes within a natural population are diverse and the genotypes obtained by natural mechanisms are also diverse. Inbred lines can be developed in allogamous species by artificial manipulations and used in breeding; typically contrasting inbred lines are combined to recreate a highly heterozygous progeny.

Implications of reproductive mode in mutation breeding. Meiosis and fertilization are important genetic processes that allow the re-assortment and recombination of genes. The manipulation of these processes enables the production of mutant populations and the refinement, selection and development of desired mutations. Through these processes, the chimeras present in M_1 plants disappear, homozygous mutant loci are generated, targeted mutant alleles are separated from other unwanted non-target mutations and the background mutational load reduced. With some exceptions such as plants with *Taraxacum* type diplospory (which involves meiosis but without subsequent fertilization), asexual reproduction (via embryos in seed-like propagules) does not result in a change in the genetic constitution due to the lack of meiosis (**Figure 24.2**). The direct consequence of this type of reproduction is that no homozygous mutant loci are formed and the mutated traits are not readily expressed due to the recessive nature of most induced mutations. Since it is also not possible to segregate the desired mutations from other non-desirable background mutations, the latter may also affect the final use of the induced mutant as a new variety.

Seed production through sexual reproduction is *via* the union of male and female gametes. For practical crop breeding purposes sexual reproduction is classified into autogamy (self-pollination) and allogamy (cross-pollination). There is a huge diversity of mechanisms in sexual reproduction: bisexual, hermaphroditic, monoclinous, perfect flowers by male and female expression in one flower, monoecious, dioecious, polygamous, etc. The classification of types is very closely related to self-pollination and cross-pollination. Methods in dissociating chimericas are therefore dependent on the type of breeding system of the species involved (**see Chapter 15**).

Many important food crop species such as rice, wheat, barley and soybean are self-pollinated species. Varieties of self-pollinated population, in general, are pure lines as out-crossing is completely restricted by natural systems such as cleistogamy (stigma is receptive and pollen shed inside closed flowers) or by artificial isolation and controlled self-pollination. Most self-pollinated species may hybridize with other genotypes and even in highly homozygous self-pollinating crops such as barley there is a low percentage of out-crossing, which can be influenced by environmental factors,

particularly stresses such as cold, drought and salinity as well as mutational stress. Japanese lawngrass, *Zoysia japonica*, and pearl millet, *Pennisetum glaucaum*, exhibit protogyny, in which the stigma appears first and matures in a flower before anthers appear, thus reducing the chance of self-pollination. In this case, plants are mostly pollinated by their neighbours. If neighbouring plants have a different genotype, heterozygous progeny will be produced. Out-crossing can be increased significantly in M_1 plants, because pollen fertility is often reduced after mutagenic treatment. For this reason it is important to isolate M_1 plants from other genotype plants.

Maize and rye and many forage grasses such as orchard grass, tall fescue and ryegrass, and legumes such as clovers, alfalfa, *Brassica* and fruit trees are cross-pollinated species. Cross-pollination is encouraged by self-incompatibility systems. Various genotypes and hence phenotypes are produced from cross-pollinations. Self-pollinated plants in cross-pollinated species can be produced by controlled pollinations, however, many cross-pollinated species exhibit inbreeding depression and the production of highly homozygous lines is made difficult if not impossible by repeated

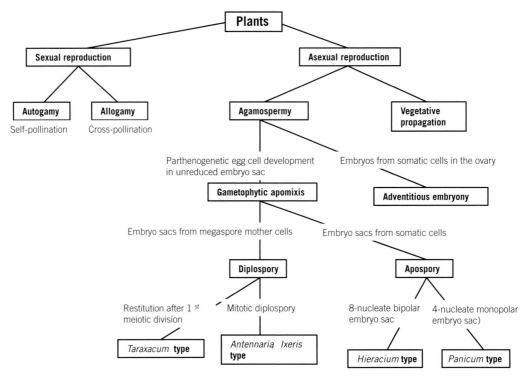

The Joint FAO/IAEA Programme

Figure 24.2 Different types of reproductive methods in plants (modification of Asker, S.E. and Jerling, L. 1992. *Apomixis in Plants*. CRC Press).

rounds of selfing, e.g. by single seed descent. However, in some cross-pollinating species such as rye which has a strong self-incompatibility system and suffers from inbreeding depression, viable 100% homozygous lines can be produced *via* doubled haploidy (Maluszynski *et al.,* 2003).

Effective mutant selection methods established for self-pollinated species are not applicable to cross-pollinated species. In a cross-pollinated population, mutated genes are present in a heterozygous condition and thus cannot be distinguished from normal plants by phenotype. The mutant can be followed and maintained in progenies by DNA diagnostics but at some stage a homozygous mutant that expresses the mutant trait is required. Therefore, the production of homozygous mutants from heterozygous material becomes a critical step in mutation breeding of such crops (methods for doing this are provided in **Chapter 18**).

2.3. Gene and Genome Biology

Gene biology and genome biology are well-established scientific disciplines and both impact on mutation breeding. As discussed in **Chapter 3**, a single gene is composed of several distinct DNA sequences and the protein product derived from exon regions may be composed of several functional domains. Screening mutations in exon regions would therefore be beneficial in increasing the efficiency of developing mutants with phenotypic effects. Furthermore, if the specific functional sequence domains are known for a target gene, these domains should be the target region, e.g. functional sequences in herbicide resistance genes (**see Chapter 31**).

Genes often work in concert, and therefore a mutation in one gene may affect the expression of other gene. This is the case for mutations in developmental genes, such as height and flowering time that have strong pleiotropic effects on a wide range of other characters, it is also important where a mutated gene is part of a linked biosynthesis pathway. In general, a mutation in a gene downstream of a pathway has less of an effect than mutations in upstream genes. For example, there are several genes involved in starch synthesis in cereal crops, mutations of the *Wx* gene only affect the synthesis of amylose, and they do not affect the overall synthesis of starch. However, mutations of other genes, e.g. the gene that produces ADP glucose pyrophos-

phorylase can affect not only starch content but also starch composition. Therefore, the *Wx* gene is a better targeted in breeding of waxy starch crops.

All living beings, according to the degree of relatedness, have many homologous genes – genes with similar structure and function. Related species share most of the genome of a common ancestor, from which they diverged millions of years ago at the beginnings of speciation. Geneticists discovered a striking synteny among plant species in which chromosomes of various species carry homoeologous (similar) genes in the same order. Synteny has relevance to mutation breeding and genetics. First, it indicates that one mutation identified in one crop species can also be induced in another crop species. This was established at the phenotype level, for example, semi-dwarf mutations can be generated in most crops, but synteny provided a genetic explanation and a platform for targeted mutation in other species. Second, knowledge of synteny is useful in identifying the candidate region of a mutated gene before starting a genetic mapping programme, and after being delimited to a chromosome region through gene mapping, the candidate gene of the mutation may be identified by searching homologues of a known gene with similar function.

Different plant species and genotypes of a given species differ in their sensitivity to mutagens; which is particularly evident in the fertility of M_1 plants (**see Chapter 14**). This can now be explained by genome and gene biology, as both are related to the survival of cells. Since some genes are essential for the stability of the plant genome and cell survival, the loss of function of such genes can directly affect the functionality of a cell. Since diploid plant species have only one copy of the genome in their gametic cells, their gametes are very sensitive to the loss of function, especially of essential genes. Similarly, embryos, seeds and plants with homozygous mutant alleles may not grow normally and may die at various growth stages. Similar mutations in polyploid plants can be tolerated due to genetic buffering of non-mutated homologues (in autopolyploids) and homoeologues (in allopolyploids). Thus, polyploid plants often possess high mutation densities relative to diploid plants when screened at DNA level (**see Chapter 20**). However, at the phenotypic level, this is not the case since mutations (very often recessive) in one genome are often masked by the wild type alleles in other genomes, hence the mutation frequency assessed at the trait level is often less in polyploid species than in diploid ones.

3. Strategies in Mutation Breeding

3.1. Approaches in Mutation Breeding

Mutation techniques can be used in different ways in crop improvement. It is therefore important to identify the most suitable approach in achieving desired objectives.

3.1.1. Direct Use of Induced Mutants

Much excitement was generated in the early days of induced mutagenesis as novel mutants overcame major obstacles in crop improvement and/or produced new and valuable variants. New forms such as semi-dwarfism, early maturity, disease resistance, etc. met immediate market demands and were often released directly as commercial varieties without recourse to refinement through cross-breeding. The development of direct mutants into commercial varieties is still a common practice in vegetatively propagated crops (VPCs) and is a major, if not the only, method of developing new varieties in some VPCs (**see Chapter 26**). For seed propagated crops (SPCs) there are many examples of direct mutants being released as new varieties, especially in diploid cereals such as rice and barley, however in general the current trend is to enter mutant lines into a breeding programme.

Breeding of traditional and local crops. Certain crop species are grown and consumed locally and have not been subject to much breeding, these are sometimes referred to as under-utilized crops, e.g. taro, amaranth and bambara groundnut. In such cases, mutation breeding is often a method of choice for the direct improvement of certain important traits, such as disease resistance. In some cases traditional varieties are preferred over "improved" varieties introduced from other countries, for example, traditional rice varieties are preferred by many farmers in certain regions of Asia, even though rice has been improved extensively elsewhere, and traditional sorghum varieties (landraces) are grown by many farmers in Sierra Leone to meet local demand. In such cases, mutation breeding can work well and elite mutant lines may be used directly as new varieties. **Chapter 33** provides an excellent example of mutation breeding as an efficient and only available technique for improving

traditional Basmati rice varieties.

Breeding for single traits and adaptability. In certain cases, a few single traits/genes play an important role in determining quality for end-user needs, either the end product itself or ease of processing. Mutation techniques can be applied to develop new varieties with such desired traits from otherwise excellent varieties. For example, many waxy rice varieties with desirable starch quality are induced mutants of high yielding non-waxy rice varieties. With respect to adaptability, quite often, a variety introduced from region A (e.g. of low latitude) performs very well in contrasting region B (e.g. of high latitude) with high yielding and desirable agronomic traits, but it is too late for commercial production. In such cases, early maturing mutants can be developed and often used as a new variety in region B. Early maturing (daylength insensitive) mutants of barley allowed the crop to be adapted to equatorial latitudes of Tanzania from the northern hemisphere (long day adapted) varieties. Mutants in genes controlling flowering time, especially vernalization and photoperiod response genes have played a major role in adapting crops to new environments. Since disease is often race specific, a resistant variety in region A may become susceptible to the same disease in region B when challenged by a different race; mutation breeding can be an effective method to improve the disease resistance in such cases. Numerous varieties of such kind have been developed in the past, such as the black spot disease in pear and apple in Japan (**see Chapter 26**), and this will remain a useful tool in breeding for adaptation.

Fine-tuning of elite lines. In cross-breeding, it is not uncommon that an elite line does not meet the requirement of one specific criterion that is needed for its release, production or increases its market value. For example, it may not have the optimal height, in cereals for example a tall variety is more prone to lodging (fall or lean over) in certain fields and environmental (weather) conditions making it more difficult to harvest. Height is a polygenic character controlled by many genes, some with large effects and others with smaller effects. Altered plant height, particularly height reduction, is easily achieved by induced mutation of major genes (**see Chapter 25**). Mutation breeding can be deployed in the tailoring of elite lines for many traits (agronomic, disease and quality).

The value of a mutant line, i.e. its potential as a new variety or use in a breeding programme will depend on the level of background mutations (mutation load). Since mutations are produced at random there will always be additional mutations across the genomes carried, some with negative effects that may or may not be immediately apparent. The following measures are recommended to minimize the effects of background mutations:

1. A relatively low dose of mutagen (e.g. lethal dose $(LD)_{20}$, **see Chapter 14**) is applied to reduce the mutation load;
2. A large M_1/M_2 population is developed and dozens of mutants for the target trait are selected, hence mutants with less negative background mutations exist;
3. Non-linked background mutations are removed by several rounds of self-pollination.

All these methods are designed to dilute the mutation load and as a consequence normally they involve the production and handling of more plant materials which is costly, takes up more storage and growing space and takes more time. A balance is therefore required between effort and expected gain.

3.1.2. Indirect Use of Induced Mutants

As mentioned above mutants are used directly for commercial production in VPCs, often with no knowledge for the genes involved. For SPCs the situation is somewhat different. New cultivated varieties are required to meet minimal standards (national and international) and provide financial benefits to growers (usually by possessing high yields and good quality). A mutant possessing a desired trait may fall short of standards. This can be due to one or several of the following reasons:

- The mutant has a heavy mutation load that cannot be removed. This is particularly common in polyploid crops that can tolerate deleterious mutations and which are often subjected to higher mutagenic doses than diploids and hence usually have relatively high mutational loads (**see Chapter 20**).
- The mutant source may be in an out-dated variety, or an unadapted line which requires major breeding upgrading to meet contemporary standards and may also be poorly adapted to the target environment.
- Breeding methods and pure line production, particularly for annual crops, are well developed and there are demands in meeting varietal DUS standards (distinctiveness, uniformity and stability) and in improving yield, quality and disease characteristics over standard varieties, thus

changes of a few traits by mutation breeding might not be sufficient to become a new variety.

- Some novel mutations may have negative effects on other traits that may not become apparent until large scale trials/commercial production are conducted, further breeding may be required to overcome such shortcomings.

Although a mutant may not be used directly as a new variety, it may have potential in breeding. In many cases the mutant line is crossed with other breeding lines, but a cross between the mutant and its parent can also be used to reduce the background mutations in certain circumstances (on average 50% of the mutations will be shed at each generation). When several mutants for different traits are developed, it is also advisable to inter-crosses these to combine (pyramid) the mutated traits into a single breeding line, in so doing the background mutational load is simultaneously reduced. Molecular markers can be used both aggressively in selecting for the desired mutant recombinants and defensively to select lines with minimal background mutation.

Once a mutant variety is developed, it can be used as a donor parent of the mutated trait in cross-breeding. This is especially the case where a mutant sets a new standard for a particular trait. For example, the barley mutant variety Golden Promise set a new standard for Scottish malt whisky. Hundreds of varieties have been developed in a wide range of commercial plants by cross-breeding using mutant varieties with traits of economic importance, such as semi-dwarf in rice and barley, disease resistance in barley (**see Chapters 1 and 2**), and herbicide tolerance in many crop species (**see Chapter 31**). Recently, mutants for novel traits, fatty acid composition (**see Chapter 32**), and low phytate content, are used in breeding new varieties for specific purposes.

Further crossing and selection is an effective way to minimize the negative pleiotropic effects of important mutations. The maize *opaque 2* (*o2*) mutation is an excellent example. The *opaque*-2 mutant contains an increased amount of lysine and has been used to improve the protein quality of maize. The *o2*-maize mutants are known to contain less α-, β-, γ-, and δ-zein, which results in increased amount of other proteins, including globulins and albumins. Despite the improvement in the amino acid composition and nutrition value, most *o2*-maize lines are not well-adapted because of their low grain yields, soft and chalky endosperm, and high susceptibility to kernel breakage and to pest and mould damages. Therefore, the *o2* could not be utilized until modifier genes to produce a vitreous (starch) endosperm while maintaining the nutritional advantages of *o2* maize were identified, which led to the development of quality protein maize (QPM). The vitreous endosperm, containing an elevated level of γ-zein, increases the resistance of *o2* maize to diseases and insect damages, increases the grain yield and improves grain appearance.

3.1.3. Other Approaches

Mutation techniques are also used in specific ways to achieve breeding objectives.

Facilitated chromosome translocation. Wide hybridization is used in many crops to introgress useful genes from wild species into cultivated forms. One mechanism used to achieve this is through chromosome translocation in which part of a chromosome from a wild species genome attaches to a chromosome of the cultivated species. This type of chromosome engineering can be greatly enhanced by mutagenic treatment (**see Chapter 19**).

Treatment of heterozygous material. Some breeders have observed wider genetic variation in progenies derived from highly heterozygous material such as F_1 or F_2 seeds treated with mutagens compared to non-treated seed. This method has been successful in producing dozens of mutant varieties. The enhanced variation is unlikely to have resulted from induced mutations since the mutation frequency is in general very low compared to the genetic variation arising from gene recombination. A possible explanation is that the mutagenic treatment enhances recombination. Since many genes of similar function or controlling a complex trait are often clustered on chromosomes, their linkage cannot be broken easily through recombination during normal meiosis. Mutagenic treatment can result in DSBs, which are repaired through homologous recombination using homologous DNA, e.g. those of sister chromosomes used as templates (**see Chapter 6**). Consequently, tight gene linkages may be broken up by mutagenic treatment.

Non-mutagenic applications. At specific doses, mutagenic irradiations can destroy the function of male gametes and thereby affect their ability to fertilize ovules, but the growth of the pollen after pollination may be unaffected. The embryos produced by aberrant pollination are often haploid and form a useful resource from which to develop doubled haploid plants. This is a popular method in developing breeding lines in citrus and other fruit crops (**see Chapter 30**).

X-rays have long been used as a tool to examine density differentials within and between tissues, especially in human medicine. In plants, X-rays have also been used as a tool to identify haploid embryos in seed batches, e.g. haploid melon embryos are less dense than normal diploid embryos and can be selected for using soft X-rays as a mass screening tool (Sauton *et al.,* 1989).

3.2. Techniques Related to Mutation Breeding

The advance of mutation breeding is much dependent on several biotechnologies, which include techniques of mutagenesis, dissociation of chimeras, production of homozygous mutants in seed crops and technologies for efficient screening and evaluating mutants.

3.2.1. Mutagenesis

Mutagenesis induced by various physical and chemical mutagens is described in detail **in Chapters 7–13**. The availability of these different mutagens and suitable treatment methods form a foundation for mutation breeding. Breeders can now choose a suitable mutagen to generate various types of mutations such as sequence changes, loss or altered gene function, chromosomal and genome rearrangements.

Generally it is the propagating material that is targeted for mutagen treatment. For SPCs this is normally the seed, for VPCs this can be *in situ* or *in vitro* vegetative propagules (whole plants, cuttings, buds, callus, etc.). The cells, tissues and organs of these propagules belong to the sporophytic generation of the plant. The male (pollen) and female (embryo sac) gameophytic phases may also be targeted. In addition to the vegetative cell, the male gametophyte generates the two sperm cells that fertilize the egg cell and the fused central cells of the female gametophyte (double fertilization) giving rise to the embryo and endosperm respectively. Mutation induction in the sperm and egg cells that generate embryos after fertilization will pass on mutations to the next generation and are thus attractive targets for mutagenic treatments in SPCs. The resulting mutant plants can be handled normally as described in relevant chapters. Treatment of pollen grains has become an efficient method for maize mutation induction.

More novel methods in gametophytic irradiation include the treatment of microspores; these are formed after meiosis, are uninucleate and develop into pollen grains in normal plant development. However, *in vitro* culture of microspores with various stress- and hormone treatments can interrupt and divert normal development and induce embryogenesis in these single cells. Microspore culture is used routinely in many species (asparagus, brassicas, barley, maize, rye, tobacco, triticale, wheats, etc.) to produce haploids and thereby doubled haploid plants. Doubled haploidy is the fastest route to total homozygosity and is of immense value to plant breeders and geneticists. The attraction in mutation breeding is that any mutation induced in a cultured microspore will be fixed in a homozygous condition immediately on developing a doubled haploid plant *via* spontaneous or induced chromosome doubling (**see Chapter 29**).

3.2.2. Production of Solid/Homo-Histont Mutants and Homozygous Mutants

Since only solid, homo-histont (non-chimerical) mutants are of practical use, technologies for producing them become critical. In SPCs, this is not difficult since the sexual reproduction process can largely solve the problem of eliminating chimeras. VPCs are different and various methods have been developed to produce homo-histont mutants (**see Chapter 26**).

In SPCs, it is generally routine to produce homozygous mutant plants and lines; since many mutations are recessive they do not express a change in phenotype when they are in a heterozygous state. Thus phenotypic selection is useless until homozygous mutants are made available. Different methods are needed to produce homozygous mutants for out-crossing and self-pollinating crops (**see Chapter 18**). Double haploid technologies are also used as efficient means for producing homozygous lines, particularly for out-crossing plants, however these technologies are currently routine for only a small number (about 30) of crop species (**see Chapter 29**).

3.2.3. Mutant Screening and Selection

From the very beginnings of induced mutagenesis it has been realized that a major constraint in mutant deployment is the ability to select rare mutant individuals. It was argued that since desired mutants occurred at such small frequencies efficient mass screening methods would be needed for their selection. The exploitation of mutants in plant breeding and genetics has always been dogged by difficulties in selecting the rare useful mutants in populations that are largely of little interest. Stadler, one of the pioneers in plant mutagenesis pointed this out at the onset of induced mutagenesis (**see Chapter 1**) and argued against induced mutagenesis as a viable plant breeding tool unless mass screening methods could be deployed. Von Sengbusch is credited with developing the first efficient mass screening method. In the 1920s he discovered 'sweet' (alkaloid-free) spontaneous mutants in various bitter and poisonous lupin species, but more significantly he developed a rapid method for their selection. Many thousands of seed were screened and the "sweet" mutants (along with other spontaneous mutants for non-shattering and water-permeable seed) paved the way for the domestication of lupins and their development as a fodder crop. The mass screening method of von Sengbusch was based on simple mecurimetry (precipitation of seed hot water extracts with potassium tetraiodomercurate) and because of its commercial value was kept secret until 1942 (von Sengbusch, 1942). The problem for plant breeding and genetics is that the ultimate goal is the production of desired phenotypes. Therefore, development of suitable high-throughput phenotyping methods will remain an important task in mutation breeding and genetics.

Visual screening and selection. Some mutant traits, e.g. plant stature, leaf colour and fertility are obvious and can be easily detected and selected by simply growing out and looking at a mutant population. Many single mass screening methods have been developed for specific traits. The simplest of these is a "shot gun" approach in which a mutant population is grown out in an area subject to a selection pressure, e.g. fields with high/low fertilizer application, fields treated with a herbicide, fields with a high incidence of disease, temperature, drought, salinity, etc. Thus mutants for fertilizer use efficiency, herbicide resistance and resistance to biotic and abiotic stresses, etc. can be selected simply by walking the field and picking out promising plants. These can then be subjected to more rigorous testing for validation and in determining inheritance. For plant breeding purposes any screening method is useful if it is cost effective and efficient. The ability to screen at seed or seedling stages is particularly beneficial as this saves time and space and removes unwanted genotypes early. Trait screening that fits into the normal procedures of plant breeding (seed production, seed storage, germination, trialling, etc.) are relatively easy (**see Chapter 25**). Others may require more specialized screens, for example, disease resistance, might be revealed through *in vitro* inoculation of leaf segments.

Screening mutants of biochemical traits. Qualities, including human health and nutrition traits have become important targets in modern breeding programmes. The analysis of quality traits normally requires specially equipped laboratories. Theoretically, mutation breeding can be used to generate novel alleles of target traits or to improve a variety for such traits; however, since tens of thousands of plants need to be screened to identify a desirable mutant, it is not always feasible for the following reasons. First, the content of chemical components is often affected by micro- and macro-environmental factors; there are often great variations among plants grown even in the same field with "uniform" agronomic practices. Therefore, it is sometimes very difficult to distinguish mutants from natural variants by chemical analysis. Second, with respect to economics it is not feasible to analyse tens of thousands plants, if the cost per unit test is expensive, this is not uncommon. For example, if it costs 10 dollars per sample (labour plus chemicals), it will cost at least 100,000 dollars to screen the plants of a reasonably sized population. The effort and costs involved in testing will therefore depend on the value of the trait sought and it is often not viable. Third, although the running of chemical analyses is often automated and efficient the whole process is often time-consuming due to lengthy sample preparation times and data analysis. This is compounded by scaling, for example, if one person can analyse 50 samples per day, then it would take 200 days to complete 10,000 samples. This is not compatible with population sizes produced nor with breeding programmes where decisions need to be made within months after harvest and before the next season's planting.

Some rapid, simple and indirect methods are helpful for pre-screening in this regard. Many quality traits for example oil, protein, amylose content can now be analysed using near infrared spectrometry (NIRS). NIRS can be used for single kernels (grains) analysis for some traits. Although there has yet been no example of NIRS being used for screening mutants, theoretically and technically it has the potential to be utilized. Other methods, such as colourimetric analysis of inorganic phosphorus (as an indicator of low phytate content) and enzymatic analysis of trypsin inhibitors are also useful for particular traits as far as they are cost- and time-effective.

High-throughput phenotyping (phenomics). The current biggest advancement in high-throughput phenotyping is the development of phenomics platforms able to measure phenotypic traits either throughout the life cycle of a plant or on specific organs such as fruits, seeds and leaves. Phenomics is emerging as an exciting new phenotypic data capture biotechnology that can keep pace with rapid advances in high-throughput genotyping. This is achieved by the development of sophisticated equipment able to measure physical and chemical parameters in a large number of samples. These methods are usually computerized and normally non-destructive. For example the MARVIN digital seed analyser/counter is able to image, weigh and process large numbers of seed in one batch rapidly (e.g. about 100 barley seed/batch/minute) and can be used to select mutant seed traits for shape, size and weight. The machine is simple to operate and one person can screen many thousands of seeds in a day. Likewise data on seed chemical composition can be gathered using near and far infra-red spectroscopy and may be exploited in selecting seed varying in quality characteristics. Additionally seed density determined by X-ray analysis may be used as another measure of quality or to screen for haploids. These methods have until recently been stand-alone platforms, but there is the potential to integrate these into a series of non-destructive tests gathering as much information as possible on large numbers of seed. This therefore would be especially attractive as a pre-screen in mutation breeding of seed crops.

Phenomics facilities are also being developed to perform non-destructive tests on plants throughout their life cycle. Some such as the Australian Plant Phenomics Facility follow plant development from seed sowing to germination to vegetative and reproductive growth to harvest (**see Chapter 41**).

Importance of uniform field trialling. The failure in the application of mutation techniques to some important traits can partially be ascribed to the non-availability of a screening method that can be applied to large mutated populations. An important consideration for field selection of mutants is homogeneity of field plots with respect to environmental conditions. Subtle environmental changes can affect trait expression and subsequently, the selection process. This is especially important in a mutation breeding programme because the screening step within the M_2 population normally aims to select a very small number of mutants from a large number of plants and should therefore minimize the selection of false positives. This is particularly relevant in screening mutants tolerant to biotic and abiotic stresses (such as drought), and mutants of altered chemical components (**see Chapter 41**).

Molecular screening of mutations. As described in **Chapter 20**, until recently the vast majority of mutants have been selected, described and catalogued based on phenotypic attributes. The explosion of genomics data and fast development of various tools for detection of DNA variation in recent years, coupled with bioinformatics, has enabled the establishment of genotypic methods for screening mutations.

3.3. Mutation Breeding for Different Traits

The strategies utilized to select and improve desirable and specific characteristics in a plant breeding programme vary with plant species, the environment where the plant will be cultivated, the farmers' cultivation method, and the utility of the product. The objectives of a mutation breeding programme are basically the same as those of any other breeding method and involve the same technologies.

3.3.1. Traits Improved trough Mutation Breeding

Based on literature and data in the FAO/IAEA Database of Mutant Variety and Genetic Stock, mutant varieties with the following improved traits have been developed.

High yield. Stable and high yield potential in specific environmental conditions is probably the most important objective of most of plant breeding programmes. Yield is a complex trait and influenced by other breeding objectives, such as plant architecture, maturity, nitrogen utilization efficiency, resistance to biotic and abiotic stresses, etc. It is difficult to use mutation breeding to improve the yield potential of crops that are well established and which have been subject to intense and refined breeding. However species new to agriculture (e.g. Jatropha) or which have received limited breeding (e.g. medicinal, spice and herb crops) can be taken to new yield levels by breeding for an ideotype, e.g. by changing plant stature.

Resistance to biotic stress. Biotic stresses are primarily diseases caused by fungi, bacteria and viruses, and the damages induced by insects, animals, nematodes,

weeds etc. Historically, mutation breeding methods have been quite successful in improving disease resistance, but not for insect resistance.

Diseases involve a complex inter-play between a host plant and a pathogen. The resistance/susceptibility response can involve several components. This is generally called a gene-for-gene relationship. Induced mutations may change the interaction and inhibit certain steps in the mechanism of infection. Numerous mutants have been developed through mutation induction, showing enhanced resistance to various diseases (virus, bacterial and to some extent fungi). For example, several authors have induced mutations at the same locus (*ml-o*) located on the short arm of chromosome 4H in barley for resistance to powdery mildew, and mutations conferring resistance to barley yellow mosaic virus have been induced in Japan. The *ml-o* mutant is interesting as it has not broken down and has provided unprecedented resistance to mildew for many decades, it is assumed that this is due to gene knock-out as several mutation methods can be used to induce this resistance. In other cases where resistance to specific pathotypes is conferred by a specific host gene allele, which might have only a single nucleotide difference from susceptible one, chemical mutagens may be deployed to provide subtle mutations in the target gene sequence.

As discussed above, simple and easy screening methods are of paramount importance for successful screening of disease-resistant mutants. In field screening, this is often performed in disease hotspots. If no suitable hotspot fields are available, artificial infections become necessary for example growing plants of a highly susceptible variety for the pathogen to multiply on and act as an inoculum source in spreading spores to test plants. For some diseases, toxins produced by the disease might be used in *in vitro* screening of resistant mutants. For example, the toxin of *Alternaria alternata* (Fr.) Keissier (= *Alternaria kikuchiana Tanaka*), which causes a serious black spot disease of Japanese pear can be efficiently used for *in vitro* screening of resistant mutants (Sanada, 1988).

Unlike diseases, there is little interaction between host plants and their insect pests, as one pest may attacks other plant species or even different plant genera. This predator-to-host interaction may explain why there are fewer examples of mutant varieties carrying induced pest resistance. Although the mechanisms of insect resistance are often not clearly defined, the mechanisms that influence the plants resistant against insects are, and include: 1) non-preference ; 2) antibiosis; and 3) tolerance.

Non-preference is a characteristic of plants whereby insects do not infest. This includes various characteristics such as colour, odour, taste, pubescence, latex production, etc. Aphids prefer greener cabbage and leafhoppers prefer soybeans with non-pubescence. These characteristics may be induced by mutagenesis. Antibiosis includes the production of substances by plants that inhibit, kill or prolong the physiological maturity of insects. Since it is difficult for mutagenesis to generate a new defence gene, it may be difficult to use this strategy in a mutation breeding programme. Tolerance towards insects is a quantitative reaction and may include characteristics such as plant vigour, the ability to produce many shoots and many roots, as well as strength of stem tissue and avoidance (little or no vegetative growth when insect pests are prevalent). Recently, there has been some progresses in identifying resistance gene(s) to brown plant hopper in rice (Fujita *et al.*, 2008), as such, it may be possible to induce resistant rice mutants efficiently by targeting the identified gene(s) related to the resistance mechanism. In this respect transgenic breeding for pest resistance has been more successful, e.g. GMO crops carrying the BT toxin gene.

Tolerance to abiotic stress. Abiotic stresses include unfavourable environmental conditions such as soil salinity, extreme pH, water deficits and flooding, and weather. The approaches utilized within these programmes are sometimes quite simple. As an example, the M_2 plants of rice were sown in high saline paddies and the M_1 calli are transplanted on the medium with high salt concentration (Lee *et al.*, 2003). From this experiment, Lee *et al.* (2003) selected two promising salinity-tolerant lines from regenerated plants. Recently, two candidate salt tolerant rice lines were induced by carbon and neon ion beams (Hayashi *et al.*, 2008)

Though many of the mechanisms of abiotic tolerance are unknown, the threat of global warming and climate change forces new approaches in adapting crops to changing environments. Huge numbers of potentially useful genotypes have been generated from mutation breeding research and breeding programmes and are available in germplasm collections, e.g. tolerance to cold, heat, daylength and drought.

Quality, nutrition and functionality. Quality usually refers to the composition of organic compounds produced and stored by plants, such as starch, protein, fatty acid, vitamins, etc. The nutritious value of harvestable products is the most important objective for the plant breeding after yield. This involves the elimination of undesired substances such as anti-nutritional factors. Raising or lowering the concentration of specific substances such as fatty acids and changing their compositional profile have been major targets for mutation breeding. Biosynthesis of such substances is controlled by the presence of a suite of enzymes. The simplest idea for modifying these compounds through mutation is by inducing knock-outs in genes involved in the metabolic pathways, thus increasing the synthesis of upstream substances and decreasing the production and concentration of downstream substances. In addition to changing the concentration of quality traits the components themselves can be changed qualitatively. A prime example here is the altered fatty acid composition of many oil crops *via* mutating genes. Canola for example is a form of rapeseed that has high quality edible oil, this was achieved by lowering the levels of toxins (glucosinolates) and the fatty acid, erucic acid by gene knock-outs induced by gamma ray irradiation. The removal of anti-nutritional factors has been key to domestication of wild plants for edible purposes for which spontaneous or induced mutations have been key, e.g. sweet lupins mentioned above.

Proanthocyanidins of barley cause problems in brewing as they produce hazy beer, the removal of proanthocyanidins was attempted by NaN_3 treatment to induce mutations that block the flavonoid pathway. Recently, mutation breeding has been used for enhancing bioavailability of important nutrients in certain crops. For example, crops with low phytic acid content are preferred because the bioavailability of mineral elements and phosphorus can be significantly increased. In this regard, two barley mutant varieties have recently been released for commercial production (see review, Raboy, 2009).

Mutation breeding has also been used for breeding crops with special functionality. In Japan, rice mutant varieties with low glutelin content, such as LGC-1 and its derivatives, have been developed for patients who must restrict protein intake, such as kidney disease patients. These varieties are now used for this type of diet therapy. Rice varieties with high content of resistant starch are being developed in China for dietary therapy of patients with type 2 diabetes. These varieties have about 10 times higher resistant starch than normal rice varieties and preliminary tests have shown they are effective for controlling the glycemic index when substituted into diets, though more studies are needed (Shu *et al.*, 2009). It is expected that more foods with such novel functions will be developed through mutation breeding.

Plant architecture. Mutation breeding is frequently used for inducing mutants affecting plant architecture. Plant architecture includes phenotypic traits such as plant height (culm length of cereals), plant type, branching character (number of tillers), size, number and shape of leaves, stolon characters, size and number of flowers, etc. Among them, mutants conferring reduced plant height, which generally has had a positive effect on increasing plant yields by reducing lodging under high fertilizer condition and high tillering. This characteristic has been frequently utilized across cereal crops and legumes. One of the most successful utilization of plant height mutations is the application of the semi-dwarfness gene, *sd* of rice. The *sd1* gene which controls short and stiff stems in rice, also confers increased grain yields through a reduction in lodging. Semi-dwarf rice varieties have been a leading example of breeding progress as defined during the "Green revolution". The first semi-dwarf mutant variety of rice var. Reimei was induced through gamma irradiation (Futsuhara, 1968). This variety's height was reduced by at least 15 cm when compared to the original var. Fujiminori, and carries the same *sd1* (semi-dwarf) allele of the spontaneous mutant var. Dee-Geo-Woo-Gen and var. Calrose 76. All of these varieties are widely utilized across Asia (Rutger, 1992). Since identification of the *sd1* gene, in Japan, 80 of 229 registered, indirect use mutant varieties represent descendants of var. Reimei. In the USA, greater use has been made of the semi-dwarf trait from cv. Dee-Geo-Woo-Gen derivatives; however, few additional induced or identified spontaneous mutations have been as successful as *sd1* (Rutger and Mackill, 2001). Similar achievements have also been made in oats, barley and other crops.

Mutants have been used to study the basic architectural building blocks of plants, the phytomer. Phytomer mutants can be used to study and predict the type of organ produced by meristems at various positions (apical or side) at various stages in the life cycle. The most basic phytomer unit consists of a stem segment with a leaf attachment which can be replicated in an apical or side direction. Studies of phytomer mutants can be used in taxonomical studies in defining what type of structure is developed from a meristem in a given species. For example a classic taxonomic difference between wheat (*Triticum*) and barley (*Hordeum*) genera is that floret production in wheat is indeterminate whereas it is determinate in barley. This leads to wheat having multiple florets per spikelet whereas barley is restricted to one. Wheat therefore produces far more seed per spike (ear) than barley. However, a phytomer mutant in barley, known as the "wheat mutant" exhibits indeterminate floret production and is therefore of interest in increasing yield (Forster *et al.*, 2007); it is also of taxonomic interest as it indicates that this particular taxonomic descriptor is governed by a single gene. Close to 100 phytomer mutants have been described in the model species barley (Forster *et al.*, 2007). Many of these altered traits are of interest to plant breeders, some examples are: *ari* – short awn, *blf1* – broad leaf, *brc1* – branched spike, *cul* – uniculm (single stem), *dub1* – double seeds, *eam* – early maturity, *ert* – short rachis internodes, *flo* – extra floret, *gig* – gigas organs, *hcm1* – short stem, *int* – intermedium spike, Kap – hooded lemma, *lax* – lax spike, *lel1* – leafy lemma, *lnt1* – low tiller number, *mnd* – many noded dwarf, *mov1* – multi-ovary, *nld* – narrow leaf dwarf, *ovl* – ovary-less, *pyr* – pyramide spike, raw – smooth awn, *sid1* – single elongated internode dwarf, *trd* – additional bract on spike, *ubs4* – unbranched style 4, *uzu1* – shortened organs, *viv* – viviparous, *vrs* – six-row spike, *wnd* – winding dwarf and *Zeo* – short stem.

Maturity. Early and late maturing mutants are frequently induced through mutagenesis and are easily identified. Early maturity in cereal and legume crops is one of the most useful characteristics for cultivation in cool temperate regions, offering the opportunity to flower in frost free conditions, harvest prior to frost and, in drought-prone regions, the ability to produce a viable crop prior to drought conditions. The flowering time and maturity of cereals and other crops is controlled by the plant's ability to sense season *via* temperature and daylength sensors, controlled by vernalization and photoperiodic sensitivity genes. Sweden was a pioneering country for mutation breeding and developed the barley variety Mari, a mutant variety exhibiting early maturity and semi-dwarfism. This was developed by radiation in 1960 and many valuable indirect-mutant varieties were developed from it. Early flowering and maturity mutants have been induced in a number of crops, other notable examples include banana, cotton, pearl millet, rice, soybean, etc.

Agronomic and related traits. The semi-dwarf trait is generally related to modern production systems. These have two major advantages: 1) they change the source sink relationship so that more energy is directed towards seed production and less to vegetative growth, and 2) amenability to mechanical harvesting. Some other characteristics of plant architecture, such as increased branch number in cereal and legume crops, more compact growth habit such as high plant density, shorter stolon length of potato, etc. can also affect the efficiency of growth and harvest, particularly in modern, mechanized agriculture. There is one unique mutation with a lower number of axillary buds in chrysanthemum, which was induced by ion beams in Japan. This mutant is very useful for the flower industry because in non-mutant chrysanthemum large flowers can only be produced by removing (hand picking) axillary flower buds from the plant. This mutation negates the need for axillary bud removal. Notable successful varieties carrying this mutation include var. Imajin (Imagine) and var. Alajin (Aladdin).

Another important plant character induced by mutagenesis is non-shattering rice. This mutation, induced by gamma-ray irradiation utilizing a shattering Indica variety, was used to release the forage rice variety "Minami-yutaka" in Japan (Kato *et al.,* 2006). The mutation, which may be induced by gamma ray irradiation in gramineous species, will be useful for improving forage crops with high levels of seed shattering in their inflorescence. To make agriculture greener, traits such as super-nodulation and more efficient use of water and other mineral elements in soils will become more and more important. The success in breeding and release of a super-nodulating soybean variety is a significant advance in this field (Takahashi *et al.,* 2005).

For species new to agriculture, e.g. blueberries, jatropha and those that have received little attention from plant breeders, e.g. medicinal plants and culinary herbs and spices, there is an urgent need to identify, develop and establish domestication traits. For species that have been in domestication for thousands of years many agronomic traits have been provided through spontaneous mutants and have been incorporated into the crop as they arise. The small grain cereals, rice, wheat, barley, etc.

provide a good example. The grassy wild progenitors of these crops possess natural seed dispersal mechanisms whereby the seed head shatters into pieces at maturity and individual units carrying seed fall to the ground and are dispersed by hooking onto passing animals. Such dispersal mechanisms are inappropriate to agriculture, and cereal crops were not established until mutants for non-shattering were found. Interestingly the barbed awn trait useful for natural dispersal has persisted, however, barbs are now thought to be associated with the dust produced during mechanical harvesting that causes 'Farmers lung' and smooth awn mutants, e.g. *raw* of barley are now of interest in addressing this modern domestication/health issue.

Post harvest traits. Post harvest traits are often overlooked but have significant value in terms of transport and storage or harvested product, processing of raw and secondary products, marketing, shelf life and ultimately delivery of affordable products to the end-user (the consumer).

4. Theoretical Considerations and Planning of a Mutation Breeding Project

The urgency in plant breeding is to produce new improved varieties as rapidly and economically as possible. Plant breeders will therefore adopt the most efficient methods to achieve this. With the availability of different techniques, mutation breeding can be applied in various situations. However, mutation breeding is not necessarily the best choice. Therefore, it is important to first assess whether mutation breeding is the right choice in achieving the objectives. Other methods include cross-breeding and transgenic breeding. However, these methods are not mutually exclusive and can be combined pragmatically by breeders in reaching goals. For all breeding approaches planning is important and requires knowledge of the biology of the crop (inbreeding or outbreeding, diploid or polyploid, etc.) in the careful utilization of all available knowledge, technology and resources.

4.1. Mutation Breeding *versus* other Breeding Methods

Mutation breeding differs from other breeding methods in several aspects. Consequently, it has advantages and disadvantages and considerations should be comprehensive, taking into account the technical feasibility of achieving the breeding objectives, access to available germplasm, access to mutation facilities, transport of M_0 and M_1 seed (or other propagation material) between mutation facility and breeding location, the human resources required (trained breeder and supporting staff), the economic resources (budget for laboratory and field work), capacity (available field space, seed storage, selection methods) and the urgency of a new variety, for example for combating a newly emerged disastrous disease. Additionally issues such as the ethical concerns of producers and end-users, intellectual property (IP) and the formalities of official release of mutant crops should also be considered, though these latter three are generally of more concern to transgenic (GMO) crops.

4.1.1. Technical Assessment of the Suitability of Mutation Breeding

Under what circumstances should mutation breeding be deployed? This is often is the first general question that is asked by a breeder.

Breeding varieties with novel traits. In certain cases, the desired trait does not exist in germplasm collections of a crop, and hence mutation breeding becomes a valuable method to generate and develop new varieties with such desired characteristics. This often happens in the following situations.

1. For plant species that are not widely used and cultivated, for example, traditional vegetable crops that are grown locally often have very limited or no available germplasm repositories and many desired traits may not exist in the locally grown material. In such cases, cross-breeding is severely limited, but mutation breeding can generate desired traits and develop new varieties in such crops.
2. There are continual demands from growers for material with new traits required for more efficient crop production. This issue becomes exacerbated by climate change. Small changes in the growing environment, e.g. 1°C increases in temperature, or extreme heat stress, and changes in precipitation are expected to require new better adapted varieties. In addition to adapting crops to the physical environment, new pests and diseases are expected to emerge. These abiotic and biotic stresses are new challenges for crop production and breed-

ing. These traits may not exist in current varieties or extant germplasm and may require mutation breeding to generate novel alleles.

3. Other examples are end-user traits, such as novel quality and nutritional/functional traits that could bring benefits to particular market sectors. For example as mentioned earlier: the low phytate trait for enhancing bioavailability of micro-mineral nutrients and high content of resistant starch for lowering the glycemic index of diabetes patients and altered starch for industrial uses (e.g. bio-ethanol).

Reducing time needed for developing a new variety. Where the required trait is available in germplasm collections of the crop species, the breeder has a dilemma, whether to cross this material with elite lines, followed by backcrossing if necessary, or to attempt to induce the variation directly into elite lines *via* mutagenesis. The advantage of mutation breeding in such cases resides on the time saved to breed a new variety. For example, homozygous lines can be obtained in the M_4 generation (four crop cycles) at the latest through mutation breeding; it would take at least 8–10 generations to produce a similar line through cross-breeding, in annual crops this translates into a saving of up to 5 years, in perennial crops this is considerably longer.

4.2. Advantages and Disadvantages of Mutation Breeding

The successful breeding of crops through traditional breeding procedures requires a wealth of genetic diversity from which the plant breeder makes appropriate selections. This diversity can be found within a species or in some instances, identified in a distant relative. Genetic variation results over time through evolutionary changes in genes. Such changes are exhibited as the natural variation in a species and this variation can be conserved by man as seed reserves, plant tissue stocks and gene banks of national and international agricultural centres. When one identifies specific accessions having characteristics appropriate to the targeted breeding objectives, the breeder can initiate the hybridization of the promising accessions and select plants with the improved characteristic(s).

If the available genetic resources do not exhibit the desired characteristics or meet the targeted breeding objectives, the inducement of genetic variation by

Box 24.2: Advantages and disadvantages of mutation breeding

Advantages

Possible to achieve instant progress in elite material.

Single trait improvements can be made to an established variety preferred by producers, processors and/or consumers.

Direct mutant varieties are possible, or limited breeding effort required.

Novel variation can be produced.

Single gene mutants with no negative pleiotropic effects are possible.

Production of environmental responsive traits (e.g. flowering, fertility under various day lengths).

For some mutagenic treatments such as gamma and X-ray, there is neither residual radiation nor chemical contamination of the treated material. The treated material is safe to handle.

Specific genes/traits can be targeted.

Possible to calculate chances of success (mutation frequency).

Disadvantages

The process is generally random and unpredictable.

Useful mutants are rare and predominantly recessive.

A heavy mutational load (mutation density) may require intensive breeding to reduce background mutations and eliminate chimeras.

Large population sizes and effective mass screening methods are required to select rare mutants.

Mutants can have strong negative pleiotropic effects on other traits, e.g. high lysine/protein in barley and low lignin/high digestibility in maize are associated with low yields (pleiotropic effects however are an issue in all forms of breeding).

Unknown interactions with environmental factors, performance may vary significantly in different environments (this is an issue in all forms of breeding).

Health risks: careful handling and disposal of waste chemical mutagens; background radiation in fast neutron treatments needs to decay before handling treated material.

Most mutants are of no use to breeding even if a large number of mutants can be produced.

The number of lines/families in a generation can mushroom after the M_1 generation.

Field trialling and germplasm storage can be expensive and require a lot of space and careful management if large mutant populations are handled. However, effective screening and selection for desired mutants at an early stage can negate this problem.

mutation is a logical and appropriate approach. One advantage to a breeder in utilizing mutation breeding is that a reduced breeding time can be achieved. If a breeder attempts to transfer a useful trait into an economically important variety, this will necessitate the crossing of the economically important variety with perhaps a less desirable variety possessing a single advantageous agronomic trait. It will then be necessary to take the F_1 hybrid and backcross it to the superior parent variety or top-cross with the latest best variety for at least seven or eight generations, with continued selection of individuals possessing the particular useful trait. If one attempts to apply mutation breeding to the same breeding objective, one may be able to induce a mutation of the desired trait in the economically important variety directly. If the trait is successfully captured in the M_2 generation, purification of the mutant requires only one or two additional generations and this may be further accelerated using doubled haploid technologies (Maluszynski *et al.*, 2003).

The advantages and disadvantages of mutation breeding are listed in **Box 24.2**. A central feature of mutation breeding is the generation of new mutated alleles which present both opportunities and challenges for the plant breeder. Advantages include the creation of new alleles that do not exist in the elite germplasm pool. Mutation induction in elite lines is very attractive to breeders as refinement into a commercial line can be achieved quickly with little further breeding effort. In some cases the induction of desired mutant alleles in a commercial line can produce an immediate 'new improved' variety. A disadvantage in mutation induction has been a limited capacity to produce dominant or co-dominant alleles, the vast majority being recessive. However, this situation is changing with the ability to generate functional mutants. Also low mutation frequencies require growing out large populations to screen for desired and rare mutants. This can be an expensive process. The advantages of mutation breeding have been exploited by plant breeders throughout the history of mutation breeding who have devised ways and means of overcoming problems.

Polyploid crops have additional issues. Mutation detection in polyploids is difficult as any homozygous mutant locus, which is easily detected in diploids, can be masked by wild type genes present at homologous (same gene in another location) and homoeologous (similar gene in another genome) loci. One of the effects of ploidy is to provide the plant with a genetic buffer, thus polyploids have a greater tolerance of mutations. The number of targeted loci is also higher (proportional to the ploidy level/number of homologous and homoeologous loci) which necessitates more intensive mutation treatments, and as a consequence more backcrossing to relieve the induced mutational load. One means of overcoming these problems is to carry out the mutation induction at a lower ploidy level and use this as a bridge to introduce mutations into the polyploid crop. Polyploids do however have an advantage in that genomes and chromosomes can be moved around, added and replaced using cytogenetic manipulations (e.g. see Law *et al.*, 1987 for methods in wheat). It is therefore possible to carry out large scale mutation sweeps in polyploids by replacing entire chromosomes.

4.3. Planning a Mutation Programme

In a mutation breeding approach it is important for the breeder to understand the mode of reproduction (sexual or asexual, self-pollinated or cross-pollinated, annual, perennial) of the crop under study. In addition, understanding how the ploidy level (diploid or polyploid) may effect trait expression is also an important consideration. Once a basic understanding of the biology of a particular species is known, the next step involves the selection of promising phenotypes and genotypes. In a mutation breeding approach, seed treatment with a mutagen, followed by selection in the M_2 is the principal approach for identifying desirable mutants. Calculating the probable number of M_1 plants is very important for the effective screening and identification of mutants in the M_2. An efficient screening approach is essential in a mutation breeding approach since a high mutation induction rate, followed by a poor screening approach could result in inefficiency and ultimately project failure. The strategy and screening methods applied by the breeder will be dependent on the breeding objectives. As an example, the method deployed for selection of a trait, such as resistance to biotic or abiotic stresses will differ from that applied to selection of traits for quality, biomass yield, etc. A well-conceived and efficient screening approach, with appropriate numbers evaluated for a targeted genotype, will offer the highest level of efficiency and project success.

4.3.1. General Considerations

It is of paramount importance to understand the biological and technological basis of mutation breeding in designing an effective mutation breeding project as described above. This includes the selection of starting materials, use of an appropriate mutagen, and deployment of efficient and affordable selection methods.

Planning and selection of a particular targeted variety for mutation induction represents an important first step. An appropriate choice of a high performing variety with selection for an appropriate targeted mutation will ultimately determine the success or failure of the mutation breeding programme. An economically successful variety is adopted for mutation induction, usually by X-ray or gamma ray irradiation, or treatment of a chemical mutagen. The selected mutagenic treatment typically results in only single or simple trait modification and aims to maintain other agronomically useful characteristics. Since the mutation rate at one locus is typically very low, the probability for simultaneous induction of two or more desirable mutation is negligible. Moreover, the accumulation of multiple, possibly interacting mutations in one individual could lower the viability of the mutant. Usually, the choice of target variety or breeding materials is determined by the needs of processors, consumers and farmers. For example, if there is a variety that possesses processor and consumer popularity, but which is susceptible to a particular disease, then the approach would be to induce and identify a mutation in that variety for disease resistance. In certain circumstances, another approach may be recommended. For example, there are two ways to develop new varieties with salinity tolerance, one is to mutate a variety that is high yielding but susceptible to salinity, the other is to mutate a variety that is low yielding but tolerant to salinity, such as a traditional local variety grown in the affected area. In such a case, because it is much easier to improve agronomic performance than to enhance salinity tolerance, the latter is the better option.

The next step is to select an appropriate mutagen. A relatively small number of mutagens are known to induce mutations effectively and efficiently. In addition, different mutagens create different types of mutation. In general, chemical mutagens induce DNA base substitutions and radiations such as gamma rays and ion beams induce small or very large deletions (**see Chapter 20**) with minor exceptions. There are also different effects on mutation induction between radiations with low linear energy transfer (LET), such as gamma rays and X-ray, and high LET radiations, such as neutrons and ion beams (**see Chapter 9**). Therefore, it is important to use a suitable mutagen to achieve the specific breeding goal.

The third step is to deploy an efficient method for mutant screening and validation. Following selection, it is important to verify if the selected trait is heritable. To assess the heritability of a newly identified mutation, it will be necessary to conduct multi-location trials across several locations for a minimum period of 3 years for annual crops, more for perennials. Following successful performance trials, data must be evaluated and a decision for variety designation and release will be pending following a variety review by committee. Data collection, storage and analysis (bio-informatics) are also important considerations.

4.3.2. Mutation Breeding in Sexually Propagated Crops

Induced mutation and breeding in sexually propagated crops is relatively straightforward as the propagating material, the seed, can be treated directly and mutant lines refined by successive rounds of meiotic recombination/breeding and selection. Furthermore seeds are easily transported from e.g. a seed store to a mutation facility and onward to a mutant screening/selection site. However, whole plants, tissues and gametes can also be treated with mutagens with selection usually operating in subsequent generations (M_2 and above). Sexually propagated crops can be divided into two types, inbreeders (self-pollinating species) and outbreeders (cross-pollinating species). Different breeding strategies are use for inbreeders and outbreeders but they share some common features. For practical considerations these include:

1. Consideration of crop biology (inbreeder, outbreeder, ploidy level, etc.).
2. Consideration of the genetic control of target traits (single gene, polygenic, dominant/recessive, pleiotropy).
3. Choice of starting material (elite genotypes, varieties, homozygous lines, etc.). Homozygous lines have an advantage as they safeguard against the selection of false positives which can arise by contamination of seed stocks (or other propagation material).
4. Material to be treated (seed, whole plants, tissue cultures, pollen, etc.).
5. Mutagen (physical, chemical, biological, somatic) and dose rate.
6. Mutation rate and thereby the population size required to recover desired mutants. Aim to pro-

duce specific mutants or a range of mutants. A population size of 100,000 M_2 plants or M_3 families is not uncommon.

7. Mass selection method, which generation should be screened? Because the vast majority of mutants are recessive their phenotypic expression cannot be observed until the mutated genes are homozygous and therefore the M_2 generation is the earliest possible generation in which to screen for mutants. For a quantitative trait, normally M_3 families and subsequent generations are used for phenotypic screening.
8. For advancing mutant generations, harvesting seed from single plant selections is recommended.
9. Additional breeding, depending on mutation density is required to clean up background mutations.
10. Costs: financial, time, space and labour.
11. Advancement (bulking and purifying) of promising lines for entry into official tests.
12. Statutory (national and industrial) checks to meet required standards of distinctiveness, uniformity and stability and food/health legislation.
13. Release and marketing.

The above list is simplistic as plant breeding is not generally a linear progression, but a continuous and progressive cyclical and rolling process. In practice any new mutant (novel variant) in a sexually propagated crop will be vectored into an ongoing programme at the time of crossing. Therefore seed must be available at sowing times to coordinate flowering. The methods are easiest for diploid inbreeding species such as rice and barley. These species have simple genetics and are naturally homozygous, it is therefore relatively easy to produce and detect homozygous (predominantly recessive) mutant lines. Obligate outbreeders lie at the other extreme. Mutant selection in outbreeding crops is more difficult due to their highly heterozygous nature. These species have natural mechanisms that promote and maximize the maintenance of heterozygosity, therefore recessive alleles play a minor role in determining phenotype. Furthermore, on (artificial) selfing of outbreeders it is difficult to distinguish between mutant recessives and normal recessives. Selfing can be difficult or impossible and the development of lines carrying homozygous recessive mutant alleles is hampered further by inbreeding depression.

A strategy for an annual seed propagated crops is as follows:

1. Year 1. Produce a large mutant population in elite germplasm using a relatively mild mutagenic treatment, grow out M_1 plants to produce M_2 seed. The aim here is to provide a large enough M_2 population in which a reasonable number of desired mutants will be present (about 10) in a population (of several thousand), and where the mutation rate per genome is small. Here the numbers will vary depending on the species, ploidy and target tissue.

2. Year 1/Year 2. The M_2 seedlings and plants (and M_3 seeds for seed traits) are screened. Apply high-throughput phenotyping to screen out candidate lines. This would be carried out in specialized facilities such as laboratories or environmentally controlled glasshouse facilities. The aim here is to reduce the population size dramatically from several thousand to about 10 candidates in the early generations. Thus subsequent breeding focuses on target traits and high value genotypes. As phenomics facilities become more advanced several traits can be assayed simultaneously and/or sequentially.

3. Year 2/Year 3. Confirm that the selected phenotype is heritable and is reproduced in field conditions.

4. Year 2/Year 3. Entry validated lines into the breeding schedule and begin crossing. Progress lines through normal breeding practices.

The scheme above can be applied with little or no knowledge of the genetic controls of the trait being targeted. Genetic in-puts however can be greatly beneficial. Knowledge of the number of genes controlling a trait can help to determine the mutagenic dose treatment and the size of the population required, the more genes involved the less the dose and/or population size. Once desired phenotypes are produced the inheritance of the trait needs to be established by fidelity (heritability) of the phenotype to the next generation. Allelism testing can be performed simply and in parallel with the breeding crossing programme and F_1s assessed to confirm or otherwise allelism to known genes controlling a trait. Thus, a breeder will know quickly whether the induced trait is a duplication of existing mutant genes (and thereby redundant) or novel.

4.3.3. Vegetatively Propagated Species

The method for mutation breeding in vegetatively propagated crops (VPCs) is similar to that of SPCs, but with some important differences. In VPCs the main harvested product is usually a vegetative organ and not the seed. As such, flowering and fertility are not major breeding goals, on the contrary in many cases non-flowering genotypes (e.g. in sugarcane) are preferred, as are seedless fruits (e.g. pineapple). Since VPCs are highly heterozygous the main mode of action of induced mutation is presumed to be the unmasking of recessive traits by knocking out the wild type (dominant) allele at a heterozygous locus. However, for those species that are asexual due to meiotic instability caused by odd ploidy levels (e.g. triploidy of banana), breeding (and mutation induction) may be performed at the diploid level after which the commercial ploidy level can be re-created.

There are two kinds of VPC plants: 1) Those that are propagated through hybridization and seed such as Japanese lawngrass (*Zoysia japonica* L.); and 2) those that are not propagated by hybridization (due to the absence of flowers or sterility).

If both vegetative and cross-breeding approaches are available within a selected species, the breeder must consider which method would have the greatest impact towards inducing the appropriate mutation. This could involve irradiation of seed or irradiation of the vegetative portions of the plant. Both methods have been successful in a wide array of mutation breeding programmes of flowers, such as in orchid and chrysanthemum. Since it takes several years for seedlings to produce flowers in fruit trees and the flowers often exhibit self-incompatibility which restricts pure line selection, mutation breeding has an advantage over hybridization in these species. In Japan, a mutant Japanese pear variety "Gold Nijisseiki" resistant to black spot disease was developed from a sport generated by planting seedlings of susceptible "Nijisseiki" in the Gamma Field (**see Chapter 26**).

In a mutation breeding programme involving a vegetatively propagated species, the possible existence of a chimera must be considered (**see Chapter 16**). Chimera formation can sometimes be avoided if adventitious buds generated from a single cell are used. If the characteristics of mutants are easily identified as flower and fruit colours and/or shapes, the propagated plants possessing only mutant cells can be obtained by cutting the stems, twigs or plant parts and propagating and screening the mutation. This can be repeated indefinitely until the pure mutant plant is identified. If *in vitro* culture of the species is available, it is often feasible to induce mutant in *in vitro* cultures followed by regeneration since an original mutated cell line is used to regenerate a plant (**see Chapter 28**). Since the techniques of *in vitro* culture of tissues and single cells have been recently developed and refined in various species, utilization of cell and tissue cultures in obtaining mutations is important.

5. Mutation Breeding in Modern Crop Improvement

The objectives of today's breeding programmes are quite different from those of earlier times. The rapid advances in molecular genetics and genomics have provided a host of DNA technologies available for crop improvement and have transformed breeding and brought about molecular plant breeding methods. As a consequence mutation breeding has thrived and gained momentum over the past decades.

5.1. Mutation Breeding for New and Novel End-Products

The continual changes in social and economic environments drive the demand for new, crop products; for food (especially health and nutrition), fuel (e.g. "greener" fuel), fibre, medicines, ornaments, industrial processes, etc. Changes in demand can alter the primary use of a crop, for example the development of "biomass crops" for the biofuel industry, rather than "food production". Breeders often possess inadequate germplasm resources for such radical changes. In such instances, mutation breeding can offer some hope for success. Recently, rice breeders have focused their attention toward the breeding of very high-yielding forage rice varieties. A few high-yielding *indica* landraces were available; however, these were not free of an undesirable seed-shattering character. Gamma ray mutagenic treatment was successful in inducing non-shattering mutants which resulted in the rapid production of new varieties such as var. Minami-yutaka (Kato *et al.*, 2006).

Another important function that mutation breeding has played and will still play in the future is the development of crop products with novel properties. New cereals and legume mutant varieties with low phytate content, or with high resistant starch content, as described earlier, are good examples. Breeding anti-nutrient less/free new varieties, or varieties with high content of micro-nutrient,

vitamins, anti-oxidants, etc. are actively pursued in many mutation breeding programmes of food crops.

5.2. Molecular Mutation Breeding

The term 'molecular mutation breeding' is defined as mutation breeding in which molecular biological knowledge, techniques and tools are used. As discussed above experimental mutagenesis of the early 20th century led to unprecedented breakthroughs in plant breeding, this however was based largely on phenotypic selection and without much knowledge of the genetic controls of target traits. As mutation science developed, some basic genetic studies (determination of dominance, mapping, allelism testing, etc.) were carried out, but it wasn't until the advent of molecular biology that the molecular basis of mutant events could be investigated and exploited in gene discovery and breeding. Techniques in molecular genetics are rapidly evolving and methods in handling the large data sets produced (bioinformatics) are expected to have a massive impact on molecular mutation breeding.

5.2.1. Utilization and Manipulation of DNA Damage and Repair Mechanism

The natural and artificial environment subjects DNA of living cells to continual damage. Cells respond to this by switching on an array of repair mechanisms, but occasionally the repair process lacks fidelity and leads to mutation. Knowledge gained from molecular genetics can now explain many phenomena emanating from classical mutation studies, e.g. differential sensitivities of plant species, genotypes and tissues to mutagenic treatments and the consequences of various repair mechanisms. For example gamma irradiation often leads to DNA DSBs, ion beams are associated with deletions of DNA fragments, ultraviolet radiation results in covalent dimerization of adjacent pyrimidine bases of DNA and chemical mutagens often cause mis-pairing or nucleotide excision etc. (**see Chapter 12**).

Furthermore two DSB repair pathways are known; homologous recombination (HR) and non-homologous end joining (NHEJ). The former is precise and results in few errors (fewer mutations) whereas the latter is error prone and can produce a high frequency of mutations. Knowledge of DNA repair can therefore be exploited in designing mutagenic treatments that maximize the frequency and type of the desired mutation. For example, plant lines, tissues or cells defective in HR repair will produce relatively higher frequencies of mutations after a mutagenic treatment than normal material and chemical mutagens are expected to produce a high frequency of amenable single nucleotide polymorphisms (SNPs). The production of SNPs, rather than fragment deletions may have the additional benefit of producing functional, dominant or co-dominant alleles rather than recessive gene knock-outs. Thus in cases where a functional mutation is required, such as herbicide resistance (**see Chapter 31**), a chemical treatment may provide the optimal effect, but where a recessive mutation is required, such as many semi-dwarfs, a physical mutation treatment may be preferred.

5.2.2. Making Use of Knowledge of Trait Genetic Control

In all cases basic knowledge of the genetics controlling the target trait is required to develop an effective molecular mutation breeding strategy. In addition to its importance for choosing a suitable mutagen, it is important to generate an appropriately sized mutant population. That is, the more the genes controlling a given trait, the more the chance to identify a mutant in a mutant population. Of course, these should be genes of major effect on the trait, since mutation of genes of minor effect cannot be identified efficiently at the phenotypic level in many circumstances.

If the gene controlling a trait is already known, mutants may be identified directly using molecular tools that target that gene.

5.2.3. Molecular Screening and Identification of Induced Mutations

TILLING approaches can be exploited in cases where the gene of interest has been sequenced. Thus a TILLING population can be designed with the expectation of recovering 10–20 or more mutations in a specific gene. Details of the TILLING technology and its application in mutation breeding are described in **Chapters 21, 22 and 27**. The ability to screen for mutated genes using molecular approaches has not merely been a technological advance, it has resulted in a revolution in mutation breeding as it has greatly enhanced the power of mutation breeding in crop improvement.

Mutation breeding of polyploidy crops is often hampered by the fact that induced mutations (often recessive) of a single gene do not have an effect at the phenotype level due to non-mutated homologous or homoeologous genes. For example, hexaploid wheat has three sets of genomes, namely A, B and D. If a mutation occurs in one chromosome of the A genome, 1A for instance, although this becomes homozygous after self-crossing in M_2 generation, the existence of non-mutated homoeologous genes in chromosomes 1B and 1D will mask the mutation on 1A and the mutant lines cannot be detected *via* phenotyping methods. By using molecular screening methods, it is now possible to identify mutants of homoeologous genes, and by combining mutated genes across a homoelogous family (e.g. a gene on 1A, 1B and 1D) into a single breeding line, the mutant phenotype is generated. As such, mutation breeding can now be applied more effectively for breeding novel varieties in polyploidy crops (**see Chapter 27**).

The power of mutation breeding of VPCs is rather limited compared with SPCs. This is because no self-crossing is involved in VPCs and hence recessive mutations cannot be revealed. Therefore, mutants are often revelations of existing heterozygous mutant alleles. Once molecular methods are used, it is possible to screen mutants heterozygous at the target locus, another round of mutation induction and screening can then reveal double (recessive) mutants. It is possible to develop a mutant line with desired phenotypes from a homozygous wild type variety. Such mutant alleles do not exist in wild type germplasm and these induced mutations are completely novel to these crops, and may bring great benefits to the product or production.

As discussed earlier, mutation breeding is often characterized by its limited power for improving the overall performance of a variety, since a mutant line often has very few mutated genes of interest. This limitation can now be overcome by either pyramiding desired mutations through crossing between complementary mutant lines derived from a single variety (for SPCs), or through repeated mutation and selection (VPCs). That is, one population can be screened for several genes of interest and mutant lines with individual desired mutations can then be crossed to produce recombinant lines with the desired mutations through marker-assisted selection. For VPCs, it is not possible to make crosses, therefore a second round of mutation induction is performed on the selected mutants, after which new lines carrying additional mutations are identified.

In summary, molecular mutation breeding has the potential to increase both the efficiency and power of plant breeding as compared to classical methods.

6. References

6.1. Cited References

Forster, B.P., Franckowiak, J.D., Lundqvist, U. *et al.* 2007. The barley phytomer. *Annals of Botany*. 100: 725–733.

Fujita, D., Doi, K., Yoshimura, A. *et al.* 2008. Molecular mapping of a novel gene, Grh5, conferring resistance to green rice leafhopper (Nephotettix cincticeps Uhler) in rice, *Oryza sativa* L. *Theor Appl Genet.*, 113(4): 567–573.

Futsuhara,Y. (1968). Breeding of a new rice variety "Reimei" by gamma-ray irradiation. *Gamma Field Symposia*. 7: 87–109.

Hayashi, Y., Takehisa, H., Kazama, Y. *et al.* 2008. Effects of ion beam irradiation on mutation induction in rice. *Proc. 18th International Conference on Cyclotrons and their Applications* (Cyclotrons), p. 237.

Kato, Y., Yoshioka, H., Iba, R. *et al.* 2006. A newly bred rice variety "Minamiyutaka" for whole-crop silage. *Bull. Miyazaki Agric. Exp. Stn.* 41: 51–60 (In Japanese with English summary).

Law, C.N., Snape, J.W. and Worland, A.J. 1987. Aneuploidy in wheat ant its uses in genetic analysis. *In*: F.G.H. Lupton (ed.) Wheat Breeding. Its Scientific Basis. London and New York: Chapman and Hall, pp. 71–127.

Lee, I.S., Kim, D.S., Lee, S.J. *et al.* 2003. Selection and characterization of radiation-induced salinity-tolerant lines in rice. *Breeding Science*. 53: 313–318.

Maluszynski, M., Kasha, K., Forster, B.P. *et al.* 2003. Doubled in: Haploid Production in Crop Plants. A Manual. Kluwer Academic Publishers, pp. 428.

Raboy, V. 2009. Approaches and challenges to engineering seed phytate and total phosphorus. *Plant Science*. 177: 281–299.

Rutger, J.N. 1992. Impact of mutation breeding in rice: a review. Mutation Breeding Review No. 8, Vienna: FAO/IAEA, pp. 1–23.

Sanada, S. 1988. Selection of resistant mutants to black spot disease of Japanese pear by using host specific toxin. *Jpn. J. Breed.* 38: 198–204.

Sauton, A., Olivier, C. and Chavagnat, A. 1989. Use of soft X-ray technique to detect haploid embryos in immature seeds of melon. *Acta Horticulturae*. 253: 131–135.

Shu, X.L., Jia, L.M., Ye, H.X. *et al.* 2009. Slow digestion properties of rice different in resistant starch. *J. Agric. Food Chem.* 57: 7552–7559.

Takahashi, M., Shimada, S., Nakayama, N. *et al.* 2005. Characteristics of nodulation and nitrogen fixation in the improved supernodulating soybean (*Glycine* max L. Merr.) variety 'Sakukei 4'. *Plant Prod. Sci.* 8: 405–411.

Vasal, S.K., Villegas, E., Bjarnason, M. *et al.* 1980. Genetic modifiers and breeding strategies in developing hard endosperm *opaque-2* materials. *In:* W. G. Pollmer and R.H. Philips (eds.) ImproVement Quality Traits Maize Grain Silage Use, London: Martinus Nijhoff, pp. 37–71.

von Sengbusch, R. 1942. Süsslupinen and Öllupinen. *Landwirtschaftliches Jahrbuch*. 91: 723–880.

von Wettstein, D., Nilan, R.A., Ahrenst-Larsen, B. *et al.* 1985. Proanthocyanidin-free barley for brewing: Progress in breeding for high yield and research tool in polyphenol chemistry. *Technical Quarterly of the Mater Brewery Association of America*. 22: 41–52.

6.2. Websites

Australian Plant Phenomics Facility:
http://www.plantphenomics.org.au

The IAEA/FAO mutant variety genetic stock database:
http://mvgs.iaea.org

Lemna Tec high-throughput plant phenotyping systems:
http://www.lemnatec.com

Australian Plant Phenomics Facility:
http://www.plantphenomics.org.au

Lamnatec imagine processing for biology:
http://www.lemnatec.com.

MARVIN digital seed analyser/counter:
http://www.hoopman-equipment.nl

TraitmillTM, CropDesign's high throughput phenotyping platform:
http://www.cropdesign.com/tech_traitmill.php

UK CropNet database:
http://www.ukcrop.net/db.html

6.3. Further Reading

Micke, A., Donini, B. and Maluszynski, M. 1990. Induced mutations for crop improvement, *Mutation Breeding Review*, 7: 1–41.

Fehr, W.R. (1987). Principles of Variety Development. Volume 1. New York: Macmillan Publishing Company, p. 536.

Fehr, W.R. 2007. Breeding for modified fatty acid composition in soybean. *Crop Sci.* 47: S72–S87.

Gustafsson, Å. 1986. Mutation and gene recombination principal tools in plant breeding. *In*: G. Olsson (ed.) Svalöf 1886–1986. Research and results in plant breeding. Stockholm: LTs förlag, pp. 76-84.

Sleeper, D.A. and Poehlman, J.M. 2006. Breeding Field Crops. 5ᵗʰ Edition. Oxford: Blackwell Publishing.

Ukai, Y. 2006. Effectiveness and efficiency of mutagenic treatments. *Gamma Field Symposia*. 45: 1–4.

C25

Mutant Phenotyping and Pre-Breeding in Barley

B.P.Forster[a,b,*] J.D.Franckowiak[c], U.Lundqvist[d], W.T.B.Thomas[a], D.Leader[a], P.Shaw[a], J.Lyon[a] and R.Waugh[a]

[a] SCRI, Invergowrie, Dundee DD2 5DA, UK
[b] Present address: BioHybrids International Ltd, P.O. Box 2411, Reading RG6 5FY, UK
[c] Agri-Science Queensland, Department of Employment, Economic Development and Innovation, Hermitage Research Station, Warwick QLD 4370 Australia
[d] Nordic Genetic Resource Center, P.O. Box 41, SE-230, 53 Alnarp, Sweden
* Corresponding author, E-MAIL: BrianForster@biohybrids.co.uk

1. Introduction

Barley has and continues to be a model species for mutation genetics and breeding. Mutagenesis in barley, and other species, is an effective method of generating phenotypic variation. Mutants have been used as tools to help understand plant development, physiology, evolution and domestication; they have also been exploited in crop improvement. Classic studies in barley have produced new genetic resources, including isogenic lines for individual mutant genes, morphological marker maps and databases on mutant stock description. Advances in bioinformatics allow phenotypic and genotypic data to be acquired and handled on a much bigger scale, with the massive data sets produced being housed and interrogated at publicly accessible websites. These advances have promoted the development of large scale multi-mutant populations and have simultaneously revitalized old mutant collections. The general applicability of mutagenesis allows the rapid generation of structured populations from germplasm of interest to scientists and breeders. Using barley as an example, practical protocols for generating, screening and selection are described for all stages in the life cycle, from caryopsis, to germination through to maturity traits. Many of the phenotypes generated are of direct interest to plant breeders and geneticists. Methods in maintaining and advancing selected lines as well as genetic characterization are also described.

Random mutation followed by the selection of beneficial traits is a natural driving force in evolution and speciation. Man has exploited this natural variation to select traits useful in the domestication and improvement of crop plants, and to adapt cultivated plants to a range of environments and uses. The production of mutations can be accelerated by the use of specific chemical or physical agents (**see Chapters 7–13**).

Induced mutants have played a major role in plant breeding and classical genetics. For example, genetic mapping populations can be developed from crosses with mutant lines to identify the specific lesion responsible for the observed mutant phenotype. Since induced mutagenesis is generally applicable to all germplasm of a given crop it is possible to generate novel mutant phenotypes in elite material. Mutants have had a significant impact on crop production worldwide with the release of over 3,000 varieties (IAEA/FAO mutant variety genetic stock database) that are either the direct products of mutation or possess a mutated ancestor (**see Chapter 1**). In cereals key mutant domestication traits include: non-dormant seed, plant stature, modified inflorescences, non-shattering spikes, free-threshing grain and large grain. Alterations in the timing of flowering have been critical in the adaptation of crops to new environments, especially adjustments in photoperiodic and vernalization responses. The use of mutant lines in plant breeding has also improved resistance to biotic and abiotic stresses, yield and quality. Perhaps the most significant example of mutation breeding is the deployment of semi-dwarf cereals in the 'Green Revolution', which resulted in dramatic yield increases in wheat (*Triticum aestivum* L.) and rice (*Oryza sativa* L.).

2. Single Gene Mutations

Simple Mendelian genetics is based on two principles: 1) the inheritance of independent genetic units, and 2) recombination of genetic units that are then distributed *via* gametes to progeny. A concept embedded in simple Mendelian genetics is that the units of inheritance, the genes, are constant entities and that variation is a result of recombination among them. However, variation *via* recombination cannot explain the sudden genetic changes that occur in nature that have been a driving force in evolution. These genetics changes are called mutations. Various mutations occur and range from gross changes to the genome, e.g. a change in the number of chromosomes to subtle single base changes in a DNA sequence (these and the methods used to induce them are discussed in greater detail in **Chapters 4 and 20**). Large genome mutations (aneuploidy, chromosome duplications, translocations and deficiencies, etc.) have been of interest in cytology studies from the onset. However, changes to plant traits are of greater interest to plant breeders and geneticists as these may be manipulated for crop improvement. Many of the traits of interest turned out to be single gene traits, followed the rules of Mendelian genetics and were thereby relatively easy to study and manipulate.

2.1. Classic Mutations in Barley

Barley, *Hordeum vulgare* L. is a small grain cereal. It is a member of the subtribe Triticeae, of the grass family,

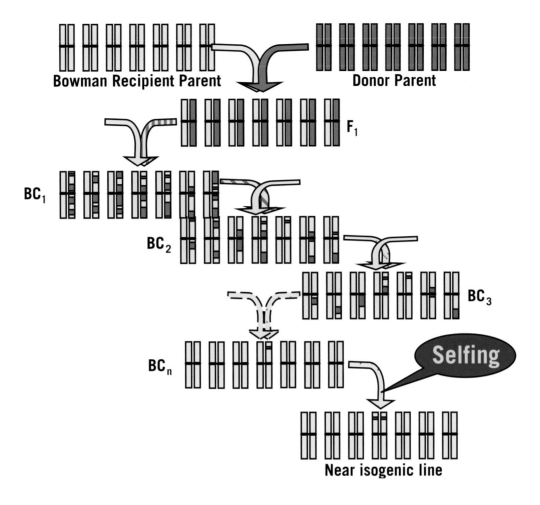

The Joint FAO/IAEA Programme

Figure 25.1 Production of mutant near isogenic lines.

the Poaceae. Barley is closely related to other important crops species; bread wheat, macaroni wheat, rye and the wheat/rye hybrid, triticale, all of which are in the *Triticeae*. The biology of barley makes it a particularly useful species for mutation studies. It has a relatively short life cycle and simple genetics (2n = 2x = 14). It is an inbreeding diploid and therefore mutants are relatively easy to detect and fix compared to polyploid and out-crossing species. Barley is also a major crop grown worldwide and is closely related to other important small grain cereals, notably bread and durum wheat, rye and the wheat rye hybrid, triticale. Barley was domesticated from the wild species *Hordeum vulgare* L. subsp. *spontaneum* (C.Koch) (2n = 2x = 14) in the Fertile Crescent some 10,000 years ago. Morphological features selected for cultivation included non-shattering spike, kernel size, spike type and adaptation to a range of environments. Mutants for spike type allowed the development of six-

rowed as well as two-rowed barley, the genes involved (*vrs1* and *int-c*) were among the first morphological mutants to be mapped (on chromosomes 2H and 4H, respectively). In the 1980s a backcrossing project was initiated in the USA to develop near isogenic lines by placing all recorded barley mutant genes individually into a common genetic background. Once mutants were isolated in a common genetic background their effects could be more easily studied. Other objectives included providing genetic stock descriptions, mapping mutant genes and providing a genetic resource. Initially 30 mutants were collected from Japan, Sweden and the USA and these were crossed and backcrossed to the locally adapted variety Bowman (**Figure 25.1**). The project has now grown to over 900 individual mutants collected from around the world with backcrossing complete for about 600. These mutants are recorded and described in the barley genetic stocks database.

The classic studies in barley resulted in new genetic resources for both research and exploitation in crop improvement. Some notable mutant traits used to improve and expand the range of the barley crop included:

- Semi-dwarfism, the *ari-e*.GP and *sdw1.d* mutants from Golden Promise and Diamant, respectively, were used to produce a number of semi-dwarf varieties, but other *ari, ert* and *brh* mutants were also exploited. Semi-dwarf cereals became important in the 1950s with the introduction of the combine harvester. Harvesting was made easier in cereal crops with short, stiff and erect straw. Semi-dwarfs were also associated with increased yield as more energy resources were available for grain rather than vegetative tissues.
- Early maturity (*eam8* from Mari). Early maturity mutants allowed the development of barley and other cereals to tropical and semi-tropical regions with short days.
- Mildew resistance (*mlo9* from Diamant). The *mlo9* mutant has been an important source of mildew resistance since Alexis was first released in 1986 (www.crpmb.org).

3. Multiple Gene Mutations

The advent of bioinformatics and developments in computational biology have revitalized interest in induced mutation as they provide a means of bringing together large data sets. Massive databases are available that describe hundreds of thousands of expressed gene sequences that can be interrogated for gene/function associations. Such data are set to grow further with the advent of the 'sequencing revolution'. However, information on phenotype, the 'phenotype gap' severely limits our ability to capitalize on the wealth of genomics data. One way to overcome this is to develop multiple mutations in structured families such as in mutation grids and Targeting Induced Local Lesions in Genomes (TILLING) populations (**see Chapters 21, 22 and 27**).

By combining mutagen dosage and population size structured families can be developed which, in theory, contain one or more (usually 10–20) mutations in every gene of the genome. These populations provide a wealth of mutated genes and phenotypes that can be exploited in forward and reverse genetics. In forward genetics the starting point is an altered phenotype, in reverse genetics the starting point is a mutated gene sequence. Both approaches aim to link phenotype to genotype. (**The reverse genetics approach is exemplified in Chapter 38**). Here we describe the forward approach using morphological mutants found in a barley mutation grid (Caldwell *et al.,* 2004). Phenotypic deviants were sought at all stages in plant development, from quiescent seed, to germination, to vegetative and reproductive growth and to maturity.

3.1. Development of a Structured Mutation Population

3.1.1. Plant Material and Methods

A two-rowed, semi-dwarf, spring barley var. Optic was chosen as it was a top quality malting variety adapted to local, Scottish, conditions. Optic, however did not possess an effective resistance gene to prevalent races of mildew in the UK and developing a mildew resistant mutant was of interest. Grain of Optic was mutagenized with ethyl-methanesulphonate (EMS) at two dosage levels, 20 and 30 mM, which produced an estimated average of 5,000 mutations per genome (Caldwell *et al.,* 2004). The treated grain is referred to as the M_0 generation and on germination produces M_1 plants. No more than two M_2 grains per M_1 plant were grown on to produce a structured population of over 21,000 M_3 families, described by Caldwell *et al.* (2004). The M_1 generation was grown up in glasshouse conditions. M_2 individual plants, M_3 rows and plots of subsequent generations were generally grown in the field. DNA was sampled from M_2 plants for genotypic analysis (mutation detection) and M_3 and subsequent generations were inspected for phenotypic variation and this was done at all stages in development. Harvested seed or ears were labelled and stored. A schematic of these procedures is given in **Figure 25.2**.

3.1.2. Maintenance of Family Structure

It is important to maintain family structure as once mutant sequences and/or phenotypes are revealed they need to be traced back to a common origin and traced forward to advanced generations. Population structure was maintained in each season by harvesting each family row separately. In order to maintain the genetic diversity within a family all plants (or a selection of plants) should be threshed as individual family bulks. Alternatively a random selection of ears along each row

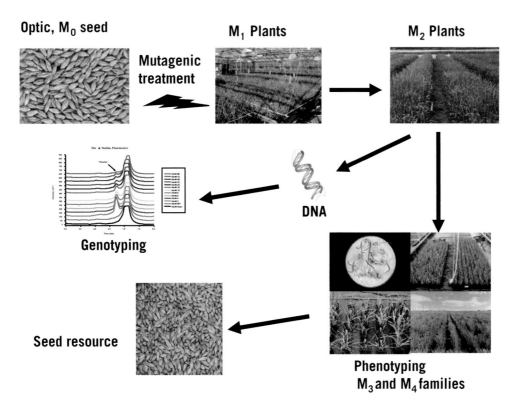

Figure 25.2 Scheme for developing and exploiting a mutation grid for forward and reverse genetics in barley.

can be selected and stored to maintain a high degree of genetic variation of the family.

Ears from mutant plants identified within family rows can be included in the bulk, but can also be harvested separately to isolate the mutant. These can then be grown on in subsequent years as progeny from single ears, and provide a true breeding genetic resource of mutants where the phenotype can be validated in the next generation.

After sampling spikes from family rows, selecting ears from single mutant plants and removing any control plants/guard rows, the remaining plants in the mutation field can be combine harvested to provide a random bulk of mutant seed. Random bulks can be utilized in large scale mutant screening, e.g. disease resistance and responses to fertilizer and herbicide treatments, but in the case of bulks the family structure is lost.

4. Phenotypic Screening and Selection

The general scheme for mutation induction, development of a structured population and the connections between phenotyping and genotyping are given in

Figure 25.2. The sheer size of mutant populations (the Optic mutation population consisted of 21,000 M_3 families) often means that phenotyping in the field is undertaken over more than one season. In the case of the Optic populations this was done over 3 years on M_3 or M_4 families grown in standard 1.5-m-long rows consisting of 15–20 plants per family. Rows were organized into plots (four rows per plot) and plots organized into beds, each family was identified by a bar-coded label and a field coordinate (bed/plot/row). Laboratory screens were used to detect mutant phenotypes in the caryopsis and germination and glasshouse tests were used to identify some seedling trait mutants. In the Optic work data were recorded electronically by handheld devices linked to a central server and transferred to the publicly accessible database. A system was developed utilizing wireless technology to electronically record data. This process ensured that recordings were immediately available to other users who were involved in the scoring process and reduced errors in transferring hand written notes. This information along with photographs of mutant phenotypes was then transferred to the publicly available database for further analysis.

Total Recording Within Defined Ranges

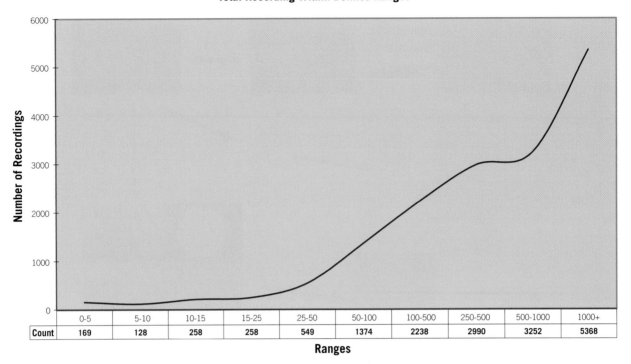

Count	169	128	258	258	549	1374	2238	2990	3252	5368
Ranges	0-5	5-10	10-15	15-25	25-50	50-100	100-500	250-500	500-1000	1000+

The Joint FAO/IAEA Programme

Figure 25.3 Frequency distribution of phenotypic mutant classes within the Optic mutation grid.

Material was examined at all stages in plant development in order to identify useful genetic diversity data for subsequent applications such as map-based gene cloning or breeding of improved lines. During the course of the work it was noticed that short stature, late flowering and light green appearance were very common mutant phenotypes. These common phenotypes were often associated with other less common mutants, i.e. rare mutants were usually present in families which exhibited one of the common mutant phenotypes. The step up in mutation frequencies provided by these three common mutant classes (**Figure 25.3**) can be exploited. Thus an alternative phenotyping strategy might be to select in the first instance for these common phenotype classes (a fourth common class which may be included is partial fertile) and to make more detailed assessments among selected families in subsequent generations. Data indicated that selection for late flowering alone represents 65% of the mutant phenotypes in the population. A benefit of a common phenotype pre-screen would be a much reduced population size in which to make more detailed observations in subsequent generations. This will however produce a bias towards late, short and light green phenotypes, and whereas semi-dwarfs have had

a major impact in the past, the structured population would be relatively unattractive to breeders as it may require several rounds of backcrossing to reduce the mutational load while isolating desired traits.

5. Mutant Phenotypes

Phenotypes were scored by the naked eye of experts familiar with barley morphological mutants or experts in specific traits such as agronomy, pathology, plant development and physiology (see Acknowledgements). An understanding of the botany of the species is also required. Phenotypic screening was not carried out until the M_3 generation at the earliest; this is because M_1 plants exhibited chimeras and physiological disorders carried over from the mutagenic treatment (**see Chapters 15 and 16**). Since the vast majority of mutants are recessive, many phenotypes are not expressed until plants become homozygous for the mutation. Phenotypes were scored continually throughout the life cycle and it was common to build phenotypic descriptions for single families/plants over the growing period and over the three year assessment period.

Table 25.1: Mutant classes

Primary class	Secondary classes	Tertiary classes
Caryopsis	Size, shape and colour	Weight Length and width Pigmentation Hull appearance
	Shrunken endosperm	Genetic Xenia
	Maternal tissues	Pericarp Testa
	Endosperm	Shrunken (genetic or xenia) Aleurone (pigments) Starch types
	Embryo	Embryo number Size and shape
	Pest reactions	Diseases Insects
Seedling development	Germination	Percentage Seedling vigour Lethality
	Seminal roots	Number Root hairs Root morphology (coiling) Root angle (geotropism) Growth rate (length)
	Shoot	Growth rate Coleoptile morphology First leaf morphology Chlorophyll pigmentation Xanthophyll pigmentation Sub-crown internode
	Pest reactions	Diseases Insects
Vegetative growth (phytomer development)	Crown development	
	Leaf blade and sheath	Blade length Blade width Coiling Folding Denticulation or serration Pubescence Surface waxes Auricles and ligules
	Leaf pigmentation and necrosis	Chlorophyll development Anthocyanin pigmentation Chlorophyll bleaching Necrotic spots Striping or streaking
	Tiller number and vigour	Prophyll morphology Tiller number Tiller morphology Vigour
	Pest reactions	Diseases Insects
	Transition to reproductive growth	

Table 25.1: Mutant classes

Reproductive growth	Leaf blade and sheath	Number and size Blade length Blade width Curling Blade folding Denticulation or serration Pubescence Surface waxes Cuticle (sheath) Auricles and ligule
	Leaf pigmentation and necrosis	Chlorophyll Pigmentation Bleaching Necrotic spots Stippling and streaking
	Culm	Number of elongated internodes Crown shape Pigmentation Thickness (vascular bundle number) Stem strength and lodging Surface waxes
	Pest reactions	Diseases Insects
	Lethality	Timing Rate
Inflorescence formation	Transition to spike development	Timing Rate Uniformity Suppression of leaf structures Abortion
	Peduncle	Length Surface waxes
	Flag leaf	Shape Size Auricles and ligule
	Collar	Suppression Size Shape
Spike development	Spike morphology	Branch suppression Rachis internode length Rachis pubescence Spike determinancy Fertile rachis internodes Spikelet number and position Spikelet fertility Surface waxes
	Spikelet morphology	Glumes Floret number Rachilla Lemma Palea Lodicules Stamens Pistils
	Floral part transition	
	Floral part abortion	

Table 25.1: Mutant classes

Meiosis and flowering (anthesis)	Microsporogenesis	Male sterility Pollen fertility
	Megasporogenesis	Female fertility Structural aberrations
	Fertilization	Anther maturation Stigma position Stigma hairs
	Seed set	Embryo development Endosperm development
	Disease reactions	
Adult plant ripening	Chlorophyll loss	
	Maturation	Rate (stay green) Uniformity
	Seed dispersal	Rachis disarticulation Spikelet adherence Threshability
	Straw strength	
	Disease reactions	

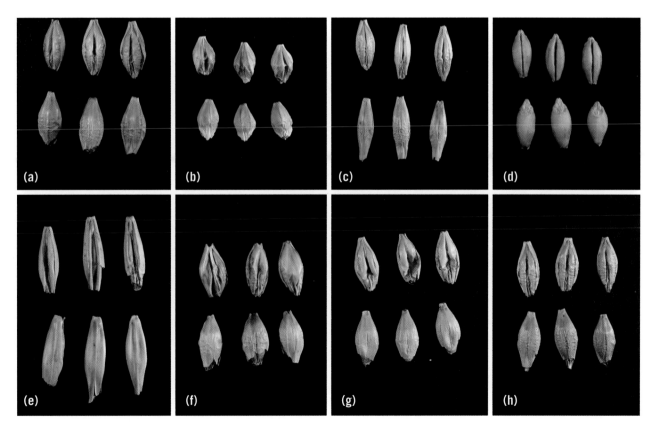

The Joint FAO/IAEA Programme

Figure 25.4 Examples of caryopsis mutants observed in stored grain of Optic grid M₃ families. Age: 0: (a) six normal caryopses abaxial sides (top) and adaxial sides (bottom) compared to mutant phenotypes: (b) small globe-shaped, (c) very thin, (d) naked, (e) large and thin, (f) double caryopses formed by fused florets, (g) defective endosperm, (h) orange lemma.

5.1. Caryopsis Mutants

The grain (also known as the kernel) of barley is a one-seeded fruit, known botanically as a caryopsis. Caryopsis mutants were generally detected by inspecting stored, quiescent grain, i.e. phenotyping is partially independent of seasonal effects and such screening can be carried out at any time as it is not season dependent. Examples of mutant phenotypes observed are given in **Figure 25.4**.

5.2. Germination Mutants

Twelve disease-free seed of each family were placed onto filter paper with 2.5 ml distilled water in 9-cm Petri dishes. Petri dishes were incubated for 3 days at 20°C in the dark and assessed for germination, seminal root and coleoptile variants. The Petri dishes were then given an additional 2.5 ml distilled water and incubated in the light for a further 3 days and assessed again, the second assessment included colour mutants of the shoot. Germination tests used stored grain and could be carried out at any time. A range of germination mutants is given in **Figure 25.5**.

5.3. First Leaf Mutants

A standard first leaf glasshouse screen was utilized: 16–20 grain were sown into 4″ pots filled with standard compost and watered. Pots were inspected after 10–14 days (first leaf) and also at four weeks (tillering). This test can be carried out at any time in a lit, temperature controlled glasshouse. Examples of first leaf mutants are given in **Figure 25.6**.

5.4. Seedling Mutants

Assessments of seedling traits were made from glasshouse tests and field grown material. Some seedling mutants are shown in **Figure 25.7**.

5.5. Tillering Mutants

Assessments of tillering characteristics were made using glasshouse and field grown material. Some examples are given in **Figure 25.8**.

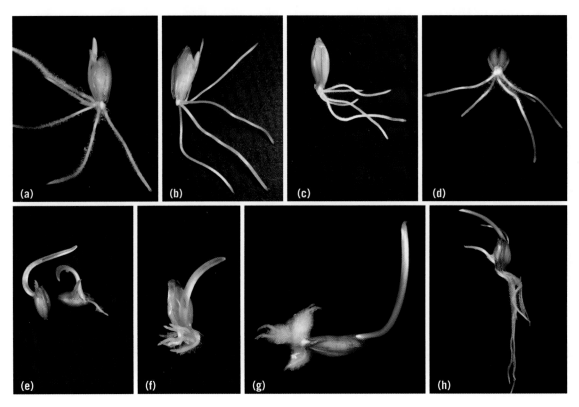

The Joint FAO/IAEA Programme

Figure 25.5 Examples of germination mutants in Optic. Age: 3–6 days: (a) normal phenotype compared to: (b) hairless roots, (c) curly roots, (d) highly geotropic roots, (e) few roots, (f) multiple short roots, (g) short and hairy roots (root hairs adhered to filter paper), (h) twin embryo.

The Joint FAO/IAEA Programme

Figure 25.6 Examples of first leaf mutants in Optic. Age: 10–14 days, mutant phenotype (right) compared to normal (left): (a) slender, (b) glossy leaf, (c) chlorina (light green), (d) albina with green tip, (e) mottled (lethal), (f) collapsing.

The Joint FAO/IAEA Programme

Figure 25.7 Examples of seedling mutants in Optic. Age: 2–3 weeks from sowing, (a) albina (lethal), (b) mottled, (c) light green (normal row to left), (d) zonata (banded mottled).

5.6. Leaf Mutants

Characterizations of leaf mutants were made using glasshouse and field grown material. Examples are given in **Figure 25.9**.

5.7. Culm Mutants

Phenotyping of stem traits was carried out primarily in the field. Some examples are given in **Figure 25.10**.

5.8. Spike and Spikelet Mutants

Variations in spike and spikelet development were examined using field grown plants. Some examples are given in **Figure 25.11**.

6. Mutant Classes

Table 25.1 lists primary, secondary and tertiary phenotypic classes that are commonly recorded in barley. Knowledge of botanical descriptors of the species under study is essential in phenotypic assessments as is knowledge of how the plant grows; its physiology, agronomy, pests and diseases. An awareness of traits of interest to scientific and breeding communities is also desirable. It is therefore advisable to have experts, in various disciplines, assess the material. Because of their size, it is not always possible to assess a whole mutant population in one season as the work is labour intensive. In the case of the Optic mutation grid assessments were made over three field seasons.

The vast majority of phenotypic data are scored on the sporophyte, although some fertility variants may be attributable to mutations in the mega- (female) and

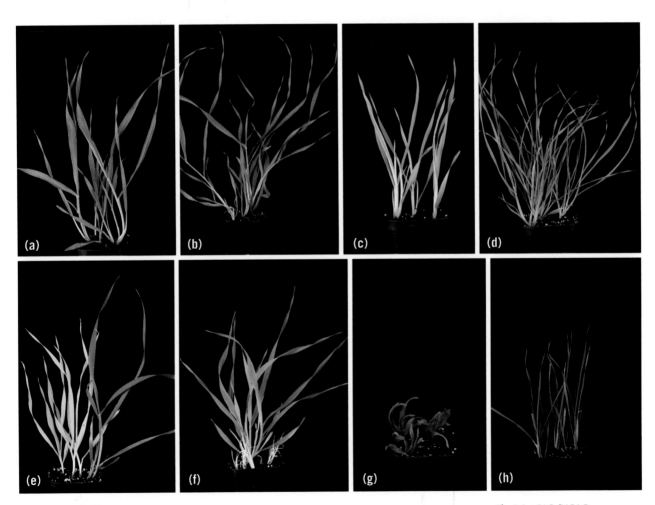

Figure 25.8 Examples of tillering mutants in Optic. Age: 4 weeks: (a) normal compared to; (b) early maturing, (c) rolled leaf, (d) thin leaf, (e) yellow (lethal), (f) albino, short with several leaves (lethal), (g) curly dwarf, (h) thin leaf and necrotic.

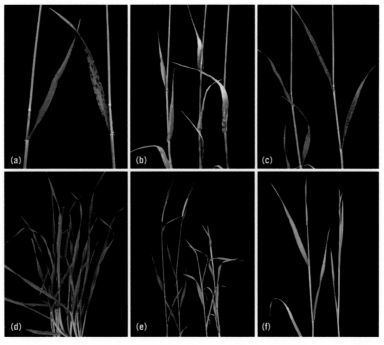

Figure 25.9 Examples of leaf mutants in Optic. Age: 4–10 weeks: (a) broad necrotic spots or flecks, (b) bleached (two examples to right), (c) small necrotic spots, (d) serrated or dented, (e) light green with necrotic flecks (two examples to right), (f) auricleless and liguleless. In (a), (b), (c), (e) and (f) the normal phenotype is shown to the left.

Figure 25.10 Examples of culm mutants in Optic. Age: 4–10 weeks: (a) orange stem of an orange lemma mutant, (b) twisted, curved and trapped leaf and stem (four stems from a single plant to right), (c) fragile stem, (d) rich wax coating, (e) glossy sheath, (f) many-noded. Wild type is shown to the left in each case.

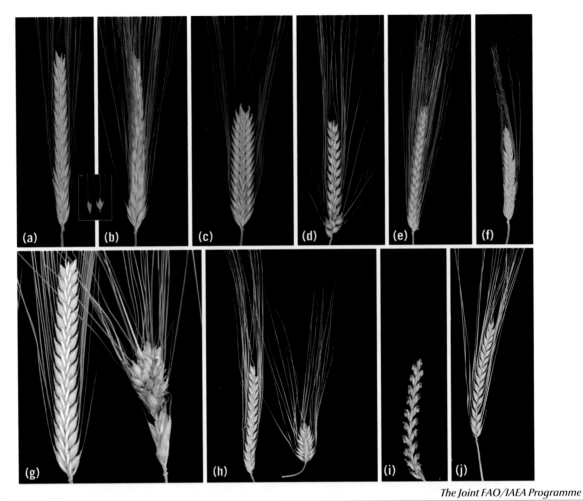

Figure 25.11 Examples of spike and spikelet mutants in Optic. Age: 8–10 weeks: (a) normal immature spike and spikelet compared to: (b) elongated outer glume (inset of spikelet), (c) dense spike, (d) multiflorous at the base, otherwise lax, (e) six-rowed, (f) curly awn, (g) branched with semi-ripe normal spike to left, (h) pyramid shaped with normal mature spike to left, (i) hooded or calcaroides, (j) small lateral spikelets.

The Joint FAO/IAEA Programme

micro- (male) gametophytes. Data can be collected throughout the life cycle (grain, seedling, vegetative and reproductive growth, floral development, flowering and ripening).

Phenotypic data were gathered over three seasons (2003 to 2005) in M_3 and M_4 rows. Among the 21,000 mutant families, 7,204 exhibited one or more visible mutant phenotypes (approx. 34%). A large number of families (4,324) exhibited more than one visible morphological mutant plant in an M_3 or M_4 row, 259 families contained lethal mutants. These include mutations effecting seed viability, embryo lethal mutants, developmentally non-viable mutants and sterility mutants. Lethal mutants can be maintained in the population as heterozygotes. Some lethal mutants are environmentally induced and seed stocks can be obtained only in certain seasons/conditions. It was possible to classify mutants into 204 phenotypic classes which

varied in representation from rare to common classes; a frequency distribution is given in **Figure 25.3**. Rare mutants, with as few as 10 representatives in the entire population included single elongated internode, albino lemma and yellow stripe mutants. At the other extreme, the most common classes were: late flowering (2,629 entries), light green (1,349) and short stature (1,356).

Families exhibiting a common mutant phenotype (late flowering, light green or short) were often observed to exhibit additional phenotypes (either in the same individual or in other plants of that family); these included other common phenotypes and less common mutant classes (**Table 25.2**). For example, of the 2,629 families scored as late flowering, 1,975 exhibited one or more additional phenotypes, the most frequent classes are given in **Table 25.2**. Taken together the three most common classes capture 55% of the observed phenotypes in the entire population.

Table 25.2: Comparison of the three common mutant classes; the total number of families exhibiting late flowering, light green and short stature phenotypes and the number of families among these that exhibit additional mutant phenotypes

Common classes	Total number families	Late	Light green	Short stature	Extreme short	Necrotic
Late flowering	2,629		562	808	216	209
Light green	1,349	562		263	52	141
Short stature	1,356	808	263		139	118

Common classes	Lethal	Thin leaf	Yellow leaf	Short awn	Weak plant	Erect growth
Late flowering	202	196	90	84	80	20
Light green	127	72	120	34	47	15
Short stature	101	176	48	66	65	12

Common classes	Early flowering	Prostrate growth	Others
Late flowering	7	12	1,101
Light green	11	5	505
Short stature	20	6	356

Web portals or databases are commonly developed for public access to data and materials The address for the Optic mutation grid is: http://bioinf.scri.ac.uk/barley/mutants and was set up to act as a gateway to the work being carried out on barley mutant populations at the Scottish Crop Research Institute (SCRI) as well as to offer tools to allow querying of phenotypic data. Experimental data can be queried in a number of ways such as by phenotype description, by year, by botanical structure (grain, leaf, stem, spike, etc.) or by the total number of families that display a particular phenotype. Access to image data associated with many of the plant families is provided. The portal also provides information for researchers who may be interested in requesting and exploiting the resources described.

The phenotypes observed in the Optic mutant population were compared to standard phenotypic descriptions from the Barley Genetic Stocks AceDB Database (BGSAD) and in several cases it was possible to assign candidate genes. These are attached to the descriptions given in the website. Comparisons between the Optic mutant phenotypic data and those listed in the BGSAD showed that about 10% of those listed in the BGSAD are not recorded in the Optic mutation grid. Missing phenotypes in the Optic mutant collection are largely in categories that require specialized screening methods, e.g. reaction to pathogens, desynapsis, and caryopsis colour and composition which could not be determined as Optic is a hulled variety. Conversely, novel mutants with no direct counterparts in BGSAD listings were found and these were found at all stages in the life cycle, notably for seminal root morphology which is an area that has been studied infrequently.

7. Germplasm Storage

Space is a major issue in storing seed and secure, compact methods need to be devised, storage of air is wasteful and promotes the spread of pests and diseases. It is also advisable that storage containers are air-tight and see-through so that seed and/or ears can be inspected without disturbance. Threshed seed or intact spikes can be stored; spike storage has advantages in that spike morphology can be inspected/verified and subsequent generations can be traced back to a single ear. Transparent screw cap 15–50 ml tubes are useful storage containers for seed or spikes harvested from single plants and single row (multiple plant) selections. Seed longevity is directly related to moisture and

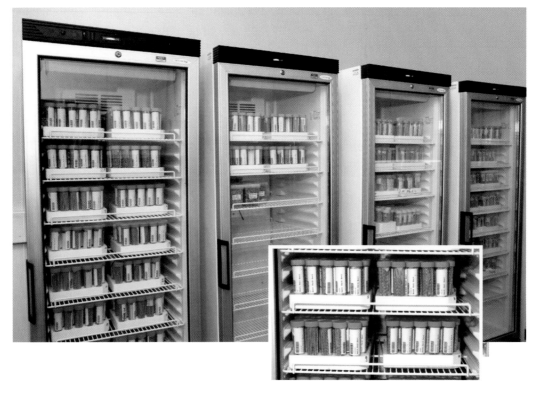

The Joint FAO/IAEA Programme

Figure 25.12 Mutant seed storage.

temperature and a compromise is often required for logistical reasons. For viability over a 10–20 year period storage in airtight containers at 4°C in the presence of a drying agent is recommended for small grain cereals such as barley. Tubes can be labelled with a bar code, family code, date of harvest, field coordinates (bed/plot/row) and phenotypic descriptors (e.g. dwarf, pale green, leaf stripe etc., **see Figure 25.12**). For larger quantities, e.g. bulked seed, larger containers are required.

8. Exploitation

8.1. Access to Information and Material

Optic mutant phenotypes are described in the database and seed can be provided on request, this is normally a common feature for various TILLING and mutation grid populations of a range of species. Various barley mutant seed stocks are available:

1. seed/spikes from family rows, the aim here is to maintain the mutational load of each family.

2. seed/spikes from selected plants, these have been selected as a first step in mutant line purification.

3. bulked seed (from all mutant families grown in the field in a particular season), which provides useful material for large scale "shot gun" phenotyping, e.g. sowing out a large area of bulked seed to select mutants for nutritional, pesticide or climatic responses.

8.2. Genetics

Mutants have played a major role in basic and classic genetic studies. Mutant lines exhibiting similar phenotypes can be crossed together to determine allelism (**see Chapter 4**), dominance relationships, the number of loci controlling a trait, the number of gene families and the number of alleles involved in trait expression. Genetic mapping populations can be developed from crosses to mutant lines and used to find the chromosomal region associated with specific mutants and to develop isogenic lines (**Figure 25.1**). A catalogue of barley genetic stocks is available at the BGSAD, which provides information on genes, gene symbols, inheritance, phenotype, origin of mutant, chromosomal location and relevant references for over 900 genes. DNA from mutants can be used to detect DNA sequence mutations in candidate genes, and lines that share a phenotype compared for commonality. The strategy is feasible with the advent of high-throughput genotypic detection systems such as TILLING (**see Chapters 21, 22 and 27**).

8.3. Advancing Selections

Once mutants have been identified from a population, they can follow two main deployment routes (**Figure 25.13**), although these are not mutually exclusive. The first step is to check whether the observed phenotype is truly new or merely another allele at an already categorized locus. The BGSAD (http://ace.untamo.net/) contains information and images of most of the mutant loci that have been reported in barley. One can search this for similar phenotypes to those identified in one's mutant population. This then provides a list of potential allelic mutants which can then be crossed to the new mutant and the F_1s can be checked to see if they are mutant or wild type. If the new mutant is not allelic to any previously known mutant, then it can be crossed to a reference genotype to begin to locate the mutant locus on the genome. A range of options are available here but all involve creating a mapping population but, as one is trying to locate one locus on the genome, it is most efficient to use Bulk Segregant Analysis (Michelmore *et al.*, 1992) with mapped markers to derive an approximate genomic location. The next step is to identify genetic markers in the region and use them to fine map the mutant locus in an extended segregating population. It is not necessary to genotype the whole population as one can use an initial screen with flanking makers to identify the potentially informative individuals for more intensive genotyping. If markers such as single nucleotide polymorphisms (SNPs) are utilized, then one can use the sequence information used to derive the SNPs to identify syntenic regions in model species such as Brachypodium and rice and thus identify a list of potential candidate genes that can be refined to a few potential targets using bioinformatics tools. The final stage is to validate candidate genes by knocking-out and/or over-expressing and evaluating

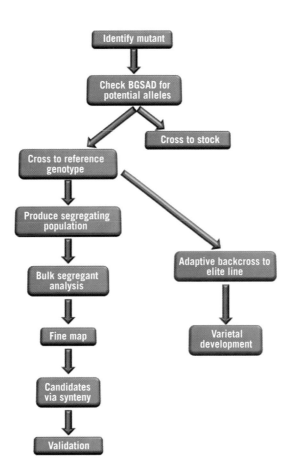

The Joint FAO/IAEA Programme

Figure 25.13 Schematic of deployment routes for mutants in barley research and improvement.

whether or not the effect on phenotype is similar to that observed in the original mutant.

Some mutants have potential application in variety development but other mutations carried elsewhere in the genome mean that direct mutants are frequently unsuitable for direct deployment as cultivated varieties. In such cases, a backcrossing programme can be used to "clean-up" the genetic background associated with the mutant whilst maintaining positive selection for the mutant phenotype. The problem with this strategy, particularly for annual crops such as barley, is that the mutant parent will probably have been surpassed by breeding progress and will thus not be competitive with other lines, unless the mutant trait confers such an end-user advantage that it attracts a market premium. An adaptive backcrossing procedure can be used to counter this situation. In such a scheme, the best available current variety is used as the "recurrent" parent so that the mutant is introgessed into a contemporary genetic background. The number of backcrosses needed depends, to some extent, upon the mutation frequency in the experiment but, given that background marker selection is not an option, at least two back-crosses should be conducted before starting a selfing or doubled haploid programme to develop a pure line.

8.4. Barley Mutations in Breeding

Classic examples of induced mutation in barley include early maturity (*eam8* induced in var. Mari), semi-dwarfism (*sdw1* and *ari-e.GP* induced in vars. Diamant and Golden Promise, respectively) and mildew resistance (*mlo9* in Diamant to produce the var. Alexis). In diploid inbreeding cereals such as barley and rice, it is not uncommon for mutant lines to become varieties directly. Typically however, mutation breeding attempts to improve a single trait, e.g. disease resistance, in otherwise elite lines: the elite line is subject to a mutagenic treatment and desired phenotypes are selected in the M_2 and subsequent generations. Although desired traits can often be found within the primary gene pool of the crop species, induced mutations have an advantage over the use of more distant gene pools (out-of-date varieties, landraces and wild relatives) in that the trait can be selected for directly within contemporary breeding germplasm.

Several of the mutants detected in the Optic mutation grid are of interest in barley breeding. Examples include mildew resistance, altered grain shape and size, naked caryopsis and orange lemma mutants. It is unlikely that any of the Optic mutants will become varieties directly as it is likely that the mutational load induced by EMS is large (1 mutation every 500,000 nucleotides), and Optic itself will soon be superseded as a variety as it was first recommended in 1995. There is however great potential in using this material for barley breeding to improve standard traits and introduce novel characters.

8.5. Other Uses

Mutants such as leaf spot and lesion mutants can be compared to disease and mineral deficiency/toxicity symptoms to investigate genes that are common or different in elucidating the mechanisms involved in biochemical responses (recognition, response, uptake, transport, etc.) to biotic and abiotic stresses.

Classically, barley mutants have played an important role in developing a better understanding of genetic systems. They have, for example, been used to elucidate phytomer models (Bossinger *et al.*, 1992). The Optic mutants include several new phytomer mutants that shed more light on the barley phytomer model and have necessitated revision (Forster *et al.*, 2007).

Root mutants are an under-represented class. Although simple tests can be devised to assess roots at germination and early seedling development there is a lack of knowledge of mutants in roots of older plants. In the Optic work several root mutants were observed at germination and it is clear that a wealth of mutants exist in older roots systems, including the nodal root system. The detection of root lodging variants is an indication that these occur. Since roots are underground organs they are difficult to observe. Simple root screening methods such as electrical capacitance designed to estimate root system size (Chloupek *et al.*, 2006), may be useful as a first screen in selecting mutants that should be studied in more detail with more sophisticated screens.

Fertility/sterility mutants are composed of a large group of variants and have potential in the discovery of the numerous genes involved in the development of reproductive organs, meiotic cell division and gametophytic differentiation. Male sterile mutants have some potential as a breeding tool. In rice male sterile mutants, particularly those that can be turned on and off by altering the environment, e.g. photoperiod are of interest in hybrid rice seed production (Rutger, 1992). Similar reversible mutants may be found after growing selected barley mutants in different environments, which can then be selected for hybrid seed production.

The large number of visual mutants that can be maintained as homozygous lines can be used as training tools for students and barley researchers to observe variants and understand the range of traits that can be altered by mutation. Some of these educational opportunities may include knowledge transfer exercises; various collections can be compiled to demonstrate genes involved in evolution and adaptation, breeding targets, chromosome location, allelic variation and to demonstrate laws of inheritance.

8.6. Limitations

Although a vast amount of phenotypic data can be collected there remain many limitations in phenotypic data of mutants. In barley the primary harvested product is the grain and therefore a collection of grain mutants would be highly desirable. Although a large mutant collection was obtained in Optic, it is limited by the fact that Optic is a semi-dwarf variety having a covered caryopsis (the lemma and palea adhere to the caryopsis, termed hulled). Many caryopsis mutations are therefore masked; this can be addressed by the development of another mutant population in a naked barley genotype. The semi-dwarf 1, *sdw1*, gene in Optic may also limit the range and expression of plant height mutants that can be observed, again this may be overcome by developing yet another population in a tall genotype. Normally only the sporophytic mutants are observed and there is little attempt made to investigate the male and female gametophytic stages, these diminutive and endosymbioitc life forms mature into pollen (male) and embryo sacs (female) and require specialized observation techniques. However, many mutants expressed during gametophytic development are likely to results in reduced fertility, and sterile and partial fertile mutant classes would provide a good starting point in which to screen for gametic phase mutants. Variation in germination percentages can be observed, but the causes require further investigation. These may, for instance be caused by biochemical mutants (isozymes, nutritional factors, malting, brewing and feed traits). The mutant descriptions are also often deficient for responses to abiotic and biotic stresses as the material is grown in normal and relatively benign environments. It is highly desirable to grow mutant populations in a range of envi-

ronments as this will reveal responses to temperature, light intensity and photoperiod as well as a range of pests and diseases. Barley, like other small grain cereals, has two root systems; seminal (or primary, derived from the seed or embryo node) and nodal (or adventitious, derived from the nodes of the culm). Since observations are normally done on aerial parts, with perhaps the exception of germination, root traits are greatly under-represented and tests are required to investigate the normally hidden underground plant parts.

9. Acknowledgements

SCRI receives grant-in-aid from the Scottish Government Rural and Environment Research and Analysis Directorate (RERAD). Jerome D. Franckowiak received an International Scientific Interchange Scheme (ISIS) fellowship from the UK Biotechnology and Biological Sciences Research Council (BBS) to work at SCRI for 3 months in 2005. We would like to thank Adrian Newton (pathologist, SCRI), Richard Keith (agronomist, SCRI), Gordon Simpson and Sandie Blackie (floral physiologists, SCRI and University of Dundee at SCRI), Jill Alexander (germination, SCRI), Nick Harberd (gibberellins expert, JIC, Norwich), David Laurie and Mary Byrne (flowering, JIC, Norwich) for help in screening parts of the Optic mutant population.

10. References

10.1. Cited References

Bossinger, G., Rohde, W., Lundqvist, U. *et al.* 1992. Genetics of barely development: mutant phenotypes and molecular aspects. *In*: P.R. Shewry (ed): Barley: Genetics, Biochemistry, Molecular Biology and Biotechnology. Wallingford: CAB International. pp. 231-264.

Caldwell, D.G., McCallum, N., Shaw, P. *et al.* 2004. A structured mutation population for forward and reverse genetics in barley (*Hordeum vulgare* L.). *The Plant Journal.* 40: 143–150.

Chloupek, O., Thomas, W.T.B., Forster, B.P. *et al.* 2006. The effect of semi-dwarf genes on root system size in field-grown barley. *Theoretical and Applied Genetics.* 112: 79–786.

Forster, B.P., Franckowiak, J.D., Lundqvist, U. *et al.* **2007.** The barley phytomer. *Annals of Botany.* 100: 725–733.

Michelmore, R.W., Paran, I., Kesseli, R.V. *et al.* **1991.** Identification of markers linked to disease resistance genes by bulked segregant analysis – a rapid method to detect markers in specific genomic regions by using segregation. *Proceedings of the National Academy of Sciences.* 88: 9828–9832.

Rutger, J.N. 1992. Impact of mutation breeding in rice – a review. *Mutation Breeding Review.* 8: 3–23.

10.2. Websites

IAEA/FAO mutant variety genetic stock database: http://mvgs.iaea.org

Barley genetic stocks database (BGSAD): http://ace.untamo.net/bgs

10.3. Further Reading

Davis, M.P., Franckowiak, J.D., Konishi, T. *et al.* **1997.** 1996 special issue. *Barley Genetics Newsletter.* 26.

Franckowiak, J.D. 1995. The brachytic class of semidwarf mutants in barley. *Barley Genetics Newsletter.* 24: 56–59.

Hagberg, A., Åkerberg, E. *et al.* **1961.** Mutations and polyploidy in plant breeding. Stockholm: Scandinavian University Books, Svenska Bokförlaget.

Khush, G.S. 2001. Green revolution: the way forward. *Nature Reviews Genetics.* 2: 815–822.

McCallum, C.M., Comai, L., Greene, A. *et al.* **2000b.** Targeted screening for induced mutations. *Nature Biology.* 18: 455–457.

C26

Mutation Breeding of Vegetatively Propagated Crops

P.Suprasanna[a,*] and H.Nakagawa[b]

[a] Nuclear Agriculture and Biotechnology Division,Bhabha Atomic Research Centre, Trombay, Mumbai 400085 India
[b] Institute of Radiation Breeding, National Institute of Agrobiological Sciences, Kami-Murata, Hitachi-Ohmiya, Ibaraki 319-2293, Japan
* Corresponding author, E-MAIL: prasanna@barc.gov.in

1. Introduction

A large number of plant species are asexually or vegetatively propagated and are known as vegetatively propagated crops (VPCs). They include many ornamentals, root and tuber crops, woody perennial and forest trees, fruit crops, and other crops such as peppermint, sugarcane, tea and many grasses. Cross-breeding of VPCs is often difficult due to various biological limitations, e.g. their long vegetative phase, high heterozygosity and polyploidy, incompatibility and other cross barriers, apomixis and sterility. Mutation techniques can overcome many of these barriers and can be used for the improvement of many VPCs.

The tulip (*Tulipa* sp.) mutant variety Estella Rijnveld developed by X-ray irradiation and officially released in 1954 in the Netherlands may be the first mutant VPC variety. Since then, there have been almost 700 VPC mutant varieties released worldwide (FAO/IAEA Database of Mutant Variety and Genetic Stocks, http://mvgs.iaea.org/. 2009.09.01).

The possibilities of mutation breeding in vegetatively propagated crops depend on many factors, such as the genetics of the characters involved, the mutagen to be used, the handling of the material after treatment, the availability of a selective screening method, etc. It is important to define both the objectives of the programme and to outline the strategy very carefully. The general principles, commonly used techniques and protocols for VPC improvement using mutation techniques are described in this chapter; *In vitro* techniques are very useful in mutation breeding of VPCs, particularly for the production of solid, non-chimera mutants and for *in vitro* selection of certain traits, they are described in **Chapter 28**.

2. Modes of Vegetative Propagation

Vegetative reproduction is a form of asexual reproduction and is also known as vegetative propagation, vegetative multiplication, or vegetative cloning. In addition to a natural vegetative mode of propagation, some VPCs can also (or can be forced to) propagate *via* seeds in certain conditions; they can also be multiplied *via in vitro* culture in commercial production.

2.1. Mode of Natural Propagation

Different plant species have evolved to asexually reproduced progenies through different mechanisms through the formation of specialized vegetative structures. **Table 26.1** presents different modes of vegetative propagation with examples in a range of plant species. In vegetative propagation, parts of the shoot are usually able to root anew and to develop into a plant. For example, strawberries develop specialized runners that serve propagation and colonization of new areas, some species form specific fruit bodies or brood buds, the potato forms tubers.

Table 26.1: Modes of vegetative propagation

Mode	Plant part	Examples
Adventitious	Roots	Asparagus and sweet potato
	Root cuttings	Raspberries, pears and apples
	Leaves through adventitious buds	*Bryophyllum* and Begonia
Stem	Runner	Lawn grasses
	Sucker	Mint
	Bulb	Onion
	Tuber	Potato and Canna
	Rhizome	Ginger
Special parts	Bulbils	Lilies, tulips, hyacinths, daffodils, *Oxalis*, pineapple and onion
	Corm	Gladioli and banana
	Lateral shoots	Strawberry and violet

2.2. Commercial Propagation

VPCs include a large number of economically important plant species such as cash crops (ornamentals, timber and fibres) and as food crops (fruits, tuber and root crops). For certain VPCs, such as some bananas, oranges and grapes, clonal propagation is necessary for the multiplication of plants of varieties that produce no viable seeds. For others propagation may be easier, more rapid, and more economical by vegetative methods than by seed.

Depending on the plant species, various plant organs are used as propagules in clonal propagation (**Table 26.1**). For example, propagation can be initiated by removing a section of a plant (bud, scion, cutting or other vegetative structures such as tuber and bulb) through *in vivo* or *in vitro* culture. Buds, scions and cuttings are very common in the multiplication of fruit trees, plantation crops and ornamental crops (e.g. citrus, rubber and rose, respectively), while tubers and bulbs are used in herbaceous vegetable and ornamental crops (e.g. potato, garlic, shallot, dahlia and lily). Vegetative or clonal propagation produce plants that have the same genetic information as the mother plants. This process is very important in horticulture since the genotype of most commercial fruit and ornamental varieties is highly heterozygous, and the unique characteristics of such plants are immediately lost if they are propagated by seeds, i.e. through the processes of meiosis.

Beside tubers, crops such as potato can be propagated through cuttings. Cuttings taken from juvenile parts of many woody perennial plants can be produced easily by adventitious roots, whereas cuttings taken from the adult part of the same plants are not able to produce adventitious roots easily. An advantage of taking cuttings is the acceleration of mass propagation of plant materials. Some plants such as pineapples and bananas, for example, are vegetatively propagated because it is not possible to produce seeded fruits. In pineapple, leaf budding has been shown to be a good way of mass propagating *in vivo*, an average of 15 leaf-buds can be obtained from a single crown. Suckers have also been a traditional method to propagate banana. However, the rate of *in vivo* propagation through suckers or leaf budding is not sufficiently rapid and does not have growth uniformity for commercial scale production, which can be solved in some species through micro-propagation using tissue culture techniques.

2.3. *In Vitro* Propagation and Propagation in Breeding Programmes

During the past decades, the ability to culture and manipulate a large number of totipotent cells has been exploited for both commercial production of propagules and for breeding programmes. Numerous protocols have been developed for different kinds of plants, for example dozens of ornamental plants. *In vitro* techniques such as micro propagation and somatic embryogenesis provide an efficient method for propagating disease-free, genetically uniform plants on a large scale compared to propagation by conventional means such as cuttings and seeds. There is also an additional benefit in that micro propagation cultures are disease free and can be easily transported. Micro propagation can be achieved by direct organogenesis *via* shoots or by somatic embryogenesis. Organogenesis *via* shoots is considered to be one of the most widely used commercial methods of regeneration. Once plants have been regenerated, they are hardened-off; then pass through gradual acclimatization in a glasshouse (or shade house depending on the crop) and then to the field. In breeding programmes of VPCs, tissue culture techniques are extremely useful; they allow mutagenic treatments of large populations, *in vitro* selection of certain traits, and cloning of selected variants in a small area, short time periods, and on a year round basis. The details are described in **Chapter 28**.

Although VPCs are multiplied asexually in commercial production, seeds are produced in conventional breeding programmes of a few plant species, such as potato, rubber and cassava. The plants produced through hybridization of two varieties always exhibit a range of genotypes and can be subject to selection of new varieties. However, asexual propagation is still used for these crops in mutation breeding programmes.

3. The General Procedure of Mutation Breeding in Vegetatively Propagated Crops

3.1. General Aspects

Both scientific and technical issues need to be thoroughly analysed before initiating a mutation breeding programme in VPCs. Mutation breeding might not be the method of choice for certain breeding objectives, or not applicable for certain plant species. Therefore, knowledge of the target crop, mode of propagation and traits to be improved are needed for the choice of starting material and the ways of eliminating chimeras.

Box 26.1 presents critical considerations that should be thought through before initiating any mutagenesis programme. The decision as to which breeding strategy (mutation *versus* cross-breeding) is appropriate for a specific situation is usually economic: which method is the easiest, the fastest, and the least costly. Mutation breeding in some cases may offer the best prospects of success.

Mutations breeding of VPCs involves the processes of mutation induction, and separating and selecting useful mutations. Effective methods to develop mutants that express phenotypic variation on an individual plant level include: adventitious bud techniques, continuous pruning, grafting and cutting-back techniques and *in vitro* culture techniques. Generally, stable mutants are not produced until after several vegetative generations.

3.2. Mutagenic Treatments

3.2.1. Target Materials

For SPCs, treatment of seeds with a mutagen is a common procedure to induce mutation, while in VPCs, meristematic buds are usually used as target material for mutation induction. Many plants possess natural systems of vegetative propagation, especially by tubers, bulbs, rhizomes, stolons, apomictic seed, etc. (**Table 26.1**). In addition, many new *in vivo* methods, such as stem or leaf cuttings or grafts, and *in vitro* methods such as cell or tissue culture are also used in commercial production. Both natural and artificial propagules can potentially be employed as the target materials of mutagenic treatment.

In practice, the ease of dissociating chimeras after mutagenic treatment (**see Chapter 16**) plays an important role in the choice of target material. If *in vitro* techniques are employed either for chimera dissociation or *in vitro* mutagenesis, consideration of their suitability for *in vitro* culture may prevail and different target materials may be used (**see Chapter 28**).

The use of heterozygous starting material (*Aa*) will be more practical for mutation work than homozygous material (*AA* or *aa*), because induced mutations are mostly recessive hence they can be expressed only in plants with the original genotype *Aa* where the dominant allele is mutated. Mutations from recessive to dominant occur at a very low frequency, i.e. probably less than 5% of all mutational events.

Box 26.1: General considerations for initiating a mutation breeding programme of VPCs [a]

- The trait for which the variation is sought, and its commercial value/potential, end-user demands.
- Genetics of the trait (dominance, recessiveness, pleiotropy, linkage).
- The crop, variety, mode of propagation, degree of heterozygosity and ploidy level.
- Need for mutation induction as an alternative to existing conventional or modern methods.
- Plant material to be used for treatment and methods to handle chimerism.
- Available information on the above aspects and limitations, if any.

[a] adapted from Broertjes and van Harten (1988).

Box 26.2: Steps in the mutation breeding in VPCs

Year 1:
1. Explants (for example, shoot meristems or axillary buds) are treated with mutagens with optimal doses (for determination of a suitable dose range, see Chapter 14).
2. Shoot growth (M_1V_1 generation) is initiated and assessed for the occurrence of chimera (in the apical and axillary bud meristems. Vegetative propagation of M_1V_1 is then followed.

Year 2:
1. Assessment of M_1V_2 shoots (shoot grown from buds of M_1V_1 plants) for possible occurrence of periclinal or homohistont mutated parts. Vegetative propagation of M_1V_2 shoots for the isolation of induced mutations.

Year 3:
1. Genetic uniformity is checked through growth of the mutant (M_1V_3) and preliminary evaluation of the mutants is taken up.

Years 4 to 9 [a]:
1. Assessment of growth of the mutant is done throughout the vegetative and reproductive stages and evaluation of mutant's performance for agronomic traits is completed.

Year 10 [a]:
1. Final assessment is made in M_1V_{10} generation and processing of improved clone/mutant for release as a mutant variety.

[a] The exact number of years needed varies from plant species to species.

Box 26.3: Some practical *in vivo* and *in vitro* techniques used for chimera dissociation

Method	Description	Plant species
Adventitious buds	Regeneration of plants through adventitious buds from leaf cuttings	Chrysanthemum
Induced axillary branching	Induced axillary shoot production from a chimera cultivar by cytokinin application	*Cordyline terminalis* L.
Leaf cuttings	Leaf cuttings regenerate shoots and the plants are mostly non-chimerical (possibly arising from single cells or cells descended from a single apical layer)	Peperomia
Adventitious roots from stem cuttings.	Generation of shoots from root cuttings (non-chimerical, originating mostly from L3)	Potato, Pelargonium
Disbudding of tubers	Regeneration of shoots from disbudded tubers (stems)	Potato
In vitro organogenesis	Regeneration through organogenesis from sectored chimeras	Begonia and several dicots and monocots
Shoot proliferation through micropropagation	Shoot cultures of chimera plants are proliferated on high cytokinin medium	Banana
Fragmentation of shoot apices	Shoot apices of periclinal chimeras are fragmented during the initiation of the culture	Grape
Shoot-tip culture technique	Shoot tips (consisting of a meristematic dome with 2–5 leaf primordia) are cultured on cytokinin medium	Banana
Multi-apexing culture technique	A combination of a high proliferation rate and production of adventitious buds is used. The absence of apical dominance and a higher proliferation rate could be responsible for the absence of mix-ploidy	Banana
Corm slice culture technique	Shoot tip explants isolated from *in vitro* plantlets with a pseudostem are isolated and necrotic tissues are removed before making 3–5 transverse sections below the meristem tip are made. Slices of the corm tissue 0.5–1 mm thick are then placed on the semi-solid multiplication medium.	Banana

3.2.2. Mutagens and Treatment

Both chemical and physical mutagens have been used, but more than 90% of released mutant varieties are developed using physical mutagens in VPCs (FAO/IAEA Database of Mutant Variety and Genetic Stocks, http://mvgs.iaea.org/). All types of ionizing radiations such as X-rays, gamma rays, fast neutrons and ion beams are effective for the induction of mutations. However, in practice, only X- and γ-rays and recently heavy ion beams are most commonly used for mutation induction in VPCs. The high penetration capacity and the fact there is no need for post-treatment handling may explain the preference of these radiations over chemical mutagens. There have also been many reports stating that chemical mutagenesis is less effective because of poor uptake and less penetration of the chemical (however, see **Chapter 12** for current views on chemical mutagenesis). Large-sized materials such as rooted cutting, bulbs, stolons and scions for grafting are difficult to treat in a reproducible manner with chemical mutagens. Chemical mutagens can be applied to tissue culture materials but the procedure requires sterilization of mutagen solution by the ultra-filtration method.

The next step is to identify a suitable dose after the target material and mutagen have been chosen. The general procedure outlined in **Chapter 14** should be followed in this process. Compared with seed mutagenesis, the moisture content of different batches of propagules of the same species may vary significantly, therefore, it is strongly recommended that previously reported doses should be used only as reference doses, and pre-trials be performed to determine the dose suitable for the material to be treated in each experiment.

3.3. Dissociation of Chimeras and Selection of Mutants

3.3.1. Production of Solid Mutated Plant

Since natural and artificial vegetative propagules are multicellular organs and several meristem cells develop into a plant or a branch, therefore, mutagenic treatment of such entities can result in chimeras (**see Chapter**

16). Mutations occurring in apical or axillary shoot meristems will lead to a lineage of daughter cells and thereby a lineage of mutated cells and then to large mutated sectors. A mutant plant replete for a specific trait could result from a single mutated cell. Both solid mutants and periclinal chimeras can be used as new varieties in vegetatively propagated plants. In situations where mutations occur outside the shoot meristem, the cells have to be stimulated to develop further shoots in order to encourage maintenance of the mutation. For example, mutagenized callus can be multiplied and stimulated to produce shoots and whole plants can be regenerated for evaluating the mutant. In chrysanthemum, gamma ray irradiated florets (small flowers on the peripheral rings of a composite flower) can be cultured for direct regeneration, a technique that has proved useful in avoiding unstable chimeras by forcing the regeneration of buds from single cells.

If chimerism cannot be avoided, it would be useful to know the chimeric pattern of M_1 plants or for a systematic dissolution of chimeric structures in vegetative organs. The mutated sectors of chimerical plants can then be grown and isolated from the non-mutated sectors through a process known as "chimera dissociation". This is achieved through a series of vegetative propagation procedures either *in vitro* or *in vivo*; the underlying principle and general protocol are described in **Chapter 16**. In practice, different methods are employed in different plant species (**Box 26.3**). Additionally, *in vitro* culture techniques are powerful means for chimera dissociation (see **Chapter 28**). For example in banana, Roux *et al.* (2001) assessed colchicine treatment-induced ploidy chimerism (mixoploidy), and chimera dissociation using three different propagation systems (shoot-tip culture technique – ST, multi-apexing culture technique – MA and corm slice culture technique – CS). The average percentage of cytochimeras was reduced from 100% to 36% after three sub-cultures using shoot-tip culture, from 100% to 24% when propagating by the corm slice culture technique and from 100% to 8% after the same number of sub-cultures using the multi-apexing technique.

An important practical consideration with respect to mutation breeding in VPCs is the availability of an existing method or one that can be developed for propagation by adventitious buds. The most important advantage should be the speed and relative ease of production of solid mutants. Adventitious bud technique is a remarkably simple way to avoid unstable chimera by forcing regeneration of buds from single cells. The proportion of mutants with uniformly altered inflorescences increase considerably in the second vegetative generation (M_1V_2), through regeneration from leaf explants. For example, in chrysanthemum, Broertjes *et al.* (1980) estimated that after X-irradiation it took only 1–2 years to produce the same range of flower colour mutants in a specific variety as at present would arise as spontaneous sports in about 10–20 years. In *Gerbera*, following X- or gamma ray treatment of *in vitro* grown leaves and subsequent regeneration of adventitious shoots, 1,250 M_1V_2 plants were raised, of which 187 mutants were isolated and only six were confirmed to be solid mutants (Laneri *et al.*, 1990). A technique has also been developed for the management of chimeric tissues through direct shoot regeneration from flower petals of chrysanthemum (Datta *et al.*, 2005). This method of chimera isolation and establishment of solid mutant has two steps: first, *in vivo* mutagen treatment of ray florets (small flowers on the peripheral rings of a composite flower) and second, *in vitro* regeneration of viable plants from mutated sectors. This method represents a quick method (M_1V_2) for the creation of solid mutants.

In Gladiolus, normally dormant corms are irradiated in springtime with about 75 Gy of gamma rays. When plants in M_1V_1 (the first vegetatively propagated generation after mutagenesis) being "sectorial" chimeras, are propagated in the usual way, that is by corms (stem tubers from orthotopic subterranean stems), the resulting M_1V_2 consists almost completely of either solid mutants or of solid normal plants.

3.3.2. Selection of Mutants

Selection for interesting mutants can take place 3–4 years after irradiation. From these mutants, clones can be made for further assessment. In this way some commercial mutants have been obtained (Broertjes and van Harten, 1989). Visual selection of the mutant phenotypic variation is the most effective method of selection in VPCs. It can be used effectively for identifying common traits and characters such as colour changes, plant morphology, earliness, resistance to pests and diseases, etc. Selection usually starts from M_1V_2 and continues for confirmation in the M_1V_3 or M_1V_4 generation. Since the probability of identifying desired mutants in a homogenous M_1V_2 population is quite low, selection is necessary in the M_1V_3 generation or in some cases in the M_1V_4 generation. In the case of ornamental plants such as tulip, selection is continued for at least four years to allow mutated cells to grow out and to produce either a solidly mutated bulb or at least a larger visibly mutated sector that will be expressed in further generations. In crops like sugarcane, stem cuttings are multiplied for three or four generations before selection is applied on the irradiated plant population.

Although each mutation breeding programme has its unique objectives, most mutations to be identified in VPCs can be categorized into the following: 1) compact type (shorter stems or shorter internodes), 2) fruit or flower colour, 3) self-fertility, 4) seedlessness, 5) time of fruit maturity and 6) disease resistance. In VPCs, phenotypic selection is made directly on plants in the field, nursery or glasshouse. In the case of selection for pest and disease resistance, challenging and/or exposure to the causative agents of the disease or its pathotoxin may be required or the plants to be tested can be grown in a disease-infected area or 'hot spot'. With ornamental plants, selection for changes in flower colour or plant morphology is very much easier, this can be done directly in the glasshouse or in the field. If a cutting back technique is to be applied, the newly grown shoots are removed after a few vegetative stages and observed for desired characters.

3.4. Evaluation of Mutant Clones and Release of New Commercial Varieties

Induced mutations have been widely used for commercial breeding of VPCs, especially with ornamental plants. After isolation of mutants with desirable traits that meet the objective, they are propagated through either *in vivo* or *in vitro* methods. The clones are then assessed in official field trials as required for the release of a new commercial variety This process is not different from cross-breeding and readers can learn more about this by reading specialized breeding books for pertinent crops.

4. Theoretical Aspects of Mutation Breeding in Vegetatively Propagated Crops

Mutations are the ultimate source of genetic variability in VPCs. In general, induced variability is no different from that generated from spontaneous mutation. The direct use of induced mutation is important for breed-

ing of VPCs, especially when the objective is to improve one or two characters, without altering the remaining genotype in a well-established variety. Therefore, the main advantages are that the basic genotypes of the established variety are only slightly altered and that the time taken to improve an outstanding variety with desired traits can be shortened as compared to traditional methods of hybridization. Inheritance in VPCs is often complex owing to high levels of heterozygosity and ploidy, which make genetic analysis difficult. Other limitations in mutation breeding in VPCs include the damage caused after radiation or chemical mutagen treatment. These include non-genetic effects that are not restricted to the first generation, their effects will only disappear after several subsequent vegetative generations.

4.1. Genetic Features of VPCs and Induced Mutations

Why can induced mutations be observed in M_1 generation in VPCs? Most VPCs are highly heterozygous, having genotypes such as *Aa, AAa* or *Aaa* at many loci (diploid or triploid or higher ploidy). This is an advantage for mutation breeding, because most induced mutations are recessive, therefore, mutants with genotype α'α and *a'aa* (*a'* represents the mutated allele) are easily manifest in M_1 generations (in chimeras such as M_1V_1, M_1V_2 or in solid mutants such as M_1V_s).

New genotypes are not necessary induced mutations. In some cases, a new genotype observed in the progeny may not necessarily result from induced mutation. Mutagenic treatment can cause re-arrangement of the meristem structure. Tetraploid cells are likely to be more resistant to irradiation than diploid cells, because doubled chromosomes in tetraploid cells compensate for chromosomal damages associated with irradiation. As a result of such effects, diploid cell layers may be replaced by tetraploid cell layers when shoot apices of periclinal chimeras in which the layers differ in ploidy level are irradiated. There have been many reports stating that irradiation can be used to transfer a mutation that is present only in the cells of one layer, to another layer.

Maintenance of chimerical mutants. In mutation breeding, the mutated sector should be developed into a solid mutant or a periclinal chimera. In ornamental plants, the flower colour depends mainly on the genetic constitution of the L1 layer of the flower petals and, in chrysanthemum; a mutated flower colour is often obtained in different genetic constitutions between L1 and L2 layers. In such periclinal mutants, it is important that periclinal chimeras are maintained as long as vegetative propagation is continued. In general, mutated cell lines are stable in periclinal chimeras, but not in the mericlinal chimeras (**see Chapter 16**).

4.2. Determining an Optimal Treatment Dose

Mutation frequency usually increases linearly with increasing dose of mutagenic treatment, but survival and regeneration capacity decreases with increasing dose. When determining an optimal dose for treatment, these two aspects should be considered.

In VPCs no meiosis is experienced throughout the process of mutant development, therefore, the genetic constitution of a mutated cell is inherited to its lineage cell lines and plants unchanged. Therefore, it is impossible to separate useful mutations from unwanted ones that occurred in the same cells. This is in sharp contrast to SPCs, where unwanted mutant genes can be separated from desired ones through self-crossing or back-crossing. Therefore, a dose lower than LD_{50} is preferred in mutation breeding of VPCs.

4.3. Features of Mutation Breeding *versus* Cross-Breeding

4.3.1. Mutation Breeding *versus* Cross-Breeding
Mutation induction techniques can greatly increase the gene mutation frequency and create new germplasm, new materials and new varieties in a relatively short period of time. Compared with other types of crops (for example, SPCs) and other breeding methods, mutation breeding of VPCs has unique advantages. The genetic variability increases considerably, by hundreds or a thousand times higher than the natural mutation frequency. The variation spectrum is wide and various, among them useful variation increases significantly, including some rare mutations that are not easy to be observed in nature or by crossing methods. Second, once the target mutation appears, it can be stabilized rapidly by vegetative propagation methods to speed up the selection process. In a situation where a desired character is controlled by a single gene, and a desirable complex character such as quality trait has to be

retained, induced mutagenesis in combination with tissue culture technique greatly enhance the efficiency of genetic improvement in VPCs.

4.3.2. Usefulness of Mutation Breeding
Mutation breeding is the most suitable method for the breeding of VPCs, since the new mutant varieties and the original ones have the same genetic background except for the mutated genes. Therefore, new mutant varieties can be used for cultivation under the same conditions as their parental varieties, which is not always the case for new varieties developed through cross-breeding where new agronomies have to be developed for new forms. In VPCs that do not produce any seed (for example banana), mutation breeding becomes one of the few available options (other than transformation) since cross-breeding is not possible.

5. Successful Examples

5.1. Japanese Pear Varieties

5.1.1. Gold Nijisseiki
The variety Nijisseiki is one of the most popular Japanese pear varieties. It occupied 28% of the total cultivated area of Japanese pear in Japan in 1990. However, var. Nijisseiki is highly susceptible to the black spot disease, *Alternaria alternate* (Fr.) *Keissier* (*Alternaria kikuchiana Tanaka*). By using chronic gamma ray radiation in the Gamma Field, a black spot-resistant Japanese pear variety, Gold Nijisseiki, was developed and released as the "pear Norin 15" by the Institute of Radiation Breeding, NIAR and Tottori Horticulture Experiment Station in 1991. The resistance is intermediate and less than that of the natural resistant germplasm compared by a developed screening test using AK-toxin to cut leaves (Sanada, 1988) probably because the periclinal chimera resistance mutation occurred only in the L2 cell layer, however black spot symptoms seldom appear in the field condition. Since the variety is very useful, it was registered in Australia in 2004 (Certificate No. 2533), too. The breeding process was as follows:

1962 Vegetatively propagated nursery plants of Nijisseiki susceptible to black spot disease were planted and grown in the Gamma Field at 53 m (30 × 10^{-2} Gy/day) – 93 m (4 × 10^{-2} Gy/day) with the interval of 8 m) from the ^{60}Co source. Since there was no effective screening method for resistant branches at the time no mutants were identified.

1981 After the application of fungicides was stopped branches were screened for resistance primarily by natural infection. One resistant mutant branch named as "γ (gamma)-1-1" was selected from the chronically irradiated trees for about 20 years without showing symptoms.

1986–1989 Regional adaptability tests at experiment stations of 27 prefectures were conducted.

1990 Since the characteristics of "γ-1-1" were identified as being the same as var. Nijisseiki except for the resistance to black spot disease the mutant line was registered as a national variety, Gold Nijisseiki (Japanese pear Norin 15).

(Adapted from Kotobuki et al., 1992).

5.1.2. Osa Gold

After the mutant branch resistant to black spot disease was identified, a rapid screen based on treatment of leaf discs with AK-toxin was established (Sanada, 1988). When the discs (7 mm in diameter) of 1st to 5th leaf of a twig were cut and placed on the filter paper soaked with the AK-toxin for 2–3 days, susceptible leaves turned black and resistant leaves remained green (**Figure 26.1**). By combining chronic gamma irradiation in the Gamma Field with this rapid screen, another intermediate resistant Japanese pear variety, Osa Gold, was bred from the self-compatible variety, Osa Nijisseiki, which is highly susceptible to black spot disease, by the Tottori Horticultural Experiment Station (THES) and Institute of Radiation Breeding, National Institute of Agrobiological Resources (NIAR). The breeding process was as follows:

1986 Vegetatively propagated nursery plants of var. Osa Nijisseiki were planted at a distance of 40, 50, 60, and 70 m from the ^{60}Co source in the Gamma Field and chronic gamma ray irradiation was applied to the growing plants.

1988 Screening tests using a bioassay using AK-toxin, (a host-specific toxin produced by *Alternaria alternata*) were conducted on the leaves of stems.

1991 Candidate resistant branches (IRB 502-13T) that were clearly more resistant to the disease than var. Osa Nijisseiki were selected. Parts of the branch were grafted to a Japanese pear tree and the resistance of the subsequent growth was identified by the bioassay method.

1992 Vegetatively propagated branches were grafted to var. Suisei of Japanese pear.

1993 Self-compatibility of the flowers and fruit characteristics of the plants were evaluated. As a result, it was demonstrated that the mutant line was completely similar to var. Osa Nijisseiki except for the resistance to black spot disease.

1997 The mutant variety Osa Gold was registered. **(Adapted from Masuda et al., 1998).**

5.1.3. Kotobuki Shinsui

Since the screening method using AK-toxin was effective, acute irradiation to scions for the induction of resistant mutants of a Japanese pear variety Shinsui (highly susceptible to black spot disease) was attempted by Tottori Horticultural Experiment Station (THES) and Institute of Radiation Breeding, National Institute of Agrobiological Resources (NIAR). The breeding process was as follows:

1987–1988 One thousand and eighty-eight dormant scions were irradiated with 60Gy of acute gamma-ray irradiation and 926 dormant scions irradiated with 80Gy (dose rate: 2.5Gy/hr), these were top-grafted to the mature tree of Japanese pear varieties, Yakumo and Hakko in the field of THES.

1988–1989 The 4th leaf of the newly developed 5,736 branches (3,568 branches from the branches irradiated with 60Gy and 2,168 with 80Gy) were screened for resistance to black spot disease using the AK-toxin bioassay.

1989 One branch (var. Nashi Houiku 2 gou) clearly exhibited the same intermediate resistance to the black spot disease as var. Gold Nijisseiki.

1993–1995 Regional adaptability tests at 27 institutes and experiment stations including National Institute of Fruit Tree Science were conducted

1996 The var. Nashi Houiku 2 gou was registered as a new variety Kotobuki Shinsui (Registered Number is No. 5436; Japanese pear var. Norin 12).

(Adapted from Kitagawa et al., 1999).

5.2. Apple

5.2.1. Houiku Indo

The AK-toxin bioassay for black spot disease screening in Japanese pear was adapted for the Alternaria blotch disease in apple by using a host-specific toxin of apple, AM-toxin. A new *in vitro* selection method was developed (Tabira *et al.*, 1993), including repeated *in vitro* propagation of shoots with weak reaction to the AM-toxin for eliminating chimeras. A new apple mutant variety, Houiku Indo, was developed by the Institute of Radiation Breeding,

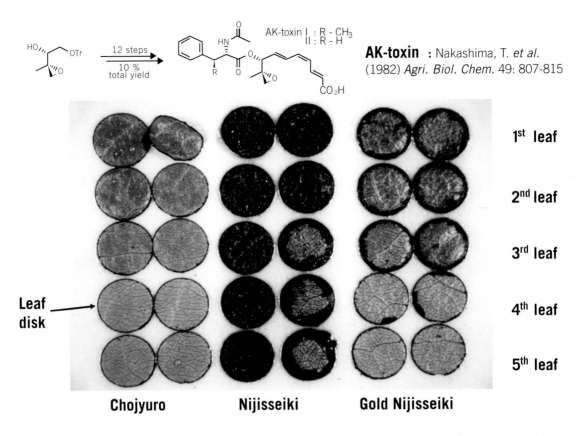

AK-toxin I : R - CH₃
II : R - H

AK-toxin : Nakashima, T. *et al.*
(1982) *Agri. Biol. Chem.* 49: 807-815

1st **leaf**

2nd **leaf**

3rd **leaf**

4th **leaf**

5th **leaf**

Leaf disk

Chojyuro   **Nijisseiki**   **Gold Nijisseiki**

The Joint FAO/IAEA Programme

Figure 26.1 Bioassay of resistance to black spot disease by *in vitro* test of AK-toxin Chojyuro: a resistant variety; Nijisseiki: a susceptible variety; Gold Nijisseiki: a mutant resistant variety.

NIAR. The breeding process was as follows:

1992 Shoots of the apple var. Indo, which is hypersensitive to Alternaria blotch were micropropagated on MS medium (Murashige and Skoog medium supplemented with 1.0 mg/l BAP and 0.75 % Bacto Agar at 25°C in a 16-hour photoperiod. Shoots cultured *in vitro* for 5–6 weeks were irradiated with gamma rays by a total exposure of 80 Gy at a dose rate of 2.5 and 5.0 Gy/h. After irradiation, a single cut shoot was placed in a 40 × 120 mm test tube containing 30 ml of micro-propagation medium among MV₁–MV₃ generation. Selection of shoots screened by using AM-toxin and propagation were repeated.

1994 One AM-toxin-insensitive clone was selected in the MV₆, irradiated by a total exposure of 80 Gy at a dose rate 5.0 Gy/h.

1995 The propagated resistant clones were top-grafted onto the apple variety, Sansa.

1999 The flowering of the grafted branch started.

2001-2003 Evaluation tests of fruits and related characteristics, as well as resistance to Alternaria blotch were conducted for 3 years and identified a resistant line, but with the quality characteristics of Indo.

2004 The line was registered as a new variety, Houiku Indo (registered number is No. 17754).
(Adapted from Yoshioka *et al.,* 2000).

5.3. Chrysanthemum

Mutation induction of chrysanthemum through a combined technique of gamma ray chronic irradiation in the Gamma Field and *in vitro* culture of irradiated plant tissues, and through ion beam irradiation were conducted and mutant varieties without chimeras have been successfully developed and released, notably in Japan.

5.3.1. Haeno-Eikou

1987 Small seedlings of a popular chrysanthemum variety Taihei were planted at various places in a Gamma Field in June with the dose of gamma rays 0.25–1.5 Gy/day until November (the total doses are 25–150 Gy for 100 days). Petals of the irradiated plants were cultured to produce callus and regenerated plants were obtained from the calli.

1988 Regenerated plants derived from the calli were transplanted to a field nursery in June, and a plant, IRB88-49, was selected from the plant population of 50 Gy of gamma ray irradiation dose.

1989-2000 The selected line was vegetatively propagated by cuttings and the characteristics of the plants were evaluated.

2001 Practical field trials were conducted in the Wadomari-cho Experiment Field, Kagoshima, and the unique colour and the shape were identified.

2002 The line, IRB88-49, was registered as the new variety, Haeno-eikou (Register No. 14563).

5.3.2. Haeno-Myojou

1987 Small seedlings of a popular chrysanthemum variety Taihei were planted at various places in the Gamma Field in June with the dose of gamma rays 0.25–1.5 Gy/day until November (the total doses were 25–150 Gy for 100 days). Petals of the irradiated plants were cultured to produce callus and regenerated plants were obtained from the calli.

1988 Regenerated plants derived from the calli were transplanted to a field nursery in June, and a plant, IRB88-30, was selected from the plant population of 50 Gy of gamma ray irradiation dose. The petals of the line were cultured and regenerated.

1989 A regenerated plant, IRB89-12 was selected.

1990 The characteristics of the IRB89-12 were evaluated.

2001 Practical field trials were conducted in the Wadomari-cho Experiment Field, Kagoshima, and the unique colour and the shape of the flowers of the mutant were identified.

2002 The line, IRB89-12, was registered as a new chrysanthemum variety, Haeno-Myojyou (Register No. 14563).

By mostly the same method using *in vitro* culture of same or different tissues from irradiated plants, another eight new varieties were developed and released by IRB.

5.3.3. Aladdin, Imagine and Aladdin 2

The variety Jimba is the most popular chrysanthemum in Japan, which is commonly used for Japanese ceremony such as funerals. Jimba has good features such as having pure white and large flowers, but has defective characteristics such as possessing many axillary flower buds and late flowering. The production of large numbers of axillary flower buds results in labour-intensive de-budding, which is thought to be 1/3 to 1/4 of total farmer's labour. Late flowering is also expensive in term of longer cultivation under heating in autumn and winter seasons in Japan. In order to overcome these agronomical constraints the Kagoshima Biotechnology Institute and Japan Atomic Energy Agency initiated mutation improvement of Jimba using ion beams. Ten thousand mutated chrysanthemum lines were produced from carbon or helium ion treatments. Two good mutants, named Aladdin and Imagine, successfully achieved the requirement for reduced axillary flower bud number. At present, Aladdin is one of the most popular varieties in Japan. As Aladdin showed no defects such as reduced DNA content after irradiation, Aladdin was used for re-irradiation to another final target trait; to improved low temperature flowering. By using ion beam irradiation, Aladdin 2 was created as the "perfect" chrysanthemum in Japan.

2000 Tissue cultures of flowers or leaf discs of Jimba were irradiated with carbon ions (320 MeV and 220 MeV) or helium ions (100 MeV and 50 MeV). Irradiation effects such as growth and regeneration rate were estimated. The best conditions for mutagenesis had been established.

2002 Among more than 10,000 mutated and regenerated plants, two good mutants, which had reduced axillary flower bud numbers compared to Jimba, were selected. At the same time, re-irradition with ion beams was initiated to improve lower temperature flowering by exposure to mutants possessing reducing axillary flower buds.

2003 Aladdin and Imagine were submitted for variety registration.

2007 Aladdin 2, which is mutated from Aladdin, was submitted for variety registration.

2009 At least 35 domestic and one foreign cooperative requested licences for using Aladdin and Aladdin 2. More than 40 million flowers have been produced commercially.

5.4. Rose

5.4.1. Hitachi Smile

1987 Small seedlings of a popular rose variety Samantha were planted in various places in a Gamma Field in June with the dose of gamma rays at 0.25–1.5 Gy/day until September 1988 (total doses were 98.4–1590.6 Gy for 16 months).

1988 The flowers of the plants were observed and the scions with mutant flower were cut and vegetatively propagated in September.

1989 The cuttings of the mutant scion were vegetatively propagated to generate non-chimeric plants and planted in the field in May.

1990 A mutant line, IRB90-4, was selected from the population with a total gamma ray dose of 492.1 Gy (1.25 Gy/day) and evaluated until 2001.

2001 The line, IRB90-4, was registered as a new rose variety, Hitachi smile (Register No. 15714). In the same way, a new variety Hitachi peony was also developed and registered.

6. References

6.1. Cited References

Broertjes, C. and Van Harten, A. M. (1988). Applied Mutation Breeding for Vegetatively Propagated Crops. Amsterdam: Elsevier.

Broertjes, C., Koene, P. and Van Heen, J.W.H. (1980). Irradiation of progressive radiation induced mutants in a mutation breeding program with *Chrysanthemum morifoium* Ram. *Euphytica.* 29: 525–530.

Datta, S.K., Mishra, P. and Mandal, A.K.A. 2005. *In vitro* mutagenesis – a quick method for establishment of solid mutant in chrysanthemum. *Curr. Sci.* 80(1): 155–158.

Laneri, U., Franconi, R, and Altavista, P. 1990. Somatic mutagenesis of *Gerbera jamesonii* hybrid: irradiation and *in vitro* cultures. *Acta Hort.* 280: 395–402.

Roux, N., Dolezel, J., Swennen, R. *et al.* 2001. Effectiveness of three micropropagation techniques to dissociate cytochimeras in *Musa* spp. *Plant Cell, Tissue* and *Organ Culture.* 66: 189–197.

Sanada, T. 1988. Selection of resistant mutants to black spot disease of Japanese pear by using host-specific toxin, Japan. *J. Breed.* 38: 198–204.

Tabira, H., Shimonaka, M., Kawakami, K. *et al.* (1993). Selection by AM-toxin of Alternaria blotch-resistant mutant from gamma rays-irradiated *in vitro* apple plants. *Ann. Phytopath. Soc. Japan.* 59: 753–754 (in Japanese).

Van Harten, A.M and Broertjes, C. 1989. Induced mutations in vegetatively propagated crops. *Plant Breeding Reviews.* 6: 55–91.

6.2. Websites

FAO/IAEA Database of Mutant Variety and Genetic Stocks:

http://mvgs.iaea.org/

FNCA. 2004:

http://www.fnca.mext.go.jp/english/mb/mbm/pdf/08

6.3. Further Reading

Van Harten, A.M. 1998. Mutation Breeding: Theory and Practical Applications. Cambridge: Cambridge University Press, pp. 1-353.

Van Harten, A.M 2002. Mutation breeding of vegetatively propagated ornamentals. *In*: A. Vainstein (ed.) Breeding for Ornamentals: Classical and Molecular Approaches. Kluwer Academic Publishers, pp. 105–127.

C27

Uses of TILLING For Crop Improvement

A.J.Slade* and C.P.Moehs

Arcadia Biosciences, 410 W. Harrison St, Ste 150, Seattle, WA 98119, USA
* Corresponding author, E-MAIL: ann.slade@arcadiabio.com

1. Introduction

TILLING, Targeting Induced Local Lesions in Genomes, is a method to uncover genetic diversity at the single nucleotide level that has widespread application in both plants and animals. This reverse genetic method facilitates research ranging from investigation of gene function to modification of traits for crop improvement. TILLING provides a means to discover induced and existing genetic diversity and can therefore be exploited in mutation breeding. This chapter discusses TILLING in a plant breeding context; the theory and protocols used in TILLING are given in **Chapters 21 and 22**. More specifically this chapter deals with 1) how TILLING can be exploited to advance crop improvement; 2) how TILLING can be used to identify useful allelic variants in genes underlying traits of interest to breeders in both diploid and polyploid crops; and 3) how comparative genomics will help expand the impact of TILLING on crop improvement in the future.

2. Mutagenesis, Breeding and Tilling

2.1. Mutagenesis and Breeding

Mutagenesis of plants with physical or chemical agents is a well-established technique in crop improvement. Over 3,000 varieties that have induced mutants in their pedigrees have been released worldwide in the past 70 years (IAEA/FAO mutant variety genetic stock database). Typically plant breeders have used mutation breeding to improve a single trait in an elite background. Not surprisingly, the traits targeted by this approach have been those that are easy to identify such as disease resistance, height or maturity date. The reason for this is readily apparent: simplicity and cost. The majority of the mutants used in these varieties were induced by physical agents, mainly γ-rays, and the mutations selected had major effects on the traits in question and could be readily distinguished from undesirable mutations. Mutation breeding is also readily applied to vegetatively propagated crops and floricultural crops in which it is easy to maintain a new and novel trait, such as flower colour, from cuttings or other vegetative tissue (**see Chapters 26 and 28**).

As useful as mutagenesis has been for increasing genetic variation in crops for breeding purposes, it has been equally important for basic plant science research. With the wide adoption of *Arabidopsis* as a model system in plant biology, induced mutants have played a prominent role in elucidating the genetics and biochemistry of a wide range of biological traits, e.g. photorespiration, the response of plants to various biotic and abiotic stresses, as well as the elucidation of amino acid biosynthesis and catabolism, to name just a few examples.

The contrasting uses of plant mutants are highlighted by the fact that the very mutants that are likely to be discarded by breeders, because of their deleterious phenotypes, are the ones of interest to researchers who wish to gain an understanding of the molecular bases of the phenotypic lesions. A classical example of this is the study of the xantha mutants of barley deficient in chlorophyll biosynthesis (von Wettstein *et al.*, 1974). A more recent example is the cloning of the rice monoculm mutant (Li *et al.*, 2003). Plants homozygous for this recessive mutation produce far few tillers than normal plants and are of no immediate value for plant breeding. Nevertheless, the cloning of the gene underlying this mutation gives new insights into the control of tillering in rice and other monocots. Interestingly, transgenic over-expression of the wild type monoculm gene, which is a GRAS-family transcription factor, in the mutant background resulted in rice plants with two to three times the numbers of tillers of wild type plants (Li *et al.*, 2003). As the authors of this paper indicate, since tiller number is related to grain yield, manipulation of this gene, whether using transgenic or non-transgenic approaches, may ultimately play a role in practical breeding efforts to increase yield in monocots.

The above examples highlight the fact that mutagenesis for crop improvement and mutagenesis for basic research are converging in ways that will benefit pure and applied plant genetics. Combining mutagenesis, TILLING and functional genomics research is expected to bridge the divide between crop improvement and basic science as genes underlying traits of interest to plant breeders are uncovered both in crops and in model plant systems like *Arabidopsis*, *Brachypodium* and rice.

2.2. TILLING for Research and Breeding

The first papers describing the concept of TILLING appeared in 2000. Eight years later TILLING has been widely adopted in the academic community for functional genomics research in both animals and plants. TILLING projects have been initiated in numerous crops including barley, maize, rapeseed, soybean, rice and wheat among others. The *Arabidopsis* TILLING project is a good example of the rapid adoption of TILLING for basic plant science research (http://tilling.fhcrc.org:9366/), which reports the discovery of 7,788 mutations in over 600 user-submitted gene targets as of January 2008.

The genetic diversity discovered by eco-TILLING or artificially generated through a TILLING programme may lead to the development of gene-specific molecular markers that can be further used in marker-assisted breeding, as other molecular markers (**Figure 27.1, see Chapter 23**).

Despite the optimism engendered by the advent of functional genomics, there is still an insufficient understanding of the genes underlying traits of interest to plant breeders. This is particularly true for traits such as yield and quality that are influenced by multiple genes each with a relatively small contribution to the trait. Plant scientists face this situation whether they wish to use non-transgenic or transgenic means to increase crop yield, quality and other traits. Recently, when a quantitative trait locus (QTL) enhancing rice yield was recently cloned and shown to be a null allele of a cytokinin oxidase, it was found that plant breeders had already selected this allele in some high yielding varieties; plant breeders are often ahead of the underpinning science as they simply select for performance. These results lead to the question: what are the candidates for TILLING that breeders might target? In addition, before embarking on a TILLING project, one must ask the question: is the hypothesized beneficial effect of new alleles of the candidate gene identified by TILLING sufficiently large to justify the costs associated with TILLING? In light of these questions, the following sections discuss the current uses of TILLING and the situations in which it offers advantages over other approaches.

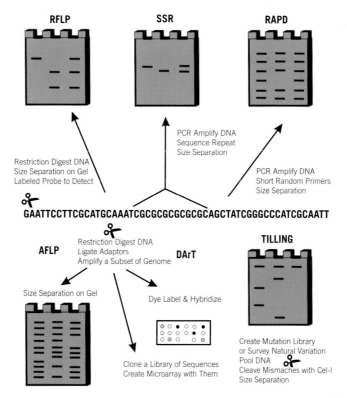

The Joint FAO/IAEA Programme

Figure 27.1 *Strategies for identifying genetic diversity.* Some commonly used strategies for developing molecular markers using differences in genetic composition between samples are illustrated. Restriction fragment length polymorphisms (RFLP) and simple sequence repeats (SSRs) identify differences between samples at a defined locus. SSRs can sometimes be multiplexed during PCR or multiplexed post-PCR prior to size separation. As an alternative to gel electrophoresis, size separation can also be performed by automated capillary electrophoresis. Random amplification of polymorphic DNA (RAPD), amplified fragment length p[olymorphisms (AFLP) and Diversity Arrays Technology (DArT) markers identify undefined regions of the genome, but markers can be cloned and sequenced to define the sequences (DArT markers are already cloned and can be most easily sequenced). TILLING identifies differences at a defined locus in natural populations or in mutation libraries. Molecular markers can then be designed to follow SNPs of interest.

3. Practical Use of TILLING for Breeding

3.1. Potential Uses of TILLING for Crop improvement

Frequently the exploitation of a particular gene for crop improvement has followed a stereotypical sequence of events: initially a particular plant is selected either in a plant breeder's field or in the laboratory because it exhibits a desired phenotype, then the plant is used by breeders to deploy the trait conferred by the gene, or the plant is characterized in the laboratory, and finally the gene underlying the trait is cloned. It is only at this point that the breeder or the academic scientist is able to ask whether new alleles of the identified gene (or orthologues in related crops) are needed for breeding or research purposes and whether such alleles are likely to be identified by TILLING or eco-TILLING.

Currently, despite the fact that TILLING populations have been developed in many important crops, there are still very few published examples in which the above sequence has been followed to the point of development of an improved crop variety. The following therefore highlights some examples that illustrate the concept of utilizing novel alleles for crop improvement. Even when the new alleles were not identified *via* TILLING, the examples discussed represent cases in which the cloning of the genes underlying the traits raise interesting questions that could be addressed by TILLING.

Recently, a major determinant of submergence tolerance in rice was cloned. The majority of rice varieties are killed by complete submergence for 14 days, while an unimproved *indica* variety FR13A is able to withstand this stress. When the submergence tolerance (*Sub1*) locus was cloned, it was found to contain a cluster of two or three members of *Apetala2*-like transcription factors similar to ethylene response factors (ERFs). In the submergence-tolerant line from which the *Sub1* locus was cloned, the three ERFs present at the locus were named *Sub1.A, Sub1.B and Sub1.C*. Submergence tolerance was correlated with the presence of a serine in a putative mitogen activated protein kinase (MAPK) binding site of *Sub1.A*; the intolerant allele of *Sub1.A* possessed a proline at the equivalent position. A further intriguing finding was that the entire *Sub1.A* coding region was absent from five of 17 *indica* varieties surveyed, and from all four of the *japonica* varieties analysed (Xu *et al.*, 2006). *Sub1.A* appears to mediate submergence-tolerance, at least in part, *via* down-regulation of one of the other ERFs present at the *Sub1* locus, *Sub1.C. Sub1.C* was consistently expressed at a higher level under submergence in submergence-intolerant varieties of rice than in submergence-tolerant rice, and transgenic ectopic expression of *Sub1.A* in a rice cultivar lacking it resulted in reduced expression under submergence of Sub1C (Xu *et al.*, 2006). In addition to providing an outstanding demonstration of the role that a single nucleotide SNP (altering a proline to a serine) in a key transcription factor can have in rice submergence tolerance, this research also raises the interesting question of the role played by Sub1.C in submergence tolerance/intolerance. This question can be addressed by TILLING to identify inactive alleles of *Sub1.C* in a *japonica* rice background that lacks *Sub1.A*.

TILLING to verify the role of particular candidate genes identified during a QTL mapping project has previously been suggested to be a particular strength of TILLING. This is indeed one of the uses to which available TILLING resources in rice are being put: Till *et al.* (2007) report finding a truncation mutation in OsDREB, a transcription factor with a role in abiotic stress tolerance (Dubouzet *et al.*, 2003). The availability of a knock-out mutation in this gene offers a new opportunity to study its function *in vivo*.

A number of variants of the TILLING/eco-TILLING technique, e.g. using agarose gels have been proven to be useful (**see Chapter 20**). These authors demonstrate the usefulness of their technique for identifying SNPs that distinguish the alleles of candidate genes of interest between the parents of a cross in rice. They also show how their technique can be used for selecting particular alleles during marker assisted backcrossing using a MYB1 transcription factor and oxalate oxidase genes involved in blast resistance as examples.

Once a gene of interest has been identified, eco-TILLING provides an easy method to assess the level of genetic diversity at that locus across a broad selection of germplasm. A proof of principle experiment has been published in which diversity at the mlo and Mla powdery mildew resistance loci of barley was assessed (Mejlhede *et al.*, 2006). Eco-TILLING of these loci offers the possibility of developing barley breeding lines containing novel allele combinations at the mlo and Mla loci. This could be especially useful in the pyramiding of resistance genes in an already resistant background.

Perhaps the best example of the use of TILLING for crop improvement to date is described in a recently published paper in which TILLING in tobacco is used to reduce nornicotine content (Julio *et al.,* 2008). Nornicotine is derived from nicotine by N-demethylation. Its presence is undesirable in cultivated tobacco because it is a precursor of the carcinogen N'-nitrosonornicotine. Nornicotine is a minor alkaloid in most commercial tobacco varieties, however, a phenomenon known as "conversion" results in the progeny of low nicotine to nornicotine converting parent plants having a much higher conversion rate of nicotine to nornicotine (Siminszky *et al.,* 2005). Using a microarray approach in which labelled cDNAs from converter and non-converter tobacco genotypes were hybridized to a custom-made tobacco expressed sequence tag (EST) microarray, several genes that were consistently up-regulated in the converter genotype compared to the closely related non-converter germplasm under conditions that favoured the conversion of nicotine to nornicotine were identified (Siminszky *et al.,* 2005). These genes were closely related cytochrome P450s, only one of which appeared to function as a nicotine N-demethylase. The next logical step was taken by Julio *et al.* (2008) who developed a TILLING population in a burley-type tobacco cultivar known to exhibit strong nicotine to nornicotine conversion ability and screened for mutations in the cytochrome P450 shown to be responsible for this activity by Siminszky *et al.* (2005). Julio *et al.* (2008) showed that plants homozygous for a null mutation in this cytochrome P450 do not accumulate nornicotine under field conditions.

The results in tobacco are a powerful demonstration of the benefits of TILLING for target selected mutation breeding. The preceding example also shows that TILLING is particularly useful for quality traits and recessive genes that are difficult for breeders to select in the field. In the coming years, numerous other examples of the utility of TILLING for crop improvement are likely to appear.

3.2. TILLING in Polyploid Wheat

Polyploidization is an important evolutionary process in which doubling of an endogenous genome or hybridization between distinct genomes permits novel combinations of genes. However, the additional complexity in the genome due to this process can complicate trait improvement. An estimated 70% of all flowering plants are polyploids including many important crops such

as wheat, cotton and rapeseed. Other crops such as maize and soybean are paleopolyploids, which are ancient polyploids with diploidized genomes. Polyploid genomes are dynamic, with duplicated genes retaining their original functions, developing new functions, or becoming silenced or lost. Bread wheat (*Triticum aestivum*), one of the world's most important food crops, is an allopolyploid with combined A, B and D genomes derived from two polyploidization events. The first polyploidization event occurred approximately 500,000 years ago to create tetraploid (AABB) wheat. The second polyploidization event occurred when ancestral tetraploid wheat (AABB) hybridized with diploid goat grass, *Triticum tauschii* (DD), around 8,000–10,000 years ago. The large size of the wheat genome (16,000 Mb) and its allopolyploid nature present additional challenges to molecular marker development and breeding because many wheat genes are present in three closely related and often functionally redundant copies. To be effective, molecular markers must distinguish between the closely related genomes. In wheat, the assignment of molecular markers to particular genomes and chromosomes has been facilitated by the development of aneuploid lines based on the capacity of the polyploid A, B and D genome chromosomes to substitute for loss of their homoeologous chromosomes (Gupta *et al.,* 2005). These lines have been useful for mapping ESTs and molecular markers onto physical maps. In addition, a smaller number of functional molecular markers have been developed for some of the important agronomic and quality traits in wheat that can be used to precisely select for these characteristics (**Box 27.1**).

In order to harness genetic diversity for the development of new traits, diversity must be first identified and accessible. A recent study found that there has been a "massive loss of nucleotide diversity" in wheat since domestication (Haudry *et al.,* 2007). The authors determined that diversity at the nucleotide level was low at the 21 loci sequenced in tetraploid wild emmer wheat and even more drastically reduced by 69% in cultivated bread wheat and by 85% in durum wheat. The limited nucleotide diversity in wheat is likely due to the small number of plants contributing to the formation of the new polyploid species and to the further restriction of domestication bottlenecks in genetic diversity. Although diversity may be limited in wheat at the single nucleotide level, the process of polyploidization results in genomic rearrangements and deletions and these

Box 27.1: Functional molecular markers in wheat

Trait	Locus	Genes
Gluten strength	HMW Glutenin	*Glu-B1, Glu-D1, Glu-A1*
& elasticity	LMW Glutenin	*Glu-A3, Glu-B3 Glu-D3*
Starch properties	Waxy	*Wx-B1, Wx-A1, Wx-D1*
Grain hardness	Puroindoline	*Pina-D1, Pinb-D1*
Semi-dwarf growth	Giberellin insensitive	*Rht-B1, Rht-D1*
Leaf rust resistance	Resistance genes	*Lr10, Lr21, Lr1*
Vernalization	AP1	*Vrn1*
	Flowering repressor	*Vrn2*
	Flowering time FT	*Vrn3*
Yellow endosperm	Phytoene synthase	*YP*
Browning & discoloration	Polyphenol oxidase	*Ppo-D1,Ppo-A1*
Grain protein content	Grain protein content	*Gpc-B1*
Photoperiod	Photoperiod insensitive	*Ppd-D1*

have been observed in polyploid wheat. The allopolyploid genome can absorb these deletions without loss of viability because many genes are present in duplicate or triplicate homoeologous copies. For example, loss of Gpc-B1 in hexaploid wheat is associated with reduced grain protein content, larger seeds and a delay in maturation, whereas RNAi suppression of GPC in diploid rice causes almost complete sterility (Dubcovsky and Dvorak, 2007). A study of the hardness locus (Ha) in wheat indicated that independent deletions of the locus have occurred recurrently in polyploid wheat, but no deletions have been found in over 200 accessions of diploid wheat (Li *et al.,* 2008).

When available, deletion (null) alleles allow novel gene combinations to be made in modifying traits. For example, combinations of deletion alleles of the waxy loci allowed the creation of the first complete waxy wheat variety (Nakamura *et al.,* 1995). Waxy has been a locus of considerable interest for plant breeders in cereal crops due to its role in the production of amylose, a primary component of starch. Loss of the enzyme, granule bound starch synthase, encoded by the waxy locus, leads to the production of starch primarily composed primarily of amylopectin. High amylopectin or "waxy" starch has unique physicochemical properties such as high peak viscosity and greater freeze thaw stability. Loss of a single homoeologous copy of the waxy gene decreases levels of amylose from about 25% to 22%, but

does not cause complete loss of amylose because of the activity of the other homoeologous genes. Even so, partial waxy wheat lacking the Wx-B1 allele is prized for Japanese Udon noodle production demonstrating that even a small reduction in amylose levels can effect starch alterations beneficial for certain applications.

In order to develop full waxy wheat, efforts were focused on identifying and combining genetic diversity for all the waxy genes in wheat. Protein isozyme markers were the first molecular markers to be employed to access diversity in the waxy loci. Wheat varieties lacking one or two of the waxy protein from the A or B genomes were identified using a modified two-dimensional gel electrophoresis that separated the closely related proteins based on size and charge. Subsequently, molecular markers based on DNA sequences were developed and thousands of wheat varieties were surveyed for their waxy alleles. Although 16% of wheat varieties from around the world were found to contain deletions in Wx-A1 or Wx-B1 genes, a deletion affecting the Wx-D1 gene was determined to be a rare occurrence. Wx-D1 null alleles were identified in only three regionally adapted landrace varieties out of thousands of wheat varieties.

3.3. Combining Multiple Alleles in Polyploids

TILLING technology provides a means to address the limited nucleotide diversity of wheat and other crops especially when favourable deletion alleles are not available or when genetic alterations other than knock-out alleles are desired. As a reverse genetic strategy, TILLING allows identification of variation in a gene of interest to be uncovered without the need for an immediately identifiable phenotypic effect. This is a considerable benefit for uncovering useful genetic variation in polyploids where similar genes may compensate functionally for the loss of a single gene, masking the effects of a mutation or gene deletion. When TILLING is applied in a polyploid, each of the multiple homoeologous genes can be targeted separately, and then favourable mutant alleles of each homoeologue in the same genetic background can be combined into one plant to finally assess a phenotype. An example of using TILLING in a polyploid to identify novel diversity is the identification of 246 new alleles of the waxy loci in bread and durum wheat (Slade *et al.*, 2005). Using TILLING, 106 novel point mutation alleles of *Wx-A1* and 90 alleles of *Wx-D1* were discovered in a mutagenized population of approximately 1,100 hexaploid wheat individuals. The wheat variety used in this study contained the null allele at the Wx-B1 locus, so no additional mutations were obtained for this gene in hexaploid wheat. A wheat

line that had a knock-out point mutation in the *Wx-D1* gene and another mutation predicted to severely affect protein function in the *Wx-A1* gene were advanced to homozygosity. The resulting line stained lightly with iodine compared with controls, demonstrating that the amylose levels were greatly reduced in the triple mutant. This general strategy of using TILLING technology to identify genetic diversity in homoeologous genes is exceptionally well suited to polyploids.

Targeting genes of interest through TILLING allows the development of novel traits and functional markers simultaneously because the mutation serves as a landmark that can be converted into a molecular marker. The use of TILLING also allows surveys of naturally occurring variation to examine thousands of individuals in a mapping population, or thousands of different varieties at a time. This approach is especially useful when no naturally occurring or very limited nucleotide diversity is available. For example, due to the rarity of detectable variation in the *Wx-D1* gene, new variants were created by starting with wheat containing A and B deletion alleles and then using mutagenesis to create new point mutation alleles that could then be detected in the background of the other two null alleles

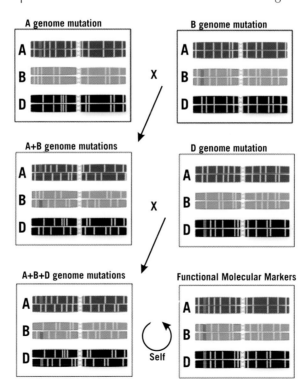

Figure 27.2 *Combining multiple alleles in a polyploid.* For many wheat genes, a phenotype will not be detectable from the alteration of only one homoeologue due to functional redundancy. Combining multiple mutations discovered through TILLING or other method is illustrated. A, B and D refer to mutations in the different homoeologous chromosomes. Mutations are shown as yellow bars and the mutation of interest as a red bar. The mutation of interest can be followed in subsequent crosses using a number of detection methods.

(Yanagisawa *et al.,* 2001). In contrast, TILLING allows identification of new alleles that don't necessarily have a knock-out gene function and that can be identified in the same genetic background. Because an allelic series of point mutations can be identified with different potential effects on protein function ranging from null to reduced function to dominant effects, a range of phenotypes may be possible. Null alleles and point mutations can have quite different phenotypic effects and are not always equivalent. For example, *Rht-B1* mutations that reduce wheat height are semi-dominant and gibberellin-resistant, but a null allele at that locus is associated with tall, gibberellin-responsive plant (Peng *et al.,* 1999).

After desirable alleles are identified through TILLING in a polyploid such as wheat from what is typically the M_2 generation, M_3 plants containing these alleles can be crossed together to produce a plant with combined mutant alleles in all three homoeologues. This process starts with crossing two mutant lines together followed by crossing the progeny of that cross to a plant containing a mutation in the third homoeologue (**Figure 27.2**). Functional markers developed to precisely identify each novel allele allow progeny with the desired combination of alleles to be rapidly selected. Siblings from these crosses lacking the desired mutations may serve as controls in early phenotyping efforts. Alternatively, backcrosses can be performed before combining mutations. This may be particularly advantageous in combining mutations in diploids, because deleterious effects of homozygous background mutations can complicate phenotypic evaluation.

3.4. Cleaning Up Background Mutations

One of the consequences of mutation breeding using chemical mutagenesis, including TILLING, is that the background genome contains high levels of random unwanted mutations. Background mutations can complicate phenotypic analysis of the effects of the modified gene or gene combinations. For example, some background mutations may interfere with plant growth or affect other aspects of the plant making it difficult to clearly determine the effects solely attributable to the desired mutations. Polyploid wheat, especially, can withstand high levels of mutation evident by the high mutation load measured in mutagenized TILLING populations of hexaploid and tetraploid wheat compared

to barley, maize, rice and *Arabidopsis* (**see Chapter 20**). The redundancy of genes in the multiple A, B and D genomes is likely the reason that wheat can withstand such a high mutation load. It has also been suggested that the time since polyploidization may play a role in allowing the relatively younger polyploid wheat genome to withstand high levels of random mutations compared with more ancient polyploids, such as maize, that have reduced redundancy in the genome.

Background random mutations can be eliminated through a backcross process while retaining the mutations of interest. In a backcross programme, the wheat line containing all the mutations of interest is crossed to an unmutagenized wild type parent line (**Figure 27.3**). The F_1 progeny is then backcrossed to the recurrent parent and the BC_1 progeny are obtained and genotyped for the alleles of interest. Only progeny with the allele assortment desired are carried on to the next backcross with the recurrent parent (BC_2). This process can be repeated as many times as needed (BC_{3-5}). This process could also be used to transfer the novel alleles into a different genetic background, all the while following the mutations with functional molecular markers developed to follow the mutations of interest.

4. Future Applications of TILLING to Plant Breeding

Recent progress in the sequencing of several complete plant genomes and EST sequencing from many diverse plant species provides new opportunities to apply knowledge gained from functional genomics in rice and *Arabidopsis* to the improvement of many different crops. Genes identified from functional genomics approaches as well as genes known to affect traits of interest in other crops become candidates for trait alteration. To capitalize on these advances and transfer information gathered in *Arabidopsis* and rice to other crops, nucleotide and amino acid sequence comparisons as well as gene synteny, the conserved order of genes in related species, can be used to identify potential orthologous genes (genes derived from a common ancestor) between crops. Identification of candidate orthologous genes is made easier by the considerable colinearity between the genomes of grass species. However, gene synteny between grass genomes is complicated by gene duplications, inversions, deletions, translocations and differences caused by transposable elements. As a result, large regions of the various grass genomes are rearranged relative to each other, and these alterations can complicate the interpretation of comparative data. Smaller scale microcolinearity between the genomes, if available, is most useful in the transfer of positional information to the cloning of candidate orthologous genes.

An increasing number of candidate genes for crop improvement have been identified using comparative approaches. For example, the Rht genes of wheat responsible for the increased yield associated with the Green Revolution were identified as orthologues of maize *Dwarf8* and *Arabidopsis GAI (Gibberellin Insensitive)* genes using sequence homology and gene synteny (Peng *et al.*, 1999). The identification of the grain protein content gene *GPC-6B1* is another example of this approach in which the wheat region on chromosome 6B syntenic to the rice region on chromosome 2 was used to define additional molecular markers to refine mapping of the GPC-6B1 locus (Distelfeld *et al.*, 2004). Armstead *et al.* used synteny between *Festuca* and rice to identify an orthologue of the stay green cotyledon colour I gene

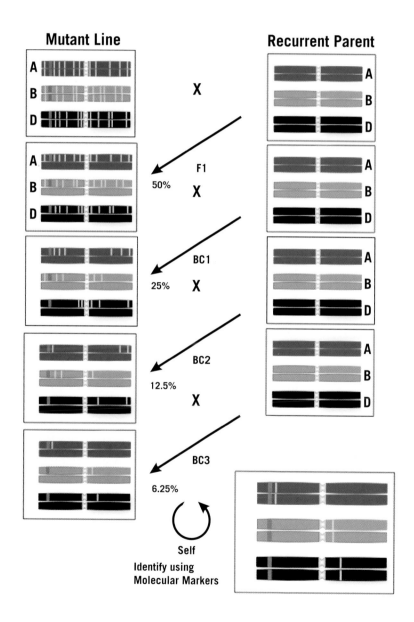

The Joint FAO/IAEA Programme

Figure 27.3 *Cleaning up background mutations.* Before or after alleles of interest are combined, the background mutations can be eliminated using a backcross process. Each backcross eliminates a certain percentage of background mutations. The mutations of interest can be retained using functional molecular markers designed to identify them.

originally described by Gregor Mendel (Armstead *et al.,* 2007). In another example, the identification of the vernalization gene, *Vrn3*, was facilitated using colinear regions between rice and barley (Yan *et al.,* 2006). Rice and wheat colinearity has also helped narrow the search for the gene responsible for tolerance to boron toxicity in wheat (Schnurbusch *et al.,* 2007). All of these examples demonstrate the utility of gene synteny and sequence homology in candidate gene identification.

Using gene-specific primers, TILLING can be applied to identify genetic diversity in candidate genes that can then be employed to generate novel traits. For polyploids especially, extension of wheat sequence information from ESTs to include genomic DNA containing introns and flanking untranslated regions helps define unique regions that distinguish between the A, B and D homoeologous genes for further study. EST sequences of wheat genes can be compared to rice gene structure to identify likely positions of intronic regions for development of gene specific primers. A programme that aligns wheat ESTs with similar rice genes facilitates candidate gene identification (Mitchell *et al.,* 2007). Additional information from ongoing projects will expand access to more and more plant genomes. And, as more genes and gene combinations that control agronomic and quality traits are defined, the utility of TILLING to identify novel genetic diversity in targeted candidate genes for breeding in many crops will increase.

5. References

5.1. Cited References

Armstead, I., Donnison, I., Aubry, S. *et al.* **2007.** Cross-species identification of Mendel's I locus. *Science.* 315: 73.

Distelfeld, A., Uauy, C., Olmos, S. *et al.* **2004.** Microcolinearity between a 2-cM region encompassing the grain protein content locus Gpc-6B1 on wheat chromosome 6B and a 350-kb region on rice chromosome 2. *Funct Integr Genomics.* 4(1): 59–66.

Dubcovsky, J. and Dvorak, J. 2007. Genome plasticity a key factor in the success of polyploid wheat under domestication. *Science.* 316(5833): 1862–1866.

Dubouzet, J.G., Sakuma, Y., Ito, Y. *et al.* **2003.** OsDREB genes in rice, *Oryza sativa* L., encode transcription activators that function in drought-, high-salt-, and cold-responsive gene expression. *Plant Journal.* 33: 751–763.

Gupta P.K., Kulwal, P.L. and Rustgi, S. 2005. Wheat cytogenetics in the genomics era and its relevance to breeding. *Cytogenet Genome Res.* 109:315-327.

Haudry, A., Cenci, A., Ravel, C. *et al.* **2007.** Grinding up wheat: a massive loss of nucleotide diversity since domestication. *Mol Biol Evol.* 24: 1506–1517.

Julio, E., Laporte, F., Reis, S. *et al.* **2008.** Reducing the content of nornicotine in tobacco *via* targeted mutation breeding. *Molecular Breeding.* DOI 10.1007/s11032-007-9138-2.

Li, W., Huang, L. and Gill, B.S. 2008. Recurrent deletions of puroindoline genes at the grain hardness locus in four independent lineages of polyploid wheat. *Plant Physiol.* 146(1): 200–212.

Li, X., Qian, Q., Fu, Z. *et al.* **2003.** Control of tillering in rice. *Nature.* 422: 618–621.

Mejlhede, N., Kyjovska, Z., Backes, G. *et al.* **2006.** EcoTILLING for the identification of allelic variation in the powdery mildew resistance genes *mlo* and *Mla* of barley. *Plant Breeding.* 125: 461–467.

Mitchell, R.A., Castells-Brooke, N., Taubert, J. *et al.* **2007.** Wheat Estimated Transcript Server (WhETS): a tool to provide best estimate of hexaploid wheat transcript sequence. *Nucleic Acids Res.* 35(Web Server issue): W148–151.

Nakamura, T., Yamamori, M., Hirano, H. *et al.* **1995.** Production of waxy (amylose-free) wheats. *Mol Gen Genet.* 248: 253–259.

Peng, J., Richards, D.E., Hartley N.M. *et al.* **1999.** 'Green revolution' genes encode mutant gibberellin response modulators. *Nature.* 400(6741): 256–261.

Schnurbusch, T., Collins, N.C., Eastwood, R.F. *et al.* **2007.** Fine mapping and targeted SNP survey using rice-wheat gene colinearity in the region of the Bo1 boron toxicity tolerance locus of bread wheat. *Theor Appl Genet.* 115: 451–461.

Siminszky, B., Gavilano, L., Bowen, S.W. *et al.* **2005.** Conversion of nicotine to nornicotine in *Nicotiana tabacum* is mediated by CYP82E4, a cytochrome P450 monooxygenase. *Proceedings of the National Academy of Sciences U.S.A.* 102: 14919–14924.

Slade, A.J., Fuerstenberg, S.I., Loeffler, D. *et al.* **2005.** A reverse genetic, nontransgenic approach to wheat crop improvement by TILLING. *Nature Biotechnology.* 23: 75–81.

Till B.J., Cooper, J., Tai, T.H. *et al.* **2007.** Discovery of chemically induced mutations in rice by TILLING. *BMC Plant Biol.* 7: 19.

von Wettstein, D., Kahn, A., Nielsen, O.F. *et al.* **1974.**

Genetic regulation of chlorophyll synthesis analyzed with mutants in barley. *Science*. 184: 800–802.

Xu K., Xu, X., Fukao, T., Canlas, P. *et al.* **2006.** Sub1A is an ethylene-response-factor-like gene that confers submergence tolerance to rice. *Nature*. 442: 705–708.

Yan, L., Fu, D., Li, C. *et al.* **2006.** The wheat and barley vernalization gene VRN3 is an orthologue of FT. *Proc Natl Acad Sci U.S.A.* 103: 19581–19586.

Yanagisawa, T., Kiribuchi-Otobe, C. and Yoshida, H. **2001.** An alanine to threonine change in the Wx-D1 protein reduces GBSS I activity in waxy mutant wheat. *Euphytica*. 121: 209–214.

5.2. Websites

Arabidopsis **TILLING project:**
  http://tilling.fhcrc.org:9366/

IAEA/FAO mutant variety genetic stock database:
  http://mvgs.iaea.org

PubMed: US national library of medicine service:
  http://www.ncbi.nlm.nih.gov

5.3. Further Reading

Till B.J., Colbert, T., Codomo, C. *et al.* **2006.** High-throughput TILLING for *Arabidopsis*. *Methods Molecular Biology*. 323: 127–135.

Till B.J., Zerr, T., Comai, L. *et al.* **2006.** A protocol for TILLING and Ecotilling in plants and animals. *Nature Protocols*. 1: 2465–2477.

Henikoff, S. and Comai, L.. **2003.** Single-nucleotide mutations for plant functional genomics. *Annual Review of Plant Biology*. 54: 375–401.

Henikoff, S., Till, B.J. and Comai, L. **2004.** TILLING. Traditional mutagenesis meets functional genomics. *Plant Physiology*. 135: 630–636.

C28

Applications of *In Vitro* Techniques in Mutation Breeding of Vegetatively Propagated Crops

P.Suprasanna[a,*], S.M.Jain[b], S.J.Ochatt[c], V.M.Kulkarni[a] and S.Predieri[d]

[a] Nuclear Agriculture & Biotechnology Division, Bhabha Atomic Research Centre, Mumbai – 400 085, India
[b] Department of Applied Biology, University of Helsinki, PL-27, Helsinki, Finland
[c] INRA CR Dijon, UMRLEG, PCMV, Dijon, France
[d] IBIMET-CNR, Area Ricerca CNR, Via Gobetti, 101, 40129 Bologna, Italy
* Corresponding author, E-MAIL: prasanna@barc.gov.in

1. Introduction

In vitro techniques, e.g. protoplast-, cell-, tissue- and organ culture, can be used in various steps and for different purposes in mutation breeding of vegetatively propagated crops (VPCs), and have become an integral part of such programmes (**Box 28.1**). The use of *in vitro* culture for chimera dissociation has been described in **Chapter 16** and the micropropagation of selected mutants is in principle not different from conventional materials, hence they are not further elaborated.

The principles and key steps of *in vitro* mutagenesis and *in vitro* selection are described, and examples are given in the section of case studies. The application of *in vitro* culture techniques in mutation breeding of VPCs relies heavily on reproducible and efficient tissue culture techniques, and readers should first check whether such techniques are available for the plant species concerned. A generalized flowchart for developing a mutant variety using the above approach is outlined in **Figure 28.1**. It is imperative that such an approach will provide advancement only if it includes investigations into different phases, from the laboratory to actual field performance testing. The time required from the induction treatment to the variety release is also very long, especially in some woody plant species, and can be 10–15 years. The results obtained will be valuable only if the choice of market stable varieties to mutate and basic traits to modify is well planned in advance.

2. *In Vitro* Mutagenesis

In vitro mutagenesis refers to mutation induction through treating explants or *in vitro* cultures (protoplasts, cells, tissues and organs) with a mutagen, followed by mutant screening/selection and characterization (**Box 28.1**). *In vitro* mutagenesis can be applied to both seed crops (e.g. haploid mutagenesis, **see Chapter 29**) and VPCs. *In vitro* mutagenesis is particularly advantageous compared with *in vivo* mutagenesis of VPCs. First, a large number of homogenous individuals can be prepared for mutagenic treatment in a relatively small laboratory space; second, the developmental patterns of cultured cells, organs and tissues can be synchronized allowing for replication of treatment, mass *in vitro* selection and, as a consequence, provide consistent and reproducible results; third, the chimeras issue is largely diminished since mutants often originate from single somatic cells. For seed crops, use of haploid cultures may provide additional benefits (**see Chapter 29**).

The whole process of *in vitro* mutagenesis involves a series of important steps: selection of proper target material (explants or cultures), determination of a mutagen and a proper dose, post-treatment handling and sub-cultures; and regeneration of plants for mutant selection. Therefore, the success of any *in vitro* mutagenesis depends on established and reproducible *in vitro* plant regeneration procedures, optimization of mutagenic treatments and efficient screening of the mutagenized populations for desired variations. Much of the variation observed *in vitro* (particularly in undifferentiated systems like callus or suspension cultures) is considered epigenetic, which implies that the variation is not caused by a mutation but by a change in gene activity caused by environmental factors during development and also by culture process. Contrary to mutations, epigenetic effects occur at much higher frequency and cannot pass through meiosis. This should be clearly distinguished from *in vitro* mutagenesis. One can also undertake extensive recombination by crossing, but this method is unsuitable in most VPCs. **Table 28.1** presents some cases of *in vitro* mutagenesis in vegetatively propagated plants reported in the literature.

Box 28.1: *In vitro* techniques in mutation breeding of vegetatively propagated crops

1. *In vitro* dissociation of chimeras. Separation of chimeras through *in vitro* culture techniques like adventitious buds or somatic embryogenesis, in which the genetic background of parent genotype is retained.
2. *In vitro* mutagenesis. Mutation induction through treating explants or *in vitro* cultures (protoplasts, cells, tissues and organs) with mutagen, followed by mutant screening/selection and characterization.
3. *In vitro* selection of mutants. Selection of *in vitro* cell lines or tissue cultures for a desired trait using a selection agent, e.g. medium with a high concentration of salt or phytotoxin.
4. Micropropagation of selected mutants. Propagation of selected mutants through *in vitro* techniques of node culture, shoot meristem culture, somatic embryogenesis, etc.

2.1. Selection of Target Material

Mutagenic treatments can be applied to explants or *in vitro* cultures (e.g. multiple shoot cultures, calli, cell suspensions, protoplasts, microspores, etc. (**Table 28.1**). A variety of explants is available, e.g. apical meristems, axillary buds, roots and tubers. A crucial factor in the choice of explants, either used for direct mutagenic treatment or for *in vitro* culture and mutagenic treatment at a late stage, is the ease of culturing. Since *in vitro* culture is dependent on plant health and the ability to produce sterile, uncontaminated cultures, certified virus-free donor plants are often used as starting material for tissue culture. The plant parts are generally washed under running tap water to remove dirt and then in mild detergent solution followed by surface sterilization with hypochlorite solution. For herbaceous plants it is often convenient to take explants (shoot tips, axillary buds, meristems and roots) from donor plants grown in a growth chamber, *in vitro*, or glasshouse; it is therefore no problem to have sufficient numbers of explants for mutagenic treatment. For other plant species, multiplication is needed to have

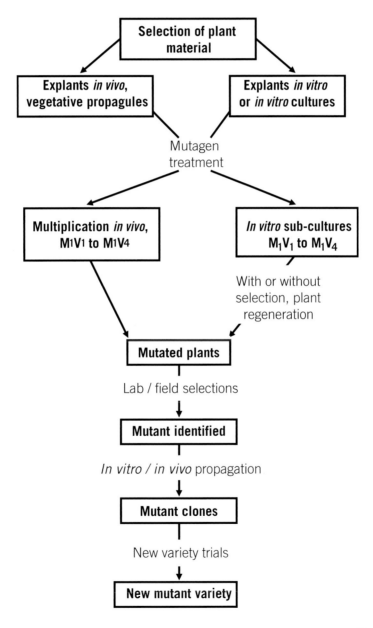

The Joint FAO/IAEA Programme

Figure 28.1 A schematic diagram of mutation breeding and use of *in vitro* techniques in vegetatively propagated crops.

a sufficient number of explants, since it is not easy to get them directly from plants.

Mutagenesis of single individual cells would be the preferred method of *in vitro* mutagenesis; however, it is practically very difficult to obtain genuine single-cell suspensions due to the tendency of cells to form clumps.

Hence, the *in vitro* plant materials can be mutagenized at any stage in the somatic embryogenesis pathway provided there is sufficient subsequent proliferation to screen out chimeras. Callus and cell suspension cultures that show good regeneration potential offer another attractive target. Protoplast cultures are ideal

Table 28.1: Reported *in vitro* mutagenesis of vegetatively propagated crops

Crop species	Treated material	Mutagen and dose (LD$_{50}$ or applied dose)	Plant regeneration route	Selected mutants/ lines	Reference
Chrysanthemum morifolium	Rooted cuttings	γ-rays (25 Gy)	Direct shoot organogenesis	Yellow flower mutants (from white and red flower varieties)	Ahloowalia *et al.* (2004)
Musa spp. banana	Shoot tips	Carbon ion beam (0.5 to 128 Gy)	Direct regeneration	Disease-resistant lines	Reyes-Borja *et al.* (2007)
Musa spp. banana	Shoot tips	γ-rays (60 Gy)	Direct regeneration	Mutant Novaria; earliness	FAO/IAEA mutant database
Banana var Lakatan Latundan	Shoot tips	γ-rays (40 Gy)	Direct regeneration	Height reduction, Larger fruit size	Roux (2004)
	Shoot tips	γ-rays (25 Gy)	Direct regeneration	Mutant variety Klue Hom Thong KU1	
Pineapple var. Queen (*Ananas comosus* (L.) Merr.)	Crowns	γ-rays	Axillary bud regeneration	Lines with reduced spines	FNCA
Begonia rex	*In vitro* cultured leaflets	(30–45 Gy)	Adventitious bud regeneration	Leaf colour and shape mutants	Buiatti (1990)
Hybrids of *Begonia x hiemalis*	*In vitro* cultured leafs	γ-rays (100 Gy)	Adventitious shoot regeneration	Leaf colour, size and form and flower mutants	Roest *et al.* (1981)
Weigela		X-rays (30 Gy)	Bud neoformation *in vitro*	Leaf and flower colour mutant varieties 'Courtadur'	Duron and Decourtye (1986)
Potato	Callus cultures	γ-rays (30–50 Gy)	Adventitious bud regeneration	Tuber colour mutants	Ancora and Sonnino (1987)
Sugarcane	Buds/callus cultures	γ-rays (20–25 Gy)	Organogenesis/ embryogenesis	Mutants for agronomic traits	Patade and Suprasanna (2008)
Cassava	Somatic embryos	γ-rays (50 Gy)	Embryogenesis	Morphological mutants; mutants with storage root yield, altered cyanogen	Roy *et al.* (2004)
Sweet potato	Embryogenic suspensions	γ-rays (80 Gy)	Embryogenesis	Mutants for salt tolerance	He *et al.* (2009)
Pear	*In vitro* shoots	γ-rays (3.5 Gy)	Microcuttings from shoots	Mutants for russeting, fruit shape and size	Predieri and Zimmerman (2001)
				Small tree size, wide branch angle and short internodes	Predieri *et al.* (1997)

Table 28.2: Some examples of radio-sensitivity of tissue culture of fruit crops [a]

Species	Mutagen	Plant explants	LD$_{50}$ (Gy)
Prunus	X-rays	Shoots	22–29
Potato	X-rays	Rachises, petiole and leaflets	15–20
Begonia	γ-rays	Adventitious buds	100
Pear	X-rays	Shoots	30
	γ-rays	Leaves	20–30
Grapevine	γ-rays	Shoots	20–40
Banana	γ-rays	Shoot tips	20–30
Strawberry	γ-rays	Shoot clumps	50
Kiwi fruit	γ-rays	Leaves	40–60
Apple	γ-rays	Leaves	10–20
Japanese plum	γ-rays	Shoots	30

[a] Adapted from Predieri (2001) and modified.

targets for mutagenesis and selection as these are in the G1 phase of cell division and therefore have potential in yielding non-chimeric mutant lines. The major disadvantage of protoplasts is their common recalcitrance to plant regeneration for a number of species, hence its actual application has so far been limited.

Besides the choice of material, considerations on the genetic constitution and ploidy may influence the success of *in vitro* mutagenesis programmes. For example, genetically heterozygous (*Aa, Aaa, Aaaa*) material can be useful in the production of recessive mutants by knocking out the "*A*" allele.

2.2. Choice of Mutagen and Doses

The principle of mutagen choice for *in vitro* mutagenesis is not different from *in vivo* mutagenesis. The most commonly used mutagens so far are physical mutagens such as γ-rays and X-rays (**Table 28.2**). The limited use and success of chemical mutagens can be attributed to both the poor uptake and penetration of the compound by cultured tissues and the difficulty in applying a chemical mutagen on vegetative parts. In reality, the choice of mutagen is also dependent on the availability and accessibility of physical facilities.

Successful *in vitro* mutagenesis is dependent upon the ability to: 1) induce target mutations at a detectable frequency with limited background mutations, and 2) regenerate plants readily from the mutagen treated plant material. Therefore, in addition to the choice of mutagen, another important step in mutagenesis experiments is to determine the appropriate mutagen dose and dose rate, which vary according to the plant species, level of ploidy, developmental stage and physiological condition. In general, since actively dividing cells are more sensitive to mutagenic treatment than non-dividing ones, the doses used for *in vitro* mutagenesis are usually lower than for *in vivo* treatment.

The common procedure to assess appropriate dose is based on understanding the sensitivity of the tissues and/or cells to a given mutagen. It is usually measured by developing a sensitivity curve based on dose and survival percentage. The dose killing about 50% of the treated materials is named LD$_{50}$, which could be obtained from a sensitivity curve (**see Chapter 14**).

Radio-sensitivity varies with the species and the genotype, with the physiological condition of plants and organs and with the manipulation of the irradiated material before and after mutagenic treatment. Reported LD$_{50}$ are given in **Table 28.2** for some species. In some cases, low dose of radiation may enhance shoot growth, e.g. in citrus species. In date palm, low doses of gamma radiation prevent loss of somatic embryogenesis for up to three years, which is normally lost within a year. The physiological status of plants and their radio-sensitivity are often correlated to water content of the tissue, since the most frequent primary target of ionizing radiation is the water molecule (**see Chapters 11 and 14**).

After the LD_{50} is determined, the most suitable dose can be chosen for treatment. LD_{50} is often chosen as a suitable dose, but the actual dose to be applied is largely determined by experience. Heinze and Schmidt (1995) suggest a LD_{50} ±10% as a starting point for the experiments. Doses lower than LD_{50} favour post-treatment plant recovery, while the use of higher doses increases the probability of inducing mutations but with a high mutational load. Therefore, LD_{20} or LD_{30} have also been used as the optimum dose as these levels of mutagens are not highly toxic to *in vitro* tissues. Although the information on radio-sensitivity and lethal effects of chemical mutagens can be obtained from previous reports, the results should not be blindly accepted. Ideally, at least a few simple experiments should be performed to generate relevant data and compare with published reports, before arriving at conclusions.

Predieri and Fasolo (1989) suggested the following optimized protocol in determining radio-sensitivity for *in vitro* leaf cultures of apple (*Malus pumila* L.). Six leaves were used per treatment in a Petri dish and cultured on a regeneration medium. The cultures were then exposed to acute irradiation (^{60}Co) with doses of 0, 10, 20, 30 or 40 Gy (42.7 Gy/min). The use of acute irradiation allows rapid treatment of the plant material. The cultures were kept in the dark and allowed to regenerate (45 days). The LD_{50} was calculated as the dose of γ radiation that reduced the number of shoots regenerated per irradiated leaf to 50% compared to un-irradiated controls.

When shoot apices are used as the target material the aim is to induce proliferation and multiplication after treatment. Shoot tips trimmed to a minimum size (as in the case of banana or sugarcane) from an actively proliferating shoot culture are placed horizontally in Petri plates with few drops of sterile water to protect the shoot tip from drying during radiation exposure. Immediately after irradiation, shoots are placed onto shoot multiplication media and the multiplication rate is recorded after four weeks of culture. LD_{50} is then calculated as the dose of γ radiation (or other treatment) that reduces the proliferation rate of irradiated shoots to 50% of un-irradiated control shoots.

2.3. Post-Treatment Handling

After the *in vitro* mutagenic treatment is given it is often necessary to provide a medium for the mutated cells to grow and develop into a cell mass, or to differentiate into embryos, shoots and ultimately plants. After transfer to fresh regeneration medium, irradiated tissues can be subjected to standard regeneration procedures, already optimized for shoot differentiation and growth. Following mutagenic treatment, undesired physiological effects can be observed at various stages up to the rooting stage. Where micro-cuttings are used in irradiation treatment, post-irradiation culture should be planned to induce meristem growth and shoot development as soon as possible. This can be done through repeated cuttings to avoid apical dominance effects, with a suitable plant growth regulator (e.g. cytokinins) to support shoot growth. In the initial period, cytokinins can be used at low concentrations (compared to the normal level used in *in vitro* cultures of the plant species), and can be increased incrementally in subsequent sub-cultures.

Tissue-cultured woody fruit plants axillary buds generally include more than one meristem. However, the outgrowth of lateral buds in growing shoots is regulated by apical dominance, a process by which the apex exerts an inhibitory action on buds situated below it (Muleo and Morini, 2006). The removal of the first developed axillary shoot allows more irradiated meristems, potentially carrying mutations, to develop from the originally treated shoots (**Figure 28.2**).

Undesired negative primary effects of irradiation, e.g. tissue browning, necrosis or chlorosis, are sometimes unavoidable, but may be reduced by incorporating antioxidants in the culture media. Cultures irradiated either during proliferation or regeneration phases should be transferred rapidly to fresh media, to reduce the formation of toxic compounds. The basal part of irradiated shoots (e.g. 1 cm in the case of banana) should also be removed for the same reason. Culturing in the dark (avoiding light exposure) may also help in limiting negative physiological effects induced by irradiation.

The lowered regeneration capability of cultures after radiation treatment has been attributed to toxic effects of gamma radiation on cells and tissues resulting in reduced competitiveness of these cells and their progenies. Intra-somatic competition which discriminates against mutagen affected cells potentially causing a loss of their cell progenies may be modified by *in vitro* conditions (medium composition or some other factors) resulting in a better competitiveness of mutant cells. In this respect partial desiccation treatment has been found to be a useful and simple method in stimulating

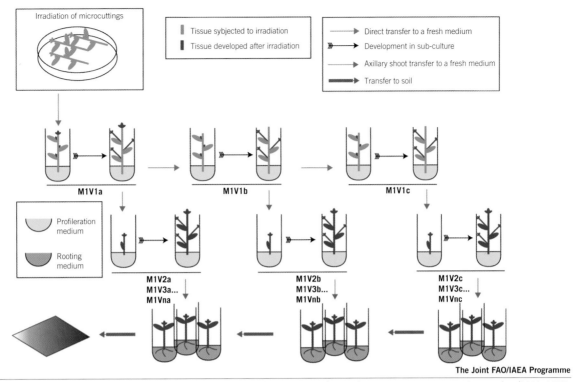

Figure 28.2 Post-irradiation culture management. Irradiated/*in vitro*/shoots are subjected to successive propagation cycles (M_1V_1, M_1V_2, M_1V_3, M_1V_n). At each cycle axillary shoots developed from irradiated meristems (blue) are cut and transferred to a fresh medium. Originally irradiated shoots (green) after the removal of the first axillary shoots (a) are submitted to two more cultures (b, c) to have more axillary shoots to develop.

the regeneration response in the case of gamma-irradiated cultures. This has proved effective in sugarcane though the mode of action is unclear.

Mutagenic treatment can also affect rooting capacity and plant development ex vitro. Maximum care should be taken at the rooting stage and during the acclimatization phase to ensure complete recovery of useful mutants without any loss. A short, acute auxin treatment can often help in increasing rooting performance. Time in culture is also associated with epi-genetic effects such as methylation.

The putative multiplied mutant plantlets may respond differentially to glasshouse, shade house or other hardening facility (depending on species and climate) and field transfer. Considerable mortality can be expected at both these stages. Hence, only experienced staff should carry out hardening and subsequent field transfer. This step normally requires 70–90% relative humidity (glasshouse, poly house, shade house, etc.) otherwise *in vitro* plants would wither and die (misting and water are usually automated for large scale production of field-ready plants). Normally in *in vitro* culture conditions, the relative humidity is over 90%. Hardened plants may be field evaluated following standard procedures depending on the trait to be evaluated either in a single or in multi-location trials.

Well-adapted *in vitro* mutant plantlets can be subjected to the selection pressure for the selection of solid mutants before transfer to the field evaluation. The selection pressure could be applied two or three times to make sure that the selected mutant lines are solid mutants.

2.4. *In Vitro* Culture and Population Development

Whether explants or *in vitro* cultures are used for mutagenic treatment, all mutagenized cultures need to undergo a series of sub-cultures to dissociate chimeras and produce plants for selection of desired mutants (**Figure 28.2**). The time required for sub-cultures depends on the type of species, the material mutagenized and the method of plant regeneration. It is not advisable to screen for mutations in the first vegetative generation after mutagenesis (M_1V_1) as mutations may remain masked or remain undetectable due to chimeras in this generation. If useful mutants are found early, these should be monitored for stability in subsequent generations, up to M_1V_4 or M_1V_6 generations.

If explants are treated for mutagenesis, once the dose is chosen, more information is needed to determine how many explants should be treated to establish a population with a reasonable expectation of containing desired mutant lines. This is dependent on both the expected mutation frequency and the tissue culture process.

The tissue structures from which plants originate are either multi-cellular or of single cell origins. Normally chimeras are major problems in plants regenerated after mutagen treatment of multi-cellular structures such as shoot tips or axillary buds (**see Chapter 16**). However, chimeras can be easily dissociated in *in vitro* culture by repetitive sub-culturing, normally about four generations (M_1V_4). Compared to *in vivo* bud mutagenesis this translates into a lower frequency of chimeric plants and a higher probability of mutant phenotype expression. It also facilitates rapid completion of propagation cycles (sub-culturing) aimed at separating mutated from non-mutated sectors. *In vitro* selection and cloning can therefore capitalize on early evaluation and large-scale multiplication systems.

2.4.1. Somatic Embryogenesis

In vitro plant regeneration occurs *via* somatic embryogenesis and organogenesis. Somatic embryogenesis is an excellent system for clonal propagation and mutation induction. The fact that somatic embryos originate from a single cell prevents chimeras among regenerated *in vitro* plants and makes them an ideal subject for mutagenesis. Also, direct mutant somatic embryos can be developed from mutagenized somatic embryo cells and germinated into whole mutant plants. This system is suitable for vegetative propagated crops, including fruits, citrus, date palm, mangos and others. Therefore, if plants can be regenerated directly through somatic embryogenesis, the rounds of sub-culture could be minimized.

The bipolar structure of the somatic embryo contains both shoots and root meristems. Somatic embryos develop and progress through distinct developmental stages, these are specific to the species, but include: globular, heart, torpedo, cotyledon and mature stages. Direct somatic embryogenesis is the formation of somatic embryos or embryogenic tissue directly from explants without the formation of an intermediate callus phase. The trend is to develop procedures that are directly embryogenic as time in culture (and particularly callus culture) is proportional to the extent of somaclonal variation in regenerated plants, and this is often undesirable (see later sections).

While the induction of secondary embryogenesis or repetitive embryogenesis can further increase the number of plants, it should not be used for *in vitro* mutagenesis programmes after mutagenic treatment, since it will increase the number of plants of the same genetic make-up. This should be taken into consideration when somatic embryogenesis occurs in callus cultures. Callus is a coherent, but unorganized and amorphous tissue, formed by the vigorous division of plant cells. In this case, limited numbers of regenerative plants from the same callus should be sampled in making up the population for mutant screening. On the other hand, embryogenic cells

---

Box 28.2: Embryogenic cells for *in vitro* mutagenesis

Advantages:

* Somatic embryos originate from a single cell and minimize or eliminate chimeras.

* Somatic embryogenic cell suspension is ideal for mutation induction due to the production of mutant somatic embryos.

* Somatic embryos can be produced in a bioreactor for their large-scale production for selection.

* Embryogenic cells are suitable for long-term storage by cryopreservation to enable preservation of mutants.

Disadvantages:

* Somatic embryogenesis is highly species- and genotype-dependent and therefore requires specific culture media formulation.

* Germination rate of somatic embryos is very poor in most crops.

* Somatic embryogenic cultures can lose their embryogenic property if they are not sub-cultured regularly.

Table 28.3: Agronomic traits that can be selectable in cultured plants

Disease resistance	Medium with pathotoxin/crude toxin/culture filtrate (CF)
Herbicide tolerance	Medium with herbicide
Salt tolerance	Medium with high salt (NaCl) concentration or sea water
Metal tolerance	Medium with high concentration of toxic metals
Cold tolerance	Cultures subject to cold temperature
Drought	Medium with polyethylene glycol or other high osmoticum
Enhanced amino acid accumulation	Medium with amino acid analogues or high concentrations of amino acids
Flooding tolerance	Culture in anaerobic conditions

are particularly preferred for *in vitro* mutagenic treatment, though they have some disadvantages (**Box 28.2**).

2.4.2. Organogenesis

Organogenesis is used for plant multiplication mainly by shoot tip culture, which is a common practice adopted by commercial laboratories. With this method, mutant plants can be multiplied in large numbers. Normally, the germination rate of somatic embryos is very low, which makes it difficult to exploit commercially for large-scale plant production. The combination of somatic embryogenesis and organogenesis would therefore be suitable for mutation induction and subsequent large-scale multiplication of the mutant plants, respectively. If shoot tips are used in gamma radiation treatment, this would generate chimerical plants depending on the occurrence of mutations in L1, L2 and/or L3 meristematic layers. The regenerated mutant plants will be unstable due to segregation in subsequent vegetative propagated generations. Therefore, it is highly desirable to dissociate chimeras by multiplying plants up to the M_1V_3 – M_1V_4 generations. This is usually done by axillary bud culture, as this represents the simplest type of *in vitro* plant propagation system, extensively applied in several ornamental plants and woody species. Axillary bud proliferation can yield an average tenfold increase in shoot number in a four to sixweek culture passage. For annual and other fast-growing species a large population of plants can be generated within a year. Adventitious shoot proliferation is the most frequently used multiplication technique in micropropagation systems.

2.5. Determination of the Number of Plants Required for Selection

The number of plants (P) to be developed for mutant selection is calculated based on the expected frequency of induction of the desired trait. Expected mutation frequencies for a single trait can be expected to have a frequency of 0.1–1.0%. For example, frequencies ranging from 0.14 to 1.93% for a number of variations in fruit traits in different varieties of pear have been reported (Predieri and Zimmerman, 2001). As an example, these authors postulated that an expected frequency of 0.5% would require 1,000 plants in order to produce five individuals that carry the desired trait. Considering three to five sub-cultures and with an average success of rooting of around 80% and survival of plants to be 90%, the total initial plant population would be 1,000 (see below for details).

Mutagenic treatments should be set to obtain a high number of plants. Less than 500 plants can limit the possibility of successful selection and 500 to 1,000 plants are a reasonable range to start with. It is also advisable to have information on the regeneration and proliferation rates of the irradiated material, number of sub-cultures post-irradiation, rooting response of the shoots or other propagules and finally the survival rates of plants post radiation exposure. The expected proliferation rate can be determined with the same regression equation as used for estimating LD_{50}. It may increase in the following sub-cultures, but it is advisable to be on the safe side, to produce more plants to compromise on the possible contamination rate, rooting and/or survival problems often encountered. The number of sub-cultures that are essentially required during the post-irradiation phase can vary from a minimum of three to a maximum of five depending on the occurrence of chimeras. The rooting percentage is often estimated to be about 80% of the regular frequency obtained for non-treated material or more directly obtained by observing cultures treated for LD determination. Plant survival must be calculated based on experience with the specific plant material. The following formula (Predeiri and Virgillo, 2007) is used to calculate the number of plant parts (X) to be treated:

$$X = P/((\alpha * b)*c)*d$$

For example: where P (the number of plants planned for field selection) is 1,000 and α (expected proliferation rate) is 3.3 and b (number of sub-cultures) is 4 and c (expected rooting percentage) is 0.85 and d (expected plant survival) is 0.90; X = 1000/((3.3*4)*0.85)*0.90 = 99.

3. *In Vitro* Selection

During *in vitro* culture, it is possible to exercise selection of mutants for agronomically useful, genetically determined traits. For selection, one can make culture medium by adding a certain amount of herbicide, salt or aluminium or exposing to physical stress such as cold or heat. Such a stress situation would kill those cells not possessing the required tolerance or resistance, and allow tolerant or resistant and surviving cells to grow. These can be picked up, multiplied through sub-cultures and regenerated into plants.

3.1. Advantages and Limitations of *In Vitro* Selection

In comparison to methodologies involving treatment of *in vivo* material, *in vitro* cultured explants provide a wider choice of controlled selection following mutagenic treatment. Screening performed *in vitro* allows for handling of large populations, avoiding the problem of working with a low number of individuals as is often the case of *in vivo* plant material. Mutagenized cell suspensions, microspores and protoplast cultures have the greatest potential for *in vitro* selection owing to their uniformity. *In vitro* selection of mutated cells has

involved exposure to herbicides, salts, toxins, culture filtrates and low or high heavy metals and temperatures. The trait of interest should be selectable at the cellular level and expressed in the regenerated plants. Not all traits can be selected at the cellular level, for example, yield, seed colour or plant height. Disease resistance, abiotic stress tolerance (particularly to salt and drought), enriched nutritional quality and herbicide tolerance are some of the traits that have been successfully selected and isolated in crop plants including VPCs. *In vitro* selected variants should always undergo specific *ex vitro* testing to confirm the existence of improved selected traits and to exclude the possible emergence of other undesired traits such as instability and non-uniformity.

3.2. *In Vitro* Selection Methods

During *in vitro* selection, two types of selections viz. single-step selection and multi-step selection are practised. Generally, an inhibitor or an anti-metabolite is added into the culture medium at a level that will either kill or inhibit the growth of most wild type cells. Desired mutated cells will be able to grow in the presence of the inhibitor. The concentration of the inhibitor depends on the sensitivity of the cells/tissues used in the experiment. It is often desirable to establish a dose–response curve with respect to growth inhibition. In single-step selection, the inhibitor is added to the culture medium, at least two or three times the level of inhibitory concentration that can kill the maximum cell population and cultures are maintained for several sub-cultures. Examples include selection for herbicide tolerance (maize), amino acid enrichment (maize, rice), disease resistance (maize, rice) and salt tolerance (rice, alfalfa). In a multi-step selection method, a sub-lethal concentration is added into the medium for cultures to grow and in the subsequent sub-cultures a gradual increase in inhibitor level is given. With this method, it has been suggested that the selected mutant trait will be more stable and more expressive, since the variant cells are in constant exposure to the increasing levels of the inhibitor.

Once mutagenesis and selection are performed, the selected population is allowed to grow, proliferate and regenerate into complete individuals. In some cases, the selected cells may have reduced efficiency for growth in the presence of a large number of other killed cells. This may be improved by providing optimal culture conditions, ideally those that can regenerate a plant from a single cell or embryogenic cell suspension culture. Nevertheless, in the absence of an optimized *in vitro* method, propagation through adventitious bud formation in which the buds develop from one or a restricted number of de-differentiated cells, is a useful option for producing mutant plants. These can then be propagated vegetatively for further field assessment of the mutant character.

4. Examples

4.1. Banana

Bananas (*Musa* spp.) are one of the world's most important fruit crops. Over 400 million people throughout the developing countries of the tropics and subtropics depend on this crop as a staple food and an important commodity for local and international trade. Several pathogens throughout the tropical regions of the world have become a threat for production, and among these, the most serious pathogen is *Fusarium oxysporum* f. sp. *cubense* race 4. There is no known chemical or cultural method available for the control of this disease, and hence, resistant germplasm needs to be developed for sustainable banana production. Development of an early, reliable and reproducible selection strategy such as the use of pathotoxin or culture filtrate can speed up the selection procedure. The selected mutant plants are transferred to a "hot spot" and the performance of the mutant plants is evaluated on the basis of survival rate and other agronomic traits including yield. Such methods have been used in efforts to either develop or screen for resistance to *Fusarium oxysporum* f. sp. *cubense* (Foc) in banana (Smith *et al.,* 2006). The following is a method adopted from their study to screen for *Fusarium* wilt resistance and superior agronomic traits using gamma-irradiated and micropropagated plants of banana.

- Shoot apices obtained from micropropagated plants were isolated and exposed to gamma radiation (^{60}Co) at 10, 20, 30, 40 and 50 Gy to check radio-sensitivity.
- About 500 explants of var. Dwarf Parfitt were irradiated at 20 Gy gamma rays and after two or three cycles of *in vitro* multiplication on cytokinin supplemented MS medium, shoots were singled out and induced for root initiation on hormone-free medium.

- The rooted plantlets were carefully removed from the test tubes and were hardened *ex vitro* for field planting.
- The micropropagated plants were planted in an experimental site uniformly infested with race 4 *Foc*. Plants were grown under normal agropractices of cultivation. When plants showed emergence of the bunch, height and circumference of pseudostem, date of bunch emergence and number of green leaves were noted.
- External symptoms such as sign of wilting, yellowing of leaves, splitting of pseudostem and petiole buckling were noted on a scale of 1–3, 1–no symptoms, 2 mild and 3 severe symptoms, plants were classified.
- Thirty-five plants showed improved agronomic performance from which pieces of rhizome with growing points were removed from plants with no wilt symptoms and replanted in a Foc infested site during November 1991.
- The plants were then maintained through M_1V_4 beginning 1993 until 1997 during which five lines (DPM 2, 25, 15, 22 and 16) showed resistance without loss in productivity.
- The five lines were micropropagated and then field planted in the Foc infested site for further evaluation.
- One line (DPM 25) was selected and evaluated along with commercial cultivar. Williams. Rhizome pieces were isolated from these agronomically superior plants and were then used for raising plants which were maintained in the Foc infested site during 1998–2003.
- DPM 25 was identified as a mutant line for further evaluation and improvement.

4.2. Sugarcane

Sugarcane (*Saccharum officinarum* L.) is an important agro-industrial sugar crop of the world; it is cultivated in more than 20 million hectares. Being a typical glycophyte, salinity in the root zone of sugarcane decreases sucrose yield, through its effect on both biomass and juice quality. Somaclonal variation in combination with *in vitro* mutagenesis and selection has been applied for the isolation of agronomically useful mutants including tolerance to biotic and abiotic stresses in the improvement of sugarcane

(Patade and Suprasanna, 2008). A methodology is presented below for the isolation of agronomically useful mutants following radiation-induced mutagenesis of embryogenic callus cultures of sugarcane (Suprasanna *et al.*, 2009).

4.2.1. Embryogenic Callus Induction and Gamma Irradiation

Embryogenic callus cultures of widely grown sugarcane var. Co 86032 were established from spindle leaf discs on MS (Murashige and Skoog, 1962) basal salts supplemented with 100 mg/l malt extract, 100 mg/l L-glutamine, 1000 mg/l casein hydrolysate, 2.0 mg/l 2,4-D, 30 g/l sucrose and 2.0 g/l gelrite. The cultures were maintained on callus induction medium (CIM) under a 16-h photoperiod (30 μmol m²/s PFD) at 25 +2°C and 70% RH by regular sub-culture on fresh callus induction medium. Embryogenic calli were then subjected to gamma radiation using Gamma Cell 220 at dose rate of 7.0 Gy/min. The irradiation doses were 10, 20, 30, 40 or 50 Gy. Radiation treated calli were immediately cultured on CIM to eliminate the radiolysis hazards and sub-cultured at least thrice, at 30-day intervals, on CIM before using for further studies.

The callus growth (mg) was determined in terms of relative growth rate (RGR) after 30 days of culture as follows: RGR = (final fresh weight–initial fresh weight)/initial fresh weight. Gamma irradiation reduced the relative growth rate of sugarcane calli with increasing doses (10–50 Gy). Significant reduction in growth rate by almost 85% was noted in 40–50-Gy treatment, whereas the 20-Gy irradiated calli exhibited almost 50% reduction of growth rate as compared to the control calli, and hence, 20 Gy was considered as the LD_{50} dose for sugarcane.

4.2.2. Plant Regeneration

Plantlets were regenerated on regeneration medium, i.e. CIM without 2,4-D. About 5-cm-long individual shoots were transferred on ½ MS medium with 2 mg/l 1-naphthaline acetic acid (NAA) for rooting. The regeneration efficiency was expressed in terms of number of plantlets regenerated in a particular treatment of gamma irradiation and salt stress. The rooted plantlets were hardened in the greenhouse.

4.2.3. Mutant Population Growth and Selection

Mutant populations were derived from *in vitro* mutagenesis of sugarcane var. CoC-671, Co 94012 and Co 86032 and field planted. At maturity, data were collected on various agronomic traits including number of millable canes, stool weight, number of internodes, cane weight, cane diameter and H.R. Brix of sugarcane mutant and control plants. The variants performing better in checks were field evaluated in the M₃ generation under the Rod Row Trial.

A wide range of agronomically desirable mutations was observed for morphological, quality- and yield-contributing characters. The spectrum of mutations was broader in the Co 94012 for morphological characters whereas, for quality and yield characters, variety Co 86032 exhibited a wider range of mutations. Mutations for early maturity (based on BRIX) were observed in Co 86032 (0.17%) and for mid late maturity in CoC 671 (0.16%). Range of mutations obtained for cane yield attributing characters was 0.09 to 0.38% at different doses of gamma rays and varieties.

4.3. Chrysanthemum

Chrysanthemum is an important ornamental plant and a leading cut flower in the world with the Netherlands, South Africa, Spain and Israel being the major world producers. Chrysanthemum is also a good model representing vegetatively propagated crops, particularly ornamentals and, the varieties are mostly hexaploid and heterozygous, thus mutational experiments become rather easier and more interesting, developing mutants with varied flower colours (Mandal *et al.*, 2000; Datta and Chakraborty, 2005).

4.3.1. Starting Materials and Mutagen Treatment

Cuttings are generally used for irradiation either with X-rays or gamma rays at doses of 15 Gy, or slightly higher (25–30 Gy). One cutting of about 10 cm in length may have ten or more axillary buds if fully grown. It is suggested that before starting experiments, preliminary mutagenic experiments with high doses should be made and the material studied to show potential for colour change. The colour sector or patches may be only a few millimetres or even microscopic in size. If a high dose of the mutagen does not give favourable colour spot changes, it may be better to give up this approach in favour of others. For acute irradiation mutagenesis, at least 100 cuttings may be used in order to obtain desired mutants. After irradiation, the growing plants are cut back including their out-growing axillary

Box 28.3: Tips for *in vitro* mutagenesis

- Use clean, actively growing tissues as the source of initial explants.
- When cuttings with dormant buds are taken, place them at 20°C, treat with fungicides and keep them in a beaker filled with water for two weeks until buds open and develop shoots.
- When shoots are 1.5–2.0 cm long, remove apical part (1.0 cm), rapidly rinse in tap water and transfer in the laminar flow cabinet. To avoid tissue browning, these steps must be followed rapidly.
- Dip explant in sodium or calcium hypochlorite (0.12–0.25% active Cl_2) solution for 10 to 20 min maximum to avoid bleaching or browning of tissues. It is better to standardize the hypochlorite concentration and treatment time for better explant response to shoot regeneration.
- After treatment, remove explants from the sterilizing solution and rinse two to three times in sterile distilled water for 3 min.
- Cut off the basal part (0.3 cm) and developed leaves of explants and culture initial explants (0.5–0.8 cm) in test tubes filled with culture medium. If the cleaning protocol is successful, explants do not brown. In some cases explants may turn light brown and start responding within 7–10 days of culture.
- In case of endogenous contamination of the explant, reducing its size and keeping it together with the sterilizing agent under vacuum for better penetration may be helpful to get rid of endogenous contaminants.
- Culture media can be based on full or half-strength MS (Murashige and Skoog, 1962) or SH (Schenk and Hildebrandt, 1972) supplemented with MS or LS (Linsmaier and Skoog, 1965) vitamins and plant growth regulators.
- NAA or 1-indole-3-butyric acid (IBA) (≥ 0.1 mg/l) are commonly used for callus induction 6-benzylaminopurine (BAP) (0.1–0.2 mg/l) stimulates growth, in some cases 0.1–0.5 mg/l giberellic acid (GA3) is effective in sustaining apex growth and shoot elongation.
- After 3 weeks, transfer explants to fresh medium amended with (0.3–1.0 mg/l) BAP, and select the most appropriate BAP concentration to yield the highest production of regularly shaped, vigorous shoots. Full strength MS is most efficient at this step.
- Determine the appropriate number of sub-cultures to obtain the highest number of shoots for mutagen treatment as well as for large-scale multiplication of mutant plants.

Box 28.4: *In vitro* physical mutagenesis

- Excise organs e.g. shoot tips from pre-sterilized plant material grown in the glasshouse, greenhouse, shade house, etc., or use *in vitro* plants, and transfer them to Petri dishes and seal dishes with parafilm.
- Irradiate at least 300 shoot tips or other organs with an optimal radiation (e.g. using LD_{50} dose). This dose can be reduced to LD_{25} or LD_{30} depending on the radio-sensitivity of crop used.
- Culture irradiated plant material on a nutrient medium containing plant growth hormones to aid recovery from the radiation shock for at least 72–96 h.
- Culture on a direct shoot regeneration medium containing auxins, cytokinins and selection agent. For example for selecting disease resistance, a crude extract of a fungal toxin for *Phytophthora cactorum* can be used with strawberry explants.
- Remove dead tissues from the Petri dishes at regular intervals to avoid any harmful effects on shoot regeneration rate.
- Regularly sub-culture regenerated shoots onto fresh culture media containing the fungal toxin selection agent at 3–4-week intervals depending on the crop. This should be continued to the M_1V_4 generation to dissociate chimeras and produce stability of mutant traits.
- After this step, multiply shoots in large numbers of the selected clone and induce rooting. Harden *in vitro* produced plantlets in a glasshouse or shade-house (depending on the crop) under 70–90% relative humidity.
- Perform initial screening of plantlets, 100–200 at a time, in the hardening house by inoculating each plant with a crude fungal extract.
- Conduct field testing on a 'hot spot' for the disease.

Box 28.5: *In vitro* chemical mutagenesis

Mutagen treatment of shoot tips

Cassava shoot tip treatment with ethyl methane sulphonate (EMS) is given as an example. This procedure can be used with organs (shoot tips, meristems, axillary buds) with some modifications depending on the crop species. The procedures are carried out aseptically, in a laminar flow cabinet.

- Remove the leaves from the plants (from liquid or semi-solid growth medium) and cut the stems into explants each with two nodes. Keep these nodal segments in a sterile plastic or glass Petri dish containing sterile distilled water, and seal them with parafilm to avoid contamination.
- Using sterile forceps, transfer the explants from the water into the homogeneous EMS solution under aseptic conditions in the air flow cabinet. As a guide, 200 ml EMS solution can used to treat 50 to 100 explants (the volume depends on the size of the explants, but it is crucial that the explants are immersed completely in the solution).
- Leave the explants immersed in EMS solution for the desired predetermined time. In order to enhance the viability of the explants, the set up should ideally stand on a gyratory shaker.
- After treatment, wash the explants in sterile distilled water by passing the explants onto sterile sieve and transferring into a conical flask or beaker containing sterile water and shaken. The process, of transferring to a fresh sterile sieve and washing through shaking in sterile water, is repeated at least three times to remove all traces of EMS.
- Collect the EMS and the wash solutions for appropriate disposal as hazard wastes.
- Transfer the washed explants into conical flasks containing liquid growth medium (6–10 explants/10 ml medium). Incubate on a horizontal gyratory shaker at 10 g under continuous light at 26°C. After 24 h, transfer the explants into new liquid medium under aseptic conditions. This change of media (effecting additional washing) may be repeated at least twice to ensure the removal of any residual mutagen, thus avoiding continuous exposure to EMS during growth and development of the plants.
- If the explants are not to be established in liquid medium (e.g. for shipment), after the last washing in growth media, transfer the explants to sterile Whatman filter paper to soak up excess liquid growth medium, and transfer to semi-solid MS basal growth medium.

buds two or three times so as to obtain shoots with larger areas of mutated sectors.

Following irradiation treatment, the fresh/dry weight is measured as an indication of retarded growth. As for the gamma ray dose, the use of only 250 rad (= 2.5 Gy) can be suitable. This dose is far below the RD_{50} (reducing dose for 50%). Such a small dose may require the production of large populations in the M_1V_1, M_1V_2 generations etc., but the incidence of accompanying unfavourable simultaneously induced mutant genes will be reduced. In order to overcome this, it is suggested to opt for low dose treatment to reduce accompanying mutant gene, and select as many (independent) mutants as possible even with the same phenotype. For example, if a yellow colour is wanted from a pink flower line, select two or three or more independent yellow colour mutant lines, e.g. yel-1 from cutting No3, yel-2 from cutting No24 and so on. Then the best one as the best performing mutant line is taken.

4.3.2. Handling the Mutant Population Through M_1V_1 to M_1V_3 and Selection

In chrysanthemum, selection is directly made in the field as the plant/flower colour, shape/type of plant or flower may be selected. The selected plant may still show chimerism in the M_1V_1 generation in the field, but these are expected to be reduced in M_1V_3 or later populations. To screen for a given trait (flower color), petal tissue of a new mutant colour sector can be cultured *in vitro* to obtain single re-differentiated mutant plant lines.

In commercial chrysanthemum production, vegetative generations advancing from M_1V_1 to M_1V_3 or more is inevitable. To shorten this process, tissue culture method is used for obtaining chimera-free mutant tissues and plants. Even if mutant tissue can be selected in the M_1V_3 or later generations, the structure of a meristem (growth corn) may still be unstable and give rise to variation.

5. Practical Protocols and Tips

During the *in vitro* mutagenesis, several steps as outlined in the chapter are important and will require optimization to establish a successful protocol for mutation induction. This can be useful in subsequent mutant isolation for agronomically useful traits. In the following sections, practical protocols and tips are provided to guide the researcher. A general protocol is outlined in **Box 28.3**.

Physical mutagen treatment of organs (shoot tips, meristems, axillary buds, etc.) for mutation induction is mainly done by gamma radiation. The protocol is more or less similar for radiation treatment of organs (**see Box 28.4**).

Chemical mutagens can also be applied to *in vitro* systems either before culture or during culture. EMS mutagen treatment is the most commonly used chemical mutagen for *in vitro* mutation induction. Methods for EMS treatment are described in **Box 28.5** though similar steps can be used for other chemical mutagen with some minor modification in the protocol.

6. References

6.1. Cited References

Ahloowalia, B.S., Maluszynski, M. and Nichterlein, K. 2004. Global impact of mutation-derived varieties. *Euphytica.* 135: 187–204.

Ancora, G. and Sonnino, A. 1987. *In vitro* induction of mutation in potato. *In*: Y.P.S. Bajaj (ed.) Biotechnology in Agriculture and Forestry, Vol 3. Berlin: Springer, pp. 408–424.

Buiatti, M. 1990. The use of cell and tissue cultures for mutation breeding. *In*: P. Parey (ed.) Science for Plant Breeding. Gottingen: EUCARPIA, pp. 179–201.

Datta, S.K. and Chakraborty, D. 2005. Classical mutation breeding and molecular methods for genetic improvement of ornamentals. *In*: S.K. Datta (ed.) Role of Classical Mutation Breeding in Crop Improvement. New Delhi: Daya Publ. House, pp. 260–265.

Duron, M. and Decourtye, L. 1986. Effets biologiques des rayons gamma appliqués à des plantes de Weigela cv. 'Bristol ruby' cultivées *in vitro*. Nuclear techniques and *in vitro* culture for plant improvement. Proc. Symp. Vienna, 1985. Vienna: IAEA, pp. 103–111.

He, S., Han, Y., Wang, Y. *et al.* **2009.** *In vitro* selection and identification of sweetpotato (*Ipomoea batatas* (L.) Lam.) plants tolerant to NaCl. *Plant Cell Tiss Organ Cult.* 96: 69–74.

Heinze, B. and Schmidt, J. 1995. Mutation work with somatic embryogenesis in woody plants. *In*: SM. Jain, K. Gupta and J. Newton (eds.) Somatic Embryogenesis in Woody Plants, Vol 1. Dordrecht: Kluwer Acad. Pub., pp. 379–398.

Linsmaier, E. and Skoog, F. 1965. Organic growth factor requirements of tobacco tissue cultures. *Physiol. Plant.* 18: 100–127.

Mandal, A.K.A., Chakrabarty D. and Datta, S.K. 2000. Application of *in vitro* techniques in mutation breeding of chrysanthemum. *Plant Cell. Tiss. Org. Cult.* 60: 33–38.

Muleo, R. and Morini, S. 2006. Light quality regulates shoot cluster growth and development of MM106 apple genotype *in vitro* culture. *Scientia Hort.* 108: 364–370.

Murashige, T. and Skoog, F. 1962. A revised medium for rapid growth and bioassays with tobacco tissue cultures. *Physiol. Plant.* 15: 472–497.

Patade, Y.V. and Suprasanna, P. 2008. Radiation induced *in vitro* mutagenesis for sugarcane improvement. Sugar Tech. 10(1): 14–19.

Predieri, S. 2001. Mutation induction and tissue culture in improving fruits. *Plant Cell Tiss. Org. Cult.* 64: 185–210.

Predieri, S. and Fasolo, F. 1989. High-frequency shoot regeneration from leaves of the apple rootstock M26 (*Malus pumila Mill.*). *Plant Cell Tiss. Org. Cult.* 17: 133–142.

Predieri, S. and Virgillo, N. 2007. *In vitro* mutagenesis and mutant multiplication. *In*: S.M. Jain and H. Häggman (eds.) Protocols for Micropropagation of Woody Trees and Fruits. Berlin: Springer, pp. 323–333.

Predieri, S. and Zimmerman, R.H. 2001. Pear mutagenesis: *In vitro* treatment with gamma-rays and field selection for productivity and fruit traits. *Euphytica.* 117:217–227.

Predieri, S., Magli, M. and Zimmerman, R.H. 1997. Pear mutagenesis: *in vitro* treatment with gamma-rays and field selection for vegetative form traits. 90: 227–237.

Reyes-Borja, W.O., Sotomayor, I., Garzón, I. et al. 2007. Alteration of resistance to black Sigatoka (*Mycosphaerella fijiensis Morelet*) in banana by *in vitro* irradiation using carbon ion-beam. *Plant Biotech.* 24: 349–353.

Roest, S., van Berkel, M.A.E., Bokelmann, G.S. et al. 1981. The use of an *in vitro* adventitious bud technique for mutation breeding of *Begonia x hiëmalis*. *Euphytica.* 30(2): 381–388.

Roux, N.S. 2004. Banana improvement: cellular, molecular biology and induced mutations. *In*: S.M. Jain and R. Swennen (eds.) Enfield, NH: Science Publishers, Inc.

Roy, J., Yeoh. H.H. and Loh, C.S. 2004. Induced mutations in cassava using somatic embryos and the identification of mutant plants with altered starch yield and composition. *Plant Cell Rep.* 23: 91–98.

Schenk, R.U. and Hildebrandt, A.C. 1972. Medium and techniques for induction and growth of monocotyledonous and dicotyledonous plant cell cultures. *Can. Jour. Bot.* 50: 199–204.

Smith, M.K., Hamill, S.D., Langdon, P.W. et al. 2006. Towards the development of a Cavendish banana resistant to race 4 of fusarium wilt: gamma irradiation of micropropagated Dwarf Parfitt (*Musa spp.*, AAA group, Cavendish subgroup). *Austr. Jour. Exp. Agri.* 46: 107–113.

Suprasanna, P., Patade, V.Y., Vaidya, E.R. et al. 2009. Radiation induced *in vitro* mutagenesis. Selection for salt tolerance and characterization in sugarcane. *In*: Q.Y. Shu (ed.) Induced Plant Mutations in the Genomics Era. Rome: Food and Agriculture Organization of the United Nations, pp. 159–162.

6.2. Websites

FAO/IAEA Database of Mutant Varieties and Genetic Stocks:
http://mvgs.iaea.org/

FNCA Mutation Breeding Manual:
http://www.fnca.mext.go.jp/english/mb/

6.3. Further Reading

Broertjes, C. and Van Harten, A.M. 1978. Application of mutation breeding methods in the improvement of vegetatively propagated crops. London: Elsevier Sci. Publ., p. 316.

Jain, M.S. 2005. Major mutation-assisted plant breeding programs supported by FAO/IAEA. *Plant Cell Tissue and Organ Culture.* 82: 113–123.

Van Harten, A.M. 1998. Mutation breeding: Theory and Practical Applications. Cambridge: Cambridge University Press, pp. 1–353.

Haploid Mutagenesis

I.Szarejko*

Department of Genetics, University of Silesia, Jagiellonska 28, 40-032 Katowice, Poland
* Corresponding author, E-MAIL: iwona.szarejko@us.edu.pl

1. Introduction

The term 'haploid' refers to plants that contain a gametic number of chromosomes (n) in their somatic tissues. Haploids can arise spontaneously from unfertilized egg cells with a very low frequency in some species, normally considered too low for practical purposes. For practical use in plant breeding and genetics, male or female gametes can be induced *in vitro* to develop haploid embryos or calli, which are subsequently regenerated into haploid plants. Having only one set of homologous chromosomes, haploid plants can not undergo normal meiosis and are sterile. However, the chromosome number of haploids can be doubled, either spontaneously or by application of a special agent (usually colchicine). The resulting individuals are called 'doubled haploids' (DHs), have two sets of identical chromosomes, and hence are fertile and completely homozygous. Therefore, 'doubled haploidy' provides the shortest route to produce completely homozygous lines from heterozygous plant material.

The first haploid plants were obtained from anthers of *Datura innoxia* cultured *in vitro* (Guha and Maheshwari, 1964). Since then, various techniques of DH production have been developed for many crop species. DH systems have been integrated into breeding programmes of many crops, including major cereals, oil crops, fruit crops, vegetables, medicinal plants and ornamentals. Application of DH systems not only saves many generations needed to produce homozygous breeding lines, but also enhances the effectiveness of selection, especially when quantitative traits are evaluated. The same advantages are apparent when haploid systems are used for mutant induction and selection. When haploid cells are subjected to mutagenic agents and then induced to develop DH plants, recessive mutations, which are the predominated form of mutations, can be recovered in the first generation after mutagenic treatment. In this way, the time needed for the development of homozygous mutants can be significantly shortened and selection of mutants becomes more efficient (**Figure 29.1**). Additionally, if selection can be carried out *in vitro*, the haploid cells or embryos provide an extremely large mutagenized population, increasing the probability of identifying even a rare mutation event.

The successful application of haploid mutagenesis depends on certain conditions. The most important is an efficient method of doubled haploid production. Although efforts to obtain DH plants have been undertaken in more than 250 plant species (Maluszynski *et al.*, 2003a, b), the efficient and reproducible DH protocols are available only for less than 30. Another important factor in the application of DH technology for mutant induction and selection is a proper choice of mutagen

Box 29.1: Terms and techniques used in haploid mutagenesis

Haploid: plants that contain a gametic number of chromosomes (n) in somatic tissues.
Doubled haploid (DH): plants that have two sets of identical chromosomes obtained through spontaneous or induced chromosome doubling of a haploid cell, embryo or tissue.

Approaches to doubled haploid production:
* *Androgenesis*: haploid plants are produced from male gametophytes through *in vitro* culture of anthers or isolated microspores.
* *Wide crosses and chromosome elimination*: haploid plants are obtained from inter-species or inter-genera crosses followed by chromosome elimination of the pollinating parent of a cross. The haploid plant has a set of chromosomes from a female parent.
* *Gynogenesis*: haploid plants are produced from cells of the embryo sac through *in vitro* culture of unfertilized ovules, ovaries or flower buds.

Critical factors that affect efficiency of androgenesis:
* *Stage of microspore development*: microspores in mid to late uni-nucleate or early bi-nucleate stage are used for *in vitro* culture, depending on the species.
* *Stresses applied prior to culture to generate a developmental pathway switch*: cold pretreatment, heat shock, osmotic stress, carbon and nitrogen starvation, colchicine, other chemicals, i.e. 2-hydroxy nicotinic acid.
* *Induction and regeneration medium*: a variety of media are used for embryo/callus induction and plant regeneration. The choice of medium depends on the species and method of culture (anther or isolated microspore cultures) and media vary in composition of nutrients and growth regulators.
* *Other factors*: genotype and growth condition of donor plants, physical conditions of anther or microspore incubation, method of chromosome doubling.
Haploid mutagenesis: mutagenic treatment applied to the haploid cells, embryos, calli in culture or plant organs containing male or female gametophytes prior to culture.
Killing curve: a dose–response curve created on the basis of a pilot test that estimates the sensitivity of treated explants to the applied mutagens.

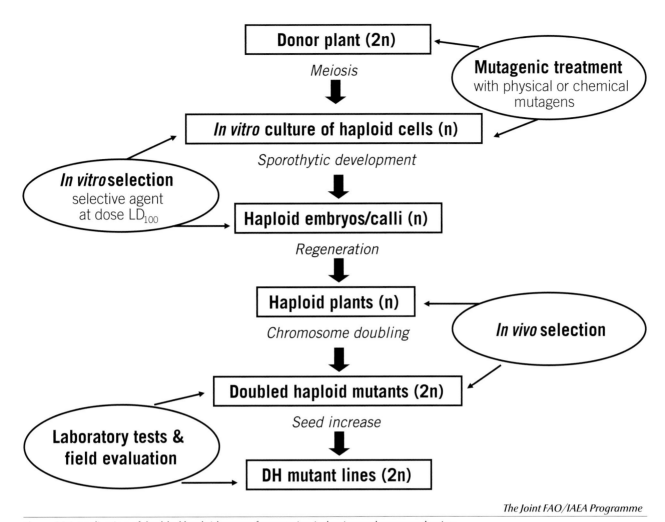

Figure 29.1 Application of doubled haploid system for mutation induction and mutant selection.

and a reliable mutagenic procedure. The existing techniques of DH production and their applicability to the major crop species are discussed below, followed by methods of applying mutagenic treatments at the haploid level. A few successful examples of *in vitro* mutagenesis and selection are presented as case studies.

2. Methods of DH Production

Three major approaches to DH production have been developed and improved upon over the past few decades (**Box 29.1**).

For many crops, DHs can be produced at a frequency enabling their efficient utilization in breeding programmes. There are a few species for which all three approaches can be efficiently used for production of DH plants, however for the majority of crops, one or two methods are predominant.

2.1. Androgenesis

The process in which microspores can be switched from their normal gametophytic development towards a sporophytic pathway is called androgenesis. Under appropriate culture conditions, microspores can undergo multiple cell divisions and form embryos or calli that can be regenerated into whole plants. The re-programming towards the sporophytic development can be induced by subjecting the microspores to various stresses. There are two main approaches used in androgenesis – anther culture and isolated microspore culture.

2.1.1. Anther Culture

Anther culture is the most widely used technology for DH production. In this technique, anthers containing microspores at the appropriate stage of development, usually mid to late uni-nucleate, are excised from spikes, panicles or flower buds under sterile conditions, plated onto induction medium and incubated at an appropriate temperature for four to six weeks. During this period, the responsive microspores undergo mitotic divisions and develop into embryos or calli. Then, microspore-derived structures are transferred onto regeneration medium where embryos can germinate into plantlets. Haploid plants can also be regenerated from microspore-derived calli through organogenesis. In many cases both embryos or embryo-like structures co-exist in the same culture together with microspore-derived calli.

The switch of microspores from the normal gametophytic (pollen) development into a sporophytic pathway is a critical step. This can be induced by a variety of stresses applied to plant material prior to culture. The detachment of anthers from the donor plant provides a stress in itself, but additional stress treatments can greatly enhance the androgenic response of microspores. Most often, temperature stresses, i.e. heat shock (32°C for a few hours to a few days) or cold pretreatment (4°C for two to four weeks); nitrogen and carbohydrate starvation or osmotic stress are applied prior to culture to promote re-programming of microspore development.

There also are many other factors that can affect anther culture efficiency. Among the most important are: growth conditions of donor plants, composition of induction medium, physical environment of culture incubation and plantlet regeneration, and genotype of plants used as donors of anthers. The proper conditions of donor plant growth are among the key factors ensuring good culture response. The material is usually cultivated in growth chambers with controlled temperature, light and humidity, although plants grown in less controlled environments (glasshouses, greenhouses, etc.) are also utilized. It is very important that plants should be properly watered, fertilized and kept free of diseases and insects. Anther cultures set up from field-grown plants have also been reported, but such material has much less synchronous microspore development and results in high variation in culture response. Among media components, the source and level of carbohydrates and nitrogen belong to the most important factors. For many species, maltose has become the preferable source of carbon in induction medium. Some species require addition of specific amino acids. The majority of anther culture media include a composition of growth regulators, although in some cases the auxin has been omitted.

Depending on the species, the population of microspore-derived embryos and germinated plants is either all haploid or consists of haploid and DH plants. The latter derive from the spontaneous chromosome doubling which can occur early in culture, usually through incomplete cytokinesis after the first microspore division, followed by fusion of what in normal development would be the vegetative and generative nuclei. When the frequency of spontaneous chromosome doubling is high, as it is in the case of barley or wheat, the haploid plants can be discarded and only DH plants are harvested without additional treatments. For most species, however, haploid plants are in the majority in microspore-derived populations; therefore the young plants are often treated with colchicine. There have also been successful attempts to apply colchicine or other chromosome doubling agents, e.g. oryzaline during embryo/calli development in vitro.

It should be pointed out that there is no single anther culture protocol available that can be used for different species. Each crop has different requirements for donor plant growth, pretreatments and media. Even within a species, different genotypes vary greatly in their response to the same applied procedure. Some recalcitrant genotypes may not respond to culture at all. This genotype dependence is one of the major factors limiting the routine use of anther culture in breeding.

2.1.2. Isolated Microspore Culture

Sporophytic development of microspores can also be achieved through in vitro culture of microspores released from anthers prior to culture. Usually, flower buds, florets or pieces of inflorescences containing microspores at a responsive stage are mechanically homogenized in a blender and microspores are separated from the debris through filtration. After a few rounds of washing and centrifugation, the microspores are finally re-suspended in an induction medium and incubated under appropriate conditions. The development of isolated microspores is more rapid and synchronous than in anther culture, and plants are generally derived through embryogenesis rather than from callus. When established, androgenesis via isolated microspore culture is the most effective system of DH production, it is however technically demanding and the development of a procedure giving similar response across genotypes requires a lot of effort.

All factors affecting anther culture efficiency are also important in isolated microspore culture with the role of stress applied to microspores prior to culture being even more pronounced. Besides temperature, osmotic and starvation shocks, treatments with various chemicals such as 2-hydroxynicotinic acid have been successful in inducing microspore embryogenesis in some species. The type of stress treatment and its duration is species- and genotype-dependent. Other critical factors that can influence the embryogenic potential of culture are: separation of viable and dead microspores before culture and determining an optimal density of microspores in the induction medium. To remove dead microspores, the microspore filtrate is centrifuged in a density gradient, e.g. maltose and mannitol, and after separation re-suspended in the induction medium to the optimal density.

The induction media used for microspore culture have the same or very similar composition to the media used for anther culture of the same species, but they are usually liquid. To ensure the proper aeration of microspores and developing structures, the cultures are often gently shaken during incubation. The media used for induction of microspore embryogenesis are either free of growth regulators or supplemented with low concentration of cytokinines and weak auxins. This is contrary to somatic tissues which require media with a strong auxin, such as 2,4-D in inducing somatic embryogenesis.

As microspore-derived embryos are the pre-dominant product of isolated microspore culture, they can be relatively easily germinated into plantlets after transfer onto regeneration medium. Thus, the regeneration potential of microspore culture is much higher than that of anther culture, where both embryos and calli are induced but the latter are more difficult to regenerate into plantlets. Another advantage of microspore culture comes from the very low level of gametoclonal variation present among embryo-derived DH plants compared to the regenerants obtained from calli. One limitation in using this technology in monocots is the high level of albino plantlets observed among microspore-derived regenerants in some genotypes, similarly to anther culture. Nevertheless, taking into account the high efficiency and other advantages of the system, it should be a method of choice for DH production, when a routine protocol of isolated microspore culture is available.

2.2. Wide Crossing Followed by Chromosome Elimination

This method of haploid production was first described by Kasha and Kao (1970) who reported the recovery of

haploid barley plants after crossing cultivated barley, *Hordeum vulgare* with the wild relative *H. bulbosum* used as a male parent. The pollen of *H. bulbosum* fertilizes the egg cell of *H. vulgare* and a hybrid zygote is formed. Then, the chromosomes of *H. bulbosum* are lost during subsequent cell divisions that create the embryo. In the first few days after pollination, only *H. vulgare* chromosomes remain in cell nuclei and the embryo becomes haploid. Due to the accompanying failure of endosperm development, the embryo is not able to mature within the seed (caryopsis) and has to be rescued through *in vitro* culture. This is usually done 12–14 days after pollination. During the next few weeks, haploid plants are developed, colchicine-treated and transferred to soil. This system, initially created for barley and often referred to as the 'Bulbosum method', has been adopted for other cereal species, including bread and durum wheat, triticale and oat in which pollen from a range of species is used.

In wheat, intergeneric crosses with maize as a pollinating parent are applied to produce haploid embryos and plantlets. The development of embryos is supported by plant hormone treatment, usually 2,4-D, applied to wheat florets and tillers immediately after pollination. The rescue of embryos onto culture medium and colchicine treatment of haploid plantlets is always necessary. The efficiency of embryo formation may be influenced by genotype of the maize parent. Some other species, e.g. sorghum and pearl millet have also been successfully applied as pollinators for bread and durum wheat. The advantages of wide crosses as a method of DH production are: very low genotype dependence, lack of albinism and absence of gameoclonal variation. It should be pointed out that haploid plants that develop from interspecific or intergeneric crosses originate from female gametic cells.

2.3. Gynogenesis

This approach to DH production is based on parthenogenic development of the unfertilized egg cell of an embryo sac under *in vitro* conditions. The mechanism stimulating the sporophytic development of the egg cell or other haploid cells of the embryo sac (antipodals or synergids) are yet unknown, but haploid embryos have been produced from unfertilized ovaries, ovules or floret buds cultured *in vitro* under appropriate conditions. The haploid embryos, after transfer onto regeneration medium, are germinated into plants. The efficiency of the method is affected by similar factors to those that influence androgenesis, i.e. donor plant growth conditions, composition of culture media, culture conditions and genotype of donor plants. However, stress pretreatments have not been successful in enhancing gynogenic response. The recovered plantlets are always haploid, therefore chromosome doubling is required. The advantage of gynogenesis is genetic stability of DH lines and a lack of albino regenerants in cereals.

Due to the potentially much lower efficiency of gynogenesis as compared to androgenetic systems, this approach to DH production is applied only for species that fail or respond very poorly in anther and microspore culture. Additionally, to make this method cost-effective, a high number of female gametic cells per one inflorescence should be available for culture, thus most efforts to improve gynogenesis have been undertaken in such species as onion (*Allium cepa*) or sugar beet (*Beta vulgaris*). Gynogenesis has been the main route to doubled haploid production in other species of Chenopodiacae and some members of Cucurbitacae family.

Gynogenesis can also be induced through pollination with irradiated pollen grains in some fruit crops (**see Chapter 30**).

2.4. Spontaneous Occurrence of Haploids

Haploids can occur spontaneously by parthenogenetic development of an unfertilized egg cell. Occasionally haploid embryos can survive to seed maturity and may germinate. This has been recognized by plant breeders for many decades, especially in seed propagated crops where rare haploids can be distinguished in fields of normal plants by their poor vigour, lack of fertility and other morphological traits. The natural production of haploid seedlings has generally been regarded as an inefficient method of generating haploids. Many approaches have been attempted to increase the frequency of parthenogenesis, including selection of seeds with twin embryos, the use of alien cytoplasm, genetic stocks carrying haploidy-inducing genes, lines with visible dominant markers or pollination with irradiated pollen (Kasha and Maluszynski, 2003). However, the frequency of haploids has remained too low for breeding purposes.

Because of their rarity, haploid seedling selection is dependent upon the availability of large seedling populations, and high-throughput detection methods such as morphology (visual screening), genotyping (homozygosity testing with DNA markers) and ploidy determination (e.g. flow cytometry). After haploid selection, the chromosome complement is doubled. Recently, the protocol for haploid seedling screening has been proposed for oil palm (Nelson *et al.*, 2008). Even though only 83 haploid plants were recovered from screening of more than 20,000,000 seedlings, haploid and DH oil palm production by this method represents a major breakthrough as oil palm has been recalcitrant to other methods of haploid production, despite intense efforts. Other advantages of haploid seedling selection include: it is a natural process and there are no issues in somaclonal variation or albinism. The applicability of high-throughput haploid seedling screening to other crops, in which haploid production is difficult or impossible by other methods, remains to be proven.

3. Efficient DH Methods Available for Major Species

The detailed protocols of DH production for many major plant species have been collected in a manual *Doubled Haploid Production in Crop Plants* (Maluszynski *et al.*, 2003a). The recent developments in DH technology and utilization are summarized in a special issue of Euphytica (2007, vol. 158) and two books: *Biotechnology in Agriculture and Forestry, Vol. 56, Haploids in Crop Improvement II* (2005) and *Advances in Haploid Production in Higher Plants* (2009) – see Further reading. **Table 1** presents species for which the effectiveness of DH production is sufficient to be utilized in breeding programmes and genetic studies.

In barley, DHs can be obtained through all possible approaches: anther culture, isolated microspore culture, wide crosses and gynogenesis, although the latter method is much less efficient than the others and is not practicable. Isolated microspores, due to their uniformity and high efficiency, are the method of choice for large commercial breeding companies, they may be, however, difficult to implement in less equipped laboratories or breeding stations. The success rate of microspore culture is very much dependent on the availability of facilities required for controlled growth of donor plants and on their genotype. Anther culture, although less effective, can be used in barley when sophisticated facilities are not available. Anther culture is also the only or the main method used for DH production in rice, maize, rye, pasture grasses, potato and some fruit and woody trees. In contrast, the effective protocols of isolated microspore culture developed for rapeseed and other *Brassica* species have made this method less amenable to plant growth conditions and applicable for large-scale mutation breeding projects.

In bread and durum wheat, wide crosses with maize and anther culture are the two most frequently utilized DH technologies. The chromosome elimination method is much less genotype dependent and can be carried on in any plant breeding station supplied with very basic *in vitro* culture equipment. It requires, however, the colchicine treatment of haploid plants and additional generations to obtain sufficient numbers of seeds. Anther culture is less labour-consuming and does not require chromosome doubling. On the other hand, high genotype dependence and albinism of many

Table 29.1: Major species with DH production protocols applicable in breeding and/or research

Crop group	Species	Method of DH production [a]
Cereals	*Hordeum vulgare* (barley)	IMC, AC, WC ('Bulbosum')
	Oryza sativa (rice)	AC
	Secale cereale (rye)	AC, IMC
	Triticum aestivum (bread wheat)	WC (wheat x maize), AC, IMC
	Triticum durum (durum wheat)	WC (wheat x maize), AC, IMC
	x *Triticosecale* (triticale)	WC (wheat x maize), AC, IMC
	Zea mays (maize)	AC, IMC
Grasses	*Lolium perenne* (ryegrass)	AC
	Lolium multiflorum (Italian ryegrass)	AC
	Phleum pratense (timothy)	AC
Oil and industrial	*Brassica napus* (rapeseed, canola))	IMC. AC
	Linum usitatissimum (linessed/flax)	IMC, AC
	Elaeis guineensis (oil palm)	SS
	Nicotiana tabacum (tobacco)	IMV, AC
Tuber and vegetables	*Allium cepa* (onion)	G (unfertilized flower culture)
	Beta vulgaris (sugar beet)	G (ovule culture)
	Brassica campestris (Chinese cabbage)	IMC. AC
	Brassica carinata (Ethiopian mustard)	IMC. AC
	Brassica juncea (Indian mustard)	IMC. AC
	Brassica oleracea (cabbage, broccoli)	IMC. AC
	Brassica rapa (cole rape, turnip)	IMC. AC
	Daucus carrota (carrot)	AC
	Solanum tuberosum (potato)	AC, WC (*S. tuberosum* x *S. phureya*)
Trees	*Citrus ssp.* (citrus)	AC
	Malus x domestica (apple tree)	AC
	Populus ssp. (poplar)	AC
	Quercus ssp. (oak)	AC

[a] Methods of DH production are listed in the order of their current usage in breeding programmes of a species. IMC – isolated microspore culture, AC – anther culture, WC – wide crosses, G – gynogenesis, SS – seedling screening.

regenerants may limit the efficiency of this method. Isolated microspore procedures are also available for wheat but this technique is much more technically demanding than two other methods of DH production.

For many important crops, among them most legumes, there is still a lack of effective methods of DH production, despite many efforts and studies undertaken. The progress achieved recently in understanding the molecular basis of microspore embryogenesis in cereals and Brassicaceae may bring new insights and indicate new directions in DH technology development also in these recalcitrant species.

4. Main Approaches in DH Mutagenesis

There are two main approaches to apply DH systems in plant mutagenesis: *in vitro* haploid mutagenesis and the production of DHs from mutated plants. In haploid mutagenesis, mutagenic treatment is applied to haploid cells, most often uni-nucleate microspores or plant organs containing male or female gametes, e.g. anthers, spikes, panicles, flower buds prior to culture. In addition, microspore-derived haploid embryos, calli or protoplasts can also be the subject of mutagenic treatment. In the second approach, M_1 plants derived from mutagenic treatment of dormant seeds are used as donors for DH production. After meiosis, M_1 plants produce gametes carrying mutations in the hemizygous stage. Anthers or isolated microspores from M_1 plants serve as explants for DH production. In both approaches, haploid cells, among them cells carrying mutations, are induced to enter the sporophytic development *in vitro* and after chromosome doubling, homozygous plants are produced (including mutant lines) and can be subjected to selection.

4.1. *In Vitro* Haploid Mutagenesis

The following factors should be considered before implementation of mutagenic treatment at the haploid level: choice of efficient method of DH production;

Table 29.2: Physical treatments applied to haploid material in Brassicaceae [a]

Species	Mutagen	Material treated	Applied doses
Brassica napus	Gamma rays	Isolated microspores	5 Gy
			5–65 Gy; LD_{50} = 15 Gy
			10–40 Gy; LD_{50} = 13 Gy
		Flower buds	10–50 Gy
		Anthers	10–60 Gy; optimal 60 Gy[b]
		Secondary embryoids	80–240 Gy; LD_{50} = 120–240 Gy
	X-rays	Isolated microspores	10–40 Gy; LD_{50} = 10.5 Gy
	Nf	Anthers	10–16 Gy; optimal 16 Gy[b]
	UV	Isolated microspores	10–60 s[c]; LD_{50} = 20 s
			15–30 s; optimal 15 s[b]
			10–120 s[c]; LD_{50} < 10 s
			5–20 s; LD_{50} = 7 s
			LD_{50} = 90 s
			LD_{50} = 75–90 s
Brassica campestris	UV	Isolated microspores	5–20 s[d]; LD_{50} = 12 s
Brassica carinata	UV	Isolated microspores	30 s – 40 min[e]; LD_{50} = 8 min

[a] When available, LD_{50} and/or an optimal dose yielding useful mutants were indicated. UV – ultra violet light; LD_{50} – dose giving 50% reduction in embryogenesis; [b] doses where significant, stimulation of embryo formation was observed; incident dose rate: [c] 33 erg $mm^{-2}s^{-1}$, [d] 2.0 W $m^{-2}s^{-1}$; [e] 3.0 J $m^{-2}s^{-1}$s.

Table 29.3: Chemical treatments applied to haploid material in Brassicaceae

Species	Mutagen	Material treated	Applied doses
Brassica napus	EMS	Isolated microspores	0.1–0.5%, 1.5 h; optimal 0.1–0.2%
		Hypocotyl segments from microspore-derived embryos	0.2, 0.25, 0.5%, 3–10 h; optimal 0.2 and 0.25%, 3–4 h
		Microspore-derived calli	0.1–0.5%, 72 h; optimal 0.15%
	ENU	Isolated microspores	5–100 μM, 5 days; optimal 20 μM
	MNU	Isolated microspores	0.2–1.5 mM, 3 h; optimal 0.2 mM, 3h
			50–200 μM, 22 h; optimal 100 μM
Brassica carinata	EMS	Isolated microspores	0.1–0.5%, 30 min; LD_{50} = 0.03%
Brassica juncea	EMS	Isolated microspores	0.1–0.5%, 1.5 h; optimal 0.1–0.2%
		Microspore-derived embryos	0–20 μM, 10–50 h; LD_{50} = 5 μM, 20 h; optimal 1–2.5 μM
	ENU	Microspore-derived embryos	0–20 μM, 10–50 h; LD_{50} = 5 μM, 20 h; optimal 2.5–5 μM
Brassica rapa	EMS	Isolated microspores	0.1–0.5%, 1.5 h; optimal 0.1–0.2%

[a]When available, LD_{50} and/or optimal doses yielding useful mutants were indicated. EI – ethylenimine; EMS – ethyl-methane-sulphonate; ENU – N-ethyl-N-nitroso urea; MNH (MNU) – N-methyl-N-nitroso urea; MNNG – N-methyl-N'-nitro-N-nitrosoguanidine: NaN_3 – sodium azide.

choice of explant to be treated; mutagen; doses and treatment procedures which can be applied for *in vitro* mutagenesis, including mutagen sensitivity test.

4.1.1. Choice of DH Production Method

When an efficient system of regenerating plants from isolated microspore culture is available, microspores at the uni-nucleate stage are the best targets for *in vitro* mutagenesis. Isolated microspores provide a plentiful population of haploid cells that are capable of rapid, uniform and synchronous development into embryos. The embryos can be easily germinated into plantlets, in contrary to the much slower and less efficient process of callus regeneration. Additionally, a high frequency of spontaneous chromosome doubling occurs in some crops at the early stage of microspore culture which results in completely fertile DH plants without the need for doubling agents.

The mutagenic treatment of isolated microspores, followed by *in vitro* selection for particular traits has been exploited most widely in rapeseed and other *Brassica* species. For crops such as *Brassica napus, B. campestris, B. carinata, B juncea* or *B. rapa*, very efficient protocols for isolated microspore culture have been developed and utilized in breeding programmes. These protocols have been adjusted for application of mutagenic treatments

both with physical and chemical mutagens (**Tables 29.2 and 29.3**). In cereals, barley is the only crop for which microspore culture techniques have been integrated with *in vitro* mutagenic treatments.

In species where microspore cultures have not been established, anther culture technique is used predominantly for DH production as a routine procedure. Mutagenic treatment of plant organs containing male gametophytes, particularly anthers, has been reported in major cereals, such as rice, wheat and barley (**Tables 29.4 and 29.5**) and some other species, among them apple, tobacco and potato (**Table 29.6**).

The 'Bulbosum method', with treatment of *H. vulgare* plants before meiosis, has been attempted to produce DH mutants in barley. The method of wide crosses followed by chromosome elimination is also routinely employed in wheat breeding, it has not, however, been used yet for wheat mutant production. Similarly, gynogenesis *via* ovary or ovule culture has not been employed for this purpose in any species, although it should be considered as the method of choice for *in vitro* mutagenesis in onion or sugar beet.

4.1.2. Choice of Explants for Mutagenic Treatment

To exploit fully the advantages of haploid mutagenesis, the mutagenic treatment must be applied at

Table 29.4: Physical treatments applied to haploid material in cereals [a]

Species	Mutagen	Material treated	Applied doses
Hordeum vulgare	Gamma rays	Spikes with microspores at uni-nucleate stage	1–10 Gy
Oryza sativa	Gamma rays	Panicles with microspores at uni-nucleate stage	5–50 Gy
		Anther cultures after 30–35 days of incubation	10–50 Gy; optimal 20 Gy[b])
		Anthers after inoculation and anther-derived calli	5–10 Gy (acute radiation) 200–300 Gy (chronic radiation)
		Microspore-derived cell cultures	30, 50, 70, 90, 120 and 150 Gy
Triticum aestivum	Gamma rays	Spikes before anther inoculation	1–10 Gy; optimal 1–5 Gy[b])

[a] UV – ultra violet light; LD_{50} – dose giving 50% reduction in embryogenesis; [b] doses where significant stimulation of embryos formation was observed.

Table 29.5: Chemical treatments applied to haploid material in cereals [a]

Species	Mutagen	Material treated	Applied doses
Hordeum vulgare	NaN_3	Isolated microspores	10^{-4}, 10^{-5} M, 1 h
		Anthers	10^{-3}, 10^{-4} M, 6 h
	MNU	Haploid seeds obtained by Bulbosum method	1 mM, 2 and 3 h
		Spikes of *H. vulgare* before meiosis (Bulbosum method)	0.1, 02 and 0.4 mM[b]
Oryza sativa	EMS	Anther culture at early stage	2–4 ml/l, 12 h[c]
		Anther culture at 0, 10[d]) and 20 days after inoculation	0.5%, 6 h
	EI	Anther culture at early stage	0.5–1 ml/l, 20 h[c]
	MNNG	Anther culture at early stage	25–100 mg/l, 15 h[c]

[a,b] Injection of mutagenic solution into the uppermost internodes of *H. vulgare* tillers at the stage just before meiosis; [c] mutagens incorporated into a liquid medium; [d] culture stage when application of mutagenic treatment significantly increased calli formation; EI – ethylenimine; EMS – ethyl-methane-sulphonate; MNH (MNU) – N-methyl-N-nitroso urea; MNNG – N-methyl-N'-nitro-N-nitrosoguanidine: NaN_3 – sodium azide.

the uni-nucleate stage of microspore development. When isolated microspore cultures are used for DH production, the treatment is usually applied soon after microspore isolation, before the first mitotic division of microspore nucleus in culture. Soon after immature pollen grains become bi-nucleate, in species such as barley and wheat, the majority of them undergo spontaneous chromosome doubling through fusion of two haploid nuclei. Usually, there is a narrow time-window, when microspores are still at the uni-nucleate stage that mutagenic treatment should be applied. When treatment is postponed to further stages of micro-

spore development *in vitro*, the advantages of haploid mutagenesis are lost. The treatment of microspores at the uni-cellular stage is also advantageous for species where the spontaneous chromosome doubling is low (such as *Brassica* sp. or rice), as it reduces the chance of chimerism in regenerative plants.

Mutagenic treatment is often applied to plant organs, such as panicles, spikes, flower buds or inflorescences containing microspores at appropriate stages for androgenesis. After treatment, anthers are excised and inoculated onto culture medium. In this case, the mutagenic treatment is applied to haploid cells, which results in

Table 29.6: Physical treatments applied to haploid material in other species [a]

Species	Mutagen	Material treated	Applied doses
Datura innoxia	Gamma rays	Anthers	10 Gy
	X-rays	Anthers	2.5–15 Gy
		Haploid protoplasts	2.5–15 Gy
Malus x domestica	Gamma rays	Anthers and flower buds	5, 10 and 20 Gy
Nicotiana plumbaginifolia	Gamma rays	Haploid protoplasts	13–23 J/kg
			1–10 Gy
	UV	Haploid protoplasts	12.5 erg mm^{-2}s^{-1}, 40 s
			25 erg mm^{-2}s^{-1}
Nicotiana sylvestris	UV	Haploid protoplasts	32 erg mm^{-2}s^{-1}
Nicotiana tabacum	Gamma rays	Anthers	10 Gy
	UV	Haploid protoplasts	32 erg mm^{-2}s^{-1}
Solanum tuberosum	Gamma rays	Inflorescences with PMC	2–30 Gy
	X-rays	Inflorescences with PMC	10 Gy

[a] PMC – pollen mother cells; LD$_{50}$ – dose giving 50% reduction in embryogenesis; [b] doses where significant stimulation of embryos formation was observed.

the recovery of completely homozygous plants after spontaneous or induced chromosome doubling. There are, however, examples of the application of mutagenic agents to anthers cultured for different periods, e.g. 10, 20 and 30 days. These examples involve rice anther cultures where the frequency of spontaneous chromosome doubling is relatively low. The majority of developing rice calli remain haploid, but as mutagenic treatment is applied to the multi-cellular structures, the regenerated plants can be chimeric and therefore segregate in the progeny. Examples of other multi-cellular haploid structures used as material for treatment include haploid embryos and their segments derived from microspore culture in rapeseed, as well as embryos produced by the 'Bulbosum method' in barley (**Tables 29.3 and 29.5**).

Haploid protoplasts or haploid cell cultures originating from microspore-derived calli have been also used as material for mutagenic treatments. The majority of reports on treatment of haploid protoplasts come from studies on tobacco and other *Nicotiana* species performed in the 1980s (**Tables 29.6 and 29.7**). Haploid *Nicotiana* protoplasts, similarly microspore-derived calli in rice, were used to generate cell suspension cultures that were utilized subsequently for *in vitro* selection of cell lines resistant to various amino acids.

Only in *Nicotiana* species where regeneration of plants from protoplasts is relatively well established, were DH mutants obtained from this procedure.

4.1.3. Applied Mutagens

Both physical and chemical mutagens have been applied for treatment of haploid explants *in vitro*. Among physical factors, gamma rays and UV light have been utilized most often, although treatments with fast neutrons and ion beams have also been attempted (**Tables 29.2, 29.4 and 29.6**). Among chemical mutagens, alkylating agents: ethyl methane sulphonate (EMS), N-methyl-N-nitrosourea (MNU), N-ethyl-N-nitrosourea (ENU), N-methyl-N¹-nitro-N-nitrosoguanidine (MNNG) and ethyleneimine (EI) have been applied most often (**Tables 29.3, 29.5 and 29.7**).

The choice of mutagenic agent depends on the facilities available and the explants to be treated. Irradiation with gamma rays is the most feasible method of treatment for all types of plant material, but it can be performed only when an *in vitro* laboratory is located in the proximity of a gamma source. In contrast to dormant seeds, cells in culture or living plant organs containing microspores are not convenient material for irradiation. UV light is a relatively cheap and easy-to-operate source of non-ionizing radiation, however due to its low penetration into plant

Table 29.7: Chemical treatments applied to haploid material in other species

Species	Mutagen	Material treated	Applied doses
Datura innoxia	MNNG	Haploid protoplasts	5–50 mg/l, 0.5 h
Nicotiana tabacum	EMS	Anthers	0.005–0.01%, 1–2 h
Solanum tuberosum	MNNG	Inflorescences with PMC	0.05–2 mM, 3 h
	MNH	Inflorescences with PMC	0.13–2 mM, 3 h

EMS – ethyl methane sulphonate; MNH (MNU) – N-methyl-N-nitroso urea; MNNG – N-methyl-N′-nitro-N-nitrosoguanidine.

tissues it can be applied only for treatment of single cell layers in culture, e.g. to the isolated microspores or haploid protoplasts (**see Chapter 7**).

Alkylating agents have proven to be very potent mutagens in inducing (predominantly) point mutations (**see Chapters 12 and 20**). When treatment with chemical agents is applied, both, mutagen concentration and time of treatment should be taken into account. Chemical mutagens, such as EMS, MNU or ENU can be dissolved easily in the culture medium used for anther or microspore culture. Usually, isolated microspores are treated directly after isolation for a relatively short time (0.5–1.5 h), although longer treatments with lower mutagen concentrations have also been applied (**Table 29.3**). When anthers or anther-derived calli and embryos are treated, the mutagen is often incorporated into the induction medium for different periods of time, ranging from several hours to a few days. Treatment of haploid explants with chemical agents has one disadvantage in comparison to the physical mutagens as it requires a thorough cleaning of treated material to remove residues of the chemical mutagen. This is especially critical for isolated microspores, as at time of treatment they are extremely fragile and can be damaged easily by the additional rounds of washing and centrifugation.

4.1.4. Assessment of Critical and Optimal Doses

Before a large scale experiment with mutagenic treatment of haploid cells is carried out, a smaller scale pilot test for estimating the sensitivity of treated explants to the applied mutagens should be performed. Treatment of isolated microspores prior to culture can drastically decrease their survival, ability to undergo divisions, develop into embryos and to germinate into plantlets. Additionally, in cereals, the mutagenic treatment can increase the proportion of albino plantlets among regenerants.

To assess the sensitivity of microspores to a mutagenic agent, a range of doses is tested and a dose–response curve, or 'killing curve' is generated. The frequency of microspore-derived embryos is estimated for each dose and the LD_{50} is indicated as a dose which reduces embryo survival by 50%. This dose is often suggested as suitable for large-scale treatment. However, it should be noted that mutagenic treatment can not only reduce the ability of microspores to develop into embryos, it can also significantly decrease their regeneration potential. Additionally, various mutagens can affect the microspore culture in different ways. In a study performed in rapeseed, UV light reduced embryo formation but not the regeneration ability of surviving embryos, whereas gamma radiation decreased both the frequency of surviving embryos and the frequency of regenerated haploid plants (MacDonald *et al.*, 1991).

It should also be considered that high doses of mutagenic treatment generate many mutations per genome, among them undesired and deleterious mutations. The presence of deleterious mutations in haploid cells is usually lethal. In contrast to the conventional mutagenesis of diploid seeds, where the desired mutations can be sorted out from the deleterious ones after meiosis in the M_1 generation, treatment of microspores in culture directly affects their survival ability. However, even in conventional mutagenesis, the dose giving 50% of survival reduction, called 'critical', is too high for breeding purposes as the high frequency of favorable mutants is accompanied by a high load of undesired mutations at other important loci. The decrease in mutation frequency can be recompensed by enlargement of mutagenized populations. Here, *in vitro* mutagenesis offers a great advantage over conventional seed treatment as extremely large populations of cells (the targets for mutagens) can be handled *in vitro* (e.g. 100,000–200,000 per 1 ml of medium). For example,

Table 29.8: Mutagens used for DH mutant production from M_1 plants [a]

Species	Mutagen	Dose applied to dormant seeds
Hordeum vulgare	Gamma rays	120 Gy
	MNH	2 × 0.5 mM, 3 h[b]
		0.1, 0.2 and 0.4 mM, 24 h
	NaN$_3$ and MNH	1.0 mM, 3 h and 0.5 mM, 3 h[b]
	EMS	10 and 20 mM, 24 h
	ENU	0.1, 0.2 and 0.4 mM, 24 h
	NaN$_3$	1 mM, 24 h
Oryza sativa	Gamma rays	100–400 Gy
		250 Gy
		300–450 Gy
	EMS	0.025 M EMS, 4 h
		94.2 mM (1%)[c]
	MNU	1 mM[d]
Solanum nigrum	EMS	10 and 20 mM, 24 h
	MNU	0.2 and 0.4 mM, 24 h
	NaN$_3$	10 and 20 mM, 24 h
Triticum aestivum	Gamma rays	150 Gy

[a] EMS – ethyl methane sulphonate; ENU – N-ethyl-N-nitroso urea; MNH (MNU) – N-methyl-N-nitroso urea; NaN$_3$ – sodium azide; [b] double treatment with 6 h inter-incubation germination period; [c] treatment of fertilized egg cells (injection of mutagenic solution into panicles at flowering) ; [d] treatment of fertilized egg cells (soaking of panicles in mutagenic solution at flowering).

the large-scale *in vitro* mutagenesis experiments of Polsoni *et al.* (1988) comprised 200 plates, each containing 1,000,000 rapeseed microspores that yielded over 2,000,000 mutagenized embryos. Dealing with such large M_2 populations is not feasible in the field. Screening for mutants in these enormously large *in vitro* populations increases the probability of identifying even rare mutation events after treatment with doses much lower than LD_{50}.

Tables 29.2–29.7 present doses of physical and chemical mutagens applied to isolated microspores, anthers, haploid calli and other explants in *Brassica* species (**Tables 29.2 and 29.3**), major cereals (**Tables 29.4 and 29.5**) and other species for which *in vitro* mutagenesis at the haploid level has been performed (**Tables 29.6 and 29.7**). When available, the data include a range of doses applied in establishing a dose–response curve, the LD_{50} dose and doses called "optimal" for which the useful mutants have been reported. It should be noted, that doses applied to haploid cells, embryos, calli, etc. are several times lower than the doses applied

to dormant seeds of the same species. Among treated materials, isolated microspores are the most sensitive to mutagens, followed by anthers before or directly after inoculation and inflorescences (spikes, panicles, etc.) with uni-nucleate microspores. Anthers after a period in culture and microspore-derived calli appear to be a little less sensitive than other explants. The finally selected doses of mutagens should be adjusted according to the results of sensitivity tests and the objective of the mutation programme. If the recovered mutants are to be used directly in breeding programmes of a particular crop, the lower doses should be applied as compared to the induction of mutants for genomic projects.

4.2. Production of DH Mutant Lines from M_1 Plants

DH systems can be incorporated into conventional mutagenesis where mutagenic agents are applied to dormant seeds. In this case, M_1 plants derived from seeds treated with physical or chemical mutagens are used as donors of gametic cells for DH production.

Depending on the species, different methods of DH production can be applied, i.e. androgenesis through anther or isolated microspore culture, wide crosses followed by chromosome elimination and gynogenesis *via* ovary or ovule culture. In practice, only anther cultures have been utilized so far for DH production from M_1 plants with examples in barley, rice, wheat and *Solanum nigrum* (**Table 29.8**).

Using M_1 plants for mutant production has some advantages and some limitations compared to *in vitro* haploid mutagenesis. First, when DH plants are regenerating from segregating gametes of M_1 plants where mutations are present at the haploid stage, all mutants are completely homozygous and do not segregate in the progeny. Second, the possibility to obtain stable, non-segregating mutants is greatly reduced when a mutagenic agent is applied after anther inoculation, i.e. to multicellular structures developing from microspores *in vitro*. Third, isolated microspores at the uni-nucleate stage are the best targets for *in vitro* mutagenesis, they are however, very sensitive to mutagenic treatments and post-treatment manipulations. The negative somatic effect of mutagenic treatment, often affecting culture efficiency drastically, is reduced greatly when the mutagen is applied to seeds, i.e. many cell generations before meiosis of the M_1 donor plants. Using dormant seeds for mutagenic treatments it is possible to employ much higher doses of mutagens than using microspores, anthers or inflorescences. Usually, the doses applied for producing the M_1 donor plants that serve as a source of gametic cells for culture are within the range of doses utilized in conventional mutagenesis. However, before a large-scale experiment is set up, the viability of microspores produced by M_1 plants should be evaluated, as too high doses of mutagens can result in a large reduction in microspore survival.

The limitations in utilizing M_1 plants for mutant production are related to dealing with much smaller populations of mutagenized cells, as compared to mutagenic treatments *in vitro*. The zygotic embryos present in dormant seeds at the time of treatment contain only a few initial cells that give rise to the cell lineages forming gametes and subsequently, the next generation. The number of these, so called 'genetically active' cells differs between species and even between varieties – there are two initial cells in *Arabidopsis thaliana* and six to twelve in barley (**see Chapter 15**). For this reason it is necessary to ensure an adequate size of the M_1 population used as a source of gametic cells for DH production. Whereas even a few donor plants can supply hundreds of thousands of microspores, each of them being a potential target for mutagenic action of *in vitro* mutagenesis, many more M_1 plants are needed to facilitate the selection of desired mutants. Due to the chimeric structure of the M_1 plants, microspores from different spikes or panicles can carry different mutations, but phenotypically similar mutants regenerated from the same floret are very likely to be identical in genetic nature. On the other hand, DH populations represent a gametic segregation ratio of 1:1 for each allele pair as compared to a 3:1 ratio in the M_2, and this makes the selection for mutants more efficient. It is difficult to estimate the exact number of M_1 plants that should be used for *in vitro* culture and very few studies address this question. In the experiment performed by Szarejko *et al.* (1995), about 50 M_1 plants for each mutagenic combination, i.e. double treatment with MNU and combined treatment with sodium azide and MNU, were used as donors of anthers. The frequency of pure mutant lines found among anther culture regenerants reached 25%.

4.3. Stimulation of *In Vitro* Culture Response by Mutagenic Treatment

While the decrease of culture ability after application of mutagens *in vitro* is a commonly observed fact, the opposite phenomenon has been reported in some cases. In rice and wheat, treatment of spikes or anthers with specific, usually low doses of physical or chemical mutagens significantly increased anther culture response measured by number of induced calli and percentage of calli regenerated plants. For example, Chen and co-workers (1996) observed a 12–38% increase in green plant regeneration after irradiation of 30–35 day-old anther cultures of six rice varieties with 20-Gy gamma rays. A similar stimulatory effect was reported in rice and barley when M_1 plants derived from mutagenized dormant seeds were used as donors for anther culture (Vagera *et al.*, 2004). The promoting effect of mutagenic treatment was especially pronounced in recalcitrant varieties that can not regenerate plants from non-treated material.

In rapeseed, in the majority of experiments, culture response declined after physical or chemical treatments. There are only a few reports indicating that treatment of anthers or isolated microspores with mutagens can significantly increase culture efficiency in this species. Nevertheless, it should be noted that the enhancement of culture response in any species is genotype-, mutagen- and dose-specific.

5. Mutant Selection

Selecting desired mutants is the ultimate goal of any mutation programme. When DH techniques are integrated with mutagenesis, the selection of mutants is somewhat different from conventional mutagenesis. In conventional mutagenesis of seed propagated crops, selection of mutants is usually initiated in the M_2 generation. Due to the chimeric structure of M_1 plants, there is usually a deficit of recessive mutant phenotypes in the segregating progeny of M_1 plants, and the vast majority of mutants are recessive. When screening for a quantitative trait, such as yield, disease resistance, abiotic stress tolerance or quality, mutant selection should be postponed until the M_3 generation. In any case, the homozygosity test must be undertaken in the selfed progeny of a selected mutant. On average, the isolation of a homozygous mutant line in seed propagated crops takes three or four generations or longer depending on the breeding system of the crop. Application of a DH system can significantly shorten this process. In haploid or DH cells and tissues, mutated traits, though recessive in nature, are expressed; therefore, selection of qualitative traits can be started immediately at the M_1 generation, which is not possible in convention mutagenesis. The selection for DH mutants with changes in quantitative traits should be carried out one generation later, similar to the conventional mutagenesis. DH mutant lines selected on the basis of their phenotype are genetically fixed and will not segregate in the progeny. Another advantage in selecting mutants using a DH system arises from the fact that the genotypes and genetic segregation ratios in DH populations are equivalent to those found in gametes. This results in simplified genetic segregations and an increased efficiency of selection for desired mutants.

5.1. *In Vivo* Selection

Although early selection for certain characters can be carried on *in vitro* (as presented below), in most cases the screening for desired mutants after haploid mutagenesis is postponed to the later stages of DH development. In this approach, haploid plants are produced directly from embryos or are regenerated from calli on medium free of any selection factor. After transfer to soil, plants are grown to maturity. If they set enough seeds, which is often high in the case of spontaneous chromosome doubling, the harvested seeds can be used directly for evaluation of DH lines and screening for mutants in a laboratory, growth room, glasshouse or field conditions. For species that have a low rate of spontaneous chromosome doubling and require colchicine treatment, one more generation may be needed for seed increase.

There are several examples of large scale microspore mutagenesis experiments in rapeseed and other Brassicas, and a smaller number of experiments in other species such as rice and barley where many valuable mutants were selected in the progeny of microspore-derived DH plants. In large-scale studies conducted in Canada, about 80,000 haploid/doubled haploid plants were obtained from microspore mutagenesis of three *Brassica* species: *B. napus*, *B. rapa* and *B. juncea*. About 1,700 DH lines of *B. napus*, 7,000 DH lines of *B. rapa* and 3,600 lines of *B. juncea* were evaluated in a two-year experiment for agronomic performance and fatty acid composition (Ferrie *et al.*, 2008). The objective was to produce DH mutant lines with altered fatty acid profiles, including lines with elevated levels of oleic acid, reduced levels of α-linolenic acid or low levels of saturated acids. The experiment appeared to be successful as many DH lines of *B. napus* and *B. juncea* were obtained with the level of saturated fatty acids reduced to 5% and 5.4%, respectively. Canola grade oil can be labelled as "low saturate" when the content of saturated fats among all fatty acids is below 7.1%. For *B. rapa*, of 7,000 DH lines evaluated for fatty acid composition, 197 lines had a level of oleic acid elevated above 55%, 67 lines had α-linolenic acid levels reduced to less than 8% and 157 lines showed low saturated acid content (less than 5%).

Similar experiments, however on a smaller scale, were performed for *Brassica carinata* with the objective to produce DH lines with modified contents of erucic acid and glucosinolates (e.g. Barro *et al.*, 2002). In barley, a relatively small mutant population of DH lines (616) was regenerated after mutagenic treatment with sodium azide applied to microspore and anther culture, but it was effective in providing various morphological and developmental mutants: dwarf, semi-dwarf, early and late heading, head type, *eceriferum*, chlorophyll seedling, partially male sterile (Castillo *et al.*, 2001). The frequency of mutated DH lines ranged from 3.8 to 15.8%, depending on the genotype. Using M_1 plants as donors for anther culture Szarejko *et al.* (1995) obtained DH lines carrying mutations in morphological, biochemical and quantitative traits at a frequency of 25%. Double treatment with sodium azide and MNU was used to produce the M_1 donor plants in this experiment. Similar treatments were applied by Jimenez-Davalos (1999) to seeds of two barley breeding lines cultivated in Peruvian highlands 3,500–4,000 m above sea level. M_1 plants served as donors for anther culture. Several true-to-type mutant lines that exceeded the yield of the parental line in these conditions were identified.

In rice, M_1 plants derived from EMS and MNU treatment of fertilized eggs were used for anther culture which produced a small number (68) of DH plants (Lee *et al.*, 2003). However, the treatment was very effective providing 31% of stable mutants for such characters as semi-dwarfism, early maturation, glabrous seed, smooth leaf, hulled seed and reduced shattering resistance. After yield evaluation, several of these mutants were used as new breeding materials in Korea. A similar approach was applied by Kyin San Myint and co-workers (2005) who obtained stable rice mutants flowering significantly earlier than the parent variety from anther culture derived from gamma-treated M_1 plants. One of these lines, with yield parameters similar to the parent, but maturing 19 days earlier, also had improved protein and amylose content and was recommended for large scale evaluation trials in Myanmar. In another study in rice, DH lines with improved agronomically important characters were selected in a DH population derived from gamma irradiation of cultured anthers (Chen *et al.*, 2001). One of the early flowering DH lines has been recommended for regional trials in the Zhejiang Province of China.

5.2. Traits Suitable for *In Vitro* Selection

One of the most important advantages of combining haploid systems with mutagenesis is the possibility of *in vitro* mutant selection, which can exploit exceedingly large populations. Such selection is carried out *in vitro* following the application of mutagenic treatment to haploid explants. Traits amenable to such selection include tolerance/resistance to disease, herbicides and abiotic stress, and altered content and/or composition of seed compounds. Examples are given below.

5.2.1. Tolerance to Herbicides and Toxins

Active components of herbicides are directly used for selection of tolerance to herbicides, while toxins excreted by pathogens are used for selection of resistance to diseases. Usually, a selective agent is included at a certain point of *in vitro* culture, at the concentration LD_{100} that kills all microspores, developing embryos, calli or plantlets except those which carry a mutation conferring resistance/tolerance to the applied factor. The earliest examples of *in vitro* selection include selection for herbicide- and disease-resistant mutants where biologically active components of herbicides or toxins produced by pathogens were added to the induction or regeneration medium at dose LD_{100} or above (**Table 29.9**). Most often, the selection agent is added at a certain stage in the culture process, and developing embryos/calli are kept in the selection medium for a longer period of time (3–4 weeks). Only surviving embryos are then germinated. Selection factors can also be incorporated into the medium used for plantlet development. After transfer to soil and subsequent chromosome doubling, the progeny of putative DH mutants are again screened for resistance/tolerance to the applied selection factor in laboratory tests and finally DH lines are evaluated in the field. Besides herbicide and disease resistance, successful selection of mutants or mutated cell lines resistant to various amino acids have been reported. Contrary to the selection applied at the whole plant level, the *in vitro* selection system allows screening of extremely large populations in a small space and with little labour.

5.2.2. Tolerance to Abiotic Stresses

The growing knowledge on biochemical pathways controlling plant responses to abiotic stresses makes it possible to explore the possibility of *in vitro* selection for mutants with elevated levels of key signalling molecules that play an important role in plant defence pathways. There are examples of cold-tolerant rapeseed DH mutant lines selected on media containing high levels of salicylic acid, jasmonic acid or proline analogues (McClinchey and Kott, 2008). All these compounds are key defence signalling molecules in cold-induced oxi-

Table 29.9: *In vitro* selection applied in haploid systems

Species	Selection objective	Selection factor	Concentration
Brassica napus	Tolerance to imidazolinone herbicides	Herbicide pursuit	40 µg/l
	Improved resistance to *Sclerotinia sclerotiorum*	Oxalic acid (OA)	3 mM
	Salt tolerance	NaCl	0.6 and 0.7%
	Cold tolerance	Salycilic acid (SA)	0.1–0.2 mM[a]
		p-Fluoro$_{-D,L}$-phenyl alanine (FPA)	0.125–0.25 mM[a]
		Jasmonic acid (JA)	0.05–0.3 mM[a]
		Hydroxyproline (HP)	5.0–10.0 mM)
		Azetidine-2-carboxylate (A2C)	0.175–0.5 mM[a]
		3,4-dehydro$_{-D,L}$-proline (DP)	0.1–0.5 mM[a]
	Reduced levels of saturated fatty acids	Heat 35°C during embryo maturation, HPLC analysis of embryo fatty acid composition	-
Brassica campestris	Improved resistance to *Erwinia carotovora*	Culture filtrates (OD$_{680}$ ~ 1.8–2.0)	30–40% (w/v)
Nicotiana plumbaginifolia	Salt tolerance	NaCl and KCl	200 mM
	Water stress tolerance	Polyethylene glycol (PEG)	25%
	Resistance to amino acid valine	L-valine	5 or 10 mM
Nicotiana sylvestris	Resistance to amino acid valine	L-valine	4 or 5 mM
Nicotiana tabacum	Resistance to amino acid valine	L-valine	8 mM
Oryza sativa	Cell lines tolerant to cyhalofop butyl (CHB) herbicide	Herbicide cyhalofop butyl (CHB)	5 mg/l
	Cell lines resistant to amino acid tryptophan	5-Methyltryptophan (5MT)	250 mol
	Cell lines resistant to amino acid lysine	S-(2-aminoethyl)-L-cysteine (AEC)	1.0 mM

[a] only concentrations of selection agents in the induction medium (NLN) are presented here.

dative and osmotic stresses. Similarly, selection for salt-tolerant rapeseed mutants was carried out on medium containing high levels of NaCl in rapeseed (Rahman *et al.*, 1995). It should be noted, however, that tolerance to a selection factor expressed by embryos and especially by calli, is not always confirmed at the whole plant level. It is necessary to evaluate the putative mutants not only during all stages of *in vitro* development but also in laboratory tests and in the field conditions. In the studies of Rahman *et al.* (1995), of 834,000 embryos developed after ENU treatment, 10 embryos survived high levels of salt in the medium, but none of the regenerated DH plants performed better than the control in field experiments. Only two lines showed improvement of some biochemical markers related to salt tolerance, such as favourable Na to K ratio or elevated levels of proline. These features may have been responsible for embryos surviving on salt-containing medium but did not result in enhanced salt tolerance of DH plants.

5.2.3. Altered Content and/or Composition of Seed Compounds

The *Brassica* microspore system provides another opportunity for *in vitro* selection. Microspore-derived haploid embryos reflect a pathway of fatty acid biosynthesis and accumulation similar to the processes occurring during *in vivo* seed maturation. Rapeseed microspore embryos express similar fatty acid composition to zygotic embryos in seeds. Thus, a small piece of tissue, usually one cotyledon of a late cotyledonary microspore embryo can be used for a non-destructive test to evaluate fatty acid composition. The remaining part, even an embryo axis without both cotyledons, can be regenerated into a plant. When embryos are produced from a mutagenized microspore population, it is possible to screen for mutated embryos with an altered oil composition and to discard all others before germination. The non-destructive method of analysis allows isolation of homozygous rapeseed mutants with an increased level of oleic acid and with an accompanying reduction of linoleic acid. Besides fatty acids, other compounds, including glucosinolates and sinapines, accumulate in microspore-derived embryos providing a possibility for *in vitro* selection of mutants with improved seed meal quality (Kott *et al.*, 1996).

5.3. Successful Examples of *In Vitro* Mutagenesis/ Selection in *Brassica* Species

The system of *in vitro* mutagenesis and selection has been best developed in rapeseed and other *Brassica* species where efficient microspore culture protocols are available. The development of microspores into embryos is uniform and rapid. Examples of large scale *in vitro* mutagenesis/selection experiments performed in *B. napus* or *B. campestris* microspore culture in order to produce herbicide-tolerant, disease-resistant, salt- or cold-tolerant DH mutant lines are presented below.

5.3.1. Rapeseed Mutants Tolerant to Imidazolinone Herbicides (Swanson *et al.*, 1989; Senior and Bavage, 2003)

In vitro microspore mutagenesis combined with a selection system was utilized to produce rapeseed mutants tolerant to the imidazolinone family of herbicides (Pursuit, Assert and Scepter). The enzyme acetohydroxyacid synthase (AHAS), the first enzyme common to the biosynthesis of leucine, izoleucine and valine is the primary target of these herbicides. AHAS is also the primary site of action for the inhibitory effects of sulphonylureas and triazole pyrimidines. Mutants tolerant to the herbicides containing these compounds were isolated in many plant species (**see Chapter 31**).

Mutagenic treatment and *in vitro* selection in microspore culture

Microspores of *Brassica napus* var. *Topas* were isolated and mutagenized with 20 μM EMS. Immediately after mutagenic treatment, the microspores were cultured in induction medium (NLN) containing the selective agent–herbicide, Pursuit. The LD_{100} concentration, 40 μg/l of Pursuit was used. The surviving embryos were transferred to solid B5 medium for 5 days and then subcultured to the same medium containing 50 μg/l Pursuit. Ten putative Pursuit-tolerant embryos were recovered from two experiments, involving 6,000 and 50,000 potential embryos. All ten embryos were regenerated into plants and transferred to soil. When plants established a good root system, they were colchicine-treated and re-potted. Seeds were collected from fertile sectors of putative mutants and increased by two generations of selfing in a glasshouse. Five of these lines produced progeny with higher tolerance to herbicide than the parent variety, but another five showed very low or no tolerance at the whole plant level. Two Pursuit-tolerant mutant lines (P1 and P2) were used for further analysis.

Genetic and biochemical analysis of mutants

Mutants P1 and P1 were backcrossed to the parent variety Topas and to each other. Tolerance in each mutant was controlled by a single semi-dominant gene and was mediated by alteration of the herbicide action to the AHAS gene. One of the mutants produced the AHAS enzyme 500 times more tolerant to Pursuit than the parent variety, Topas. The segregants from F_2 progeny of the P1 × P2 cross tolerated higher levels of herbicide than either homozygous mutant alone.

Glasshouse and field herbicide trials

Plants of variety Topas were killed by 20 g/ha Pursuit both in glasshouse and field experiments and were seriously damaged by much lower doses of herbicide. The mutant line P1 tolerated two times higher levels of Pursuit than the 50 g/ha recommended for field use, and the P2-tolerant line withstood the rates of herbicide up to 10 times the recommended dose. Both mutant lines were tolerant to 500 g/ha (P1) and 12,000 g/ha (P2) of herbicide Assert in glasshouse spray tests.

Mutant commercialization

The double mutant line harbouring both genes has been commercialized in Canada under the name

SmartCanola. The variety expressed tolerance to imid-azoline and sulphonylurea herbicides without deleterious effect on plant development, yield, maturity, quality and disease resistance.

5.3.2. Rapeseed Mutants Tolerant to Stem Rot (*Sclerotinia sclerotiorum*) (Liu *et al.*, 2005)

Plant material and microspore culture

Twenty varieties and breeding lines expressing different levels of resistance to *Sclerotinia sclerotiorum* were used as donor plants for microspore culture. Microspores were isolated, cultured in NLN medium and incubated in darkness for 1 month. For regeneration, the cotyledonary embryos were transferred onto B5 medium. After 20–25 days, the regenerated embryonic seedlings were cut into 2-mm-long segments and planted onto MS medium for callus induction. After four weeks the developing calli were sub-cultured and maintained for further experiments.

Mutagenic treatment

Calli (2 mm) of one line were used to develop a dose-response curve after EMS treatment. Calli were grown on MS medium containing the following EMS concentrations: 0, 0.1, 0.15, 0.3 and 0.5% for 72 hours. The surviving calli were transferred on MS medium and after a four-week recovery their growth was determined. For a large-scale treatment, the dose 0.15% EMS for 72 h was applied. For each of eight *B. napus* lines, 1,800 to 2,200 calli were cultured on MS medium with this dose. The calli that survived the mutagenic treatment were recovered on MS medium without EMS.

In vitro selection

Oxylic acid (OA) is a toxin produced by *S. sclerotiorum*. To find an appropriate concentration of OA for selection of calli with improved resistance, calli from four lines were cultured on MS medium with different concentrations of OA (0, 1, 3, 5 and 8 mM) for four weeks. After this time period, the surviving calli were counted and weighed. For routine screening, a concentration of 3 mM was used. After the initial selection with this dose, the surviving calli underwent a three-week phase of re-growth on MS medium without OA, after which they were cut into smaller pieces and subjected to the second cycle of screening with 5 or 8 mM OA incorporated into MS medium.

Plant regeneration and production of DH lines

After the second cycle of testing, a total of 1,600 surviving calli from 17 donor lines were transferred onto an MS regeneration medium. The regenerated plantlets were removed to soil, vernalized and colchicine treated for chromosome doubling. In total, 242 plants after OA selection were regenerated and finally 54 DH lines were obtained and used for evaluation of stem rot resistance in laboratory and field experiments.

Glasshouse and field evaluation of DH lines

The resistance of the obtained DH lines to stem rot (*S. sclerotiorum*) was evaluated in a greenhouse by artificial inoculation of detached leaves with mycelia of *S. sclerotiorum* and by feeding OA at the seedling stage. Based on resistance assessment and observation of agronomic characters, 22 DH lines were selected for field trials. Further analysis led to the selection of two DH lines, named M083 and M004, which expressed 50% less disease indices than the control and performed better than the parents in several agronomic parameters.

5.3.3. Mutants Tolerant to Soft Rot Disease (*Erwinia carotovora*) in Chinese Cabbage (Zhang and Takahata, 1999)

Plant material and microspore culture

Plants of Chinese cabbage (*Brassica campestris*) var. Ho Mei were used as donors for microspore culture. Microspores were isolated and cultured in the modified ½ NLN medium for 1 day at 32°C, followed by incubation at 25°C, in darkness. After three weeks of culture, the cotyledonary embryos were transferred to B5 regeneration medium and incubated at 25°C under a 16 h photoperiod for three to four weeks.

Mutagenic treatment

Immediately after isolation, microspore suspensions were subjected to UV radiation. The open dishes were placed on the flow bench at a distance of 65 cm from a UV tube (incident dose rate of 2.0 W mm^{-2}s^{-1}) for various lengths of time, from 5 to 80 s. Additionally, microspores cultured for various time (0–96 h) after isolation were UV irradiated for 40 s. LD$_{50}$ was determined, corresponding to UV irradiation for 12 s prior to culture. For large-scale experiments, irradiation for 20 s was applied.

In vitro selection

In the initial experiment, culture filtrates of a soft rot (*Erwinia carotovora*) isolate were used to examine their toxicity and usefulness for *in vitro* selection. Microspore-derived cotyledonary embryos were transferred to B5 medium containing 0–40% (w/v) culture filtrates (OD_{680}~1.0) for three to four weeks after which period their survival was evaluated. For further experiments, B5 medium containing 30–40% (w/v) soft rot filtrate (OD_{680}~1.8–2.0) were used. After incubation on selection medium for three to for weeks, the surviving embryos were sub-cultured on B5 medium for two to three weeks and then transferred to soil.

Production of DH lines resistant to soft rot disease

Out of 6,657 cotyledonary embryos grown on medium containing 30–40% bacterial filtrate, 250 plantlets were regenerated and 152 plants developed to the flower stage. One hundred and two plants underwent spontaneous chromosome doubling and produced seeds. The resistance to *E. carotovora* was evaluated in the progeny of 46 fertile microspore-derived plants in a glasshouse test. Twelve DH lines described as resistant or moderately resistant were detected among all analyzed.

5.3.4. Cold-Tolerant Mutants in Canola (McClinchey and Kott, 2008)

Plant material and microspore culture

The five spring canola (*Brassica napus*) breeding lines were used as donors for microspore culture. Microspores were isolated and cultured following the standard procedure for rapeseed microspore culture, including incubation in NLN medium, application of heat stress at 30°C for the first two weeks to induce embryogenesis, followed by incubation at 24°C for the next three weeks on a gyratory shaker and transfer of cotyledonary embryos on solid B5 medium for germination. After 30 days, the plantlets were transplanted to soil.

Mutagenic treatment

Directly after plating, microspores were exposed to 280-nm short-wave UV light delivered by the model C81 lamp. The duration of exposure varied between 75 and 90 s for a particular canola line and was determined on the basis of dose–response curve in the preceding pilot experiments. The doses causing 50% embryogenesis reduction were chosen for large-scale mutagenesis.

Development of *in vitro* chemical selection

The objective of this study was to develop canola lines with increased tolerance to cold. The applied strategy of *in vitro* selection was aimed at detecting mutated microspores or embryos with changes in various biochemical pathways leading to the elevated levels of key defence signalling molecules, such as salycilic acid (SA), *p*-fluoro$_{-D,L}$phenyl alanine (FPA) and jasmonic acid (JA). All these molecules are known to play a significant role in the frost-induced oxidative stress pathway. Another approach was to select mutants over-producing proline, which is known to protect plant tissues in the cold-induced osmotic stress pathway. In this case, *in vitro* selection of mutated microspores and embryos was carried out by using three proline analogues: hydroxyproline (HP), azetidine-2-carboxylate (A2C) and 3,4-dehydro$_{-D,L}$proline (DP).

All chemicals were first tested on non-mutagenized microspore cultures to evaluate the proper concentration of a specific compound and the time of its application giving an adequate selection pressure. The concentration used ranged from 0.05 to 10 mM. On the basis of these experiments, the concentrations of selection agents for 1–3-day-old microspore cultures (**Table 29.9**) and developing plantlets were established. The concentrations used were found to be chemical- and genotype-dependent. The mutagenized microspores were subjected to the selection chemicals in NLN medium until the surviving embryos were identified and transferred onto B5 germination medium. For SA, FPA, A2C and DP, the *in vitro* selection was continued, by flooding the solid B5 medium with an appropriate concentration of a chemical agent when the plantlets were 24 and 40 days old.

Characterization of DH lines tolerant to cold

Haploid mutant plants were colchicine treated at the beginning of flowering. A total of 329 microspore-derived DH lines were obtained from the *in vitro* mutagenesis/selection experiments. They showed a wide range of cold tolerance in laboratory freezing tests. All chemical selective agents, except for FPA, enabled selection of lines with improved cold tolerance compared to the parent lines. Seventy-four DH lines expressed significantly enhanced tolerance in −6°C freezing test and 19 lines had better winter survival than the winter canola checks. Out of four mutants tested for proline content, two mutant lines selected on medium with the proline analogues, accumulated more proline than the parent lines, when exposed to 4°C environment. All cold-tolerant mutant lines performed similarly to the parents with regard to seed quality and agronomic parameters.

5.3.5. Canola Mutants with Reduced Level of Saturated Fatty Acids (Beaith *et al.*, 2005)

Plant material, microspore culture

Four rapeseed (canola) varieties and breeding lines were used as donors for isolated microspore cultures. The microspore plating, embryo initiation and germination were conducted according to the procedure outlined above, except for the elevated temperatures applied during culture and the excision of embryo cotyledons for HPLC analysis.

Mutagenic treatment

For mutation induction, the isolated microspores were exposed to 254-nm short wave UV light for 90 s. This dose caused 50% microspore embryogenesis reduction.

In vitro selection

The objective was to produce DH mutants with reduced levels of saturated fats, palmitic and stearic acids. The heat treatment during embryo maturation was applied in order to (artificially) increase the saturated fat levels, this enables the clear identification of putative mutated embryos with the lowest and highest content of saturated fatty acids. Heat (35°C) was applied during embryo maturation for 18 days, following the microspore incubation at 30°C for 17 days. After 35 days of culture, when embryos had reached maturity, they were subjected to HPLC analysis of fatty acid composition. For this purpose, the cotyledons were excised and the remaining embryo axes were transferred onto B5 solid medium. Although the weight of embryos matured at 35°C was drastically reduced, the cotyledons provided enough tissue for the HPLC analysis while the embryo axes were able to develop into plantlets. Only embryos with positive changes in fatty acid profile were maintained in culture, those showing a lack or negative alterations were discarded. After germination, 30-day-old haploid plants were transferred to soil conditions, and at the bolting stage these were colchicine treated to induce chromosome doubling.

Characterization of DH mutant lines

In total, 389 mutated embryos that were able to maturate

at 35°C were screened for fatty acid composition. These embryos showed a range of fatty acid composition in both directions. Fifty-two per cent of them had reduced palmitic acid content and 46% had lower stearic acid level. The reduced level of saturated fatty acids was confirmed in the progeny of DH lines grown at normal temperatures. The lowest level of palmitic acid was 3.9% and stearic acid 0.9%, compared to 5.5% and 1.7% in the controls, respectively. Several DH mutant lines with major saturates levels below 5.5% were selected.

6. Other Applications of DH Systems in Connection with Mutation Techniques

The advantage of DH can be utilized in plant mutagenesis not only during mutant induction and early selection. Given that DH is the shortest route to complete homozygosity, DH systems can be employed during other stages of a mutation breeding programme. Often, a desired mutation in a selected mutant line is accompanied by other mutations which negatively affect plant yield or other agronomically important traits. In such cases, the procedure called 'mutant cleaning', based on a few rounds of backcrossing to the parent variety, can be proposed. The selection of recombinants carrying a desired mutation and 'clean' from the undesired ones can be facilitated by implementation of a DH system to produce homozygous lines from these crosses. This can speed up the production of desired true-to-type mutant lines and increase the efficiency of their selection in a desired genetic background.

DH technology can also be applied when a mutant line is used in a cross-breeding programme. The application of DH systems for production of pure lines from F_1 progeny of the crosses between a mutant line and other varieties or breeding lines gives the same advantages as its use in conventional breeding programmes. Of special value is the rapid fixation of recombinant lines in the homozygous stage when a mutant line carries changes in quantitatively inherited traits. DH populations obtained from such crosses are 'immortal' (can be maintained indefinitely) and can be evaluated over many years in replicated and multi-locational trials. In contrast to generations obtained by selfing or backcrossing, DH lines do not segregate in the progeny, therefore the potential phenotypic variation within a line should be attributed only to environmental factors.

Genetic analysis of mutants including mapping mutated genes is another area for combining DH and mutation technology. Over years, DH mapping populations derived from crosses between polymorphic varieties have been used to produce molecular marker maps in barley, wheat, rice, rapeseed and other species. If a similar DH population is created from a cross between a mutant line and the variety being a parent of the mapping population, it can be used for mapping of genes underlying the mutated phenotype. Such approaches facilitate chromosomal location of a gene of interest as it is possible to utilize information on the position of DNA markers linked to the gene from genetic reference maps. Dealing with DH populations instead of F_2 generations is especially recommended when quantitative traits are to be analyzed.

Microspore mutagenesis provides a possibility to create a large population of DH lines carrying mutations in a homozygous state. Such a population can serve as a TILLING platform for detecting mutations in genes of interest (**see Chapter 21**) when DNA is isolated from each line and seeds deposited in a seed bank. Creation of TILLING populations from mutagenized microspores can save one generation compared to the traditional TILLING population based on the development of M_1 and M_2 progeny. It also allows the direct recovery of a pure mutant line, instead of looking for a homozygous mutant in the M_3 progeny. All lethal mutations will be eliminated from the DH population, which should be considered as the advantage in applying TILLING for breeding purposes. Such a population could be used for comparative studies with the traditional TILLING platform.

7. References

7.1. Cited References

Barro, F., Fernandez-Escobar, J., De La Vega, M. *et al.* **2002.** Modification of glucosinolate and eruic acid contents in doubled haploid lines of *Brassica carinata* by UV treatment of isolated microspores. *Euphytica.* 129: 1–6.

Beaith, M.E., Fletcher, R.S. and Kott, L.S. **2005.** Reduction of saturated fats by mutagenesis and heat selection in *Brassica napus* L. *Euphytica.* 144: 1–9.

Castillo, A.M., L. Cistue, M.P. Valles, J.M. Sanz, I. Romagosa, J.L. Molina-Cano. **2001.** Efficient production of androgenic doubled-haploid mutants in barley by the application of sodium azide to anther and microspore cultures. *Plant Cell Rep.* 20: 105–111.

Chen, J. L. and Beversdorf ,W.D. **1990.** Fatty acid inheritance in microspore-derived populations of spring rapeseed (*Brassica napus* L.). *Theor. Appl. Genet.* 80: 465–469.

Chen, Q.F., Wang, C.L., Lu, Y.M. *et al.* **1996.** Rice induced mutation breeding by irradiated anther culture. *Acta Agric Zhejiangensis.* 8: 202–207.

Chen, Q.F., Wang, C.L., Lu, Y.M. *et al.* **2001.** Anther cuture in connection with induced mutations for rice improvement. *Euphytica.* 120: 401–408.

Guha, S., and Maheshwari, S.C. **1964.** *In vitro* production of embryos from anthers of Datura. *Nature.* 204: 497.

Ferrie, A.M.R., Taylor, D.C., MacKenzie, S.L. *et al.* **2008.** Microspore mutagenesis of *Brassica* species for fatty acid modifications: a preliminary evaluation. *Plant Breeding.* 127: 501–506.

Jimenez Davalos, J.E. **1999.** Development of doubled haploid mutant lines of barley (*Hordeum vulgare* L.) through *in vitro* anther culture and evaluation of DH lines in field conditions. MSc Thesis, La Molina National Agrarian University, Lima, Peru, pp. 76 (in Spanish).

Kasha, K.J. and Kao, K.N. **1970.** High frequency haploid production in barley (*Hordeum vulgare* L.). *Nature.* 225: 874–876.

Kasha, K.J. and Maluszynski, M. **2003.** Production of doubled haploids in crop plants. An introduction. *In*: M. Maluszynski, K. J. Kasha, B. P. Forster and I. Szarejko (eds.) Doubled Haploid Production in Crop Plants. A Manual. Dordrecht: Kluwer Academic Publishers, pp. 1–4.

Kott, L., Wong, R., Swanson, E. *et al.* **1996.** Mutation and selection for improved oil and meal quality in *Brassica napus* utilizing microspore culture. *In:* S.M. Jain, S.K. Sopory and R.E. Veilleux (eds.) *In Vitro* Haploid Production in Higher Plants. Vol. 2. Dordrecht: Kluwer Academic Publishers, pp. 151-167.

Kyin San Myint, Khine Oo Aung and Khin Soe. 2005. Development of a short duration upland rice mutant line through anther culture of gamma irradiated plants. MBNL&R 1: 13–14.

Lee, S.Y., Cheong, J.I. and Kim, T.S. 2003. Production of doubled haploids through anther culture of M₁ rice plants derived from mutagenized fertilized egg cells. *Plant Cell Rep.* 22: 218–223.

Liu, S., Wang, H., Zhang, J. *et al.* 2005. *In vitro* mutation and selection of doubled haploid *Brassica napus* lines with improved resistance to *Sclerotinia sclerotiorum*. *Plant Cell Rep.* 24: 133–144.

MacDonald, M.V., Ahmad, I., Menten, J.O.M. *et al.* 1991. Haploid culture and *in vitro* mutagenesis (UV light, X-rays, and gamma rays) of rapid cycling *Brassica napus* for improved resistance to disease. *In*: Plant Mutation Breeding for Crop Improvement. Vol. 2. Vienna: IAEA, pp. 129–138.

Maluszynski, M., Kasha, K.J., Forster, B.P. *et al.* 2003a. Doubled Haploid Production in Crop Plants. A Manual. Dordrecht: Kluwer Academic Publishers, pp. 428.

Maluszynski, M., Kasha, K.J. and Szarejko, I. 2003b. Published protocols for other crop plants. *In*: M. Maluszynski, K. J. Kasha, B. P. Forster, and I. Szarejko (eds.) Doubled Haploid Production in Crop Plants. A Manual. Dordrecht: Kluwer Academic Publishers, pp. 309–335.

McClinchey, S.L. and Kott, L.S. 2008. Production of mutants with high cold tolerance in spring canola (*Brassica napus*). *Euphytica.* 162: 51–67.

Nelson, S.P.C., Dunwell, J.M., Wilkinson, M.J. *et al.* 2008. Methods of producing haploid and doubled haploid oil palms. PCT, International publication number WO 2008/114000 A1.

Polsoni, L., Kott, L.S. and Beversdorf, W.D. 1988. Large-scale microspore culture technique for mutation-selection studies in *Brassica napus*. *Can. J. Bot.* 66: 1681–1685.

Rahman, M.H., Krishnaraj, S. and Thorpe, T.A. 1995. Selection for salt tolerance *in vitro* using microspore-derived embryos of *Brassica napus* cv. Topas, and the characterization of putative tolerant plants. *In Vitro Cell Dev. Biol.–Plant.* 31: 116–121.

Senior, I.J. and Bavage, A.D. 2003. Comparison of genetically modified and conventionally derived herbicide tolerance in oilseed rape: a case study. *Euphytica.* 132: 217–226.

Swanson, E.B., Herrgesell, M.J., Arnoldo, M. *et al.* 1989. Microspore mutagenesis and selection: canola plants with field tolerance to the imidazolinones. *Theor. Appl. Genet.* 78: 525–530.

Szarejko, I. and Forster, B.P. 2006. Doubled haploidy and induced mutation. *Euphytica.* 158: 359–370.

Szarejko, I., Guzy, J., Jimenez Davalos, J. *et al.* 1995. Production of mutants using barley DH systems. *In*: Induced Mutations and Molecualar Techniques for Crop Improvement. Vienna: IAEA, pp. 517–530.

Vagera, J., Novotny, J. and Ohnoutkova, L. 2004. Induced androgenesis *in vitro* in mutated populations of barley. *Hordeum vulgare. Plant Cell Tiss. Organ Cult.* 77: 55–61.

Zhang, F. and Takahata, Y. 1999. Microspore mutagenesis and *in vitro* selection for resistance to soft rot disease in Chinese cabbage (*Brassica campestris* L. ssp. pekinensis). *Breed. Sci.* 49: 161–166.

7.2. Further Reading

Forster, B.P., Heberle-Bors, E., Kasha, K.J. *et al.* 2007. The resurgance of haploids in higher plants. *Trends in Plant Sci.* 12(8): 368–375.

Forster, B.P. and Thomas, W.T.B. 2005. Doubled haploids in genetics and plant breeding. *Plant Breeding Reviews.* 25: 57–88.

Maluszynski, M., Szarejko, I. and Sigurbjörnsson, B. 1996. Haploidy and mutation techniques. *In*: S.M. Jain, S.K. Sopory and R.E. Veilleux (eds.) *In Vitro* Haploid Production in Higher Plants. Vol. 1. Dordrecht: Kluwer Academic Publishers, pp. 67–93.

Maraschin, S.F., de Priester, W., Spaink, H.P. *et al.* 2005. Androgenetic switch: an example of plant embryogenesis from male gametophyte perspective. *J. Exp. Botany.* 56 (417): 1711–1726.

Palmer, C.E., Keller, W.A. and Kasha, K.J. (eds.) 2005. Biotechnology in Agriculture and Forestry, Vol. 56 Haploids in Crop Improvement II. Berlin: Springer-Verlag, p. 318.

Pauls, K.P., Chan, J., Woronuk, G. *et al.* 2006. When microspores decide to become embryos – cellular and molecular changes. *Can. J. Bot.* 84: 668–678.

Shariatpanahi, M.E., Bal, U., Heberle-Bors, E. *et al.* 2006. Stresses applied for the re-programming of plant microspores towards *in vitro* embryogenesis. *Physiologia Plantarum.* 127: 519–534.

Szarejko, I. 2003. Doubled haploid mutant production. *In*: M. Maluszynski, K.J. Kasha, B.P. Forster and I. Szarekjo (eds.) Doubled Haploid Production in Crop Plants: A Manual. Dordrecht: Kluwer Academic Publisher, pp. 351–361.

Touarev, A., Forster, B.P. and Jain, S.M. (eds.) 2009. Advances in Haploid Production in Higher Plants. Berlin: Springer Science + Business Media.

Use of Irradiated Pollen to Induce Parthenogenesis and Haploid Production in Fruit Crops

M.A.Germana*

Dipartimento DEMETRA. Facoltà di Agraria. Università degli Studi di Palermo. Viale delle Scienze, 11. 90128 Palermo, Italy
* Corresponding author, E-MAIL: agermana@unipa.it

1. Introduction

Worldwide, the most cultivated fruit trees are citrus, banana, grape, apple, peaches, pear, plum, apricot and kiwifruit. World fruit production amounted to almost 500 million metric tonnes in 2007 (FAOSTAT, Database). The awareness of the necessity of a healthier food intake will probably increase the demand for fruit and the variety on offer in the near future. In addition, market demands for quality of the product and of the plants that produce them are increasing, especially with regard to a higher level of resistance or tolerance to biotic and abiotic stresses and the development of more sustainable agricultural practices.

Fruit breeding focuses on numerous goals and increasingly employs biotechnological tools, alongside more conventional approaches such as hybridization, selection and mutation induction. For example, among biotechnological methods, haploid and DH production through gametic embryogenesis allows the single-step development of complete homozygous lines from heterozygous parents, increasing the efficiency of perennial crop breeding programmes.

Haploids (sporophytic plants with the gametophytic chromosome number), originate from a single gamete and DHs are obtained by the spontaneous or induced doubling of their chromosome numbers. They have aroused the interest of breeders since the discovery of the first natural haploid in *Datura stramonium* and *Nicotiana* in the 1960s because of their potential use in mutation research, selection, genome mapping, production of inbred parental lines required in hybrid production (capitalizing on heterosis), genetic transformation and increasing the efficiency of crop breeding programmes. "Gametoclonal variation" is the variation often observed among plants regenerated from cultured gametic cells and, consists of genetic, morphological and biochemical differences that can contribute to crop improvement. It results from both meiotic and mitotic division and can be induced by cell and tissue culture. Chromosome doubling procedures are often required and the segregation and independent assortment is observed for selection purposes (**see Chapter 29**). DHs can be very useful also for genome mapping, providing reliable information on the location of major genes and quantitative trait loci (QTLs) for economically important traits. New superior varieties produced *via* gametic embryogenesis have been reported for rice, wheat, tobacco, maize and pepper, and doubled haploidy is used routinely in breeding programmes for variety development in many crops, e.g. aubergine, pepper, barley and rapeseed. Furthermore, haploid cells are ideal for mutation induction and mutant selection. Mutation induction played an important role in fruit breeding, especially in the middle of the 20th century. DH production is, among the *in vitro* tissue culture methods, an efficient system to increase the range of applications of mutation techniques, facilitating the screening for both recessive and dominant mutants, avoiding chimerism and shortening the time taken up by breeding programmes. The main aspects of haploid mutagenesis are the immediate expression of recessive mutations and the possibility of obtaining, by doubling chromosomes, complete homozygous diploids. In addition to mutagenic treatment of haploid cells, mutation treatments of gametophytes can be used for the induction of haploids.

The characteristics of woody plants, like a long reproductive cycle, a high degree of heterozygosis, large size and, often, self-incompatibility, make it impossible to perform several generations of selfing to obtain homozygous lines using conventional inbreeding methods. Therefore, methods to produce DHs are particularly attractive for fruit trees. Considerable research has been carried out since the 1970s to obtain haploids for fruit tree breeding through gametic embryogenesis, not always successfully. Haploids can be induced by regeneration from the male gamete (pollen embryogenesis) or from the female gamete (gynogenesis). *In vitro* anther or isolated microspore culture techniques are usually the most effective and widely used methods of producing haploids and DHs. Haploid production through pollen embryogenesis has been reported in about 200 species, but mostly in a limited number of families, such as *Solanaceae, Cruciferae* and *Gramineae*; many other families (particularly woody plants and *Leguminosae*) remain recalcitrant. With respect to haploid production *via* gynogenesis, several methods can be used: selection of seedlings, *in situ* or *in vitro* parthenogenesis induced by pollen from a triploid plant followed by *in vitro* embryo culture and *in situ* parthenogenesis induced by pollination with irradiated pollen. The most effective of these in the major fruit crops is haploid production *via* parthenogenesis induced through pollination with irradiated pollen.

2. Haploid Development *via* Irradiated Pollen Pollination

Parthenogenesis (the production of an embryo from an egg cell without the participation of the male gamete) or apogamy (the production of an embryo from a gametophytic cell other than the egg) can lead to haploid embryos. Parthenogenesis induced *in vivo* by irradiated pollen, often followed by *in vitro* culture of embryos, is exploited in obtaining haploids in fruit crops. This method is commonly used in those species in which *in vitro* pollen embryogenesis has not been applied successfully, and consists of rescuing immature seeds or embryos obtained through pollination with irradiated pollen. Irradiation to destroy pollen or egg cell nuclei was among one of the first applications of nuclear techniques in plant breeding. Haploid embryo production after pollination with irradiated pollen was first discovered by Blakeslee *et al.* in 1922 to obtain gynogenic haploids in *Datura stramonium*; later in 1985 it was reported by Raquin in *Petunia*, and it has been successful in fruit trees such as apple, pear, kiwifruit and citrus. Production of haploids *via* irradiated pollen-induced parthenogenesis is based on the 'Hertwig effect'.

2.1. The 'Hertwig Effect'

The 'Hertwig effect' is a phenomenon originally described by O. Hertwig (1911, see Pandey and Phung, 1982) in frog and consists of the production of maternal individuals with the use of male gametes treated with high doses of ionizing radiation (radium). On the other hand, subjecting the sperm to low doses of radiation prior to fertilization caused massive mortality of the embryos. The explanation of the Hertwig effect is that high doses cause complete inactivation of the sperm and permit the development of haploid embryos, while low doses do not totally destroy the sperm genome, resulting in aberrant embryos.

2.2. Discovery of 'Hertwig Effect' in Plants

Pandey and Phung (1982), carrying out studies in four *Nicotiana* species, showed that the 'Hertwig effect' occurs in plants as well as in animals. They reported that a parthenogenetic haploid and diploid maternal individual may develop after pollination using male gametes treated with high doses (50 and 100 Kr) of ioniz-

ing radiation, and that they can incorporate certain genes from the irradiated pollen. At lower doses (10–20 Kr), fewer seedlings were produced and many of them were abnormal and died soon after germination or before reaching maturity. Moreover, low level radiation may damage only part of the generative nucleus while maintaining its capacity to fertilize the egg cell, and lead to hybridization. One explanation for this phenomenon is that growth promoting substances inducing parthenocarpic fruit and seed testa development are released from the pollen coat. The genetic constitution of both maternal and paternal parents, particularly of the latter donor parents, are significant factors in the successful induction of parthenogenetic embryos through the use of irradiated pollen.

2.3. Fruit and Seed Development After Pollination With Irradiated Pollen

Irradiation does not hinder pollen germination, but prevents pollen fertilization, stimulating the development of haploid embryos from ovules. In fact, several research studies have shown that irradiated pollen can germinate on the stigma, grow within the style and reach the embryo sac, but cannot fertilize the egg cell and the polar nuclei (that produce the embryo and endosperm, respectively). However, it can stimulate the division of the egg cell, inducing parthenogenesis and development of parthenocarpic fruit. Pollen response to irradiation is genotype- and dose-dependent. Usually, low doses induce conventional mutational damages, while higher doses increase the frequency of parthenogenetic embryos.

Pollen irradiation can include genetic and development phenomena, such as selective gene transfer, 'egg transformation' *via* incorporation of fragments of male DNA after high pollen irradiation doses and the mentor pollen effect use to overcome incompatibility barriers.

Autonomous endosperm is sometimes observed after irradiated pollination in apple (e.g. Zhang and Lespinasse, 1991) and cacao (e.g. Falque, 1994). The development of autonomous endosperm is a very rare event in sexually reproducing angiosperms.

3. Steps in Doubled Haploid Production

The practical steps of DH production *via* parthenogenesis are described below. The methods are based on the *in vitro* culture of immature seeds or embryos obtained from pollination with pollen irradiated (usually by gamma rays from cobalt 60). The success of this technique is dependent on several factors such as the choice of radiation dose, the developmental stage of the embryos at the time of culture, the media requirements and the culture conditions. Methods include various steps: selection of parents and emasculation, pollen irradiation, pollination, seed or embryo rescue, ploidy level determination, chromosome doubling and identification of homozygous plants by molecular markers (**Figure 30.1**).

3.1. Selection of Parents and Emasculation

One of the main factors affecting parthenogenesis induced by irradiated pollen is the genotype of the plant. Research involving large numbers of different genotypes showed that the genotype of both the pollen donor and female recipient influences the number of parthenogenetic seed obtained. In kiwifruit, moreover, both the genotype of the pollen and female parents influences the ploidy of the regenerated plants. Some pollen donors appear to be more efficient in producing parthenogenetic haploids compared to others and different female lines can respond differently to the irradiated pollen of the same genotype. However, genotype has a less marked effect on *in situ* parthenogenesis induced by irradiated pollen than on *in vitro* pollen embryogenesis.

Emasculation is performed in flowers of the female parent, to make flowers unattractive to pollinating insects and to carry out controlled pollination. Flowers of the pollen donor are harvested just before anthesis. Anthers are removed and placed in a dessicator until dehiscence, and the dry pollen is then exposed to radiation treatment.

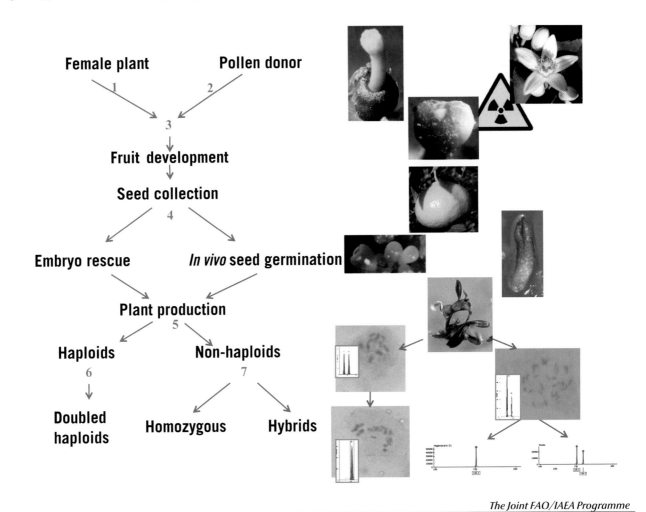

The Joint FAO/IAEA Programme

Figure 30.1 Scheme of doubled haploid production *via* parthenogenesis induced by irradiated pollen in fruit crops. 1: selection of parents and emasculation; 2: pollen irradiation; 3: pollination; 4: seed or Embryo rescue; 5: ploidy level determination; 6: chromosome doubling; 7: identification of homozygous plants by molecular markers.

3.2. Pollen Irradiation

Radiation is a physical mutagen. Irradiation has been used in fruit crop breeding to inactivate one nucleus in protoplast fusion to produce cybrids or alloplasmic hybrids (diploid somatic hybrid plants containing the nuclear genome of one parent and the cytoplasmic genome either of the other parent or a combination of both parents). Pollen radiation has been used also in the "mentor (or pioneer) pollen" technique to overcome self-incompatibility, particularly in apple and pear.

The types of radiation available for mutagenesis are ultraviolet radiation (UV, 250–290 nm) and ionizing radiation (X-rays, gamma-rays, alpha and beta particles, protons and neutrons), the latter having the capacity to penetrate deeper into tissue. Among ionizing radiation, X-rays and above all gamma rays, have been the most used, particularly for fruit trees. Gamma rays are generally obtained from isotope sources (e.g. ^{60}Co or ^{137}Cs), while the most commonly used source of far–UV radiation is the germicidal lamp. Optimizing the dose (intensity and amount) to apply is one of the most important steps in the mutagenic induction, because it affects the percentage of survival of the treated material and the mutation frequency. The radiation energy is expressed in Gy (Gray) and the sensitivity to radiation depends on: 1) the species and genotypes, 2) the plant's physiological status, and 3) the treatment of the material before and after the mutagenic treatments. Regarding the rate at which the dose is applied to plant material it is possible to distinguish between acute and chronic doses. **Table 30.1** gives commonly used treatments for pollen irradiation in major fruit crops.

3.3. Pollination

Irradiated pollen is used to pollinate the female parent 24–48 h after emasculation. Controlled pollination technique consists of applying irradiated pollen to the apical part of the stigma of a gynoecium in a emasculated flower, with a brush or by a fingertip. *In vitro* pollination of the stigma of an excised gynoecium implanted in solid culture medium has also been applied in *Citrus clementina* Hort. ex Tan. to produce haploids. A mature stage of the pistils is usually necessary for gynogenic embryo regeneration.

Several research studies have shown that irradiated pollen can germinate on the stigma, grow within the style and reach the embryo sac, but cannot affect normal double fertilization (fertilize the egg cell and the polar nuclei). However, it can stimulate the division of the egg cell, inducing parthenogenesis and development of parthenocarpic fruit.

3.4. Seed or Embryo Rescue

The combination of pollination with irradiated pollen and *in vitro* culture of immature embryos has improved the efficiency of haploid production in many fruit crops. The timing of embryo rescue is critical and it varies among species. Some authors suggest in apple and pear that embryos should be removed and cultured no later than six weeks after pollination, but this time can depend upon genotype and temperature during the preceding period. Postponing excision will decrease the relatively small number of viable embryos, though the larger size of older embryos facilitates their *in vitro* survival and propagation. Apple embryos from either normal or irradiated pollen pollination are normally cultured at about 50 days after pollination; 30-day-old or younger embryos fail to develop. In pear embryos are normally rescued from 80-day-old developing seed.

Embryo development can be improved by manipulation of the numerous factors affecting their rescue and growth. The culture medium, with all its components (macro and micronutrients, carbon source, plant growth regulators, activated charcoal; other substances; pH, osmotic potential and solidifying agents) can determine the success of embryo rescue. Different genotypes show very diverse medium requirements in promoting the development of parthenogenetic embryos. The incubation environment (temperatures, light and culture vessels) can affect the success of the culture, but usually they do not receive much attention in embryo rescue experiments (especially light quality, irradiance and photoperiod).

3.5. Ploidy Determination

Different methods can be used to determine ploidy level of regenerants obtained. These include: chromosome counting, flow cytometry analysis, stomatal size measurement, chloroplast number of the guard cells and nucleus size. The chromosome counting method can be laborious and cumbersome, and requires actively growing tissues (usually root tips). Chromosome counting is further handicapped in many woody species because of their small size. Although flow cytometry has a high start-up cost (requiring expensive equipment) it allows rapid analysis of a large number of samples and for this reason it is gaining more importance than other methods. Haploid, but also diploid, triploid or hexaploid regenerants have been obtained *via* parthenogenesis induced by radiated pollen in fruit crops (**Table 30.1**).

3.6. Chromosome Doubling

Spontaneous doubling can occour during the gametic embryogenesis process, but with different frequencies, depending on the method of haploid production used. Particularly, in pollen embryogenesis the rates of spontaneous doubling are much higher than in gynogenesis. During adventitious *in vitro* regeneration, it is also possible to obtain spontaneous chromosome doubling, without chemical treatments. Moreover, the frequency of spontaneous doubling depends on the genotype and when it is too low, efficient chromosome doubling protocols have to be developed. Duplication agents (anti-microtubule agents or spindle inhibitors) can be applied to the regenerated plantlets or at the early stages of gametic embryogenesis; furthermore, it is possible to apply them *in vivo* or *in vitro*. Colchicine is the most commonly used anti-microtubule agent, but oryzalin, trifluralin, pronamide and amiprophosmethyl (APM) are also used. A compromise between toxicity towards the treated tissues and the efficiency in genome doubling has to be found to maximize success.

3.7. Identification of Homozygous Plants

Because of the spontaneous diploidization of the haploids, ploidy level determination cannot always identify gametic plants, and molecular markers (as well as isozyme analyses) have been employed to decide the gametic origin of regenerants; since diploids could arise from un-wanted normal pollination these need to be identified and eliminated. This is relatively easy to do with genetic markers as haploids and DHs will be homozygous whereas normal embryos will exhibit heterozygosity. Random amplification of polymorphic DNA (RAPD) and microsatellite analyses have been employed to assess homozygosity and to characterize regenerants obtained *via* parthenogenesis induced by irradiated pollen in fruit crops. It is expected that faster genotyping based on DNA sequence will predominate in the future.

4. Practical Examples in Major Fruit Crops

4.1. Kiwifruit

Both male and female gametic methods of producing haploids, that is by pollen embryogenesis or by gynogenesis, were attempted in kiwifruit (*Actinidia deliciosa* A. Chev.) in the 1980s, but only the second approach has been successful. Pandey, the pioneer in this field, pollinated several female kiwifruit varieties with lethally irradiated pollen of different male and inconstant male genotypes (Pandey *et al.*, 1990). Inconstant males are male genotypes that occasionally develop small ovaries with viable ovules. Irradiation doses of 500, 700 and 900 Gy did not affect pollen germination and seeds were set regularly. Seeds were germinated *in vitro*, after three to six months storage at 4°C. Ploidy level reduction was estimated by measuring the size of nuclei of both embryos and endosperm cells with an image analyser. Two plants from Pandey's breeding programme with half the chromosome number, that is 87 chromosomes (kiwifruit is hexaploid: 2n = 6x = 174) were produced. A similar experiment was successfully carried out a few years later in France by Chalak and Legave (1997), who produced parthenogenetic tri-haploids by pollinating the female var. Hayward with pollen from two male genotypes. The M_1 and M_2 pollen was irradiated with gamma rays at a dose of 200–1,500 Gy. The pollen irradiation treatment had little effect on pollen germination, but decreased fruit set, fruit growth and the number of viable seeds produced per fruit. Seeds were sown in a glasshouse; seedling ploidy was assessed by flow cytometry and plantlets were analysed using isozymes and RAPD markers. The analyses revealed a reduction in the chromosome set, from 6x to 3x in many seedlings. These nevertheless displayed a high level of heterozygosity, compatible with the polyploidy and the heterozygosity of the original seed plants. At the lowest irradiation dosage (200 Gy), hexaploid seedlings were also obtained (derived from complete fertilization), whereas tri-haploid doubling was easily obtained by treating scarified leaves removed from tissue-cultured shoots with oryzalin, a herbicide with a strong binding affinity to plant tubulin which is capable of depolymerization of the cell microtubules' structure resulting in chromosome doubling.

4.2. Citrus Species

Citrus species represent the largest fruit production worldwide, with over 115 million tonnes produced during 2007 (FAOSTAT database). All cultivated forms of *Citrus* and related genera (*Poncirus, Fortunella,* etc.) are diploid with a monoploid number of chromosomes (n = x = 9). Triploid and tetraploid forms of Citrus also exist. With regard to haploids and homozygous plants obtained from female gametes, in *Citrus natsudaidai* haploid seedlings were first obtained by the application of gamma rays. One haploid embryo was obtained from an immature seed from a diploid (Clementine mandarin) x diploid (Pearl tangelo) cross (Esen and Soost, 1972) and a haploid plant was obtained from the cross between 'Banpeiyu' pummelo and "Ruby Red" grapefruit (Toolapong *et al.,* 1996).

In Citrus, *in situ* or *in vitro* parthenogenesis can be induced by triploid pollen followed by *in vitro* culture of embryos. In particular, three haploid plants were obtained from *in vivo* crosses of two mono-embryonic diploids (clementine and "Lee") x a triploid hybrid of "Kawano natsudaidai" (*Citrus natsudaidai* Hayata; Oiyama and Kobayashi, 1993). Triploidy of pollen, like irradiation, does not hinder pollen germination, but prevents double fertilization and stimulates the development of haploid embryoids from ovules. Haploid and diploid embryoids did not show any difference in size, however, haploid seedlings grew very slowly in the soil. Restriction endonuclease analyses of nuclear ribosomal DNA and of chloroplast DNA determined the maternal origin of these haploids.

Haploid plantlet regeneration through gynogenesis in *Citrus clementina* Hort. ex Tan., var. Nules, has been induced by *in vitro* pollination with pollen from a triploid plant (Gemanà and Chiancone, 2001). The pollen source chosen was 'Oroblanco', a triploid grapefruit-type citrus obtained in 1958 through a cross between an acidless pummelo (*Citrus grandis* Osbeck) and a seedy, tetraploid grapefruit (*Citrus paradisi* Macf.). The *in vitro* stigmatic pollination technique consisted of applying pollen to the apical part of the stigma of an excised gynoecium implanted in solid culture medium. Some ovaries developed into brownish and friable callus, sometimes breaking to reveal ovules. From this kind of ovary, the gynogenic embryoids emerged four to five months after *in vitro* pollination, which is practically the same time required for haploid plant regeneration *via* anther culture. Pollination of mature pistils was necessary for gynogenic embryoids regeneration. Although, unlike clementine anther culture, embryogenic calluses were not obtained, parthenogenesis induced *in vitro* by triploid pollen can be employed usefully in attempting to obtain haploids in mono-embryonic genotypes of Citrus for which pollen embryogenesis has not yet been successful in haploid production.

With regard to the *in situ* parthenogenesis induced by irradiated pollen followed by *in vitro* embryo culture, the production of nine haploid plantlets, which did not survive, and two embryogenic callus lines have been obtained in clementine (*Citrus clementina* Hort. ex Tan.), var. SRA 63 after *in situ* parthenogenesis induced by pollen of Meyer lemon (*Citrus meyeri* Y. Tan.) irradiated at 300, 600 and 900 Gy at the rate of 6 Gy/min from a cobalt 60 source (Ollitrault *et al.,* 1996). A haploid plant produced through *in vivo* induced gynogenesis by pollination of Nules clementine with irradiated pollen of Fortune mandarin, followed by *in vitro* embryo rescue, has been selected for citrus genome sequencing by the International Citrus Genome Consortium (ICGC) (http://int-citrusgenomics.org/).

Flowers of clementine SRA 63 were pollinated in the field with the irradiated pollen; fruits were picked at maturity and small seed selected. Embryos were recovered from the small seeds and cultured *in vitro* on a medium enriched with sugar and plant growth regulators (30 g/l sucrose, 1 mg/l GA_3 or 1 mg/l Kinetin).

While the *in vitro* pollen germination rate was not affected by irradiation up to 900 Gy, seed production and size were reduced by this treatment dose. Ploidy level of the obtained plantlets was evaluated by flow cytometry and isozyme analyses have been carried out to characterize the regenerants. Haploid induction has been obtained in the mandarins Fortune (*C. clementina* Hort. ex Tan. x *C. tangerina* Hort. ex Tan.) and Ellendale (*C. reticulata* Blanco x *C. sinensis* L. Osb.) by Froelicher *et al.* (2007) through *in situ* gynogenesis by pollination with irradiated pollen of "Meyer" lemon. Pollination was carried out for three genotypes of mandarin with four levels of gamma ray-irradiated pollen (150, 300, 600 and 900 Gy). Again small seed were selected from which embryos were rescued and cultured. Ploidy levels of plantlets, determined by flow cytometry, were haploid, diploid and triploid. Microsatellite analysis showed that diploid and triploid plants were the result of crosses between mandarin and lemon.

4.3. Apple

Apples (*Malus domestica* (L.) Borkh.), with more than 64 million tonnes produced during 2007 (FAOSTAT database), are among the most produced fruits in the world, alongside citrus, bananas and grapes. Haploid production has been obtained in *Malus domestica* (L.) Borkh, 2n = 2x = 34, using several methods. Several authors have reported pollen-derived plants from anther culture in this species. The induction of embryogenesis from cultured apple anthers is still low and highly genotype-dependent. Pollen embryogenesis and plant formation from isolated apple microspores has been reported for several apple varieties (Höfer, 2004).

James at al. (1985) reported that 50 days after pollination with pollen of 'Baskatong' irradiated at 50 and 100 kr, about 50% of the seeds had developed endosperm, while a proportion of these seeds also contained an embryo. The 'stimulative parthenocarpy' may be attributed to auxins produced by the developing endosperm.

Haploid plants have been obtained through *in situ* parthenogenesis induced by pollination of var. Erovan with pollen irradiated at 500 to 1,000 Gy, followed by *in vitro* embryo culture on MS medium containing 0.1 mg/l NAA, 1 mg/l BA and 1 mg/l gibberellic acid (GA$_3$). Trees chosen as female parents were isolated in a cage before flowering, and three or four flowers per cluster were pollinated after manual emasculation at the balloon stage of flower development. This technique has been successfully applied to other apple varieties with different gamma rays doses from Cobalt 60 (e.g. Zhang *et al.*, 1992).

A study on the influence of pollination techniques, plant growth regulators for fruit set and *ex* vitro germination on homozygous plant production by parthenogenesis *in situ* has been carried out by De Witte and Keulemans (2000). They reported beneficial effects of bee-booster pollination (only for strongly self-incompatible genotypes, such as Elstar and Granny Smith), by reduction of fruit drop through daminozide applications and sowing well-developed seeds after stratification instead of *in vitro* embryo germination. Positive effects are produced by repeated pollination (the first one with pollen irradiated at a high dose, 500 Gy, and the second one with pollen irradiated at low dose, 100 Gy) on embryo yield (although not in all the genotypes in all years). Furthermore, these authors reported that the pollination of one stigma with 'low dose pollen (50 Gy)' and the other four with 'high dose pollen (500 Gy)' resulted in a higher embryo yield compared to pollination of all the stigmas with 'high dose pollen'. The king flowers produced more embryos compared to lateral flowers in the cluster.

Pollination with irradiated pollen often results in the formation of seeds with just endosperm or with endosperm and embryos. Pollen tubes of irradiated pollen grains at high dose treatments may only carry a single sperm viable nucleus into the embryo sac, allowing only single rather than double fertilization, either of the egg cell or of the fused polar nuclei.

In situ parthenogenesis has been induced in the female parent Lodi and Erovan by irradiating the pollen of a clone RR homozygous for the marker gene coding for anthocyanin production by gamma rays from Cobalt 60 at 45 Gy/min (500, 1,000 and 1,500 Gy) (Zhang *et al.*, 1990). The young fruits were collected at different times after pollination and ovules and immature seeds were cultured *in vitro* (immature seeds were given a cold treatment to break dormancy). Six embryos were recovered from Erovan and three from Lodi but only two plants from Erovan survived and exhibited a haploid chromosome number.

Zhang and Lespinasse, (1991) pollinated four apple varieties (Erovan, Golden Delicious, R1-49 and X6677) with gamma-irradiated pollen of the clone RR at the doses from 125 to 1,000 Gy. Pollination by irradiated pollen affected fruit set, seed number and seed content and induced the formation of parthenogenetic embryos, especially when the irradiation doses were over 125 Gy. There was an influence of the female genotype on the percentage of empty seeds, of embryo-containing seeds or seeds with only endosperm. Immature embryos were extracted two to three months after pollination and cultured *in vitro* after two months of cold treatment (3°C). Eleven haploid plants (detected by chromosome count) were obtained in all genotypes and the best radiation doses were from 200 to 500 Gy. The authors concluded that pollination with gamma-irradiated pollen in apples can promote: 1) the production of parthenocarpic fruits, and 2) the development of parthenogenesis with the production of haploid and diploid embryos as well as endosperms with or without an accompanying embryo. They stated also that the optimum conditions for inducing *in situ* parthenogenesis are: 1) pollen irradiation doses of 200 and 500 Gy; 2) collection of fruits three months after pollination; 3) *in vitro* culture of immature embryos and 4) cold treatment (3°C) of embryos for two months.

Emasculated flowers of variety Idared were pollinated with gamma-irradiated Baskatong pollen to stimulate the egg cells to develop into haploid embryos. The embryos obtained were dissected out of the immature seeds (70 and 140 days after pollination) and cultured *in vitro* (Verdoodt *et al.,* 1998). The shoots obtained were analyzed using the S-alleles of the two parent genotypes in order to select the true homozygotes.

Genotype has a much greater effect on *in vitro* pollen embryogenesis than on *in situ* parthenogenesis induced by irradiated pollen. In addition, this technique is relatively easy to perform and therefore useful in fruit tree breeding.

The cytological effects of pollen gamma irradiation at 50 and 100 kr on both embryo and endosperm development were studied in *Malus × domestica* by Nicoll *et al.* (1987). They showed that the fruit and seed set was reduced by increasing doses of pollen irradiation, while embryo sacs resulting from the treatments differed in number and morphology of endosperm nuclei and in the presence or absence of an embryo. Nuclear abnormalities included enhanced numbers of polyploid restitution nuclei, bridges between nuclei, excluded metaphase chromosome fragments and disrupted mitotic synchrony. Generally, a high dose of pollen irradiation (100 kr) generated an all-or-nothing response in the embryo sac, either creating highly abnormal embryos and/or endosperms which aborted, or showing relatively normal development. The callus obtained from excised endosperm derived from 100 kr pollen showed a smaller average genome size than one from control (un-irradiated) pollen.

4.4. Sweet Cherry

The regeneration of four homozygous lines has been obtained by *in situ* parthenogenesis induced by pollination with irradiated pollen, followed by immature embryo and cotyledon culture in sweet cherry (*Prunus avium* L.), variety Altenburger (Höfer and Grafe, 2003). The pollen parent used was *Prunus cerasifera* var. Atropurpurea, carrying a marker gene R encoding anthocyanin synthesis in the whole plant. Pollen was irradiated by gamma rays from Cobalt 60 at doses of 250, 500 and 750 Gy at 1.3 or 4 Gy/min for the first year and only 750 and 1,200 Gy at 4 or 10 Gy/min for

the second year experiments. The fruit set markedly decreased when pollination was performed by irradiated pollen, in contrast to *in vitro* germination of pollen which was not significantly decreased compared to the control.

4.5. Plum

Peixe *et al.* (2000) carried out a study reporting the effect of irradiation with gamma rays (0, 100, 200, 500 and 1,000 Gy) of pollen of the var. Stanley on its viability, fruit setting and seed development after pollination of a European plum (*Prunus domestica* L.), var. Rainha Clàudia Verde. Results showed that gamma-irradiated (200 Gy) pollen can induce the formation of 2n endosperm and abnormal embryo development and that it is possible to obtain haploid induction in this genotype by *in situ* parthenogenesis.

4.6. Pear

Untreated and 50 kr irradiated pollen of var. Bonne Louise d'Avranches was used to pollinate flowers at the balloon stage of the pear (*Pyrus communis* L.) varieties Conference, Doyenné du Comice and Gieser Wildeman. The irradiated pollen stimulated a rather large "parthenocarpic" fruit set, presumably associated with endosperm rather than embryo development during the first four to six weeks after pollination (Sniezko and Visser, 1987). The irradiated pollen-induced pear embryos showing limited viability, because of their presence in immature fruit and their absence in mature fruits. The *in vitro* culture of immature embryos was suggested to rescue the putative haploids.

Haploids and DHs have been obtained by *in situ* parthenogenesis induced through irradiated pollen (P) or by seedling selection (S) from three pear varieties: Doyenné du Comice (P and S), Williams (S) and Harrow Sweet (S) (Bouvier *et al.,* 2002). Particularly, among more than 10,000 seedlings obtained from 12 different crosses, 17 were selected because of their 'haploid phenotype' and 12 resulted, by chromosome counting, as haploids (2n = x = 17). With regard to the induction of haploid plants by irradiated pollen, four genotypes were chosen as female parents: Doyenné du Comice, Williams, President-Hèron and Harrow Sweet. The male parent was TNR12.40, a selected clone carrying a homozygous dominant marker gene for the colour red and the pollen was treated with gamma irradiation (from cobalt 60 at 0, 200, 250 and 500 Gy). The percentage of pollen germination was only somewhat affected by irradiation, whereas the higher the irradiation dose, the lower the number of embryos per fruit. Immature embryos were removed from seeds three months after pollination and they were cultured *in vitro* at 3°C for three months to break embryo dormancy and later at 24°C. Three plants from var. Doyenné du Comice and one from Williams were from maternal origins. Although some spontaneous chromosome doubling occurred, most of the doubled haploids were obtained by oryzalin *in vitro* treatment (200–300 μM).

4.7. Cocoa

Cocoa (*Theobroma cacao*) pollen was treated with 0, 50, 100, 200, 500 and 1,000 Gy gamma rays (Falque *et al.,* 1992). Pollen viability and *in vitro* germination were not affected by irradiation. Irradiated pollen tubes penetrated into the styles and reached the ovules 20 hours after pollination. With irradiation over 100 Gy no fruit set was obtained. Further experiments have been carried out in order to induce parthenogenesis in *Theobroma cacao* L. (Falque, 1994). Pollen of the clone K5 was gamma irradiated at 0, 50, 70 and 90 Gy and used to pollinate flowers of the clones IMC67 and T85/799. Irradiation affected pod size, ripening time and number of beans. No haploid was obtained. Morphological mutants were obtained from IMC67 pollinated with 50 Gy irradiated K5 pollen, and some of them had inherited paternal alleles of enzymic markers without showing the axil-spot dominant character. Accidental self-fertilization of the two self-incompatible clones, through mentor pollen effect, was only observed in rare cases.

5. Acknowledgements

The Author wishes to thank Raffaele Testolin for critically reading the manuscript.

6. References

6.1. Cited References

Blakeslee, A. F., Belling, J., Farnham, M. E *et al.* **1922.** A haploid mutant in the Jimson weed, *Datura stramonium. Science.* 55: 646–647.

Bouvier, L., Guérif, P., Djulbic, M. *et al.* **2002.** Chromosome doubling of pear haploid plants and homozygosity assessment using isozyme and microsatellite markers. *Euphytica.* 123: 255–262.

Chalak, L. and Legave, J. M. 1997. Effects of pollination by irradiated pollen in Hayward kiwifruit and spontaneous doubling of induced parthenogenetic trihaploids. *Sci. Hort.* 68: 83–93.

De Witte, K. and Keulemans, J. 2000. Influence of pollination techniques, plant growth regulators for fruit set and *ex vitro* germination on homozygous plant production by parthenogenesis *in situ* in apple. *In*: M. Geibel, M. Fischer and C. Fischer (eds.). Proceedings of Eucarpia, Symposium on Fruit Breeding and Genetics. Acta Hort. 538, ISH. pp. 309–314.

Esen, A. and Soost, R. K. 1972. Unexpected triploids in citrus: their origin, identification, and possible use. *J. Hered.* 62: 329–333.

Falque, M. 1994. Pod and seed development and phenotype of the M_1 plants after pollination and fertilization with irradiated pollen of cacao (*Theobroma cacao*). *Euphytica.* 75(1-2): 19–25.

Falque, M., Kodia, A. A., Sounigo, O. *et al.* **1992.** Gamma-irradiation of cacao (*Theobroma cacao*) pollen: Effect on pollen grain viability, germination and mitosis and fruit set. *Euphytica.* 64: 167–172.

Froelicher, Y., Bassene, J.-B., Jedidi-Neji, E. *et al.* **2007.** Induced parthenogenesis in mandarin for haploid production: induction procedures and genetic analysis of plantlets. *Plant Cell Report.* 23: 937–944.

Germanà, M.A. and Chiancone, B. 2001. Gynogenetic haploids of *Citrus* after *in vitro* pollination with triploid pollen grains. Plant Cell Tissue Organ Culture. 66: 59–66.

Höfer, M. 2004. *In vitro* androgenesis in apple – improvement of the induction phase. *Plant Cell Report.* 22: 365–370.

Höfer, M. and Grafe, C. 2003. Induction of doubled haploids in sweet cherry (*Prunus avium* L.). *Euphytica.* 130: 191–197.

James, D. J., Passey, A.J., Mackenzie, K.A.D. *et al.* 1985. The effect of pollen irradiation on the development of the post-fertilization ovule of apple (*Malus pumilia* Mill.) *In*: G.P.C. Chapman *et al.* (eds.) Experimental Manipulation of Ovule Tissue. London: Longman, pp. 210–224.

Nicoll, M.F., Chapman, G.P. and James, D.J. 1987. Endosperm responses to irradiated pollen in apples. *Theor Appl. Genet.* 74: 508–515.

Ollitrault, P., Allent, V. and Luro, F. 1996. Production of haploid plants and embryogenic calli of clementine (*Citrus reticulata Blanco*) after *in situ* parthenogenesis induced by irradiated pollen. *Proc. Intern. Soc. Citriculture. South Africa.* 2: 913–917.

Oiyama, I.I. and Kobayashi, S. 1993. Haploid obtained from diploid X triploid crosses of Citrus. *J. Japan. Soc. Hort. Sci.* 62(1): 89–93.

Pandey, K.K. and Phung, M. 1982. 'Hertwig effect' In plants: induced partenogenesis through the use of irradiated pollen. *Theor. Appl. Genet.* 62: 295–300.

Pandey, K.K., Przywara, L. and Sanders, P.M. 1990. Induced parthenogenesis in kiwifruit (*Actinidia deliciosa*) through the use of lethally irradiated pollen. *Euphytica.* 51: 1–9.

Peixe, A., Campos, M.D., Cavaleiro, C. *et al.* 2000. Gamma-irradiated pollen induces the formation of 2n endosperm and abnormal embryo development in European plum (*Prunus domestica* L., cv. "Rainha Clàudia Verde"). *Sci. Hort.* 86: 267–278.

Raquin, C. 1985. Induction of haploid plants by *in vitro* culture of *Petunia* ovaries pollinated with irradiated pollen. *Z. Pflanzenzucht.* 94: 166–169.

Sniezko, R. and Visser, T. 1987. Embryo development and fruit-set in pear induced by untreated and irradiated pollen. *Euphytica.* 1: 287–294.

Toolapong, P., Komatsu, H. and Iwamasa, M. 1996. Triploids and haploids progenies derived from small seeds of 'Banpeiyu' pummelo, crossed with 'Ruby Red' grapefruit. *J. Jpn. Soc. Hort. Sci.* 65: 255–260.

Verdoodt, L., Van Haute, A., Goderis *et al.* 1998. Use of the multi-allelic self-incompatibility gene in apple to assess homozygosity in shoots obtained through haploid induction. *Theoretical and Applied Genetics.* 96: 294–300.

Yehia, M.M. 1985. Shoot regeneration from callus derived from nucellar explant and immature embryo tissue of Williams Bon Chrétien and Conference pear. *Rep. E. Malling Res. Stn.* For 1984: 245–250.

Zhang, Y.X. and Lespinasse, Y. 1991. Pollination with gamma-irradiation pollen and development of fruits, seeds and parthenogenetic plants in apple. *Euphytica.* 54: 101–109.

Zhang, Y.X., Lespinasse, Y. and Chevreau, E. 1990. Induction of haploidy in fruit trees. *Acta Hort.* 280: 293–304.

Zhang, Y.X., Bouvier, L. and Lespinasse, Y. 1992. Microspore embryogenesis induced by low gamma dose irradiation in apple. *Plant Breed.* 108: 173–176.

6.2. Websites

International Citrus Genome Consortium (ICGC)
http://int-citrusgenomics.org/

FAO/IAEA, Mutant Varieties Database. Plant Breeding and Genetic Section. Vienna, Austria. 2006.
http://www-infocris.iaea.org/MVD/.

FAOSTAT Database.
http://faostat.fao.org/default.aspx

6.3. Further reading

Forster B.P., Herberle-Bors, E., Kasha, K. J. *et al.* 2008. The resurgence of haploids in higher plants. *Trends in Plant Science.* 12: 602–609.

Forster, B.P. and Thomas, W.T.B. 2005. Doubled haploids in genetics and plant breeding. *Plant Breeding Reviews.* 25: 57–88.

Germanà, M.A. 2006. Doubled haploid production in fruit crops. *Plant Cell Tissue Organ Culture.* 86: 131–146.

Germanà, M.A. 2007. Haploidy. *In*: I. Khan (ed.) Citrus. Genetics, breeding and biotechnology. Wallingford: CABI, pp. 167–196.

Germanà M.A. 2009. Haploid and doubled haploids in fruit trees. *In*: A. Touraev, B. Forster and M. Jain (eds.) Advances in Haploid Production in Higher Plants. Springer, pp. 241-263.

Maluszynski M., Kasha, K.J., Forster, B.P. *et al.* 2003. Doubled Haploid Production in Crop Plants: A Manual. Dordrecht: Kluwer Academic Press.

Sestili, S. and Ficcadenti, N. 1996. Irradiated pollen for haploid production *In:* S.M. Jain, S.K. Sopory and R.E. Veilleux (eds.) *In vitro* Haploid Production in Higher Plants. Vol. 1. Dordrecht: Kluwer Academic Publishers, pp. 263-274.

Van Harten, A.M. 1998. Mutation Breeding Theory and Practical Applications. Cambridge: Cambridge University Press.

C31

Herbicide-Tolerant Crops Developed from Mutations

S.Tan* and S.J.Bowe

BASF Corporation, 26 Davis Drive, Research Triangle Park, NC 27709-3528, USA
* Corresponding author, E-MAIL: siyuan.tan@basf.com

423

1. Introduction

A herbicide is a chemical substance or cultured biological organism used to kill plants or suppress the growth of plants. Herbicide-tolerant crops (HTCs) have been widely and increasingly adopted around the world because they have several advantages over other weed management systems. HTCs in combination with their corresponding herbicides are able to control many weeds that cannot be, or are less effectively, controlled by other means. It is particularly effective when the weed is a closely related species to the crop. For instance, the imidazolinone-tolerant rice (*Oryza sativa*) and wheat (*Triticum aestivum*) in combination with imidazolinone herbicides can control red rice (*Oryza sativa*) and jointed goatgrass (*Aegilops cylindrica*) which no other herbicide can selectively control in cultivated rice and wheat. The imidazolinone-tolerant maize (*Zea mays*) and sunflower (*Helianthus annuus*) in combination with imidazolinone herbicides are able to control parasitic weeds such as witchweed (*Striga* spp.) and broomape (*Orobanche* spp.) very effectively. HTCs can also be good rotational crops for fields previously treated with the herbicide for which the HTC is designed because the herbicide residue carried over from previous years in the soil will not injure the HTC. Another reason for the popularity of HTCs is the increasing difficulty of discovering a new herbicide with a known or a novel mode of action. As a result, there is a strong demand to extend the utility of existing herbicides to crops that are not naturally tolerant to these herbicides.

A common and effective approach to allow farmers to use herbicide in a herbicide-susceptible crop is to change the crop genetically with a traditional, mutational or transgenic technique to make the crop tolerant to the herbicide. Transgenic technique inserts a foreign gene into an organism, while mutational technique mutates a native gene. After the genetic change, the enzyme, a protein that catalyzes a biochemical reaction and encoded by the herbicide-tolerant gene, becomes insensitive or less sensitive to the herbicide than the normal enzyme, or the modified plant can degrade the herbicide before the herbicide reaches the target enzyme, thus making a crop tolerant to the herbicide (**Table 31.1**). The herbicide tolerance trait that results from the genetic modification can be used not only for the development of a HTC but also as a selectable marker.

Many HTCs have been commercialized, and they can be classified into several categories depending on their herbicide tolerance, method of development and mechanism of tolerance (**Table 31.1**). Commercial HTCs developed from mutations include imidazolinone-tolerant maize, rice, wheat, rapeseed (*Brassica napus*), sunflower and lentil (*Lens culinaris*); sulphonylurea-tolerant soybean (*Glycine max*) and sunflower; cyclohexanedione-tolerant maize; and triazine-tolerant rapeseed (**Table 31.1**). All commercial, non-transgenic

Table 31.1: Commercial herbicide-tolerant crops, their mechanisms of tolerance, and development methods

	Mechanism of tolerance		
Development method	Insertion or mutation of a gene that encodes a tolerant enzyme targeted by the herbicide	Insertion or mutation of a gene that encodes an enzyme to degrade herbicide	Reference
Mutation of a native gene (non-transgenic)	Imidazolinone-tolerant maize, rice, wheat, rapeseed, sunflower and lentil. Sulphonylurea-tolerant soybean and sunflower. Cyclohexanedione-tolerant maize. Triazine-tolerant rapeseed.	None	Tan *et al.*, 2005; Slinkard *et al.*, 2007; Green, 2007; Gabard and Huby, 2004; Zagnitko *et al.*, 2001; Sundby *et al.*, 1993
Insertion of a foreign gene (transgenic)	Glyphosate-tolerant maize, soybean, cotton, rapeseed, alfalfa, and sugar beet.	Glyphosate-tolerant rapeseed. Glufosinate-tolerant maize, cotton, and rapeseed. Bromoxynil-tolerant cotton[a] and rapeseed.[a]	Duke, 2005; Green, 2009

[a] Not commercially available anymore.

HTCs were developed from mutations of genes that encode altered enzymes targeted by herbicides. Commercial HTCs developed from transgenes include glyphosate-tolerant maize, soybean, cotton (*Gossypium* spp.), rapeseed, alfalfa (*Medicago sativa*), and sugar beet (*Beta vulgaris*); glufosinate-tolerant maize, cotton and rapeseed; and bromoxynil-tolerant cotton and rapeseed (**Table 31.1**). Genes conferring tolerance to most herbicide classes are available, but many of them have not yet been developed into commercial HTCs.

2. Target Genes and Enzymes for Commercial Herbicide-Tolerant Crop Development

Herbicides control weeds generally through binding and inhibiting the function of key enzymes or proteins of vital biochemical pathways and physiological processes. For instance, imidazolinone and sulphonylurea herbicides, glyphosate and glufosinate inhibit critical enzymes of amino acid biosynthesis; aryloxyphenoxypropionate and cyclohexanedione herbicides inhibit a key enzyme of fatty acid biosynthesis; and triazine herbicides and bromoxynil bind to D1 protein and block the electron transport of photosynthesis. A crop with a herbicide-insensitive enzyme targeted by the herbicide or with the ability to degrade the herbicide quickly can tolerate the herbicide with little or no injury. Three enzymes or proteins and their genes have often been the targets for mutation discovery of herbicide-tolerant variants. They are acetohydroxyacid synthase, acetyl-CoA carboxylase and D1 protein of photosynthesis (**Table 31.2**).

2.1. Acetohydroxyacid Synthase (AHAS) and *AHAS* Gene Mutations

AHAS, also called acetolactate synthase (ALS), is a key enzyme of branched-chain amino acid biosynthesis that catalyzes the condensation of two pyruvate molecules to acetolactate and the synthesis of 2-acetohydroxybutyrate from pyruvate and 2-ketobutyrate (**Box 31.1**). Leucine, valine, and isoleucine are the products of this biochemical pathway and are important amino acids and building blocks of proteins. The AHAS can be inhibited non-competitively by five major chemical families including imidazolinones, sulphonylureas, triazolopyrimidines, pyrimidinylthio(or oxy)-benzoates, and sulphonylamino-carbonyltriazolinones (**Table 31.2**). The inhibition blocks the biosynthesis of leucine, valine, and isoleucine and causes plant death (**Box 31.1**). The examples of imidazolinone herbicides are imazamox, imazaquin, imazapic, imazapyr,

Table 31.2: Target enzymes of herbicides and their gene mutations used in developing commercial herbicide-tolerant crops

Enzyme or protein	Acetohydroxyacid synthase	Acetyl-CoA carboxylase	D1 protein of photosynthesis II
Catalytic function	Biosynthesis of branched-chain amino acids	Biosynthesis of fatty acids	Facilitation of electron transport in photosynthesis II
Inhibitor or herbicide	Imidazolinones, sulphonylureas, triazolopyrimidines, pyrimidinylthio(or oxy)-benzoates, sulphonylamino-carbonyltriazolinones	Aryloxyphenoxypropionate, cyclohexanedione	Triazines, bromoxynil, many other herbicides
Type of inhibition	Non-competitive	Non-competitive	Competitive
Coding gene	*AHAS* gene	*ACCase* gene	*psbA*
Gene location	Nucleus	Nucleus	Chloroplast
Common mutation point for herbicide tolerance	A122, P197, A205, W574, S653	I1781, W2027, I2041, D2078, G2096	V219, S264
Reference species	*Arabidopsis thaliana*	*Alopecurus myosuroides*	*Arabidopsis thaliana*
Inheritance of the mutant	Incompletely dominant	Incompletely dominant	Maternal
Method for mutant discovery	Selection or mutagenesis followed by selection	Selection	Selection

Box 31.1: Catalytic function of susceptible and resistant *AHAS* enzymes

Schematic diagrams showing catalytic function of susceptible and resistant AHAS enzymes without a herbicide present (A and B) and inhibition of the herbicide (C and D). The herbicide can bind to susceptible AHAS (C), inhibits the AHAS catalytic activity, and results in plant death. In contrast, the herbicide cannot bind to resistant AHAS (D) because of the altered AHAS structure resulting from amino acid substitution led by *AHAS* gene mutation. As a result, the herbicide does not inhibit catalytic activity of the resistant AHAS, and the resistant plant is not affected by the herbicide.

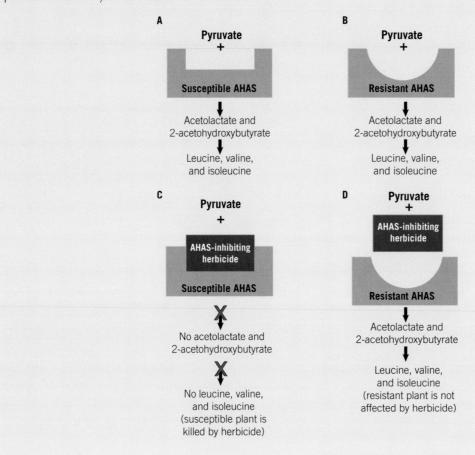

and imazethapyr; and examples of sulphonylurea herbicides are chlorsulphuron, rimsulphuron, and thifensulphuron-methyl.

The AHAS enzyme consists of large catalytic subunits and small regulatory subunits encoded by nuclear genes. Since maize, rice, soybean, sunflower and lentil are diploid; rapeseed allotetraploid; and bread wheat hexaploid; the number of *AHAS* gene loci is expected to be different from crop to crop. The polypeptide of the AHAS large subunit has about 670 amino acids. Each amino acid in the protein is determined by a set of three successive nucleotides in mRNA called a coding triplet or codon. Five commonly occurring mutation sites in the gene of the AHAS large subunit that confer tolerance to AHAS-inhibiting herbicides have been found in plants. The wild type amino acid and codon position of these mutation sites are A122, P197, A205, W574 and S653 in

reference to the *AHAS* gene of *Arabidopsis thaliana* (All codon or amino acid positions in this chapter are in reference to *Arabidopsis thaliana* unless stated otherwise) (**Table 31.2**). The five amino acids are folded in the adjacent area of the AHAS enzyme quaternary structure and this folding pocket has been proposed as the binding site of the AHAS-inhibiting herbicides where herbicides contact and interact with the AHAS enzyme.

Substitution of the amino acid at any of these five critical positions may prevent a herbicide from binding and inhibiting the AHAS enzyme because of protein structural change resulting from the amino acid substitution (**Box 31.1**). The altered AHAS enzyme resulting from the *AHAS* gene mutation is less sensitive or no longer sensitive to the herbicide. Thus the plant with the resistant AHAS has normal biosythesis of leucine, valine and isoleucine and is not affected by the herbicide

Figure 31.1 Sunflower plants with imidazolinone-tolerant A205V mutation of *AHAS* gene (left pot) were not affected by foliar spray of imazamox herbicide, but sunflower plants without the tolerant mutation (right pot) were killed by the same herbicide spray.

(**Box 31.1 and Figure 31.1**). Many *AHAS* gene mutants have been used to develop crops tolerant to AHAS-inhibiting herbicides. There have been no significant agronomical pleiotropic effects noted with the commercial *AHAS* gene mutations. To date, commercially available, mutated *AHAS* gene alleles in all crops are incompletely-dominant (partially-, semi- or co-dominant).

2.2. Acetyl-CoA Carboxylase (ACCase) and *ACCase* Gene Mutations

ACCase is a key enzyme of the fatty acid biosynthesis and catalyzes carboxylation of acetyl-CoA to form malonyl-CoA (**Table 31.2**). Phospholipids are the products of the fatty acid biosynthesis and are critical to the building of new membranes required for cell growth. The ACCase in the cytosol of all plants and also in the plastids and mitochondria of grass species is a homomeric enzyme that has all functional domains on a single polypeptide. In contrast, all other ACCases in the plant are heteromeric enzymes composed of four distinct subunits. The homomeric ACCase can be inhibited non-competitively by two chemical families of herbicides, aryloxyphenoxypropionates such as haloxyfop and cyclohexanediones such as cycloxydim and sethoxydim (**Table 31.2**).

The ACCase enzyme is encoded by a nuclear gene and consists of about 2,300 amino acids depending on the species. Five commonly occurring mutation positions have been found in the homomeric *ACCase* gene that confer tolerance to ACCase-inhibiting herbicides in plants (**Table 31.2**). The wild type amino acids and codon numbers of the five positions are I1781, W2027, I2041, D2078 and G2096 in reference to blackgrass (*Alopecurus myosuroides*), and they are located within the active site cavity of the ACCase's carboxyl transferase to which herbicides bind.

The ACCase enzyme with an amino acid substitution at position I1781 or D2078 tends to be resistant to both aryloxyphenoxypropionate and cyclohexanedione herbicides, but the altered ACCase from the mutation at position W2027, I2041 or G2096 tends to be more resistant or only resistant to aryloxyphenoxypropionate herbicides. The I1781L mutation of the *ACCase* gene discovered in maize has been used to develop cyclohexanedione-tolerant maize. The trait is incompletely dominant. There have been no significant agronomical pleiotropic effects noted with this mutation.

2.3. D1 Protein of Photosynthesis and *psbA* Gene Mutations

The D1 protein of photosynthesis II (PS II) facilitates electron transport from plastoquinone A (Q_A) to plastoquinone B (Q_B) in the PS II system, one of the two functional photosynthesis assemblies that is activated at shorter light wavelength and is responsible for oxygen evolution of photosynthesis (**Table 31.2**). The D1 protein can be bound by herbicides from several chemical families including triazines and bromoxynil. The herbicides bind the Q_B-binding niche on the D1 protein competitively and block the electron transport from Q_A to Q_B in PS II. The D1 protein is encoded by a chloroplastic gene called *psbA*. Several mutation sites of the *psbA* gene have been reported in plants. The two commonly-occurring mutation sites are V219 and S264 in reference to *Arabidopsis thaliana*. The S264G mutation of *psbA* gene has been used to develop triazine-tolerant rapeseed. Because the *psbA* gene is located in the chloroplast, the triazine-tolerance trait is inherited maternally. The triazine-tolerance trait in rapeseed tends to be associated with low yield and oil content and poor seedling vigor. This is the only herbicide-tolerance trait commercialized from a mutation that shows any agronomic penalty from the mutation.

3. Development of Herbicide-Tolerant Mutants

Spontaneous mutations and variants of the *AHAS*, *ACCase* and *psbA* genes occur commonly in natural plant populations. For instance, the spontaneous mutation frequency for resistance to AHAS-inhibiting herbicide in *Arabidopsis thaliana* was estimated to be of the magnitude of 10^{-9}. Imposing selection pressure by spraying inhibitors of AHAS, ACCas and D1 protein onto the plant population or adding the inhibitor in tissue culture medium will kill the plants or tissues that are susceptible to the inhibitors. In contrast, a plant or cell with a herbicide-tolerant variant of *AHAS, ACCase* or *psbA* genes resulting from spontaneous mutation will survive the selection pressure and can be easily identified with the herbicide treatment.

Several herbicide-tolerant mutants have been discovered through this simple selection method from either cell culture or farm field (**Table 31.3**). Cyclohexanedione-tolerant maize and several mutants of imidazolinone-tolerant maize and rice were selected from cell culture using a culture medium containing a cyclohexanedione or AHAS-inhibiting herbicide. Similarly, one of the imidazolinone- or sulphonylurea-tolerant sunflower mutants and the triazine-tolerant trait in rapeseed were discovered in farmer's fields from natural populations after herbicide spray.

Chemical mutagenesis followed by selection can increase mutation frequency and therefore improve the chance of discovering herbicide-tolerant mutants when compared with the method of selecting mutants solely from spontaneous mutation. Many breeders use chemical mutagens to create herbicide-tolerant mutants beyond spontaneous mutation. After the chemical mutagenesis, mutants that are tolerant to a herbicide are generally selected by treating the progeny of the mutagen-treated plants with the herbicide. In fact, most of the herbicide-tolerant mutants were created and selected through this two-step method of chemical mutagenesis followed by herbicide selection (**Table 31.3**).

Table 31.3: Creation and selection of herbicide-tolerant mutants in crops

Plant materials for mutagenesis or selection	Seed	Cell	Pollen	Microspore	Plant
Direct selection with herbicides					
Spontaneous mutation		Maize, rice			Sunflower, rapeseed
Mutagensis followed by herbicide selection					
Chemical mutagensis					
Ethyl methane sulphonate	Rice, wheat, lentil, soybean, sunflower		Maize		
Ethyl nitrosourea	Soybean			Rapeseed	
N-nitroso-N-methylurea	Soybean				
Sodium azide	Wheat				
Gamma irridation	Rice				

Several chemical mutagens including ethyl methane sulphonate (EMS), ethyl nitrosourea, N-nitroso-N-methylurea and sodium azide have been used in mutagenesis for treating seeds, pollen and microspores of a plant to develop herbicide-tolerant mutants (**Table 31.3**). Among the chemical mutagens used in the development of herbicide-tolerant mutants, EMS is the most popular. Treating dose, time and procedure of chemical mutagens varied depending on types of mutagens, crops and plant organs; and details about the mutagen treatment and subsequent selection of each herbicide-tolerant mutant will be discussed in the following sections of this chapter.

Imidazolinone-tolerant rice, wheat and lentil; sulphonylurea-tolerant soybean; and one of the imidazolinone- or sulphonylurea-tolerant sunflower mutants were developed by soaking seeds with chemical mutagens followed by herbicide selection (**Table 31.3**). In comparison, several imidazolinone-tolerant maize mutants were developed by soaking pollens with a chemical mutagen followed by pollinating maize plants with the mutagen-treated pollens and spraying M_1 plants with the herbicide. Similarly, imidazolinone-tolerant rapeseed was developed by treating microspores with a chemical mutagen followed by culturing the microspores in the medium containing an imidazolinone herbicide (**see Chapter 29**).

Gamma irradiation was also used in enhancing gene mutation frequency in herbicide-tolerant mutant development (**Table 31.3**). For example, rice seeds were treated with gamma rays to increase mutation frequency in the development of imidazolinone-tolerant rice. Gamma irradiation was also used in creating sulphonylurea-tolerant lentil. However, mutagenesis with gamma irradiation was not as popular, nor as successful as chemical mutagenesis in the development of herbicide-tolerant mutants. In contrast to gamma irradiation, chemical mutagenesis needs less expensive equipment and can be carried out more easily. Many breeders have access to a chemical mutagen, but not a gamma ray source. To date there has been no commercial herbicide-tolerant crop created and developed through gamma irradiation.

Mutagenesis including chemical mutagenesis may cause mutation of desirable genes at the same time of creating herbicide-tolerance mutation, resulting in undesirable characteristics of a mutant. However, if the mutated gene conferring undesirable phenotype is not closely linked to the gene conferring herbicide tolerance, it is easy to introduce the latter without the former into an elite variety through backcrossing and imposing selection pressure on each backcrossed generation with herbicide spray. After several generations of herbicide selection, any undesirable mutation from mutagenesis can be separated from the herbicide tolerance mutation.

4. Commercial Herbicide-Tolerant Crops Developed from Mutations

4.1. Herbicide-Tolerant Maize

4.1.1. Imidazolinone-Tolerant Maize

Cell culture and pollen mutagenesis were used in *in vitro* selection to develop imidazolinone-tolerant maize (**Table 31.4**). The cell calluses induced in culture of the maize A188 x B73 cross were cultured in the medium containing imazaquin to select imazaquin-tolerant cells. Several imidazolinone-tolerant lines were selected including XI12, XA17, QJ22 and XS40. An important feature of this type of *in vitro* selection is that since it is based on biochemical cell metabolism the whole plant is not required for selection purpose; hence selection can be carried out efficiently in the laboratory and not in the field during a limited growing season. The XI12 and XA17 lines have been used subsequently to develop commercial IT and IR maize hybrids. The imidazolinone-tolerant maize was commercialized in 1992 and is marketed as CLEARFIELD® maize.

The second method was pollen mutagenesis by using a chemical mutagen followed by selection with the herbicide. Fresh pollen grains of the maize line UE95 were added to 1% ethyl methane sulphonate solution and were shaken vigorously for 30 seconds and then shaken four or five times every three minutes for 40 minutes. The treated pollen was then brushed onto the silks of de-tasselled UE95 for pollination. The M_1 seeds were planted and sprayed with imazethapyr. Several imidazolinone-tolerant maize lines including Mutant 1 and Mutant 2 were successfully selected. Subsequently, imidazolinone-tolerant maize developed from this source has been also marketed.

Maize has two *AHAS* gene loci, als1 on the long arm of chromosome four and als2 on the long arm of chromosome five. XI12 and XA17 are from mutations of the

Table 31.4: Mutations of the *AHAS* gene conferring tolerance to AHAS-inhibiting herbicides

| Crop and mutant | Mutation | | | Mutagenesis and selection with herbicide | | Reported herbicide tolerance | Reference |
	Amino acid substitution[a]	Nucleotide substitution	Locus	Chemical mutagen	Plant material and method		
Maize (*Zea mays*)							
XI12	S653N	G to A	als2	No	Cell culture	Imidazolinone	Dietrich, 1998
QJ22	S653N	G to A	als1	No	Cell culture	Imidazolinone	Dietrich, 1998
Mutant 2	S653N	G to A	No data	EMS	Pollen mutagenesis	Imidazolinone	Bright *et al.*, 1992
XA17	W574L		als2	No	Cell culture	Imidazolinone, other AHAS inhibitors	Bernasconi *et al.*, 1996
Mutant 1	A122T	G to A	No data	EMS	Pollen mutagenesis	Imidazolinone	Bright *et al.*, 1992
Rice (*Oryza sativa*)							
93AS3510	G654E	G to A	No data	EMS	Seed mutagenesis	Imidazolinone	Croughan, 2005
Several	S653N	G to A	No data	EMS	Seed mutagenesis	Imidazolinone	Croughan, 2005
Kinmaze mutant	W574L + S653I		No data	No	Cell culture	Pyrimidinyloxy-benzoate and other AHAS inhibitors	Shimizu *et al.*, 2006
Wheat (*Triticum aestivum*)							
Several	S653N	G to A	ALS1	Sodium azide, EMS	Seed mutagenesis	Imidazolinone	Pozniak *et al.*, 2004
TeaIMI 11A	S653N	G to A	ALS2	EMS	Seed mutagenesis	Imidazolinone	Pozniak *et al.*, 2004
Rapeseed (*Brassica napus*)							
PM1	S653N	G to A	AHAS1	Ethyl nitrosourea	Microspore mutagenesis	Imidazolinone	Barnes and Vanstraelen, 2009
PM2	W574L	G to T	AHAS3	Ethyl nitrosourea	Microspore mutagenesis	Imidazolinone and sulphonylurea	Barnes and Vanstraelen, 2009
Sunflower (*Helianthus annuus*)							
IMISUN	A205V	C to T	AHAS1	No	Whole plant	Imidazolinone	Kolkman *et al.*, 2004
CLHA-PLUS	A122T	G to A	AHAS1	EMS	Seed mutagenesis	Imidazolinone	Sala *et al.*, 2008
SURES	P197L	C to T	AHAS1	No	Whole plant	Sulphonylurea	Kolkman *et al.*, 2004
Several	No data		No data	EMS	Seed mutagenesis	Sulphonylurea	Gabard and Huby, 2004
Lentil (*Lens culinaris*)							
RH44	No data		One of two loci	EMS	Seed mutagenesis	Imidazolinone	Slinkard *et al.*, 2007
Soybean (*Glycine max*)							
W20	P197S		als1	NMU	Seed mutagenesis	Sulphonylurea	Sebastian, 1992; Green, 2007
W4-4	Two mutations at two loci			NMU	Seed mutagenesis	Sulphonylurea, Imidazolinone	Sebastian, 1992

[a] In reference to *Arabidpsis thaliana*.

als2, but QJ22 and XS40 are from mutations of the als1 (**Table 31.4**). A single nucleotide change or substitution in DNA can lead to gene mutation, subsequently codon change in mRNA and consequently amino acid sequence change of a protein. The X112 and QJ22 have an altered AHAS enzyme with S653N amino substitution (from serine to asparagine or S to N at codon 653) encoded by an *AHAS* gene mutation resulting from a single nucleotide substitution from AGT to AAT at codon 653 (**Table 31.4**). Similarly, the Mutant 2 from the pollen mutagenesis has the same S653N *AHAS* gene mutation resulting from a base pair substitution from G to A at codon 653. In contrast, the XA17 from the cell selection has the W574L mutation of the *AHAS* gene (**Table 31.4**). Mutant 1 from the pollen mutagenesis has the A122T mutation resulting from a single nucleotide substitution from G to A at codon 122 (**Table 31.4**). The amino acid substitution of A122T, W574L or S653N modifies the herbicide-binding pocket of the AHAS enzyme and prevents the herbicide from binding and inhibiting the enzyme.

4.1.2. Aryloxyphenoxypropionate- and Cyclohexanedione-Tolerant Maize

Similar to imidazolinone-tolerant maize, aryloxyphenoxypropionate- and cyclohexanedione-tolerant maize lines were also discovered *in vitro* from herbicide selection of maize cells (**Table 31.5**). Several sethoxydim-tolerant callus lines including S1, S2, S3 and S4 and haloxyfop-tolerant callus lines including H1 and H2 have been selected from a maize tissue culture of A188 x B73 cross in a medium containing sethoxydim or haloxyfop. Cyclohexanedione-tolerant inbred lines were developed by backcrossing public inbred lines with plants regenerated from the S2 callus cultures. The materials were transferred to maize breeding companies in 1990, and sethoxydim-tolerant maize hybrids were developed from the trait and were first marketed in 1996. Subsequently cycloxydim-tolerant maize hybrids have been also developed using the same mutation, they are marketed in Europe as DUO System and are increasingly adopted by farmers.

Two distinct ACCase isoforms, ACCase I and ACCase II, have been found in maize. ACCase I is sensitive to herbicide, while ACCase II is relatively insensitive to or less inhibited by sethoxydim or haloxyfop than ACCase I. The DNA analysis of the sethoxydim- and haloxyfop-tolerant maize DK592 showed that the *ACCase* gene, presumably the *ACCase I* gene, has a single nucleotide substitution at codon 1781 in reference to blackgrass

or at codon 1769 in reference to wheat (**Table 31.5**). Codon nucleotides TTA replaced ATA, and as a result, the encoded ACCase enzyme has the I1781L amino acid substitution. The alteration of the ACCase prevents some aryloxyphenoxypropionate and cyclohexanedione herbicides from binding and inhibiting the enzyme and makes the plant tolerant to the herbicides.

4.2. Imidazolinone-Tolerant Rice

The commercial imidazolinone-tolerant rice was developed from two seed mutagenesis projects (**Table 31.4**). In the first project, 1.5 kg seeds of the rice variety AS3510 were treated with 0.5% EMS solution for 16 hours immediately prior to planting. The M$_2$ plants were treated with imazethapyr to select imidazolinone-tolerant plants. A single rice plant survived the herbicide treatment and was designated as 93AS3510. The imidazolinone-tolerance trait of 93AS3510 was used to breed CLEARFIELD® rice varieties CL121 and CL141, which were commercialized in 2001. Exposing rice seeds to gamma rays was also attempted in this project but did not result in any commercial herbicide-tolerant mutant.

In the second project, 136 kg seeds of the rice variety Cypress were treated with 0.175% EMS aqueous solution for 23 hours. The M2 plants were sprayed post-emergence with imazapyr or imazapic, and seven tolerant lines including PWC16, PWC23, CMC29, CMC31, WDC33, WDC37 and WDC38 were selected. Several commercial, imidazolinone-tolerant rice varieties and hybrids such as CL161 and XL8 have been bred by using the mutation from this source. Crop breeders in the US often apply for plant variety protection for their new rice varieties, readers can get information on more CLEARFIELD® rice varieties from the USDA plant variety protection office website listed in reference section of this chapter.

Imidazolinone-tolerant rice mutants used for commercial variety development have a single nucleotide substitution in their *AHAS* gene that is responsible for the imidazolinone tolerance. The line 93AS3510 has the nucleotide substitution from GGG to GAG at codon 654, resulting in the G654E amino acid substitution of the AHAS enzyme (**Table 31.4**). In comparison, the lines PWC16, PWC23, CMC29, WDC33 and WDC38 have the S653N amino acid modification of the AHAS enzyme resulting from the nucleotide substitution of AGT with AAT at codon 653 (**Table 31.4**). Both G654E and S653N mutations confer good tolerance to imidazolinone herbicides.

Besides the commercial imidazolinone-tolerant rice mutations, a double mutation of the *AHAS* gene has also been reported in rice. This double mutation was discovered from a cell culture of the rice cultivar Kinmaze in medium containing bispyribac-sodium, a pyrimidinyloxy-benzoate herbicide. The *AHAS* gene has double mutations of W574L and S653I and confers higher resistance to bispyribac-sodium than to imazaquin or chlorsulphuron (**Table 31.4**).

4.3. Imidazolinone-Tolerant Wheat

Similar to imidazolinone-tolerant rice, imidazolinone-tolerant wheat was developed from two seed mutagenesis projects using chemical mutagens (**Table 31.4**). In the first project, 5,000 seeds of the winter wheat variety Fidel were treated with 1 mM sodium azide solution for two hours after being soaked in water for 24 hours without and then with air bubbling. The M_2 seeds were seed-treated with the herbicide imazethapyr and then followed by a soil spray of imazethapyr. Four plants, FS1, FS2, FS3 and FS4, survived the imidazolinone treatments; their phenotype and enzyme tolerance profiles were nearly identical. The tolerance mutation from this source has been used as a trait donor for the imidazolinone-tolerant wheat varieties that have been marketed as CLEARFIELD® wheat since 2001.

In the second seed mutagenesis project, approximately 40,000 seeds of the spring wheat variety CDC Teal were treated with 0.3% EMS solution for six hours after being soaked in distilled water for four hours. The M_2 plants were sprayed with herbicide imazamox as the selection agent. Several plants including TealIMI 11A and 15A survived the herbicide treatment, and a new mutation was discovered from TealIMI 11A. The tolerance trait from this source has also been used to develop imidazolinone-tolerant wheat. Several dozens of CLEARFIELD® winter and spring wheat varieties have been commercialized such as ORCF-102, Infinity CL and AP604 CL. Readers can get information about more CLEARFIELD® wheat varieties in the US from the USDA plant variety protection office website listed in reference section of this chapter.

Wheat has three *AHAS* gene loci, ALS1, ALS2 and ALS3, and they are located on the long arm of chromosomes 6D, 6B and 6A, respectively. The FS and TealIMI 11A mutants have an *AHAS* gene mutation on genomes D and B, respectively, and the mutation causes a single nucleotide substitution from G to A at codon 653, resulting in the amino acid substitution of S653N of the AHAS enzyme (**Table 31.4**). Similar to the S653N mutation in maize and rice, the S653N mutation in wheat also confers good tolerance to imidazolinone herbicides.

4.4. Herbicide-Tolerant Rapeseed

4.4.1. Imidazolinone-Tolerant Rapeseed

Imidazolinone-tolerant rapeseed was developed from chemical mutagenesis of microspores (**Table 31.4**). Microspores of the rapeseed variety Topas were isolated and treated with 20 µM ethyl nitrosourea, a chemical mutagen. The treated microspores were cultured in the medium containing imazethapyr to produce embryos, and the survived embryos then grew into plantlets. The plantlets were transferred to soil, and the chromosomes of the plantlets were doubled with colchicine. Two lines exhibited excellent tolerance to imazethapyr and were designated as P1 and P2, also referred to as PM1 and PM2. Both PM1 and PM2 have been subsequently used as trait donors for breeding imidazolinone-tolerant rapeseed that was commercialized in 1995 and is marketed as CLEARFIELD® canola.

There are five *AHAS* gene loci in rapeseed. AHAS1 and AHAS5 are located on genome C, but AHAS2, AHAS3 and AHAS4 are located on genome A. The PM1 has a single nucleotide modification from G to A at codon 653, resulting in the S653N amino acid substitution of the AHAS1 enzyme (**Table 31.4**). In contrast, the PM2 has a single base-pair change from G to T at codon 574, resulting in the W574L amino acid substitution of the AHAS3 enzyme. Similar to the case in maize, both S653N and W574L mutations of the *AHAS* gene in rapeseed confer good tolerance to imidazolinone herbicides and are responsible for the imidazolinone tolerance observed with PM1 and PM2 mutants. However, PM2 is much more tolerant to imidazolinone herbicides than PM1.

4.4.2. Triazine-Tolerant Rapeseed

Triazine-tolerance in rapeseed was derived from spontaneous mutations found in bird's rape (*Brassica rapa* or *B. campestris*). A triazine-tolerant bird's rape biotype was originally discovered in two maize fields in Quebec, Canada. Repeated backcrossing of the rape-

Table 31.5: Gene mutations conferring cyclohexanedione and triazine tolerance

| Gene | Crop | Mutant | Mutation | | | Mutagenesis and selection with herbicide | | Reported herbicide tolerance | Reference |
			Amino acid substitution	Nucleotide substitution	Locus	Chemical mutagen	Material and method		
ACCase gene	Maize	S2	I1781L [a]	A to T	ACCase I	No	Cell culture	Aryloxy-phenoxy-propionate, cyclohexa-nedione	Zagnitko *et al.,* 2001
psbA	Rapeseed	ATR-5Tw	S264G [b]	A to G	in chloro-plast	No	Inter-specific crossing with Brassica rapa [c]	Triazine	Sundby *et al.,* 1993

[a] In reference to blackgrass (*Alopecurus myosuroides*) or I1769L in reference to wheat (*Triticum aestivum*).
[b] In reference to *Arabidopsis thaliana*.
[c] The triazine-tolerant *Brassica rapa* (*B. campestris*) was discovered from maize fields in Quebec, Canada.

seed cultivar Tower (*B. napus*) with the resistant *B. rapa* biotype resulted in a triazine-tolerant *B. napus* germplasm referred to as ATR-5Tw. The ATR-5Tw germplasm was released and used to develop triazine-tolerant rapeseed cultivars that have been marketed since the 1980s. Triazine-tolerant rapeseed varieties tend to have lower yield and oil content than conventional varieties. The farmer has therefore to consider the tradeoff between good weed control with triazine herbicides and the potential reduction of yield or oil content by using triazine-tolerant varieties.

The mutation conferring triazine tolerance in *B. napus* occurred at the chloroplastic *psbA* gene that encodes the D1 protein of photosynthesis II. The nucleotide substitution from AGT to GGT at codon 264 resulted in the S264G alteration of the D1 protein (**Table 31.5**). Because of the substitution, triazines are no longer able to bind to the D1 protein and compete with Q_B, and consequently cannot interrupt the electron transport of PS II and injure the plant.

4.5. Sunflowers Tolerant to AHAS Inhibitors

4.5.1. Imidazolinone-Tolerant Sunflower

Two imidazolinone-tolerance traits have been discovered in common sunflower. One was from a field whole plant selection, and the other from seed mutagenesis (**Table 31.4**). An imidazolinone-tolerant wild sunflower biotype with a spontaneous *AHAS* gene mutation was discovered in a soybean field in Kansas, USA. The tolerant wild sunflower was crossed with several domestic sunflower lines. As a result, several imidazolinone-tolerant germplasm lines including IMISUN-1, IMISUN-2, IMISUN-3 and IMISUN-4 were created. The seeds of these lines were distributed to sunflower breeders for developing imidazolinone-tolerant sunflower hybrids. Several dozens of imidazolinone-tolerant sunflower hybrids carrying this trait have been developed and commercialized as CLEARFIELD® sunflower since 2003.

Three *AHAS* genes, *AHAS1*, *AHAS2* and *AHAS3*, have been identified in common sunflower. The IMISUN mutant has a point mutation on the *AHAS1* gene with the base-pair substitution from GCG to GTG at codon position 205 and the amino acid substitution of A205V of the encoded AHAS1 enzyme (**Table 31.4**). Similar to the mutant discovered in Kansas, another imidazolinone-tolerant wild sunflower biotype was found in South Dakota, USA and also has the A205V *AHAS* gene mutation.

Surprisingly, selfed-progeny of an intermediately tolerant phenotype of the Kansas mutant was observed not to segregate further, hinting that more than one gene may contribute to the total tolerance. The inheritance of the imidazolinone-tolerance trait from Kansas appears to be additively controlled by two genes. It has been also proposed that the tolerance is controlled by the partially dominant A205V *AHAS* gene and a modifier gene that confers no tolerance by itself but can enhance the former. However, the second additive or modifier gene is yet to be found.

For the seed mutagenesis project, approximately 60,000 seeds of sunflower line BTK47 were treated with 0.25% EMS for 15 hours, then soaked in 1.5% sodium thiosulphate solution for 0.5 hours, and finally rinsed continually with tap water for 1 hour. After M_2 was sprayed with herbicide imazapyr, eight resistant plants were identified. After several generations of screening, one of the resistant lines closely resembling BTK47 attributes was selected and designated as CLHA-PLUS (**Table 31.4**). CLHA-PLUS was found to have an *AHAS1* gene mutation with the nucleotide substitution of GCG to ACG at codon 122, resulting in A122T amino acid substitution of the AHAS1 enzyme (**Table 31.4**). The mutated *AHAS1* gene allele in CLHA-PLUS is partially dominant and confers a higher level of imidazolinone tolerance than the allele in IMISUN.

4.5.2. Sulphonylurea-Tolerant Sunflower

Similar to imidazolinone-tolerant sunflower, two sulphonylurea-tolerance traits were discovered in common sunflower. One was from a field whole plant selection, and the other from seed mutagenesis (**Table 31.4**). A sulphonylurea-tolerant wild sunflower biotype was discovered in Kansas, USA, and the tolerance trait was introduced into domestic sunflower lines and resulted in the germplasm lines SURES-1 and SURES-2. Tolerance is conferred by a mutation on the *AHAS1* gene, and codon position is 197 with a codon base-pair substitution from CCC to CTC and an amino acid substitution from proline to leucine in the encoded AHAS1 enzyme (**Table 31.4**).

Seed mutagenesis of sunflower was carried out by treating the seeds of sunflower line HA89B with 20–35 mM EMS for 18 hours under continuous stirring (**Table 31.4**). The M_2 plants were subjected to selection pressure by being grown in a hydroponic screening system containing thifensulphuron-methyl or rimsulphuron. As a result, several sulphonylurea-tolerant lines including M7 were selected. Subsequent AHAS assays showed that the AHAS enzyme of all mutants was tolerant to sulphonylureas. Apparently, the sulphonylurea

tolerance is also derived from a mutated *AHAS* gene. However, there is no published information available about the DNA sequence of the mutated *AHAS* gene and its locus in M7. Sulphonylurea-tolerant sunflower has been developed and commercialized from this source.

4.6. Imidazolinone-Tolerant Lentil

The imidazolinone-tolerant lentil trait was developed from seed mutagenesis (**Table 31.4**). Five kilogrammes of mixed lentil seeds were treated in EMS solution and then planted in the field. An imidazolinone-tolerant line designated as RH44 was discovered from a population of approximately one million M_3 seeds using imidazolinone herbicides as the selection agent. The RH44 line showed an increased imidazolinone resistance as compared with the conventional lentil variety CDC Richlea, but showed little resistance to sulphonylureas. The RH44 line has been used as an imidazolinone-tolerance trait donor to develop imidazolinone-tolerant lentil varieties that were first commercialized as CLEARFIELD® lentil in 2006. Besides chemical mutagenesis, gamma irradiation was also reported to generate sulphonylurea-tolerant lentils.

An enzyme assay showed an increased tolerance of RH44 AHAS enzyme to imazapic or imazethapyr when compared with the conventional variety CDC Richlea, indicating that the tolerance is a result of *AHAS* gene mutation. However, there is no published information available about the DNA sequence of the mutated *AHAS* gene. Lentil is believed to have two *AHAS* genes, and one of them mutated in RH44. The tolerance trait is incompletely dominant.

4.7. Sulphonylurea-Tolerant Soybeans

Sulphonylurea-tolerant soybeans have been developed through chemical seed mutagenesis followed by herbicide screening (**Table 31.4**). Seeds of soybean varieties Williams, Williams 82 and A3205 were soaked with mutagen EMS or N-nitroso-N-methylurea (NMU) to increase mutation frequency. Among the sulphonylurea-tolerant soybean mutants selected, W20 was derived from Williams treated with NMU and has been used as a trait donor to develop sulphonylurea-tolerant soybeans. Sulphonylurea-tolerant soybean varieties have been marketed as STS® soybeans since 1994.

The tolerance of W20 is conferred by a mutation on

als1 locus at codon 197 with an amino acid substitution from proline to serine in the encoded AHAS enzyme (**Table 31.4**). Backcrossing of W20 with Williams 82 and several other soybean lines showed that the mutated allele (*Als1*) is inherited dominantly at the whole-plant level. However, AHAS enzyme assays showed clear evidence for a dosage effect of the mutated allele, indicating that the tolerance trait is semi-dominant.

In addition to the mutation in W20, a second sulphonylurea-tolerant *AHAS* gene mutation named *Als2* was also found in the second-round NMU mutagenesis of the soybean seeds that already possessed one mutation for herbicide resistance, and the "second generation" mutant is designated as W4-4 (**Table 31.4**). The *Als2* resides at a second locus, segregates independently from the first mutation, and is inherited semi-dominantly. The W4-4 carries two mutations, has a higher level of sulphonylurea resistance than W20, and is cross-resistant to an imidazolinone herbicide. Besides *AHAS* gene mutations, four non-target based and sulphonylurea-tolerant soybean mutants were also discovered from seed mutagenesis of variety Williams, through using ethylnitrosourea as the mutagen. The tolerance is conferred by a single recessive gene in all four mutants, and each mutation resides at one of three loci.

In summary, spontaneous and mutagen-induced mutations have been successfully selected and utilized in developing crops tolerant to several classes of herbicides. All herbicide-tolerant mutants used in commercial HTCs were derived from a single nucleotide substitution of genes that encode tolerant enzymes or proteins targeted by herbicides. The alleles of all commercial herbicide-tolerant mutations are incompletely dominant and not pleiotropic, except for the triazine-tolerant mutation that is inherited maternally and linked with several agronomic traits. To confer commercial tolerance to herbicides, some herbicide-tolerant alleles can be heterozygous, others need to be homozygous, and the rest must be stacked with another tolerant gene. The herbicide-tolerance trait can be incorporated in an elite variety through crossing a trait donor with the elite variety.

5. References

5.1. Cited References

Barnes, S. and Vanstraelen, S. 2009. Assay for imidazolinone resistance mutations in *Brassica* species. US patent 7595177. http://www.uspto.gov/patft/index.html.

Bernasconi, P., Woodworth, A.R., Rosen B.A. *et al.* 1996. A naturally occurring point mutation confers broad range tolerance to herbicides that target acetolactate synthase. *J. Biol. Chem. correction 271: 13925 and original. 270:* 17381–17385.

Bright, S.W.J., Chang, M.T., Evans, I.J. *et al.* 1992. Herbicide resistant plants. *Patent Application of World Intellectual Property Organization.* WO92/08794.

Croughan, T.P. 2005. Resistance to acetohydroxyacid synthase-inhibiting herbicides. US patent 6943280. http://www.uspto.gov/patft/index.html.

Dietrich, G.E. 1998. Imidazolinone resistant AHAS mutants. US Patent 5767361. http://www.uspto.gov/patft/index.html.

Duke, S.O. 2005. Taking stock of herbicide-resistant crops ten years after introduction. *Pest Manag. Sci.* 61: 211–218.

Gabard, J.M. and Huby, J.P. 2004. Sulphonylurea-tolerant sunflower line M7. US patent 6822146. http://www.uspto.gov/patft/index.html.

Green, J. 2007. Review of glyphosate and ALS-inhibiting herbicide crop resistance and resistant weed management. *Weed Technol.* 21: 547–558.

Green, J. 2009. Evolution of glyphosate-resistant crop technology. *Weed Sci.* 57: 108–117.

Kolkman, J.M., Slabaugh, M.B., Bruniard, J.M. *et al.* 2004. Acetohydroxyacid synthase mutations conferring resistance to imidazolinone or sulphonylurea herbicides in sunflower. *Theor. Appl. Genet.* 109: 1147–1159.

Pozniak, C.J., Birk, I.T., O'Donoughue, L.S. *et al.* 2004. Physiological and molecular characterization of mutation-derived imidazolinone resistance in spring wheat. *Crop Sci.* 44: 1434–1443.

Sala, C.A., Bulos, M., Echarte, M. *et al.* 2008. Molecular and biochemical characterization of an induced mutation conferring imidazolinone resistance in sunflower. *Theor. Appl. Genet.* 118: 105–112.

Sebastian, S.A. 1992. Soybean plants with dominant selectable trait for herbicide resistance. US patent 5084082. http://www.uspto.gov/patft/index.html.

Shimizu, T., Nakayama, I., Nagayama, K. *et al.* 2006. Gene encoding acetolactate synthase. US patent 7119256. http://www.uspto.gov/patft/index.html.

Slinkard, A.E., Vanderberg, A. and Holm, F. 2007. Lentil plants having increased resistance to imidazolinone herbicides. US patent 7232942. http://www.uspto.gov/patft/index.html.

Sundby, C., Chow, W.S. and Anderson, J.M. 1993. Effects on photosystem II function, photoinhibition, and plant performance of the spontaneous mutation of serine-264 in the photosystem II reaction center D1 protein in triazine-resistant *Brassica napus* L. *Plant Physiol.* 103: 105–113.

Tan, S., Evans, R.R., Dahmer, M.L. *et al.* 2005. Imidazolinone-tolerant crops: history, current status and future. *Pest Manag. Sci.* 61: 246–257.

Zagnitko, O., Jelenska, J., Tevzadze, G. *et al.* 2001. An isoleucine/leucine residue in the carboxyltransferase domain of acetyl-CoA carboxylase is critical for interaction with aryloxyphenoxypropionate and cyclohexanedione inhibitors. *PNAS.* 98: 6617–6622.

5.2. Websites

ACCase and ALS mutation tables:

http://www.weedscience.com

Canadian Food Inspection Agency. Plants evaluated for environmental and livestock feed safety:

http://active.inspection.gc.ca/eng/plaveg/bio/pnt-vcne.asp

Center for Environmental Risk Assessment. GM crop database:

http://www.cera-gmc.org/?action=gm_crop_database

Health Canada. Novel food decisions:

http://www.hc-sc.gc.ca/fn-an/gmf-agm/appro/index-eng.php

Herbicide Resistance Action Committee. Classification of herbicides according to mode of action:

http://www.hracglobal.com/

USDA. Certificate status database of plant variety protection office:

http://www.ars-grin.gov/cgi-bin/npgs/html/pvplist.pl

5.3. Further Reading

Duggleby, R.G., McCourt, J.A. and Guddat, L.W. 2008. Structure and mechanism of inhibition of plant acetohydroxyacid synthase. *Plant Physi. Biochem.* 46: 309–324.

Duke, S.O. (ed.). 1996. Herbicide Resistant Crops, Boca Raton, Fl: CRC Press.

Gressel, J. 2002. Molecular biology of Weed Control. London: Taylor & Francis.

Mallory-Smith, C.A. and Retzinger, Jr E.J. 2003. Revised classification of herbicides by site of action for weed resistance management strategies. *Weed Technol.* 17: 605–619.

Newhouse, K., Wang, T. and Anderson, P. 1991. Imidazolinone-tolerant crops. *In*: D.L. Shaner and S.L. O'Conner (eds.) Imidazolinone Herbicides. Boca Raton, Fl: CRC Press. pp. 139–150.

Powles, S.B. and Holt, J.A. 1994. Herbicide Resistance in Plants: Biology and Biochemistry. Boca Raton, Fl: CRC Press.

Preston, C. and Mallory-Smith, C.A. 2001. Biochemical mechanisms, inheritance, and molecular genetics of herbicide resistance in weeds. *In*: S.B. Powles and D.L. Shaner (eds.) Herbicide Resistance and World Grains. Boca Raton, Fl: CRC Press. pp. 23–60.

Reade, J.P.H. and Cobb, H. 2002. Herbicides: modes of action and metabolism. *In*: R.E.L. Naylor (ed.) Weed Management Handbook. Oxford: Blackwell Science. pp. 134–170.

Tan, S., Evans, R. and Singh, B. 2006. Herbicidal inhibitors of amino acid biosynthesis and herbicide-tolerant crops. *Amino Acids.* 30: 195–204.

C32

Mutation Breeding for Fatty Acid Composition in Soybean

Y.Takagi[a], T.Anai[a] and H.Nakagawa[b,*]

[a] Laboratory of Plant Genetics and Breeding, Faculty of Agriculture, Saga University, Honjyo-machi 1, Saga 840-8502, Japan
[b] Institute of Radiation Breeding, National Institute of Agrobiological Sciences, Kami-Murata, Hitachi-Ohmiya, Ibaraki 319-2293, Japan
 Present address: Biomass Research & Development Center, National Agriculture and Food Research Organization, 3-1-1 Kannondai, Tsukuba 305-8517, Japan
* Corresponding authors, E-MAIL: sukeyuki@po.bunbun.ne.jp, anai@cc.saga-u.ac.jp, ngene@affrc.go.jp

1. Introduction

Soybean is one of the most important oil crops providing about 30% of the world's oil production. It is widely used for cooking and numerous industrial processes. Differences in the fatty acid composition of soybean oil affect the physical, chemical and nutritional characters of its products. Soybean oil of common varieties is composed of, on average, 25.0% oleic acid, 52.0% linoleic acid and 8.0% linolenic acid. **Table 32.1** shows the fatty acid composition of different vegetative oils and their relative market prices. The fatty acid composition of soybean oil is characterized as low oleic acid content (with high linoleic acid) and high linolenic acid content. Due to this composition, the price of soybean oil is typically lower than other oil crops.

The relative ratio of saturated and unsaturated fatty acids determines the property of seed oils. Oils with a high content of the mono-unsaturated fatty acid, oleic acid, have greater stability during high-temperature heating in cooking than oils with a normal oleate content. Soybean oil with a high content of poly-unsaturated fatty acids (linoleic acid and linolenic acid) produces an undesirable flavour and odour when maintained for long periods at high room temperature. Hydrogenation treatment is used to improve the oil stability by reducing its linolenic acid content; however, during this process, trans fatty acids are produced, which have been associated with an increased risk of coronary disease in humans. The United States Food and Drug Administration (1999) expressed the need for soybean oil that exhibits a total content of linolenic acid of less than 1% in order to reduce the total levels of *trans*-type fatty acid generated by the hydrogenation treatment. As a result, oils from sunflower, safflower and olive with high oleic acid content are considered to be of superior quality.

Various breeding programmes have been set up during the past two decades, which aim to modify soybean fatty acid composition, including a mutation breeding programme. Seed X-ray irradiation of the var. Bay has produced several mutants with varying fatty acid composition and the underlying genes and mutations have been investigated. A summary of mutants that express unique fatty acid composition is provided along with other achievements in fatty acid improvement in soybean research.

2. Diversity of Fatty Acid Composition and its Genetic Control

2.1. Fatty Acids Biosynthesis

Every oil crop produces unique oils with a specific fatty acid composition (**Table 32.1**). In the seed of traditional rapeseed, the levels of eicosenoic acid and erucic acid are 14.5% and 41.0%, respectively, and these levels are higher than those found in the seed of other oil seed crops. The fatty acid composition of oil crops such as soybean, maize, traditional sunflower and traditional safflower varieties exhibit higher levels of linoleic acid (50–80%), and lower levels of oleic acid (10–30%). Linolenic acid is not found in the oils of maize, traditional sunflower and traditional safflower, but it is as high as 8.0% in soybean. Olive oil is characterized by a higher content of oleic acid (76.2%) and a lower level of linoleic acid (5.5%). Palm oil is characterized by higher levels of palmitic acid (43.0%) and oleic acid (44.2%). Essentially, fatty acid composition is quite diverse within and across oil crops and between the traditional (non-improved) and improved varieties. Notable examples are zero erucic acid rapeseed, high oleic sunflower and high oleic safflower.

The biosynthesis of fatty acids, common in all the plant species, involves a carbon elongation process: palmitic acid (16 carbons:0 double bonds) → stearic acid (18:0), and a desaturation reaction process, that is stearic acid (18:0) → oleic acid (18:1) → linoleic acid (18:2) → linolenic acid (18:3) (**Figure 32.1**). Oleic acid (18:1) can be further extended to eicosenoic acid (20:1) and erucic acid (22:1) through a carbon chain elongation process that is unique to traditional rapeseed. The fatty acid composition of a particular variety or species is largely determined by the genes coding for enzymes involved in the biosynthesis of individual fatty acids.

The incorporation of "target genes", known to modify the fatty acid components of oil crops, were initially identified in natural populations of a specific species, and then incorporated into breeding materials. For example, high levels of erucic acid were controlled by 2 pairs of genes ($E_1E_1E_2E_2$) acting additively in the biosynthesis of eicosenoic acid (20:1) and erucic acid (22:1) in traditional rapeseed. In a screen of a diverse array of world germplasm collections, a gene was identified and eventually used to develop "zero erucic" rapeseed, which has no erucic acid. In "zero erucic" rapeseed vari-

Table 32.1: Fatty acid composition (% of total oil) of different vegetable oils, and their market prices

Seed oil	Fatty acids							
	Palmitic (16:0)	Stearic (18:0)	Oleic (18:1)	Linoleic (18:2)	Linolenic (18:3)	Eicosenoic (20:1)	Erucic (22:1)	Price (US$/18 L) [a]
Rapeseed (traditional)	4.0	1.5	17.0	13.0	9.0	14.5	41.0	42
Rapeseed (zero erucic)	9.2	1.6	59.1	19.4	10.2	0.2	0.3	26
Soybean	11.5	3.9	24.6	52.0	8.0	–	–	25
Maize	12.1	2.3	28.7	56.2	0.7	–	–	42
Sunflower (traditional)	6.4	4.8	19.6	68.7	<0.5	–	–	25
Sunflower (mid oleic)	4.5	4.4	54.3	36.3	<0.5	–	–	33
Sunflower (high oleic)	3.6	4.2	81.4	10.3	<0.5	–	–	50
Safflower (traditional)	7.2	2.1	9.7	81.0	0.0	–	–	58
Safflower (high oleic)	5.4	2.3	80.0	12.3	0.0	–	–	58
Olive	14.6	3.1	76.2	5.5	0.6	–	–	92
Palm	43.0	4.1	44.2	8.7	0.0	–	–	21

[a] The price is based on the data in 2004.

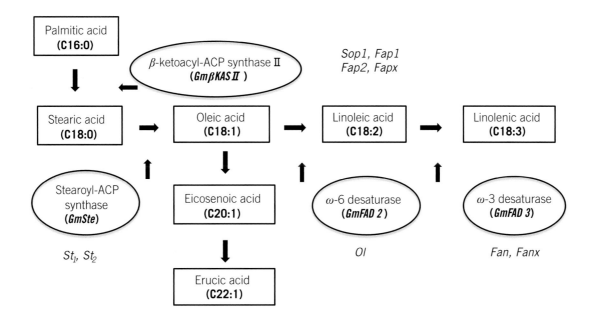

Figure 32.1 Genes responsible for different fatty acid biosynthesis.

eties, the recessive alleles $e_1e_1e_2e_2$ are present and the carbon elongation reactions (oleic acid → eicosenoic acid→ erucic acid) are blocked.

In a second example, a gene regulating high oleic acid content of safflower was discovered in a landrace from Pakistan and transferred to commercial varieties.

The alteration of the fatty acid composition of safflower is provided and compared to other oil seed crops in **Table 32.1.** The 10 to 80% increase in oleic acid content of safflower was considered a breeding success. As a result, the linoleic acid content in a newly derived safflower variety was reduced to approximately 10%. In this

Table 32.2: Fatty acid composition of different soybean mutants, and their genotypes

Mutant lines	Fatty acid (% of total oil)					Genotype
	Palmitic (16:0)	Stearic (18:0)	Oleic (18:1)	Linoleic (18:2)	Linolenic (18:3)	
J3	6.1	3.4	26.5	55.4	8.6	*sop1*
C1726	8.5	3.9	22.1	56.0	9.6	*fap1*
LPKKC-3	4.4	3.2	27.1	55.5	9.8	*sop1, fap1*
KK7	14.2	5.0	23.3	49.5	7.9	*fapx*
J10	17.2	5.0	19.7	48.4	9.7	*fap2*
HPKKJ10	21.6	5.7	18.0	45.6	9.0	*fapx, fap2*
KK2	10.3	7.2	23.1	51.6	7.9	st_1
M25	9.5	21.2	15.6	44.0	9.8	st_2
M25KK2	9.4	31.4	12.4	38.4	8.4	st_1, st_2
M11	10.3	4.3	35.8	41.4	8.2	ol^{α}
M23	8.9	4.4	48.6	29.5	8.5	*ol*
M24	10.0	5.3	27.3	51.5	5.9	$fanx^{\alpha}$
M5	9.8	4.9	25.5	55.3	4.9	*fan*
LOLL	10.3	4.5	27.7	54.7	2.9	$fan, fanx^{\alpha}$
B739	7.9	4.3	14.3	55.1	18.4	lin^h
Bay (wild type)	10.6	4.2	25.2	51.7	8.3	

biosynthetic process, the desaturation step of converting oleic acid (18:1) to linoleic acid (18:2) was blocked due to the presence of the *ol* gene allele.

2.2. Inter-Varietal Variation of Fatty Acid Composition in Soybean

In order to modify the fatty acid composition, 251 soybean varieties from the USA and 172 landrace lines from Japan and Korea were evaluated to identify inter-varietal variation. The results indicated that linoleic and linolenic acid contents were controlled in a polygenic manner. Genetic studies showed that the contents of oleic acid and linolenic acid content appeared to be controlled by two and one genes, respectively. But, specific quantification of both fatty acids from soybean accessions did not reveal any genotypes with substantially high oleate and low linolenate. It is thought that the effects of each gene on fatty acid content are small, indicative of a quantitative trait, and are easily modified by environmental effects, although the genetic variation of fatty acid contents occurs in the natural population of soybean.

Unlike prior successes in rapeseed, and sunflower and safflower germplasm to identify fatty acid variations, a similar evaluation of soybean germplasm failed to identify any major genes that would dramatically alter the fatty acid composition of soybean. As a consequence, the induction of mutants to enhance the diversity of fatty acid composition of soybean was initiated.

3. Comparison of Fatty Acids in Mutant and Natural Populations

X-ray induction of mutations in soybean var. Bay was initiated in 1987. Preliminary evaluation and screening of M_2 mutants exhibited alterations in fatty acid composition. Finally, 46 mutant lines were isolated out of an M_2 population of 12,266 individuals. Fatty acid compositional changes of the mutant lines are listed in **Table 32.2**. These include the low palmitic acid content mutant J3 (6.1%), the high palmitic acid content mutant J10 (17.2%), and other mutant lines, e.g. M25 with a high stearic acid content (21.2%), M23 with a high oleic acid content (48.6%), M5 with a low linolenic acid content

Table 32.3: Mean and range of fatty acid composition (% of total oil) in different mutants and natural varieties of soybean (1992)

Materials	No. of lines	Palmitic acid (16:0)	Stearic acid (18:0)	Oleic acid (18:1)	Linoleic acid (18:2)	Linolenic acid (18:3)
Mutants of Bay [a]	46	10.6 (6.3–16.7)	5.2 (2.9–16.1)	26.5 (17.5–48.2)	49.8 (32.2–60.5)	7.9 (4.6–12.6)
Var. Bay [b]	(30 plants)	11.6 (11.3–11.9)	2.9 (2.7–3.1)	23.3 (20.8–25.9)	54.1 (52.6–55.7)	8.2 (7.9–8.5)
Varieties [c] Mean (range)	99	12.1 (10.0–15.5)	2.9 (1.6–4.1)	25.1 (14.2–44.3)	51.9 (36.6–61.1)	8.0 (5.2–12.6)

[a] Fatty acid mutant lines generated from cv. Bay;
[b] Fatty acid composition was analyzed on an individual plant basis;
[c] Landrace varieties from Japan and Korea. Data were shown in mean (range).

(4.9%) and B739 with a high linolenic acid content (18.4%). **Table 32.3** shows the comparison of fatty acid composition of the 46 mutant lines and 99 soybean landraces; it clearly shows that the 46 individual mutant lines have a wider range of fatty acid composition than the landraces. These results indicated that mutation induction is a useful approach for enhancing diversity of fatty acid composition in soybean.

4. Mutant Genes Affecting Fatty Acid Content

Traditional breeding, selection, characterization and experimentation have provided plant breeders with a cornucopia of mutant genes. The basic studies on plant fatty acid biosynthesis of soybean have also been propelled by advances in genomics, especially in model species such as *Arabidopsis thaliana*. **Figure 32.1** illustrates the fatty acid biosynthetic pathway of soybean seed oil. Palmtoil-ACP (16:0) is synthesized from maronil-ACP (4:0) through the condensation (fatty acid elongation) cycle, which involves three types of β-ketoacyl-synthase (βKASs). Palmitoil-ACP is synthesized by βKASI and βKASIII, then steroil-ACP (18:0) is synthesized by βKASII. Following this process, stearoil-ACP is converted into oleoil-ACP by stearoil-ACP desaturase. Acyl-ACP thioesterase catalyzes the acyl-ACPs, and palmitic acid, stearic acid and oleic acid are produced in the same pathway. Oleic acid is then converted to linolenic acid through linoleic acid by two distinct microsomal fatty acid desaturases, omega-6 and omega-3. These fatty acid biosynthesis enzymes are encoded by *GmβKASII*, *GmSte*, *GmFAD2* and *GmFAD3* gene families, respectively.

4.1. Mutant Genes for Reduced and Elevated Palmitic Content

Palmitate content in ordinary soybean oil is about 11%. C1726 was isolated as a mutant with low palmitic acid following chemical treatment with EMS and its underpinning allele is designated as *fap1*. Additional mutant alleles, *fap3* and *sop1*, were identified as conferring low palmitic acid content. A double mutant line carrying homozygous alleles of *fap1* and *fap3* had reduced palmitic acid content as low as 3.5% (Horejsi *et al.*, 1994). Several low palmitic acid mutants have been isolated and a double mutant line (*fap1* plus *sop1*), LPKKC-3, derived from a cross between J3 and C1726, exhibited a reduced palmitic acid level of 4.4% (**Table 32.2**).

High palmitic acid content mutants have also been isolated. For example, line C1727 carrying the *fap2* gene exhibits a palmitic acid content as high as 17.3%. A double mutant line, HPKKJ10, was selected from the progeny of a cross between KK7 (14.2%) possessing *fapx* and J10 (17.2%) possessing *fap2*; it exhibits a high palmitic acid content of 21.6% (**Table 32.2**). Several additional genes have been reported to confer elevated palmitic acid content such as the *fap4, fap5, fap6 and fap7* genes (Narvel *et al.*, 2000). Pyramiding these genes into a single line can further increase palmitic acid content, i.e. a triple mutant line with *fap-2b, fap4, fap6* has a palmitic acid content as high as 39.9%.

4.2. Elevated Stearic Acid Content

Stearate content in ordinary soybean oil is about 3.5%. Three mutant lines which exhibit elevated stearic acid contents were isolated from soybean populations fol-

Table 32.4: Genetic basis of phenotypes and reduced linolenic acid mutant lines of soybean

Isoform/Locus	Line	Allele	Mutation type	Reference
GmFAD3-1a	M24	fanx$^{\alpha}$	frame shift	Anai et al., 2005
(GmFAD3B)	A5	fan	large deletion	Bilyeu et al., 2005
	C1640	fan	premature stop codon	Chappel and Bilyeu, 2006
GmFAD3-1b	M5	fan	frame shift	Anai et al., 2005
(GmFAD3A)	J18	fan	large deletion	Anai et al., 2005
	A29	fan	missplicing	Bilyeu et al., 2006
GmFAD3-2a	A29	fan	amino acid substitution	Bilyeu et al., 2006
(GmFAD3-2b)				

lowing EMS or sodium azide treatment. The mutant line A6 with the fas$^{\alpha}$ gene allele exhibits the highest stearic acid content up to 30.4%. Allelism tests were made between A6 and four newly isolated high stearic acid mutants, ST1, ST2, ST3 and ST4. The results from hybrids between the ST2 line (st2 gene) and A6 (fas$^{\alpha}$ gene) indicated the mutants were allelic and exhibited a higher stearic acid content (Bubeck et al., 1989). X-ray induced mutants have also been isolated and some hybrids exhibited elevated stearic acid content. A double recessive line (st$_{r}$ st$_{2}$) derived from a cross between KK2 (7.2%) and M25 (21.2%) exhibits a very high content of stearic acid up to 31.4% (**Table 32.2**).

4.3. Elevated Oleic Acid Content

Oleic acid content, which is of interests to consumers who purchase health food, ranges from 14.2–44.3% among traditional soybean varieties (**Table 32.3**). This range is wider than the other fatty acids. However, since oleic acid content is highly affected by environment, especially by the temperature during the stage of seed ripening, it is thought to be an unstable characteristic. Two high oleic acid mutants, M23 (48.6%) and M11 (35.8%), have been developed through X-ray irradiation of the var. Bay (25.2%). Evaluation of the offspring from the cross between M23 and M11 suggested that each mutant carried a different allele (ol and ol$^{\alpha}$, respectively) at the same locus for oleic acid content. The wild type allele Ol for low oleic acid content (var. Bay) was partially dominant to ol and completely dominant to the allele ol^{a}, and ol^{a} was completely dominant to ol.

There was a negative relationship between oleic and linoleic acid contents in both mutants, which suggests

that the mutant alleles, ol and ol^{a}, may control the linoleic acid content by blocking the oleic acid desaturation step (**Figure 32.1**). Presumably, multiple genes are involved in the fatty acid synthesis of oleic acid. In order to increase oleic acid content, the ol gene allele and other gene mutations that act additively are necessary.

4.4. Reduced and Elevated Linolenic Acid Content

Linolenic acid is thought to be a cause of quality deterioration of soybean oil by oxidation, and hence breeding projects aim to reduce its content to less than 1%. Among traditional varieties, linolenic acid content varies from 5.2–12.6% (**Table 32.3**), and is highly affected by environmental conditions. Hammond and Fehr (1983) isolated a mutant line named A5 from an EMS-treated population. It has a low linolenic acid content of ~3.4% and was later identified to possess a mutant allele of the fan gene. Two double mutant lines A16 and A17 (fan, fan2) were developed from the cross between low linolenic acid mutants A23 and A5, and exhibited a lower linolenic acid content than either of their parents. Another double mutant line, LOLL, which was developed from a cross between M5 (4.9%) and M24 (5.9%) and carries two recessive mutant genes of fan and fanx$^{\alpha}$, has a linolenic acid content as low as 2.9% (**Table 32.2**).

Mutants exhibiting high levels of linolenic acid content were also developed. For example, the mutant line B739 expressed a linolenic acid content of 18.4%. This high linolenic acid trait was identified to be controlled by the lin^{h} gene.

Table 32.5: Soybean lines with altered fatty acid composition developed by combination of induced mutations

Line	Palmitic (16:0)	Stearic (18:0)	Oleic (18:1)	Linoleic (18:2)	Linolenic (18:3)	Genotype and Reference
HPS	23.1	11.4	13.3	44.6	8.6	sop2, fap2/St$_1$, st$_2$/Ol/Fan, Fanx
MHPDHL	19.3	2.2	31.8	43.4	3.3	Sop2, fap2/St$_1$, st$_2$/ol/fan, fanx$^\alpha$
HPLOLL	25.2	2.8	18.5	50.9	2.6	sop2, fap2/ St$_1$, st$_2$ /Ol/fan, fanx$^\alpha$
HSO	11.2	11.8	20.4	48.2	8.4	Sop1, Fap2/ St$_1$, st$_2$ /ol/Fan, Fanx
DHL	11.7	2.8	50.6	32.2	2.7	Sop1, Fap2/ St$_1$, St$_2$/ol/fan, fanx$^\alpha$
N98-4445A	9.6	4.0	58.9	25.0	2.6	USDA-ARS/North Carolina State University

5. Genetic Characterization of Mutant Genes

Genetic analysis of the fatty acid content in soybean mutants suggests that three or more genes may be involved in each step of fatty acid synthesis. For example, four microsomal GmFAD3 isoforms, GmFAD3-1α GmFAD3-1b, GmFAD3-2α and GmFAD3-2b desaturase genes have been identified in soybean, all of which may affect linolenic acid content. The genetic bases of some mutant genes corresponding to three major low-linolenic acid loci have been identified (Anai et al., 2005). It was found that the low linolenic lines M24, A5 and C1640, which were designated as possessing the fanx$^\alpha$ or fan allele, had a deletion in the GmFAD3-1a gene, and M5, J18 and A29, which also possess the fan allele, had a deletion in the GmFAD3-1b gene. In each mutant line, single base or large deletions were identified in the sequence of the responsible gene (**Table 32.4**). Other deletion mutants have been found in the GmβKASII gene of palmitic acid, the GmSte gene of stearic acid and the GmFAD2 gene of oleic acid gene. However, only limited information is available concerning additional fatty acid mutants and the vast majority of fatty acid mutants of soybean have yet to be characterized. Identification of the genetic basis of these mutant alleles enables the development of allele-specific molecular markers, which can be used in marker-assisted selection programmes. This will greatly enhance the effectiveness of fatty acid composition improvement because the fatty acid content is highly influenced by environmental conditions, e.g. location, temperature, solar radiation, etc. A superior understanding of the relationships between the corresponding gene(s) and the phenotype of individual mutants will be valuable for future soybean oil breeding programmes.

6. Soybean with New Fatty Acid Composition

Soybean varieties with altered fatty acid composition can be useful for unique use of soybean oil. Some newly developed genetic recombinant lines associated with three or more mutant genes are provided in **Table 32.5**. High oleic acid and low linolenic acid content in soybean are of potential value for increasing the market price of the product as well as meeting end-user demands. To attain this goal, a DHL line with high oleic acid and low linolenic acid content, which was developed through a cross between M23 with high oleic acid (ol gene) and LOLL with low linolenic acid (double mutant carrying fan and fanx$^\alpha$ alleles), was developed. The DHL has extremely high oleic acid content (50.6%) and reduced linolenic acid (2.7%), compared to the original var. Bay (25.2% oleic acid and 7.6% linolenic acid).

In the USA, an experimental line designated as N98-4445A exhibits high oleic acid (~ 59%) and low linolenic acid (~2.6%) contents. The extremely high oleic acid line, (N98-4445A x M23)-1, obtained from the cross between N98-4445A and M23 attained a 73% oleic acid content. Two or more genes of N98-4445A and with complementary gene act ol of M23 believed to be responsible for this extremely high oleic acid level.

Multiple genes are presumed to be involved in each fatty acid biosynthetic pathway and various interactions are likely to occur among them. Major goals are to continue the development of soybean germplasm that exhibits higher than 70% oleic acid and lower than 1% linolenic acid content. In order to accomplish the allele-specific genotypic selection for these traits, the utilization of genetic markers for particular gene families will provide a more efficient approach than phenotypic selection (e.g. gas–liquid chromatography) that is subject to environmental influences. Multiple sources of alleles for each candidate gene will provide further benefits in the breeding for superior fatty acid profiles, and assist in minimizing the fixation of alleles linked to target genes.

7. References

7.1. Cited References

Alt, J.L., Fehr, W.R., Welke, G.A. *et al.* 2005. Transgressive segregation for oleate content in three soybean populations. *Crop Sci.* 45: 2005–2007.

Anai, T., Yamada, T., Kinoshita, T. *et al.* 2005. Indication of corresponding genes for three low-α-linolenic acid mutants and elucidation of their contribution to fatty acid biosynthesis in soybean seed. *Plant Sci.* 168: 1615–1623.

Bilyeu, K.D., Palavalli, L., Sleper, D.A. *et al.* 2003. Three microsomal omega-3 fatty aid desaturase genes contribute to soybean linolenic acid levels. *Crop Sci.* 43: 1833–1838.

Bubeck, D.M., Fehr, W.R. and Hammond, E.G. 1989. Inheritance of palmitic and stearic acid mutants of soybean. *Crop Sci.* 29: 652–656.

Hammond, E.G. and Fehr, W.R. 1983. Registration of A5 germplasm line of soybean (Reg. No. GP44). *Crop Sci.* 23: 192.

Horesji, T.F., Fehr, W.R., Welke, G.A. *et al.* 1994. Genetic control of reduced palmitate content in soybean. *Crop Sci.* 34: 331–334.

Narvel, J.M., Fehr, W.R., Ininda, J. *et al.* 2000. Inheritance of elevated palmitate in soybean seed oil. *Crop Sci.* 40: 635–639.

7.2. Further Reading

Acquaah, G. 2007. Principles of Plant Genetics and Breeding. Oxford: Blackwell Publishers.

Cardinal, A.J. 2008. Molecular genetics and breeding for fatty acid manipulation in soybean. *In*: J. Janick (ed.) Plant Breeding Reviews. Oxford: John Wiley & Sons, Inc. pp. 259–294.

Fehr, W.R. 2007. Breeding for modified fatty acid composition in soybean. *Crop Sci.* 47: S72–S87.

Kinney, A.J. 1997. Genetic engineering of oilseeds for desired traits. *In*: J.K. Setlos (ed.) Genetic Engineering, Principles and Methods. Vol. 19. New York: Plenum Press. pp. 149–166.

Ohlrogge, J.B., Browse, J. and Sommerville, C.R. 1991. The genetics of plant lipids. *Biochem. Biophys. Acta* 1082:1-26.

Winter, P.C., Hickey, G.I. and Fletcher, H.L. 1998. Instant Notes in Genetics. BIOS Scientific Publishers, Berlin: Springer.

Genetic Improvement of Basmati Rice Through Mutation Breeding

G.J.N.Rao*, A.Patnaik and D.Chaudhary

Division of Crop Improvement, Central Rice Research Institute, Cuttack 753006, India
* Corresponding author, E-MAIL: raogjn@yahoo.com

1. Introduction

The passion for exquisite types of food and drink in the world calls to mind names like Champagne wine, Scotch whisky, which with their rich flavour and unique taste bestow pleasure to connoisseurs. These premium names in international trade are recognizable, set themselves apart from others in their category and are best recognized by their branding, specific ingredients and processing. For wines, their quality and character are determined by using grapes of specific varieties and produced in particular areas. Basmati, the Champagne of rice, the unique aromatic rice from the Indian sub-continent is one of the handful of rice varieties in the world that are internationally traded with a specific name. Basmati derives its name from *Bas* (aroma) and *mati* (already ingrained).

Also known as "Queen of Fragrance", "Prince of Rice", "Crown Jewel" of South Asian rice, Basmati is romanticized as nature's gift to the Indian sub-continent and had been favoured by Emperors and praised by poets for hundreds of years (Thakrar and Ahuja, 1990). Basmati rice is defined by its origin (geographical location) and by genotype. Historical and archeological findings indicate that varieties with such unique morphological and quality attributes are not present in traditional rice-growing areas anywhere in the world, signifying the emphasis on the place of origin of Basmati. Therefore, the names of the varieties were often derived from their originating localities and authentic Basmati rice cultivation is confined to the Indo-Gangetic plains of India (1 m.ha) and Pakistan (0.75 m.ha).

Basmati rice varieties constitute a distinctive group with their pleasant aroma, exclusive grain and cooking quality. Basmati's matchless characteristics include long slender kernels with a high length/breadth ratio with the grain having a slightly twisted tip. During cooking, the kernels elongate linearly by nearly twice their original length, remain separate, non-sticky and possess a soft fluffy texture with visible hoops or corrugation on the cooked kernels (**Figure 33.1**). Appreciated for its exquisite aroma and taste, Basmati rice commands a premium price in both domestic and international markets (about three times higher than non-Basmati rice) and traditional Basmati labels also receive duty exemption in some markets. The demand for Basmati rice is strong and increasing.

The traditional Basmati rice varieties are low yielding with tall and weak culms, light green leaves (**Figure 33.1**) and respond poorly to fertilizer application. They are characterized by photo-thermo sensitivity, late maturity and lodging susceptibility, which makes it difficult to fit them into cropping systems. To address these undesirable features, Basmati breeders of the sub-continent have long been attempting to breed high-yielding Basmati rice of shorter duration coupled with lodging resistance.

2. Types of Basmati Rice and Breeding Approaches

2.1. Recognition of a Basmati Rice Variety

To protect the unique Basmati quality, there are strict criteria for defining a rice variety as a Basmati rice (**Box 33.1**). Therefore only a handful of Basmati rice varieties have been approved for general production (**Box 33.2**).

In India, Basmati rice varieties are classified into two categories: Traditional Basmati and Evolved Basmati (**Box 33.2**). All varieties derived from cross-breeding can only be ranked as Evolved Basmati, while those from pure line selection or selection from natural mutation can be recognized as Traditional Basmati.

2.2. Genetic Improvement of Basmati Rice

In the efforts to breed high-yielding, semi-dwarf, non-lodging Basmati varieties, two main breeding approaches have been used for improvement of Basmati rice over the past century.

Box 33.1: Definition of Basmati rice

- Long kernel (≥6.61 cm), which increases substantially on cooking (≥1.7 X).
- The cooked grain has high integrity and high discreteness and distinctive aroma, taste and mouth feel.
- Either a traditional or evolved variety where, a traditional Basmati variety means land races or varieties of rice of uniform shape size and colour traditionally recognized as Basmati, an evolved Basmati variety generally meaning a variety derived from a cross of which one of the parental varieties is a traditional Basmati variety.

Source: http://www.eicindia.org/eic/qc&i/enotfn-rice-68.htm

Figure 33.1 Crop, grain, polished kernel and cooked kernel of Basmati 370.

2.2.1. Pure-Line Selection Breeding

Pure-line selection breeding has been adopted to improve locally adapted land races by selecting and multiplying desirable plants identified in natural populations, which has led to the release of a number of Basmati varieties (**Box 33.2**). For example, Basmati 370 was a pure-line selection of Dehradooni Basmati identified at Kala Shah Kaku research station in Pakistan (Ahuja *et al.*, 1995); it has become a bench mark for quality and has brought an export boom to both India and Pakistan.

2.2.2. Cross-Breeding

Breeders have also employed cross- and backcross breeding approaches in their efforts to develop semi-dwarf Basmati varieties. These methodologies are useful in breaking undesirable linkages between preferred quality traits and undesirable agronomic characters. For example, Haryana Basmati-1, a semi-dwarf aromatic rice variety for Haryana (India) was developed from a cross between Sona and Basmati 370 (Panwar *et al.*, 1991). Using this strategy, a number of evolved Basmati rice varieties have been developed (**Box 33.2**).

2.3. Mutation Breeding of Basmati Rice

Government constrained definitions mean that Basmati breeding involves only traditional breeding methods and this severely limits the options for improvement of Basmati rice. In this regard, mutation breeding is one of the few viable options of improving the overall performance of Basmati while keeping its status. New varieties developed through mutation breeding can potentially be classified as traditional Basmati, as long as they retain the quality characteristics of their parents (Patnaik *et al.*, 2006).

Mutation breeding has been highly successful in other rice improvement schemes (Futsuhara *et al.*, 1966; Rutger *et al.*, 1977) and similar successes can be expected for Basmati. However, notable successes are rare though dozens of mutants have been induced from traditional Basmati varieties, most did not show significant yield increases and did not retain quality characteristics. In Pakistan, a Basmati 370 mutant was released as Kashmir Basmati, but failed to become popular. However, the recent success in developing a

---

Box 33.2: Basmati rice varieties approved for general production

India
Traditional: Basmati 370, Basmati 386, Type 3, Taroari Basmati, Basmati 217, Ranbir Basmati.
Evolved: Pusa Basmati 1, Punjab Basmati, Haryana Basmati-1, Kasturi, Mahisugandha, Super Basmati, Pusa Basmati 1121.

Pakistan
Traditional: Basmati 370, Pakistan Basmati.
Evolved: Basmati 385, Super Basmati, Basmati 198.

high-yielding semi-dwarf, non-lodging Basmati mutant at the Central Rice Research Institute, India may provide a good example of how to improve traditional specialty varieties using mutation breeding.

3. Breeding of a New Basmati Mutant Variety 'CRM 2007-1'

3.1. The Breeding Process

3.1.1. Growing and Harvesting of M_1 Populations

Grain of two traditional Basmati rice varieties, Basmati 370 and Pakistan Basmati and an evolved Basmati, Pusa Basmati 1, were treated with γ-rays (100, 150, 200 Gy). The objective of the programme was to induce non-lodging, semi-dwarf Basmati varieties from the traditional varieties; while yield improvement was the major objective in the case of evolved Basmati. Since the Basmati improvement programme was acknowledged to be a difficult task, around 150–200 g of grains for each treatment (dose) were irradiated so as to get around 7,000–8,000 plants per treatment in the M_1 generation. The treated materials were sown immediately after irradiation in shallow trays along with the control (non-irradiated parent). Care was taken to eliminate contamination of grains from other genotypes and 25–30-day-old seedlings were transplanted in a well puddled field at one seedling/hill with extremely close spacing ($\leq$5 cm × 5 cm) to curtail tillering in the plants (a practice used to obtain only one or two tillers/plant). The control was planted similarly. As expected, the plants produced shorter panicles with fewer grains than normal. During harvest, the panicles from all plants were pooled for each dose and highly sterile panicles were all discarded. It is important to note that this is a pragmatic method aimed at producing desired phenotypes quickly and therefore, in contrast to TILLING procedures (**see Chapter 21**), no effort was made to develop a structured population.

3.1.2. Growing, Selecting and Harvesting of M_2 Plants

In the M_2 generation, around 100,000 plants per variety were raised from the bulked seeds and were transplanted at one seedling/hill with normal spacing (20 cm × 15 cm in wet season; 15 cm × 15cm in dry season). Application of recommended doses of fertilizer and need-based plant protection measures were undertaken. During harvest,

first, all tall plants (in traditional varieties) were harvested as they tended to camouflage semi-dwarf plants in the field and lodging was common in wet season in the parent varieties as they did not possess a stiff culm. Of the remaining, plants with non-lodging phenotypes were selected as it was difficult to ascertain the nature of short stature (genetic or environment-shading due to the effect of neighbouring tall plants in the population). From the populations of Pusa Basmati 1, plants possessing good panicle types without awns were selected. The flowering date was recorded and plants that flowered earlier or later than the parent were tagged so that they could be harvested separately. As it was not possible to evaluate quality at the M_2 generation, the plants with grains similar to Basmati were selected. In addition, in order to avoid loss of potentially promising single plants, a relatively higher number of plants (16,860) was collected from the M_2 generation.

3.1.3. M_3 and Higher Generations

A few kernels from each plant selected at the M_2 stage were checked for aroma and grain chalkiness and several lines without the desirable quality (aroma and clear translucent endosperm without any chalkiness) were discarded. The grains from the primary panicles of selected M_2 plants were grown as an M_3 family line. From the M_3 generation onwards, panicles from four or five single plants were collected from each line and were grown separately in the next generation with the same line number. From the identified semi-dwarfs, several lines were rejected as they were not true breeding. The lines with desired levels of grain quality, non-lodging and semi-dwarf stature were advanced. Many lines were rejected as they had poor or undesirable panicle characters, e.g. size, exertion, spikelet number per panicle or low grain weight, sterility and awn, etc. Another criterion was to select plants that were similar to their parent (traditional varieties) except for short stature, and earnest efforts were made in this regard. As a consequence, the number of lines reduced drastically with generation advancement, which concluded at M_7. After M_7, only three, seven and ten mutant lines from Basmati 370, Pusa Basmati 1 and Pakistan Basmati, respectively, were isolated. Only these lines were evaluated in Observational (OYT) and Replicated Yield Trials (RYT) before nomination to multi-location evaluation trials in the Basmati growing area (North West India). Though all three mutant lines of Basmati 370 recorded higher yields at the station

Table 33.1: Performance of mutant line CRM 2007-1 and its parent Basmati 370

Traits	CRM 2007-1	Basmati370
Plant height (cm)	115	135
Number of days from sowing to 50% flowering	92	104
Number of panicle bearing tillers/m^2	302	344
Panicle length (cm)	29.4	27.3
Number of grains per panicle	116	91
Panicle weight (g)	3.2	2.6
Yield (kg/ha)	3720	2870
Harvest index	0.41	0.28
1000 grain weight (g)	23.6	23.1

Table 33.2: Yield performance of CRM 2007-1 and two Basmati varieties at the Central Rice Research Institute research farm in wet and dry season (yield in kg/ha)

Variety	2001		2002		2003		2004	
	Wet	Dry	Wet	Dry	Wet	Dry	Wet	Dry
CRM 2007-1	5038	4900	4233	4938	4955	4878	3729	4261
Pusa Basmati 1	-	3800	3876	3950	-	4128	3555	4163
Taroari Basmati	2667	-	2482	2267	2026	2017	-	2276

trials, one line i.e. CRM 2007-2 was discarded since it did not attain the desired purity level, which might be due to out-crossing-derived segregation.

Of the three mutant lines, CRM 2007-1 is a semi-dwarf and early, with reduced plant height of ~20 cm and shortened growth duration of ~10 days compared to its parent (**Table 33.1, Figure 33.2**). It showed great promise by out-yielding both parent and Pusa Basmati 1, the yield control in the station trials (**Table 33.2**). The yield increase in the mutant over the parent was accounted for by the increase in number of grains per panicle and panicle weight.

3.2. Performance of Mutants

3.2.1. Basmati Growing Areas

In the multi-location trials, the mutant line CRM 2007-1 performed well in the traditional Basmati growing states of Punjab and Haryana. The mutant recorded significantly higher yields over the two Basmati controls i.e. Pusa Basmati 1, the yield control (6.96%) and Taroari Basmati, the quality control (40.24%; **Table 33.3**). CRM

2007-1 had very similar growth duration (113 days from sowing to 50% flowering) and plant height (108 cm) to Pusa Basmati 1, but about one week earlier and 24 cm shorter than Taroari Basmati.

The grain quality tests and the Basmati panel tests on CRM 2007-1 established conclusively its Basmati quality and good acceptability of the produce (**Tables 33.4 and 33.5**).

The three-year mandatory multi-location trials clearly demonstrated the superiority of CRM 2007-1 over Pusa Basmati 1 in yield; its grain quality also meets the Basmati standards. Therefore, CRM 2007-1 was identified for release as a Basmati variety in the North Western Region of India by the Variety Identification Committee in 2006. CRM 2007-1 has also performed well in the farmers' fields of Punjab and Haryana and the yield levels recorded up to 6.2 t/ha indicates its high yield potential in the Basmati zone.

3.2.2. Non-Basmati Areas

In Orissa, India, a non-traditional Basmati area, CRM 2007-1 performed well and gained wide consumer

Table 33.3: Yield performance of CRM 2007-1 comparative in multi location trials

Year	Grain yield (kg/ha)	Yield advantage (%) over controls	
		Pusa Basmati 1	Taroari Basmati
2001	3358	8.57	31.07
2002	3536	--	25.61
2003	4485	5.36	64.04

Table 33.4: Grain quality characters of CRM 2007-1 (IET 17276) and CRM 2007-3 (IET 18987)

Character	CRM 2007-1	CRM 2007-3	Pusa Basmati 1	Taroari Basmati
Milling yield (%)	70.4	68.9	68.7	69.5
Head rice recovery (%)	56.4	33.2	57.3	50.6
Kernel length (mm)	7.47	7.20	7.62	7.22
Kernel breadth (mm)	1.89	1.78	1.76	1.84
Length/breadth	3.95	4.04	4.32	3.92
Kernel length after cooking (mm)	13.0	11.9	14.1	14.0
Elongation ratio	1.74	1.65	1.85	2.02
Alkali spreading value (1-7)	5.0	4.0	7.0	5.0
Amylose content (%)	25.9	25.0	24.4	23.3
Gel consistence (mm)	70.0	56.0	61.0	70.0
Grain chalkiness	Absent	Absent	Very occasionally present	Absent

acceptance. Possessing the Basmati quality, CRM 2007-1 fetched a premium price in the local market thus ensuring higher economic returns to the farmers in the region. Due to good performance in Orissa and ready acceptance by farmers, CRM 2007-1 was released as "Geetanjali" in 2005 and was notified by the Government of India in 2006.

Another mutant line, CRM 2007-3, also derived from Basmati 370, performed well in the multi-location trials. It had out-yielded the control variety Pusa Basmati 1 by 6.47% and Taroari Basmati by 29.40%. However, this mutant could not be recognized as a Basmati due to its low head rice recovery and low kernel length after cooking (KLAC) value as per Basmati standards (**Table 33.4**).

3.3. Performance of Mutants of Pusa Basmati 1 and Pakistan Basmati

Two semi-dwarf mutants from Pakistan Basmati and four semi-dwarf, awnless mutants from Pusa Basmati 1 were also evaluated in the multi-location trials conducted by AICRIP in different years. The mutants of Pusa Basmati 1,

i.e. CRM 2203-1, CRM 2203-2, CRM 2203-3 and CRM 2203-4 showed 7.91, 7.01, 15.39 and 21.24% yield increase over Pusa Basmati 1, the yield control respectively. Both the mutants from Pakistan Basmati also displayed high yields. The mutant CRM 2202-117 showed 14.9% higher yield than Pusa Basmati 1 while the other mutant CRM 2202-118 displayed a 7.5% higher yield over Dubraj, the yield control, in another AFGON trial.

However, none of these mutant lines could be classified as Basmati due to deficiencies in one or more quality traits. Both mutant lines of Pakistan Basmati did not have the required kernel length and other traits; although most mutant lines of Pusa Basmati 1 had the required kernel length, they failed to meet the requirement for kernel length after cooking (**Table 33.6**).

4. Experiences and Implications

The success achieved in the development of a high yielding, semi-dwarf, non-lodging Basmati has dem-

Table 33.5: Panel test scores[a] of CRM 2007-1 (IET 17276)

Variety	Appearance	Cohesiveness	Tenderness on		Taste	Aroma	Elongation	Overall acceptability
			Touching	Chewing				
CRM 2007-1	4.1	4.0	3.5	4.2	2.8	3.8	2.8	3.4
Pusa Basmati 1	4.2	4.2	4.1	4.5	3.6	3.8	3.8	4.0
Taroari Basmati	4.1	3.9	3.9	4.3	3.5	4.2	3.6	3.9

[a] Values are in the scale of 1–5

Table 33.6: Gain quality of mutant lines of Pusa Basmati 1 and Pakistan Basmati

Mutant (IET No.)	Mill (%)	HRR (%)	KL (mm)	KB (mm)	L/B Ratio	Grain Type	Grain Chalk	KLAC	AC (%)
CRM2203-1 (IET 18416)	73.8	53.5	6.62	1.79	3.69	LS	VOC	11.5	24.97
CRM2203-2 (IET 18988)	72.9	42.3	6.60	1.72	3.83	LS	VOC	9.4	22.64
CRM2203-3 (IET 18417)	73.1	49.5	6.63	1.76	3.76	LS	VOC	11.0	24.33
CRM2203-4 (IET 18008)	73.0	69.6	6.31	1.75	3.61	LS	A	10.6	23.32
CRM2202-117 (IET 18007)	67.0	40.3	5.95	1.77	3.36	SS	A	10.2	23.49
CRM2202-118 (IET 18364)	63.0	42.1	6.25	1.76	3.55	LS	VOC	11.0	21.65

Mill – milling percentage (%), HRR – head rice recovery (%), KL – kernel length (mm), KB – kernel breadth (mm), LS – long slender, SS – short slender, A – absent, VOC – very occasionally present, KLAC – kernel length after cooking (mm), ER – elongation ratio, AC – amylose content (%)

onstrated clear potential of mutation breeding in the genetic improvement of Basmati rice. From a pure line selection like Basmati 370, a traditional low yielding Basmati, significant enhancement in yield (29.6%) was obtained using mutagenesis, without altering its famous quality traits. The success in yield enhancement in CRM 2203-4 (21.2%) over Pusa Basmati 1, an improved high-yielding Basmati clearly demonstrates the potential existing in Basmati rice for improvement through mutation.

However, many selected mutant lines with improved yield or other agronomic traits failed to meet the stringent criteria of a Basmati rice. These results implied that mutant lines derived from a speciality variety like Basmati rice might not attain its original quality criteria. Therefore, it is important to isolate many mutants with targeted traits, e.g. semi-dwarf, non-lodging, as carried out in this study, and then subject them to quality analysis to discard those that fail to meet the criteria.

Some of the experiences that can be shared from the program on Basmati rices are:

1. Requirement of a large M_2 population size (>100,000 per variety), a large population has to be raised to select mutants with the desirable traits.

2. The second and key element is that production of initial large populations in the M_1 and M_2 should be followed by specific and stringent phenotypic selection which minimizes the number of lines taken forward to the next generation. This is important in terms of saving space and costs.

3. Maintaining the grain quality is the major and the most difficult aspect of Basmati breeding. The basic selection criteria should be grain quality first as these speciality rices are known by their quality rather than by yield. Though careful selection was followed over generations among large populations at each stage, very few semi-dwarf mutants

having Basmati quality could be isolated and out of these, only CRM 2007-1 (Geetanjali) possesses the set quality standards of Basmati.

4. The number of traits to be selected should be restricted to one or two as the present example is a clear demonstration of the non-recovery of all the grain quality traits of Basmati in the mutants.

5. Single plant-based selection should be practised as line purity is critical for the success of the breeding programme.

6. Seasonal variation – if the objective is to breed a Basmati for both wet and dry seasons, selection of plants for wet and dry seasons should be performed separately to obtain line uniformity as lines selected in the wet season tend to behave differently in the dry season.

5. Acknowledgements

The authors wish to thank the director, C.R.R.I. for the facilities and also duly acknowledge the Directorate of Rice Research, Hyderabad, India for the data on the AICRIP trials for the years 2001, 2002, 2003 and 2004 in its annual progress Reports.

6. References

6.1. Cited References

Ahuja, S.C., Panwar, D.V.S. and Uma A. 1995. Basmati rice – the scented pearl. CCS Haryana Agricultural University, Hisar, Haryana. 63.

Futsuhara, V., Toriyama, K. and Tsunoda, K. 1967. Breeding of a new rice variety 'Reimei' by gamma-ray irradiation. *Jap. J. Breed.* 17: 85.

Panwar, D.V.S., Gupta, K.R. and Battan, K.R. 1991. HKR 228. A semi dwarf aromatic rice strain for Haryana, India. *Intl. Rice Res. Newsl.* 16: 16–17.

Patnaik, A., Chaudhary, D. and Rao, G.J.N. 2006. Genetic improvement of long grain aromatic rices through mutation approach. *Plant Mutation Reports.* 1(1): 11–16.

Rutger, J.N., Peterson, M.L. and Hu, C.H. 1977. Registration of Calrose 76 rice. *Crop Sci.* 17: 978.

Thakrar, R. and Ahuja, S.C. 1990. Potential prospects for export of Basmati rice. *In*: K. Muralidharan and E.A. Siddiq (eds.) New Frontiers in Rice Research. Hyderabad: DDR. pp. 382–387.

6.2. Websites

Rice variety notification in India:
http://www.eicindia.org/eic/qc&i/enotfn-rice-68.htm

Rice Association, 2007. Rice Market Briefing 65.A.001:
http://www.riceassociation.org.uk/

Mutation Breeding of Sweet Cherry (*Prunus avium* L.) var. 0900 Ziraat

B.Kunter[a,*], M.Bas[b], Y.Kantoglu[a] and M.Burak[c]

[a] Turkish Atomic Energy Authority (TAEK), Sarayköy Nuclear Research and Training Center (SANAEM), Division of Agriculture, Ankara, Turkey
[b] Atatürk Horticultural Central Research Institute (AHCRI), Yalova, Turkey
[c] Republic of Turkey, Ministry of Agriculture and Rural Affairs, General Directorate of Agricultural Research
* Corresponding author, E-MAIL: burak.kunter@taek.gov.tr

1. Introduction

Turkey is the motherland and first producing country of sweet cherries (*Prunus avium* L.). It is also one of the largest exporters of sweet cherry and accounts for 20% of the global export market. The ancient Roman historian Pliny the Elder mentioned that the cherry tree was taken by the Roman General Lucius Licinius Lucullus when he defeated Mithridates VI, King of Pontus, around 70 BC in ancient Cerasus or Cerasion of modern day Turkey, which is a major cherry cultivation area and ranks either as first or second in worldwide cherry production with an annual production amount of 200,000 tons–35,000 tons of this amount is exported annually. Due to Turkey's geography, plantations are located at different altitudes, which results in a wide harvest period making it possible to export cherries with a similar quality for almost five to six weeks with different maturation periods. Nearly all of Turkey's export cherries consist of 0900 Ziraat, and the vast majority of these cherries are marketed in Europe.

0900 Ziraat is self-incompatible and tends to grow vigorously and is low yield on standard rootstocks. Although these traits are deleterious they are outweighed by the huge demand from external markets for the 0900 Ziraat sweet cherry variety.

In **Table 34.1**, the Mutant Variety Database of International Atomic Energy Agency (FAO/IAEA), nine mutant cherry varieties are listed (Mutant Variety Database and Donini, 1980). Apart from the literature, there is a rich diversity of mutant cherry varieties which are not registered but used as breeding parents.

Because of the economic importance of variety 0900 Ziraat and the success of mutant cherry varieties, gamma ray irradiations have been used to induce "compact" and "self-fertile" types in this popular variety.

2. Mutagenic Treatment and Growing Out the M_1V_1 Generation

This project has been carried out in collaboration with the Turkish Atomic Energy Authority, Sarayköy Nuclear Research and the Training Center and Yalova Atatürk Horticultural Central Research Institute (AHCRI).

In 2000, in order to determine the "effective mutation dose", dormant scions of 0900 Ziraat were irradiated with 25, 30, 35, 40, 45, 50, 55 and 60 Gy of gamma rays with ^{60}Co as a irradiation source. Fifty scions (20 cm long) were used for each irradiation dose. Subsequently, irradiated buds were immediately grafted on *Prunus avium* L. rootstock using T-budding. Grafted seedlings were grown in a glasshouse with 50% shading.

On the 60th day following bud break and sprouting, the effective mutation dose was calculated by linear regression on the basis of shoot length (Donini, 1980, Saamin and Thompson, 1998) and determined as 33.75 Gy (**Figure 34.1**).

In 2001 approximately 2,000 buds were irradiated at 33.75 Gy. In early spring of 2002, when shoots reached suitable thickness (1–1.5 cm diameter of basal scion), young trees were transplanted to orchard conditions, with a spacing of 3 × 5 m (**Figure 34.2**). The varieties Bigarreau gaucher and Starks gold, which are known to be compatible with 0900 Ziraat, were used as pollinators.

Table 34.1: Recorded mutant sweet cherry varieties in IAEA Mutant Variety Database (MVD)

MVD ID	Mutant Variety Name	Mutagen	Country	Year	Main Character
251	Compact Lambert	X rays	Canada	1964	Compact growth
252	Compact Stella 35B11	X rays	Canada	1974	Compact growth
253	Lapins	Mutant cross	Canada	1983	Fruit size
254	Stella	Mutant cross	Canada	1968	Self-fertile
255	Sunburst	Mutant cross	Canada	1983	Fruit size
274	Burlat C1	Gamma rays	Italy	1983	Compact growth
275	Nero II C1	Gamma rays	Italy	1983	Compact growth
276	Ferrovia spur	X rays	Italy	1976–1992	Shortness
2305	Sumste samba	Mutant cross	Canada	2000	Fruit size

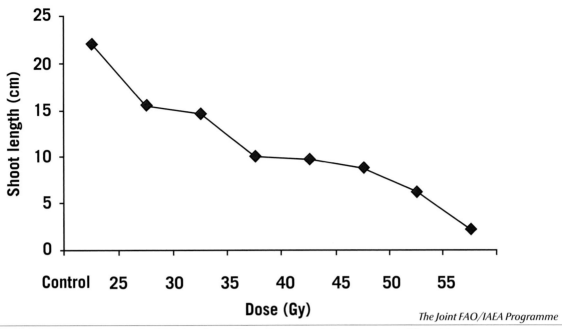

The Joint FAO/IAEA Programme

Figure 34.1 Shoot length of sweet cherry on the 60th day after grafting with treatment of gamma rays at different doses.

The Joint FAO/IAEA Programme

Figure 34.2 Growing stages of mutated sweet cherry plantations in AHCR, Yalova, Turkey.

3. Procedures for the M_1V_2 and Following Generations before Fruit Set

In 2002 during the growth season (M_1V_2) various morphological growth characters were noted, e.g. growth vigor, branching pattern, leaf and bud structures. At this stage mutation frequencies (%) were calculated from observed abnormalities (short internodes, rent leaf, chlorophyll mutation, different leaf size) on 150 randomly selected shoots (**Figure 34.3**). The mean mutation frequency was estimated to be about 4%, which is similar to the value obtained in the cherry var. Bing

(Saamin and Thompson, 1998). At this stage the main objective was to produce a reasonably sized and varied mutant population exhibiting variation throughout the vegetative period (bud sprouting to leaf fall). Growth of each year was considered as a new mutant generation (M_1V_2, M_1V_3, M_1V_4, etc.). In each generation general habit was noted along with detailed morphological descriptors of every branch. All data and abnormalities (if any) were recorded up to the fruit setting stage.

After the young trees had completed their juvenile phase (subsequent to flower formation), fruit setting, fruit quality, fruit pomology (fruit weight, width, height,

The Joint FAO/IAEA Programme

Figure 34.3 Some leaf mutants in the M_1V_2 generation. A: abnormal leaves; B: extra large leaf, C: chlorophyll mutation (indicated by the arrow).

The Joint FAO/IAEA Programme

Figure 34.4 Flowering trees (left) of sweet cherry; one branch is covered with a nylon net tied at both ends for self pollination (shown on the left).

peduncle length and seed weight), fertility, cracking rate and soluble solid content rate (brix) were determined in samples of 30 fruits through three repeated rounds. Most trees completed their juvenile phase at M_1V_5 or M_1V_6. At this stage there were 371 trees surviving from the approximately 2,000 grafted buds at the M_1V_1 generation. The decrease in numbers was caused by a latent effect of the irradiation treatment, poor grafting success, weakness of shoot growth and susceptibility to winter cold.

The trees and fruits of all mutant candidates were photographed and archived during the M_1V_5 or M_1V_6. Self-compatibility was assessed by observing fruit set in 100 randomly selected and isolated flowers to prevent pollen contamination before flowers blossom (**Figure 34.4**). At the end of petal fall stage, fruit numbers were

counted. The time and methods scheme of the experiment is below (**Figure 34.5**)

4. Performance Criteria and Production of Mutants for Registration

By the year 2008, eight mutant candidates were identified among 371 living trees in the plantation, these were selected according to morphological and pomological characteristics such as growth habit, fruit shape, fruit set, fruit weight, fruit height, peduncle length, seed weight, soluble solid content rate (brix) and self-fertility. Measurement of minimum 9 g fruit weight, 5 cm peduncle length, 28 mm fruit size, 15% self-fruit and 16% soluble solid content were accepted as selection criteria. In

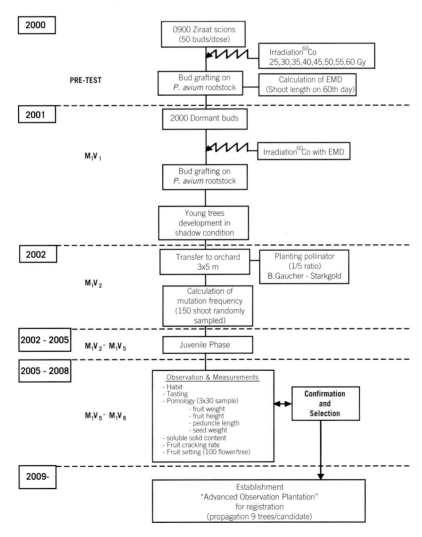

The Joint FAO/IAEA Programme

Figure 34.5 A schematic diagram illustrating the process of mutant production in sweet cherry.

addition the total yield over the first three years was taken into account. Some of these data are presented in **Table 34.2** and shown in **Figure 34.6**.

Dwarf mutants were among the selected as these are of economic significance (**Table 34.3**). Other candidates were observed to be highly fertile, semi-dwarf, additionally possessing low cracking rates. However, no self-fertile types have been observed.

According to the US packing and sales standards on cherry (Anonymous, 2006), fruits smaller than 10 ROW size (equivalent. 26.6 mm diameter and 8.7–10.6 g fruit weight) are not suitable for export. The mutant candidates have proven to be economically valuable considering their average fruit size of 29 mm even in dry years; and while the control trees produced a yield of 18–20 kg/tree, the yield of the candidate trees was more than 29 kg.

A plantation established for advance observations and registration work by the Ministry of Agriculture for Turkey will commence on the basis of the findings gathered from this plantation.

5. References

5.1. Cited References

Anonymous, 2006. Agricultural Marketing Resource Center, Commodity Profile: Cherries, Sweet and Tart. pp. 1–9.

Donini, B. 1980. Mutagenesis Applied to Improve Fruit Trees. Induced Mutations in Vegetatively Propagated Plants. Vienna: IAEA Panel Proceeding Series.

Saamin, S. and Thompson, M.M. 1998. Radiation-induced mutations from accessory buds of sweet cherry, *Prunus avium* L. cv 'Bing'. *Theoretical and Applied Genetics*. 96: 912–916.

5.2. Websites

FAO/IAEA Database of Mutant Variety and Genetic Stocks:

http:// mvgs.iaea.org

The Joint FAO/IAEA Programme

Figure 34.6 Pictures showing the tree, fruit and branches of some dwarf and semi-dwarf mutants. Mutant 25/1, 25/55 and 30.10 are about 50%, 20% and 40% smaller in tree size than their wild type parent.

Table 34.2: Observations of some mutant variety candidates with dwarf, semi-dwarf and high-yield traits

Type	Year	Fruit weight (g)	Fruit width (cm)	Fruit height (cm)	Peduncle length (cm)	Seed weight (g)	Soluble solid content (brix) (%)	Yield (kg)	Cracking (%)
Mutant 25/55	2005	10,4	2,8	2,4	5,7	0,56	14,8	0,115	-
	2006	10,7	3,0	2,5	5,9	0,46	17,4	1,772	48,0
	2007	9,4	2,7	2,4	5,8	0,40	15,5	4,700	0,0
	2008	8,9	2,5	2,4	5,5	0,30	19,6	5,200	0,0
	Mean	9,9	2,8	2,4	5,7	0,43	16,8	11,79 [a]	-
Observations	Semi-dwarf, (20–30% shorter than parent), high yield								
Mutant 35/25	2005	-	-	-	-	-	-	-	-
	2006	9,2	2,7	2,3	5,4	0,49	19,6	0,602	33,2
	2007	10,1	2,7	2,5	5,9	0,43	12,8	4,550	0,0
	2008	9,3	2,6	2,4	4,9	0,34	18,7	3,200	0,0
	Mean	9,7	2,7	2,4	5,4	0,42	17,0	8,350 [a]	-
Observations	Dwarf (50–60% shorter than parent), high yield								
Mutant 50/28	2005	-	-	-	-	-	-	-	-
	2006	12,7	3,5	2,7	5,7	0,52	16,4	3,580	43,4
	2007	10,4	2,8	2,6	5,9	0,40	19,1	13,000	0,0
	2008	8,3	2,5	2,4	5,0	0,37	19,0	12,900	0,0
	Mean	10,5	2,9	2,6	5,5	0,43	18,2	29,480 [a]	-
Observations	Vigorous (15–20 % shorter than parent), grand high yield (50% more than control in tree basis)								
Control	2005	-	-	-	-	-	-	-	-
	2006	10,3	3,0	2,4	5,0	0,50	15,7	0,400	23,7
	2007	9,8	2,7	2,3	5,6	0,44	12,9	8,400	0,0
	2008	8,0	2,4	2,1	5,1	0,37	18,3	10,400	0,0
	Mean	9,3	2,7	2,3	5,0	0,43	15,6	19,200 [a]	-

[a] Cumulative yield for first four year.

Table 34.3: Comparative yield analysis in unit area basis

Subject	Dwarf	Dwarf	Control
Planting space (m)	2.5 × 2.5	3 × 3	6 × 6
Area/tree (m²)	6.25	9	36
Tree number/ha	1,600	1,100	270
Yield/tree (kg) [a]	8	8	20
Yield/ha (kg)	12,800	8,800	5,400
Improved productivity (%)	137	62	0

[a] Cumulative total yield.

Section 5

Mutations in Functional Genomics

C35

Cloning Genes for Mineral Uptake: Examples Using Rice Mutants

J.F.Ma*

Institute of Plant Science and Resources, Okayama University, Chuo 2-10-1, Kurashiki 710-0046
* Correspomding author, EMAIL: maj@rib.okayama-u.ac.jp

1. Introduction

The recent completion of sequencing the rice genome has propelled genomics studies towards the identification of genes and understanding their functions. It is estimated that rice contains 37,000–38,000 genes (International Rice Genome Sequencing Project, 2005), but the functions of the vast majority of these genes are unknown. Many approaches have been suggested in cloning and analyzing a gene, one of these is to use mutants in a forward genetic approach.

Plants require 17 elements for their growth and development, of which 14 are minerals (others are provided by water, carbon dioxide and oxygen). When a mineral is taken up by the roots from the soil it is translocated to the shoots, and then distributed within the plant. Many genes are believed to be involved in this process. On the other hand, plants also have to overcome excessive levels of toxic minerals, such as Cd and Al, in problem soils. Therefore, tolerant genes to these stresses must exist. However, only a few genes to date have been identified in these processes. Most mineral-associated genes remain to be cloned; however, two examples are given of how to isolate mutants and functional genes related to plant nutrition.

2. Example 1: Identification of a Si Transporter Gene

Silicon is a beneficial element for plant growth. Its main function is to help plants to overcome multiple biotic and abiotic stresses. Silicon is especially important for optimal growth and sustainable production of rice. Rice is able to accumulate up to 10% Si in the shoot. This high accumulation has been attributed to the ability of the roots to take up Si rapidly. Physiological studies have shown that Si uptake by rice roots is mediated by a transporter; however, the genes responsible for the high Si uptake are unknown. Rice mutants with a low Si content were first isolated to identify genes responsible for Si uptake.

2.1. Screening Mutants Defective in Si Uptake

Tolerance to germanium (Ge) was used as an indicator for screening rice mutants defective in Si uptake. Germanium is an analogue of Si and previous studies have shown that plant roots take up Si and Ge similarly. However, once it is taken up, Ge is toxic to plants, which is easily characterized by brown spots on the leaf blades (**Figure 35.1**). One mutant showing tolerance to Ge toxicity was isolated from 64,000 M_2 plants obtained from seed mutagenized with NaN_3. This was named *low silicon mutant 1* (*lsi1*).

Short-term uptake experiments showed that Si uptake of the mutant was much lower than that of wild type (WT) rice at both low and high Si concentrations. When the two lines were cultured in a nutrient solution containing 1.5 mM Si for one month, the Si content of the mutant was 30% that of the WT. In contrast, there was no difference in the P and K concentrations between the mutant and WT (**Table 35.1**). These results indicate that *lsi1* is a mutant defective in Si uptake.

2.2. Cloning and Characterizing the Si Transporter Gene, *Lsi1*

An F_2 mapping population was derived from a cross between *lsi1* and WT and used in genetic analyses.

The Joint FAO/IAEA Programme

Figure 35.1 Spot symptoms of Ge toxicity in rice leaf blades (control above).

Table 35.1: Shoot concentrations of Si, P and K of WT and *lsi1* cultured in a nutrient solution with 1.5 mM Si for 1 month

Line	Shoot concentration (%)		
	Si	P	K
WT	4.62	0.51	3.18
lsi1	1.43	0.57	3.23

Among 89 F$_2$ seedlings, 71 seedlings showed a high Si uptake similar to that of WT, while 18 seedlings showed a low Si uptake similar to that of *lsi1*. This segregation ratio fitted a 3:1, suggesting that the low Si uptake of mutant is controlled by a single recessive gene.

Genetic mapping of the mutant gene was performed by segregation analysis in F$_2$ seedlings generated from a cross between *lsi1* and the indica rice variety Kasalath. A total of 128 EST-based polymerase chain reaction (PCR) markers and 58 microsatellite markers, which are scattered around the rice genome, were used for rough mapping of the gene. As a result, the gene was located at about 19 cM between the microsatellite maker RM5303 and an EST-based PCR marker on chromosome 2. For fine mapping of *Lsi1*, new markers were developed that narrowed the genomic region of *Lsi1* to 13.9 kb between the two flanking markers (**Figure 35.2**). A gene candidate within this region was identified using rice GAAS software and the gene sequences of WT and *lsi1* compared. A sequence mutation in the candidate gene, a base change from G in the WT to A in the mutant was revealed. This resulted in an amino acid change from alanine in the wild type to threonine in the mutant at the position of 132 aa. The gene consists of five exons and four introns. The cDNA of this gene was 1,409 bp long and the deduced protein consisted of 298 amino acids. The gene is predicted to encode a membrane protein similar to water channel proteins, aquaporins. The predicted amino acid sequence has six transmembrane domains and two Asn-Pro-Ala (NPA) motifs, which are well conserved in typical aquaporins. Blast search and ClustalW analysis revealed that *Lsi1* belongs to the Nod26-like major intrinsic protein (NIP) sub-family.

An RNAi technique (to suppress the expression of Lsi1) was used to confirm *Lsi1* involvement in Si uptake. In RNAi transgenic lines, Si uptake was significantly reduced compared with that of the vector control plants, and the expression of *Lsi1* in the RNAi transgenic lines was significantly suppressed correspondingly (**Figure 35.3**). The RNAi transgenic lines also showed higher resistance to Ge toxicity. In contrast, there was no difference in water uptake between the RNAi lines and the vector control lines (Ma *et al.*, 2006). Taken together, the results indicate that the *Lsi1* gene is responsible for a high Si uptake in rice.

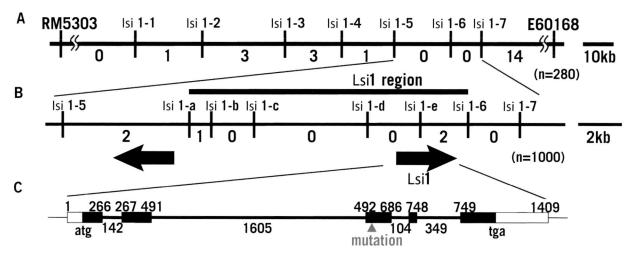

The Joint FAO/IAEA Programme

Figure 35.2 Mapping of *Lsi1* and gene structure. (A,B) *lsi1* mutation was mapped on the long arm of chromosome 2 between markers RM5303 and E60168. C) *Lsi1* gene structure at genomic sequence. Red arrow shows the *lsi1* mutant position.

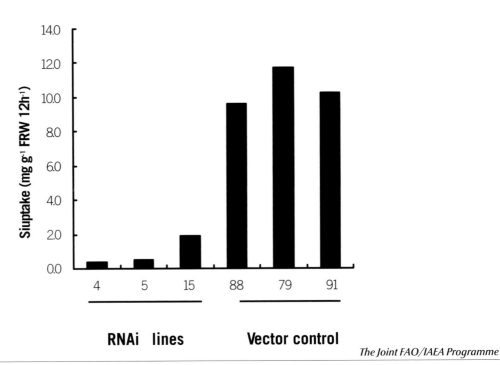

Figure 35.3 Si uptake of the RNAi transgenic lines and the vector control lines.

The function of *Lsi1* in terms of expression and localization was also investigated. Quantitative RT-PCR analysis revealed that *Lsi1* was mainly expressed in the root, but not in shoots nor in panicles (**Figure 35.4A**). The expression was regulated by Si supply; the expression was decreased to one fourth in plants continuously supplied with Si. The Lsi1 protein was localized in the basal region of seminal roots, on the distal side of both the exodermis and endodermis (**Figure 35.4A**). In both the exodermis and endodermis, the Casparian strips are formed during root maturation in rice, which are a barrier to apoplastic permeation of water and solutes. These apoplastic barriers apparently restrict silicic acid permeability. Therefore, for transport of Si into the cells, Lsi1 is required to be expressed on both the exodermis and endodermis.

By using the same approach, recently a gene (*Lsi2*) encoding an efflux transporter of silicon has been

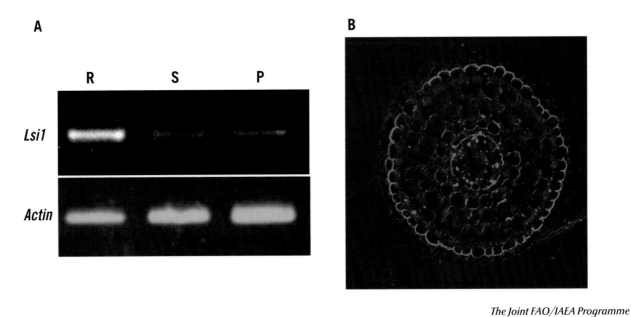

Figure 35.4 Expression of *Lsi1* and localization of the Lsi1 protein. A) expression of Lsi1 in the roots (R), shoots (S) and panicle (P) grown in a paddy field. The sample was taken at week 10 after transplanting. B) Lsi1 immunolocalization stained with an anti-Lsi1 polyclonal antibody.

identified in rice (Ma *et al.,* 2007). This transporter is also localized at the exodermis and endodermis like Lsi1. However, in contrast to Lsi1, Lsi2 is localized at the proximal side (Ma *et al.,* 2007).

3. Example 2: Identification of Al-Tolerant Genes

Soil acidity is a major problem limiting crop productivity throughout the world. Although crop productivity on acid soils is limited by multiple abiotic stress factors such as deficiency of phosphorus, calcium and magnesium, toxicity of manganese, and so on, toxicity of aluminum (Al) has been regarded as a major factor limiting crop productivity on acid soils. Soil Al dissolves at low pH in an ionic form (mainly Al^{3+}), which inhibits root elongation at a very low concentration, and subsequently the uptake of water and nutrients, leading to reduced plant growth and increased susceptibility to environmental stresses.

There are large variations in Al resistance among plant species and varieties within a species. Rice is the most Al-resistant species among small-grain cereal crops, although genotypic variation also exists in this species (Foy, 1988). However, the physiological mechanism responsible for the high Al resistance in rice has not been elucidated. Unlike other cereal species such as rye, wheat and barley, secretion of organic acid anions does not seem to be a mechanism for Al resistance (Ma *et al.,* 2002). Several genetic studies on Al resistance in rice have been carried out and three to ten quantitative trait loci (QTLs) have been identified in different genetic backgrounds and experimental conditions. One major QTL for Al resistance was detected on chromosome 1 in all populations (Wu *et al.,* 2000; Nguyen *et al.,* 2001, 2002, 2003; Ma *et al.,* 2002). This QTL has been recently reported to be orthologous to Alt$_{SB}$, a major Al resistance gene in sorghum on chromosome 3 (Magalhaes *et al.,* 2004). Another major rice Al resistance QTL has been identified on chromosome 3 (Wu *et al.,* 2000; Nguyen *et al.,* 2003), which is homeologous to the Triticeae group 4 chromosomes. These studies provided an important starting point for isolating genes responsible for high Al resistance in rice. However, no Al-resistance genes have been cloned from rice so far. One possible reason is that the contribution of each QTL detected is not large (less than 30%), giving rise to difficulty in discriminating between phenotypes when the map-based cloning technique is applied. As an alternative strategy to QTL mapping, mutants sensitive to Al were exploited.

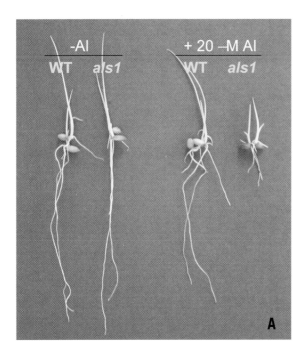

The Joint FAO/IAEA Programme

Figure 35.5 Phenotype of wild type rice (WT, var. Koshihikari) and an Al-sensitive mutant (*star1/als1*). A) seedlings (4 days old) were exposed to a 0.5-mM CaCl$_2$ solution (pH 4.5) containing 0 or 20 μM Al for 72 h; B) seedlings (25 days old) were cultured in an acid Andosol (pH 4.8) or a near-neutral alluvial soil (pH 6.5) for 20 days.

3.1. Screening Al-Sensitive Mutants

M_3 populations of an Al-resistant variety of rice (*Oryza sativa* L. cv. Koshihikari) were generated by γ-ray treatment and were subsequently screened for Al-sensitive mutants. The screening test was done with 50 μM Al and relative root elongation was used as an index for evaluation of Al resistance (**Figure 35.5**). In the first screening, 60 putative mutants were selected from a total of 560 lines. The second round of screening resulted in 14 putative mutants. Finally, two mutants were isolated in the third screening of M_4 seedlings; the most Al-sensitive mutant (*star1*, sensitive to Al rhizotoxicity, formerly named as *als1*, Al sensitive 1) is described here.

The morphology and growth rate of both the root and shoot were similar between the WT plant and *star1* mutant in the absence of Al (**Figure 35.5A**). However, in the presence of Al at 20 μM, root elongation of the WT plant was not significantly inhibited, whereas that of the mutant was severely inhibited. Furthermore, when both lines were grown in an alluvial soil without Al toxicity, the growth of both roots and shoots of the mutant were similar to those of the wild type (**Figure 35.5B**), whereas in the acid soil at pH 4.8, the growth of the root and the shoot of the mutant decreased compared to the WT by 85% and 70%, respectively. Reduction of shoot growth is probably a consequence of root growth inhibition, which decreased the uptake of nutrients and water.

The Al content of the root apices (0–1cm) was significantly higher in the mutant than in the WT plant irrespective of Al concentration in the culture solution (**Figure 35.6**). This result suggests that Al-exclusion mechanisms rather than internal detoxification are disrupted in the mutant. There were no significant differences in the sensitivity to La and Cd toxicity between the WT plant and the mutant, indicating that the mutation is specific to Al toxicity.

3.2. Genetic Analysis and Molecular Mapping of *STAR1/Als1*

Genetic analyses were performed on progeny from crosses between the *star1/als1* mutant and a WT plant (var. Koshihikari). The F_1 seedlings were highly resistant to Al like the WT plant, suggesting that *als1* is a recessive mutant. Further analysis with the F_2 progeny resulted in 70 Al-resistant seedlings (>30% relative root elongation) and 20 Al-sensitive seedlings (<20% relative root elongation). Thus the Al-resistant and the Al-sensitive seedlings segregated at a 3:1 ratio ($\chi^2 = 0.37$, $0.5 < P < 0.75$), suggesting that the sensitivity to Al in *als1* is controlled by a single recessive gene.

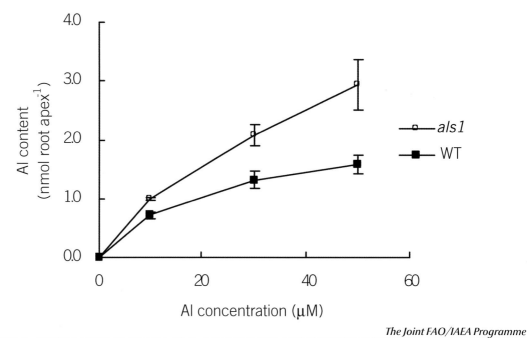

The Joint FAO/IAEA Programme

Figure 35.6 Al content in the root apices of wild type rice (var. Koshihikari) and the mutant (*als1/star1*). Four-day-old seedlings were exposed for 24 h to a 0.5-mM $CaCl_2$ solution (pH 4.5) containing 0, 10, 30 or 50 μM Al. Root apices (0–1 cm) were excised and the Al concentration was determined by graphite furnace atomic absorption spectrophotometry. Error bars represent ± SD ($n = 3$).

The gene (*STAR1/Als1*) responsible for Al resistance in rice was mapped using a segregating F_2 population from the cross:mutant (*star1/als1*) X Kasalath. Bulked segregant analysis (Michelmore *et al.*, 1991) was performed by pooling equal amounts of DNA from 10 Al-sensitive plants and comparing this with a bulk from 10 Al-resistant F_2 plants. Among 80 indel markers examined, 59 markers showed polymorphism between the mutant and Kasalath and were then used to examine polymorphism among the Al-sensitive and Al-resistant bulks. As a result, the marker R6M44 on the long arm of chromosome 6 was found to be associated with Al sensitivity. Co-segregating analysis using the 27 Al-sensitive F_2 plants showed that R6M44 was located 24.7 cM from *STAR1/Als1*. Further fine mapping of the gene exploited two polymorphic indel markers MaOs0615 and MaOs0619. Linkage analysis showed that *STAR1/Als1* was located between MaOs0615 and MaOs0619, with a distance of 3.8 cM and 1.9 cM, respectively (**Figure 35.7**). For fine mapping of *STAR1*, 716 homozygotes with

low Al tolerance was used. Three polymorphic indel markers between the above two markers was developed and this gene was mapped between MaOs0624 and MaOs0617 on the two overlapping PAC clones (AP003770 and AP003771) (Huang *et al.*, 2009). With further development of two new indel markers and three CAPS markers on the two PACs, the gene was finally mapped to an 88-kb candidate region on PAC clone AP003771. There are 14 predicated genes within this region based on rice genome annotation on the TIGR website (http://www.tigr.org), including seven putative retrotransposon genes. Sequencing the other seven candidate genes revealed that one gene has a 15-bp deletion in the second exon region in the mutant compared to the WT, while there were no mutations in other genes. This gene consists of four exons and three introns, encoding a 291-amino-acid protein. The gene (*STAR1/Als1*) is predicted to encode a nucleotide-binding domain (NBD) of a putative ATP-binding cassette (ABC) transporter protein, which has all typical

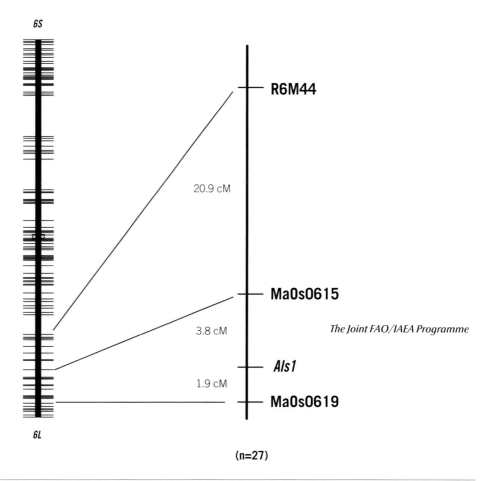

Figure 35.7 Linkage relationship between indel markers and a gene for Al resistance (*STAR1/Als1*) on chromosome 6 in rice.

motifs conserved in NBD, Walker-A, Q-loop, ABC signature, Walker-B, D-loop and H-loop. According to new nomenclature, STAR1 was categorized into the NO subgroup of the subfamily I, which includes NBD proteins with low similarity and unknown functions. In rice genome, there is no homologue of STAR1. Database searches led to finding one homologue of STAR1 each in maize, Arabidopsis, grape and moss (*Physcomitrella patens*), but their function is unknown. Further work showed that STAR1 interacts with STAR2 to function as a bacterial-type ABC transporter, which is required for Al tolerance in rice (Huang *et al.,* 2009).

Since the map-based cloning technique has been established, successful isolation of a mutant is a key step in the process of gene cloning. For this, a reliable, less labour-intensive and time-consuming, and a more convenient screening method is important. Integrated knowledge of plant nutrition, plant physiology, chemistry and biochemistry is required to help to develop a successful screening method.

4. Acknowledgements

The study was partly supported by a grant from the Ministry of Agriculture, Forestry and Fisheries of Japan (Rice Genomics for Agricultural Innovation (IPG-006 and QT-3006) to J.F.M.) and a Grant-in-Aid for Scientific Research from the Ministry of Education, Culture, Sports, Science and Technology of Japan (Nos. 21248009 and 22119002). I thank Fangjie Zhao for his critical reading of this manuscript.

5. References

5.1. Cited references

Foy, C.D. 1988. Plant adaptation to acid aluminum-toxic soils. *Commun Soil Sci. Plant Ana.* 19: 959–987.

International Rice Genome Sequencing Project 2005. The map-based sequence of the rice genome. *Nature.* 436: 793–800.

Huang, C.F., Yamaji, N., Mitani, N. *et al.* 2009. A bacterial-type ABC transporter is involved in aluminum tolerance in rice. *Plant Cell.* 21: 655–667.

Ma, J.F., Shen, R., Zhao, Z. *et al.* 2002. Response of rice to Al stress and identification of quantitative trait loci for Al tolerance. *Plant Cell Physiol.* 43: 652–659.

Ma, J.F. and Furukawa, J. 2003. Recent progress in the research of external Al detoxification in higher plants: a mini review. *J. Inorg. Biochem.* 97: 46–51.

Ma, J.F., Tamai, K., Yamaji, N. *et al.* 2006. A silicon transporter in rice. *Nature.* 440: 688–691.

Ma, J.F., Yamaji, N., Mitani, N. *et al.* 2007. An efflux transporter of silicon in rice. *Nature.* 448: 209–211.

Magalhaes, J.V., Garvun, D.F., Wang, Y. *et al.* 2004. Comparative mapping of a major aluminum tolerance gene in sorghum and other species in the Poaceae. *Genetics.* 167: 1905–1914.

Michelmore, R.W., Paran, I. and Kesseli, R.V. 1991. Identification of markers linked to disease resistance genes by bulked segregant analysis – a rapid method to detect markers in specific genomic regions using segregating populations. *PNAS.* 88: 9828–9832.

5.2. Further Reading

Kochian, L.V., Hoekenga, O.A. and Piñeros, M.A. 2004. How do crop plants tolerate acid soils? Mechanisms of aluminum tolerance and phosphorous efficiency. *Annu Rev Plant Biol.* 55: 459–493.

Ma, J.F. and Takahashi, E. 2002. Soil, Fertilizer, and Plant Silicon Research in Japan. Amsterdam: Elsevier.

Ma J.F. and Yamaji, N. 2008. Functions and transport of silicon in plants. *Cellular and Molecular Life Sciences.* 65:3049-3057.

Molecular Genetics of Symbiotic Plant–Microbe Interactions in a Model Legume, *Lotus japonicus*

H.Kouchi*

National Institute of Agrobiological Sciences, Tsukuba, Ibaraki 305-8602, Japan
*Corresponding author, E-MAIL: kouchih@nias.affrc.go.jp

1. Introduction

Legume plants are able to form root nodules by symbiotic interactions with soil bacteria of genera, *Rhizobium*, *Sinorhizobium*, *Bradyrhizobium*, *Mesorhizobium* and *Azorhizobium* (referred to as *Rhizobium* hereafter). The micro-symbionts in the nodules fix atmospheric nitrogen and supply the fixed nitrogen to the host legumes, thus making legume–*Rhizobium* symbiosis highly beneficial in agricultural practice.

Identification of rhizobial lipochichin oligosaccharide signal molecules (Nod factors) in the early 1990s has led to great progress in understanding the functions of bacterial genes that are responsible for symbiosis. However, identification of host legume genes that are essential for symbiosis has been very limited. Two legume species, *Lotus japonicus* and *Medicago truncatula*, are diploid with relatively small genome sizes and are capable of molecular transfection, and are thus model legumes for molecular genetic studies (**Box 1**).

The accumulation of molecular genetics capabilities, such as expression sequence tags (ESTs) (Asamizu *et al.,* 2000; Kouchi *et al.,* 2004), high-density linkage maps and genome sequencing (Sato *et al.,* 2008) has allowed a number of host legume genes required for symbiosis to be isolated. Recent developments of molecular cloning of host legume symbiotic genes using a model legume, *L. japonicus* is described here. Available resources and databases for *L. japonicus* are listed (see websites in References).

2. Symbiotic Mutants of *Lotus japonicus*

2.1. The Nodulation Process

In general, the interactions between legume plants and *Rhizobium* bacteria initiate from the colonization of rhizobia on legume root hairs, followed by invasion of rhizobia into root epidermis and then cortical cells through infection threads. Concomitantly with these bacterial infection processes, cell division is induced in the root cortex to form a nodule primordium, which develops into a highly organized symbiotic organ, the root nodule. Rhizobia reside in nodule cells and are enclosed by a plant-derived symbiosis-specific membrane (peribacteroid membrane, PBM), differentiate

into bacteroids, and finally fixation of atmospheric nitrogen takes place (**Figure 36.1**).

2.2. Types of Symbiotic Mutants

Lotus japonicus mutants were mostly generated by chemical (ethyl methane sulphonate: EMS) mutagenesis, T-DNA insertion and/or somatic mutation through re-generation of plants from de-differentiated calli or cultured cells (Schauser *et al.,* 1998; Szczyglowski *et al.,* Kawaguchi *et al.,* 2002). The most comprehensive list of the symbiotic mutants identified so far is given by Sandal *et al.* (2006) together with gene map positions. These symbiotic mutants were categorized into three groups according to the developmental stages of symbiosis in which the defects are attributed, i.e. non-nodulating (Nod⁻), defective in cooperative histogenesis (Hist⁻), and formation of ineffective nodules (Fix⁻). Nod⁻ mutants are attributed to defects in very early steps of symbiotic interactions, and hence neither bacterial infection nor nodule primordium formation occurs. Hist⁻ mutants are characterized by defects in infection thread formation and/or its growth, and accompanied by incomplete nodule organogenesis. Fix⁻ mutants form apparently normal nodules containing endosymbiotic bacteria, but their nodules show no or very low nitrogen-fixing activity. Development of nitrogen-fixing symbiosis is depicted in **Figure 36.1** together with representatives of genes and/or genetic loci belonging to these categories.

Besides these mutant categories, there is another category that has defects in regulation of nodule numbers. This category contains so-called hyper-nodulation (Nod⁺⁺) mutants, which show excessive nodulation owing to defects in "autoregulation" (Oka-Kira and Kawaguchi, 2006).

Box 36.1: Characteristics of *Lotus japonicus*

Genome size	470–490 Mb, 0.5 pg/haploid
Generation time	2–4 months
Seed production	~6,000/plant, 10–20 seeds/pod
Plant height	~30 cm
Transformation	Hypocotyl infection by *A. tumefaciens*

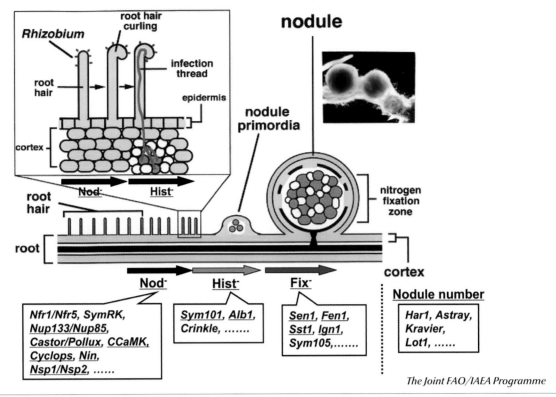

Figure 36.1 Nodulation process of *Lotus japonicus* and representatives of symbiotic mutants (revised from Kawaguchi *et al.,* 2002).

3. Early Symbiotic Gene Cascade

3.1. Perception of Nod Factors and Common Symbiotic Pathway

Nod⁻ mutants are attributed to the defects in very early steps of symbiotic interactions, such as Nod factor perception and/or the immediate downstream signal transduction pathways that precede infection and nodulation processes. About half of Nod⁻ mutants identified so far are defective in symbiosis with arbuscular mycorrhiza (AM) fungi (Myc⁻), indicating that both symbioses share common mechanisms. Since the symbiosis with AM fungi is distributed through more than 80% of plant species and is shown to have an evolutionary origin far preceding legume–*Rhizobium* symbiosis, it is thought that legume plants have acquired the ability of symbiosis with *Rhizobium* bacteria on the basis of pre-existing mechanisms for mycorrhizal symbiosis (Kistner *et al.,* 2005).

Figure 36.2 shows the current working model of a symbiotic gene cascade in *Lotus japonicus*. *NFR1* and *NFR5* encode putative Nod factor receptors and the defects in these genes lead to complete lack of the responses to Nod factors and *Rhizobium* inoculation. However, those mutants are able to achieve symbiosis with AM fungi, thus showing Nod⁻Myc⁺ phenotype

(Radutoiu *et al.,* 2003). Mutants with a Nod⁻Myc⁻ phenotype are placed on a "common sym pathway" (CSP) and their causal genes are termed "common sym genes". The most prominent physiological event on the CSP is calcium spiking, which appears as an oscillation of Ca²⁺ concentration in root hair cells in response to Nod factor application (**Figure 36.2**). Calcium spiking is thought to play a pivotal role in early symbiotic signalling. The CSP genes are classified into two groups according to their positions before and after calcium spiking. *SymRK* encodes a receptor-like protein kinase, and *LjNup133* and *LjNup85*, encode nucleoporins, are positioned upstream of calcium spiking. CASTOR and POLLUX are both putative calcium-gated ion channel proteins and are also crucial for induction of calcium spiking. The mutants defective in these genes fail to induce calcium spiking upon bacterial inoculation. The exact biochemical functions of these genes are still to be elucidated. CCaMK is a calcium- and calmodulin-dependent protein kinase and is postulated to work as a decoder of calcium spiking. It was first identified as the causal gene of a symbiotic mutant *Ljsym72*, which showed the Nod⁻Myc⁻ phenotype but was able to induce calcium spiking in response to bacterial inoculation. *Ljsym72* was shown to be a loss-of-function mutant of LjCCaMK. Interestingly, the *snf1* mutant, which forms

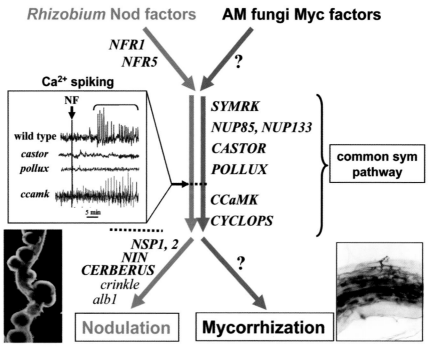

Figure 36.2 A working model of the symbiotic gene cascade in *Lotus japonicus* at early stages of symbiosis with Rhizobia and AM fungi. The loci of which causal gene was isolated are bold faced. (courtesy of Dr. H. Imaizumi-Anraku).

spontaneous nodules without *Rhizobium* bacteria, has been demonstrated to be due to the gain-of-function mutation of the same gene. In the *snf1* mutant, CCaMK exhibits kinase activity constitutively in an independent manner to calcium spiking by a single amino acid substitution at the autophosphorylation site in its kinase domain. These findings demonstrate the key regulatory role of CCaMK in activating the downstream signalling cascade(s) that lead to nodule organogenesis (Tirichine *et al.*, 2006). Recently, a novel CSP component CYCLOPS was identified, which was shown to interact directly with CCaMK (Yano *et al.*, 2008).

3.2. Nodulation-Specific Pathway

Nod⁻ (Myc⁺) mutants downstream of the common sym pathway are allocated to the nodulation-specific pathway. A putative transcription factor, NIN, and GRAS family transcription factors, NSP1 and NSP2, have been identified as the components of the earliest step of the nodulation-specific pathway. Further genetic approaches such as analyses of double and/or triple mutants, together with biochemistry of isolated gene products, will reveal more insights of early symbiotic signalling pathways. Since transformation with gain-of-function CCaMK^{T265D} is demonstrated to induce spontaneous nodulation, transfection of CCaMK^{T265D} into

the background of the other Nod⁻ mutants would be a powerful approach to investigate interrelationships of individual symbiotic genes.

Host plant signalling specifically responsible for mycorrhization is totally unknown at present. This is mainly due to the fact that symbiosis with AM fungi is obligatory which makes it difficult to analyze the mycorrhization process in great detail. Key to overcoming these difficulties is the identification of the signal molecules (Myc factors) produced by AM fungi and the isolation of Nod⁺Myc⁻ mutants.

4. Infection Process and Symbiotic Nodule Formation

4.1. Infection and Nodule Organogenesis

Hist⁻ mutants represent genetic loci that have effects on the infection process. The mutants of this category are characterized by defects in infection thread formation and/or its growth, as well as incomplete nodule organogenesis with absence of bacterial infected cells. The *alb1* and *crinkle* mutants are typical examples of this category. In *crinkle*, infection thread development is arrested upon penetration into the epidermal cells (**Figure 36.3**) and nodule organogenesis is blocked at very early stages, resulting in the formation of small bumps. Very thick infection threads mostly aborted in root hair cells are formed in *alb1* mutants. Even when the infection threads develop in root cortical cells the bacteria fail to be released into host cell cytoplasm and are tightly packed in the infection threads. The nodulation process in *alb1* appears to be more advanced than *crinkle*, but the nodules are irregularly shaped with undeveloped vascular bundles.

Recently, a novel Hist⁻ mutant, *Ljsym101* (*cerberus*) has been identified and the *CERBERUS* gene was cloned. The *cerberus* mutant forms many small bumps, with no successful infection of rhizobia. In this mutant, rhizobia colonize at curled root hair tips, but formation of infection threads are severely aborted. Even when the infection threads are formed on rare occasions, the growth is arrested near the root hair tip or at the epidermal layer, and fails to penetrate the cortical cells. These phenotypes strongly suggest that the *CERBERUS* gene is required for very early steps of infection thread formation, thus it is positioned upstream of *CRINCLE* and *ALB1*.

4.2. Genes Involved in Bacteroid Differentiation and Regulation of Nitrogen Fixation

In contrast to Nod⁻ and Hist⁻ mutants, genetic loci in Fix⁻ mutants are attributed to much later stages of symbiotic nodule development. This involves the differentiation of rhizobia into bacteroids followed by induction of nitrogenase, and the organization of metabolic functions required for nitrogen fixation in the host nodule cells. Fix⁻ mutants form morphologically normal nodules with infected cells containing endosymbiotic bacteria, but exhibit no or very low nitrogen-fixing activity. Therefore, the genetically controlled specific interactions between legumes and rhizobia are not restricted only to the early stages of symbiosis as mediated by Nod factors, but also exist at rather later stages of nodule development.

Phenotypes of representative Fix⁻ mutants are summarized in **Table 36.1**. Among these Fix⁻ mutants *Ljsym75* (*sen1*) and *Ljsym81* (*sst1*) are the most well characterized. The nodules formed on the *sen1* mutant are morphologically comparable to those on wild type plants, but exhibit no nitrogen fixation activity. Nitrogenase proteins in the endosymbiotic bacteria are hardly detectable. Electron microscopy suggested that in *sen1* nodules, differentiation of rhizobia into bacteroids is incomplete, and thus the bacteria are unable to induce nitrogenase genes. The *Sen1* gene has now been cloned (**Table 36.1**).

The *sst1* mutant forms nodules with very low nitrogen fixation activity (**Figure 36.4**), and bacteroid differentiation resembles that in wild type nodules rather than *sen1* nodules. The gene was identified as encoding nodule-specific sulphate transporter (SST1) (Krusell *et al.*, 2005). SST1 protein is a component of peribacteroid membrane (PBM) and transport SO_4^{2-} to bacteroids from plant cell cytosol (**Figure 36.4**), thus meeting the demand for sulphur by intracellular rhizobia to support continuous synthesis of nitrogenase and the related electron transfer proteins.

Bacteroid differentiation is one of the central subjects in the studies of Fix⁻ mutants. Biochemical and molecular mechanisms underlying differentiation of *Rhizobium*

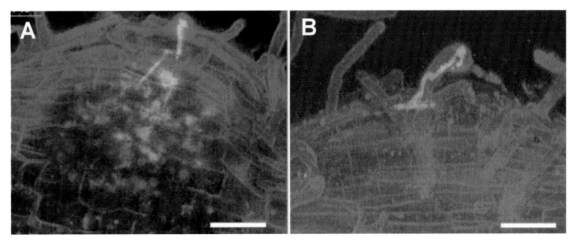

The Joint FAO/IAEA Programme

Figure 36.3 Infection process in wild type (A) and *crinkle* (B) in which the infection thread is arrested. Infection thread can be seen by green fluorescence from GFP-labelled *Mesorhizobium loti*. Bars = 50 μm (from Tansengco *et al.*, 2003)

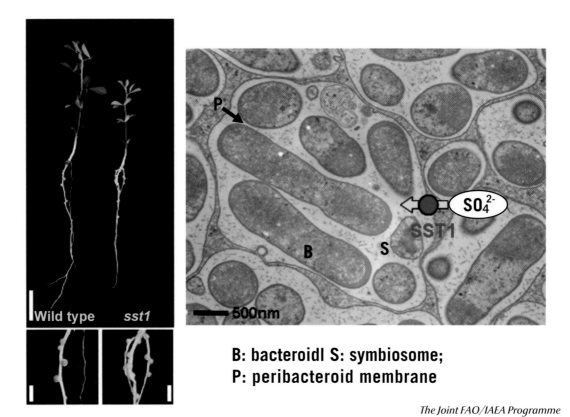

B: bacteroidl S: symbiosome;
P: peribacteroid membrane

The Joint FAO/IAEA Programme

Figure 36.4 Left: Phenotype of *sst1* mutant. Nodules formed on the mutant are pale pink and nitrogen fixing activity is very low. Bars = 2 cm (upper panel) and 1 mm (lower panels) (From Krusell *et al.,* 2005). Right: Possible function of SST1. (courtesy of Prof. N. Suganuma)

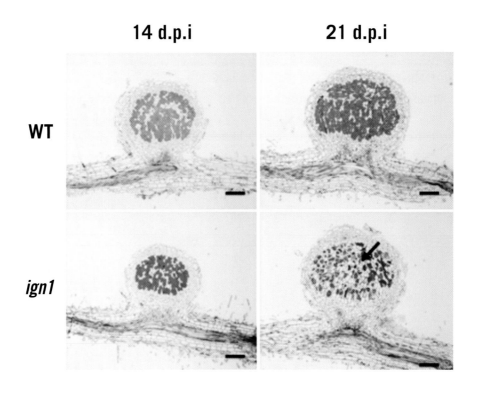

The Joint FAO/IAEA Programme

Figure 36.5 Premature senescence of *ign1* nodules. dpi = days post *M. loti* inoculation. Bars = 20 μm.

Table 36.1: Representatives of Fix- mutants of *Lotus japonicus* and their phenotypes[1]

Locus	Endocytosis	Nitrogenase	N$_2$ fixation	Early senescence	Gene cloning status
sen1	+	–	–	+	Cloned
fen1	+	+	+/–	+	Cloned
sst1	+	+	+/–	+	Sulphate transporter
ign1	+	+	–/+	+++	Ankyrin repeat membrane protein
Ljsym105	+	+	+/–	+++	Underway

[a] A more comprehensive list is given by Sandal *et al.,* 2006. + and – mean positive and negative for each phenotype, respectively. +/– in N$_2$-fixation means presence of activity at low levels.

bacteria into the symbiotic form, bacteroids, has not been studied well from the host plant side nor from the bacterial side. A recent transcriptome analysis using a DNA array constructed on the basis of whole genome sequencing of *Mesorhizobium loti* revealed that bacteroid differentiation involves global and drastic alterations in gene expression in rhizobia (Uchiumi *et al.,* 2004). Differentiation to nitrogen-fixing bacteroids is strictly controlled by interactions with the host nodule cells. Fix⁻ mutants provide a powerful tool to investigate such interactions and the host plant factors required for bacteroid differentiation as well as its maintenance and persistence.

Nodules formed on Fix⁻ mutants exhibit premature senescence in common, that are characterized by abnormal and/or excessive vacuolation of infected cells, enlargement of symbiosomes and disintegration of infected cell cytoplasm. The temporal patterns and extent of these phenotypes of premature senescence is considerably different between each Fix⁻ mutant. The premature senescence of nodules is not simply due to nitrogen deficiency caused by ineffective nitrogen fixation. It is more likely that it reflects the activation of a kind of plant defense response to exclude non-fixing or inefficient endosymbionts which confer no benefit to the host plants. For instance, a Fix⁻ mutant, *ign1*, exhibits very rapid disintegration of symbiosomes of the infected cells as compared with the other Fix⁻ mutants identified so far (**Figure 36.5**). IGN1 was demostrated to be a novel plasma membrane protein containing ankyrin repeats, but its biochemical functions remain to be elucidated (Kumagai *et al.,* 2007). Investigations on molecular mechanisms of premature senescence of nodules in these Fix⁻ mutants may lead us to understand how intracelular rhizobia are stably maintained in the

host legume cells without invoking defense responses against microbial invasion.

5. Acknowledgements

The author thanks Drs. H. Imaizumi-Anraku and Y. Umehara of NIAS, and Prof. N. Suganuma of Aichi University of Education for providing information on unpublished data.

6. References

6.1. Cited References

Asamizu, E., Nakamura, Y., Sato, S. *et al.* **2000.** Generation of 7137 non-redundant expressed sequence tags from a legume. *Lotus japonicus. DNA Res.* 7: 127–130.

Kawaguchi, M., Imaizumi-Anraku, H., Koiwa, H. *et al.* **2002.** Root, root hair, and symbiotic mutants of the model legume *Lotus japonicus. Mol. Plant-Microbe Interact.* 15: 17–26.

Kistner, C., Winzer, T., Pitzschke, A. *et al.* **2005.** Seven *Lotus japonicus* genes required for transcriptional reprogramming of the root during fungal and bacterial symbiosis. *Plant Cell.* 17: 2217–2229.

Kouchi, H., Shimomura, K., Hata, S. *et al.* **2004.** Large-scale analysis of gene expression profiles during early stages of root nodule formation in a model legume, *Lotus japonicus. DNA Res.* 11: 263–274.

Krusell, L., Krause, K., Ott, T. *et al.* **2005.** The sulfate transporter SST1 is crucial for symbiotic nitrogen fixation in *Lotus japonicus* root nodules. *Plant Cell.* 17: 1625–1636.

Kumagai, H., Hakoyama, T., Umehara, Y. *et al.* 2007. A novel ankyrin-repeat membrane protein IGN1 is required for persistence of nitrogen-fixing symbiosis in root nodules of *Lotus japonicus*. *Plant Physiol.* 143: 1293–1305.

Oka-Kira, E. and Kawaguchi, M. 2006. Long-distance signaling to control root nodule number. *Curr. Opin. Plant Biol.* 9: 496–502.

Radutoiu, S., Madsen, L.H., Madsen, E.B. *et al.* 2003. Plant recognition of symbiotic bacteria requires two LysM receptor-like kinases. *Nature.* 425: 585–592.

Sandal, N., Petersen, T.R., Murray, J. *et al.* 2006. Genetics of symbiosis in *Lotus japonicus*: Recombinant inbred lines, comparative genetic maps and map position of 35 symbiotic loci. *Mol. Plant-Microbe Interact.* 19: 80–91.

Schauser, L., Handberg, K., Sandal, N. *et al.* 1998. Symbiotic mutants deficient in nodule establishment identified after T-DNA transformation of *Lotus japonicus*. *Mol. Gen. Genet.* 259: 414–423.

Szczyglowski, K., Shaw, R.S., Wopereis, J. *et al.* 1998. Nodule organogenesis and symbiotic mutants of the model legume *Lotus japonicus*. *Mol. Plant Microbe Interact.* 11: 684–697.

Tansengco, M.L., Hayashi, M., Kawaguchi, M. *et al.* 2003. *Crinkle*, a novel symbiotic mutant that affects the infection thread growth and alters the root hair, trichome, and seed development in *Lotus japonicus*. *Plant Physiol.* 131: 1054–1063.

Tirichine, L., Imaizumi-Anraku, H., Yoshida, S., Murakami, Y., Madsen, L.H., Miwa, H., Nakagawa, T., Sandal, N. *et al.* 2006. Deregulation of a Ca^{2+}/calmodulin dependent kinase leads to spontaneous nodule development. *Nature.* 441: 1153–1156.

Uchiumi, T., Ohwada, T., Itakura, M., Mitsui, H., Nukui, N., Dawadi, P., Kaneko, T., Tabata, S., Yokoyama, T., Tejima, K. *et al.* 2004. Expression islands clustered on symbiosis island of *Mesorhizobium loti* genome. *J. Bacteriol.* 186: 2439–2448.

Yano, K., Yoshida, S., Müller, J. *et al.* 2008. CYCLOPS, a mediator of symbiotic intracellular accommodation. *Proc. Natl. Acad. Sci. USA.* 105: 20540–20545.

6.2. Websites

Genome sequences, linkage map and markers:
> http://www.kazusa.or.jp/lotus/

EST database:
> http://www.kazusa.or.jp/en/plant/lotus/EST/

Gene expression database:
> http://www.kazusa.or.jp/en/plant/lotus/EST/

***Mesorhizobium loti* Genome:**
> http://www.kazusa.or.jp/rhizobase/

Legume Base (Japan National BioResource Project):
> http://www.shigen.nig.ac.jp/legume/legumebase/index.jsp

EMS mutant library (Riken):
> http://grs.psc.riken.jp/index.html

TIGR *Lotus japonicus* Gene Index:
> http://rice.tigr.org/tigr-scripts/tgi/T_index.cgi?species=l_japonicus

Lotus TILLING homepage:
> http://www.lotusjaponicus.org/tillingpages/homepage.htm

6.3. Further Reading

Hayashi, M., Miyahara, A., Sato, S. *et al.* 2001. Construction of a genetic linkage map of the model legume *Lotus japonicus* using an intraspecific F-2 population. *DNA Res.* 8: 301–310.

Imaizumi-Anraku, H., Takeda, N., Charpentier, M. *et al.* 2005. Plastid proteins crucial for symbiotic fungal and bacterial entry into plant roots. *Nature.* 433: 527–530.

Murakami, Y., Miwa, H., Imaizumi-Anraku, H. *et al.* 2006. Positional cloning identifies *Lotus japonicus* NSP2, a putative transcription factor of the GRAS family, required for NIN and ENOD40 gene expression in nodule initiation. *DNA Res.* 13: 255–265.

Nishimura, R., Hayashi, M., Wu, G.-J. *et al.* 2002. HAR1 mediates systemic regulation of symbiotic organ development. *Nature.* 420: 426–429.

Schauser, L., Roussis, A., Stiller, J. *et al.* 1999. A plant regulator controlling development of symbiotic root nodules. *Nature.* 402: 191–195.

Mutational Dissection of the Phytochrome Genetic Systems in Rice

M.Takano*, X.Xianzhi and N.Inagaki

Photobiology and Photosynthesis Unit, Division of Plant Science, National Institute of Agrobiological Sciences, Tsukuba, Ibaraki 305-8602, Japan
* Corresponding author, E-MAIL: mtakano@nias.affrc.go.jp

1. **Introduction**
2. **Mutant Isolation**
    2.1. *Tos17* Retrotransposon Mutants
    2.2. Mutants Induced by Gamma Rays
3. **Functional Analysis of Phytochrome Mutants**
    3.1. Genetic Basis of Phytochrome Mutations
    3.2. Effects on Photomorphogenesis
    3.3. Effects on Light-Inducible Genes
    3.4. Effects on Flowering Time
4. **Acknowledgements**
5. **References**

1. Introduction

Light is one of the most important environmental stimuli, playing a pivotal role in the regulation of plant growth, development and metabolic activities. The perception of environmental light by plants is achieved by a family of plant photoreceptors which are capable of detecting a wide spectrum of wavelengths ranging from UV to far-red light: these include the phytochromes, cryptochromes, phototropins and several others. Phytochromes are the most extensively characterized photoreceptors in plants. However, their diverse functions in the regulation of plant development have been characterized mainly in dicots, and very little information in this regard is available in monocots, principally due to the lack of phytochrome mutants. Phytochrome mutants have been developed in rice in order to advance knowledge of phytochrome function in plants, especially in terms of exploring potential differences between monocots and dicots. Therefore in addition to studying phytochromes in a monocot, the development of mutants in rice would allow for comparisons of these photoreceptors between monocots and dicots.

Phytochromes in higher plants are encoded by small gene families. Rice has only three members, *PhyA, PhyB* and *PhyC* (Kay *et al.,* 1989; Dehesh *et al.,* 1991; Basu *et al.,* 2000), while *Arabidopsis* has five phytochromes, *PhyA* to *PhyE* (Clack *et al.,* 1994). The relatively small number of phytochrome genes in rice is an advantage in studying the functions of individual phytochromes by the isolation and analysis of mutants. A reverse-genetics strategy was adopted to identify rice phytochrome mutants as a forward approach was not possible because the phenotype of rice deficient in phytochrome was unknown. The Rice Genome Research Program (http://rgp.dna.affrc.go.jp/) has generated a gene knock-out system using the retrotransposon *Tos17* and has isolated mutants for many genes of interest. In this system, the *Tos17,* which is activated in tissue culture, can be used to generate a large number of transposon-tagged mutant lines. DNA isolated from mutant plants is organized into 'mutant panels' (Hirochika, 1999), which may be screened efficiently by the polymerase chain reaction (PCR) by combining a primer for a gene of interest with a primer for *Tos17.* The PCR products can be exploited in the isolation of mutant alleles of the gene. This chapter describes the isolation and characterization of rice phytochrome mutants and experiments designed to study their effects on photomorphogenesis, light-inducible genes and flowering time.

2. Mutant Isolation

2.1. *Tos17* Retrotransposon Mutants

Large populations of rice (*Oryza sativa* L. var. Nipponbare) mutants generated by *Tos17*-mediated mutagenesis were made available by the Laboratory of Gene Function at the National Institute of Agrobiological Sciences (Tsukuba, Japan). Details of mutagenesis with *Tos17* have been described (Hirochika, 1999). In order to identify mutants for specific genes from a large mutant population (around 50,000 mutant lines), the population was divided into 52 groups, and within each group the lines were aligned into two- or three-dimensional matrices. A particular group of mutant lines was termed a 'mutant panel'. For example, a mutant panel of 625 lines consisted of a matrix in which all of the mutant lines were aligned in 25 rows and 25 columns. Plant samples were collected and mixed to make a pool for DNA extraction from all of the individual mutants in a particular row and column. The pooled DNA was subjected to PCR analysis so that a maximum of 50 (25 + 25) PCR reactions were enough to survey 625 mutant lines.

Mutant lines with an insertion of *Tos17* in the *PhyA* gene could be identified by using primers specific to *PhyA* and to *Tos17.* If a desired mutant exists in the panel, DNA pools from one row and one column must give amplified DNA fragments of the same size. The mutant line can be located on the matrix by the row/column identifier. After extensive screening of mutant panels, five different mutant alleles of *phyA* were isolated (**Figure 37.1A**). In addition, eight *phyA* mutant alleles were found in the FST (flanking sequences tagged) database (**Figure 37.1A**). Attempts were also been made to isolate *phyB* and *phyC* mutants from the mutant panels, but only one mutant line for *phyC* was identified (**Figure 37.1C**) and no candidates have been obtained for *phyB* mutants (Takano *et al.,* 2005). As a result there is a bias for the transposition of *Tos17* at a sub-chromosomal level and the sub-chromosomal region where *PhyA* resides appears to be a focus for the transposition of *Tos17.*

The insertion sites of *Tos17* in the *PhyA* (**A**) and *PhyC* (**C**) genes, and mutation sites found in the *PhyB* (**B**) gene, are schematically depicted. Exons are represented as

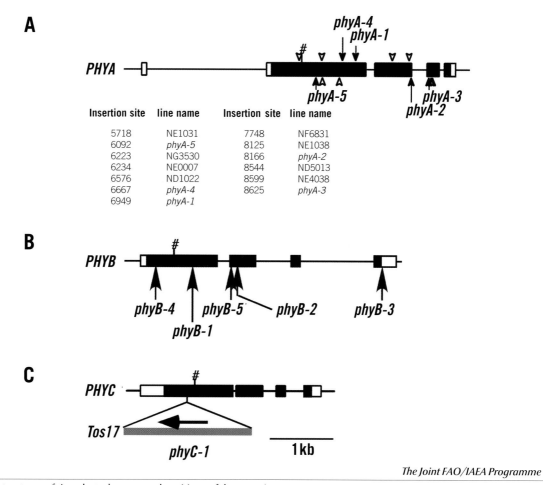

The Joint FAO/IAEA Programme

Figure 37.1 Gene structures of rice phytochromes and positions of the mutations.

black boxes and open boxes are 5'- and 3'-untranslated regions. The chromophore binding sites are indicated by #. Thirteen different alleles of rice *phyA* mutations were identified (A). *Tos17*-insertion sites are indicated by black arrows (isolated by our group) or open arrowheads (found in the FST database). *Tos17* sequences were inserted in both orientations; those with forward orientation are presented above the bar and those with reverse orientation are below. Nucleotide numbers are based on the rice *Phy1B* gene sequence (Accession No. X14172). The horizontal arrow in **C** shows the orientation of the inserted *Tos17* (grey bar).

2.2. Mutants Induced by Gamma Rays

Since no lines were found with a *Tos17* insertion in *PhyB* among the mutant panels, other mutation methods were used to obtain this class of mutants. This was achieved by phenotypic screening of an M$_2$ generation of a rice population mutagenized by γ-ray irradiation. Seedlings of rice *phyB* mutants are expected to

have long coleoptiles under red light. About 3,000 M$_2$ seedlings were screened from γ-ray mutagenized Nipponbare for the elongated coleoptile phenotype (referred to as *elc* hereafter) under continuous red light (Rc). The screen yielded one *elc* mutant (*elc-1*) in which the coleoptile was relatively straight and long compared to control seedlings (Takano *et al.*, 2005).

Rice phytochromes function as repressors of flowering induction under long day (LD) conditions. Thus, 37 lines of early flowering mutants which had been isolated in the Institute of Radiation Breeding for coleoptile lengths were grown under Rc irradiation. This screen yielded four additional mutants at the *elc* locus (alleles *elc-2, elc-3, elc-4 and elc-5*). The background variety of *elc-2* is Nipponbare, and that of *elc-3, elc-4* and *elc-5* is Norin 8.

Sequencing of each of the *elc* mutants was done to test whether the mutant phenotypes were caused by lesions in this gene. As expected, either a deletion or an insertion in the coding region of the *PhyB* gene was detected in all *elc* mutants isolated (**Figure 37.1B**). The insertion or deletions induced frame shifts and subsequently created new stop codons, resulting in a truncated PhyB protein in *elc-1, -2, -4* and *-5* mutants (**Figure 37.2**). These mutations are large enough to abolish the normal function of *PhyB*. However, the *elc-3* mutation appeared to be less drastic: a deletion of 33 base pairs removed the stop codon along with a sequence corresponding to the C-terminal 9 amino acid residues and added 62 irrelevant residues at the C-terminus of *PhyB* (Takano *et al.*, 2005). From these results, it was concluded that all *elc* mutants are alleles of *phyB*, they were therefore renamed from *elc-1, elc-2, elc-3, elc-4* and *elc-5* to *phyB-1, phyB-2, phyB-3, phyB-4* and *phyB-5*, respectively.

Details of mutations detected in the *elc* mutants and their consequent aberrant ORFs are shown along the *PhyB* cDNA sequence (Accession No. AB109892) and its deduced amino acid sequence.

3. Functional Analysis of Phytochrome Mutants

3.1. Genetic Basis of Phytochrome Mutations

In order to confirm that the mutants isolated lacked phytochrome function, Western blot analyses were carried out on soluble proteins extracted from seedlings of *phyA-2, phyB-1* and *phyC-1* mutants using a monoclonal antibody against rye PhyA and polyclonal antibodies raised against rice PhyB and PhyC proteins. The results indicate that each antibody is specific to the individual phytochrome and that the mutant lines tested are null. It was noted that the content of PhyC protein was greatly reduced in the *phyB* mutant, a phenomenon which has also been observed in *Arabidopsis* (Monte *et al.*, 2003). Double mutants were obtained by crossing mutant lines together and selfing the F_1, i.e. crosses were made

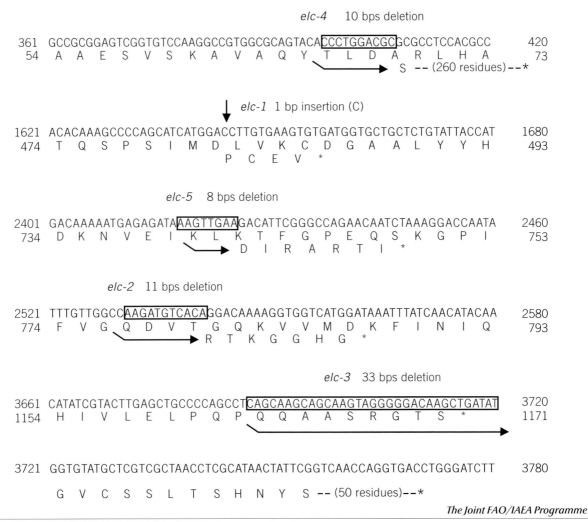

The Joint FAO/IAEA Programme

Figure 37.2 Mutations in the PhyB gene in *elc* mutants.

with *phyA-2* or *phyA-4, phyB-1* and *phyC-1* mutants with eachother and selection made *phyA+phyB, phyA+phyC,* and *phyB+phyC* double mutants in the F$_2$.

3.2. Effects on Photomorphogenesis

The distinct roles of phytochromes in the photomorphogenesis of rice seedlings were examined by comparing single- and double-phytochrome mutants and Nipponbare (wild type, WT) grown under various light conditions (**Figure 37.3**), including continuous irradiation with FR (FRc) and red light (Rc) for nine days. No apparent differences were observed between WT and phytochrome mutants grown in the dark.

Under FRc, coleoptiles of *phyA* mutants were longer than those of WT but still shorter than those of dark-grown seedlings (Dark). The *phyC* single mutation showed no effect, but *phyA+phyC* double mutants had long coleoptiles, as long as those of dark-grown seedlings (**Figure 37.3**, FRc). Therefore, phyA is a dominant photoreceptor for FRc and the function of phyC is dispensable when phyA is functional. But in the absence of phyA, phyC partially compensates for the phyA function on the FR-mediated coleoptile growth inhibition (the difference between *phyA* and *phyA+phyC*). It was also found that *phyA* is involved in the inhibition of mesocotyl

elongation and the induction of the gravitropic response in crown roots under FRc (Takano *et al.,* 2001). When grown under natural light conditions, phyA mutants develop normally and display a vegetative phenotype and a flowering time that are indistinguishable from the WT. Therefore, loss of phyA function in rice appears to cause phenotypic changes that are mainly restricted to the de-etiolation process.

Phenotypic characterization of *phyB* mutant seedlings showed that phyB is responsible for the response to Rc in rice (**Figure 37.3**, Rc). However, it was revealed that phyB is not the sole photoreceptor for Rc, because coleoptile elongation was significantly inhibited by Rc even in *phyB* mutants, although the extent of inhibition was less than that in the WT (**Figure 37.3**, Rc). Such an incomplete effect is not observed in *Arabidopsis*, where etiolated *phyB* seedlings display a marked insensitivity to Rc with respect to almost all aspects of seedling de-etiolation. Thus, phyB plays a main role in responding to Rc in *Arabidopsis*, whereas in rice, phyB does not seem to be a predominant player in the R-mediated response. Such a difference was also observed in the expression modes of light-inducible genes, mentioned below.

Seedlings were grown in darkness (A), or under FRc (B), or Rc (C) for nine days at 28°C. The fluence rates of FRc and Rc are 15 µmol photons m^{-2} s^{-1}. Two seedlings

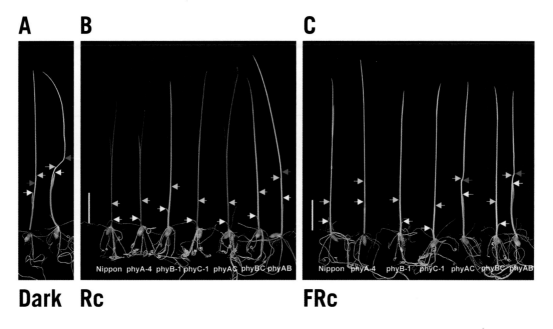

The Joint FAO/IAEA Programme

Figure 37.3 Nine-day-old seedlings of WT and phytochrome mutants grown under FRc or Rc.

of Nipponbare (WT) grown in the dark are in A. In B and C, WT and single- and double-phytochrome mutants are aligned for comparison from left to right: WT, *phyA, phyB, phyC, phyA+phyC, phyB+phyC* and *phyA+phyB*. White and yellow arrows indicate apexes of coleoptiles and first leaves, respectively. Red arrows indicate second nodes. All pictures are the same magnitude, and scale bars at the left are 10 mm.

In *Arabidopsis*, most phytochrome mutants have been isolated by forward genetics screens designed to identify lines with reduced sensitivity to light. However, such screens have not yielded any *phyC* mutants, suggesting the marginal contribution of phyC to photomorphogenesis. Loss of phyC function did not cause clear phenotypic differences in seedling or vegetative growth in rice, indicating that phyC also plays a minor role in photomorphogenesis in this species. The light-dependent inhibition of coleoptile and mesocotyl elongation under Rc or FRc were examined, but no differences were observed between the WT and *phyC* mutants (**Figure 37.3**).

3.3. Effects on Light-Inducible Genes
In order to elucidate the distinct roles of phytochromes in gene expression, the expression levels were examined in *Lhcb* (*Light-harvesting chlorophyll a/b binding protein*) and *RbcS* (*Ribulose bisphosphate carboxylase/oxygenese small subunit*) genes in seedlings of WT and phytochrome single and double mutants grown in darkness or under FRc or Rc. Light-inducible genes such as *Lhcb* and *RbcS* were slightly induced by FRc even in the *phyA* mutant, but this expression was completely abolished in the *phyA+phyC* double mutant (**Figure 37.4A**). These results indicate that phyC is able to perceive FR to induce at least *Lhcb* and *RbcS* expression among light-inducible genes, although the contribution is minor (**Figure 37.4A**).

A. Nipponbare (WT) and phytochrome-single (*phyA, phyB, phyC*) and double (*AB, phyA+phyB; AC, phyA+phyC; BC, phyB+phyC*) mutants were grown in the dark (D) or under FRc for four days. *LhcB* and *RbcS* gene expression was analyzed by Northern hybridization with gene specific probes. Ribosomal RNA (rRNA) was stained with Methylene Blue on the same blot as a quantity control.

B. Rice seedlings from Nipponbare (WT) and phytochrome-single (*phyA, phyB, phyC*) and double (*phyA+phyC, phyB+phyC, phyA+phyB*) mutants were

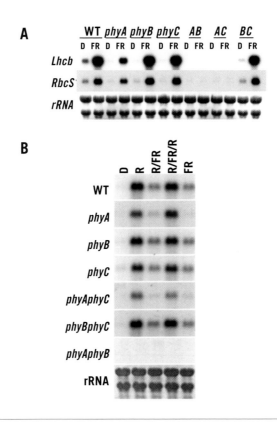

Figure 37.4 Induction of *Lhcb* and *RbcS* genes by continuous or pulses of R or/and FR light in phytochrome mutants.

grown in complete darkness for seven days (D) and then treated with a single pulse of red light (R), a pulse of FR immediately after a red pulse (R/FR), a train of R-FR-R pulses (R/FR/R) or a single FR pulse (FR). Seedlings were harvested 3 h after the pulse irradiation and *LhcB* gene expression was analyzed by Northern hybridization. Ribosomal RNA (rRNA) was stained with Methylene Blue on the blot of WT as a quantity control.

In *Arabidopsis*, phyA is responsible for *Lhcb* gene induction in the photo-irreversible very low fluence response (VLFR) mode, while phyB mediates *Lhcb* expression in the R/FR photo-reversible low fluence response (LFR) mode. As a result, *Lhcb* expression induced by R did not appear to be reversed by subsequent FR irradiation in the WT and the *phyB* mutant (Hamazato *et al.* 1997). In contrast, etiolated seedlings of WT and *phyB* mutant rice did show R/FR photo-reversibility in the induction of *Lhcb* gene expression. Unexpectedly, even *phyB+phyC* double mutants, in which phyA is the only active phytochrome, showed R/FR reversibility (**Figure 37.4B**). These results clearly indicate that phyA is involved in *Lhcb* induction, not only in VLFR mode but also in R/FR reversible LFR mode. Therefore, the molecular properties of phyA seem to differ between *Arabidopsis* and rice, being more specialized for FR perception in *Arabidopsis*, while perceiving both R and FR in rice. A model comparing phytochrome function in these two species is shown in **Figure 37.5**.

3.4. Effects on Flowering Time

Light is a crucial factor in determining flowering time. WT and phytochrome mutant plants were grown in paddy field conditions with natural day-length, which resembled experimental long day (LD) conditions, and the time to flowering was measured (**Figure 37.6A**). WT plants flowered in about 100 days after germination and so did the *phyA* mutants. However, *phyB* and *phyC* mutants flowered about two weeks earlier than WT. The flowering time of the *phyB+phyC* double mutant was the same as that of *phyB* or *phyC* monogenic mutants. The results obtained suggest that cooperative functioning of *phyB* and *phyC* is necessary to suppress flowering under LD conditions. Interestingly, *phyA+phyC* double mutants flowered very early, more than 20 days ahead of *phyC* monogenic mutants. The *phyC* monogenic mutation caused a moderate phenotype of early flowering under LD conditions. However, the *phyC* mutation in combination with the phyA mutation accelerated flowering time in LD conditions, while the *phyA* mutation by itself had little effect on flowering time. These results suggest that *phyC* and *phyA* affect flowering time at separate points in the same signalling pathway.

In short day (SD) conditions (**Figure 37.6B**), *phyC* mutants flowered at the same time as the WT. On the other hand, *phyA* mutants showed slightly late flowering compared to Nipponbare or *phyC* mutants, and

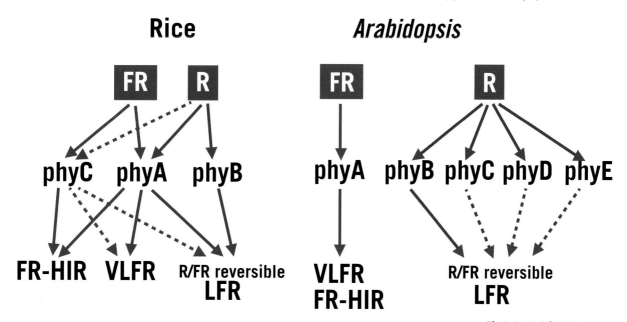

The Joint FAO/IAEA Programme

Figure 37.5 Schematic drawing of photoperception modes of distinct phytochromes in rice and *Arabidopsis*. Solid arrows indicate a predominant contribution and dotted arrows a minor one.

the same extent of delay of flowering was observed in *phyA+phyC* double mutants. The *phyB* monogenic mutants were early flowering compared with the WT or *phyC* mutant even in SD conditions, and the flowering time of *phyB+phyC* double mutants was the same as that of *phyB* monogenic mutants. These observations indicate that phyC is dispensable for flowering time determination in SD conditions.

In order to evaluate the differences in flowering times between LD and SD conditions more precisely, double mutants were grown in the same growth chamber under LD conditions (**Figure 37.6B**, shaded bars). Floral induction in the *phyB+phyC* double mutants was greatly enhanced with the reduction in day length and this behaviour matched that of a *phyB* single mutant. Interestingly, *phyA+phyC* double mutants showed the same flowering times in both SD and LD conditions. Furthermore, the flowering time of the *phyA+phyB* double mutants was earlier in LD than in SD conditions. These results suggest that in rice the light signals mediated by the phytochromes promote flowering in response to SD conditions, while they delay flowering in long days. Among the phytochromes, phyB alone seems to be involved in SD signal transduction but not in LD signalling, while phyA mediates the signals in response to both photoperiods.

Nipponbare (WT) and phytochrome-single (*phyA, phyB, phyC*) and double (*phyA+phyC, phyB+phyC, phyA+phyB*) mutants were grown in the paddy field (**Figure 37.6A**) or in a growth chamber set as SD (10L/14D, **Figure 37.6B**: open bars) or LD (14L/10D, **Figure 37.6B**: shaded bars), and their flowering times were measured. The means ± SE obtained from 20 plants are displayed.

By comparing these results in rice with those obtained in *Arabidopsis* (Neff and Chory, 1998), it can be concluded that phyB and phyC act to delay flowering under inadequate photoperiod for the plants in both rice and *Arabidopsis* (LD for rice and SD for *Arabidopsis*). The contributions of phyB and phyC are the same in rice, while phyB function is dominant over phyC in *Arabidopsis* (Monte *et al.*, 2003). The phyA functions, especially in combination with other phytochromes, are quite different between rice and *Arabidopsis*. In rice, the *phyA* mutation greatly accelerated flowering in the *phyB* and *phyC* background, but no significant effects of the *phyA* mutation were observed in the *phyB* and *phyC* background in *Arabidopsis*.

4. Acknowledgements

We thank Dr. H. Hanzawa for providing us with the monoclonal antibody against rye *PhyA*. We are grateful to Dr. M. Tahir for critical reading of the manuscript. We also thank K. Yagi, M. Sato and Y. Iguchi for their technical assistance. This work was partly supported by the Rice Genome Project (IP-1006).

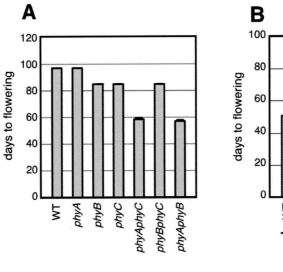
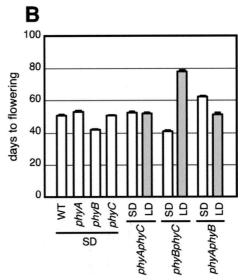

The Joint FAO/IAEA Programme

Figure 37.6 Comparison of flowering times between WT and phytochrome mutants under natural day lengths and SD conditions.

5. References

5.1. Cited References

Basu, D., Dehesh, K., Schneider-Poetsch, H.J. *et al.* **2000.** Rice PHYC gene: structure, expression, map position and evolution. *Plant Mol. Biol.* 44: 27–42.

Clack, T., Mathews, S. and Sharrock, R.A. 1994. The phytochrome apoprotein family in Arabidopsis is encoded by five genes: the sequences and expression of PHYD and PHYE. *Plant Mol. Biol.* 25: 413–427.

Dehesh, K., Tepperman, J., Christensen, A.H. *et al.* **1991.** phyB is evolutionarily conserved and constitutively expressed in rice seedling shoots. *Mol. Gen. Genet.* 225: 305–313.

Hamazato, F., Shinomura, T., Hanzawa, H. *et al.* **1997.** Fluence and wavelength requirements for *Arabidopsis CAB* gene induction by different phytochromes. *Plant Physiol.* 115: 1533–1540.

Hirochika, H. 1999. Retrotransposons of rice as a tool for forward and reverse genetics. *In*: K. Shimamoto (ed.) Molecular Biology of Rice. pp. 43–58.

Kay, S.A., Keith, B., Shinozaki, K. *et al.* **1989.** The sequence of the rice phytochrome gene. *Nucleic Acids Res.* 17: 2865–2866.

Monte, E., Alonso, J.M., Ecker, J.R. *et al.* **2003.** Isolation and characterization of *phyC* mutants in Arabidopsis reveals complex crosstalk between phytochrome signaling pathways. *Plant Cell.* 15: 1962–1980.

Neff, M.M. and Chory, J. 1998. Genetic interactions between phytochrome A, phytochrome B, and cryptochrome 1 during Arabidopsis development. *Plant Physiol.* 118:27-35.

Takano, M., Inagaki, N., Xie, X. *et al.* **2005.** Distinct and cooperative functions of phytochromes A, B and C in the control of de-etiolation and flowering in rice. *Plant Cell.* 17: 3311–3325.

Takano, M., Kanegae, H., Shinomura, T. *et al.* **2001.** Isolation and characterization of rice phytochrome A mutants. *Plant Cell.* 13: 521–534.

5.2. Websites

The Rice Genome Research Program:
http://rgp.dna.affrc.go.jp/

5.3. Further Reading

Briggs, W.R., Beck, C.F., Cashmore, A.R. *et al.* **2001.** The phototropin family of photoreceptors. *Plant Cell.* 13: 993–997.

Briggs, W.R. and Huala, E. 1999. Blue-light photoreceptors in higher plants. *Annu. Rev. Cell Dev. Biol.* 15: 33–62.

Hirochika, H., Sugimoto, K., Otsuki, Y. *et al.* **1996.** Retrotransposons of rice involved in mutations induced by tissue culture. *Proc. Natl. Acad. Sci. USA.* 93: 7783–7788.

Hirochika, H. 1997. Retrotransposons of rice: their regulation and use for genome analysis. *Plant Mol. Biol.* 35: 231–240.

Lin, C. 2002. Blue light receptors and signal transduction. *Plant Cell.* 14:S207-S225.

Whitelam, G.C., Patel, S. and Devlin, P.F. 1998. Phytochromes and photomorphogenesis in Arabidopsis. *Philos. Trans. R. Soc. Lond. B. Biol. Sci.* 353: 1445–1453.

C38

T-DNA Insertion Mutagenesis

R.S.Sangwan[a,*], S.Ochatt[b], J.-E.Nava-Saucedo[a] and B.Sangwan-Norreel[a]

[a] Unité de Recherche EA3900 "Biologie des Plantes et Contrôle des Insectes ravageurs", Université de Picardie Jules Verne, Laboratoire Androgenèse et Biotechnologie, Faculté des Sciences, 33 rue Saint-Leu, 80039 Amiens, France
[b] Laboratoire de Physiologie Cellulaire, Morphogenèse et Validation (PCMV) Unité Mixte de Recherches en Génétique et Ecophysiologie des Légumineuses à Graines (URLEG), Centre de Recherches INRA de Dijon, B.P. 86510, 21065 Dijon Cedex, France
* Corresponding author, E-MAIL: Rajbir.Sangwan@u-picardie.fr, Tel: 0033 3 22827651 Fax: 0033 3 22827612

1. Introduction

The last two decades have been very rewarding for plant scientists as they have capitalized on recombinant DNA technology and induced mutations, created by mutagens such as transposable elements and insertional mutagenesis (**see Glossary, Box 38.1**) to explore many aspects of gene and chromosome behaviour. Most mutations in plants have been induced at random, either by using chemicals, ionizing radiation, or by disrupting the gene through the insertion of a DNA fragment. The inserted DNA can be either a mobile genetic element, such as a transposon, or a piece of bacterial DNA, known as T-DNA, that is integrated in the plant genome following co-cultivation with *Agrobacterium tumefaciens*, gene bombardment or direct gene transfer *via* electroporation, microinjection or polyethylene-glycol treatment of protoplasts. T-DNA-induced mutants/genes could be cloned because they are tagged by the T-DNA insertion. Thus, tagging an unknown gene with a transposon or with DNA greatly simplifies the subsequent isolation of the gene. The lack of movement of T-DNA is both an advantage and a disadvantage. It is advantageous during gene cloning, but makes it impossible to generate multiple alleles by reversion. Several large-scale T-DNA and transposon insertional mutagenesis projects using these strategies are underway to support gene discovery projects (Pan *et al.,* 2003). The mutagenic potential of transposable elements has been widely exploited in research on many organisms to clone genes using transposon sequences as molecular tags (transposon tagging). Although, application is limited to species with active, well-defined transposon systems, such as maize and *Antirrhinum*, it has been demonstrated that it may be possible to clone any gene from any plant species by the application of transposon tagging (**see Chapter 39**).

Since many plant scientists use *Arabidopsis* as a model plant for mutational experiments, extensive collections of lines of mutants, as well as comprehensive genetic maps based on biochemical and morphological mutations and on polymorphic DNA markers, are now available. However, one of the best methods for generating the tagged mutants is based on insertional mutagenesis with the T-DNA from *Agrobacterium tumefacians* (e.g. Feldmann, 1991; Radhamony *et al.,* 2005); these inserts do not move and hence are not analogous to transposons, but many of the cloning strategies exploited in

transposon tagging can be applied to T-DNA generated mutants as well. A T-DNA insertion may reveal gene function *via* a gene knock-out and a gene knock-up (over-expression and mis-expression) or through expression patterns revealed by modified insertion elements. Moreover, T-DNA tagged mutants are central to reverse genetics in *Arabidopsis* and are used in a similar way to conventional mutants in crosses, phenotypic assessments and molecular analyses. The T-DNA tag typically causes a loss of gene expression and results in a monogenic recessive mutation in the gene. This chapter highlights the significance of insertional mutagenesis in functional genomic studies and summarizes T-DNA mutagenesis schemes, including methodologies for creation of T-DNA mutant collections, T-DNA vectors, isolation of the tagged gene and comparisons of the frequency of mutant recovery using various strategies. In addition, concrete examples are given in cloning genes using T-DNA insertion lines, step by step.

2. Creation of Mutant Collections by T-DNA Insertion in Plants

Agrobacterium tumefaciens-mediated T-DNA transfer has been the method of choice for insertional mutagenesis in plants. **Figure 38.1** illustrates the T-DNA tagging system, including the process of T-DNA integration and its structure. Any sequences inserted between the T-DNA borders can be transferred into the plant cell; however, the integration process consists of several steps involving bacterial proteins and plant enzymes responsible for DNA replication and repair. Using the *Agrobacterium* system, large collections of T-DNA transformants of *Arabidopsis* have been reported by several groups and have been used for reverse genetics (Pan *et al.,* 2003). Users can access this information through a range of databases, described by Pan *et al.,* 2003. Interestingly, T-DNA integration seems to involve mostly transcriptionally active regions and it has been suggested that >80% of T-DNA insertions are related to functionally active genes.

Moreover, to create a T-DNA mutant gene bank or collection, i.e. a population of transgenic plants with gene-tagged individuals and isolating the tagged genes, involves many well-defined steps. **Figure 38.2** shows a flow chart for T-DNA mutant generation in *Arabidopsis*. It describes the protocols used to gener-

Box 38.1: Glossary

50Kan medium	Culture medium containing 50 mg/l of Kanamycin.
Ac-Dc	The first transposon system described by Barbara McClintock. Ac ('activator') is an autonomous element, and Dc ('dissociation') refers to a non-autonomous element of this system.
Allele	One of the variant forms of a gene at a particular locus, or location, on a chromosome. Different alleles produce variation in inherited characteristics.
Binary vectors	A modified *Agrobacterium* T-DNA is introduced into the plant on a separate-coli compatible plasmid; the necessary virulence functions are provided in trans from a Ti plasmid devoid of T-DNA.
Cointegrate vector	Is produced by integrating the recombinant intermediate vector (containing the DNA insert) into a disarmed pTi. The co-integration of the two plasmids is achieved within *Agrobacterium* by homologous recombination. The genes of interest to be transferred into plants are initially cloned in *E. coli*.
Forward genetics	This process begins with a mutant phenotype but says nothing about the nature of the gene.
Hemizygous	If there is only one copy of a gene for a particular trait In a diploid organism, the organism is hemizygous for the trait.
Heterozygous	Having two different alleles at a given locus on a pair of homologous chromosomes. A genotype consisting of two different alleles of a gene for a particular trait (Aa). Individuals that are heterozygous for a trait are referred to as heterozygotes.
Homologous recombination	Specific gene disruptions are made by replacing the endogenous gene with an altered version of that gene multimer.
Homozygous	Having two identical alleles of a particular gene (e. g. GG, or TT).
Mini-Ti	The plasmid containing disarmed T-DNA is called mini-Ti or micro-Ti, e.g., pBinl9, and has the origins for replication in both *E. coli* and *Agrobacterium*. The mini–Ti approach simplifies procedures for exploiting T–DNA as a gene vector for plants.
Null alleles	An allele that makes no gene product or whose product has no activity.
Phenotype	The characteristics of the organism, usually with respect to the traits a particular gene controls.
Reverse genetics	This process initiates from the gene sequence and tries to generate a mutant phenotype.
Somaclonal variation	In tissue culture transformation systems, mutants can result from exposures to exogenous hormones, creating 'somaclonal variants'.
Suppressor mutation	A mutation that counteracts the effects of another mutation. A suppressor maps at a different site than the mutation it counteracts.
	T-DNA is the transferred DNA of the tumor-inducing (Ti) plasmid of some species of bacteria such as *Agrobacterium tumefaciens* and *Agrobacterium rhizogenes*.
T-DNA mutagenesis	The same procedure of T-DNA transfer can be used to disrupt genes *via* insertional mutagenesis. Not only does the inserted T-DNA sequence create a mutation but it also 'tags' the affected gene, thus allowing for its isolation. This method is used widely to study gene function in plants, such as the model plant *Arabidopsis thaliana*.
Ti plasmid	Is a circular plasmid that often, but not always, is a part of the genetic equipment that *Agrobacterium tumefaciens* and *Agrobacterium rhizogenes* use to transduce its genetic material to plants.
TILLING (Targeted Induced Local Lesions in Genomes)	A PCR mutagenesis protocol to create single-base mutations in particular genes. This method allows the creation of multiple alleles.
Transposable elements 'jumping genes'	Discrete DNA sequences able to transport directly to other locations within a genome.
Transposon systems	Include both autonomous elements and non-autonomous elements. Autonomous elements encode the functional transposase enzyme, and can thus mobilize themselves. Non-autonomous elements do not encode functional transposes, and therefore, can not transpose on their own, but they can be mobilized by the autonomous element of the same transposon system.
Wild type	An organism as found in nature; the dominant allele usually found in nature, and from which mutations produce other dominant or recessive alleles.

T-DNA Tagging System

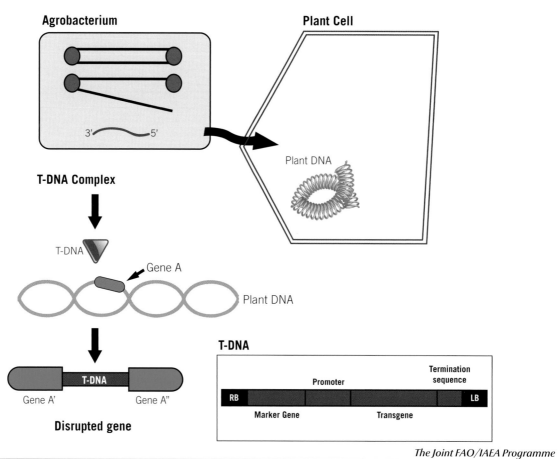

The Joint FAO/IAEA Programme

Figure 38.1 T-DNA insertion mutagenesis at a random location within a gene. A schematic representation of the T-DNA insertion and tagging system. The T-DNA, including any sequences inserted between the borders, can be transferred by the *Agrobacterium* to the plant cell, where it integrates into the genome. RB – Right border; LB – Left border.

ate a transgenic population, the genetic analysis of the transformants, the screens that have been conducted and gene isolation strategies. Genetic segregation and molecular analysis such as Southern analysis are done to confirm that the mutant contains T-DNA as well as demonstrating that the mutant phenotype is not due to somaclonal variation. Moreover, co-segregation of the genetic marker carried by the T-DNA and the isolated phenotype is necessary to show that the mutation is linked to the T-DNA insertion. For transformation, several vectors have been developed (below) however, to be able to use T-DNA effectively as a gene tag, protocols for efficient generation of a large number of transgenic individuals to screen for specific phenotypes are required. Usually the primary transformation is referred to as T_0 and the seeds from T_0 plants/lines are called T_1 (first generation after transformation), T_3 plants/lines are obtained after out-crossing/selfing the T_2 plants/lines.

Usually the mutant screening is done in T_3 lines. Because primary transformants (T_1/T_2 lines) are heterozygous for the insert (selectable marker and mutation), progeny (T_3 lines) will segregate 3:1 (resistance:recessive) for the dominant selectable marker (for a single insert), but 1:3 (mutant:wild type) for any recessive mutation.

2.1. T-DNA Tagging Vectors

Plants such as *Arabidopsis* can be transformed stably using *Agrobacterium tumefaciens*-mediated transfer of T-DNA (**Figure 38.2**). The T-DNA, which is flanked by left- and right-border (LB and RB, **see Figure 38.1**) sequences, resides on a tumour-inducing (Ti) plasmid. The Ti plasmid also carries many of the transfer functions for mobilizing the T-DNA. Important considerations when choosing a vector include the resistance marker in bacteria, the resistance marker in plants, the size of the vector, and

Tissue Culture System

Protoplast, stem, leaf, root, explants

Co-cultivation

Agrobacterium

Primary Transformants

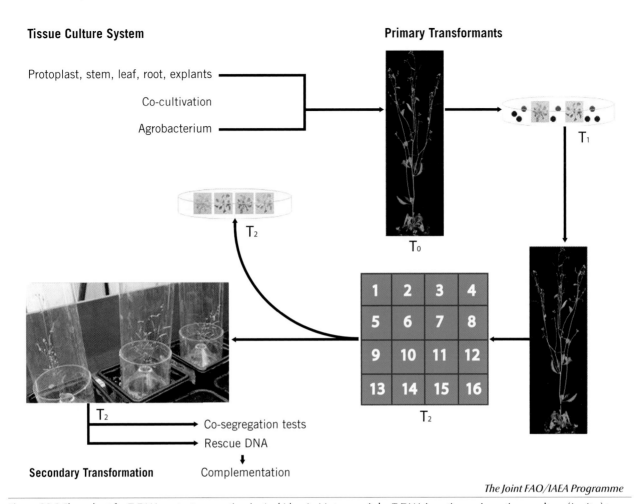

T_1

T_0

T_2

1	2	3	4
5	6	7	8
9	10	11	12
13	14	15	16

T_2

T_2

→ Co-segregation tests

→ Rescue DNA

↓

Complementation

Secondary Transformation

The Joint FAO/IAEA Programme

Figure 38.2 Flow chart for T-DNA mutant generation in *Arabidopsis*. Mutagenesis by T-DNA insertion, using a tissue culture (*in vitro*) transformation protocols. Wild type explants of *Arabidopsis* are cultivated with *Agrobacterium* harbouring a T-DNA containing a kanamycin-resistance gene. The transgenic population/seeds are germinated and the resulting plants (referred to as T_0) are grown to maturity and their progeny (T_1) sown on media containing kanamycin (kan). Preliminary phenotypic screening and Southern analysis are done at this T_1 stage. Individual resistant seedlings are grown to maturity and their progeny (T_2) collected (the box depicts 16 independent transformants). The segregation T_2 families are assayed for the number of inserts (resistant: sensitive) and screened for visible alterations in phenotype. Co-segregations tests on the resulting mutants demonstrate whether the mutation is due to a T-DNA insert. Isolated mutants are further characterized and complementation done to show that the mutant gene was responsible for the mutation.

the configuration of unique restriction sites available for cloning. The most common bacterial resistance markers are kanamycin, streptomycin or spectinomycin, gentamycin and tetracycline. With respect to plant resistance markers, several families of T-DNA vectors, such as the pSLJ, pPZP, pCAMBIA (http://www.cambia.org.au), and pGreen series, include members conferring different resistances. The most widely used plant resistance markers are probably those for the antibiotic kanamycin and the herbicide phosphinothricin or glufosinate ammonium, better known by its trade names Basta. Old protocols used *Agrobacterium tumefaciens* strain LBA4404, but this strain often does not appear to be virulent enough for the vacuum-infiltration method. Better strains are C58 derivatives such as GV3101 (pMP90) and GV3101 (pMP90RK).

2.1.1. Types of Commonly Used Vectors

Co-integrate vectors. The genes to be transferred into plants are initially cloned in *E. coli* for ease in the cloning procedures. A co-integrate vector is produced by integrating the recombinant intermediate vector into a disarmed pTi. The co-integration of the two plasmids is achieved within *Agrobacterium* by homologous recombination (**Figure 38.3A**). The critical feature is that the vector by itself will not replicate in *Agrobacterium* without first integration into the resident Ti plasmid.

Binary vectors: A binary vector consists of a pair of plasmids of which one plasmid contains disarmed T-DNA sequences (at least the left and right borders of T-DNA must be present), while the other contains *vir* region, and ordinarily lacks the entire T-DNA including the borders (**Figure 38.3B** shows a binary vector). The

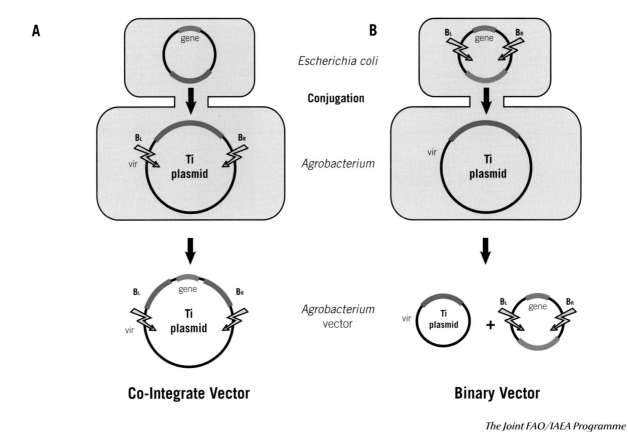

A   B

Escherichia coli

Conjugation

Agrobacterium

Agrobacterium
vector

Co-Integrate Vector   **Binary Vector**

The Joint FAO/IAEA Programme

Figure 38.3 Schematic representation of co-integrate and binary vectors. A) Co-integrate vector system. The cloned gene and a selectable marker (not shown) are cloned in pBR322. This plasmid is then mobilized to *Agrobacterium*. A single cross-over results in the co-integrate vector (black box indicating the pBR homology). The entire sequence between the T-DNA borders (wavy lines) which includes a pBR repeat will be transferred to the plant cell. B) Binary vector system. The gene is cloned in-between the T-DNA borders of a broad host range plasmid. When mobilized to *Agrobacterium* this plasmid is able to replicate autonomously. The *vir* functions are supplied in *trans* by the plasmid.

plasmid containing disarmed T-DNA is called mini-Ti or micro-Ti, e.g. pBinl9, and has the origins for replication in both *E. coli* and *Agrobacterium*. The DNA insert is integrated within the T-region of mini-Ti, and the recombinant mini-Ti is cloned in *E. coli*. Binary vectors (or autonomous vectors) are vectors which need not recombine into the resident Ti plasmid. These vectors are the most commonly used for plant transformations.

Apart from the border repeats, the T-DNA can be modified at will without interfering with the process of gene transfer to the plant cell and integration into the genome. This leads to using wide variety of tagging vectors, such as passive and active tagging vectors.

Passive vectors: Passive tagging vectors contain a marker gene either lacking a promoter sequence or one linked to a 'minimal promoter' notionally missing sequences necessary to produce normal developmentally regulated gene expression, yet sufficient to initiate transcription. The rationale of these vectors is that fol-

lowing insertion into the plant genome, flanking plant sequences able to direct gene expression will lead to the expression of the marker gene. Screenable markers, such as NPTII, Lux or GUS can be used to screen for promoters effectively as transcriptional fusions or, where the marker gene product can tolerate amino-terminal extensions, translational fusions.

Active vectors: Active T-DNA tagging vectors or tags that do not respond passively to the expression patterns of the chromosomal environment into which they insert, but actively modify it in a positive or negative, manner. In other words, this means activation of expression of flanking plant DNA sequences by the T-DNA tag following insertion into the genome. It could be envisaged that T-DNA tags bearing transcriptional enhancer sequences could be used to this effect. This concept has been used to tag plant sequences and to dissect the molecular basis of auxin synthesis, transport or action.

2.2. Genetic Transformation Systems

A special criterion (i.e. no or little somaclonal variation) has to be considered in the transformation method for T-DNA mutagenesis. For example, a root transformation system (Valvekens *et al.,* 1988) is good for introducing genes into *Arabidopsis* but it is not likely to be extremely useful for insertion mutagenesis as it generates somaclonal variation. In Arabidopsis many methods for *in vitro* plant regeneration have been developed using protoplast, cotyledons, leaf stem and root explants (Valvekens *et al.,* 1988), however, a high frequency of somaclonal variation was found in the regenerants. When *Agrobacterium*-mediated transformation was added to this regeneration protocol a concomitant increase in somaclonal variation was also observed. This is mainly because the regeneration of transgenic plants generally required long (weeks to months) exposure of plant tissues to exogenous phytohormones. As a result, somaclonal variations or mutants that are not due to the insertion of T-DNA can arise with higher frequency. These untagged mutations make it laborious to find the rare tagged mutants. Therefore, *in planta* transformation approaches (detailed below) have been developed which greatly reduce the problem of somaclonal variation. Several *in vitro* and *in planta* transformation techniques, using *Agrobacterium* for T-DNA mutagenesis are available and these are detailed below.

2.2.1. *In Vitro* Transformation Systems

Classically, *Agrobacterium*-mediated transformation has been obtained using co-cultivation of plant explants such as leaf, stem, root, etc. and then transfer to a regenerating medium containing phytohormones and antibiotics. Regeneration of transgenic plants, using *in vitro* genetic transformation methods requires careful co-cultivation of the explants (cells, callus or tissues), introduction of DNA using *Agrobacterium tumefaciens* or particle bombardment, selection of transformed cell lines and regeneration of plants. Transformation mediated by *Agrobacterium* has provided a reliable means of creating transgenics in plant species. The various methods of transformation that have been developed, and the list of plants successfully transformed, have been reviewed by Sangwan and Sangwan-Norreel (1989). In *Arabidopsis*, several *Agrobacterium*-mediated genetic transformation methods have been developed *via* explants obtained from leaves, stems, roots, cells

and protoplasts. Using these *in vitro* techniques, T-DNA insertion banks were created and screening of these transgenic families has revealed the presence of mutations affecting many different aspects of plant growth and development. Previously, it has been reported that zygotic embryos provide an alternative source of explant tissue for regeneration at high frequencies and without apparent somaclonal variations. One possible explanation of this success is that cells of zygotic embryos have undergone only a few cell divisions after completing fertilization. These cells may thus remain totipotent and therefore more responsive to *in vitro* manipulations. Taking into account the problem of somaclonal variation, a reliable and general transformation system has been developed using the *Arabidopsis* zygotic embryo as a model system. **Figure 38.4** illustrates the transformation protocol, following which no somaclonal variants were observed among a total of 5,000 primary transformants (Veyres *et al.,* 2008).

2.2.2. *In Planta* Transformation Systems

Several *in planta* transformation techniques, using *Agrobacterium* have been available since 1987 (Radhamony *et al.,* 2005). Initially, seed transformation with *Agrobacterium* was developed (Feldmann and Marks, 1987) and later this became a valuable method for insertional mutagensis and gene isolation in *Arabidopsis. Arabidopsis* seeds imbided in fresh cultures of *Agrobacterium*, were grown to maturity and progeny obtained by selfing were found to maintain a marker gene carried by T-DNA. The ability to raise a large number of independent transformants has allowed this approach to be used effectively for gene tagging. As *Agrobacterium* is capable of systematically transforming diverse cell types when infiltrated into plants, vacuum infiltration techniques in *Arabidopsis* became important and were developed (Bechtold *et al.,* 1993; Clough and Bent, 1998). *In planta* transformation has been refined further, for example, the "floral dip" method consists of simply dipping developing floral tissues into a solution containing *Agrobacterium tumefaciens* (Clough and Bent, 1998). With these *in planta* transformation methods, most transformed progeny are genetically uniform (non-chimeric) and the somaclonal variation associated with tissue culture and regeneration is minimized. Thus, the ease of this method means that, in addition to being useful for routine transformation, it can also be used to create the large number of independent transformants

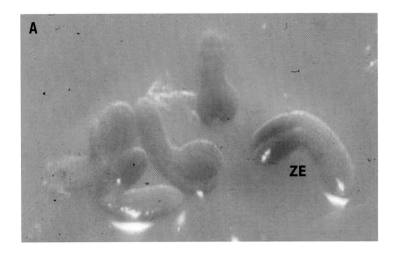
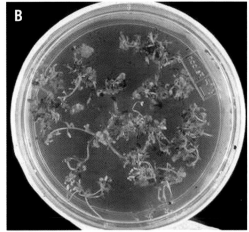

The Joint FAO/IAEA Programme

Figure 38.4 Transformation of zygotic embryos (ZE) (A) followed by regeneration of transgenic (B).

that are required for insertional mutagenesis of the genome.

Extensive work in T-DNA tagging (**see Section 3 below**) has become possible because of improvements in techniques for *Agrobacterium*-mediated transformation. For example, using transformation methods based on "floral dips" (Clough and Bent, 1998), a large population of T-DNA transformed lines of *Arabidopsis* can be generated, thus having good chances of finding a transgenic plant carrying a T-DNA insert within any gene of interest.

2.2.3. Selection of Transformants

The general steps in transformants selection from T_0, T_1 and T_2 generations are the following:

- T_0 lines/plants: are referred as the primary transformants grown on the selective (kanamycin) medium. Southern analyses are done to check the transgenic status as well as the number of copy of the T-DNA inserts. T_0 plants are grown to maturity to get the T_1 seeds.
- T_1 lines/plants: To ensure that only unique transformants are harvested, T_1 seeds are surface sterilized and plated on 50K medium. Testing of T_1 seeds on the 50K medium will indicate the average number of functional inserts per transformant. T_1 plants are grown to maturity to get the T_2 seeds.
- T_2 lines/plants: T_2 or T_3 lines are used for mutant screening and segregation analysis

To select for transformed plants, sterilized T_1 seeds are plated onto a selection (kanamycin) medium. Transformants are identified as kanamycin-resistant seedlings that produced green leaves and well-established roots within the selective medium. Transformants are grown to maturity in a glasshouse. Southern analyses are done to confirm the transgenic status of the transformants. Generally, DNA integrates at a single locus, but it usually consists of arranged multimers of the donor DNA. Once integrated into the genome, DNA is usually maintained stably. Most mutants caused by T-DNA insertion are recessive and hence not seen until the regenerants are self-pollinated, i.e. are present in a homozygous state. Chimeras can also be created during transformation, complicating recognition of novel phenotypes. As a consequence of these phenomena, only a portion of the variation observed after T-DNA transfer results from T-DNA insertional mutagenesis. Moreover, insertion appears to be random: T-DNA has no significant target-site preference and no obvious hotspots have been detected.

2.2.4. Mutant Screening and Verification

To prove that mutation is caused by a T-DNA insertion, the following criteria must be met:

- Mendelian segregation of the mutant phenotype, and
- Co-segregation of a T-DNA insertion with the mutant phenotype.

It has been shown that if a T_2/T_3 line is segregating in a Mendelian manner for both the Kan[R] marker and the mutation of interest, most probably it can be considered that the gene is tagged than when the two traits are not segregating appropriately (Feldmann, 1991; Veyres *et al.*, 2008). The best method for co-segregation of early expression of phenotypes is to plate about 200 T_2 or T_3 seeds on 50K medium and a segregation ratio of 1:2:1

(Kan^R, mut: Kan^R, WT: Kan^S, WT) most likely indicates a single insert. For further confirmation, at least 200 seeds from the line that is segregating 1:2:1 on non-selective medium should be plated. This should generate a 3:1 ration (WT: mutant). In addition, transfer about 50 mutant seedlings/plantlets from non-selective medium to 50K medium. If any of the mutant plantlets are sensitive, the plantlets will bleach with arrested root growth. If all 50 plantlets are resistant then tight linkage is established. If any of the 50 plantlets bleaches on 50K medium the insert has probably not been caused by the mutation. However, due care should be taken with this assay as some mutants that show a general bleach after transfer, such as defective in root development, may, in fact, be Kan^R. Southern analysis (on at least 7–10 plants) should be done to establish that there is not a silent T-DNA insert segregating in the family. If the mutant successfully passes both the transfer test and the Southern analysis, gene isolation and molecular complementation can proceed.

2.3. Somaclonal Variation Versus T-DNA Mutation

In vitro regenerated plants may display a different phenotype to that of the plant from which the cells originated – this is known as somaclonal variation. The basis of somaclonal variation is not understood, although chromosomal rearrangements, the activation of endogenous transposons, modifications of the nuclear DNA content, change in DNA methylation status and mutation may all be contributing factors. Among factors clearly shown to increase the extent and susceptibility of tissues to somaclonal variation are the time spent *in vitro* prior to regeneration, the hormonal content of the media used and the regeneration strategy adopted. As a result, in tissue culture systems, somaclonal variation – mutants which are not due to the insertion of T-DNA can arise with high frequency, these large numbers of untagged mutants making it laborious to find the rare tagged mutants. The frequency of somaclonal variants from such *in vitro* manipulations often reaches a few percentages of the primary regenerants, about the same frequency as the T-DNA tagged mutations, and this particularly affects the relative nuclear DNA content as was recently reported for *Tnt1* retrotransposon insertion mutants of the model legume *Medicago truncatula*. Thus, in gene tagging experiments, it is difficult to distinguish insertional gene mutations from the background of somaclonal variants, particularly in cases where they are linked. Moreover, the identification and cloning of any gene of interest is not practical using tissue culture-based transformation systems, e.g. in addition to generating the somaclonal variations, the size of the transgenic population obtained is usually small. However, somaclonal variation can be minimized if semi-organized or embryonic tissues are used as the target for transformation, particularly if the target cell can enter embryogenesis either directly or only after a short callus phase. Another way to avoid somaclonal variation is to introduce the DNA directly into the germ line – into targets such as meristematic tissues or microspores that have been re-programmed to become embryos (Touraev *et al.*, 1997). To circumvent the problem of somaclonal variations, Feldmann and Marks (1987) have devised an *Agrobacterium*-mediated seed transformation technique that avoids tissue culture *in vitro*. By this method several thousand lines carrying T-DNA-induced mutations have been isolated in *Arabidopsis* (Feldmann, 1991). However, this procedure yields only a low frequency of transformation (0.32%), chimeric transformants containing both transformed and non-transformed tissue frequently occur, and several parameters affecting its reproducibility have not yet been determined.

3. Insertion Mutant Facilitated Gene Cloning

The inserted sequence not only disrupts the expression of the gene but also tags the effected gene and as the sequence of the inserted sequence is known, therefore, the gene in which the insertion has occurred can be easily recovered, using several cloning or PCR-based strategies. As *Arabidopsis* introns are small and have little intergenic material, the T-DNA insertion (in the order of 5 to 25 kb in length) generally produces a disruption of gene function. Therefore, in a large T-DNA bank/ lines, chances of finding a transgenic plant carrying a T-DNA insert within any gene of interest are very high.

3.1. Techniques for the Isolation of Mutated Genes: Recovery of Plant DNA Flanking the T-DNA

After it has been shown that a mutation co-segregates with the kanamycin-resistance marker in the T-DNA, there are a wide range of molecular techniques available for the isolation of the mutated gene: recovering plant DNA flanking the T-DNA inserts. One can simply construct a genomic library of the mutant and isolate clones containing T-DNA sequences. Some of these clones will contain plant DNA as well. However, currently, the PCR-based techniques are more commonly used, mainly because these are simple, relatively easy, rapid and inexpensive. For example, it has also been reported that the sequences flanking the insertion can be easily identified from single or low copy number lines using inverse PCR (IPCR) and by thermal asymmetric interlaced (TAIL) PCR (detailed below). In addition, a simple and efficient PCR strategy to amplify the T-DNA flanking region known as T-linker specific ligation PCR (T-linker PCR) has also been reported.

Commonly used techniques used to isolate the mutated gene are described (from the most complex to simple ones) such as Lambda library Screening, Plasmid rescue, Inverse PCR, TAIL-PCR and Genome Walking. Generally speaking, the success rates of isolating the mutated genes are higher with the production of Lambda library and plasmid rescue than with Inverse PCR or with TAIL-PCR. In all cases, however, unequivocal proof about the function of gene is obtained by performing molecular complementation, *i.e.*, by demonstrating that the introduction of a wild type allele would eventually restore the phenotype.

Lambda library screening: Often, this has been accomplished by construction of a genomic DNA library in a bacteriophage λ vector followed by screening of the library using the T-DNA as a probe. The production of a lambda library should always generate a plant flanking DNA. However, this approach has at least two drawbacks. First, the preparation and screening of the genomic library is tedious and time consuming. Second, genomic DNA may contain sequences or covalent modifications (e.g. methylation) that render recombinant bacteriophage clones "unstable" in packaging reactions or in host bacteria, thus precluding efficient cloning of certain regions.

Plasmid rescue: A schematic representation of plasmid rescue is shown in **Figure 38.5**. It is performed by restricting a small amount of plant DNA containing a T-DNA insert, followed by self ligation and transformation of the resulting fragments into competent *E. coli* and selection of antibiotic resistant colonies and analysis of the plasmids to uncover those that contain

plant DNA. Before beginning plasmid recovery for a particular tagged gene it is important to have shown that there is only a single insert in the transformed line. If the map position of the mutation of interest is known, it is possible to recover plant flanking DNAs from lines with multiple inserts, followed by identification of a polymorphism for each, and placement on the genetic marker map to identify the correct plasmid. This protocol has been described in detail. This technique does not work for all inserts. If the correct fragment is not recovered in one or two tries, it is best to simply generate a lambda library to recover the flanking plant DNA.

The PCR-based techniques: PCR is a technique that has revolutionized molecular biology by virtue of its utility in amplifying microgram amounts of a specific, targeted gene from as little as a single copy of starting template DNA. Various modifications to the classic PCR have been carried out to identify DNA sequence flanking a known sequence.

Inverse PCR: This involves digestion of the DNA to produce a fragment containing a known sequence, followed by circularization by intra-molecular ligation. Schematic representation of inverse PCR is shown in **Figure 38.6**. The resulting circular template is then PCR amplified *via* the use of divergent primers which have been designed to anneal within the region of known sequence. This permits amplification of the flanking sequence by positioning two primers, each of which binds to the known sequence "inside out" on the circle. Therefore, this strategy maintains specificity at each primer binding site. This method is only suitable when the digested DNA can be monomerically circularized. Difficulties with inverse PCR include the requirement for two restriction sites that flank the priming region and inefficient PCR amplification of closed circular DNA. This inefficient PCR amplification occurs if a convenient restriction enzyme site is not present to linearize the circle between the 5' ends of the two amplifying primers prior to PCR amplification. Without linearization, double-stranded circular DNA is amplified much less efficiently than linear DNA. Since, at present, only regions of limited size can be enzymatically amplified using PCR, and since primers must be synthesized from known sequences, the inverse PCR approach is not amenable to assaying very long distances in the genome. However, the inverse PCR technique does

498

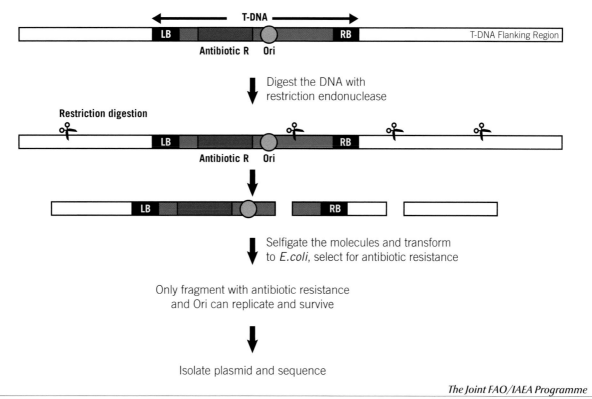

Figure 38.5 Schematic representation of plasmid rescue.

eliminate the need to construct and screen genomic libraries in order to walk hundreds, if not thousands, of base pairs into flanking regions. In theory, inverse PCR could be carried out repeatedly allowing progressively more distant flanking sequences to be determined.

TAIL-PCR: TAIL-PCR utilizes a set of nested sequence specific primers together with a shorter arbitrary primer (AD) of lower melting temperature. Schematic representation of TAIL-PCR is shown in **Figure 38.7.** PCR conditions are modified in such a manner that both primers or alternatively, only the specific primer functions well. This is achieved by use of high and reduced stringency cycles, enabling target sequence to be amplified preferentially over non-specific products. Usually the border primers are used in combination with AD primers. Three rounds of PCR are carried out, in which the primary PCR product is diluted 50-fold and used as template for secondary PCR in presence of nested specific primer. Subsequently the secondary PCR product is diluted 10-fold and used as a template for a third round of PCR with a second nested specific and AD primer. Running the PCR products side by side allows isolating the band with a 'shift' in product size between the secondary and tertiary reactions next to each other.

Genome walking: Here the starting DNA must be very clean and have a high average molecular weight, requiring a higher quality preparation than the minimum suitable for Southern blotting or conventional PCR. The DNA is digested by blunt end cutting enzymes. *DraI, PvuII, EcoRV* and *StuI* are the enzymes usually provided in the Universal Genome Walker kit. However, other enzymes that leave a blunt end could be also be used, expanding the number of libraries that can be generated. The digested DNA is ligated with the adaptors. These are referred to as "Genome Walker libraries". The next step is to design a pair of nested gene specific primers, these primers should be 26–30 nucleotides in length and have a G/C-content of 40–60%. This ensures that the primers anneal effectively to the template at an annealing and extension temperature of 67°C. The primary PCR reaction uses the outer adaptor primer (AP1) and the outer gene-specific primer (GSP1). The product of the primary walk is then diluted and used as a template for a secondary walk with the nested adaptor (AP2) and nested gene-specific (GSP2) primers. The major PCR products obtained are gel extracted and sequenced.

Panhandle PCR: This method requires the generation of a template shaped like a pan with a handle, hence the name. The template is generated by restriction

499

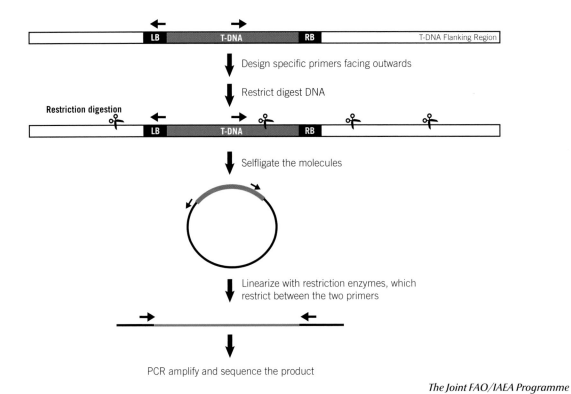

Design specific primers facing outwards

Restrict digest DNA

Restriction digestion

Selfligate the molecules

Linearize with restriction enzymes, which restrict between the two primers

PCR amplify and sequence the product

The Joint FAO/IAEA Programme

Figure 38.6 Schematic representation of inverse PCR. It involves digestion of DNA from mutant plant with an enzyme that cuts mainly once within the T-DNA followed by self-ligation. The resulting circular template is then PCR amplified using appropriate primers from the T-DNA region. The flanking plant DNA is shown.

enzyme digestion of genomic DNA followed by ligation to a single-stranded oligonucleotide. This ligated-oligonucleotide has a free 3' end which is complementary to the known region of DNA. In this method, single strands of DNA which contain the complement of the ligated-oligonucleotide undergo self-annealing, forming a stem-loop structure. The ligated oligonucleotide can then prime template-directed DNA polymerization. PCR amplification of the unknown DNA can subsequently be carried out because a known sequence now flanks both ends of the unknown DNA.

Capture PCR: In this technique only one primer is derived from the known sequence, whereas the second primer anneals to a linker, which has been ligated to one end of the digested DNA sample. The second primer is also biotinylated to enable isolation of subsequent extension products on a streptavidin support. Ligation of a linker to allow the use of a defined second primer is also the strategy employed in Ligation Mediated PCR, which is a method that has been used to generate *in vivo* footprints of transcription factors bound to DNA. In this case, the linkers are ligated to blunt ends which have been generated by random chemical cleavage of the genomic DNA tem-

plate, followed by denaturation of and primer extension from a known sequence.

Targeted inverted repeat amplification: *This* eliminates the steps and sequence artifacts associated with cloning and permits genome walking into unclonable regions of DNA. The genomic DNA is restriction enzyme digested and then ligated to the 3' end of a 5'-phosphorylated oligonucleotide using a short bridging oligonucleotide as a splint. The phosphorylated oligonucleotide is designed to create 5'-end extensions that are complementary to the known sequence. Following denaturation and reannealing under dilute conditions that promote intrastrand annealing and under high stringency, only those DNA strands that contain the known sequence will form a stem-loop structure with a recessed and phosphorylated 5' end. This stem-loop renders a substrate for a subsequent heat-stable ligation reaction to another oligonucleotide that anneals to the known sequence immediately adjacent to the phosphorylated oligonucleotide high-stringency annealing site. The oligonucleotide appended to the phosphorylated oligonucleotide by the heat-stable ligase can, when present in its free, nonligated form, prime DNA polymerase-mediated amplification of those strands modified by site-specific ligation

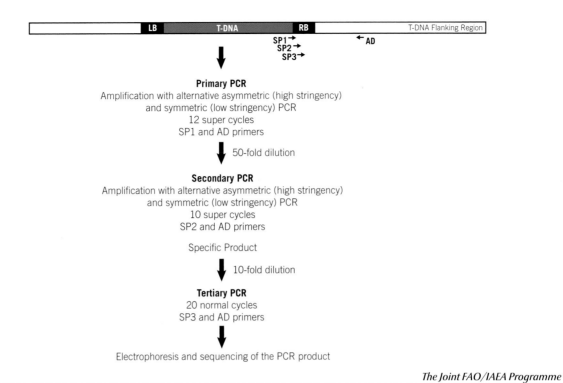

| LB | T-DNA | RB | T-DNA Flanking Region |

SP1→
SP2→
SP3→
←AD

↓

Primary PCR
Amplification with alternative asymmetric (high stringency)
and symmetric (low stringency) PCR
12 super cycles
SP1 and AD primers

↓ 50-fold dilution

Secondary PCR
Amplification with alternative asymmetric (high stringency)
and symmetric (low stringency) PCR
10 super cycles
SP2 and AD primers

Specific Product

↓ 10-fold dilution

Tertiary PCR
20 normal cycles
SP3 and AD primers

↓

Electrophoresis and sequencing of the PCR product

The Joint FAO/IAEA Programme

Figure 38.7 Schematic representation of thermal asymmetric interlaced PCR (TAIL-PCR). It involves three consecutive rounds of PCR, performed with a set of three nested T-DNA specific primers (SP1, SP2 and SP3) and a small, arbitrary primer. The positions of three nested primers in the T-DNA region are indicated by arrow heads. AD is arbitrary degenerate primer indicated by arrow and T-DNA flanking regions are also shown.

to this same oligonucleotide. This is followed by one or two nested DNA amplifications, with the final amplification primed by the phosphorylated oligonucleotide in its free, non-ligated form.

Rapid Amplification of Genomic DNA Ends (RAGE):
The Rapid Amplification of Genomic DNA Ends (RAGE) approach involves ligation of digested genomic DNA to plasmid DNA. PCR primers are designed so that one is homologous to a known target region in the genomic DNA sequence, and the other is homologous to a defined plasmid sequence. Using the RAGE method, cloning is accomplished *via* digestion of genomic DNA with a single restriction endonuclease, and the products are ligated into the linearized, phosphatased plasmid vector. PCR is performed directly on the ligated DNA employing nested primers and two rounds of amplification to increase specificity. There are a few modifications to this procedure; the genomic DNA is directionally cloned into the multiple cloning site of a plasmid vector. The ligated products are then transformed into bacteria and plated on selective media, and the plasmid DNA recovered from the pool of surviving clones is PCR amplified. These additional

straightforward steps give relative enrichment for the genomic sequence of interest and generate a simple DNA template, enabling easy amplification by PCR.

Vectorette PCR: Here synthetic duplexes are ligated to the restriction enzyme digested ends of DNA. The unique feature of this method lies in the construction of the synthetic duplexes, termed vectorette units. Vectorette units contain a bubble region of non-complementarity. The vectorette PCR primer is identical to one of the non-complementarity portions. The vectorette PCR primer, therefore, contains no region of complementarity to the end modified DNA unless polymerase extension is initiated from an upstream portion of a DNA strand. The DNA strand of interest is amplified to the extent that this initial DNA primer extension (from the non-vectorette primer that anneals to the known region) is specific for the strand of interest. A limiting factor with this method may be the specificity in the primer extension step that generates an annealing site for the vectorette primer. This is because primer extension from a site near the 5′ end of any DNA strand will create an annealing site for the vectorette primer, which results in a PCR product.

PCR walking or adaptor ligation PCR: DNA is digested with restriction enzymes yielding blunt-ended fragments that are ligated to asymmetric adaptors. Using PCR primers specific to the adaptor sequences combined with PCR primers designed for the T-DNA, it is possible to amplify unknown genomic regions flanking the T-DNA insertion site. The design of the adaptor sequences allows PCR amplification of fragments between the adaptor specific primer and the T-DNA-specific primer only. Non-specific PCR amplification between adaptor-specific primers is prevented by the presence of an amine group on the exposed 3′ end of the adaptor. This vectorette feature prevents extension of the lower adaptor DNA strand, which would result in the formation of an adaptor primer-binding site. An adaptor primer-binding site is created only by PCR extension from the T-DNA-specific primer. Furthermore, if any PCR products are generated with full-length adaptor sequences at both ends, amplification is suppressed following the denaturation step during PCR cycling because the terminal repeats that are present in these sequences form panhandle structures.

Further methods rely on the use of less specific second primers: restriction site-dependent PCR employs oligonucleotide primers specific for given restriction enzyme recognition sequences, in conjunction with a specific primer derived from the known DNA sequence, whereas targeted gene walking is based on PCR initiation with primers bearing only partial homology at the 3′ end of the target sequence. The disadvantage of this latter method is that it produces numerous non-specifically amplified species in addition to the correct PCR product.

Random primer PCR: This utilizes a combination of short random primers and one primer for a known sequence, the extra product formed due to the presence of the known primer is eluted and sequenced.

Restriction site-dependent PCR (RSD-PCR): RSD-PCR utilizes the natural restriction sites in the genomic DNA to design universal primers, and has two rounds of PCR protocol without nested process. The RSD-PCR protocols have been optimized, and successfully used for chromosome walking and new gene cloning *via* degenerated primers. RSD-PCR strategy is based on these principles: restriction sites dispersed throughout the genome are candidates for universal pairing; a universal primer is a combination of selected restriction sites and degenerate sequence at the 3′ and 5′ end, respectively. A two-round PCR protocol is designed and optimized for the RSD-PCR to amplify the single-strand target template from genomic DNA by a specific primer, and amplify the target gene by using the specific primer and one of the universal RSD primers. In the first round, the single-strand target gene from genomic DNA is amplified by a single specific primer under stringent conditions. The PCR product is diluted 50-fold and used as a template in the second round. Then, the double-strand target gene is amplified by a universal primer and the specific primer.

Targeted gene walking: This is a modification of PCR that allows the amplification of unknown sequences adjacent to a known sequence without intervening steps. The overall protocol for "gene-walking PCR" consists of three consecutive steps: A series of PCR reactions are performed in parallel, with identical components in each tube except for different "walking" primers. Later, an oligomer-extension assay is performed in a small aliquot of PCR reaction with a nested, kinased internal primer, to identify products that represent "walks" from the targeted site. Further, the labelled band is excised from the gel, re-amplified and directly sequenced.

4. T-DNA Insertion Mutant Resources, Gene Cloning and their Limitations

A number of genes involved in various metabolic and developmental pathways have been cloned from the T-DNA tagged lines (Radhamony *et al.*, 2005). For example, genes involved in the ethylene signal transduction pathway have been isolated. The engineering of these genes and others in the pathway might make it possible to generate plants that are more resistant to environmental stresses or improve senescence qualities of fruits and vegetable parts of plants.

Although, many agriculturally useful plants are the product of traditional plant breeding strategies based on quantitative genetic variation and on random and spontaneous genetic events, biotechnology is playing an increasingly important role, e.g. most European barley varieties are produced *via* doubled haploidy and marker-assisted selection is commonplace, especially for disease resistance genes. Therefore, functional genomics in combination with genetic engineering for particular commercial traits such as exploiting changes in plant architecture: height, morphology, leaf size, number and shape, root proliferation, the timing of flowering, and the number, size and shape of seeds, has enormous potential (**see Chapter 25**). By identifying the key genes known to control basic steps in plant architecture, it may be possible to change almost any aspect of plant development; this would lead to the production of a range of novel horticultural and agricultural varieties. In the past, the best-known examples have been the height reductions in wheat and rice that facilitated the "green revolution" in cereal production in the 1960s. This modification in plant architecture is the visual evidence of variation in the behaviour and timing of the expression of a large number of regulatory genes and chemical signalling pathways within and between cells of a meristem. Recently, Forster *et al.* (2007) have demonstrated the use of mutants to elucidate the phytomer model of barley, which includes for example an indeterminate floral mutant similar to that of wheat which may be exploited in increasing barley yields. In the past dwarfing and flowering time mutants have played a major role in crop improvement in cereals. Therefore, T-DNA mutagenesis represents a promising approach to the molecular isolation of genes with an essential function, e.g. during plant growth and development.

4.1. T-DNA Insertion Mutant Resources

Transformation mediated by *Agrobacterium* has been widely used for insertional mutagenesis in different plant species. Huge collections of T-DNA transformants of *Arabidopsis* have been generated independently by several groups and have been used extensively for reverse genetics (Sessions *et al.,* 2002; Krysan *et al.,* 2002; http://signal.salk.edu/cgi-bin/tdnaexpress;http://Arabidopsis.info/info/MASC_2007.pdf.;http://www.GABI-Kat.de). Recently, more than 225,000 independent *Agrobacterium-mediated* T-DNA insertion events in the genome of *Arabidopsis thaliana* have been reported (http://signal.salk.edu/cgi-bin/tdnaexpress) that represent near saturation of the gene space. However, the precise locations were determined for about 88,000 T-DNA insertions, which resulted in the identification of mutations in >21,700 of the 29,454 predicted *Arabidopsis* genes. Genome-wide analysis

of the distribution of integration events revealed the existence of a large integration site bias at both the chromosome and gene levels. The first advantage of using T-DNA as an insertional mutagen is that it directly generates stable insertions into genomic DNA and does not require additional steps to stabilize the insert. The second advantage is that T-DNA insertion appears to be completely random, as no T-DNA integration hot spots or integration preferences have been reported. The *Arabidopsis* Biological Resource Center (ABRC) at the Ohio State University and the Nottingham *Arabidopsis* Stock Centre (NASC) in UK, have been established to disseminate biological materials, including mutant germplasm, gene, DNA, data, descriptions, etc.

Aside from the advantage of tagging the disrupted locus, T-DNA has other advantages over traditional mutagenesis systems: low copy number and random insertion. A database (http://atidb.cshl.org/) has been developed for archiving, searching and analyzing insertional mutagenesis lines, in the model plant species *Arabidopsis thaliana*. For the websites that provide information see **Box 38.2**.

4.2. Major Insertion Mutants in Plants

Other than *Arabidopsis*, information on genomic sequences and physical maps are now available for many plants such as rice, maize, tomato, *Medicago* and *Brassica*. For example, the rice genome has been extensively covered because of its small size among cereal genomes, nearly complete genome sequence and economic importance as a major crop. T-DNA tagging has been used successfully for gene discovery in rice. An initial database has also been constructed using T-DNA flanking sequences. T-DNA tags have also been observed to insert preferentially into gene-rich regions. A database of insertion sites in rice is publicly available at http://www.genomics.zju.edu.cn/riceetdna. In tomato new tools for functional analysis based on insertional mutagenesis, with an activator/dissociator (Ac/Ds) system in the background of the miniature cultivar Micro-Tom has been established. The T-DNA mutational approach is now being modified successfully to tag genes in a number of economically important plant species such as, rice, tomato, *Brassica*, *Medicago* and poplar. Thus T-DNA tagging in conjunction with other mutation based techniques like transposon insertion, TILLING, etc. would not only continue to provide useful

Box 38.2: Gene disruption resources in *Arabidopsis*

Laboratory	Sequenced insertions	Source
Salk Institute	94,947	http://signal.salk.edu/cgi-bin/tdnaexpress
TMRI	100,000	http://www.tmri.org/en/partnership/sail_collection.aspx
GABI-KAT	20,764	http://www.mpiz-koeln.mpg.de/GABI-Kat/
FLAG	11,500	http://flagdb-genoplante-info.infobiogen.fr/

information in *Arabidopsis* but is likely to prove an efficient tool for functional genomics in other plants also.

4.3. Gene Cloning Using T-DNA Insertion Lines: An Example of Brassinosteroid Dwarf Mutants.

T-DNA insertion mutagenesis is a compromise between the ease of mutant generation and the ease of cloning. Therefore, by first generating a mutant that demonstrates the phenotypic consequences of a mutation in a gene, the context for understanding gene function is enhanced enormously. Moreover, the versatility of T-DNA tagging vectors is such that they can be employed to search not only for specific genes but also sequences of DNA controlling expression. In recent years there has been an explosion of cloned genes in *Arabidopsis*, rice and, to a lesser extent, other plant species (Radhamony *et al.,* 2005). Data now suggest the utility of T-DNA tagging in other plants such as rice, maize, tobacco and raise the possibility of applying this mutagen to the agronomically important crop plants. As an example of this T-DNA tagging approach, the discovery that two differentially regulated genes in rice control the rate-limiting step in brassinosteroid (BR) biosynthesis introduces several exciting possibilities for engineering the BR pathway to increase the yield in rice and other monocots.

The importance of Brassinosteroids: BRs are a new group (the sixth group) of plant growth regulators with significant growth-promoting activity. They were first isolated and characterized from the pollen of *Brassica napus* L. and are now regarded as ubiquitous in the plant kingdom. BRs are considered as phytohormones with pleiotropic effects, as they influence several developmental processes such as growth, germination of seeds, rhizogenesis, flowering, abscission, maturation and senescence. They also confer resistance to plants against various abiotic stresses. BRs are biosynthesized from campestrol, the biosynthetic pathway was elucidated and later shown to be correct through the analysis of BR biosynthesis mutants in *Arabidopsis*,

tomatoes and peas. Moreover, the studies conducted in *Arabidopsis* with brassinosteroid-deficient, sensitive and insensitive mutants provide convincing evidence that BRs are essential for plant growth and development. For example, BR biosynthetic mutants such as *bri1*, *cpd1* and *dwf1* have been identified among de-etiolated dwarfs of *Arabidopsis*. The growth of mutants was found to be restored by exogenous application of brassinolide and not by gibberellin or IAA. Thus, new discoveries of the physiological properties of BRs allow us to consider them as highly promising, environmentally-friendly natural substances suitable for wide application in plant protection and yield promotion in agriculture.

During the last decade, several groups (including ours) have isolated dwarf mutants/genes, especially the *Br* mutants, using T-DNA insertion mutants. For example, a T-DNA insertion library based on an *Agrobacterium*-mediated zygotic embryo transformation technique in *Arabidopsis* can be established and used to screen for dwarf phenotypes in F_2 plants (e.g. Veyres *et al.,* 2008). One of the mutants identified carried a recessive nuclear mutation, *boul (bul)* (**Figure 38.8**). By characterizing the dwarf mutants, several groups have shown the involvement of BRs in the cell elongation process. The *Br*-dwarf gene is the latest in a series of genes that, through their effects on plant growth and development, have potential use in crop plants. *Br* genes have been over-expressed and studied in rice with an erect phenotype and with a significant increase in photosynthesis and yield. These new genes may allow the time and duration of growth/height to be altered in transgenic crop plants. In food crops such as rice, wheat and corn the extent of plant growth is a critical factor in determining the seed number and size- for instance in rice inflorescence. It will be interesting to determine the relationship between genes controlling this aspect of determinacy and dwarfism.

Moreover, convincing evidence shows that an agronomically important trait can be enhanced by manipulating BR levels without undesirable side-effects. In

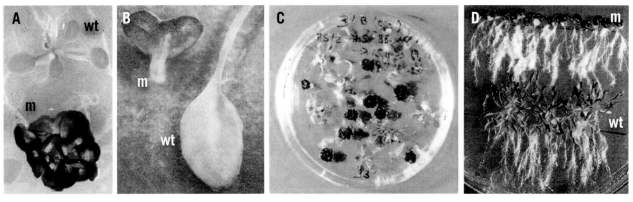

Figure 38.8 Morphological differences between wild type and *bul* mutant, a T-DNA mutant (A) Wild type (wt) and *bul* (m) plants after 15 days of culture. (B) Leaves of wt and mutant. (C) and (D) wt and mutant plants after 4 weeks of culture.

the case of the two rice C-22 hydroxylases, one plays a primary role in shoot elongation and reproductive development, whereas the other controls leaf inclination. Loss of the former gene (OsDWARF4L1) results in semi-dwarf plants with small seeds, whereas loss of the latter gene (OsDWARF4) results in normally shorter plants with erect leaves. The more erect leaf architecture most probably increases photosynthesis in the lower leaves and thus improves yields under dense planting conditions, thus indicating that increases in yield are possible for all crops. Manipulation of BR biosynthesis or other signal transduction pathways should prove a useful approach to this end. Given that loss-of-function mutants in the BR pathways are dwarfed, it was hypothesized that the over-expression of the same genes would produce taller plants with greater biomass and yield. Using T-DNA mutants, many of the growth and developmental processes, including meristem differentiation, hormone perception, flower development and morphogenesis have been studied in *Arabidopsis*, and comprehensive reviews for each exist (Radhamony *et al.*, 2005).

4.4. Limitations of Insertion Mutants for Gene Cloning

Some problems have been reported while analyzing the T-DNA tagged lines, e.g. multiple insertions, complex arrangement of T-DNA, insertion of vector backbone sequences, modifications of the DNA content per nucleus, chromosomal duplication and rearrangements or a combination of these (Radhamony *et al.*, 2005). For example, complex patterns of T-DNA integration, including the transfer of vector sequences adjacent to T-DNA borders and the large frequency of concatemeric T-DNA insertions, can complicate further PCR analysis for reverse genetics. Moreover, small and major chromosomal rearrangements induced by T-DNA integration have been frequently observed, leading to difficulties in the genetics analysis of the insertion, such as mutant phenotypes that are not correlated with the T-DNA insertion. Moreover, the efficient transformation systems in *Arabidopsis* have made it feasible for laboratories to generate thousands of independent transformants rapidly and as a result T-DNA is now a widely used insertional mutagen in *Arabidopsis*. However, the simplicity of using insertional mutations should not overshadow the power of other approaches, such as chemical and physical mutagenesis.

5. Acknowledgements

We thank Dr. Geetha Govind for her assistance in writing a portion of the work on the molecular techniques for recovery of plant DNA flanking the T-DNA.

6. References

6.1. Cited References

Azpiroz-Leehan, R. and Feldmann, K.A. 1997. T-DNA insertion mutagenesis in *Arabidopsis*: going back and forth. *Trends in Genetics*. 13: 152–156.

Bechtold, N., Ellis, J. and Pelletier, G. 1993. *In planta Agrobacterium*-mediated gene transfer by infiltration of adult *Arabidopsis thaliana* plants. *CR Acad. Sci.: Life Sci.* 316: 1194–1199.

Clough, S.J. and Bent, A.F. 1998. Floral dip: a simplified method for Agrobacterium-mediated transformation of *Arabidopsis thaliana*. *The Plant Journal*. 16: 735–743.

Feldmann, K.A. 1991. T-DNA insertion mutagenesis in *Arabidopsis* mutational spectrum. *The Plant Journal*. 1: 71–82.

Feldmann, K.A. and Marks, M.D. 1987. *Agrobacterium*-mediated transformation of germinating seeds of *Arabidopsis thaliana*. *Molecular and General Genetics*. 208: 1–9.

Forster, B.P., Franckowiak, J.D., Lundqvist, U. *et al.* 2007. The Barley Phytomer. *Annals of Botany*. 100: 725–733.

Krysan, P., Young, J.C., Jester, P. *et al.* 2002. Characterization of T-DNA insertion sites in *Arabidopsis* thaliana and the implications for saturation mutagenesis. OMICS A J. *Integrative Biology*. 6: 163–174.

Pan, X., Liu, H., Clarke, J. *et al.* 2003. ATIDB: *Arabidopsis thaliana* insertion database. *Nucleic Acids Res*. 31: 1245–1251.

Phillips, R.L., Kaeppler, S.M. and Olhoft, P. 1994. Genetic instability of plant tissue cultures: Breakdown of normal controls. *Proc. Natl. Acad. Sci. USA*. 91: 5222–5226.

Radhamony, R.N., Prasad, A.M. and Srinivasan, R. 2005. T-DNA insertional mutagenesis in *Arabidopsis*: a tool for functional genomics. *E. Journal Biotechnology*. 8: 1–20.

Sangwan, R.S. and Sangwan-Norreel, B.S. 1989. Genetic transformation and plant improvement. *In*: B.S. and R.S. Sangwan (eds.) Biotechnology in Agriculture. Dordrecht: Kluwer Academic Press, pp. 299–337.

Touraev, A., Stöger, E., Voronin, V. *et al.* 1997. Plant male germ line transformation. *The Plant Journal*. 12: 949–956.

Valvekens, D., Van Montagu, M. and Van Lijsebettens, M. 1988. A. *tumefaciens*-mediated transformation of *Arabidopsis thaliana* root explants by using kanamycin selection. *Proc. Nat. Acad. Sci. USA*. 85: 5536–5546.

Veyres, N., Danon, A., Aono, M. *et al.* 2008. The *Arabidopsis* sweetie mutant is affected in carbohydrate metabolism and defective in the control of growth, development and senescence. *The Plant Journal*. 55: 665–686.

6.2. Websites

CAMBIA:
http://www.cambia.org.au

GABI-Kat SimpleSearch (http://www.GABI-Kat.de) is a Flanking Sequence Tag (FST)-based database for T-DNA insertion mutants.

Ochatt, S.J. 2006. Flow cytometry (ploidy determination, cell cycle analysis, DNA content per nucleus). *In*: *Medicago truncatula* Handbook, Chapter 2.2.7. Online version, http://www.noble.org/MedicagoHandbook/

T-DNA Express: Arabidopsis Gene Mapping Tool:
http://signal.salk.edu/cgi-bin/tdnaexpress

The multinational coordinated Arabidopsis thaliana functional genomics project:
http://Arabidopsis.info/info/MASC_2007.pdf.

6.3. Further Reading

Azpiroz-Leehan, R. and Feldmann, K.A. 1997. T-DNA insertion mutagenesis in *Arabidopsis*: going back and forth. *Trends in Genetics*. 13: 152–156.

Catterou, M., Dubois, F., Smets, R. *et al.* 2002. hoc: an *Arabidopsis* mutant overproducing cytokinins and expressing high *in vitro* organogenic capacity. *Plant Journal*. 30:1–17.

Feldmann, K.A. 2006. Steroid regulation improves crop yield. *Nature Biotechnology*. 24: 46–47.

Fujioka, S. and Yokota, T. 2003. Biosynthesis and metabolism of brassinosteroids. *Annual Review of Plant Biology*. 54: 137–164

Koncz, C., Martini, N., Meyerhofer, R. *et al.* 1994. Specialized vectors for gene tagging and expression studies. *In*: S.B. Galvin and R.A. Schilperoort (eds.) Plant Lol. Biol. Manual, Vol. B2 . Dordrecht: Kluwer Academic Press. pp. 1–22.

Ochatt, S.J. 2008. Flow cytometry in plant breeding. *Cytometry*. A73: 581–598

Parinov, S. and Sundaresan, V. 2000. Systems for insertional mutagenesis. *Current Opinion Biotech*. 11: 157–161.

Wang, K.L.-C., Hai, L. and Ecker, J.R 2002. Ethylene biosynthesis and signaling networks. *Plant Cell*. 14: S141–S151.

Transposon Mutagenesis for Functional Genomics

Q.H.Zhu*, N.Upadhyaya and C.Helliwell

Division of Plant Industry, Commonwealth Scientific and Industrial Research Organization, GPO Box 1600, Canberra, ACT 2601, Australia
* Corresponding author, E-MAIL: qianhao.zhu@csiro.au, Tel: 61 2 6246 4903, Fax: 61 2 6246 5000

1. Introduction

With the completion of genome sequencing of *Arabidopsis* and rice, the model species for dicotyledonous and monocotyledonous plants, respectively, and the production of large expression sequence tag (EST) collections in a number of plant species, plant science has entered a new era of functional genomics that aims to unravel the biological functions of all plant genes. Mutant phenotypes provide direct links between genes and their biological functions in plant growth and development. Generation and evaluation of mutants is therefore essential to provide new impetus for genetic studies and functional genomics in plants. Chemical and ionizing radiation mutagenesis have been routinely used to create mutants for genetic analyses and plant breeding, but it is generally time-consuming, and difficult to identify the responsible gene. A number of functional genomics tools have been developed to accelerate linking mutant phenotypes to genes. Among these, insertional mutagenesis using T-DNA and/or transposable elements (TEs) is not only able to create large-scale mutant populations but also provides molecular tags for further identification and characterization of the tagged gene. Compared with T-DNA insertional mutagenesis (**see Chapter 38**), mutagenesis using TEs has several advantages. First, large-scale mutagenized populations can be produced using a relatively small number of initial TE insertion lines as many independent insertions can be generated among the progeny of a single TE insertion line. This is particularly useful for plant species for which transformation is inefficient. Second, T-DNA transformants tend to have multiple copies of the T-DNA insert at one locus while transposon insertions are usually single copy. Third, a chromosome region of interest, such as a quantitative trait locus (QTL) controlling an agronomically important trait, can be intentionally mutagenized using a TE insertion line harbouring a TE close to the QTL. Fourth, a tagged gene can be confirmed by recovery of revertants resulting from excision of the TE tag. This chapter introduces the principles and strategies of transposon mutagenesis in gene identification, discusses the approaches used for mutant identification and verification, and highlights development of enhanced tagging systems.

2. Plant Transposable Elements

Transposable elements were first discovered by Barbara McClintock in the 1940s as the causative agent of maize kernel variegation (**see Chapter 1**). Since then TEs have been found to be ubiquitous genetic elements in both prokaryotes and eukaryotes. Genome sequencing and annotation have shown that TEs are the single largest component of the genetic material of most eukaryotes. They account for 15–80% of plant genomes, for instance, ~30% of the rice genome consists of TEs (International Rice Genome Sequencing Project, 2005), and are driving forces of genome evolution due to their abilities to alter genome organization and gene sequences.

Eukaryotic TEs are categorized into two groups according to their structure (**Figure 39.1, A–C**) and transposition mechanism (**see Box 39.1 for details**): Class I elements or retrotransposons and class II elements or DNA transposons. Class I elements can be divided into long terminal repeat (LTR) retrotransposons with both LTRs in direct repeat orientation (**Figure 39.1B**), and non-LTR retrotransposons that are further divided into long interspersed nuclear elements (LINEs) and short interspersed nuclear elements (SINEs). Class II elements are characterized by short terminal inverted repeats (TIRs) and duplication of a short genomic sequence at the site of insertion, a feature conserved in each TE superfamily, for example, the *hAT* (**hobo-_Ac_-Tam3**) superfamily generates 8-bp target site duplication (TSD) upon insertion (**Figure 39.1D**). Both groups of TEs contain autonomous and non-autonomous elements with transposition of the latter controlled by the former.

3. Transposon Mutagenesis Strategies

3.1. Transposable Element as a Gene Identification Tool

Transposable elements, by their capacity to excise from the original location in the genome and reinsert into a new genomic location, have the potential to induce mutations that result in detectable mutant phenotypes due to knock-out of the gene function. The disrupted gene can then be identified by analyzing the DNA sequences surrounding the inserted TE (**Figure 39.2**). This approach of gene isolation, transposon tagging, does not require prior knowledge of the gene to be isolated, but does require the sequence of the TE to

Box 39.1: Glossary

Autonomous TEs	TEs with open reading frames (ORFs) encoding transposase – a protein required for transposition. They are active or are capable of transposition by themselves.
Flanking sequence tags (FSTs)	DNA sequences flanking a TE insert, which are usually isolated using PCR-based approaches or plasmid rescue if the TE-containing vector is engineered with a bacterial origin of replication (*ori*) and an appropriate antibiotic resistance gene.
Forward genetics	A gene characterization approach starting with a mutant phenotype followed by cloning and identification of the corresponding gene.
Germinal excision	Excision of transposon occurring in germline cells that can be passed on to the next generation.
Launching pad	The integration site of a T-DNA from which a transposon is excised and reinserted into a new location in the genome, also called the donor site.
Long interspersed nuclear elements (LINEs)	Long DNA sequences (>5 kb) contain an internal promoter and at least two genes that are responsible for transcription, reverse transcription and integration, respectively. These properties enable them to copy both themselves and other non-coding LINEs, consequently to enlarge the genome.
Non-autonomous TEs	TEs without or with only partial ORFs encoding transposase. In most cases they are derived from internal deletion of an autonomous TE. They are inactive but can be transposed in the presence of an autonomous TE.
Non-targeted or random mutagenesis	A transposon mutagenesis strategy that aims to mutagenize a gene of interest, a specific chromosomal region (regional mutagenesis) or to create a collection of mutants.
One-element system	A transposon tagging strategy in which only an autonomous transposon is used.
Regional mutagenesis	A transposon mutagenesis strategy that aims to mutagenize a certain chromosomal region based on the linked transposition characteristic of some transposons, such as associator/dissociator (*Ac/Ds*) and suppresor-mutator/defective suppressor-mutator (*Spm/dSpm*).
Retrotransposons	Class I TEs that transpose *via* an RNA intermediate by a 'copy and paste' mechanism. Copy number increases after transposition.
Reverse genetics	A gene characterization approach starting with the gene of interest aiming to understanding the gene function by the generation and analysis of mutant phenotypes.
Revertant	Wild type phenotype progeny resulting from germinal excision of a transposon insert that restores the ORF of the disrupted gene.
Short interspersed nuclear elements (SINEs)	Short DNA sequences (<500 bp) containing an internal pol III promoter at the 5′ end and other sequence of unknown origin in the 3′ half of the element. They do not encode a functional reverse transcriptase but can be transposed in the presence of active LINEs.
Somatic excision	Excision of a transposon occurring in somatic cells resulting in mosaic tissues, an early-stage excision results in large reversion sectors, whereas a late-stage excision results in a small reversion sector. These excisions are not passed on to the next generation.
Targeted mutagenesis	A transposon mutagenesis strategy that aims to generate transposon tagged alleles of a known mutant identified by conventional mutagenesis approaches to identify the gene responsible.
Transposable elements (TEs)	DNA sequences that can excise from one chromosomal location and reinsert into new chromosomal locations in the genome.
Transposons	Class II TEs that transpose *via* a DNA intermediate by a 'cut and paste' mechanism. Copy number does not change before and after transposition.
Transposon tagging	A gene identification strategy by insertion of a transposon into a genomic region that may or may not result in a detectable mutant phenotype, and the gene can be isolated by cloning the DNA sequences flanking the transposon insert that serves as a molecular tag.
TILLING (Targeted Induced Local Lesions in Genomes)	A PCR mutagenesis protocol to create single base mutations in particular genes. This method allows the creation of multiple alleles.
Two-element system	A transposon tagging strategy in which the autonomous element is immobilized by removing its *cis*-sequence, usually one of the terminal inverted repeats, and the immobilized autonomous and its corresponding non-autonomous element are transformed into two independent transgenic lines. Transposition of the non-autonomous element is initiated by crossing these two transgenic lines.

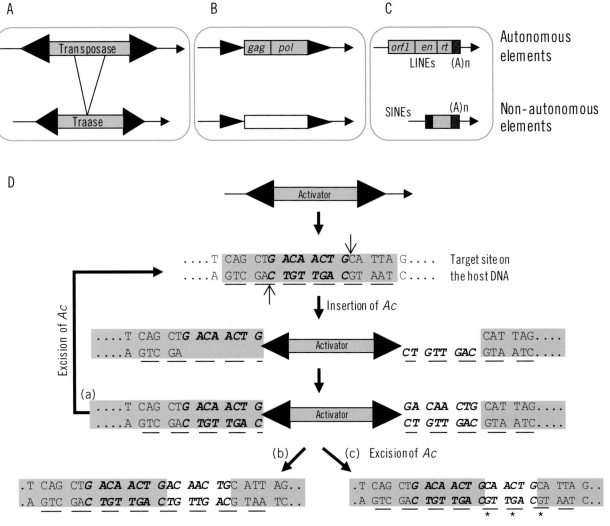

Figure 39.1 Schematic representations of structures of transposable elements. A, class II DNA transposons. The autonomous element encodes transposase and is capable of transposition by itself, whereas the non-autonomous element is a derivative of the autonomous element due to internal deletion. Both elements contain terminal inverted repeats (TIRs; black solid triangles) that are essential for transposition. B, LTR retrotransposons. These have long terminal repeats (LTRs) in direct orientation (black triangles). The autonomous elements contain at least two open reading frames (ORFs; *gag* and *pol*) encoding a capsid-like protein and a polyprotein, respectively, which are required for reverse transcription and transposition. The non-autonomous element lacks partial or all coding sequences. C, LINEs and SINEs. LINEs contain coding sequences (*orf1*, *en* and *rt*) that encode a *gag*-like protein, endonuclease and reverse transcriptase, respectively. SINEs are character-ized by an internal pol III promoter (black rectangle). Both LINEs and SINEs terminate by a simple sequence repeat, usually poly(A). D, target site duplication (TDS) and excision footprint. Upon insertion of *Activator (Ac)* in the exon of a host gene, an 8-bp TSD (bold italic letters) is generated and the ORF (highlighted in grey background and individual codons for amino acids are underlined) is interrupted, resulting in a loss-of-function mutation. Upon excision of the *Ac* element, the ORF could be restored resulting in revertant (a), but in most cases, excisions are imprecise resulting in frameshift mutation (b) or restoration of the reading frame but changed in the amino acid sequence (c) that may or may not show the wild type phenotype, i.e. revertant, depending on whether or not the changes are located at the conserved domain. Note that only two possible excision events are shown.

be used. Isolation and characterization of class II DNA transposons from maize (*Zea mays*) and snapdragon (*Antirrhinum majus*) in the 1980s (Fedoroff *et al.,* 1983 and Bonas *et al.,* 1984, respectively) laid the cornerstone for application of transposon tagging in gene identification.

After cloning of the first gene, *bronze* (*bz*), using the *Activator* (*Ac*) element in maize (Fedoroff *et al.,* 1984), a number of genes were subsequently identified using

endogenous transposons in several plant species. The importance of transposon tagging for gene identifica-tion was quickly recognized by geneticists working with plants that lack well-characterized endogenous TEs, leading to the introduction of maize transposons into a range of heterologous plant species, including tobacco (*Nicotiana tabacum*), *Arabidopsis*, tomato (*Solanum lycopersicum*), rice (*Oryza sativa*) and soybean (*Glycine*

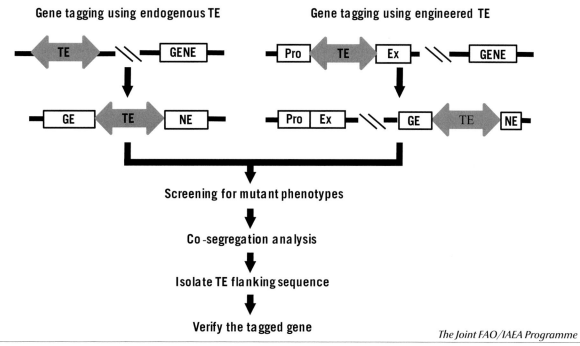

Gene tagging using endogenous TE   **Gene tagging using engineered TE**

Screening for mutant phenotypes

Co-segregation analysis

Isolate TE flanking sequence

Verify the tagged gene

The Joint FAO/IAEA Programme

Figure 39.2 General principle of transposon tagging in plants. A gene is disrupted by insertion of a TE, which results in a mutant phenotype that can be identified in the segregating population. Once the co-segregation relationship between the transposon insert and the mutant phenotype is confirmed, the genomic sequence of the interrupted gene can be isolated using inverse PCR or TAIL-PCR with primers specific to the TE. Identity of the gene can be found by Basic Local Alignment Search Tool (BLAST) search of sequence databases using the isolated genomic sequence as a query. In a heterologous system, the engineered TE is usually inserted between a constitutively-expressed promoter (Pro) such as CaMV 35S and an excision marker (Ex) such as hygromycin resistance gene (*HPT*) that will express upon excision of the element so that excision of the element can be monitored.

max). Several genes were first cloned using the transformed transposon as a molecular tag in *Arabidopsis* and petunia. These efforts demonstrated the feasibility of transposon tagging in heterologous plant species and encouraged plant biologists to further refine and develop the tagging systems to increase screening and tagging efficiencies. To date, elegant transposon tagging systems have been established in several plant species, leading to isolation and characterization of many genes. There are two tagging strategies: 1) targeted mutagenesis and 2) non-targeted or random mutagenesis. Targeted mutagenesis is mainly applied in maize, whereas non-targeted mutagenesis is applied in plants with well-characterized endogenous transposons as well as in heterologous plants.

3.2. Targeted Mutagenesis

Normally recessive homozygous mutants are available for the target gene. Plants carrying active transposon(s) are crossed to the homozygous mutants. The F_1 population is screened for rare progeny showing the mutant phenotype, which are analyzed to check the presence

of the transposon and to confirm the co-segregation of the transposon with the mutant phenotype. The flanking sequence tag (FST) of the transposon is then isolated using inverse PCR (iPCR) or thermal asymmetric interlaced (TAIL) PCR (**Figure 39.3A**). The identity of the mutant gene is determined by searching sequence databases using the isolated FST as a query. This approach cannot be applied to identify genes encoding indispensable functions for plant growth and development as no homozygous mutant will be available.

In maize, *Ac/Ds, Enhancer (En)/Inhibitor (I)* or *Spm/dSpm* and *Mutator (Mu)* have all been successfully used in targeted mutagenesis. These elements showed different tagging efficiencies or mutation rates due to differences in copy number and transposition characteristics. It has been shown that both *Ac/Ds* and *Spm/dSpm* tend to transpose to genetically linked sites. In studies of the *p* and *bz* loci in maize, ~60% of Ac transpositions were to genetically linked sites (e.g. Dooner and Belachew, 1989), with the majority of transposed *Ac* located within 10 cM of the donor site (e.g. Dooner and Belachew, 1989). *Ac/Ds* and *Spm/dSpm* have similar mutation rates at ~10^{-5}–10^{-6}/gene per generation (**Table 39.1**), but it can

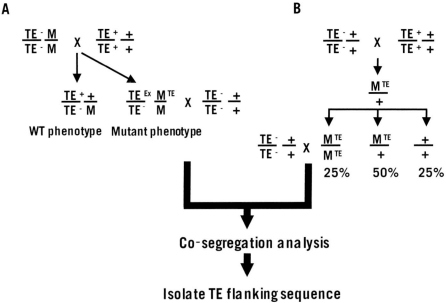

$$\text{A} \qquad\qquad\qquad\qquad\qquad\qquad \text{B}$$

Co-segregation analysis

Isolate TE flanking sequence

The Joint FAO/IAEA Programme

Figure 39.3 Schematic representations of transposon tagging strategies. A, targeted mutagenesis. A homozygous mutant without active transposon is crossed to a wild type plant with active transposon(s). Among the F_1 population, rare mutant progeny are identified from the majority of plants showing the wild type phenotype. They are crossed to wild type plants to identify the transposon co-segregating with the mutant phenotype using the segregating population. The transposon flanking sequence can then be isolated using PCR-based approaches. B, non-targeted mutagenesis. Transposon-active plants are crossed to wild type plants and the resulting F_1 progeny are self-pollinated to generate a segregating population (F_2). Recessive mutants are identified and backcrossed to wild type plants to identify the transposon co-segregating with the mutant phenotype. M: mutant allele; M^{TE}: mutant allele with a TE insertion; +: wild type allele; TE^+: with TE insertion; TE^-: without TE insertion.

be up to 100-fold higher if the gene of interest is linked to the active transposon. The mutation rate of *Mutator* is 10–100-fold higher than that of *Ac/Ds* and *Spm/dSpm* largely due to its high copy number and a transposition preference for genetically unlinked sites (Lisch *et al.*, 1995), but the characteristic of high copy number may complicate the subsequent co-segregation analysis. In addition, the *Mu* element has a low germinal excision rate and revertants are difficult to recover.

3.3. Non-Targeted Mutagenesis

Non-targeted mutagenesis is a more general tagging approach. It is not only suitable for cloning genes causing

lethality or infertility because the mutated allele can be maintained through heterozygotes, but also for generating large-scale transposon insertion populations. In this approach, plants with active transposons are crossed to wild type plants. Dominant mutations that are related to seed development might be observed in F_1 progeny due to fertilization of the mutagenized gametes, but recessive mutations can only be screened with a 3:1 segregation ratio for the normal and mutant phenotype in the F_2 population (**Figure 39.3B**). After confirmation of the co-segregation relationship between the transposon and the mutant phenotype, the corresponding gene is able to be cloned using similar approaches used in targeted mutagenesis.

Table 39.1: Mutation frequency of transposons determined by targeted mutagenesis in maize

Transposon	Copy number in the genome	Mutation frequency	Target site duplication (bp)	Transposition preference	Utilization in heterologous systems
Ac/Ds	Low	$\sim10^{-6} - \sim10^{-5}$	8	Linked > unlinked	Yes
Spm/dSpm or En/I	Low	$5.8 \times 10^{-6} - 4.6 \times 10^{-5}$	3	Linked > unlinked	Yes
Mu	Low–High	$1.7 \times 10^{-5} - 3.9 \times 10^{-4}$	9	Unlinked > linked	No

The focus of non-targeted mutagenesis can be on a single gene or genes involved in a specific metabolic pathway, but only genes functioning in a single copy manner can be effectively identified and analyzed. To ensure that the gene of interest can be identified and cloned, it is necessary to increase the size of the mutant population and the density of the transposon inserts, i.e. towards achieving saturation mutagenesis. These populations can be screened for the gene of interest based on a forward genetics approach, but they are more routinely screened using a reverse genetics approach (see below). *Mutator* is the most commonly used element in maize because of the aforementioned characteristics, whereas *Ac/Ds* and *Spm/dSpm* elements are more commonly used in other plant species.

3.4. Regional Mutagenesis

Saturating a plant genome with transposon insertion can be achieved by using transposons with high copy number, such as *Mu* elements. Several projects have been set up to mutagenize the maize genome using *Mu* elements, such as the Trait Utility System for Corn (TUSC), Maize Targeted Mutagenesis (MTM), UniformMu and RescueMu (Brutnell, 2002). Alternatively, transposons such as *Ac/Ds* and *Spm/dSpm* that preferentially transpose to genetically linked sites are first used to saturate a chromosomal region, i.e. regional saturation, and finally to achieve whole genome saturation by combining each saturated region. For example, with a population of insertion lines in which the *Ac* elements are spaced at 10–20 cM intervals, it is possible to

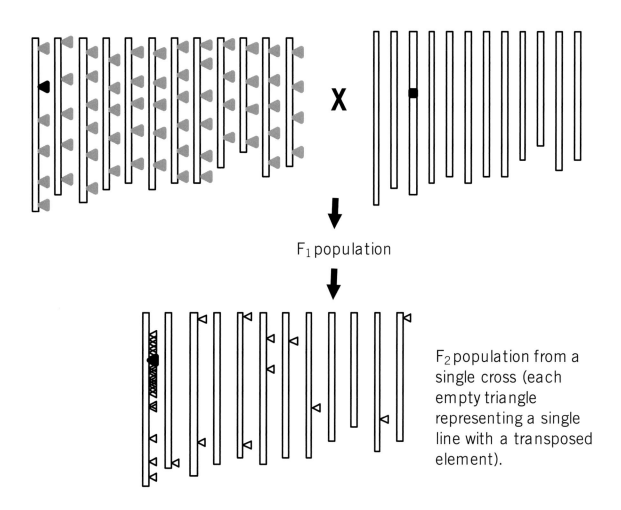

F$_1$ population

F$_2$ population from a single cross (each empty triangle representing a single line with a transposed element).

The Joint FAO/IAEA Programme

Figure 39.4 Saturating the rice genome using regional mutagenesis. Starting *Ds* or *dSpm* insertion lines (each triangle representing a single insertion line) with the transposons evenly distributed across all 12 chromosomes are selected from available resources based on the FST mapping information. Each line is crossed to *Ac* or *Spm* (represented by black square) line to re-mobilize *Ds* or *dSpm*. Because of the linked transposition characteristic of *Ds* and *dSpm*, the majority of the transposed elements (empty triangles) will be located at the flanking region of the launching pad (black triangle) although unlinked transposition events (on the same or different chromosome) are also present.

Table 39.2: Transposon tagging population resources in plants

Species	Institution	Transposon used	Mutagenized loci	FST lines available	Database web link
Rice	CIRAD-INRAIRD-CNRS, Ge'noplante, France	Tos17	100,000	11,488	http://urgi.versailles.inra.fr/
	NIAS, Japan	Tos17	500,000	34,844	http://tos.nias.affrc.go.jp
	CSIRO, Plant Industry, Australia	Ac/Ds, GT and ET	16,000	~3,000	http://www.pi.csiro.au/fgrttpub
	EU-OSTID, European Union	Ac/Ds, ET	25,000	1,300	http://orygenesdb.cirad.fr/
	Gyeongsang National University, South Korea	Ac/Ds, GT	30,000	4,820	http://www.niab.go.kr/RDS
	Temasek Lifesciences, Singapore	Ac/Ds, GT	20,000	2,000	NA
	University of California, Davis, USA	Ac/Ds, GT En/I	20,000	4,630 9,036	http://www-plb.ucdavis.edu/Labs/sundar
Arabidopsis	RIKEN, Japan	Ac/Ds	~14,000	11,800	http://rarge.gsc.riken.jp/phenome
	University of Cologne, Germany	En/I, AT	9,471	NA	http://arabidopsis.info/CollectionInfo?id=71
	CSHL, USA	Ac/Ds, GT and ET	NA	~7,700	http://genetrap.cshl.org/
	John Innes Centre, UK	Ac/Ds, En/I, GT and ET	>43,200	>1,200	http://www.jicgenomelab.co.uk/projects/transposon-tagging/sins.html
Maize	Boyce Thompson Institute, USA	Ac	NA	62	http://bti.cornell.edu/Brutnell_lab2/Projects/Tagging/BMGG_pro_tagging.html
	CSHL, USA	Mutator	43,776	NA	http://mtm.cshl.org/
	University of Florida, USA	UniformMu	31,548	~35,000	http://www.hos.ufl.edu/amsweb/research.html; http://pgir.rutgers.edu/endosperm.info/NewEndosperm.html
	NSF, USA	RescureMu	NA	14,887	http://www.maizegdb.org/rescuemu-phenotype.php
	Pioneer Hi-Bred International Inc.	Mu	~1,000,000	NA	NA

AT: activation tagging; ET: enhancer trap; GT: gene trap; NA: not available. Some of the numbers of the mutagenized loci and FST lines in rice are based on Krishnan *et al.* (2009).

mutagenize the whole maize genome using the regional mutagenesis strategy (Walbot, 2000). In *Arabidopsis*, several regional mutagenized populations have been created using *Ac/Ds* or *Spm/dSpm*.

In rice, where a large number of *Ds* and dSpm insertion lines have been created and FSTs of these lines have been or are being isolated in several laboratories (**Table 39.2**), the initial launching pad population can be generated by selecting insertion lines based on the mapping information of the FSTs. Each launching pad line is then crossed to *Ac* or *Spm* lines to saturate the surrounding region of the launching pad. Assuming that 50% of the transposed *Ds* and *dSpm* elements insert within a 1-Mb (~4 cM in rice) genomic region flanking the launching pad, ~400,000 insertion plants (~930 plants with independently transposed *Ds* or *dSpm* to be produced from each line) from ~430 *Ds* or *dSpm* launching pad lines in which the *Ds* or *dSpm* elements are evenly distributed throughout the rice genome with 1-Mb interval will most likely be sufficient to saturate the whole genome (**Figure 39.4**). More launching pad lines would establish the population more rapidly.

Regional mutagenesis is also particularly attractive for cloning and identification of QTLs that have been extensively studied due to their agronomic importance, but are hard to characterize using traditional approaches. Even for plant species with a whole genome sequence, such as rice, regional mutagenesis is more efficient than a map-based cloning approach, a time-consuming procedure. To use regional mutagenesis for QTL cloning, an insertion line with a transposon mapped to the QTL or its surrounding region is searched and retrieved from the insertion mutant libraries and crossed to a line expressing transposase to generate a transposon mutagenized population containing multiple mutant alleles at the locus of interest. Segregating populations (usually F_2 or subsequent generations) are then screened for mutants in which the expected mutant phenotype co-segregates with the transposon insertion. Once confirmed, the nature of the QTL is identified by isolation of the flanking sequence of the transposon.

Box 39.2: Application of the rice *Tos17* insertion population and reverse genetics in gene identification

Tos17 is a 4114-bp long LTR-type retrotransposon and has a low copy number in the rice genome compared with other types of retrotransposons. For instance, Nipponbare, the variety selected by IRGSP (the International Rice Genome Sequencing Project) for rice genome sequencing, contains only two copies of *Tos17* per haploid genome. After activation by tissue culture, *Tos17* copy numbers could be increased up to 30, and all of them became inactive in the regenerated plants. Several features make *Tos17* an ideal molecular mutagen for high-throughput mutagenesis in rice. First, the copy number of *Tos17* correlates with the duration of tissue culture, suggesting that it is possible to control the number of *Tos17* elements and consequently the mutation rate. Second, *Tos17* tends to transpose into unlinked genomic sites. Third, *Tos17* prefers low-copy-number sequences and genes as integration targets. Approximately 50,000 *Tos17* insertion lines containing ~500,000 mutated loci have been generated at the National Institute for Agrobiological Sciences (NIAS, Japan) and are publically available. This has become an important resource for reverse genetics in rice. There are two ways to identify insertion in the gene of interest, e.g. gene A in the following diagram, in this population. One (I) is to search the *Tos17* FST database using the sequence of gene A as a query. The FST database was established by systematically amplification of *Tos17* FSTs using PCR-based approaches followed by sequencing using DNA samples prepared from each individual line. To date, ~35,000 *Tos17* FSTs have been catalogued in the database. Another way (II) is to screen DNA pools using primers specific to gene A and *Tos17*. Seeds of *Tos17* lines are pooled in a three-dimensional matrix, and DNA samples are prepared from tissues derived from each pool of seeds. For example, if an insertion in gene A is located at plate I, row B and column 2, it can be identified by positive PCR results in all these three pools after screening of 30 DNA pools. To facilitate functional characterization of these *Tos17* lines and to benefit plant biologists, especially those working on rice and other monocotyledonous species, all *Tos17* lines have been recently phenotyped in the field through collaboration of seven laboratories, and a public database containing this phenotype information has been created, which is available at http://tos.nias.affrc.go.jp.

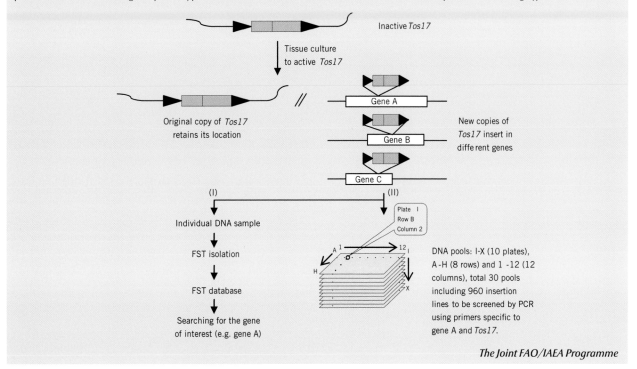

The Joint FAO/IAEA Programme

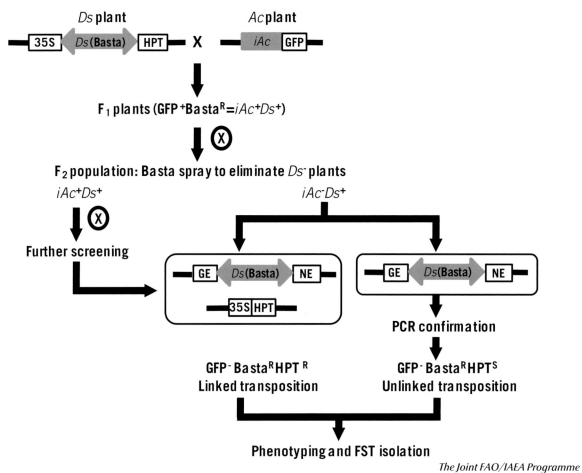

Figure 39.5 Generation and screening of stable transposon insertion plants using the two-element system. Transgenic plants expressing transposase from an immobilized autonomous element (*iAc*) are crossed to transgenic plants harbouring a non-autonomous element (*Ds*) that is inserted between a CaMV 35S promoter (35S) and the hygromycin resistance gene (HPT) to induce transposition of the *Ds* element. The *Ds* element contains the Basta resistance gene so that progeny with the *Ds* element will be resistant to Basta. F$_1$ progeny containing both *iAc* and *Ds* can be confirmed simply by checking the expression of GFP using the germinating seeds and spray of Basta at seedling stage. F$_2$ progeny are first screened with the GFP marker to separate GFP positive (with *iAc*) and negative (without *iAc*) plants. Seedlings are then sprayed with Basta to remove plants without the *Ds* element. GFP negative plants surviving after Basta selection are putative stable *Ds* insertion plants (*iAc$^-$Ds$^+$*). Plants with a transposed *Ds* linked or unlinked to the launching pad can be distinguished by hygromycin selection. The HPT^s stable *Ds* insertion plants should be further confirmed by a positive PCR result using primers located at CaMV 35S and the hygromycin gene to discard the plants with an un-transposed *Ds* element. Stable *Ds* insertion plants are subjected to phenotyping and FST isolation. F$_2$ progeny showing as GFP positive and resistant to Basta can be used for further screening after self-pollination.

3.5. Forward Genetics and Reverse Genetics

Forward genetics is a phenotype-based approach that is straightforward but is not suitable for gene identification without a detectable mutant phenotype or gene family members, as insertion in one member of the gene family does not necessary lead to a mutant phenotype due to functional redundancy. Some mutations may require specific conditions to give rise to a detectable mutant phenotype. Screening mutants using forward genetics tends to be laborious and requires much space, especially for plant species of large size. Despite these limitations, forward genetics has played and will continue to play an important role in gene identification in plants, particularly for the genes controlling morphological traits which are easily phenotyped.

Reverse genetics aims to gain insight into the function of the gene of interest, such as an annotated new gene or a member of a gene family, by generation and analysis of phenotype(s) in the corresponding mutant. RNA silencing technology using hairpin transgenes has been widely applied in plants as a complementary tool for characterization of gene function. Recent development of artificial miRNAs provides a more efficient and accurate RNA silencing technology in plants. In spite of successful application of RNA silencing in investigation

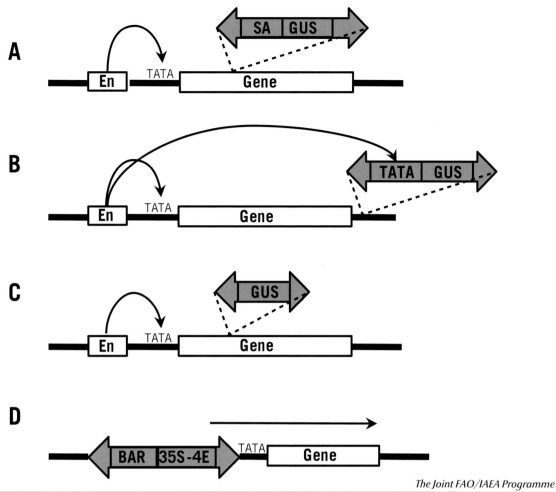

Figure 39.6 Schematic representations of gene trap, enhancer trap and activation tagging system. A, gene trap system. The modified transposon (shown in grey box with arrow heads representing TIRs) has a sequence containing a splice acceptor (SA) fused to a promoter-less reporter gene (GUS) that will express upon insertion of the transposon into an exon or intron of an endogenous gene due to generation of a fusion transcript as a result of interaction between the splice donor of the gene and the SA before the reporter gene. B, enhancer trap system. The reporter gene (GUS) is fused to a minimal promoter (TATA) that could be activated by an endogenous enhancer (En) upon insertion of the transposon within or adjacent to a gene. C, promoter trap system. The modified transposon containing a promoter-less reporter gene (GUS) that will be activated upon insertion of the transposon within exon of a gene. D, activation tagging system. The modified transposon containing a selection marker (BAR, Basta resistance gene) and a tetramer of the CaMV 35S enhancer (35S_4E) next to and pointing to one of the termini of the transposon. Transcription of an endogenous gene will be activated or enhanced because of interaction with the tetramer enhancer. Both termini of the transposon could have a tetramer enhancer so that it is capable of up-regulating endogenous genes from either border.

of gene function in plants, knock-out by RNA silencing is often not complete. In contrast, knock-out mutations resulting from T-DNA or transposon insertion are usually null mutations, leading to a more precise understanding of the functions of the gene of interest.

To apply reverse genetics effectively, a large collection of transposons induced mutant plants needs to be created to increase the chance that the gene of interest is mutagenized. To this end, a large number of transposon mutagenized populations have been generated in several plant species. An incomplete list of such populations is shown in **Table 39.2**. In addition, it should be possible to determine effectively the presence of an insertion mutant for the gene of interest in the mutagenized population. Usually, PCR is used to screen pooled DNAs using a gene-specific primer and a primer specific to the transposon tag, which has been discussed in detail in **Chapter 38**. The use of rice *Tos17* in gene identification for reverse genetics is shown in **Box 39.2**. PCR-based methods are very efficient for identification of insertions from single- or low-copy-number insertion lines. The sensitivity of the PCR technique, especially in combination with confirmation of the PCR products by hybridization with a gene-specific probe, allows detection of the gene of interest within a pool of hundreds or thousands of individuals. For

The Joint FAO/IAEA Programme

high-copy-number insertion lines, transposon display and amplification of insertion-mutagenized sites (AIMS) can be used. Once the FST is confirmed by sequencing, the loss-of-function phenotypes of the gene can then be examined after identification of homozygous mutants.

In order to facilitate gene identification and functional analysis, large-scale isolation of FSTs has been or is being performed for most of the major transposon tagging populations in various plant species. Several systematically catalogued databases of transposon FSTs have been established in *Arabidopsis*, rice and maize, most of these FSTs have been mapped to the corresponding genome and are searchable *via* web-based interfaces (**Table 39.2**). These FST databases will significantly facilitate gene characterization in plants especially for the vast majority of genes for which no mutant phenotype is available. Finally, the availability of complete genome sequences combined with reverse genetics can allow every gene to be characterized.

3.6. Verification of Transposon Tagged Mutations

In order to establish robust associations between phenotype and gene, the relationship between the mutant phenotype and the causative transposon insertion has to be confirmed by co-segregation analyses using hybridization- or PCR-based methods. TSD is created at the insertion site upon insertion of a transposon. Excision of the transposon is usually imprecise and leaves part or all of the TSD, the so called excision footprint, behind (**Figure 39.1D**). When a transposon inserts at the exon of a gene, the imprecise excision footprint may result in a frameshift mutation. In this case, a transposed transposon closely linked to the donor site could be associated with the mutation, resulting in a false relationship with the mutant phenotype. Therefore, co-segregation results obtained by hybridization using a transposon-specific probe do not always guarantee that the mutation is induced by this particular transposon. The function of the transposon tagged gene has to be supported by other independent experiments, such as alleles created by other methodologies, knock-out phenotype using RNA silencing technology or complementation experiments using the wild type copy of the tagged gene. But restoration of a wild type phenotype upon excision of the transposon provides strong evidence that the transposon was responsible for the mutation.

4. Enhanced Transposon Tagging Systems

Use of transposons as a molecular mutagen for gene isolation was initially exploited endogenously in maize and snapdragon. Although *Ac* and *Mu* elements have been utilized in large-scale mutagenesis experiments in maize, sophisticated transposon tagging systems using *Ac/Ds* or *Spm/dSpm* were established in heterologous systems by modification of the *Ds* or *dSpm* elements to incorporate selection markers and by controlling transposition of the *Ds* or *dSpm* elements using a two-element tagging system.

4.1. Development of a Two-Element System

In heterologous plant species, initially only the autonomous element (*e.g. Ac*) was transformed to generate a tagging population. To monitor transposition of the element, the *Ac* element is usually engineered between a constitutively expressed promoter such as the Cauliflower Mosaic Virus promoter (CaMV 35S) and an excision marker that will express upon excision of the element. Mutations caused by insertion of the *Ac* element are unstable as the transposability of the *Ac* element remains throughout the life of a plant. Some plants will maintain the mutant phenotype after imprecise excision of the element, but it would not be possible to isolate the corresponding gene due to loss of the molecular tag. After several generations, insertion plants will accumulate increasingly 'noisy' backgrounds due to the imprecise excision of the element, making co-segregation analysis much harder. Furthermore, screening of the transposed autonomous element largely relies on labour- and cost-consuming PCR and/or Southern blot analysis.

To overcome these problems, two-element systems were developed so that a transposed non-autonomous element can be stabilized by segregating away an immobilized autonomous element. In the non-autonomous element, a reinsertion marker such as antibiotic- or herbicide-resistance gene is incorporated to select progeny bearing the transposed elements. To monitor the excision events or to select against non-excision events, the non-autonomous element is inserted between a promoter and an excision marker gene so that excision of the element results in expression of the excision marker. This excision marker also serves as a launching pad indicator so that genetically unlinked transposition events can be separated or enriched by selection against the launching pad indicator. Progeny expressing the reinsertion marker and the excision marker have the element transposed into linked sites, whereas progeny expressing only the reinsertion marker have the element transposed into unlinked sites. Appropriate selection markers can also be engineered in the construct harbouring the immobilized autonomous element so that stable transposants can be easily selected by counter selection of this marker. An example of such a two-element tagging system is shown in **Figure 39.5**. To facilitate recovery of FSTs of the transposon inserts, components for plasmid rescue function, including the antibiotic resistance gene and plasmid origin of replication, can be engineered into the non-autonomous element (**see Chapter 38**). In other words, by judicious choice of an appropriate combination of marker genes, the efficiency of screening stable transposants can be significantly improved.

The autonomous and non-autonomous elements can also be brought together by co-transformation. Similarly, stable transposants can be obtained by segregating away the autonomous element from the double transformants. In rice, a system in which a transiently expressed transposase is used to induce transposition of *Ds* during tissue culture steps has been established so that stable *Ds* insertion lines (i.e. without integrated *iAc*) can be directly regenerated by tissue culture (Upadhyaya *et al.,* 2006). The main advantage of this system is that stable *Ds* insertion lines can be produced as primary transformants, thereby increasing screening efficiency and reducing the space required for growing the segregating population.

4.2. Gene, Enhancer and Promoter Trap Systems

Both *Ac/Ds* and *Spm/dSpm* tend to insert preferentially into or near genes. Analysis of distribution of insertion positions within genic regions in rice showed that 30–40% of insertions were in exons (Krishnan *et al.,* 2009). Obviously, insertion of transposons in exons is expected to disrupt the gene function but it may or may not lead to visible mutant phenotypes due to existence of functional redundant gene(s), lethality or insertion in non-essential genes. To overcome these difficulties in phenotyping, gene and enhancer trap systems have been developed and integrated into transposon tagging systems in plants. A gene trap contains a modified

transposon consisting of an intron with multiple splicing acceptor sites fused to the coding region of a reporter gene, usually a β-glucuronidase (GUS) or GFP gene (**Figure 39.6A**). A fusion protein including the reporter gene and the N-terminal portion of the host gene will be produced when the element is inserted into either an exon or an intron of a host gene in the same transcriptional orientation; therefore, expression pattern of the reporter gene reflects the activity and regulation of the disrupted genes. An enhancer trap contains a modified transposon harbouring a reporter gene driven by a minimal promoter (**Figure 39.6B**). Upon insertion at, or adjacent to, a host gene, the minimal promoter, may be cis-activated by enhancer elements in the host gene leading to the expression of the reporter gene. A promoter trap is a variation of the enhancer trap, in which expression of a promoter-less reporter gene relies on insertion of the modified transposon element downstream of an active endogenous promoter (**Figure 39.6C**).

A clear advantage of these trap systems is that the expression pattern of the tagged gene can be studied in detail by analysis of the GUS or GFP expression pattern during plant development. Information from such expression patterns can be very useful in the subsequent phenotypic analysis of the homozygous insertion mutants. Although functionality of the enhancer trap system may not be affected by its insertion orientation if it is close to one endogenous enhancer, expression of the reporter gene of the gene trap or the promoter trap requires the reporter gene to be in the same orientation as the tagged gene. In other words, the disadvantage of these uni-directional trap systems is that there is neither selection against insertions outside genes, nor against insertions in which the reporter gene is in the opposite orientation relative to transcription of a tagged gene. This drawback has been partly overcome by development of a bi-directional trap system (Eamens *et al.,* 2004).

4.3. Activation Tagging System

Insertional mutagenesis using T-DNAs or transposons usually generates recessive loss-of-function mutations. Genes with lethal or deleterious knock-out phenotypes are not amenable to the loss-of-function approach, and knock-out of a single gene of a gene family does not always produce detectable mutant phenotypes due to functional redundancy. These limitations of insertional mutagenesis could be partly compensated for by activation tagging which produces dominant gain-of-function mutations. Activation tagging requires a T-DNA containing a constitutive promoter or a strong enhancer, usually a tetramer of the CaMV 35S enhancer that reads out of the T-DNA terminus (**Figure 39.6D**). Integration of such a T-DNA upstream of an adjacent gene will lead to increased transcription of the gene or changed expression pattern depending on the activity of the enhancer. Weigel *et al.* (2000) reported that genes at least 3 kb from the enhancer can be activated to give a novel phenotype. Although activation tagging has been mainly applied to T-DNA tagging systems and large collections of such T-DNA lines have been created in *Arabidopsis* and rice, it has also been successfully adapted for transposon tagging systems for gene identification and characterization (e.g. Wilson *et al.*, 1996; Marsch-Martinez *et al.*, 2002; Ayliffe *et al.*, 2007 and Qu *et al.*, 2008).

The first version of the transposon-based activation tagging system was designed by Coupland and colleagues (Wilson *et al.* 1996), who modified the *Ds* element by inserting a CaMV 35S promoter, which points outwards from the element, close to one end of the element so that a dominant gain-of-function mutation is expected upon insertion of the transposed *Ds* upstream of a gene. Using this engineered *Ds* element, a semidominant mutation that increases expression of TINY, an AP2 domain containing gene in *Arabidopsis*, was identified. Later, like the T-DNA-based activation tagging systems, a tetramer of the CaMV 35S enhancer was placed near one end of the *Ds* or *dSpm* element to distribute the CaMV 35S enhancers throughout the genome using the two-element tagging strategy (e.g. Marsch-Martinez *et al.* 2002 and Qu *et al.* 2008). In *Arabidopsis* populations, 0.1-1% of the activation tagged inserts showed a dominant or semidominant inheritable phenotype (e.g. Marsch-Martinez *et al.* 2002), which was similar to or ~10 fold higher than the frequency observed in the T-DNA-based activation tagging population (Weigel *et al.* 2000), suggesting that activation tagging using transposons is more effective than using T-DNA. The 10 fold difference in activation efficiencies observed in the two *Arabidopsis* populations (e.g. Marsch-Martinez *et al.* 2002) might be caused by the difference in length of the TIR sequences used to construct the mobile elements. Shorter TIR sequences might prevent them from forming internal hairpins or double stranded RNAs, a precursor for production of small interfering RNAs

involved in *cis*-silencing of repeat sequences. No phenotypic data were available for the activation tagging population in rice thus far, but enhanced expression of rice genes adjacent to *Ds* insertions was confirmed by semi-quantitative reverse transcription PCR (Qu *et al.* 2008). Recently, a novel *Ac/Ds*-based activation tagging system in which a modified *Ds* element (*UbiDs*) contains two outwards inverted copies of the maize ubiquitin 1 promoters near each end of the element has been established in barley (Ayliffe *et al.* 2007). The majority of the transposed *UbiDs* elements activated high levels of adjacent flanking sequence transcription.

5. Prospects

One of the main objectives for the plant biologist in the "-omics" era (genomics, phenomics, metabolomics, etc.) is to identify and characterize all plant genes in order to have a thorough understanding of the physiology, morphology, genetics and development of plants. This relies on the application of many approaches including insertional mutagenesis. Transposon mutagenesis with its unique advantages over T-DNA mutagenesis has proven to be an important functional genomics tool in achieving these goals. Although the collections of T-DNA insertion lines have now outnumbered those of transposon insertion lines in *Arabidopsis* largely due to its relatively easy *Agrobacterium*-mediated *in planta* transformation, transposon-based mutagenesis is still the routine or the only approach for generation of large-scale insertion populations in other plants, especially those where large-scale transformation methods have not been developed. Recent advances further demonstrate the feasibility of using transposon mutagenesis in the study of gene functions in the majority of flowering plant species, if not all. Large-scale mutagenesis populations with transposons or retrotransposons inserted as a molecular tag have been established in a number of plant species, including several important food crops such as rice, maize and barley. With the availability of the genome sequences of *Arabidopsis* and rice and large number of ESTs in other plant species, the current focus of transposon tagging is not only to increase the size of mutagenized populations but to map the insertion sites by isolation of the transposon FSTs. To take further advantage of TE-based mutagenesis, it is necessary to increase the density of TE tags and to finally to achieve saturation mutagenesis of plant genomes. Regional mutagenesis should play an important role towards this goal. Development of other methodologies, such as high-throughput phenomics facilitates, will greatly enhance the progress on phenotyping the TE tagged populations, which has been the bottleneck in large-scale functional genomics studies thus far. Furthermore, as the majority of the transposon insertion lines do not show a mutant phenotype under normal conditions, it is important to initiate customized phenotyping in conditions such as cold, drought and/or diseases, to fully take advantage of the transposon mutagenesis populations in discovery of functional plant genes. Special facilities such as the Australian Plant Phenomics Facility are being set up for such purposes.

6. References

6.1. Cited References

Ayliffe, M.A., Pallotta, M., Langridge, P. *et al.* 2007. A barley activation tagging system. *Plant Mol. Biol.* 64: 329–347.

Bonas, U., Sommer, H., Harrison, B.J. *et al.* 1984. The transposable element *Taml* of *Antirrhinum majus* is 17 kb long. *Mol. Gen. Genet.* 194: 138–143.

Brutnell, T.P. 2002. Transposon tagging in maize. *Funct. Integr. Genomics.* 2: 4–12.

Dooner, H.K. and Belachew, A. 1989. Transposition pattern of the maize element Ac from the bz-M2 (Ac) allele. *Genetics.* 122: 447–457.

Eamens, A.L., Blanchard, C.L., Dennis, E.S. *et al.* 2004. A bidirectional gene trap construct for T-DNA and *Ds* mediated insertional mutagenesis in rice (*Oryza sativa* L.). *Plant Biotech. J.* 2: 367–380.

Fedoroff,N.V., Wessler, S. and Shurem, M. 1983. Isolation of the transposable maize controlling elements *Ac* and *Ds*. *Cell.* 35: 235–242.

Fedoroff, N.V., Furtek, D.B. and Nelson, O. 1984. Cloning of the *bronze* locus in maize by a simple and generalizable procedure using the transposable controlling element *Activator (Ac)*. *Proc. Natl. Acad. Sci. USA.* 81: 3825–3829.

International Rice Genome Sequencing Project 2005. The map-based sequence of the rice genome. *Nature.* 436:793–800.

Krishnan, A., Guiderdoni, E., An, G. *et al.* 2009. Mutant resources in rice for functional genomics of the grasses. *Plant Physiol.* 149: 165–170.

Lisch, D., Chomet, P. and Freeling, M. 1995. Genetic characterization of the *Mutator* system in maize: behavior and regulation of *Mu* transposons in a minimal line. *Genetics.* 139: 1777–1796.

Marsch-Martinez, N., Greco, R., Van Arkel, G. *et al.* 2002. Activation tagging using the *En-I* maize transposon system in *Arabidopsis. Plant Physiol.* 129: 1544–1556.

Qu, S., Desai, A., Wing, R. *et al.* 2008. A versatile transposon-based activation tag vector system for functional genomics in cereals and other monocot plants. *Plant Physiol.* 146: 189–199.

Upadhyaya, N.M., Zhu, Q.-H., Zhou, X.-R. *et al.* 2006. Dissociation (*Ds*) constructs, mapped *Ds* launch pads and a transiently expressed transposase system suitable for localized insertional mutagenesis in rice. *Theor. Appl. Genet.* 112: 1326–1341.

Walbot, V. 2000. Saturation mutagenesis using maize transposons. *Curr. Opin. Plant Biol.* 3: 103–107.

Weigel, D., Hoon Ahn, J., Blazquez, M.A. *et al.* 2000. Activation tagging in *Arabidopsis*. *Plant Physiol.* 122: 1003–1013.

Wilson, K., Long, D., Swinburne, J. *et al.* 1996. A *Dissociation* insertion causes a semidominant mutation that increases expression of TINY, an *Arabidopsis* gene related to *APETALA2. Plant Cell.* 8: 659–671.

6.2. Websites

The Australian Plant Phenomics Facility:
http://www.plantphenomics.org.au/

6.3. Further Reading

Bennetzen, J.L. 2000. Transposable element contributions to plant gene and genome evolution. *Plant Mol. Biol.* 42: 251–269.

Fedoroff, N.V. 1989. About maize transposable elements and development. *Cell.* 56: 181–191.

Gierl, A. and Saedler, H. 1989. The En/Spm transposable element of Zea mays. *Plant Mol. Biol.* 13: 261–266.

Springer, P.S. 2000. Gene traps: tools for plant development and genomics. *Plant Cell.* 12: 1007–1020.

Walbot, V. 1992. Strategies for mutagenesis and gene cloning using transposon tagging and T-DNA insertional mutagenesis. *Annu. Rev. Plant. Physiol. Plant Mol. Biol.* 43: 49–82.

Zhu, Q-H., Eun, M.Y., Han, C.D. *et al.* **2007.** Transposon insertional mutants: a resource for rice functional genomics. *In*: N.M. Upadhyaya (ed.) Rice Functional Genomics—Challenges, Progress and Prospects. New York: Springer, pp. 223–271.

Site-Directed Mutagenesis in Higher Plants

K.Osakabe, H.Saika, A.Okuzaki and S.Toki*

Division of Plant Sciences, National Institute of Agrobiological Sciences, Kannondai, Tsukuba, Ibaraki 305-8602, Japan
* Corresponding author, E-MAIL: stoki@affrc.go.jp

1. Introduction

Site-directed mutagenesis is a molecular biology technique in which a mutation is created at a defined site in a DNA molecule. Historically, this has been a method of introducing mutations into plasmid DNA. In general, site-directed mutagenesis requires sequence data of the wild type gene. Recent advancements in genome analysis have provided information on the whole genome sequence of model plants and important crops. This allows modification of the genomic sequences of plants of interest *via* homologous recombination (HR) mediated gene targeting (GT) and chimeric oligo-mediated mutagenesis. Many agronomically valuable phenotypes and natural variants have been caused by point, or only a few, mutations. Point mutations, especially can be regarded as the cleanest and most direct gene manipulation technique for future molecular breeding in plants. In addition the production of the knock-out gene mutants in plants *via* site-directed mutagenesis is a useful technique in functional studies.

2. Site-Directed Mutagenesis *via* Gene Targeting

2.1. Molecular Basis of Gene Targeting

GT is a molecular technique to introduce modifications into the endogenous genomic sequences *via* HR. HR is one of the repair systems for double-stranded DNA breaks (DSBs) in somatic cells, and is the process by which a strand of DNA is broken and joined to the end of a different DNA molecule (**see Chapter 6** for the details of HR) . This repair system is generally the more accurate pathway of DSB repair, but it requires the presence of the undamaged sister chromatid or homologous chromosome as a template. In the GT method, an exogenous DNA which includes homologous sequences with the target gene and the modification of interest is normally used as a template instead of undamaged homologous DNA.

HR is active in the late S to G2 phases of the cell cycle when the sister chromatid is available as a homologous template. HR also requires DSBs, which can not only be generated by treatment with DNA-damaging stresses such as ionizing radiation, UV or chemical mutagens, but also by cellular processes such as DNA replication. Some HR genes are activated by DSBs. Thus, these are key features for improvement of the GT efficiency.

2.2. Molecular Processes of Gene Targeting in Higher Plants

GT *via* HR enables precise modification of a target gene. Thus, gene modification by GT is a powerful tool for functional analysis of targeted genes. Moreover, GT is thought to be a 'clean' transformation technology as it involves the complete substitution of target gene. GT has been shown to be efficacious in the production of novel phenotypes in plants, suggesting that GT has potential in molecular breeding of crops. To date, successful GT events of endogenous gene in flowering plants have been reported in *Arabidopsis*, tobacco, maize and rice (**Table 40.1**), though these experiments are in their infancy.

True GT (TGT), in which the target gene is modified as expected, is explained by the double crossover model or one-sided invasion (OSI) model (**Figure 40.1**). In all GT events in higher plants except zinc finger nucleases (ZFNs) stimulated GT, an *Agrobacterium*-mediated transformation system is used to introduce donor DNA (**see Chapter 38**). It has been proposed that a single-stranded T-DNA (ssT-DNA) molecule is imported into the plant nucleus, and it converts to a double-stranded T-DNA (dsT-DNA). In the case of dsT-DNA (**Figure 40.1A**), both 3'-ended single strands from the targeted DSB sites in the plant genome invade homologous dsT-DNA sequences to initiate new DNA synthesis. The resolution of the Holiday junction (**see Chapter 6**) results in non-crossover or crossover recombinants. Non-homologous sequences with plant genome such as border sequence are eliminated in this process. In the case of ssT-DNA (**Figures 40.1B and 40.1C**), one of the 3'-ended single strands from the targeted DSB sites in the plant genome anneals to homologous ssT-DNA sequences to initiate a new DNA synthesis. If the newly synthesized 3'-ended sequence anneals to the homologous sequences at another DSB site in the plant genome, non-homologous sequences with the plant genome are eliminated (**Figure 40.1B**). However, if the newly completely synthesized 3'-ended sequence is ligated to another DSB site, the border sequences are integrated into the plant genome *via* non-homologous end joining (NHEJ) (**Figure 40.1C**). A mismatch correction occurs after DNA synthesis and resolution of heteroduplex regions.

Aberrant recombination events such as ectopic GT (EGT), in which the modified target sequence is integrated elsewhere in the genome, are frequently

Table 40.1: GT in higher plants

Plant species	Gene name	Selection marker(s)	Regenerants
Tobacco	ALS	ALS	22 EGT plant lines (1990), 403 candidate calli (2009)[a]
Arabidopsis	TGA3	nptII, GUS	1–2 calli (1995)
	AGL5	nptII	1 TGT plant (1997)
	ADH	nptII, codA	1 callus, 1 plant line (2001)
	PPO	PPO	3 TGT plants, 6 EGT plants (2001)
	ALS	ALS	1 TGT plant, 1EGT plant (2006)
	CRU3	GFP	556 candidate seeds (2005)[b]
Rice	Waxy	HPT, DT-A	6 TGT plants (2002)
	ADH2	HPT, DT-A	9 TGT plants (2007)
	MET1a	HPT, DT-A	15 candidate plants (2009)
	ALS	ALS	66 candidate plants (2007)
	OASA2	OASA2	2 TGT plants (unpublished data)
Maize	IPK1	PAT	124 calli (2009)[a]

This table is modified and updated from Iida and Terada (2005). [a] GT with ZFNs (see 3.2). [b] GT with Rad54 (see 3.1).

Box 40.1: Glossaries of site-directed mutagenesis

Random mutagenesis
Physical agents (such as X-rays, UV light) or chemical agents (base analogues, alkylating and cross-linking reagents) increase the frequency of mutations above the natural background level, and also cause changes of genetic information of the organisms. These mutations randomly occur in the genome. Methods of introducing these mutations involve random mutagenesis.

Targeted-selected mutagenesis
Even though mutations happen randomly in the genome of the target organism, a sensitive screening method such as Targeting Induced Local Lesions in Genomes (TILLING), allows directed identification of a mutation in a specific gene. Coupling of TILLING with random mutagenesis is an example of targeted-selected mutagenesis.

Insertional mutagenesis (transposon mutagenesis)
Insertional mutagenesis is a method of producing mutations in a genome by insertion of mobile genetic elements and T-DNA. In nature, retrotransposons and transposons are the major agents of insertional mutagenesis. In higher plants, T-DNA insertion *via Agrobacterium* has been used to produce a large number of mutant collections in *Arabidopsis* and rice.

Gene targeting
Gene targeting (GT) is a transformation technology that can modify a target gene using homologous recombination (HR). There are two types GT: true GT (TGT) in which the target gene was modified as expected, and ectopic GT (EGT) in which the modified target sequence is integrated elsewhere in the genome.

Single-stranded T-DNA
Single-stranded T-DNA (ssT-DNA), the T-strand corresponds to the coding strand of the T-DNA region. A T-complex is a T-strand with a single affixed VirD2 and coated with VirE2. It is formed in the plant cytosol and translocates to the nucleus where the T-DNA subsequently integrates into the plant genome.

Exogenous marker
In the universal GT system, exogenous selection markers are needed to select cells in which T-DNA integrates at random or targeted sites, or to stop growth of cells in which T-DNA integrates at random. Selection markers derived from micro-organisms are usually used.

Zinc finger nucleases
Zinc finger nucleases (ZFNs) are chimeric proteins composed of several (in general three to four) zinc finger DNA binding domains and a DNA cleavage domain from the restriction endonuclease, *Fok* I. Each zinc finger is a stretch of about 30 amino acids that binds to a particular three-base DNA sequence. The binding motif of the zinc finger domain can be changed to obtain zinc finger arrays with different sequence specificity.

Chimeric oligo
A chimeric oligo is a short strand (68–72 nt) of synthesized DNA oligonucleotides, which also contains 2'-O-methyl RNA. The chimeric oligo is self-complementary and takes the form of a double-hairpin structure. A duplex region is designed to have homology to the endogenous target locus, but with one of the middle bases so as to make a mis-match with the target sequence.

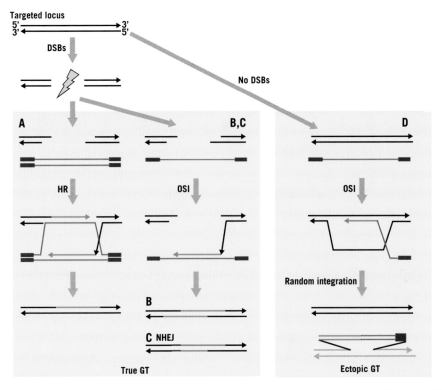

The Joint FAO/IAEA Programme

Figure 40.1 Schematic representation of DSB-induced GT process. (A) double crossover model TGT, (B) and (C) OSI model TGT, and (D) classical one-sided invasion model EGT. Black lines indicate the targeted genomic DNA with arrows 5' to 3'. Red and blue lines indicate homologous sequence with plant targeted locus and border sequences of T-DNA, respectively. Green lines indicated newly synthesized DNA. Grey lines in panel D indicate the non-targeted genomic sequence.

detected in addition to TGT. EGT is explained by OSI and NHEJ (**Figure 40.1D**). The broken 3'-ended ssT-DNA anneals to homologous sequences in the plant genome to initiate new DNA synthesis. Additional sequences in the plant genome are synthesized at the 3'-end of ssT-DNA and then the modified T-DNA is integrated at random in the genome without changing the target locus.

2.3. Selection Strategies in Gene Targeted Plants

In the *Agrobacterium*-mediated transformation system, the majority of the T-DNA molecules are integrated randomly into plant genomic DNA *via* NHEJ. GT *via* HR occurs in the order of 10^{-3} to 10^{-6} compared with random integration of T-DNA in the plant genome *via* NHEJ in higher plants. Thus, in order to select cells and plants in which GT has occurred, it is important to distinguish these from those that are non-transformed, and cells in which donor DNA is inserted at random. In general, there are two types of exogenous markers: 1) a positive marker to select cells in which T-DNA

integrates at a random site or a targeted site; and 2) a negative marker to stop growth of cells in which T-DNA integrates at random.

2.3.1. Positive–Negative Selection Using Exogenous Markers

In positive–negative selection using exogenous markers, a positive selectable marker which is inserted into a targeted sequence and negative selectable marker(s) which are placed at the ends of a targeted sequence, are used in a vector for GT. In this strategy, a positive exogenous marker remains in the targeted locus, and these are therefore not 'clean technologies'.

The *Waxy* and *ADH2* genes encode the granule-bound starch synthase and alcohol dehydrogenase, respectively, in rice; they can be knocked out by positive–negative selection. **Figure 40.2A** is a diagram illustrating the knock-out of the *Waxy* gene. Diphtheria toxin A (DT-A) protein inactivates the elongation factor 2 and inhibits protein biosynthesis; it has been demonstrated that DT-A can be used as a strong negative marker of GT in rice. In their GT vector, the hygromycin phosphotransferase (hpt)

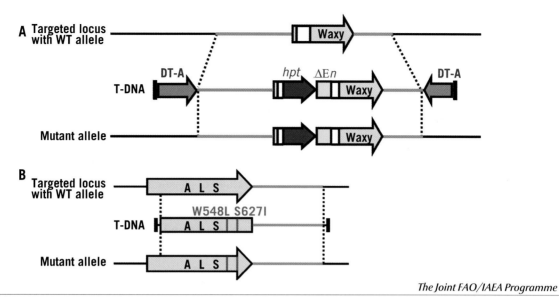

The Joint FAO/IAEA Programme

Figure 40.2 Schematic representation of rice GT. The green bars represent the homologous sequences corresponding to the *Waxy* or *OsALS* locus (Terada *et al.,* 2002; Endo *et al.,* 2007). (A) GT strategy for the disruption of *Waxy* gene. The white region in the *Waxy* box indicates the first intron sequence. In the T-DNA, a hygromycin phosphotransferase expression cassette (hpt) with maize *En* terminator (ΔEn) is used as a positive marker and two diphtheria toxin expression cassettes (DT-A) are used as negative markers. (B) GT strategy for *OsALS* gene. A sequence encoding the chloroplast-targeting signal is deleted in the T-DNA, rendering the gene non-functional. The two mutations (W548L and S627I) that confer herbicide BS tolerance are marked by solid red vertical lines.

expression cassette is used as the disruptive gene and two DT-A expression cassettes are placed at both ends of the transformation vector. They succeeded in obtaining GT plants at a frequency of 1–2 % (targeted calli per survived calli) similar to that reported in mouse embryogenic stem cells. In this system, only cells in which TGT occurs and functional DT-A cassettes are not inserted into genome DNA, as shown in **Figure 40.1A and B**, can survive.

2.3.2. Target Gene Specific Selection

In cases where the selectable marker is coincidentally also the gene of interest, e.g. a herbicide or antibiotic, target gene specific selection can be exploited in GT. Exogenous markers are not always necessary in this strategy because a target gene itself can work as a positive selectable marker.

Acetolactate synthase (ALS, see also **Chapter 31**) catalyzes the initial step common to the biosynthesis of the branched-chain amino acids and is the initial target of many herbicides such as bispyribac sodium (BS). **Figure 40.2B** is diagram illustrating the modification of the *OsALS* gene. By using target gene-specific selection strategy, rice plants tolerant to BS were obtained by introducing two base substitutions that confer tolerance to BS into the *OsALS* gene. A truncated ALS sequence, which lacks the chloroplast-targeting signal sequence

and harboring two base mutations that confer tolerance to BS, was constructed as a vector for GT of *OsALS*; the truncated ALS is thought to be non-functional when it integrates into the rice genome; it becomes functional when the two point mutations introduced by GT work as positive selection markers against BS. Using this strategy, 66 independent plants were regenerated from 1500 *Agrobacterium*-transformed calli (approximately 30 g FW). A sensitivity test of ALS enzymatic activity using GT plants showed that BS tolerance level of GT homozygous plants exceeded that of plants over-expressing the mutated *ALS* gene containing the same mutation as GT plants. This result suggested that exclusion of the wild type, BS-sensitive *OsALS* gene is important to produce novel crops, hyper-tolerant to BS.

3. Challenges in Improving Gene Targeting

3.1. Over-Expression of Homologous Recombinational Repair Genes

GT is a general method applied in yeast and mammalian cell research. In higher plants the frequency of GT events, e.g. in rice and *Arabidopsis* is still relatively low because of low HR frequency. It seems that the dominance of NHEJ

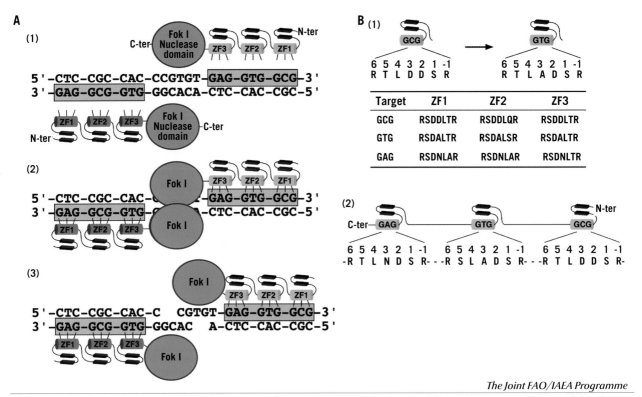

Figure 40.3 Schematic representaion of ZFNs structure and designing the custom ZF arrays. (A) Typical structure of ZFNs and the cleavage of the recognition sequence by heterodimeric ZFNs. (A-1) Multiple fingers are joined together, and are further fused to the nuclease domain of the *Fok* I restriction endonuclease. In this figure, three yellow colored ZF arrays are designed to bind with 5′-GAGGTGGCG-3′ in light blue colored box of the plus strand, and three magenta colored ZF arrays are designed to bind with 5′-GTGGCGGAG-3′ in the light blue box of the minus strand. (A-2) Zinc fingers recognize nucleotide triplets of target sequences. (A-3) When the *Fok* I monomers (the nuclease domain) are brought into proximity by DNA binding, nuclease functions to create a DSB at the spacer region between two ZFN target sequences. (B) Modular assembly to make ZF arrays. (B-1) The ZF recognizing the sequence of 5′-GCG-3′ can be exchanged with a ZF domain recognizing the sequence of 5′-GTG-3′ by modification of amino acids within only the alpha-helix sequence. Examples of ZF domains for 5′-GAG-3′, 5′-GCG-3′ and 5′-GTG-3′ are given in the table shown at the bottom of Figure 3B (1) (Liu *et al.,* 2002). (B-2) Three ZFs recognizing 5′-GAGGTGGCG-3′ were assembled according to the table in B-1.

3.2. Role of Zinc Finger Nucleases in Double-Stranded DNA Breaks

over HR in plant DNA repair, and by implication the integration of foreign DNA into the plant genome, does not permit efficient HR.

One attempt to improve the efficiency of HR is the manipulation of the expression of HR genes. Overexpression of *Escherichia coli recA* or *ruvC* in plant cells showed some gains in the HR frequency with a modest enhancement (two- to three-fold). However, these gains were not confirmed by GT experiments. Shaked *et al.* (2005) reported that over-expression of yeast *RAD54* gene in *Arabidopsis* enhances GT frequency by an average of 27-fold. However, over-expression of the *Arabidopsis RAD54* gene in *Arabidopsis* was shown to lead to no enhancement of intra-chromosomal HR, suggesting that more detailed analysis of *RAD54*-overexpressed plants may be necessary in developing an efficient GT system.

Enhanced HR can be achieved in plant cells by creating DSBs at the target site. For example, expression of *I-Sce* I, a rare-cutting restriction enzyme, has been shown to lead to a significant increase in HR-mediated GT in tobacco. Expression of rare-cutters can also lead to site-specific integration of foreign DNA molecules and to induce site-specific mutagenesis. Nevertheless, their use is limited to rarely occurring natural recognition sites or to specific artificial target sites. To overcome this problem, ZFNs have been developed. ZFNs are chimeric proteins composed of a synthetic zinc finger-based DNA binding domain and a DNA cleavage domain. ZFNs can be specifically designed to cleave virtually any long stretch of double-stranded DNA sequence by

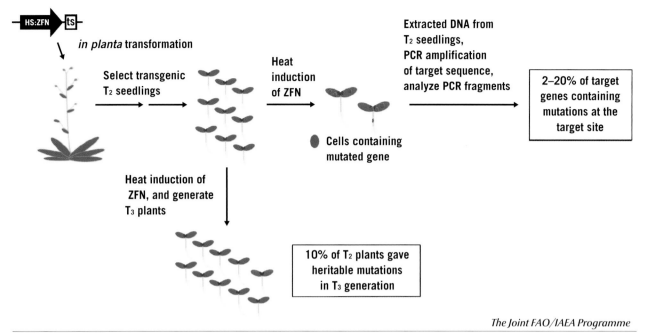

The Joint FAO/IAEA Programme

Figure 40.4 The diagram of the results by ZFN-mediated mutagenesis reported by Lloyd *et al.* (2005). HS:ZFN = the *Arabidopsis HSP18.2* promoter + ZFNQQR gene cassette, ts = the target sequence of ZFNQQR.

modification of the zinc finger DNA binding domain (for reviews see Durai *et al.*, 2005; Camenisch *et al.*, 2008).

Functional ZFNs are dimer with each monomer composed of a non-specific DNA cleavage domain from the *Fok* I endonuclease fused to a zinc finger array engineered to bind a target DNA sequence of interest. A typical structure is shown in **Figure 40.3A**. The DNA-binding domain of a ZFN is typically composed of three to four zinc finger arrays. Each zinc finger motif binds to a three-base pair sequence and as such a protein including more zinc fingers targets longer sequences. Key amino acids, at positions –1, 2, 3 and 6 relative to the start of the α-helix contribute to the specific binding to the DNA sequence. These amino acids can be changed, while maintaining the remaining amino acids as a consensus backbone, to generate ZFNs with different sequence specificities. The rules for selecting target sequences have been described in several papers (e.g. Mandel and Barbas, 2006; Sander *et al.*, 2007). This modular assembly method along with an example of the construction of the ZF array binding to the sequence of 5'-GAG-GTG-GCG-3' is shown in **Figure 40.3B**. Depending on the required specifications of the end-product, the constructed ZF arrays can be customized *via* cell-based engineering techniques.

The non-specific nuclease domain of *Fok I* is functionally independent of its natural DNA-binding domain and is therefore employed in the construction of ZFNs.

Since the domain must be dimerized to accomplish a DSB it is necessary that a nuclease is also bound to the opposite strand by virtue of another ZFN molecule bound to its target sequence as shown in **Figure 40.3A**. The two target sites need not be homologous, so long as ZFNs targeting both sites are present. In order to form a dimer, two ZFN molecules must meet with their respective recognition sites not less than four to seven base pairs apart, but also not so far apart that they cannot dimerize. To accomplish this, the length of the linker between the zinc finger DNA binding domain and the DNA cleavage domain needs to be optimized.

Specifically designed ZFNs can introduce a DSB at a target site on the gene of interest. This introduction of the DSB up-regulates HR at that site. Indeed, expression of ZFNs has been shown to enhance HR in several organisms. In fruit fly, engineered ZFNs that recognize the yellow gene were expressed in flies in the presence of donor DNA (Bibikova *et al.*, 2003). GT can occur at remarkable efficiencies. In some cases, more than 1% of the progeny flies had sustained a GT event. According to a study in human cells using a defective green fluorescent protein (GFP) reporter gene that was repaired by ZFN-mediated HR (Porteus and Baltimore, 2003), cleavage of the GFP reporter by ZFNs enhanced HR more than 2000-fold. In addition, Urnov *et al.* (2005) reported that engineered ZFNs can be used to correct human genetic defects through HR. Mutations in

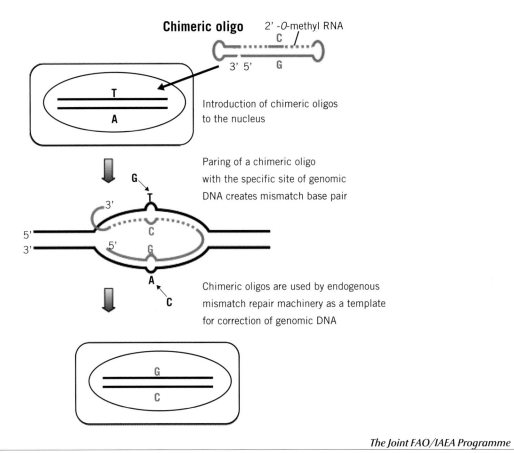

Chimeric oligo   2' -O-methyl RNA

Introduction of chimeric oligos
to the nucleus

Paring of a chimeric oligo
with the specific site of genomic
DNA creates mismatch base pair

Chimeric oligos are used by endogenous
mismatch repair machinery as a template
for correction of genomic DNA

The Joint FAO/IAEA Programme

Figure 40.5 Model of chimeric oligo-mediated mutagenesis.

the *IL2RG* gene that cause X-linked severe combined immune deficiency (SCID) were corrected at frequencies approaching 18%. Furthermore, about 7% of the cells acquired the desired genetic modification on both chromosomes.

ZFNs can also be used to increase the frequency of gene targeting in higher plants. To give a proof of principle, Shukla *et al.* (2009) and Townsend *et al.* (2009) designed ZFNs for endogenous genes of maize (Inositol-1,3,4,5,6-penta-kisphophate 2-kinase (*IPK1*) gene; Shukla *et al.*, 2009) and tobacco (*ALS*; Townsend *et al.*, 2009). They introduced ZFN expression cassettes into plant cells with plasmid DNA as double-stranded DNA *via* electroporation (Townsend *et al.*, 2009) or whisker-mediated transformation (Shukla *et al.*, 2009). Both studies revealed that over 20% of selected lines showed GT events. These works demonstrate that cleavage of a chromosomal target by ZFNs dramatically stimulates HR-mediated gene targeting in plants, and provides a basis for future experiments with ZFNs directed to any endogenous genomic locations.

4. Site-Directed Mutagenesis *via* Methods other than Gene Targeting

4.1. ZFN-Mediated Mutagenesis

ZFN-mediated DSBs at the target sequences can potentially be repaired by HR or lead to enhanced HR at the break site as described in the above section. Nevertheless, genomic DSBs in many organisms, and particularly in higher plants, are typically repaired by NHEJ machinery. NHEJ repairs DSBs by joining the two ends together and usually produces no mutations, provided that the cut is clean and uncomplicated. However, in some instances, the repair will be imperfect, resulting in deletions, insertions and/or substitutions and even capture of foreign DNA molecules at the repaired site. This phenomenon leads to the possibility that ZFNs can be used to disable dominant mutations in heterozygous individuals by producing DSBs in the mutant allele. This technology has been applied successfully in several organisms including *Arabidopsis*.

Lloyd *et al.* (2005) applied ZFN in *Arabidopsis* for site-directed mutagenesis. They used a model system with a synthetic target site for a previously reported three-finger type ZFNQQR. First, the construct composed of the synthetic target site and the ZFNQQR gene driven by the heat-shock promoter was introduced into the genome of *Arabidopsis*. After transformation, the seedlings from the transgenic lines were heat-shocked and DNA from the seedlings analyzed. The results showed that, as expected, induction of ZFN expression resulted in mutations at the chromosomal target locus at frequencies as high as 0.4 mutations per cell, and 10% of the heat-induced plants showed the heritable mutations in the next generation (**Figure 40.4**). This work demonstrates that ZFNs can work at high efficiencies in plant nuclei for site-directed mutagenesis.

4.2. Chimeric Oligo-Mediated Mutagenesis

Chimeric oligo-mediated mutagenesis was initially developed to induce single-base mutations in chromosomal genes of mammalian cells. A chimeric oligo consists of 68 synthesized oligonucleotides and takes the form of double hairpin structures, which have a DNA "mutator" region of 5 nucleotides complementary to the target site flanked by 2'-*O*-methyl RNA bridges of 10 nucleotides each. Modification of the RNA residues by 2'-*O*-methylation of ribose sugar is designed to avoid degradation by RNaseH in cells. Chimeric oligos are thought to bind to the specific-site of the target gene and form a D-loop structure, and be used as a template for correcting the targeted gene through DNA mismatch repair (**Figure 40.5**).

The utility of chimeric oligos has been demonstrated in some plant species, such as maize (Zhu *et al.*, 2000), tobacco (Kochevenko and Willmitzer, 2003) and rice (Okuzaki and Toriyama, 2004). In plants experiments, chimeric oligo were delivered into calli by particle bombardment or into protoplasts by electroporation. In mammalian cells, chimeric oligo have been delivered by lipofection or by transfection.

In all plants, the same endogenous genes, *e.g.* ALS were selected as a target so that cells having a modified *ALS* gene could be selectable with herbicides. Unlike the high frequency levels of chimeric oligo-mediated mutagenesis – up to 40%, have been reported in mammalian systems – the frequencies in plants are estimated to be much lower, such as 10^{-6} in tobacco, 10^{-4} in maize

and 10^{-4} in rice. In two cases using maize and tobacco, the regenerated plants with modified genes were fertile, and modified *ALS* genes were genetically inherited through normal Mendelian segregation and stable in subsequent generations (Zhu *et al.*, 2000; Kochevenko and Willmitzer, 2003).

However, using BY-2 cells and rapeseed protoplasts, Ruiter (2003) showed that the frequency of base changes at the target sequence was at the same level regardless of the presence of the chimeric oligo, therefore, they suggested that spontaneous mutation obscures the effect of chimera oligonucleotides in plants. Model systems have also been developed to examine the utility of chimeric oligos using transgenic cells with integrated markers such as GFP, bar and a fusion between GFP and bar as reporter genes. Other markers assays include the use of reporter plasmid DNA. Plant extracts have been reported to catalyze DNA repair (e.g. banana and maize). In addition culture media conditions can affect targeting efficiency.

In addition to plant species and genotype, there are many factors that affect the targeting frequencies among experiments, such as differences in cell type, target sequences, transcription activities of the target gene, design and quality of oligonucleotides, the procedure of inducing oligonucleotides, enzyme activity of the homologous recombination system or the DNA mismatch repairing system, selection agents or culture condition and so on. In order to improve the frequency of chimeric oligonucleotide-directed gene targeting in plants, it will be necessary to optimize the detailed conditions depending on the cell type and variety of each plant species through model assay systems. The remaining problem of this technology is how we can isolate mutated cells or tissues without using selectable marker.

Once these problems are solved, oligonucleotide-directed gene targeting will become a powerful tool for investigating gene function or producing improved crops.

5. Key Issues for Future Practical Use

There are three major issues in the improvement of GT efficiency in the context of transformation systems. First, the frequency of T-DNA integration *via Agrobacterium* should be increased because the number of cells in which GT occurs is thought to be highly dependent on the number of initial transformed cells. For example, several hundred independent transformation events in rice culture occur in small pieces of callus (5–10 mm diameter) using efficient transformation protocols. Thus, highly efficient transformation systems are required to establish highly efficient GT technology in transforming recalcitrant plants. Second, T-DNAs' integration into the plant nucleus should be easily accessed to the endogenous homologous sequence, and then HR between T-DNA and the plant genomic sequence should occur efficiently. Finally, it is important to distinguish efficiently between cells in which GT has and hasn't occurred.

Although custom-made ZFNs can drastically enhance the efficiency of GT in several organisms including plants, a bottleneck in the application of ZFNs has been the generation of highly specific engineered zinc finger arrays. A lack of specificity of the zinc finger DNA binding domains used in the construction of the ZFNs leads to recognition of secondary degenerate sites that they were not intended to target. Hence it is critical to design highly specific ZFNs. Protocols and web-based tools describing modular-based approaches to ZFN design are now available (http://www.scripps.edu/mb/barbas/zfdesign/zfdesignhome.php; http://bindr.gdcb.iastate.edu/ZiFiT/). However, high failure rates were also reported when using these tools.

Sustained expression of ZFNs in cells is likely to contribute to cellular toxicity. To overcome this problem, RNA injection approaches have been taken to express ZFNs in zebrafish and fly cells; however, it is difficult to apply this method to higher plants, hence alternative methods are needed to regulate the expression of ZFNs, e.g. using inducible promoters such as heat shock promoters or drug-inducible promoters.

6. References

6.1. Cited references

Bibikova, M., Beumer, K., Trautman, J.K. *et al.* 2003. Enhancing gene targeting with designed zinc finger nucleases. *Science*. 300: 764.

Camenisch, T.D., Brilliant, M.H. and Segal, D.J. 2008. Critical parameters for genome editing using zinc finger nucleases. *Mini Review of Medical Chemistry*. 8: 669–676.

Durai, S., Mani, M., Kandavelou, K. *et al.* 2005. Zinc finger nucleases: custom-designed molecular scissors for genome engineering of plant and mammalian cells. *Nucleic Acids Research*. 18: 5978–5990.

Endo, M., Osakabe, K., Ono, K. *et al.* 2007. Molecular breeding of a novel herbicide-tolerant rice by gene targeting. *Plant Journal*. 52: 157–166.

Iida, S., and Terada, R. 2005. Modification of endogenous natural genes by gene targeting in rice and other higher plants. *Plant Molecular Biology*. 59: 205–219.

Kochevenko, A. and Willmitzer, L. 2003. Chimeric RNA/DNA oligonucleotide-based site-specific modification of the tobacco acetolactate syntase gene. *Plant Physiology*. 132: 174–184.

Lloyd, A., Plaisier, C.L., Carroll, D. *et al.* 2005. Targeted mutagenesis using zinc-finger nucleases in *Arabidopsis*. *Proc. Nat. Acad. Sci. USA*. 102: 2232–2237.

Okuzaki, A. and Toriyama, K. 2004. Chimeric RNA/DNA oligonucleotide-directed gene targeting in rice. *Plant Cell Report*. 22: 509–513.

Porteus, M.H. and Baltimore, D. 2003. Chimeric nucleases stimulate gene targeting in human cells. *Science*. 300: 763.

Ruiter, R., Brande, V.D., Stals, E. *et al.* 2003. Spontaneous mutation frequency in plants obscures the effect of chimeraplasty. *Plant Molecular Biology*. 53: 715–729.

Sander, J.D., Zaback, P., Joung, J.K. *et al.* 2007. Zinc Finger Targeter (ZiFiT): an engineered zinc finger/target site design tool. *Nucleic Acids Research*. 35:W599-605. To access "ZiFit" program, jump to http://zifit.partners.org

Shaked, H., Melamed-Bessudo, C. and Levy, A.A. 2005. High-frequency gene targeting in Arabidopsis plants expressing the yeast RAD54 gene. *Proc. Nat. Acad. Sci. USA*. 102: 12265–12269.

Shukla, V.K., Doyon, Y., Miller, J.C. *et al.* 2009. Precise genome modification in the crop species *Zea mays* using zinc-finger nucleases. *Nature*. 459: 437–441.

Terada, R., Urawa, H., Inagaki, Y. *et al.* 2002. Efficient gene targeting by homologous recombination in rice. *Nature Biotechnology*. 20: 1030–1034.

Townsend, J.A., Wright, D.A., Winfrey, R.J. *et al.* 2009. High-frequency modification of plant genes using engineered zinc-finger nucleases. *Nature*. 459: 442–445.

Urnov, F.D., Miller, J.C., Lee, Y.L. *et al.* 2005. Highly efficient endogenous human gene correction using designed zinc-finger nucleases. *Nature*. 435: 646–651.

Zhu, T., Mettenburg, K., Peterson, D.J. *et al.* **2000.** Engineering herbicide-resistant maize using chimeric RNA/DNA oligonucleotides. *Nature Biotechnology.* 18: 555–558.

6.2. Websites

Zinc Finger Tools:
http://www.scripps.edu/mb/barbas/zfdesign/zfdesignhome.php

ZiFiT: software for engineering zinc finger proteins (V3.3):
http://zifit.partners.org/

6.3. Further Reading

Endo, M., Osakabe, K., Ichikawa, H. *et al.* **2006.** Molecular characterization of true and ectopic gene targeting events at the acetolactate synthase gene in *Arabidopsis. Plant and Cell Physiology.* 47: 372–379.

Johzuka-Hisatomi, Y., Terada, R. and Iida, S. **2008.** Efficient transfer of base changes from a vector to the rice genome by homologous recombination: involvement of heteroduplex formation and mismatch correction. *Nucleic Acids Research.* 36: 4727–4735.

C41

Phenomics in Plant Biological Research and Mutation Breeding

C.R.Schunk and M.Eberius*

LemnaTec GmbH, Schumanstr. 18, 52146 Wuerselen, Germany
* Corresponding author, E-MAIL: matthias.eberius@lemnatec.com

1. Introduction

The following chapter will try to provide an outline of the fast-evolving approach to characterizing plants by the way they grow, develop and look over a range of the electromagnetic spectra far beyond human vision. Plant phenomics (as the more general term) corresponds to plant phenotyping and provides methods and technical approaches to measure phenotypes. The chapter highlights the essential points and focuses on basic approaches. Examples are given to illustrate these points with a focus on application to mutation breeding. However, it should become apparent that most approaches and phenotyping technologies of phenomics have wide ranging applications, especially where large numbers of plants, seedlings, seeds or fruit require characterization. Furthermore, this can be achieved in a non-destructive, comprehensive and reproducible manner. The data are captured swiftly and stored for subsequent detailed analysis.

2. The Concept of Phenomics

2.1. Classical Phenotyping in Breeding and the New Opportunities Offered by Phenomics

During the course of the last millennia selection has played a crucial role in plant domestication and crop improvement. Direct visual evaluation of plants across their entire life cycle has always played an important role in the selection of relevant plant specimens. In most cases huge numbers of single plants or small plots, mostly in the field, were and are classified according to prevailing breeding targets. This classification is mostly performed in a half-quantitative manner and only leads to reproducible results because a small number of very experienced breeders with a good eye or intuitions are able to evaluate large plant numbers.

This highly successful phenotypic approach can now be augmented by genotypic selection. The enormous progress in genetic research, particularly at the DNA sequence level can be exploited by breeders in marker-assisted selection, tracing traits through pedigrees, legitimacy testing, etc., and in combination with other biotechnologies can improve breeding efficiency. However, the explosion in data acquisition platforms in genomics has out-paced that of phenotyping and the genotype–phenotype gap has widened in recent years, i.e. our ability to associate gene functions with plant phenotypes of relevance to breeding has remained largely static.

In the past 15 years developments in IT support of high-resolution, fast imaging and storage as well as large-scale data processing have led to entirely new potentials for the comprehensive characterization of large plant numbers. The development of multi-spectral recognition systems and usage of comprehensive automatization, particularly in the glasshouse domain, delivers images and measurement results for individual plants or small groups of plants in a quality, number,

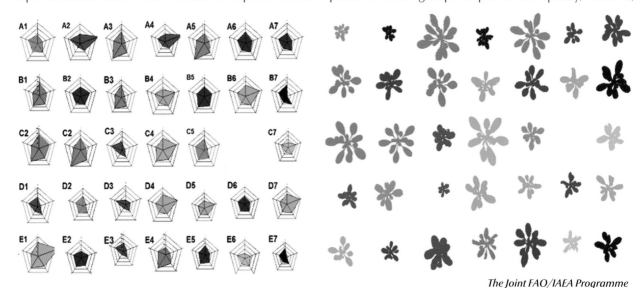

The Joint FAO/IAEA Programme

Figure 41.1 Static (one point of measurement) phenotype grouping of *Arabidopsis thaliana* plants, based on five automatically measured traits (normalized values in the polar diagram in clockwise order: growth rate, eccentricity, compactness, stockiness and surface coverage). Colours of polar diagrams correspond to plants on the right side. Such data can be analysed in QTL-studies.

frequency and reproducibility formerly unknown. One major part of the current development is to supplement the highly successful and long-standing phenotype recognition of experienced breeders in the quantitative domain by comprehensive image recognition and to evaluate the resulting data with biostatistical methods to obtain significant results. These data may then be correlated optimally with the quantitative data of the other "-omics" disciplines and, in this way, will enlarge the scope of knowledge about plants, which will in turn impact on efficient selection in plant breeding.

2.2. Static Phenotyping

Due to the huge requirements of man hours and concentrated work of phenotyping staff, classical breeding has been largely restricted to a limited number of assessment sessions. As this mostly concerns assessing differences between various lines and particularly their yield it is doable, though crude. In automated phenotyping one-off evaluation of test objects such as seeds, fruits or plants (**Figure 41.1**) has allowed huge progress, because a large number of quantitative parameters can be registered objectively, quickly and reproducibly for all single plants or plant parts (e.g. fruit or seed).

2.3. Non-Destructive, Individual, High-Frequency Phenotyping

A major advantage of non-destructive measurement techniques (particularly for optical evaluation) and automated imaging is the possibility to repeatedly image the

same plants within a week or even daily, thus achieving a high measurement density. This form of dynamic phenotyping generates completely new insights into the high dynamics of plant growth, because variability between plants as well as variability in time can be recorded quantitatively. In this way a biomass measurement (e.g. dry weight) that could formerly only be taken once per plant can now be explained by the accumulated result of a series of factors such as germination velocity and growth rate of the leaf area in various phases. Each plant is measured repeatedly and from these measurements distinct, biologically relevant parameters may be determined (**Figure 41.2**). This in turn helps in the development of growth models for genotypes integrating the influence of the environment. These model-based parameters are then far less widely distributed.

This approach of repeated individual measurement for single test objects allows the detection of significant differences between groups of varying genetic background or under changing environmental conditions, even for smaller plant numbers. It is particularly important in cases where seeds and arable land are expensive or where only single individuals are available because of crossings or mutation tests. Germination and seedling establishment are also major factors determining the success of varieties in specific environments.

2.4. High-Throughput Plant Imaging

High-throughput plant imaging can be achieved by automation by either moving plants to an imaging apparatus or by moving the apparatus to the plant. Currently

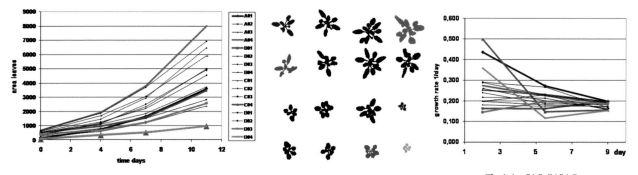

The Joint FAO/IAEA Programme

Figure 41.2 Left: Raw data in the form of growth curves of *Arabidopsis* plants (shown in the middle section). The biggest plant (A04, red), the two smallest plants (C04 light blue, D04 yellow) and some others (A01 dark blue, B01 green, D03 green) have been highlighted. Curves are difficult to interpret, a major complication being the different starting points as plants have small but relevant size differences at the first measurement due to variation in germination speed. Right: A model based on exponential growth rate curves. The model reveals that the biggest plant (A04 red) is not the fastest growing. The smallest plants (C04 light blue, D04 yellow) had extremely high growth at the start but retarded later. The slowest growing plants (B01, D03 both green) reached medium sizes due to initial rapid germination rates and the plant with the significantly highest overall growth rates reached just a medium size during the test (A1, dark blue).

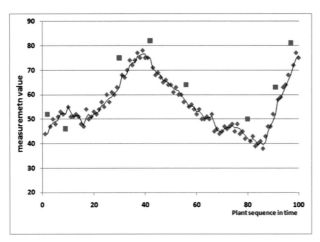

The Joint FAO/IAEA Programme

Figure 41.3 Schematic example showing the measurement pattern of a high number of control-like plants. Outliers (even under massively changing conditions) can be identified (in red) along the moving averages. Leaf temperature may be analysed in this manner.

image analysis units have the capacity to image more than 1,000 plants in pots per day during daytime or, depending on their size, analyse between 10,000 to 100,000 seedlings in trays. The high-throughput does not allow simultaneous imaging of extremely dynamic processes (such as photosynthesis, stomata behaviour and resulting leaf temperature), but it does allow the determination of relative differences between plants. In cases of measuring parameters of photosynthesis (and other parameters that vary over a day) the measurements can be timed to minimize effects of varying light conditions that can occur in glasshouses. However, moving averages of parameters and plants deviating from them can be handled and outlying individuals identified in a statistically significant way (**Figure 41.3**).

As time frames and environmental conditions, e.g. during transport from the glasshouse into the imaging unit, remain constant for all plants in automated systems, significant measurement data can still be collected, even though it is expected that the measurement as such will influence the results.

2.5. Quantitative Comprehensive Analysis

Every assessment, even when performed by highly experienced breeders or scientists, will eventually bring humans to a performance limit concerning time, and it taxes their patience all the more (at the expense of quality) when large numbers of parameters are to be recorded simultaneously. Furthermore, it needs to be stated that in most projects of targeted science and breeding the interest is often specific and limited to certain traits. This restraint is often born out of necessity rather than desire. In reality this means that in one biomass screening only aspects which are important for selection are measured quantitatively (e.g. germination, growth rate, dry weight, flowering time, yield), or that only the limited percentage of plants with the lowest (e. g. disease) infection will be analysed further.

This pragmatical approach means that large parts of potentially available data are never collected.

Raw data acquisition and analysis are often separated to avoid time-consuming performance problems. Images using various cameras in different imaging compartments are taken in parallel to maximize plant throughput. Once the images are taken, up to 100 parameters per image are extracted in the background, using high-end image analysis and computing, to be stored in powerful databases without any further human involvement (**Figure 41.4**).

Preselection or prevalidation of the parameters to be recorded is no longer necessary, providing that the imaging method in general will include the desired data.

2.6. Storage of Raw Data

As the raw data is uploaded for storage in the form of images it may be analysed immediately after imaging in the database. Or alternatively, analysed at a later time for all plants, when the test series has long been finished, or even in later years should new questions arise. Completely new research areas will present themselves with data mining, and statisticians and biologists will be able to use the comprehensive documentation to gain new insights from existing data (Edwards *et al.*, 2009). In this way costs for further tests can be significantly reduced. In addition, the plausibility of results can be checked against the background of new models. With this concept extensive screenings with highly dynamic and perishable plants or fruits can be made accessible to many scientists and breeders worldwide. A similar form of high-quality mass storage of internationally transferable data has already been established and used successfully in the area of elementary particle research at CERN , and can now be applied to phenotyping.

2.7. Glasshouses and Climate Chambers

When comparing the car development industry and phenomics, climate chambers and glasshouses are the equivalent of wind tunnels and roller rigs in car test-

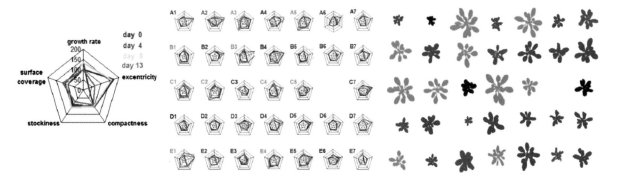

Figure 41.4 Polar diagrams representing four shapes and one growth rate parameter of rosette plants, each for four measurement intervals (left). All 25 values are shown for 34 plants (middle). Colour-coded plant silhouettes (right) represent the time-dependent development of all phenotypic parameters: yellow – growing stockiness; blue – diminishing stockiness; purple – all parameters except excentricity nearly constant; green – growth rate slowing down. In each tray some 850 data points are generated and phenotyping can process 100 trays in six hours for each measurement with a total of 85.000 data points for the full analysis. Such data generation provides important insights into the dynamic relation between gene, proteome and growth dynamics and this is not feasible by manual data capture.

ing. Environmental conditions indoors certainly differ from outdoor conditions and any plants selecting in phenomics facilities will need to be validated in the field. Nevertheless controlled environments can be well-monitored enough to allow for sufficient observation and sufficient mathematical modelling of certain plant characteristics, and have the potential to broaden biological or physical understanding. When model simulations conform to the results of the experiment, increasingly deeper insights into the practically relevant procedures are possible, even when the exact physics or biology remain (yet) unclear.

Growth conditions in the field represent a very realistic part of natural growth conditions at any one location. But this alone is not sufficient to regard them as ideal test environments for screening programmes as environmental conditions may vary between growing seasons and within the test plot. Depending on the target of a screening programme other factors may be more crucial. In order to judge the influence of different genotypes on the phenotype it is of utmost necessity to be able to access all plants easily and quickly for measurements and at the same time to keep growth conditions for all plants as similar as possible (**Figure 41.5**). If in long-term breeding programmes only a small number of plants are available for each genotype (it may be only a single specimen), light and temperature conditions need to be controlled within strict limits and watering of all plants needs to be kept comparable. This can only be achieved in glasshouses or – under even tighter control – in climate chambers, but not in the field.

It remains yet to be decided in a general discourse

if the advantages reached through better controllability and the reduced need for cooling in large climate chambers can outweigh the assumed problems with purely "unnatural" light. In any case the allocation of near to natural light conditions in growth chambers can allow a shortening of breeding times due to multiple harvest periods per year, particularly when field breeding in the other hemisphere (for annual crops) proves impossible.

Optimal access to individual plants for image analysis can only be guaranteed by screening single pot plants. At the same time this offers the incredible opportu-

Figure 41.5 Fully automated glasshouse compartment with 600 plants. Plants are transported on conveyor belts through doors (left upper edge of the image) to the imaging system. The box visible between the two doors is a pesticide spraying unit.

nity to supply each plant individually with water and keep soil humidity in all pots similar. As pots generally provide less buffering against fluctuations in humidity levels due to their smaller size, it is necessary to get the plants to a predefined weight once or twice a day. This solves the huge problems of pot water management in non-automated glasshouse tests with manual watering or flood watering namely the irregular watering or constant watering to 100 % water holding capacity.

The Joint FAO/IAEA Programme

Figure 41.6 Examples of maize plants grown without light competition (left) where the starting or any average leaf angle (red line) is of little relevance, while plants grown under light competition as in the field show clearly defined leaf angles (right)

Solely relying on "field similarity" to create a meaningful screening is not an adequate line of reasoning for or against a certain test scenario. In some cases it may even make sense to grow a plant, e.g. a maize plant in relatively large single pot units at great distances from each other (avoiding root competition), to study their maximum biomass potential and its distribution within and between roots and shoots. However, important breeding traits such as leaf angle cannot be reasonably assessed in this way, because this is heavily influenced by environmental components (especially plant-to-plant interactions, **Figure 41.6**).

A realistic leaf angle cannot be measured or deducted in single plant tests For many cereal species it has also been shown that multiple plants in a single pot can lead to significant changes in phenotypes and that phenotypes displayed in the glasshouse are lost in field grown plants. Image analysis can deal remarkably well with complex growth behaviour, even for multiple plants per pot, and quantify a large variety of growth parameters.

Growth chambers have utility in the identification of fac-

tors involved in pest and disease resistance. For example, the use of leaf discs placed on agar has proven to be a very successful method for screening for pest and disease resistance. In most cases, homogeneous, non-resistant leaves are infected with various kinds of fungi or populated with leaf sucking or eating insects, before or after spraying them with different chemicals to identify certain effects. Screenings may be performed in 6- to 96-well plates.

In testing for genetic disease resistance leaves from a range of varieties are tested without the addition of chemicals. Test routines like these can be performed manually, but for large sample numbers and tests for multiple pests and diseases image-based automated and quantitative assessment is hugely effective. This allows research into the specific resistance patterns for individual plants that may have been sampled from seed collections, or progeny of crosses or mutant populations. Various parameters can be imaged: the percentage of the infected surface, the size of spots (for example in response to a rust fungus) or the overall spreading dynamics of symptoms. These may be assessed by single or multiple imaging. Leaf feeding assays enable the monitoring of consumed leave amounts and of the mobility/survival of the feeding organisms. In tests with aphids, parameters like growth, reproduction rate, mobility or mortality rate may be measured. In addition, the movement of pest organisms can be initiated by controlled heat or UV-light. Thus leaf disc assays provide a simple and effective tool for any kind of resistance screening, as they do not require the complete plant and thus allow several resistance tests to be carried out for each plant.

3. Key Technologies Used in Phenomics

3.1. Plant Phenotyping at Medium Frequency

Whenever a daily amount of several hundred plants needs to be moved, it is sensible to employ completely automated plant transport systems. There are, however, various reasons for avoiding the long-term cultivation of plants on these transport systems, and for putting them on short conveyor belt loops or in manual loading units for imaging and watering. Short conveyor loops will offer advantages compared to complete manual loading when several detection systems are to be used, because multiple plants can be dealt with in parallel

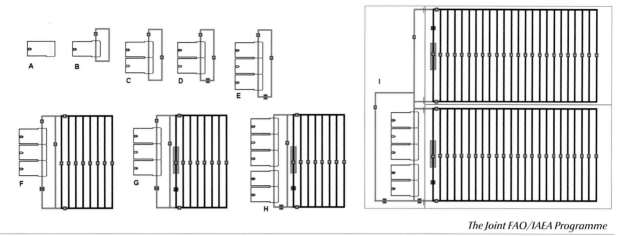

Figure 41.7 Phenotyping systems of different sizes starting from one single manually loaded imaging unit, (A) small conveyor loops with 1–3 imaging units for different cameras (B–E) up to systems with many plants standing on conveyors in the glasshouse including spray units (green) and separated glasshouse compartments (F–I).

without additional manual handling. Moreover, short storage belts can be employed as buffers for loading and unloading of the imaging system, using the intermediate times while plants are imaged to take them from or bring them back to their glasshouses or growth chambers (**Figure 41.7**).

Medium frequency in this case implies the daily phenotyping of several hundred similar objects, as the frequencies, depending on systems and evaluation complexity, may remain between 20 and 60 items per hour. Plants or trays are identified by barcode scanning or a permanently attached RFID-chip.

3.1.1. Root Columns

Root assessment in transparent pots or columns is often performed with medium throughput as the root columns need to be shaded between imaging to avoid algae growth and to keep the root system appropriately

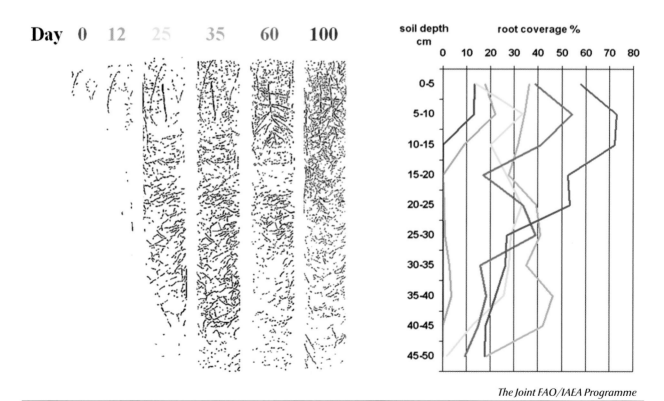

Figure 41.8 Root profiles of one column extracted from images for different points of time after the start of a test (left). Root coverage in the soil column at different points of time, dependent on depth. Root reduction after day 35 is a result of a water logging event.

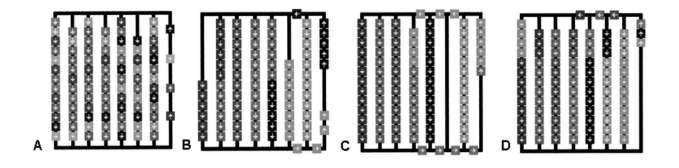

Figure 41.9 Random mode (A); Half line mode (B); Full line mode (C); First plant systematic (D).

climatized. When the root columns are constantly orientated in the same way, root growth can easily be followed on the image series (**Figure 41.8**).

If large root columns are to be scanned with high resolution in visible light or in near infrared light conditions, a large number of fused images are needed because of the large bend in the column. This in turn leads to imaging times of several minutes. Rectangular containers are much faster to image, but tend to form root accumulation in corners. However, by employing special cars and lifting units and adjusting the glasshouse accordingly, columns may also be used in high-throughput testing.

3.1.2. Agar Plates

For small seedlings such as *Arabidopsis*, where inclined agar plates are used to grow roots these can be placed flat mechanically for imaging in high-throughput systems. To achieve this, the multi-well plates are put down, imaged and then lifted, taking up to 100 at a time. Generally this is also possible for systems with manual plate loading, when these are put down for an even shorter period of time.

Special systems allow the imaging of the plates with a high frequency when a limited amount of upright plates revolved around an imaging channel. The plates will be transported automatically.

3.2. Automated Plant Phenotyping at High Frequency

3.2.1. Plant Identification and Tracking

In phenotyping it is essential that plants can be identified fast and reliably during their entire growth cycle. Barcodes may get soiled over time, but RFID-chips identify plants safely and reliably at every crucial point, be it before imaging, watering or sampling. Depending on the employed chip type and the frequency band used, the chips may be scanned within areas ranging from a few centimetres up to several metres. To identify individual plants that are in close proximity, systems with short ranges are particularly useful. RFID-chips can be attached to pots, plants, fruits or flowers or to the cars that carry pots and trays. If plants are to remain on cars for longer terms, these are commonly equipped with integral RFID-chips. As the chips are storage media, both individual plant names and additional information about tests or manual phenotype classification can be saved (e.g. with hand-held scanners) or buffered before being transferred into the database. The chips are mostly water- and weather-resistant and can even be buried in the pots or attached to plants in the field. As read-out is unproblematic, plants can be permanently tracked and their plant history retraced, even during intense randomization schemes.

3.2.2. Conveyor Belts in Glasshouses and Climate Chambers

Employing conveyor belts for plant transport in glasshouses provides a series of advantages for comprehensive phenotyping. Therefore many automated glasshouses have been equipped with conveyors, not only in research where large plant numbers need to change position frequently. The changing of location for each individual plant is essential and sensible for various reasons. Slow and almost continual rotation is highly effective in plant randomization, to minimize the effects of gradients and hotspots in temperature, light, and humidity and wind velocity in the short term. It can be achieved by following different randomization strategies (examples given in **Figure 41.9**).

A complete randomization may lead to frequent changes in adjacent plants, but a thorough distribu-

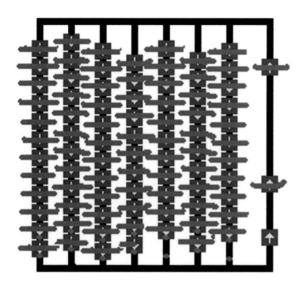

Figure 41.10 Moving Field Design for a 99:1-layout. Two stoppers (red dots) and an alternating positioning of the rows ensure minimum overlap, thus avoiding leaf damage. By always moving the first row out first (every 2nd plant) and then taking the alternating row somewhat behind, mechanical stress for the leaves is minimized.

Through cooling the air close to the glasshouse floor by blowing from the side or using pipes under the belts, cooler air can be layered at the pot level and a temperature gradient achieved in the soil similar to field conditions. Some plants need these temperature differences for their biomass production and roots normally experience lower temperature than aerial plant parts. The effect can even be increased by using suitable covers between the belts and employing transparent pots for root research observation. Choosing a specific floor color under the belts may simulate the presence or absence of competitors to the plant, depending on the screening objectives, for example, red soil color implies field-like competition.

Depending on the conveyor system weights between 1 and 40 kg per transport unit may be carried. Distances between the lines can be chosen in a way that the plants may hang over their pots, as long as they do not grow into the plants in the adjacent line. The belt motion does not lead to significant or problematic changes in phenotype, even for tomatoes that have been in the same spot and suffered no wind movement in the glasshouse for several years; at the most it will cause a slight shortening in comparison to unmoved plants, and thus lead to a stabilization that is also rather common in the field.

When it comes to choosing a belt system, low vibration belts have a significant advantage over roll systems as far as smooth running is concerned and when there is a danger that loose soil mixtures might segregate. In the very rare cases of newly planted test objects being particularly shock-sensitive, shock-absorbing buffers and controlled belt movement are options worth considering.

tion of all plants to all positions can only be achieved within a random distribution system over a long time. Also, with linear working through imaging series, control (standard) plants should re-occur regularly to be used as reference data sets. Watering combined with complete randomization may lead to plants in particularly small pots and during drought stress testing with relatively low soil humidity being exposed to hugely varying stress intensities. But block-by-block random designs similar to the field may also be deployed with modifications for conveyor systems. Here half or quarter lines of plants remain together, forming virtual trains and changing their position in the glasshouse in a strictly controlled manner to achieve an equal spatial distribution over time. In this way watering schemes can also be structured and equalized. By employing several virtual trains on one conveyor belt line, plants from the line end will regularly end up in interior positions.

One exception to permanent randomization in blocks or single plants is the moving field concept in the strictest sense, which is particularly suitable for big plants with overhanging leaves, such as maize (**Figure 41.10**). In order to arrange these plants in close proximity, but nevertheless minimize shearing during movement, they can be moved, as it were, as an entire field (with some care taken for plant leaves in end rows).

3.2.3. Plant Transport Systems

When plants do not hang over the pot rim and only rarely need to be moved for measurement, watering and imaging, crane systems such as those commonly employed in commercial gardening may be suitable. With these systems plants will be transferred out of the glasshouse and onto a conveyor belt and – after measuring, watering and weighing – be transported back into the glasshouse. Often small single pots can be placed on trays for easier transport.

These transport systems are particularly suitable when growing plants need to be selected and newly arranged on the basis of image analysis results, as is the case in resistance screening. Here a constantly increasing

The Joint FAO/IAEA Programme

Figure 41.11 One or more cameras move over multi-well plates or trays in one or more layers depending on the growth chamber design.

number of densely growing plants in trays need to be sorted out or classified according to growth velocity during their life cycle, to filter out those that are most resistant right to the very end of the test.

3.2.4. Static Multi-Well and Tray Assessment at High-Throughput – Mobile Cameras

There is sometimes a need to phenotype large numbers of relatively small plants exclusively from the top within narrow time frames and without moving them. Here the sensible approach is to deploy moving cameras (**Figure 41.11**). Systems such as these are built into climate chambers, even on several levels if it is possible to light the respective levels separately and minimize temperature gradients. The layout of shelves depends on the prevailing climate system and related specifications for maximum illumination.

Plant heights are limited by the growing vessel and generally range from 96-well multi-well plates and Japanese cultivation plates (well diameter approx. 1 cm) to trays or single pots up to 10 cm in diameter.

3.2.5. Mobile Cameras to Assess Closed Cultures in Glasshouses

In cases where moving the test objects is not efficient due to plant sizes or numbers, or when growth conditions are to be kept closer to glasshouse production scenarios, it is often extremely useful to move one or more cameras around in the glasshouse instead. Cameras and illumination plus additional sensors can be installed on an autonomous car which stops in front of every plant,

takes the relevant images after identifying the plant and then proceeds to the next test object. The specific challenges of ambient light and appropriate positioning can be overcome by shading and bright light illumination during imaging. Such imaging approaches make sense e.g. for pepper, tomato or cucumber phenotyping in glasshouse breeding programmes when often thousands of genotypes need to be assessed.

3.2.6. Mobile and *In Situ* Watering and Weighing

Comprehensively programmed and controlled watering and weighing of individual plants or replicates in small groups is only one of a series of important requirements in precise phenotyping.

Water is nearly always a limiting factor in growth and the difference of soil humidity between plants watered without full individual control very easily reduces the significance of screening results between different treatments groups by increasing variability within treatments. Bigger plants need more water and different air humidity levels at different points in the glasshouse also impacts on evapo-transpiration (**Figure 41.12**).

In addition flooding tables and surplus watering can provoke nutrient differences due to leaching, and always having 100% water holding capacity as a top level is not very realistic as it does not compare to the field environment. With individual quantitative data on water usage the comparison with imaging data, and thus a continuous calculation of the water use efficiency (WUE) of each plant, becomes possible. While fixed humidity sensors in each pot and adjustable dripping lines may tackle the problem of different soil humidity profiles in each pot, the fixed wiring for sensors and piping for water is often inhibitory to randomization and imaging. For fast, spatially resolved and non-destructive assessment of soil humidity, NIR-cameras, filtered for the specific water band, can provide a wide range of information, measuring soil humidity through transparent material or directly at the soil surface or in deeper layers (by employing capillary sticks) and making it visible through transparent windows into the whole soil profile. This information can also be used for individual watering in multi-well trays.

3.2.7. Pest and Disease Control

Highly automated phenotyping systems need to control undesirable pests and disease just like any other glasshouse system. For this purpose specific spraying

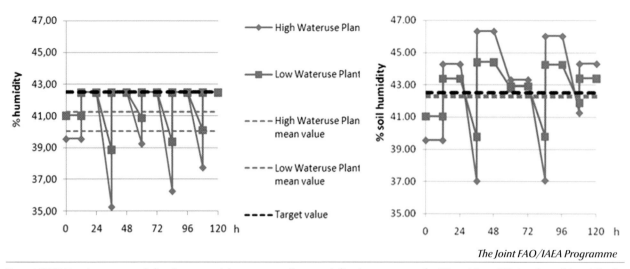

The Joint FAO/IAEA Programme

Figure 41.12 Watering to a pre-defined target weight corresponding to a defined percentage of soil humidity will bring the soil humidity for all plants to the same value only once a day. Plants that need much water will evaporate more, and thus the average value during the day will be significantly lower than that for plants needing less (left). Dynamic watering takes the higher consumption of water by some plants into account by adding a percentage offset to the water consumed relative to the target value. In most cases a 30% offset is sufficient to keep the average humidity of plants with different water demands nearly constant over time. Values depend on absolute water consumption, water holding capacities and pot sizes.

units are available which can either imitate field plant protection spraying or perform their own particularly careful version to protect the valuable plants – calculations of comprehensive cost per plate in some cases reach US$700, demonstrating the need to apply any kind of treatment cautiously.

Such cabinets contain a selected number of plants passing through the spray nozzles. Surplus spraying solution is collected for later re-use or controlled disposal.

The same units may also be used to create artificial rain, which is quite important as some test screenings need simulated rainfall. Boron tolerance screens for example strongly need the possibility to wash off excreted borate, at least for some treatment groups, to allow better assessment and prediction of how specific varieties would behave under natural conditions. Additionally, such programmed washing as a uniform starting point can make individual analysis of borate excretion much easier compared to test setups where plants are washed individually at similar or different points in time.

All these methods can be used to confine pest control to certain areas of the glasshouse where appropriate.

3.3. Controlled Stress Induction Management and Assessment

3.3.1. Watering and Weighing

As shown above precise and closely monitored watering of plants is an important requirement for any kind of

plant screening where plants are to face environmental conditions as homogeneous as possible.

A highly precise control of water supplies is hugely important particularly in screening for tolerance towards different drought stress patterns. Due to the fact that pots contain a lower absolute amount of water compared to bulk field soil, particularly under drought conditions, it is a specific challenge to keep a defined, constant, low soil humidity level or vary it according to a pre-defined plan. Like the slow reduction of soil humidity in the field, lowering the average soil humidity in pots may also be achieved by reducing the target weight of the pots on a daily basis (**Figure 41.13**).

Plants may be provided with the same target weight or the same average soil humidity over the day, irrespective of how much water they actually use due to their size or different water use efficiencies. Bigger plants will accordingly get more water as they would also extract larger amounts in the field from the bulk soil. Thus, drought stress screenings provide more reliable statistical significances if factors depending on the environment, plant size or water use efficiency of different varieties are compensated. This is independent of the question whether the aim of the test is screening for tolerance, yield stability, growth, selection, gene regulation or other parameters. Watering might also be performed during the night hours and moreover extremely regularly and faster than with any manual individual and defined watering, providing an even more homo-

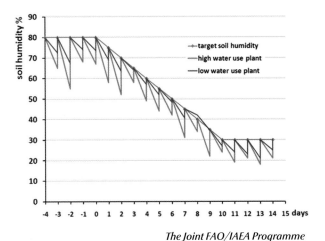

The Joint FAO/IAEA Programme

Figure 41.13 Fast, but consistent drying down of all plants from 80 to 30% soil humidity. The slope is chosen to be only slightly less shallow than the water reduction of a plant with less demand for water. But on day 8 less water is evaporated than the expected 5% per day. This form of programmed drying provides the best possible comparison between plants as the stress is nearly identical.

geneous daily water supply as well as growth improvements to all plants, and additionally reducing variability within treatments. Depending on the plants and set-up, the tests can be performed with pots closed at the bottom (watered from the top) or open pots (watering from the bottom into special saucers), to achieve the desired humidity gradients in the soil, defining specific plant or particularly root growth conditions.

3.3.2. Addition of Nutrients

Regular fertilisation of plants may be part of the normal plant cultivation process, striving for optimal or even unrestricted nutrient provision. Sometimes tests for nutrient use efficiency or very specific nutrient stress can be a part of a stress screening. In both cases it is important that all plants or at least all plants of a certain treatment group receive the same absolute amount of fertilizer. For this purpose fully automated programmable dosage systems can add nutrients and water in separate weight and watering cycles or in one cycle with multiple dosage of water and nutrient concentrates through one pipe. The immediate adding of water after nutrient addition minimizes the risk of local injuries and helps to distribute the nutrients. When various nutrients are to be provided to plants in different ratios, multiple pump systems allow sequential or parallel addition. Fertigation (the application of nutrients with water) systems may also be deployed.

3.3.3. Water Logging

While drought stress is a common stress tested, water logging or flooding also plays an important role in plant cultivation and generates massive yield losses in many crops. Fully automated phenotyping systems also provide excellent tools to test for these stressors, using closed or double walled pots with leak-proof, high outer pots. In these cases the target weight needs to be adjusted to such an amount that water levels in the pots are increased to the intended water logging or flooding level according to weight. At the planned end of the stress phase – which should be independent of the evaporation rate of the plant – the water level can be actively reduced by sucking water out of the specialized pot types. Thus the speed of water reduction can be far better controlled and does not depend on the evaporation of the plant alone, which in turn largely depends on plant size, thus generating negative control circles for stress: big plant, faster evaporation, greater loss of water, higher stress, higher need for water, plants dies faster.

3.4. Non-Destructive Multi-Spectral Imaging

3.4.1. Visible Shoot Imaging

Visible light imaging is the most common approach used in phenotyping. In the course of the last years immense progress has been made in the field of camera resolution. Nevertheless, one still has to keep in mind that resolution as such is in no way a relevant quality criterion. Instead of pixel number the chip size is much more decisive, particularly for maintaining high camera sensitivity. While high resolution in combination with

The Joint FAO/IAEA Programme

Figure 41.14 Colour classification of a maize plant. Colour classification reduces a color continuum to a statistical and human manageable dataset.

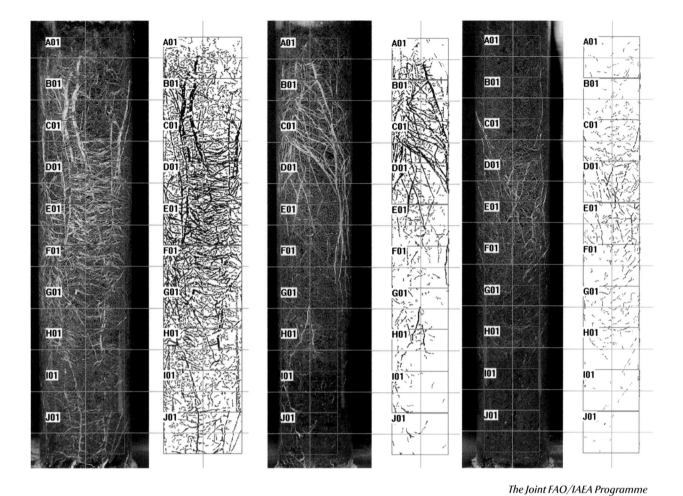

The Joint FAO/IAEA Programme

Figure 41.15 Three different maize varieties: Bangui (left), Sweet corn (middle), Prelude (right) showing three different root growth patterns.

good lens systems might be ideal to identify small details such as beards under bright light conditions, for lower light applications, e. g. in fluorescence imaging (see below), lower resolution cameras with a higher sensitivity are to be preferred.

Keeping the illumination conditions constant in well-illuminated, closed imaging cabinets allows reproducible color classification (**Figure 41.14**) as long as all other camera parameters are fully controlled. Only tightly controlled conditions such as these provide an appropriate base for using the resulting images in a measurement system.

Colour classification is an excellent way to demonstrate that classifying all green shades of an image in three or four color classes and comparing the resulting distribution between different plants provides significant information for the segregation of plants or treatment groups. In some cases different shades of green are predominant parameters of phenotypes and are especially common in mutant populations. Correlating

these with growth rates can provide additional information on photosynthetic efficiency.

3.4.2. Non-Visibility Imaging

While the spectral range around 900 nm is often already considered as near infrared, it is nevertheless visible using "normal" CCD-chips based on silicon, at least if filters are removed from the chips. In combination with 900-nm LED illumination and appropriate filtering, camera systems adapted in this way allow day and night imaging in a wavelength range invisible to plants. Thus plants can be imaged at night/in the dark without disturbing their diurnal rhythm. This is important for photobiological studies and the monitoring of plant movement (Fankhauser *et al.*, 2010).

3.4.3. Visible Root Imaging

To achieve high resolution, multiple images per root column are made by automatically shifting the position of the column during the image acquisition process.

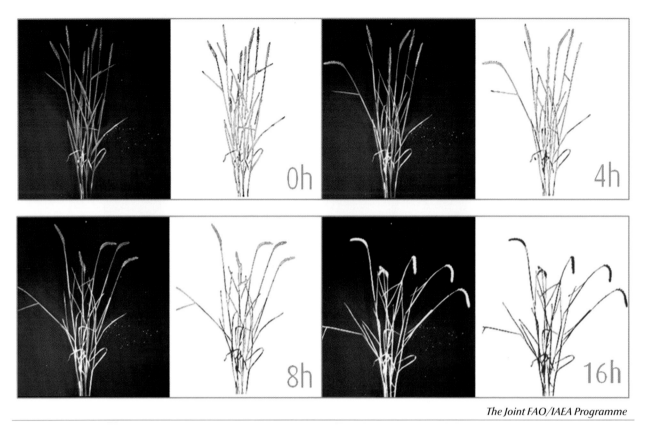

Figure 41.16 Depth dependency or root density for three different maize genotypes at the same period of development.

These images are then automatically transferred into the software and the column can be seen as a whole.

As roots may grow non-homogeneously around the column, each column is turned four times by 90° to create a complete record of all visible roots. For better spatial tracking of local changes, the columns can be specifically designed to only fit into the holder in a certain position, thus making it possible to take images that come close to time-lapse shots.

As most effects of root density and development are interesting on the depth axis, the LemnaTec software allows for the soil column to be separated in as many virtual slices as required. In this way thorough root density profiles can be compiled.

Figure 41.15 shows an analysed image of three root columns for three different maize varieties: It is evident directly from the images how root growth characters are observed with differences in density, color and width (**Figure 41.16**).

When dynamic imaging is used in combination with time resolution, the great advantages of non-destructive quantitative image analysis become more obvious. Importantly, repeated measurement allows taking both the initial and the final value of any time interval to be

taken into account, to characterize effects not only relative to the mean value of the control, but to the relative change of the same plant. This makes it possible, for example, to calculate relative growth rates for the roots, taking differences in absolute plant size into consideration, based e.g. on different germination times. Thus a huge amount of otherwise detectable "biological variability" can be attributed to different initial values for individual plants and removed from the resulting data.

This is particularly valid and important if dynamic stressors such as drought, fertilizer or herbivorous pests are brought into the system at pre-determined points of time (after several weeks) of plant growth, when plant size and root development have reached a certain distribution, e.g. due to slightly different germination times, smaller differences in growth conditions and exponential growth rates.

3.4.4. Fluorescence Imaging
Blue wide-field fluorescence. Fluorescence phenotyping of complete plants is always challenging as high and homogeneous excitation in combination with sensitive detection is required. Combining cold light fluorescence bulbs with large, blue cut-off filter plates provides the

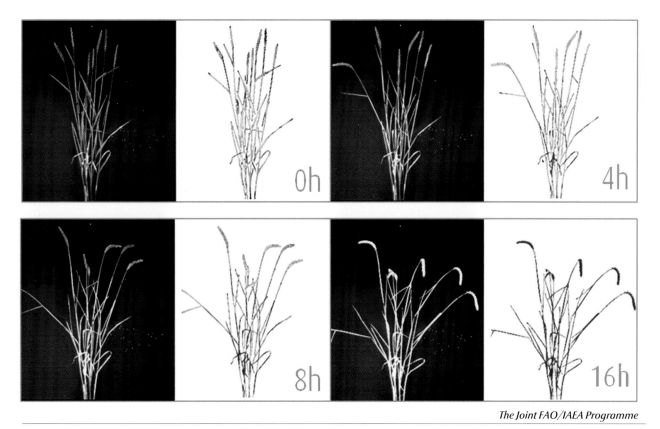

The Joint FAO/IAEA Programme

Figure 41.17 Wheat dries down in warm ambient conditions. NIR-imaging shows a strong increase in reflectance as the water in the leaves is extremely reduced. Blue/green false colors represent high water content, while yellow/red colors symbolize low water content (high reflectance).

basis for homogeneous blue excitation. Detecting all light in a range higher than 510 nm by using a sensitive color camera allows the measuring of any fluorescence that can be excited by blue light. Based on this technology complete plants, up to fully grown maize or 2.5 m Miscanthus or sugarcane, can be imaged for green fluorescent protein (GFP), red fluorescent protein (RFP) or other fluorophores as well as for continuously excited chlorophyll fluorescence. Measuring chlorophyll and GFP-like fluorescence in parallel provides an ideal basis for later image processing by e.g. normalizing GFP-signals of 3D-plants and arbitrarily oriented leaves through chlorophyll intensity. This kind of chlorophyll imaging contains without a doubt less information than PAM-like measurements, but it can be done very fast and with reproducible results even for really big plants. Depending on the aim of the screening, blue wide-field imaging can provide sufficient information on chlorophyll, particularly in comparison with other slower and more complex chlorophyll fluorescence imaging methods.

Chlorophyll PAM. Due to the restricted size of the illumination field PAM-systems are frequently used for small plot areas and two-dimensional plants like Brassica rosettes and small seedlings (www.walz.de). Taking this into account the integration of PAM-systems in high-throughput imaging with moving cameras or small imaging cabinets becomes attractive. Due to the high degree of automation even complex analysis schemes (including dark adaptation phases in the range of 20 to 60 minutes) can be easily realized by placing storage conveyor belts in dark tunnels between glasshouses and imaging cabinets.

For larger, more complex plants PAM-like illuminations are not yet available.

Chlorophyll Kautsky. A very interesting approach to chlorophyll activity measurement is the Kautsky effect measurement. Fast-flashed images are taken in a sequence and allow the calculation of Epsilon for each pixel of the leaf. However, it is not actually biologically clear what a change in Epsilon means and how to interpret it. Nevertheless, experiments have shown that adding pesticides led within minutes to extremely visualizable changes of Epsilon in leaves, whereas visible light did not reveal any differences (Jalink, 2009).

Luminescence imaging. The development of cameras with a very high sensitivity and a very good signal to noise ratio which do not need liquid nitrogen cooling opens new opportunities for luciferin testing. Test systems wherein plants are sprayed with luciferase before imaging can be completely automated to also include the spraying procedure. While these tests provide medium rather than high-throughput, the automation of long imaging times is even more important for the imaging of as many plants per day as possible.

3.4.5. NIR Imaging

Due to the limited availability of cameras with a sensitivity range between 900 and 1,700 nm (near infrared), the exploitation of this range is still in its infancy. While Fourier transform spectroscopy of homogenized and dried samples is very common in the analysis of proteins, NIR imaging is a newly emerging technology in this field.

3.4.6. Water Leaf Scanning

In the last years NIR cameras (900 to 1700 nm) have become increasingly popular as prices of this very specific camera type have reached affordable levels for deployment in phenotyping systems. The most common application is the reflective measurement of water in leaves, as water has highly absorbing bands between 1,450 and 1,550 nm. While well-watered leaves thus show a dark color, even a slight reduction in water content results in an increase in brightness, i.e. reduced absorption of the leaves – visible long before wilting or leaf rolling (**Figure 41.17**).

As a result water filling of tomato plants after watering can be made visible. Other phenomena like leaves showing temporarily increased water content at the beginning of drought stress, caused by the closing of stomata when roots start running out of water, is also detectable. Due to the fast and non-destructive measurement NIR water imaging is a very interesting alternative to the destructive assessment of leaf water potential. This is a good example of how specific information gained from destructive measurement methods, which nobody would ever want to apply to thousands of plants each day, can be substituted by a very effective non-destructive imaging method. As is usual when applying imaging methods, they should never be expected to replace or even substitute another method. Rather, it is always very effective to use the new method independently to gain more information. The final statistics can then reveal what is very important additional information for high-throughput phenotyping of plant stress.

3.4.7. Water NIR-Root Scanning

While there is a wide range of methods to assess growth, morphology and architectural features of aerial plant parts, the root system remains comparatively under-investigated. The number of already existing monitoring systems is limited, particularly when larger numbers of plants need to be analysed non-destructively under high-throughput conditions. As many essential influencing factors for root growth directly depend on the soil, it would be of special interest to assess real soil samples. This is particularly important when complex interactions between plants, roots, nutrients, drought, salinity or herbivorous insects are to be analysed and artefacts caused by artificial media need to be minimized or preferably eliminated. The quantification of root architecture requires the shape and structure of roots to be analysed. This can then be studied in relation to information on their ability and efficiency to extract water and nutrients from the soil, in addition to the static requirements. As uptake is to a large degree mediated by root hairs, many very laborious additional analyses for their quantification are necessary, a fact which is further complicated by their growth in soil. Even if assessment can be performed with soil, any form of root hair analysis can only be done on very short root parts, which bears the additional risk of not being representative for all roots concerned. Thus the full efficiency of a plant's water extraction capacity is difficult to assess, particularly if one takes into account that much of the transfer might be strongly connected with the presence of fungi, which still remain completely unquantifiable through the assessment of root architecture. The general challenge here is that some important features related to root efficiency such as hair roots and symbiosis with fungi cannot be made directly visible by imaging. To overcome this problem the LemnaTec NIR-approach is to look directly at the change of water distribution in the root column during plant growth.

By using specially designed NIR-imaging units and appropriate cameras, changes in the relative water content of the outer soil surface can be monitored. Particularly for plants needing a lot of water, the liquid will diffuse through the column more slowly than plants can actively extract it from the soil. As the soil column is much taller than wide, it can be assumed that a change

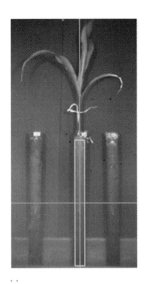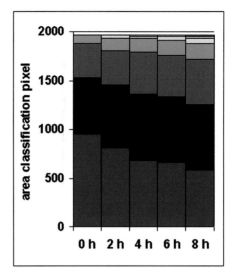

The Joint FAO/IAEA Programme

Figure 41.18 NIR imaging of a maize plant in a root column. Original image (left), color-classified image of soil water distribution with blue for highest and red for lowest water density (middle), change of areas with different moistures over time (right).

of the water content in a specific depth is more dependent on water extraction by the roots than by diffusion within the soil. This is particularly true with repeated imaging during daytime, when most water is used up, and when watering is then performed in the evening, allowing for equal dispersion during the night. Specific non-capillary top layers and the closed bottom of the columns virtually rule out evaporative losses.

A very important advantage of this method is that root efficiency not visible from the outside can now, for the first time, be quantified. Instead of destructively acquired estimates of root hair density, the global water extraction efficiency measurement allows for much larger numbers of plants to be screened, which would never be possible with detailed observation of each plant. In addition, any kind of drought stress experiment where pots are weighed for the calculation of average soil humidity is massively enhanced when the distribution of water within the soil can be monitored. **Figure 41.18** shows just such an analysis for a maize plant.

3.4.8. IR Heat Imaging

High-throughput 3D imaging systems provide a unique chance to quantify differences in leaf temperature, either within one plant or even between various plants, under highly controlled and automated conditions.

Depending on the environment in which the plants are grown and the aim of the screening, IR-imaging can be performed with different set-ups. All approaches need to consider that leaf temperature imaging is a highly dynamic process, as opening and closing of stomata may change leaf temperatures within minutes. Additionally, the absolute leaf temperature is – beside genetic or phenotypic plant-related factors – strongly dependent on actual irradiation, air temperature, air humidity, air velocity and, for thicker plant parts, the "history" of all these factors over the last hour or so.

For these reasons keeping the environmental factors as constant or at least as controlled as possible – e. g. in growth chambers – is one promising approach. Letting plants circulate fast to homogenize conditions is another helpful method, particularly in glasshouses. In addition, taking measurements for a large number of plants over a short time allows a far better normalization of results on the moving average and – based on this – identification of plants that deviate and are for this very reason more interesting. Conveyor-based systems also enable placement of the plants in well-climatized, pre-conditioning zones, where they all stay for the same pre-defined time under controlled conditions concerning light, airflow, temperature and humidity. This allows separation of the environmental factors from the phenotypic or genetic factors. Even the plants' reactions to changing conditions can thus be monitored, particularly when the imaging unit itself is also kept as a highly controlled environment.

All these features of automated measurement provide homogeneous measurement conditions where manual measurement would never be able to keep conditions similar in quite such a controlled way. This example also highlights the aspect that automating is not just a way to increase quantity of data, but can even be used to generate new types of data or data with much less measurement noise and higher reproducibility, by standardizing the plant preconditioning (or sample preparation, as it is more generally called) far better than ever before.

Such imaging offers great potential for screening mutants for surface (epidermis) properties such as hairiness and waxiness.

3.5. Image Pre-Processing

3.5.1. General

The role image processing plays in plant phenotyping is significant, but many projects tend to overestimate the specific task of developing appropriate image processing algorithms as part of the phenotyping projects.

Any reliable image processing needs reproducible imaging conditions, i.e. images taken under conditions suitable for scientific imaging to be used as a measurement method. Therefore it makes sense to invest in highly controlled imaging where camera parameters, light intensity and direction, image background and plant positioning can be controlled automatically, at best by a set of parameters taken directly from the database.

3.5.2. Requirements for Imaging Algorithms

Based on such images there are several image processing algorithms available with various functions. They help to separate the plant from the background, classify plant shape quantitatively, characterize plant parts and provide data on plant color. Again having the most complex or best performing algorithms is far less important than being able to identify the robust and reliable information of a defined set of parameters. While scientific image analysis in general works retrospectively – i.e. by having a set of images for subsequent, detailed analysis, with relevant amounts of human interference in parameter selection – phenotypic image processing for high-throughput needs algorithms which perform satisfactorily without human control, as image numbers simply get too large to control them. Data need to be available within minutes or at least hours after imaging and not only at the end of a long-term project. To reach this aim any kind of image analysis should be completely integrated into the whole automated phenotyping process. No manual interference should be needed to transfer images, apply algorithms or copy data results from file A to file B.

Consequently, fully integrated phenotyping systems include powerful, but clear-cut and transparent image analysis devices, as part of their complete routine from imaging to data presentation. As methods are generally developed at the beginning of larger screenings and then retained, it makes sense to resort to already existing expertise in development and implementation. After all, image analysis is just one component of all measurement methods, and is relatively uncoupled from the type and quality of the results. This is in clear contrast to the statistical retrospective analysis of data where they are intrinsically related to the real experiment. Therefore statistics are – in contrast to image analysis – always at the core of the research and competence of the user of any phenotyping technology.

While other books are already providing crucial information on biological image processing, some core aspects shall also be discussed here to explain the specific needs of biological image processing in contrast to the widespread technical image processing (Russ, 1999).

3.5.3. Image analysis as a measurement method

The fine art of image analysis includes a tremendous number of algorithms for pattern matching, neuronal networks and decision making devices. This all sounds very adequate for phenotyping, but it should be considered that phenotyping is a quantitative measurement method rather than a tool to obtain meaning from images. Biological meaning is after all a question of skilful statistical analysis, interpretation of data in a certain context and cross checking of image-based data with data from other sources in relevant comprehensive plant models, integrating much more than just the image analysis results. Therefore, the tendency to expect too much final interpretation from the image analysis itself should be avoided. Any kind of interpretation of non-measurable factors as part of an image analysis is scientifically very risky as all factors used as a base for a certain extrapolation may change over time, by application or with varieties, and then create mere artefacts. Therefore image analysis which solely provides data on projected leaf area should never extrapolate (even if it seems quite suggestive) leaf parts hidden by others. The human brain is working continuously with such extrapolations, but they should not be part of a scientific measurement method as it strives to identify interesting plants and not to extrapolate probabilities. If an *Arabidopsis* rosette is very compact with largely overlapping leaves, describing this by a set of three parameters such as projected leaf area, compactness and smallest non-convex shape area is much more successful than any complex approach to calculate the hidden leave parts from what is visible. Projected leaf area from top and side may even convey very efficiently the light interception, more so than any attempt to identify the inclination of each single leaf. Nevertheless, there are certainly cases where specific interpretative image analyses make more sense in answering specific, well-defined questions such as fast selection of seeds or seedlings.

3.5.4. Two-Dimensional Imaging and Three-Dimensional Data Representation

Raw images are nearly always 2D and to employ specific techniques with multiple cameras or structured light opens new options for the reconstruction of 3D objects (LIT 3D images). This is quite appealing, but one needs to consider carefully in which cases the informational value of the 3D image is actually higher than that of the combined image analysis results of several 2D images. There are cases where this is correct, but it should always be weighed against the great need to add a large amount of seemingly suggestive external information to the algorithms that produce such 3D models. If the model fits really well, this is fine, but there is the high risk of deriving artefact parameters though it is no longer clear to what degree the applied 3D plant model gets over-stretched. In general, if they are not used very carefully, 3D models just integrate various shape parameters into a seemingly pleasant visual result, disguising the individual parameters. Only very thorough data extraction from such models, while at the same time keeping the entire modelling process in mind, really allows drawing clear-cut and reproducible conclusions better than those directly obtained from a set of 2D images.

3.5.5. Image Analysis: No Need to Be Fully Comprehensive

The very pleasant effect from high volume data storage of all images taken is that only very loosely aggregated raw data is collected. As a result it is not necessary to extract all kinds of specific data, even if there is no obvious reason to do it. As long as the statisticians are happy with the accumulated data, it can make sense to store 100 values from each image right from the beginning, but having

sound backup systems for keeping the images safe and well-ordered is far more important. Storage of all relevant images allows re-analysis at any time in the future, when new needs arise or insights appear and become relevant for further biological questions.

3.6. Phenotyping Based Cross Technologies for Other –Omics

3.6.1. Sampling Procedures for Chemical Analysis: Proteomics, Metabolomics and Genomics

Two major concerns in any kind of manual plant part sampling or in making other kinds of manual measurements are: 1) the wrong assignment of samples due to incorrect labelling, and 2) the sampling time. For many sampling procedures, e.g. in gene regulation studies, it is of utmost importance to sample swiftly and at a specific time of the day or developmental stage and then label the samples as fast and as accurately as possible before freezing. With increasing throughput and efforts to lower costs for any kind of analysis, the numbers involved in such samplings will grow. To avoid mistakes, sampling vials can be pre-labelled with a reliable barcode. When using hand-held scanners, first the barcode of the vial is scanned and immediately afterwards the RFID-chip of the plant is registered to connect both datasets. The results are either stored on the hand-held or the RFID-chip or directly transferred (wireless) to the central database. Thus the sampler can concentrate on performing fast manual work without the danger of confusing or mis-naming samples. Entire QTL- or other -omics analysis studies can be compromised by faults in data handling. As conveyor belts can transport plants efficiently and on request, current procedures require about five people standing side by side in one place, taking one sample every 10 s (including freezing), can manage 1,800 samples within one hour, thus taking full advantage of automatic plant identification and vial/plant transportation on request.

3.6.2. Spectroscopy for Compositional Analysis

A growing number of technologies using small or even portable sensor systems are becoming available, measuring a wide range of signals related to specific components that can be of interest to metabolomics and quality traits in crops.

Chlorophyll is the most common component measured, but others like phenols, starch, sugar and volatile substances are evolving. Such example shows how phenotyping technology can assist other -omics as well as opening new fields of phenotyping.

4. Selected Methodologies in Phenomics

4.1. Transformation and Mining of Phenotype Data

4.1.1. Old and New Parameters – The Issue of Correlation and Character of Measurement Values

The use of phenotyping parameters is generally based on pragmatic decisions by those who need fast and efficient methods to assess sufficient data for their research or breeding. For example, the assessment of biomass in pre-imaging phenomics times, was simplified to cutting the test plants and measuring their weight. Depending on the size and number of plants, fresh or dry weight was chosen whichever was easier to handle. If for example weighing could be done immediately after cutting and the number of plants was very large, fresh weight was often preferred. Dry weight was frequently favoured for smaller plant numbers where enough drying capacity was available. Unfortunately, neither of these two measurement values represents a "true value" for biomass, but both concepts intend to provide reproducible results to be used as statistical values, to characterize and distinguish plant genotypes. Nevertheless both approaches are well-established and often used as they are based on a well-defined measurement scheme, reproducible and successful in differentiating plants. Introducing additional parameters in image-based phenotyping, such as projected leaf area or image-based volumes (Rajendran *et al.,* 2009), often involves hard work in finding correlations with "classical", manually measured parameters. If correlation remains limited, this is seen as a drawback of the new parameter. The "historical" approach tends to ignore the fact that many parameters have different physical bases of measurement and for this reason can only be correlated under very tightly controlled conditions and rarely in a scientifically reasoned way; any kind of projected leaf area measurement does not "see" the thickness of leaves (a genotype just having small but thick leaves would be an outlier in such a correlation, suggesting a fault of the parameter). Importantly in this example, a plant phenotype like this would be very interesting as the ratio between fresh weight and projected leaf area could

be seen as a parameter for leave thickness. This little example illustrates that the establishment of any kind of correlation based on a measurement hypothesis only generates more problems than it could finally ever help to solve in any productive statistical assessment. For this reason any measurement parameter as such should be seen as non-correlated to any other recent or older parameter. If statisticians do find strong correlations afterwards, this may be due to the physical redundancy of the parameters, the self-similarity of the plants or just pure chance, due to high homogeneity of the genotypes assessed. The reasons supporting a correlation should always be assessed carefully and extrapolation for future measurements (often under different conditions) should never be taken lightly.

4.1.2. Plant Shape Parameters

Any kind of shape needs more than one parameter for its characterization. Even the description of a very simple shape such as a circle needs two parameters (circumference and diameter). In this case the constant ratio factor between both (Pi = 3.14…) is specifically characterizing the shape as a circle. For more complex shapes more parameters are necessary, for example when aiming to describe plant shapes accurately. For this reason a set

Figure 41.19 The colored lines show different human approaches to measuring rosette diameter by hand. The circle is the basis for an objective automated measurement of diameter based on the smallest encompassing circle.

of different parameters is used, which have different mathematical definitions such as compactness, 2nd moments, 2nd moments ratio, ferret diameter orientation or roundness. The ferret diameter or caliper length is a good example for an essential and non-trivial transformation of the manual parameter "rosette diameter" into a mathematically reliable measurement. While diameter is precisely defined for the measurement of a circle, caliper length for an irregular leaf rosette contains a great deal of subjectivity (**Figure 41.19**).

Taking the diameter of the smallest encompassing circle is, by contrast, a highly reproducible measurement. Nevertheless many shape parameters – like diameter and circumference as such – have only a restricted shape descriptive value and an even less general biological meaning. Such meanings are often developed after data transformation. The use of top and side projected areas of plants to describe dynamic leaf rolling under drought stress for maize is an example for such a transformation – eliminating absolute sizes by using similarity models – which is quite common in other engineering sciences (compare Buckingham's Pi-theorem in engineering).

Specific analyses of maize plants identifying internode length and leaf angle are examples of highly developed and highly specialized morphological parameters. While these measurements appear to be very targeted and appropriate for biological samples, it should always be considered that for fast, flexible and efficient image analysis more general parameters like leaf orientation distribution seem rather abstract at first glance, but can be applied much more easily to a far wider range of plants, when not all of them follow a rather idealized pattern, like the very straight leaves of maize grown under field density conditions.

4.2. Advanced Statistics

Bioinformatics tailored to the appropriate purpose is a crucial tool for any kind of phenotyping, but especially high-throughout. Experience with many successful phenotyping systems shows that using less and rather well-defined basic parameters and more high-level statistics for well-designed screening tests is a very promising approach. Nevertheless, any such statistics should refer to a comprehensible biological basis and should not just hold a loose connection to physiological basics, in order to avoid over-interpretation of data and utilizing of mathematical artefacts instead of well-founded pheno-

types of biological and crop significance. All this requires a very multi-disciplinary approach, where statisticians need a good basic feeling or intuitive model of biological processes. In addition, the biologists need to design experiments in collaboration with statisticians. If the work is based on diffuse questions merely suggested in screening approaches, even sound phenotyping will only be able to identify more or less trivial answers. For this reason it is strongly recommended to make the biostatistical approach an intrinsic part of each screening, starting well in advance when planning an experiment and following all steps. Mere end-of-the-pipe statistics will not be helpful in finding biologically relevant answers (Edwards, 2009).

4.3. Phenotyping Infrastructure – Approaches and Tendencies

High-throughput plant phenotyping requires significant amounts of hard- and software and also trained staff to handle each step of the whole process. As a result of these requirements different approaches have been developed over the last decade as to how such phenotyping centres could be initiated, maintained and shaped. The following aims to highlight some developments that have contributed to the current stage in phenomics.

In agricultural fertilizer and nutrient research, for example, Professor Mitscherlich developed a conveyor systems for automated watering as early as the 1960s, and he established several of these systems in removable glasshouses in the German Democratic Republic . At the time, any kind of growth and phenotype assessment was more or less done manually.

But with the impressive automatization of genetic research and the increasing knowledge about model plant and crop genetics, the need for larger phenotyping facilities in glasshouses and growth chambers became urgent. The dramatic progress in imaging and IT has since provided a boost in the amount of affordable tools available for the performance of automated and often non-destructive high-throughput phenotyping.

For obvious reasons, the agricultural industry was among the first to realize the potential in establishing phenotyping facilities both for model plants and crops. While some were started by large breeding companies, other smaller companies made phenotyping their core business, focusing for example on *Arabidopsis* (Paradigm Genetics started in 1997) or rice (e.g. Cropdesign starting in 1998). Accelerated crop development and the patenting of genetic functions were the primary focus of this industry-driven research.

Due to the need for centres with a certain critical mass, smaller breeding companies started to operate phenotyping centres as cooperatives, helping them to develop smaller, but often focusing on high-value crops like fruits and vegetables, in many cases these were geared to specific regions or environments.

State research centres often place their focus on the characterization of seed banks for germplasm to allow better exploitation of the as-yet-untapped treasures that have been collected there over long periods of time (e.g. barley genetic stocks, including mutant stocks). At the same time such centres intend to provide services for basic and applied research groups and to help the regional breeding industry in an effort to remain competitive (e.g. in Australia). There is a good reason for the fact that phenotyping centres generally develop in close relation to regionally or nationally important crops. This is particularly true as such centres can help to assert a region's rights over their national genomic heritage. Even for similar crops growth conditions and challenges such as diseases, soil properties or climatic factors vary massively between regions, resulting in the need for regionally orientated phenotyping centres. In addition, quarantine regulations for specific diseases frequently make regional solutions a must. As a result public–private partnerships have been established, as well as combinations of local public infrastructure developments and international research initiatives.

An increasing number of phenotyping centres are now emerging in Europe, Australia, the USA, Canada, Eastern Asia, South America, India and China. Unfortunately, economically challenged regions, particularly in developing countries, but also agriculturally important states like Russia, Ukraine and White Russia are lagging behind.

5. Phenomics in Plant Mutation Breeding

There are many interesting cases where mutation breeding may take advantages of high-throughput phenotyping. A major driver for the application of phenomics to mutant populations is that the selection of mutants in a mutant population can be extremely arduous, involving manual selection often of huge populations (e.g. 20,000 M_3 families, **see Chapter 25**). Difficulties in screening for mutants are exacerbated by the fact mutants are predominantly recessive, and desired mutants are often extremely rare chance events. As such mutation breeding was considered non-scientific, a label that has dogged the uptake and release of its potential. High-throughput phenotyping sweeps this obstacle aside.

The key issue is to spread out, germinate or let grow the seeds or plants in such a way that image analysis has a chance to identify the specific phenotype based on sizes, shapes, colors, fluorescence or on multi-image parameters like growth rates. Growing plants under high-density conditions makes sense especially if they need to be grown over longer durations. High numbers of plants or computing times are never an issue if the trait is extractable from the image. If seeds or germinating plants can be handled in multi-well plates (e. g. 48-well plates) a screen of 20,000 seeds just corresponds to 420 plates being six loads of a ScanalyzerHTS system, which can be analysed in one or two days depending on the imaging resolution needed. If plants are cultivated in 20-well soil trays the analysis of 1,000 trays (20,000 plants) would take around eight hours per imaging run on a Scanalyzer3D system as imaging frequencies in the range of 25 s per tray are no problem as long as trays stand on or are fed to conveyor belts. If growth rates needed are, one day per week is enough to screen 20,000 plants in a well-organized phenotyping facility.

Analysis is not slower than imaging, which allows sorting of plants online after they leave the imaging units.

This example shows that high-throughput phenotyping is very suitable for high plant numbers, and handling size of units just depends on the sizes of the plants. If images are taken once complexity of the analysis almost does not matter. As a result the same analysis methods are used for high-throughput as well for high-content analysis where lower number of plants are analysed in greater detail. This detail will in many cases be restricted to the imaging frequency or the numbers of images per plant during each imaging (turning plants add imaging from several sides) which would take somewhat more time. Multi-frequency imaging does not need additional time as imaging units work sequentially at the same frequency as one unit would do.

5.1. Provision of Plant Management Infrastructure for Uniform Testing

To quantify relevant growth parameters in large-scale research programmes, it is very important to keep growth conditions as similar as possible for all plants, or at least to monitor the environmental factors closely enough to allow statistical or model-based compensation for environmental factors. For these various reasons randomization in the glasshouse and comprehensive water management as part of most phenotyping programmes are of special importance, particularly if 5–15% changes in a specific trait could already signify an important advantage for a specific crop.

5.2. Comprehensive Quantitative Performance Profiles of Mutated Plants

While screening of mutants implies focusing on the most interesting traits of a specific pre-determined breeding programme, high-throughput phenomics allow comprehensive observation of many phenotypic traits. This means that not only the top 10% of interesting plants get a sound data coverage, but all plants, as long as they are phenotyped. Thus the full set of information is available for each single plant, and as long as plant samples are at least stored for genetic analysis for a much greater percentage of plants showing any deviation from the wild type a gene–phenotype–correlation can be established immediately or retrospectively. This has huge implications for the efficiency of mutant phenotype selection and will also impact on the understanding of gene function and its relation to phenotype and environmental interactions. Thus the long-term value of any mutation programme can be increased massively, even for researchers with a very different focus, located on the other side of the world, and the data may be of relevance years after the original experiment.

5.3. Trait Targeting

5.3.1. Biotic Stress

Biotic stress testing for individual (pre-selected mutant) plants has to be very well organized if it is to be done fast and efficiently at an early stage of screening. Having only single plants or small numbers on hand can mean that less than one plant is available per resistance test. This might be helpful in minimizing screening costs and accelerating the process overall, but at the same time requires specific concepts on how to handle test designs. One approach in such a case – keeping the efficient plant logistics and identifications of automated phenotyping systems in mind – could be similar to the following layout: The respective plant is grown and all growth parameters are measured. At a certain plant size one leaf or parts of a leaf of minor importance for growth are collected from all specimens, cut into pieces and put on agar plates for classical leaf disc screening, after infecting different parts with various pests. These may be aphids, feeding larvae, various fungi or other plant diseases. The result of these tests can be read out within the next two weeks generally, by using a high-throughput phenotyping system for multi-well plates. As all plants and leaf discs are uniformly identified by a unique label, datasets will be immediately matched after measurement. As a result important decisions can be made well in advance, e.g. to use additional leaf material for more specific tests, to grow this plant further or even to plan targeted crossings based on the resistance screening results. This example clearly illustrates how reliable and fast information can greatly accelerate screening for mutants. An approach such as this allows for earlier decisions, as entire resistance profiles of individual plants are determined, before the need to go into the field and make single resistance screenings at hotspots.

Identifying plants with higher resistance to root diseases is far easier than any field testing as test systems with transparent pots allow the comprehensive monitoring of root growth and also to watch reactions to infections nearly as easily as with shoot development.

5.3.2. Abiotic Stress

Due to the complexity of defining and imposing specific stress for and on plants, it makes sense to control stress levels as tightly as possible. To this end, automated phenomics systems allow randomization of plants for those parameters with limited control options (light, temperature, air humidity) and, at the same time, full control over soil humidity or nutrient status. Imposing the same drying down gradients on all plants is particularly important if the plant material is commonly very heterogeneous, as is a special feature of mutant populations. Otherwise any drought stress screening would just be a water usage screen for big plants that need more water and face stress earlier than smaller plants.

The same would apply to salt stress where plants that need to take up higher amounts of water would also face higher stress levels. As water uptake is quantitatively measured based on water loss (in systems with covered soil to prevent evaporation), such factors which might otherwise compromise a screening can be integrated into the stress assessment. To avoid killing individual, valuable plants through extreme stress, imaging of leaf rolling in visible light and leaf water changes in near-infrared light or changes in chlorophyll activity under fluorescent light can be used to quantify early stress symptoms, instead of screening for symptoms that show near death also visible to the naked eye. In addition, the imaging results can be used to stop stressing plants individually by no longer reducing soil humidity or even by going back to the non-stress level soon enough to guarantee recovery. In many cases medium stress levels plus recovery monitoring might be far more relevant in any event for an informed selection of plants than their survival of high stress level situations.

5.4. Quantification and Selection

5.4.1. Trawling for Small Quantitative Differences

A large part of screenings deals with the handling of large numbers of plant data, manpower, pots, soil and glasshouse capacities. Informed decision making can help to select relevant plants earlier on in the test, thus reducing the total number of plants that need to be screened till they reach full maturity. As shown above any screening based on a certain dataset taken at a specific point in time is not comprehensive. For this reason highly automated phenomics, which provides an information chain – from image acquisition to data transformation and statistical analysis – allows an evaluation of plants for all parameters involved very close in time to the actual screening. Data for each plant is thus available quickly and only plants of interest may be taken on to the next measurement or sorting point. This is an enormous advantage compared to any system that requires huge interaction of data and manpower for the calculation of decisive results.

The fast provision and analysis of data are particularly useful for the selection of plants to be taken forward for practical plant breeding (including mutation breeding); selected plants may be used in crossing or other plant breeding technologies such as haploid/doubled haploid generation and genotyping. Selected plants may also be taken up in gene expression studies. Here again if all data have been aggregated in close temporal proximity, interesting plants can be sampled before a certain phenomenon has already passed by. As an example: if early stress symptoms measured by NIR water imaging of leaves is correlated to gene regulation, it certainly makes sense to have such stress symptoms identified within hours of imaging. Even for parameters such as growth rate, fast identification of a 10% increase could be decisive in selecting a valuable mutant.

The option to sort and select plants pre-selected at the database level while they are circulating past a sampling station, e.g. for leaf discs or other plant part phenotypes, dramatically accelerates the sampling of rare and valuable plants among a very large group. This is of special importance for mutation breeding where selection of rare phenotypes is more relevant than functional analysis of plant development and interaction of different factors.

5.4.2. Seed and Seedling Phenotyping

Seed phenotyping is becoming more and more important for various reasons. A major driver in seed phenotyping is that for many crops the seed is the primary agricultural product. Selection for high-yielding and high quality seed, including novel seed products, is therefore of interest to breeders. Seed samples are also relatively easy to handle as they are transportable, take up little space and selected seed may be grown on when and where required. Recently, Nelson *et al.* (2009) demonstrated the use of high-throughput phenotyping of seeds and seedlings in the detection of haploids and doubled haploids (and thereby F_1 hybrids) in oil palm. Since haploids occur naturally in seeds and seedlings of many crop species, albeit at extremely low frequencies (e.g. 1 in 10,000) this screen has wide potential application as it overcomes the hurdle of rarity. Any phenotypic screening needs to be able to handle large amounts of seedlings in a non-destructive manner in order to select interesting specimens. Technologies in this field are developing fast and may need hugely different approaches in the future, depending on the tested crop species. Nevertheless, such methods for selecting rare but identifiable seeds or seedlings from thousands of others is a classical example of how phenomics produces technological tools that allow the handling of well-established breeding methods in a much more efficient way. Thus technology is finally providing not only a quantitative change based on larger throughput numbers, but also allowing completely new approaches which would never have been used without adequate automation technology.

Similar arguments can be made for other plant parts of harvestable interest such as fruit-, tuber-, leaf-, flower- and wood-specific phenotyping.

5.4.3. *In Vitro* Screening

As calli and small or regenerating plants are mostly grown under sterile conditions on agar, they are ideal objects for high-throughput screening. It makes sense to grow these in regular geometrical schemes such as in multi-well plates, to have a distinctive correlation between data and plant/callus.

High-throughput phenotyping systems with moving cameras can take multiple images per plate at various image magnifications. Depending on the screening aims or available markers, a range of parameters can be analysed, including color, shape, growth rate and water content (NIR) to specific fluorescence-related to chlorophyll, anthocyanins, phenols or other components of tissues and organs. Any of these parameters might provide information to identify cells, calli, embryos or plantlets which show targeted patterns of mutagenesis or simply reveal differences compared to "normal" (wild type) background values. Such plants or calli, etc., can then be selected for further regeneration and hardening to produce complete plants. Comprehensive multiparameter patterns may be of great value in retrospective studies in identifying early markers of interesting plants, and also in minimizing losses (by not regenerating cultures of no or little potential).

In the case of cell suspensions that will provide large numbers of small calli, the early detection of unwanted chimeras is of particular value in increasing the development of useful regenerants. All these automated steps allow the handling and early screening of larger plant numbers and provide informed decisions on how to define selection criteria.

5.5. Modelling Mutation Effects

The rapidly decreasing costs of genotyping and many newly evolving bio-techniques to identify mutations will aid in characterizing mutant plants far in advance of current methods. This is particularly important for the early identification of mutants present in a heterozygous condition, i.e. where mutant phenotypes are masked. To focus such crossings on really interesting mutations, the modelling of genotype to phenotype interactions will become increasingly important in the simulation of how a final line with an identified mutation is expressed a certain genetic background. To reach this aim all available knowledge and measurement capacities are required to develop and validate optimal models of genotype-related phenotypes. The maximum control over environmental conditions is needed to separate environmental factors from genetic factors. Thus glasshouse screening can be genuinely compared with the "wind tunnel" testing for cars mentioned above: in providing an infrastructure for the highly controlled and monitored experiments, to test and separate different factors of plant development and the expression of desired phenotypes.

6. Outlook

Phenotyping and, more generally, phenomics have their basis in the human capability for plant pattern recognition and classification. This approach in the understanding of plants and breeding is several thousand years old and has proven to be highly successful. Nevertheless, the requirements of quantitative phenotyping when dealing with large numbers of plants in high frequency have dramatically increased due to the advances in genomics. The innovative deployment of automation technology and the ability to provide fast storage and effective data crunching capabilities were the prerequisites to reach today's standards.

The history of active and controlled (induced) mutagenesis is much shorter than that of plant breeding and classification in general. Nevertheless, several approaches in the enhanced development of new varieties have gained importance over the last century. All these techniques aim to improve and accelerate breeding to keep pace with rising food demands, changing growing conditions and other challenges for modern varieties such as pest and disease control and the development of novel products to satisfy changing market demands. Modern phenomics provides a wide range of new and efficient tools and approaches to make mutation breeding more effective. And thus it not only presents quantitative changes, but actually creates new opportunities in usage for specific plants or breeding targets that would not be available without the new sensors, automatization, analysis and data mining capacities of modern high-throughput plant phenotyping.

7. References

7.1. Cited References

Eberius, M. and Lima-Guerra, J. 2009. High throughput phenotyping – data acquisition, transformation and analytics. *In*: D. Edwards *et al.* (eds.) Bioinformatics – Tools and Applications. New York: Springer.

Edwards, D. *et al.* 2009. Bioinformatics – Tools and Applications. New York: Springer.

Fankhauser C. *et al.* 2010. The *Arabidopsis* Phytochrome Kinase Substrate 2 protein is a phototropin signalling element that regulates leaf flattening and leaf positioning. *Plant Physiology*. In press.

Jalink, H. 2009. Fast and non-invasive phenotyping of plant health/stress status using LED induced chlorophyll fluorescence transient imager. *EPSO workshop* 2–3 November.

Nelson, S.P.C. *et al.* 2009. Breeding for high productivity lines *via* haploid technology. *In*: PIPOC 2009: Agriculture, Biotechnology & Sustainability Conference. Kuala Lumpur: Malaysian Palm Oil Board, pp. 203–225.

Rajendran, K., Tester, M. Roy and Stuart J. 2009. Quantifying the three main components of salinity tolerance in cereals. *Plant Cell and Environment*. 32: 237–249.

Russ, J.C. 1999. The Image Processing Handbook. Boca Raton: CRC Press.

7.2. Websites

http://www.lemnatec.com
http://www.cropdesign.com
http://www.pioneer.com
http://www.ipk-gatersleben.de
http://www.vib.be
http://www.plantaccelerator.org
http://www.basf.com/plantscience
http://www.fz-juelich.de/icg/icg-3
http://www.spicyweb.eu
http://www.bayercropscience.de
http://www.scynexis.com
http://www.syngenta.de

1. Introduction to Appendices 1–3

Appendix 1 was adopted from the training manual, *Radiosensitivity of a Number of Crop Species to Gamma and Fast Neutron Radiation*, prepared by H. Brunner (1985) of the Plant Breeding Unit, FAO/IAEA Agriculture & Biotechnology Laboratory, International Atomic Energy Agency Laboratories, Seibersdorf, Vienna International Centre, Vienna, Austria.

Appendices 2 and 3 were compiled by H. Nakagawa of the Institute of Radiation Breeding, National Institute of Agrobiological Sciences, P.O. Box 3, Kami–Murata, Hitachi–Ohmiya, Ibaraki 319–2293, Japan.

The References cited in Appendices 2 and 3 are as follows:

[1] **Yamaguchi, T. 1988.** Mutation breeding of ornamental plants. *Bull. Inst. Rad. Breed.* 7: 49–67.

[2] **Miksche, J.P. and Shapiro, S. 1963.** Use of neutron irradiations in the Brookhaven Mutation Programme. Biological Effects of Neutron and Proton Irradiations. Vienna: IAEA. Vol. 1, pp. 393–408.

[3] **Yamaguchi, T. 1982.** Mutation breeding in vegetable crops. *Gamma Field Symposia.* 21: 37–53.

[4] **Ukai, Y. 1983.** *In*: Y. Yoshio Watanabe and H. Yamaguchi (eds.) Mutation Breeding Gamma. Field Symposia No. 20 Suppl. Tokyo: Yokendo, p. 306 (in Japanese).

[5] **Sparrow, A.H., Sparrow, R.C., Thompson, K.H. *et al.* 1965.** The use of nuclear and chromosomal variables in determining and predicting radiosensitivities. *Radiation Botany.* Suppl. 5: 101–132.

[6] **Bhatt, B.Y., Bora, K.C., Gopal-Ayengar, A.R. *et al.* 1961.** Aspects of irradiation of seeds with radiations. *Effects of ionizing radiations on seeds*. Vienna: IAEA, pp. 591–607.

[7] **Singleton, W.R. (ed.) 1958.** Nuclear Radiation in Food and Agriculture. p. 193.

[8] **Nishida, M. 1969.** The induced semi-dwarf mutants of apples by gamma irradiation. *IRB Technical News.* 2: 1–2.

[9] **Yamakawa, K. 1969.** Induction of male-sterile tomato mutants by gamma irradiation. *IRB Technical News.* 3: 1–2.

[10] **Nakajima, K. 1970.** Gamma ray induced sports from a rose variety (Peace). *IRB Technical News.* 4: 1–2.

[11] **Fujita, H. and Takato, S. 1970.** An Entire leaf mutant in Mulberry. *IRB Technical News.* 5: 1–2.

[12] **Yamakawa, K. 1970.** Radiation-induced mutants of chrysanthemum and their somatic chromosome number. *IRB Technical News.* 6: 1–2.

[13] **Kukimura, H. 1971.** On the artificial induction of skin colour mutation of sweet potato (*Ipomoea batatas* LAM) tuber. *IRB Technical News.* 8: 1–2.

[14] **Takagi, Y. 1973.** Radiosensitizing gene (rs-1) in soybean variety. *IRB Technical News.* 12: 1–2.

[15] **Ikeda, F. 1974.** Radiation-induced fruit color mutation in the apple var. Fuji. *IRB Technical News.* 15: 1–2.

[16] **Hiraiwa, S. and Tanaka, S. 1979.** Glabrous mutants of rice. *IRB Technical News.* 22: 1–2.

[17] **Fujita, H., Yokoyama, T. and Nakajima, K. 1980.** Re-treatment of induced mulberry mutants with gamma-rays. *IRB Technical News.* 23: 1–2.

[18] **Ukai, Y. 1983.** "Crossing-within-spike-progeny method". An effective method for selection of mutants in cross-fertilizing plants. *IRB Technical News.* 25: 1–2.

[19] **Kondo, T. and Ohba, K. 1984.** Juvenile leaf-form mutation in *Chamaecyparis obtusa*. S. et Z. *IRB Technical News*. 26: 1–2.

[20] **Yatou, O. 1985.** Radiosensitivity of callus of safflower, *Carthamus tinctorius* L. *IRB Technical News*. 27: 1–2.

[21] **Sanada, T., Nishida, T. and Ikeda, F. 1986.** Resistant mutant to black spot disease of Japanese pear. *IRB Technical News*. 29: 1–2.

[22] **Kukimura, H. 1987.** Variation of berberine content in embryoid of *Coptis* spp. irradiated by γ-ray. *IRB Technical News*. 31: 1–2.

[23] **Iida, S. and Amano, E. 1987.** A method to obtain mutants in outcrossing crops. -Induction of seedling mutants in cucumber using pollen irradiation. *IRB Technical News*. 32: 1–2.

[24] **Amano, E. and Tsugawa, H. 1985.** Effect of gamma-irradiation on garlic and nagaimo. *Jpn. J. Breeding*. 35. Suppl. 2: 216–217.

[25] **Nagatomi, S. 1987.** Personal communication.

[26] Irradiation record at Institute of Radiation Breeding (Non-publication)

[27] **Sparrow, A.H. and Schwemmer, S.S. 1974.** Correlations between nuclear characteristics, growth inhibition, and survival-curve parameters (LD_n, whole plant D_0 and D_q) for whole-plant acute gamma-irradiation of herbaceous species. *Int. J. Radiat. Biol.* 25: 565–581.

[28] **Fujii, T. and Matsumura, S. 1959.** Radiosensitivity in plants III. Experiments with several polyploid plants. *Jpn J. Breeding.* 9: 245–252. (in Japanese)

[29] **Sparrow, A.H., Schwemmer, S.S. and Bottino, P.J. 1971.** The Effects of external gamma radiation from radioactive fallout on plants with special reference to crop production. *Radiation Botany.* 11: 85–118.

[30] **Fuji, T. 1983.** Radio-sensitivity and modifying factors. *In:* Mutation Breeding, Gamma Field Symposia, 20, Suppl., Yokendo. pp. 49–62.

[31] **FAO/IAEA. 1985.** Mutation Breeding Review. 3.

[32] **FAO/IAEA. 1985.** Mutation Breeding Newsletter. 26.

[33] **FAO/IAEA. 1986.** Mutation Breeding Newsletter. 27.

[34] **FAO/IAEA. 1986.** Mutation Breeding Newsletter. 28.

[35] **FAO/IAEA. 1987.** Mutation Breeding Newsletter. 29.

[36] **FAO/IAEA. 1987.** Mutation Breeding Newsletter. 30.

Reference 8 is for "Malus pumila (apple)" of chronic irradiation.

Appendix 1: Radio-sensitivity of various plant species to fast neutrons

Genus (family)/Common name	Species (Latin name)	No. Line tested	GR50 (Gy, range)	Typical dose (Gy)
POACEAE				
Oat	*Avena sativa*	12	8–13	3–06
Barley	*Hordeum vulgare*	48	8–12	3–06
Rice	*Oryza sativa*			
	japonica	42	23–38	10–20
	indica	34	33–47	15–25
Rice	*Oryza glaberrima*	14	26–36	12–22
Bread wheat	*Triticum aestivum*	62	16–24	3–7
Durum wheat	*Triticum durum*	21	12–17	3–6
Sorghum	*Sorghum vulgare*	14	18–25	4–10
Pearl millet	*Pennisetum typhoides*	8	12–18	4–08
Maize	*Zea mays*	34	8–19	3–8
Fescue	*Festuca arundinacea*	3	8–10	2–5
FABACEAE				
Peanut	*Arachis hypogaea*	14	18–28	8–18
Pigeon pea	*Cajanus cajan*	6	25–35	10–20
Chickpea	*Cicer arietinum*	9	28–45	18–28
Lentil	*Lens esculenta*	7	9–14	5–10
Lupine	*Lupinus albus*	3	15–28	8–14
	angustifolius	2	18–28	10–15
Common bean	*Phaseolus vulgaris*	26	10–14	6–12
Lima bean	*Phaseolus lunatus*	8	16–27	7–14
Mung gram	*Phaseolus aureus*	6	50–70	25–40
Pea	*Pisum sativum*	34	5–12	3–7
Horse bean	*Vicia faba major*	4	1.2–1.8	0.5–1.0
	Vicia faba minor	9	3–4	1.5–3.0
Cowpea	*Vigna unguiculata*	8	25–45	15–30
Mungbean	*Vigna radiata*	4	50–75	20–40
Soybean	*Glycine max*	28	20–40	10–18
BRASSICACEAE				
Radish	*Raphanus sativus*	12	60–80	20–45
Arabidopsis	*Arabidopsis thaliana*	6	55–75	20–40
LILIACEAE				
Onion	*Allium cepa*	8	7–12	
Leek	*Allium scorodoprasum*	2	8–11	
MALVACEAE				
Cotton	*Gossypium arboretrum*	3	4–7	2–4
Cotton	*Gossypium hirsutum*	18	8–12	3–7
ASTERACEAE				
Safflower	*Carthamus tinctorius*	4	45–55	15–30
Niger	*Guizota abyssinica*	3	14–17	5–8
Lettuce	*Lactuca sativa*	5	10–15	4–7
TILIACEAE				
Jute	*Corchorus olitorius (2n)*	3	38–45	15–27
	Corchorus olitorius (4n)	3	34–42	15–25

GR50 = 50% seedling height (or epicotyl height) reduction after irradiation of quiescent seeds equilibrated to 12–14% moisture with fast neutrons derived from the SNIF (Standard Neutron Irradiation Facility). Precision of the applied doses: ±5%.

Appendix 2: Radio–sensitivity of plants to chronic irradiation and suggested dose for mutation induction

Genus/species	English name	Suggested dose (Gy/day)	Reference
Abelmoschus esculentus	Okra (gumbo)	<4.0	[4]
Acer rubrum (6x or 8x)	Red maple	1.0–2.0	[7]
Acer spicatum	Mountain maple	1.0–2.0	[7]
Allium cepa	Onion	<0.5	[4]
Allium cepa	Onion	4.0–8.0	[7]
Allium fistulosum	Welsh onion	<0.5	[4]
Alnus japonica	Alder	0.5–1.0	[4]
Althaea rosea (6x or 8x)	Hollyhock	4.0–8.0	[7]
Antirrhinum majus	Snap dragon	0.5–1.0	[4]
Antirrhinum majus	Snap dragon	2.0–4.0	[7]
Apium graveolens	Celery	<0.5	[4]
Arachis hypogaea	Peanuts	1.0–1.5	[4]
Asparagus officinalis	Asparagus	1.0–1.5	[4]
Avena sp.	Ooats	<0.5	[4]
Beta vulgaris	Sugar beet	1.5–2.0	[4]
Betula sp.	Birch	0.5–1.0	[4]
Brassica napus	Rapeseed	2.0–3.0	[4]
Brassica oleracea	Cabbage	0.5–1.0	[4]
Brassica pekinensis	Chinese cabbage	0.5–1.0	[4]
Canna generalis	Canna	2.0–4.0	[7]
Capsicum flutescence (2x, 4x)	Red pepper	2.0–4.0	[7]
Capsicum sp.	Red pepper	1.0–1.5	[4]
Celosia cristata (4x)	Cockscomb	4.0–8.0	[7]
Celosia sp.	Cockscomb	1.0–1.5	[4]
Chamaecyparis obtusa	Hinoki falsecypress	<0.5	[4]
Chamaecyparis obtusa S.et Z.	Hinoki falsecypress	0.076–0.112	[19]
Chamaecyparis pisifera	Sungold falsecypress	<0.5	[4]
Chenopodium album (4x)	Lamb's quarter	4.0–8.0	[7]
Chrysanthemum arcticum (8x)	Arctic daisy	8.0–16.0	[7]
Chrysanthemum ircutianum (4x)	Oxeye daisy	4.0–8.0	[7]
Chrysanthemum lacustre (22x)	Pyrenees daisy	8.0–16.0	[7]
Chrysanthemum nipponicum	Nippon daisy	2.0–4.0	[7]
Chrysanthemum yezoense (10x)	Kohamagiku (in Japanese)	8.0–16.0	[7]
Chrysanthemum sp.	Chrysanthemum	<4.0	[4]
Citrullus vulgaris	Watermelon	1.5–2.0	[4]
Coleus blumei (4x)	Coleus	2.0–4.0	[7]
Coleus sp. (4x)	Coleus	0.5–1.0	[4]

Appendix 2: Radio–sensitivity of plants to chronic irradiation and suggested dose for mutation induction

Genus/species	English name	Suggested dose (Gy/day)	Reference
Commelina coelestis	Dayflower	1.0–2.0	[7]
Coptis japonica	Ouren (in Japanese)	30.0	[22]
Cornus florida	Dogwood	0.5–1.0	[7]
Cosmos sp.	Cosmos	1.0–2.0	[7]
Cosmos sp.	Cosmos	<0.5	[4]
Criptomeria japonica	Japanese cedar	<0.5	[4]
Cucumis melo	Melon	2.0–3.0	[4]
Cucumis sativus	Cucumber	2.0–3.0	[4]
Cucurbita sp. (4x)	Pumpkin	8.0–16.0	[7]
Cucurbita sp.	Pumpkin	3.0–4.0	[4]
Cynodon dactylon	Bermuda grass	2.0–3.0	[4]
Dahlia (8x)	Dahlia	0.5–1.0	[4]
Dahlia (hybrid) (8x)	Dahlia	2.0–4.0	[7]
Datura stramonium	Jimson weed	2.0–4.0	[7]
Daucus carota	Carrot	1.5–2.0	[4]
Digitaria sanguinalis	Large crabgrass	16.0–60.0	[7]
Diospyros sp.	Persimon	<0.5	[4]
Festuca arundinacea	Tall fescue	1.0–1.5	[4]
Gladiolus (hybrid) (6x)	Gladiolus	16.0–60.0	[7]
Gladiolus sp.	Gladiolus	<4.0	[4]
Glycine max (rs–1)	Soybean	<0.5	[4]
Glycine max (Rs–1)	Soybean	0.5–1.0	[4]
Gossypium hirsutum (4x)	Cotton	2.0–4.0	[7]
Gossypium sp.	Cotton	0.5–10	[4]
Graptopetalum bartramii (2x)	Patagonia mountain leatherpetal	16.0–60.0	[7]
Graptopetalum macdougallii (22x)	English lavender	16.0–60.0	[7]
Helianthus annuus	Common sunflower	1.0–2.0	[4]
Helianthus annuus L.	Common sunflower	4.0–8.0	[7]
Hibiscus sp.	Hibiscus	4.0	[4]
Hordeum vulgare	Barley	<0.5	[4]
Ilex (4x)	Holly	1.0–2.0	[7]
Impatiens sultanii	Busy lizzie	0.5–1.0	[7]
Ipomoea batatas	Sweet potato	2.0–3.0	[4]
Ipomoea noctiflora	Moonflower	4.0–8.0	[7]
Iris sp.	Iris	2.0–3.0	[4]
Iris (hybrid) (4x)		8.0–16.0	[7]
Kalanchoe blossfeldiana	Kalanchoe	8.0–16.0	[7]

Appendix 2: Radio–sensitivity of plants to chronic irradiation and suggested dose for mutation induction

Genus/species	English name	Suggested dose (Gy/day)	Reference
Kalanchoe daigremontiana	Devil's backbone	4.0–8.0	[7]
Kalanchoe tubiflora	Chandelier plant	16.0–60.0	[7]
Kalmia latifolia	Mountain laurel	2.0–4.0	[7]
Lactuca sativa	Lettuce	1.0–1.5	[4]
Lactuca sativa	Lettuce	1.0–1.5	[4]
Lactuca sativa	Lettuce	4.0–8.0	[7]
Lenophyllum pusillum		8.0–16.0	[7]
Lenophyllum texanum		16.0–60.0	[7]
Lilium longiflorum	Easter lily	0.3–0.5	[5]
Lilium sp.	Lily	<0.5	[4]
Linum usitatissimum	Flax	3.0–4.0	[4]
Linum usitatissimum	Flax	8.0–16.0	[7]
Liriodendron tulipifera	Tulip tree	2.0–4.0	[7]
Lolium multiflorum	Italian rye grass	0.5–1.0	[4]
Lolium perenne	Perennial rye grass	0.5–1.0	[4]
Luzura acuminata (8x)	Hairy woodrush	16.0–60.0	[7]
Luzula multiflora (4x)	Heath woodrush	16.0–60.0	[7]
Luzura pallescens (4x)	Eurasian woodrush	16.0–60.0	[7]
Luzura purpurea	Purple woodrush	2.0–4.0	[7]
Lycopersicon esculentum	Tomato	0.5–1.0	[4]
Lycopersicon esculentum	Tomato	4.0–8.0	[7]
Magnolia sp.	Magnolia	1.0–2.0	[7]
Malus pumila	Apple	0.115–0.230	[8]
Malus sp.	Apple	0.5–1.0	[4]
Medicago sativa	Alfalfa	0.5–1.0	[4]
Melilotus officinalis	Yellow sweetclover	2.0–4.0	[7]
Mirabilis jalapa	Four o'clock	2.0–4.0	[7]
Mollugo verticillata (8x)	Green carpetweed	8.0–16.0	[7]
Morus rubra	Mulberry	<0.5	[4]
Morus rubra	Mulberry	0.088	[11]
Narcissus sp.	Narcissus	<0.5	[4]
Nicotiana bigelovii (4x)	Indian tobacco	2.0–4.0	[7]
Nicotiana glauca	Tree tobacco	2.0–4.0	[7]
Nicotiana glauca x N. langsdorffii (6x)		2.0–4.0	[7]
Nicotiana langsdorffii	Tobacco	2.0–4.0	[7]
Nicotiana rustica (4x)	Sacred tobacco	2.0–4.0	[7]
Oryza sativa	Rice	2.0–3.0	[4]

Appendix 2: Radio–sensitivity of plants to chronic irradiation and suggested dose for mutation induction

Genus/species	English name	Suggested dose (Gy/day)	Reference
Paeonia sp.	Peony	<0.5	[4]
Petunia sp.	Petunia	1.0–1.5	[4]
Petunia hybrida	Petunia	4.0–8.0	[7]
Phaseolus vulgaris	Kidney bean	0.5–1.0	[4]
Phaseolus vulgaris	Common bean	8.0–16.0	[7]
Phleum pratense	Timothy	1.0–1.5	[4]
Phytolacca decandra (4x)	Poke root	2.0–4.0	[7]
Pieris japonica	Japanese andromeda	4.0–8.0	[7]
Pinus densiflora	Japanese red pine	<0.5	[4]
Pinus thunbergii	Japanese black pine	(2.5)	[26]
Pisum sativum	Pea	0.5–1.0	[4]
Pisum sativum	Pea	2.0–4.0	[7]
Populus sp.	Poplar	0.5–1.0	[4]
Prunus persica	Peach	<0.5	[4]
Prunus persica	Peach	2.0–4.0	[7]
Pyrus malus	Apple	1.0–2.0	[7]
Pyrus serotina cv. Nijisseiki	Japanese pear	0.04–0.15	[21]
Pyrus sp.	Pear	0.5–1.0	[4]
Raphanus sativus	Radish	1.5–2.0	[4]
Rhododendron sp.	Rhododendron	1.0–2.0	[7]
Ricinus communis L.	Castor bean	4.0–8.0	[7]
Rosa sp.	Rose	1.0–1.5	[4]
Rosa (Hybrid Tea Rose)	Rose	4.0–8.0	[7]
Saccharinum officinarum	Sugarcane	1.5–2.0	[4]
Saintpaulia sp.	African violet	1.0–1.5	[4]
Saintpaulia sp.	African violet	4.0–8.0	[7]
Secale cereale	Rye	<0.5	[4]
Sedum acre (12x)	Sedum	8.0–16.0	[7]
Sedum aizoon	Sedum	4.0–8.0	[7]
Sedum album (16x)	White stonecrop	8.0–16.0	[7]
Setaria italica	Italian millet	2.0–3.0	[4]
Setcreasea sp.	Purple heart	0.5–1.0	[7]
Solanum melongena	Aubergine	0.5–1.0	[4]
Solanum tuberosum	Potato	1.5–2.0	[]
Sorghum sudanense	Sudan grass	1.0–1.5	[4]
Sorghum vulgare	Sorghum	1.0–1.5	[4]
Spinacia oleracea	Spinach	1.0–1.5	[4]

Appendix 2: Radio–sensitivity of plants to chronic irradiation and suggested dose for mutation induction

Genus/species	English name	Suggested dose (Gy/day)	Reference
Stachyurus sp.	Kibushi (in Japanese)	4.0–8.0	[7]
Taxus media		0.3–0.5	[7]
Thea sp.	Tea	<0.5	[4]
Tradescantia ohiensis	Bluejacket	0.3–0.5	[7]
Tradescantia paludosa	Confederate spiderwort	<0.5	[4]
Tradescantia paludosa	Confederate spiderwort	0.3–0.5	[7]
Trifolium incarnatum	Crimson clover	0.5–1.0	[4]
Triticum aestivum	Wheat	<0.5	[4]
Tulipa sp.	Tulip	<0.5	[4]
Vicia angustifolia	Vetch	2.0–4.0	[7]
Vicia faba	Broad bean	<0.5	[4]
Vicia faba	Broad bean	0.5–1.0	[7]
Vicia tenuifolia (4x)	Cow vetch	2.0–4.0	[7]
Vitis vinifera	Grape	<0.5	[4]
Xanthium sp. (4x)	Cocklebur	4.0–8.0	[7]
Zea mays	Corn (maize)	1.0–1.5	[4]
Zinnia elegans	Common zinia	2.0–4.0	[7]

Appendix 3: The radio-sensitivity of plants to acute gamma irradiation estimated by LD_{50}, RD_{50} and the suggested dose (SD) for practical application

Genus/species	English name	Organ	SD (Gy)	LD_{50} (Gy)	RD_{50} (Gy)	Reference
Abelmoschus esculentus	Okra (gumbo)	Seeds		700	500	[1]
Aberia sp.	Kai apple	Rooted cuttings	30			[1]
Abies balsamea	Balsam fir			7.5		[29]
Abies concolor	White fir			8.1		[29]
Abies grandis	Grand fir			6.2		[29]
Abies lasiocarpa	Alpine fir			6.2		[29]
Acer rubrum	Red maple		45	51.1		[5]; [29]
Acer saccharum	Sugar maple		300	47.2		[5]; [29]
Achimenes sp.	Japanese pansy	Detached leaves	30 (10–40)			[1]
Aconitum sp.	Monk's hood	Tuber	(25–200)			[26]
Adiantum sp.	Maidenhair fern	Spores	300			[1]
Aechmea fasciata	Urn plant	Seeds	300–400			[1]
Aegilops cylindrica	Jointed goatgrass	Seeds	300			[28]
Aegilops squarrosa	Goatgrass	Seeds	300			[28]
Aegilops triuncialis	Barbed goatgrass	Seeds	400			[28]
Aegilops ventricosa	Bulbed goatgrass	Seeds	300–400			[28]
Aesculus octandra	Yellow buckeye			71.1		[29]
Agave rigida	Sisal hemp				80–120	[29]
Agropyron cristatum	Crested wheatgrass	Seedling		20		[29]
Agropyron cristatum	Crested wheatgrass				20–40	[29]
Agropyron trachycaulum	Bearded wheatgrass				20–40	[29]
Agropyron intermedium	Wheatgrass				40–60	[29]
Agrostis gigantea	Red top	Seeds	300			[4]
Agrostis sp.	Bent grass	Seeds	200			[4]
Agrostis sp.	Bent grass		400			[26]
Allium cepa	Onion		100–200			[2]
Allium cepa	Onion		150			[3]
Allium cepa	Onion	Seeds	<100			[4]
Allium cepa	Onion	Seeds	(25–80)			[26]
Allium cepa	Onion			29		[27]
Allium cepa	Onion		18.9		10–20	[29]
Allium fistulosum	Welsh onion	Seeds	<100			[4]
Allium porrum	Leek				10–20	[29]
Allium sativum L.	Garlic	Bulb	4–8			[24]
Allium sativum L.	Garlic	Bulb	(4–20)			[26]
Allium sativum L.	Garlic	Cultured tissue	(10–20)			[26]

Appendix 3: The radio-sensitivity of plants to acute gamma irradiation estimated by LD$_{50}$, RD$_{50}$ and the suggested dose (SD) for practical application

Genus/species	English name	Organ	SD (Gy)	LD$_{50}$ (Gy)	RD$_{50}$ (Gy)	Reference
Allium sativum	Garlic	Bulblets	(11.2)			[26]
Allium sativum	Garlic				10–20	[29]
Allium tuberosum	Chinese chive	Seeds	(20–200)			[26]
Allium schoenoprasum	Chives				10–20	[29]
Aloe brevifolia	Aloe				15.5	[27]; [29]
Alstroemeria sp.	Lily of inca	Rhizomes	3.5–5			[1]
Alstroemeria sp.	Lily of inca		5–7			[1]
Alstroemeria sp.	Lily of inca	Rhizomes	(30)			[26]
Alstroemeria sp.	Lily of inca	Rhizomes	(4–7)			[26]
Alstroemeria sp.	Lily of inca	Seeds	200			[26]
Althaea rosea	Hollyhock	Seeds	500–700	1200	590	[1]
Amaranthus tricolor	Joseph's coat	Seeds		190	120	[1]
Amaranthus sp.	Amaranth	Seeds	(100–200)			[26]
Amaryllis belladonna	Belladonna lily	Bulbs		10.5		[29]
Amaryllis sp.	Amaryllis	Bulbs	5–10			[1]
Amorphophallus konjak	Glukomannan	Tuber	(2.5–20)			[26]
Ananas comosus	Pineapple	Crown bud		350		[25]
Ananas comosus	Pineapple	Leaf bud			250	[25]
Ananas comosus	Pineapple	Crown section		89.7	250	[29]
Andropogon gerardi	Big bluestem				160–200	[29]
Andropogon scoparius	Little bluestem				60–80	[29]
Anemone coronaria	Garden anemone	Seeds		400	–	[1]
Anemone fulgens	Flame anemone	Bulbs		20.4	–	[29]
Anemone sp.	Anemone	Seeds	100			[1]
Anemone sp.	Anemone	Small tubers	150			[1]
Anethum graveolens	Dill				60–80	[29]
Aphanostephus skirrobasis	Lazy daisy		20			[5]
Apium graveolens	Celery				60–80	[29]
Aquilegia akitensis Huth.	Columbine	Seeds		580	300	[1]
Aquilegia vulgaris	European columbine	Seeds		680	500	[1]
Arachis hypogaea	Peanuts	Seeds	200	100	40–60	[4]; [29]
Arctium lappa	Edible burdock	Seeds	<100			[4]
Armeria maritima Willd.	Common thrift	Seeds		110	50	[1]
Asparagus officinalis	Asparagus		20–80		60–80	[2]; [29]
Asplenium sp.	Hart's tongue fern	Spores	100			[1]
Aster novi-belgii	New york aster	Seeds		220	140	[1]

Appendix 3: The radio-sensitivity of plants to acute gamma irradiation estimated by LD_{50}, RD_{50} and the suggested dose (SD) for practical application

Genus/species	English name	Organ	SD (Gy)	LD_{50} (Gy)	RD_{50} (Gy)	Reference
Astilbe sp.	Astilbe		(5–50)			[26]
Astragalus sinicus	Astragol (locoweed)	Seeds	500–700			[4]
Astragalus sinicus (2x)	Astragol (locoweed)	Seeds	200–400			[28]
Astragalus sinicus (4x)	Astragol (locoweed)	Seeds	200–400			[28]
Avena sativa	Oats		150–250			[2]
Avena sativa	Oats		300	170–270	20–40	[4]; [29]
Avena 'Condor'	Spring oat	2–4 Leaves	9.2			[29]
Avena 'Orbit'	Spring oat	Seedling	19.5	34.2		[29]
Begonia x hiemalis	Begonia	Detached leaves	15–30			[1]
Begonia rex-cultorum	Rex begonia	Detached leaves	100			[1]
Begonia	Rex begonia		5–100			[4]
Begonia sp.	Begonia	Plant	(130–180)			[26]
Bellis perennis	English daisy	Seeds		300	270	[1]
Bellis sp.	Daisy	Seeds	300			[4]
Beta cicla	Swiss chard				60–80	[29]
Beta vulgaris	Sugar beet	Seeds	300			[4]
Beta vulgaris	Sugar beet	Seeds	(200)			[26]
Beta vulgaris (2x)	Sugar beet	Seeds	200			[28]
Beta vulgaris (4x)	Sugar beet	Seeds	400–700			[28]
Beta vulgaris	Sugar beet	Plant			200	[29]
Beta vulgaris	Beet				80–120	[29]
Betula lutea	Yellow birch		20	42.8		[5]; [29]
Bombax malabaricum	Cotton tree	Seeds		300	270	[1]
Bougainvillea sp.	Bougainvillea	Rooted cuttings	10–25			[1]
Bouteloua gracilis	Blue grama				120–160	[29]
Brassica campestris	Bird rape	Seeds	800<	1200	860	[1]; [4]
Brassica campestris	Bird rape				80–120	[29]
Brassica hirta	White mustard				80–120	[29]
Brassica juncea	Mustard		350			[3]
Brassica juncea	India mustard				120–160	[29]
Brassica napobrassica	Rutabaga				80–120	[29]
Brassica napus	Winter rape	Seeds	<800			[4]
Brassica napus	Winter rape				>240	[29]
Brassica nigra	Black mustard	Seeds		>>550		[6]
Brassica nigra	Black mustard				80–120	[29]
Brassica oleracea	Flowering cabbage	Seeds	500–700	960	780	[1]; [4]

Appendix 3: The radio-sensitivity of plants to acute gamma irradiation estimated by LD_{50}, RD_{50} and the suggested dose (SD) for practical application

Genus/species	English name	Organ	SD (Gy)	LD_{50} (Gy)	RD_{50} (Gy)	Reference
Brassica oleracea	Cauliflower		200–400			[2]
Brassica oleracea	Cabbage	Seeds	300	151		[4]; [27]
Brassica oleracea	Cabbage	Seedling		112.3		[29]
Brassica oleracea var. *acephala*	Kale				60–80	[29]
Brassica oleracea var. *botrytis*	Cauliflower				60–80	[29]
Brassica oleracea var. *gemmifera*	Brussel sprouts				120–160	[29]
Brassica oleracea var. *italica*	Broccoli				80–120	[29]
Brassica pekinensis	Chinese cabbage	Seeds	200			[4]
Brassica pekinensis	Chinese cabbage				80–120	[29]
Brassica rapa	Turnip				120–160	[29]
Brassica rapus	Rape seed		600–1500			[2]
Brassica rapus		Seeds	200			[4]
Brassica sp.		Seeds	(400)			[26]
Bromus inermis	Smooth brome				40–60	[29]
Bromus sp.	Brome grass		100–200			[2]
Buddleia sp.	Buddleia	Plants	20–30			[1]
Cajanus cajan	Pigeon pea	Seeds		150		[29]
Calanthe discolor	Calanthe	Bulbs	(440)			[26]
Calendura officinalis	Pot marigold	Seeds		<1200	<1200	[1]
Calendura officinalis	Pot marigold	Seeds	<800			[4]
Callistephus chinensis Nees.	China marigold	Seeds	100	210	160	[1]; [4]
Calochortus (av.of 2 spp.)	Mariposa lily	Bulbs		21.5		[29]
Calonyction aculeatum House	Common moon flower	Seeds		150	100	[1]
Campanula medium	Canterbury bells	Seeds		150–200	100–150	[1]
Canavalia gladiata		Seeds	200			[4]
Canna sp.	Canna	Rhizomes	10–30			[1]
Capsella bursa-pastoris				580		[27]
Capsicum annuum	Red pepper	Seeds		170	140–150	[1]
Capsicum annuum	Red pepper		135			[3]
Capsicum annuum var. *grossum*	Bell pepper	Seeds	100			[[4]]
Capsicum annuum (2x)	Bell pepper		200–300			[28]
Capsicum annuum (4x)	Bell pepper		400			[28]
Capsicum flutescence	Bell pepper	Seeds		240		[29]
Capsicum flutescence	Bell pepper				60–80	[29]
Capsicum sp.		Seeds	(110)			[26]

Appendix 3: The radio-sensitivity of plants to acute gamma irradiation estimated by LD_{50}, RD_{50} and the suggested dose (SD) for practical application

Genus/species	English name	Organ	SD (Gy)	LD_{50} (Gy)	RD_{50} (Gy)	Reference
Cardiospermum halicacabum	Balloon vine	Seeds		110	80	[1]
Carica papaya	Papaya		200–350	120		[2]; [29]
Cariopteris incana	Blue spirea	Seeds		160	110	[1]
Carthamus tinctorius	Safflower	Seeds			400	[20]
Carthamus tinctorius	Safflower	Seedling			60	[20]
Carthamus tinctorius	Safflower	Callus			120	[20]
Carya cordiformis	Bitternut hickory			76.9		[29]
Carya laciniosa	Shellbark hickory			41		[29]
Carya ovata	Shagbark hickory			60.3		[29]
Carya tomentosa	Mockernut hickory			76.9		[29]
Carya illinoensis	Pecan			290		[29]
Castanea denntata	American chestnut			37.7		[29]
Castanea sp.	Chestnut	Scion	(20)			[26]
Castanea sp.	Chestnut	Cuttings	25–50			[2]
Catharanthus roseus		Seeds		180	150	[1]
Cedrus libani	Cedar-of-lebanon			8.4		[29]
Celosia cristata	Feather cockscomb	Seeds		660	400	[1]
Celosia cristata	Cockscomb	Seeds		480	220	[1]
Centaurium sp.		Seeds				[6[
Cheiranthus cheiri	Wall flower	Seeds		700	500	[1]
Cheiranthus cheiri	Wall flower	Seeds	500–700			[4]
Chionodoxa luciliae	Glory-of-the-snow	Bulbs		28.1		[29]
Chlorophytum elatum			5			[5]
Chrysanthemum coccineum	Common pyrethrum	Seeds		100	50	[1]
Chrysanthemum lacustre (22n)					67.5	[27]; [29]
Chrysanthemum maximum	Shasta daisy	Seeds	400	600	360	[1]; [4]
Chrysanthemum morifolium	Chrysanthemum	Rooted cuttings	15–25			[1]
Chrysanthemum morifolium	Chrysanthemum	Cuttings	15–25			[1]
Chrysanthemum morifolium	Chrysanthemum		50–200			[12]
Chrysanthemum nipponicum		Seeds	300			[4]
Chrysanthemum nipponicum					30.2	[27]
Chrysanthemum paludosum		Seeds		400	90	[1]
Chrysanthemum yezoense (6n)					91.5	[27]
Chrysanthemum sp.	Chrysanthemum		3–4			[2]
Chrysanthemum sp.	Chrysanthemum		25			[4]
Chrysanthemum sp.	Chrysanthemum		(13–17.4)			[26]

Appendix 3: The radio-sensitivity of plants to acute gamma irradiation estimated by LD$_{50}$, RD$_{50}$ and the suggested dose (SD) for practical application

Genus/species	English name	Organ	SD (Gy)	LD$_{50}$ (Gy)	RD$_{50}$ (Gy)	Reference
Chrysanthemum sp.	Chrysanthemum		(30)			[26]
Chrysanthemum sp.	Chrysanthemum	Seeds	(200–300)			[26]
Cirsium japonicum	Japanese thistle	Seeds		200	160	[1]
Cirsium sp.	Thistle	Seeds	(200–600)			[26]
Citrullus vulgaris	Watermelon	Seeds	300	600		[4]; [29]
Citrullus vulgaris	Watermelon			>300		[7]
Citrullus vulgaris (2x)	Watermelon		200			[28]
Citrullus vulgaris (4x)	Watermelon		400			[28]
Citrullus vulgaris	Watermelon				60–80	[29]
Citrus paradisi	Grapefruit			32.7		[29]
Citrus reticulata	Mandarin orange			49.1		[29]
Citrus sinensis	Sweet orange			41.8		[29]
Citrus limonia	Lemon			41.8		[29]
Citrus sp.	Unshu orange	Scion	(40–80)			[26]
Citrus sp.		Callus	(120–200)			[26]
Citrus sp.		Scion	(80)			[26]
Clematis sp.	Clematis	Rooted cuttings	2–5; (10–15)			[1]; [26]
Coix lacryma-jobi var. *Ma-yuen*	Job's tear	Seeds	300; (200)			[4]; [26]
Colocasia esculenta	Taro		100			[4]
Conifer	Conifer	Rooted cuttings	20			[1]
Convallaria majalis	Lily-of-the-valley	Bulbs (roots)		18.6		[29]
Coptis japonica	Ouren		(200–400)			[29]
Coptis japonica	Ouren	Embryoid	(2.5–12.5)			[26]
Corchorus capsularis	Jute		50–200			[2]
Coreopsis tinctoria	Plain coreopsis	Seeds		140	30	[1]
Cosmos bipinnatus	Cosmos	Seeds		<700	<700	[1]
Cosmos bipinnatus	Yellow cosmos	Seeds		130–220	120–150	[1]
Cosmos sp.	Cosmos	Rooted cuttings	20			[1]
Crepis capillaris	Smooth hawksbeard		10			[5]
Crocus sp.	Crocus	Dormant corms	10–15			[1]
Crocus sp.	Crocus	Bulbs		9.8		[29]
Cryptomeria japonica	Cryptomeria			12.2		[29]
Cucumis melo	Cantaloupe		150–200			[2]
Cucumis melo	Cantaloupe	Seeds		500		[6]
Cucumis melo var. *cantalupensis*	Cantaloupe				80–120	[29]
Cucumis melo	Muskmelon				60–80	[29]

Appendix 3: The radio-sensitivity of plants to acute gamma irradiation estimated by LD_{50}, RD_{50} and the suggested dose (SD) for practical application

Genus/species	English name	Organ	SD (Gy)	LD_{50} (Gy)	RD_{50} (Gy)	Reference
Cucumis melo	Muskmelon	Seeds	400			[4]
Cucumis sativus	Cucumber		200–400			[2]
Cucumis sativus	Cucumber	Seeds	400			[4]; [26]
Cucumis sativus	Cucumber	Pollen in anther	20–40			[23]
Cucumis sativus	Cucumber	Young plants	(10–80)			[26]
Cucumis sativus	Cucumber				20–40	[29]
Cucurbita sp.	Squash		200–250			[2]
Cucurbita pepo	Squash	Seeds			73	[27]
Cucurbita pepo	Squash	Seedling		66.5		[29]
Cucurbita pepo	Acorn squash				60–80	[29]
Cucurbita pepo var. *medullosa*	Zucchini squash				200–240	[29]
Cucurbita maxima	Winter squash				200–240	[29]
Cucurbita moschata	Butternut squash				160–200	[29]
Cucurbita sp.	Pumpkin	Seeds	400			[4]
Cucurbita sp.		Seeds	(300–500)			[26]
Cupressus duclouxiana	Bhutan cypress			15.8		[29]
Cyanotis somaliensis	Furry kittens				83.5	[27]; [29]
Cymbidium sp.	Cymbidium	Cultured tissue	(60)			[26]
Cymbidium sp.	Cymbidium	Back bulb	(25–50)			[26]
Cynara scolymus	Globe artichoke	Seeds	500–700	700	620	[1]; [4]
Cynara scolymus	Globe artichoke				80–120	[29]
Cynodon dactylon	Bermuda grass		(30–70)			[26]
Cyclamen persicum	Cyclamen	Seeds	90–100; (150)			[1]; [26]
Cyclamen sp.	Cyclamen	Seeds	(5–10)			[26]
Dactylis glomerata	Orchard grass				40–60	[29]
Dactylis sp.	Orchard grass	Seeds	100	110		[4]; [29]
Dahlia sp.	Dahlia	Tubers	15–25			[1]
Dahlia x cultorum	Dahlia			190	150	[1]
Datura fastuosa	Double purple	Seeds		130	100	[1]
Datura quercifolium	Oak-leaf datura	Callus	(100–700)			[26]
Datura striata		Callus	(50–300)			[26]
Daucus carota	Carrot	Seeds	500–700			[4]
Daucus carota	Carrot				80–120	[29]
Delphinium ajacis	Larkspur	Seeds		190	150	[1]
Dianthus caryophyllus	Carnation		75			[2]

Appendix 3: The radio-sensitivity of plants to acute gamma irradiation estimated by LD$_{50}$, RD$_{50}$ and the suggested dose (SD) for practical application

Genus/species	English name	Organ	SD (Gy)	LD$_{50}$ (Gy)	RD$_{50}$ (Gy)	Reference
Dianthus sp.	Carnation	Cutting	(40–120)			[26]
Dianthus sp.	Carnation	Rooted cuttings	40–60			[1]
Dianthus sp.	Carnation	Unrooted cuttings	80–120			[1]
Dianthus barbatus	Sweet William	Seeds		330	240	[1]
Dianthus barbatus x D. sinensis		Seeds		900	480	[1]
Dianthus sp.	Carnation	Seeds	400			[4]
Dimorphotheca sinuata	Cape marigold	Seeds		400–700	200–660	[1]
Dimorphotheca sp.	Daisey	Seeds	500–700			[4]
Dioscorea batatas	Chinese yam	Bulbil	20–50			[24]
Dioscorea batatas	Chinese yam	Bulbil	20–50			[26]
Diospyros sp.	Persimon	Scion	20–80			[26]
Dolichos lablab	Hyacinth bean	Seeds		500		[6]
Echinacea purpurea Moench.	Purple cornflower	Seeds	<100	70	60	[1]; [4]
Echinochloa utilis	Burnyard millet	Seeds	500–700; (200–400)			[4]; [26]
Echinochloa utilis	Burnyard millet	Seeds	(300)			[26]
Echinops ritro L.	Small globe thistle	Seeds	100	230	90	[1]; [4]
Eleusine coracana	Finger millet	Seeds		300		[6]
Eriobotrya japonica	Japanese loquat	Scion	(50–100)			[26]
Eriobotrya japonica	Japanese loquat	Young plants	(40–60)			[26]
Eucalyptus obliqua	Eucalyptus			30		[29]
Euphorbia sp.	Eucalyptus	Rooted cuttings	30–50			[1]
Eustoma grandiflorum	Lisianthus	Young plants	(25–50)			[26]
Eustoma grandiflorum	Lisianthus	Seeds	(25–120)			[26]
Exacum sp.	German violet	Cultured plant	(4)			[26]
Fagopyrum sp.	Buckwheat		(50–100)			[2]
Fagopyrum esculentum	Buckwheat	Seeds	400; (150–250)			[4]; [26]
Fagopyrum sagittatum	Buckwheat				60–80	[29]
Fagus grandifolia	American beech			64.1		[29]
Festuca elatior	Meadow fescue	Seeds		190		[29]
Festuca elatior	Meadow fescue	3 week seedling			37.1	[29]
Festuca elatior	Meadow fescue	7 week seedling			24.8	[29]
Festuca elatior	Meadow fescue	3–7 week seedling			35.7	[29]
Festuca elatior	Meadow fescue				20–40	[29]
Festuca ovina	Sheep fescue				40–60	[29]

Appendix 3: The radio-sensitivity of plants to acute gamma irradiation estimated by LD_{50}, RD_{50} and the suggested dose (SD) for practical application

Genus/species	English name	Organ	SD (Gy)	LD_{50} (Gy)	RD_{50} (Gy)	Reference
Festuca rubra	Red fescue	Seeds	100			[4]
Festuca sp.	Fescue	Seeds	<100			[4]
Ficus carica	Common fig			62.1		[29]
Fragaria sp.	Strawberry		60–100			[2]
Fragaria sp.	Strawberry	Callus	(40–80)			[26]
Fragaria var. *'Takane'*	Strawberry	Stolon			65.3	[29]
Fragaria sp.	Strawberry				120–160	[29]
Fraxinus americana	White ash		35	~70		[5]; [29]
Fritillaria meleagris	Fritillary, checkered	Bulbs		6.5		[29]
Forsythia x intermedia	Forsythia	Rooted cuttings	40–80			[1]
Gaillardia pulchella	Rosering gaillardia	Seeds	500–700	<700	<700	[1]; [4]
Gazania x splendens	Pied gazania			600	460	[1]
Gazania sp.	Gentiana	Seeds	(25–120)			[26]
Gazania sp.	Gerbera	Plants	15			[1]
Gerbera jamesonii	Gerbera			200	130	[1]
Gerbera sp.	Gerbera	Seeds	100			[4]
Gerbera sp.	Gladiolus	Dormant corms	40			[1]
Gerbera sp.	Gladiolus		70–200			[2]
Gladiolus H.V. Friendship	Gladiolus		50	321		[5]; [27]
Gladiolus (av. of 4 var.)	Gladiolus			126.6		[29]
Tulipa sp.	Tulip		3–5			[1]
Glycine max	Soybean		100–200		80–120	[2]; [4]; [29]
Glycine max (rs-1)	Soybean	Seeds	<100			[4]
Glycine max (Rs-1)	Soybean	Seeds	200			[4]
Glycine max	Soybean	Seeds	100–200			[14]; [16]
Glycine max	Soybean	Seeds		110		[29]
Glycine max 'Hill'	Soybean	Early blooming			9.6	[29]
Gomphocarpus fructicosus	Cotton bush	Seeds		<700	320	[1]
Gomphocarpus sp.		Seeds	300			[4]
Gomphrena globosa	Globe amaranth	Seeds		140–250	110–190	[1]
Gossypium araboreum	Cotton	Seeds		260	220	[1]
Gossypium hirsutum	Cotton		150–400		60–80	[2]; [29]
Gossypium sp.	Cotton	Seeds	100			[4]
Gossypium sp.	Guzmania	Seeds	33			[1]
Gypsophila elegans	Common gypsophila	Seeds	300	560	400	[1]; [4]
Haemanthus katherinae	Blood lily			9.6		[27]

Appendix 3: The radio-sensitivity of plants to acute gamma irradiation estimated by LD$_{50}$, RD$_{50}$ and the suggested dose (SD) for practical application

Genus/species	English name	Organ	SD (Gy)	LD$_{50}$ (Gy)	RD$_{50}$ (Gy)	Reference
Haworthia fasciata	Zebra plant				16.5	[27]; [29]
Hedera helix	Common ivy	Plants	40			[1]
Helianthus annuus	Common sunflower	Seeds	(200–240)	460	340	[1]; [26]
Helianthus annuus	Common sunflower				40–60	[29]
Helianthus debilis	Beach sunflower	Seeds		130	110	[1]
Helichrysum bracteatum	Strawflower	Seeds	200	300	240	[1]; [4]
Helipterum manglessii	Mangles sunray	Seeds		700	660	[1]
Helipterum roseum Benth.	Mangles sunray	Seeds		<700	700	[1]
Helipterum roseum	Pink paper daisy	Seeds	500–700			[4]
Helipterum sp.	Mangles sunray	Seeds	500–700			[4]
Hibiscus cannabinus	Kenaf			64.3		[27]
Hibiscus trionum	Flower of an hour			220	–	[1]
Hibiscus sp.	Hibiscus	Rooted cuttings	50			[1]
Hibiscus sp.	Hibiscus		100–200			[1]
Hibiscus sp.	Hibiscus		5			[1]
Hibiscus sp.	Hibiscus	Seeds	(125–165)			[26]
Hordeum bulbosum	Bulbous barley	Seeds	(200)			[26]
Hordeum vulgare	Barley		50–400			[2]; [4]
Hordeum vulgare	Barley	Seeds	100			[4]
Hordeum vulgare	2-Rowed barley	Seeds	200; (100–200)			[4]; [26]
Hordeum vulgare	6-Rowed barley	Seeds	300	130–200	18	[4]; [29]
Hordeum vulgare	Barley	Callus	(180)			[29]
Hordeum vulgare	Spring barley	2–4 leaves		4.7		[29]
Hordeum vulgare	Spring barley	Ear emergence		6.2		[29]
Hordeum vulgare	Spring barley	Seedling		13.7–19.9		[29]
Hordeum vulgare	Spring barley				10–20	[29]
Humulus lupulus	Hops	Cutting	(10–40)		40–60	[26]; [29]
Hyacinthus orientalis	Hyacinth		2			[5]
Hyacinthus sp.	Hyacinth	Bulbs	2–5	10.6		[1]; [29]
Impatiens balsamina	Garden balsam	Seeds		170	150	[1]
Ipomoea batatas	Sweet potato	Tuber	200; (24–100)			[4]; [26]
Ipomoea batatas	Sweet potato	Seeds	400			[4]
Ipomoea batatas	Sweet potato		100–200			[13]
Ipomoea batatas	Sweet potato				120–160	[29]

Appendix 3: The radio-sensitivity of plants to acute gamma irradiation estimated by LD_{50}, RD_{50} and the suggested dose (SD) for practical application

Genus/species	English name	Organ	SD (Gy)	LD_{50} (Gy)	RD_{50} (Gy)	Reference
Ipomoea tricolor	Morning glory			140	110	[1]
Iris sp.	Iris	Fresh corms	10			[1]
Iris sp.	Iris	Bulbs	100–350			[2]
Juglans nigra	Eastern black walnut			38.3		[29]
Juglans regia	Persian walnut			48		[29]
Juncus effusus	Mut rush	Rootted plants	(200–500)			[26]
Juniperus communis	Common juniper			14.9		[29]
Juniperus virginiana	Eastern red cedar			13.5		[29]
Kalanchoe daigremontiana	Devil's backbone			475		[27]
Kalanchoe x hybrida				422		[27]
Kalanchoe sp.		Detached leaves	15–20			[1]
Kniphofia uvaria	Torchlily	Bulbs		8.4		[29]
Laburnum sp.	Golden chain	Plants	20–30			[1]
Lactuca sativa	Lettuce		200–400			[2]
Lactuca sativa	Lettuce	Seeds	200–400			[4]; [26]
Lactuca sativa	Lettuce			69.1	73	[27]
Lactuca sativa	Lettuce	Seedling		~50	~45	[29]
Lactuca sativa	Lettuce				40–60	[29]
Lagenaria leucantha	Bottle gourd	Seeds		210	160	[1]
Lagerstroemia indica	Indian lilac	Seeds		260	160	[1]
Larix decidua	European larch			7.7		[29]
Larix laricina	Eastern larch			6.9		[29]
Larix leptolepis	Japanese larch		7	8.5		[5]; [29]
Larix occidentalis	Western larch			8.5		[29]
Larix (2 spp.)				7–8.3		[29]
Lens culinaris	Lentil				10–20	[29]
Leonotis leonurus	Lion's tail	Seeds		170	110	[1]
Leonotis nepetaefolia	Lion's hair	Seeds		110	80	[1]
Lespedeza cuneatum	Sericea	Seeds		370–460		[29]
Lespedeza stipulacea	Korean lespedeza	Seeds		>400		[29]
Liatris scariosa	Eastern blazing star	Seeds		170	130	[1]
Liatris scariosa	Gay feather	Seeds	100			[4]
Lilium formosanum	Formosa lily	Bulbs		8.9		[29]
Lilium longiflorum	Easter lily		1.5			[5]
Lilium longiflorum	Easter lily	Bulbs		11.4		[29]
Lilium regale	Regal lily	Bulbs		9.1		[29]

Genus/species	English name	Organ	SD (Gy)	LD$_{50}$ (Gy)	RD$_{50}$ (Gy)	Reference
Lilium superbum	Turk's-cap lily			11.7		[27]
Lilium sp.	Lily	Bulb-scales	2.5			[1]
Lilium sp.	Lily		150–200			[2]
Limonium dumosum	German statice	Seeds		70	60	[1]
Limonium sinuatum	Sea lavender	Seeds		120	40	[1]
Limonium sinuatum	Statice	Seeds	<100			[4]
Limonium sinuatum	Statice	Cultured tissue	(4)			[26]
Linum usitatissimum	Flax		150–300			[2]
Linum usitatissimum	Flax	Seeds	500–700		120–160	[4]; [29]
Lolium multiflorum	Italian rye grass		300			[18]
Lolium multiflorum (2x)	Italian rye grass	Seeds	300			[4]
Lolium multiflorum	Italian rye grass	Seeds	500–700			[4]
Lolium perenne	Perennial rye grass	Seeds	200			[4]
Lolium perenne	Perennial rye grass	3 week seedling			15.9	[29]
Lolium perenne	Perennial rye grass	7 week seedling			19.3	[29]
Lolium perenne	Perennial rye grass	3–7 week seedling			19.2	[29]
Lolium perenne	Perennial rye grass				20–40	[29]
Lupinus albus	White lupine	Seeds	200			[4]
Lupinus angustifolius	Blue lupine	Seeds		>400		[29]
Lupinus hirsutus	Blue lupine		150–300			[2]
Lupinus luteus	Yellow lupine	Seeds	400	600	440	[1]; [4]
Lupinus luteus	Yellow lupine		150–250			[2]
Lupinus sp.	Lupine	Seeds	200			[4]
Lycopersicon esculentum	Tomato		100–400			[2]; [3]; [4]
Lycopersicon esculentum	Tomato	Seeds	200–300; (400)			[9]
Lycopersicon esculentum	Tomato	Seeds	(400)	130–370		[26]; [29]
Lycopersicon sp.	Mini-tomato	Seeds	200–300			[26]
Lycopersicon sp.	Tomato	Seedling		133	176	[29]
Malus sp.	Apple	Grafted plants	20–30			[1]
Malus	Apple	Scions	30–35			[1]
Malus pumila	Apple	Plant	100–282			[15]
Malus sp.	Apple	Callus	(50–100)			[26]
Malus sp.	Apple	Scion	(20–80)			[26]
Malus sp.	Apple	Cultured tissue	(96–160)			[26]
Manihot dulcis	Cassava			35		[29]

Appendix 3: The radio-sensitivity of plants to acute gamma irradiation estimated by LD$_{50}$, RD$_{50}$ and the suggested dose (SD) for practical application

Genus/species	English name	Organ	SD (Gy)	LD$_{50}$ (Gy)	RD$_{50}$ (Gy)	Reference
Manihot esculenta	Cassava		300			[4]
Mathiolaincana	Stock	Seeds	100–200	250–500	160–390	[1]; [4]
Medicago orbiculatus	Button clover	Seeds		210		[29]
Medicago sativa	Alfalfa		250–500		40–60	[2]; [29]
Medicago sativa	Alfalfa	Seeds	300	380–620		[4]; [29]
Medicago sativa	Vernal alfalfa				80–120	[29]
Melilotus sp.	Sweet clover	Seeds		590		[29]
Mentha piperita	Peppermint	Cuttings	40–60			[2]
Mentha piperita	Peppermint		120		140–200	[5]; [29]
Mentha spicata	Spearmint				120–160	[29]
Mimosa pudica	Mimosa	Seeds		740	410	[1]
Morus rubra	Mulberry	Plant	100			[17]
Musa textilis	Abaca		150–200			[2]
Muscari armeriacum	Muscari	Detached leaves	10–15			[1]
Muscari (average of 2 spp.)	Grape hyacinth	Bulbs		39.4		[29]
Narcissus pseudo-narcissus	Daffodil	Bulbs		9.3		[29]
Narcissus sp.	Narcisus	Dormant bulbs	4–10			[1]
Narcissus (average of 3 spp.)	Narcisus	Bulbs		15		[29]
Nephelium lit-chi	Lit-chi	Seeds	(40–80)			[26]
Nicotiana alata	Winged tobacco	Seeds	<200			[28]
Nicotiana debneyi		Seeds	200			[28]
Nicotiana glauca	Tree tobacco	Seeds	200			[28]
Nicotiana glutinosa	Tobacco	Seeds	400			[28]
Nicotiana gossei	Tobacco	Seeds	700			[28]
Nicotiana langsdorffii	Tobacco	Seeds	200			[28]
Nicotiana longiflora	Longflower tobacco	Seeds	<<200			[28]
Nicotiana megalosiphon		Seeds	400–700			[28]
Nicotiana paniculata	Tobacco	Seeds	400			[28]
Nicotiana rotundifolia		Seeds	400			[28]
Nicotiana rustica	Sacred tobacco	Seeds	400			[28]
Nicotiana suaveolens	Australian tobacco	Seeds	400–700			[28]
Nicotiana sylvestris	Flowering tobacco	Seeds	200			[28]
Nicotiana tabacum	Tobacco		300–400		60–80	[2]; [29]
Nicotiana tabacum	Tobacco	Seeds	300–400			[4]; [28]
Nicotiana tabacum x N.debneyi		Seeds	200–400			[28]
Nigella damascena	Nigella			25.4	22	[27]; [29]

Genus/species	English name	Organ	SD (Gy)	LD$_{50}$ (Gy)	RD$_{50}$ (Gy)	Reference
Orchis sp.	Orchid	Seeds	(60–200)			[26]
Orchisgraminifolia	Uchou-ran	Bulbs	(10–50)			[26]
Ornithogalum virens	Star-of-bethlehem	Bulbs		11.2		[29]
Ornithogalum sp.	Ornithogalum	Detached leaves	5–10			[1]
Orychophragmus violaceus		Seeds	200	300	200	[1]; [4]
Oryza alata	Rice (wild)	Seeds	200			[28]
Oryza australiensis	Rice (wild)	Seeds	<200			[28]
Oryza eichingeri	Rice (wild)	Seeds	<200			[28]
Oryza latifolia	Rice (wild)	Seeds	<200			[28]
Oryza minuta	Rice (wild)	Seeds	<200			[28]
Oryza officinalis	Rice (wild)	Seeds	<<200			[28]
Oryza sativa (sensitive)	Rice	Seeds	200			[4]
Oryza sativa	Lowland rice	Seeds	300			[4]
Oryza sativa	Upland rice	Seeds	300			[4]
Oryza sativa	Rice	Seeds	(200–300)	400	120–160	[6]; [26]; [29]
Oryza sativa f.spontanea	Rice	Seeds	400–500			[28]
Oryza sativa	Rice	Seeds		150–420		[29]
Oryza sativa	Rice	Panicle emergence			143	[29]
Panicum miliaceum	Common millet	Seeds	300			[4]
Paspalm dilatatum	Dallisgrass		100–300			[2]
Paspalm dilatatum	Dallisgrass	Seeds	200; (150–300)	>320		[4]; [26]; [29]
Paspalm dilatatum	Dallisgrass				120–160	[29]
Pastinaca sativa	Parsnip				60–80	[29]
Pelargonium sp.	Elargonium	Plants	10–12.5			[1]
Pennisetum typhoides	Pearl millet	Seeds		300		[6]
Pennisetum glaucum	Pearl millet				40–60	[29]
Perilla sp.	Siso (in japanese)	Seeds	(50–100)			[26]
Persea americana	American avocado			28.1		[29]
Petroselinum crispum	Parsley				80–120	[29]
Petunia hybrida	Petunia		100–150			[2]
Petunia sp.	Petunia	Plants	(70–150)			[26]
Petunia parodii	Petunia	Callus	(450)			[26]
Pharbitis nil Choisy	Morning glory	Seeds	(36)	140	110	[1]; [26]
Phaseolus anglaris	Adzuki bean	Seeds	100; (100)			[4]; [26]

Appendix 3: The radio-sensitivity of plants to acute gamma irradiation estimated by LD_{50}, RD_{50} and the suggested dose (SD) for practical application

Genus/species	English name	Organ	SD (Gy)	LD_{50} (Gy)	RD_{50} (Gy)	Reference
Phaseolus limensis	Lima bean		60–120		20–40	[2]; [27]; [29]
Phaseolus limensis	Lima bean	Seedling		42–62	23.9	[29]
Phaseolus limensis	Lima bean	Flower bud			4.2	[29]
Phaseolus limensis	Lima bean	Flower and pod			14.6	[29]
Phaseolus limensis	Lima bean	Pod			63.4	[29]
Phaseolus lunatus	Haricot bean		70			[3]
Phaseolus mungo	Black gram	Seeds		>500		[6]
Phaseolus vulgaris	Common bean		80–160			[2]
Phaseolus vulgaris	Common bean		3–100			[3]
Phaseolus vulgaris	Kidney bean	Seeds	200; (30–150)			[4]; [26]
Phaseolus vulgaris	Kidney bean				40–60	[29]
Phleum pratense	Timothy		100		60–80	[4]; [29]
Phlox drumondii	Drummond phlox	Seeds		400	160	[1]
Phlox drumondii	Phlox	Seeds	200			[4]
Physalis alkekengi	Chinese lantern plant	Seeds		180	140	[1]
Picea abies	Norway spruce			7.3		[29]
Picea engelmanni	Engelmann spruce			7.3		[29]
Picea glauca	White spruce		3	7.7		[5]; [29]
Picea mariana	Black spruce			8.4		[29]
Picea pungens	Colorado spruce			7.6		[29]
Picea rubens	Red spruce			5.7		[29]
Picea (4 spp.)			6.3–11.9			[29]
Pinus caribaea	Slash pine			7.7		[29]
Pinus contorta	Shore pine			7		[29]
Pinus densiflora	Japanese red pine			6		[29]
Pinus griffithii	Himalayan pine			5		[29]
Pinus lambertiana	Sugar pine			4.1		[29]
Pinus nigra	Austrian pine			6.1		[29]
Pinus ponderosa	Ponderosa pine			5.8		[29]
Pinus resinosa	Red pine			7		[29]
Pinus rigida	Pitch pine			6.7		[29]
Pinus strobus	Eastern white pine		1.5	5.2		[5]; [29]
Pinus sylvestris	Scotch pine			6.2		[29]
Pinus taeda	Loblolly pine			6.3		[29]
Pinus virginiana	Virginia pine			6.9		[29]

Appendix 3: The radio-sensitivity of plants to acute gamma irradiation estimated by LD_{50}, RD_{50} and the suggested dose (SD) for practical application

Genus/species	English name	Organ	SD (Gy)	LD_{50} (Gy)	RD_{50} (Gy)	Reference
Pinus (3 spp.)			4.7–8.2			[29]
Pisum sativum	Pea	Seeds	100	>100		[4]; [6]
Pisum sativum	Pea	Seedling		22.4	11.1	[29]
Pisum sativum	Pea	Vegetative			3.9	[29]
Pisum sativum	Pea	Flowering			2.5	[29]
Pisum sativum var. *arvense*	Field pea				20–40	[29]
Platycerium sp.	Platycerium	Spores	500			[1]
Platycodon grandiflorum	Balloon flower	Seeds		230	150	[1]
Poa pratensis	Kentucky blue grass	Seeds	100			[4]
Podophyllum peltatum	Mayapple		2.5			[5]
Pogostemon cablin	Patchouli	Cutting	(100)			[26]
Pogostemon cablin	Patchouli	Callus	(50–200)			[26]
Polyanthes tuberosa	Tuberosa	Bulbs	20			[1]
Poncirus trifoliata	Trifoliate orange	Callus	(10–30)			[26]
Populus tremuloides	Quaking aspen			48		[29]
Portulaca grandiflora	Portulaca	Cuttings	10–40			[1]
Portulaca grandiflora	Portulaca	Potted plants	100			[1]
Prunus amygdalus	Almond			31.1		[29]
Prunus armeniaca	Apricot	Cuttings	40–60			[2]
Prunus armeniaca	Apricot			30		[29]
Prunus avium	Mazzard cherry			36		[29]
Prunus x cerasus	Sour cherry			58.5		[29]
Prunus domestica	Garden plum			46		[29]
Prunus percica	Peach		100–200	46		[2]; [29]
Prunus persica	Peach	Cuttings	20–60			[2]
Prunus persica	Peach	Scions	(60)			[26]
Prunus sp.		Grafted plants	20–30			[1]
Prunus sp.	Almond	Cuttings	60–120			[2]
Prunus sp.	Almond		80–120			[2]
Prunus sp.	Cherry	Cuttings	20–40			[2]
Prunus sp.	Cherry	Scions	(20–60)			[26]
Pseudotsuga douglasii	Douglas fir			9.9		[29]
Pseudotsuga sp.				4.6		[29]
Psidium guajava	Guava	Seeds		170		[29]
Psidium sp.	Guava	Plants	(20–80)			[26]
Psidium sp.		Spores	300			[1]

Appendix 3: The radio-sensitivity of plants to acute gamma irradiation estimated by LD_{50}, RD_{50} and the suggested dose (SD) for practical application

Genus/species	English name	Organ	SD (Gy)	LD_{50} (Gy)	RD_{50} (Gy)	Reference
Pyrus communis	Pear		60–100			[2]
Pyrus malus	Apple	Cuttings	20–50			[2]
Pyrus malus	Apple			46		[29]
Pyrus serotina	Japanese pear	Scions	(60)			[26]
Quamoclit cardinalis	Cardinal climber	Seeds		700	460	[1]
Quercus borealis var. *maxima*	Eastern red oak			36.5		[29]
Quercus rubra	Northern red oak		20			[5]
Raphanus sativus	Radish		200–500			[2]
Raphanus sativus	Radish		400			[4]
Raphanus sativus	Radish	Seeds	(80–150)			[26]
Raphanus sativus	Radish			190	80–147	[27]; [29]
Raphanus sativus (2x)	Radish	Seeds	400			[28]
Raphanus sativus (4x)	Radish	Seeds	700			[28]
Raphanus sativus	Radish	Seedling		129	88.7	[29]
Ranunculus sp.	Ranunculus	Tubers	100			[1]
Rheum rhaponticum	Rhubarb		150–400		40–60	[2]; [9]
Rhododendron sp.	Rhododendron	Rooted cuttings	10–60			[1]
Rhus succedanea	Japanese was tree	Seeds	(160–220)			[26]
Ricinus communis	Castor bean	Seeds		360	280	[1]
Ricinus communis	Castor bean				80–120	[29]
Rosa multiflora	Multiflora rose		150–200			[2]
Rosa sp.	Rose		100–200			[2]; [10]
Rosa sp.	Rose	Dormant plants	40–100			[1]
Rosa sp.	Rose	Cuttings	60–80			[2]
Rosa sp.	Rose	Dormant twig	(300)			[26]
Rosa sp.	Rose	Fruits	(15)			[26]
Rosa sp.	Rose	Plants	(100–150)			[26]
Rudbeckia hirta	Gloriosa daisy	Seeds		60–190	50–150	[1]
Rumex aquaticus (14x)	Western dock				128	[27],[2]
Rumex confertus	Russian dock				96	[2]
Rumex conglomeratus (2x)	Clustered dock				160	[27],[2]
Rumex crispus (6x)	Curled dock			213	211	[27],[2]
Rumex hydrolapathum (20x)	Water dock				60	[27],[2]
Rumex maritimus (4x)	Golden dock			163	120	[27],[2]
Rumex obtusifolius (4x)	Broad-leaved dock			152	153	[27],[2]
Rumex orbiculatus (16x)	Broad-leaved dock				78	[27],[2]

Appendix 3: The radio-sensitivity of plants to acute gamma irradiation estimated by LD_{50}, RD_{50} and the suggested dose (SD) for practical application

Genus/species	English name	Organ	SD (Gy)	LD_{50} (Gy)	RD_{50} (Gy)	Reference
Rumex palustris (6x)	Marsh dock				164	[27],[2]
Rumex pseudonatronatus (4x)	Field dock			182	178	[27],[2]
Rumex pulcher (2x)					152	[27],[2]
Rumex salicifolius (2x)	Willow dock			141		[2]
Rumex sanguineus (4x)	Bloody dock			140	124	[27],[2]
Rumex scutatus (2x)	French sorrel			137		[2]
Rumex stenophyllus (6x)	Narrow-leaf dock			113	241	[2]
Rumex thyrsiflorus (2x)	Compact dock			50.8		[2]
Saccharinum officinarum	Sugarcane		40–100		60–80	[2]; [29]
Saccharinum sp.	Sugarcane	Young plant		150–200		[25]
Saintpaulia sp.	African violet	Detached leaves	30–40			[1]
Salvia splendens	Salvia	Seeds		170	150	[1]
Sambucus canadensis	Elder		9			[5]
Sanguisorba officinalis	Great burnet	Seeds		150	130	[1]
Saponaria officinalis	Rock soapwort	Seeds		<1200	<1200	[1]
Saponaria	Soapworts	Seeds	<800			[4]
Scilla sp.	Scilla	Bulbs	5–10			[1]
Scilla sp.	Scilla	Detached leaves	1–5			[1]
Scilla hispanica	Spanish bluebell	Bulbs		10.8		[29]
Scilla sibirica	Siberian squill	Bulbs		7.2		[29]
Secale cereale	Rye		80–150		10–20	[2]; [29]
Secale cereale	Rye	Seeds	100; (200)	80–160		[4]; [26]; [29]
Sedum alfredii	Sedum		75		231	[5]
Sedum oryzifolium	Sedum		75			[5]
Sedum rupifragum	Sedum		150		230	[5]
Sedum rupifragum	Sedum				210	[29]
Sedum ternatum	Sedum		120			[5]
Sedum tricarpum	Sedum		150			[5]
Sesamum indicum	Oriental sesame	Seeds		400		[6]
Sesamum indicum	Oriental sesame				80–120	[29]
Setaria italica	Italian millet	Seeds	400	140–460	280	[1]; [4]; [6]; [29]
Silene pendula	Drooping silene	Seeds	300	470	300	[1]; [4]
Solanum mammosum	Nipplefruit	Seeds		150	100	[1]
Solanum melongena	Aubergine	Seeds	200		80–120	[4]; [29]

Genus/species	English name	Organ	SD (Gy)	LD_{50} (Gy)	RD_{50} (Gy)	Reference
Solanum muricatum	Melon pear	Rooted cutting	(40–80)			[2]
Solanum pseudo-capsicum	Jerusalem cherry	Seeds	100	250	160	[1]; [4]
Solanum tuberosum	Potato		100; (40–80)			[2]; [4]
Solanum tuberosum	Potato	Shoot emergence			16.6	[2]
Solanum tuberosum	Potato	Stolon formation			22.4	[2]
Solanum tuberosum	Potato	Tuber initiation			93.3	[2]
Sorghum nervosum	Broom sorghum	Seeds		250	190	[2]
Sorghum sudanense	Sudan grass	Seeds	300			[4]
Sorghum vulgare	Sorghum	Seeds	300		40–60	[4]; [29]
Sorghum vulgare	Sorghum	Seeds		>400		[29]
Spinacia oleracea	Spinach	Seeds	100			[4]
Spinacia oleracea	Spinach	Seeds	(100–200)			[26]
Spinacia oleraces	Spinach	Seedling		118		[29]
Spinacia oleraces	Roundseed spinach				20–40	[29]
Spiraea cantoniensis	Spiraea	Seeds	(150–200)			[26]
Streptocarpus sp.	Streptocarpus	Detached leaves	30			[1]
Swertia japonica	Senburi (in Japanese)	Seeds	(100–400)			[26]
Syringa sp.	Lilac	Plants	30			[1]
Tagetes erecta	African marigold	Seeds		760	360	[1]
Tagetes patula	French marigold	Seeds		720	480	[1]
Taxus (2 spp.)	Conifer			4.8–12		[29]
Thea sp.	Tea	Seeds	<100			[4]
Thea sp.	Tea	Scions	(10–40)			[26]
Thuja occidentalis	Northern white ceder		6			[5]
Thuja sp.				9.7		[29]
Tigridia pavonia	Tigerflower	Bulbs		33.2		[29]
Tradescantia paludosa	Confederate spiderwort			14.4		[27]
Tradescantia sp.				15.7		[27]
Tradescantia navicularis	Chain plant			42.5		[27]
Trifolium incarnatum	Crimson clover			250->650		[29]
Trifolium pratense	Red clover	Seeds	500–700	350->1080		[4]; [29]
Trifolium pratense	Red clover				80–120	[29]
Trifolium repens	White clover	Seeds	400			[4]
Trifolium repens	White clover			243	123	[27]; [29]
Trifolium repens	White clover	3 week seedling			114	[29]

Appendix 3: The radio-sensitivity of plants to acute gamma irradiation estimated by LD_{50}, RD_{50} and the suggested dose (SD) for practical application

Genus/species	English name	Organ	SD (Gy)	LD_{50} (Gy)	RD_{50} (Gy)	Reference
Trifolium repens	White clover	7 week seedling			234	[29]
Trifolium repens	White clover	3–7 week seedling			140	[29]
Trifolium repens	White clover	Seedling		242		[29]
Trifolium resupinatum	Perusian clover		250–350			[2]
Trifolium squarrosum	Clover		250–350			[2]
Trifolium subterraneum	Subterranean clover		250–350			[2]
Trifolium sp.		Seeds	500–700			[4]
Trillium grandiflorum	Trillium		2			[5]
Triticale sp.	Triticale	Seeds	200			[4]
Triticale sp.	Triticale	Seeds	(50–300)			[26]
Triticum aegilopoides (AA)		Seeds	100			[28]
Triticum aestivum	Bread wheat		250–300			[4]
Triticum aestivum	Wheat	Seeds	200			[4]
Triticum aestivum	Wheat	Seeds	(150–200)			[26]
Triticum aestivum	Spring wheat				31.1	[29]
Triticum aestivum	Wheat				20–40	[29]
Triticum aestivum	Spring wheat	2–4 leaves			14.1	[29]
Triticum aestivum	Spring wheat	Ear emergence			9	[29]
Triticum aestivum	Spring wheat	Anthesis			17.8	[29]
Triticum aestivum	Spring wheat	Seedling		30.9		[29]
Triticum aestivum	Spring wheat	Seedling		3.45	20.6	[29]
Triticum aestivum	Winter wheat	Ear emergence			8.6	[29]
Triticum aestivum	Winter wheat	Anthesis			15.6	[29]
Triticum dicoccoides (AABB)	Wild emmer wheat	Seeds	100			[28]
Triticum durum (AABB)	Durum wheat	Seeds	100–200			[28]
Triticum monococcum (AA)		Seeds	100–200			[28]
Triticum Spelta (AABBDD)		Seeds	100–200			[28]
Triticum sp.	Wheat		150–200			[2]
Tritonia crocata	Tritonia (montbretia)	Bulbs		68.4		[29]
Tropaeolum majus	Nasturtium	Seeds		360–400	170–300	[1]
Tropaeolum majus	Nasturtium			115		[27]
Tsuga sp.				6.96		[29]
Tsuga sp.	Tulip	Bulbs	(20–80)			[26]
Tsuga sp.	Tulip (darwin)	Bulbs		9.9		[29]
Tulipa fosteriana	Tulip 'red emperor'	Bulbs		10.7		[29]

Appendix 3: The radio-sensitivity of plants to acute gamma irradiation estimated by LD_{50}, RD_{50} and the suggested dose (SD) for practical application

Genus/species	English name	Organ	SD (Gy)	LD_{50} (Gy)	RD_{50} (Gy)	Reference
Tulipa kaufmanniana	Waterlily tulip	Bulbs	20–80			[29]
Vaccinium sp.	Blueberry	Cuttings	60–80; (20–80)			[2]; [26]
Vicia faba	Broad bean	Seeds	<100			[4]
Vicia faba	Broad bean			9.8	14	[27]
Vicia faba	Broad bean	Vegetative			2.2	[29]
Vicia faba	Broad bean	Flowering			1.1	[29]
Vicia sativa subsp. *sativa*	Vetch	Seeds	100			[4]
Vicia villosa	Hairy vetch	Seeds	100–200	170		[4]; [29]
Vigna sinensis	Cowpea	Seeds	200	110	60–80	[4]; [29]
Viola cornuta	Horned viola	Seeds		<700	560	[1]
Viola sp.	Violet	Seeds	500–700			[4]
Viscaria viscosa	German catchfly	Seeds		300	200	[1]
Viscaria sp.	Clammy campion	Seeds	200			[4]
Vitis vinifera	Grape	Cuttings	20–60			[2]
Vitis vinifera	Grape	Scions	(20–100)			[26]
Vitis spp.	Grape	Seeds		<40–<50		[29]
Zea Mays	Corn (maize)		150–200			[2]
Zea mays	Corn (maize)	Seeds	300			[4]
Zea mays H.V.Golden Bantam	Corn (maize)		7.5			[5]
Zea mays	Corn (maize)			52.3	40	[27]
Zea mays	Corn (maize)	Seeds	(100–300)			[26]
Zea mays	Corn (maize)				40	[29]
Zea mays (hybrid)	Corn (maize)				42	[29]
Zea mays	Corn (maize)	Seeds		>150		[29]
Zea mays	Corn (maize)	2 leaves		8		[29]
Zea mays	Corn (maize)	Seedling		45–50		[29]
Zea mays	Corn (maize)				20–40	[29]
Zephyranths sp.	Zephyr lily			8.2		[29]
Zinnia elegans	Common zinia	Seeds		170	150	[1]
Zingiber sp.	Zinger	Roots	(20–30)			[26]
Zingiber sp.	Zinger	Cultured tissue	(40–60)			[26]
Zingiber sp.	Lawngrass		(400–500)			[26]
Zingiber sp.	Lawngrass		(300–500)			[26]